Waterflooding

G. Paul Willhite
Professor of Chemical and Petroleum Engineering
U. of Kansas

SPE Textbook Series, Volume 3

Henry L. Doherty Memorial Fund of AIME
Society of Petroleum Engineers
Richardson, TX USA

Dedication

This book is dedicated to my parents, family, friends, colleagues, teachers, and students who contributed to this book in ways they may never know.

Disclaimer

This book was prepared by members of the Society of Petroleum Engineers and their well-qualified colleagues from material published in the recognized technical literature and from their own individual experience and expertise. While the material presented is believed to be based on sound technical knowledge, neither the Society of Petroleum Engineers nor any of the authors or editors herein provide a warranty either expressed or implied in its application in the design, implementation, or analysis of waterflooding. Correspondingly, the discussion of materials, methods, or techniques that may be covered by letters patents implies no freedom to use such materials, methods, or techniques without permission through appropriate licensing. Nothing described within this book should be construed to lessen the need to apply sound engineering judgment nor to carefully apply accepted engineering practices in the design, implementation, or application of the techniques described herein.

ISBN 978-1-55563-005-8

11 12 13 14 15 16 / 16 15 14 13 12 11

Society of Petroleum Engineers
222 Palisades Creek Drive
Richardson, TX 75080-2040 USA
http://store.spe.org/
books@spe.org
1.972.952.9393

SPE Textbook Series

The Textbook Series of the Society of Petroleum Engineers was established in 1972 by action of the SPE Board of Directors. The Series is intended to ensure availability of high-quality textbooks for use in under-graduate courses in areas clearly identified as being within the petroleum engineering field. The work is directed by the Society's Textbook Committee, one of more than 40 Society-wide standing committees, through members designated as Textbook Coordinators. The Textbook Coordinators and the Textbook Committee provide technical evaluation of the book. Below is a listing of those who have been most closely involved in the final preparation of this book. Many others contributed as Textbook Committee members or others involved with the book.

Textbook Coordinators

L.K. Thomas, Phillips Petroleum Co., Bartlesville, OK
H.M. Staggs, ARCO Oil & Gas Co., Plano, TX

Textbook Committee (1986)

Medhat M. Kamal, chairman, Flopetrol Johnston, Houston
Theodore R. Blevins, Chevron U.S.A., Ventura, CA
John F. Evers, U. of Wyoming, Laramie
Wilmer A. Hoyer, Exxon Production Research Co., Houston
Stephen H. Neuse, Hudson Consultants Inc., Tulsa
Fred H. Poettmann, Colorado School of Mines, Golden
David S. Pye, Union Geothermal Div., Indio, CA
Philip A. Schenewerk, Ensearch Exploration, Denver

Preface

This book was written as a text for senior students in petroleum engineering. Like many books, it began as a collection of notes and papers for a course at the U. of Kansas entitled Secondary Recovery. The opportunity to consolidate these materials into a coherent, organized text was the direct result of the SPE Textbook Committee, who sought authors for a text on EOR. During the intervening years, the chapter on waterflooding in the original outline was expanded into a much-needed text to complement the widely used SPE monograph, *Reservoir Engineering Aspects of Waterflooding*, by F.F. Craig Jr., and SPE Reprint Series No. 2a, *Waterflooding*, a collection of case histories and reference papers.

In this text, I have attempted to develop waterflooding from fundamental principles because I believe that fundamental concepts apply universally to any proposed or existing waterflood. Many of these concepts carry over into all other EOR processes. Not all fundamental concepts are fully understood, particularly on a microscopic level. In these cases, I have presented the current understanding, and recognized differences of opinion. This is particularly evident in discussions on determination of reservoir wettability and viscous fingering. Numerous problems of various degrees of difficulty are included at the ends of Chaps. 2 through 6. Some can be solved with a hand calculator, but many require preparation of short computer programs. A few require the use of programs available in the published literature. In all cases, the emphasis is on understanding of fundamental concepts and/or the application of these concepts to solve waterflooding problems.

One goal of this text is the development of a systematic procedure for the design of a waterflood. The design experience is an important growth opportunity for students and engineers alike. In this experience, we have to grapple with the difference between theory and practice, the shortage of data, or the absence of a good reservoir description. Recognition that the reservoir is more complicated than many engineers and geologists once imagined is a growing part of this experience. For this reason, an extensive treatment of reservoir geology and its role in the design and operation of waterfloods is included in Chap. 7. In this regard, I am indebted to Jim Ebanks Jr., a long-time friend and geologist, who generously provided counsel, encouragement, and critique on this subject to an engineer who was trying to learn enough about geology to replace guesses with defendable assumptions.

In the preparation of this text, I have read hundreds of papers concerning all aspects of waterflooding. One cannot go through this experience without gaining a profound appreciation for the work of so many scientists and engineers who sought to understand and develop this process in an environment that is often poorly defined. Many of these papers did not fit within the context of this text but were valuable nonetheless.

Writing is a lonely and demanding task. My efforts were supported by many people. My colleagues in the Dept. of Chemical and Petroleum Engineering at the U. of Kansas taught courses while I wrote portions of this text during sabbatical leaves. My students in CPE 695 contributed over the years this text was under development by critiquing the manuscript, the problems, and the instructor. Several of my colleagues also pointed out the inevitable small errors that tend to creep into manuscripts and galley proofs. My colleagues in the industry and at other universities were quite helpful. Bill Brigham provided notes from his waterflood course, including several good problems, and Jim McCaleb and Jim Ebanks provided excellent critiques of Chap. 7. Others mentioned in appropriate sections of the text provided problems, figures, and information not available in the published literature. I was fortunate to spend my sabbatical leave in 1984–85 at Chevron Oil Field Research Corp., LaHabra, CA, where I was able to complete final details of the text.

Several other people at the U. of Kansas provided continuing support throughout the preparation of the text and deserve special recognition. Ruth Sleeper, Emily Chung, and Debbie Fowler typed portions of the manuscript. Vera Sehon did all graphic and photographic work. In Dallas, many members of the SPE publications staff worked diligently to make our publication deadline. The assistance of Christine Butcher and Holly Hargadine was exceptional. The contributions of Bud Staggs and Kent Thomas, editors from the SPE Textbook Committee, were appreciated.

G. Paul Willhite

Introduction

Waterflooding was practiced as an art for years before a scientific basis for waterflood design developed. This understanding evolved primarily in the late 1940's and 1950's from extensive research and development efforts by companies and universities combined with field experience in the 1960's and 1970's. This book presents waterflooding design by building a fundamental understanding of multiphase flow in porous media.

In developing this text, I sought to present concepts that reveal basic features for two- and three-dimensional (2D and 3D) displacement processes without resorting to numerical simulations. Chap. 1 describes the historical development of waterflooding and reviews primary producing mechanisms. Concepts of two-phase flow and microscopic displacement efficiency are reviewed in Chap. 2. Waterflooding is introduced in Chap. 3 through the study of linear displacement processes. Displacement in two dimensions is covered in Chap. 4 through analysis of waterflood performance in pattern floods. The streamtube model is introduced as a tool to design waterfloods by extending concepts developed for linear systems to areal floods. Heterogeneous reservoirs are treated in Chap. 5, first as noncommunicating layered systems. Crossflow between layers is approximated by developing the concept of pseudorelative permeability. Although numerical simulators are beyond the scope of this text, examples of numerical simulation applied to one-dimensional (1D), 2D, and 3D displacement problems are included to illustrate capabilities of numerical simulators.

Chap. 6 presents one approach to waterflood design. I emphasize the development of a model to simulate injection rates and displacement performance as an integral part of waterflood design. Models ranging from empirical methods to numerical simulators are introduced in order of increasing complexity. Field experience combined with geological interpretation is discussed in Chap. 7 to illustrate the role of geological interpretation in the design and operation of waterfloods. Many of the waterflood design procedures can be prepared as small computer programs. Selected computer subprograms are included to help the student write more complex design programs with a reasonable effort.

Contents

Chapter 1
Background

1.1 Introduction

Enhanced oil recovery (EOR) refers to any method used to recover more oil from a reservoir than would be produced by primary recovery.[1] In primary production, oil is displaced to the production well by natural reservoir energy. Sources of natural reservoir energy are fluid and rock expansion, solution gas drive, gravity drainage, and the influx of water from aquifers.

Most EOR processes involve the injection of a fluid (gas or liquid) into the reservoir. Waterflooding is the most widely applied EOR process and is the subject of this text. Other EOR processes are planned for a future volume.

This chapter is organized into two sections. First, the historical development of waterflooding is outlined. Then primary production mechanisms are reviewed to define the initial condition of the reservoir at the time waterflooding operations are being considered.

1.2 Development of Waterflooding

The discovery of crude oil by Edwin L. Drake at Titusville, PA, on Aug. 27, 1859, marked the beginning of the petroleum era. Although the first oil well produced about 10 B/D [1.6 m^3/d], within 2 years other wells were drilled that flowed thousands of barrels per day.[2] Production rates from these shallow Pennsylvania reservoirs declined rapidly as reservoir energy was depleted. Recovery was a small percentage of the amount of oil estimated to be initially in place. As early as 1880, Carll[3] raised the possibility that oil recovery might be increased by the injection of water into the reservoir to displace oil to producing wells.

The practice of waterflooding apparently began accidentally. The experience in the Bradford field, PA, is typical.[4] Many wells were abandoned in the Bradford field following the flush production period of the 1880's. Some were abandoned by pulling casing without plugging, while in other wells the casing was left in the wells, where it

corroded. In both cases, fresh water from shallower horizons apparently entered the producing interval. Water injection began, perhaps as early as 1890, when operators realized that water entering the productive formation was stimulating production. By 1907, the practice of water injection had an appreciable impact on oil production from the Bradford field. The first flooding pattern, termed a circle flood, consisted of injecting water into a well until surrounding producing wells watered out. The watered-out production wells were converted to injection to create an expanding "circular" waterfront. Many operators were against the injection of water into a sand. A Pennsylvania law requiring plugging of abandoned wells and dry holes to prevent water from entering oil and gas sands was construed as prohibiting waterflooding, so waterflooding was done secretly. In 1921, the Pennsylvania legislature legalized the injection of water into the Bradford sands.

The practice of water injection expanded rapidly after 1921. The circle-flood method was replaced by a "line" flood, in which two rows of producing wells were staggered on both sides of an equally spaced row or line of water intake wells.[4] By 1928, the line flood was replaced by a new method termed the "five-spot" because of the resemblence of the pattern to the five spots on dice.[4] Waterflooding was quite successful in the Bradford field. Fig. 1.1 shows the production history of the Bradford field for more than 100 years of producing life. The effects of waterflooding on production from this field are apparent.

Waterflooding, called secondary recovery because the process yielded a second batch of oil after a field was depleted by primary production, spread slowly throughout the oil-producing provinces. Water-injection operations were reported in Oklahoma in 1931, in Kansas in 1935, and in Texas in 1936.

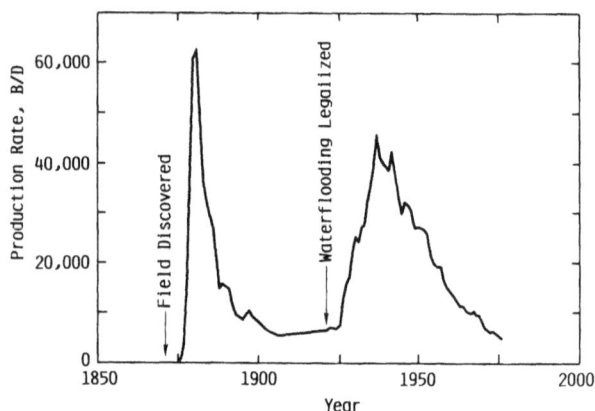

Fig. 1.1—Production history of the Bradford fields.

The slow growth of water injection was caused by several factors. In the early days, waterflooding was understood poorly. Gas injection developed about the same time as waterflooding and was a competing process in some reservoirs. Major discoveries of crude oil in the U.S. in the 1920's and 1930's led to proration in several states. The capability to produce oil was much greater than market demand. Consequently, primary depletion of many reservoirs was controlled by market demand.

In the intervening years, major oil discoveries were made throughout the world. Shut-in production capacity exceeded demand. Large supplies of low-cost imported oil also prolonged the primary life of reservoirs, delaying implementation of water injection. Interest in waterflooding developed in the late 1940's and early 1950's as reservoirs approached economic limits and operators sought to increase reserves. By 1955,[5] waterflooding was estimated to contribute more than 750,000 B/D [119 200 m³/d] out of a total production rate of 6.6 million B/D [10⁶ m³/d] in the U.S. Waterflooding is practiced extensively throughout the world. In the U.S. as much as half of the current oil production is thought to be the result of water injection.

1.3 Primary Production

The amount of oil that can be displaced by the natural reservoir energy associated with a reservoir varies with reservoir type. Reservoirs are classified into five broad categories based on the principal source of reservoir energy. These are (1) water drive, (2) solution gas drive, (3) fluid expansion, (4) gas-cap drive, and (5) gravity drainage.

Water Drive

A water-drive reservoir has a hydraulic connection between the reservoir and a porous, water-saturated rock called an aquifer. The aquifer may underlie all or part of the reservoir. Often the aquifer is at the edge of the field, as shown in Fig. 1.2.

The water in an aquifer is compressed. As reservoir pressure is reduced by oil production, the water expands, creating a natural waterflood at the reservoir/aquifer boundary. Reservoir energy is also supplied by compressibility of the rock in the aquifer. When the aquifer is large and contains sufficient energy, the entire reservoir may be "waterflooded" by proper management of fluid withdrawal rates. Recovery efficiencies of 70 to 80% of the original oil in place (OOIP) are possible in some water-drive reservoirs. Reservoir geology, heterogeneity, and structural position are important variables affecting recovery efficiency. Strong water-drive reservoirs have been found throughout the world. Examples include the East Texas field, the Arbuckle reservoirs of Kansas, and the Tensleep reservoirs in Wyoming.

Some water-drive reservoirs are connected to aquifers that have limited amounts of energy. Unless there is extensive geological information about the aquifer from drilling or other records, the extent of the aquifer and its capability to provide reservoir energy is not known until well into the primary production period. Usually, reservoir pressure is monitored with fluid withdrawal. Water influx can be calculated. The leveling of reservoir pressure at some withdrawal rate is a measure of water-drive capability. If the aquifer cannot supply sufficient energy

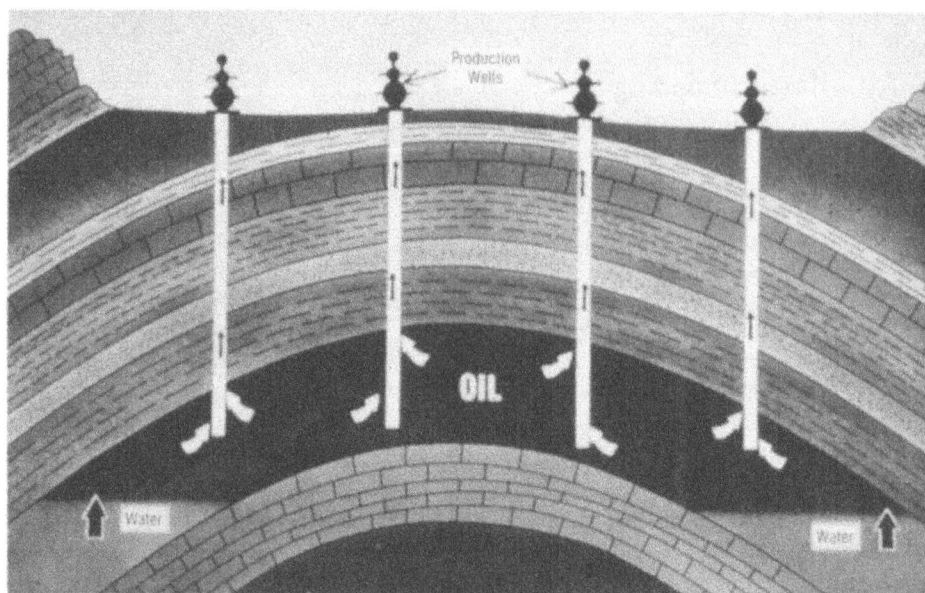

Fig. 1.2—Water-drive reservoir.

Fig. 1.3—Dissolved gas drive.

to meet desired fluid withdrawal rates while maintaining reservoir pressure, an edgewater injection program may be used to supplement natural reservoir energy. The program, called pressure maintenance, is a waterflood. It follows that reservoirs with strong aquifers are seldom waterflood candidates. However, reservoir heterogeneity may limit the effect of a natural water drive to a portion of the reservoir.

Solution Gas Drive

Crude oil under high pressure may contain large amounts of dissolved gas. When the reservoir pressure is reduced as fluids are withdrawn, gas comes out of solution and displaces oil from the reservoir to the producing wells, as shown in Fig. 1.3.

The efficiency of solution gas drive depends on the amount of gas in solution, the rock and oil properties, and the geological structure of the reservoir. Recoveries are low, on the order of 10 to 30% of the OOIP. Recovery is low because the gas is more mobile than the oil phase in the reservoir. As pressure declines, gas flows at a faster rate than oil, leading to rapid depletion of reservoir energy, which is noted by increasing gas/oil ratios (GOR's) in the field. Solution-gas-drive reservoirs are usually good candidates for waterflooding.

Undersaturated Reservoirs

A crude oil is undersaturated when it contains less gas than is required to saturate the oil at the pressure and temperature of the reservoir. When the oil is highly undersaturated, much of the reservoir energy is stored in the form of fluid and rock compressibility. Pressure declines

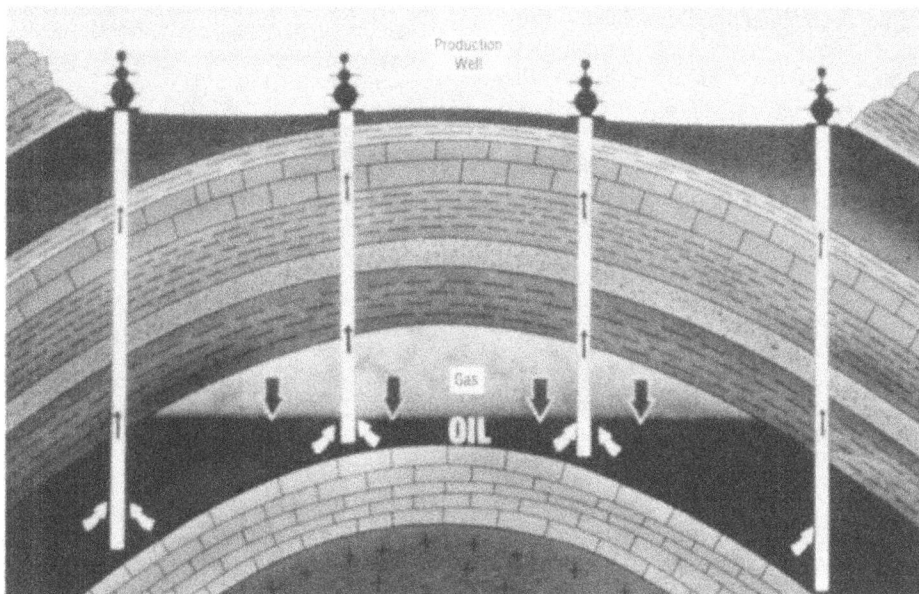

Fig. 1.4—Gas-cap drive.

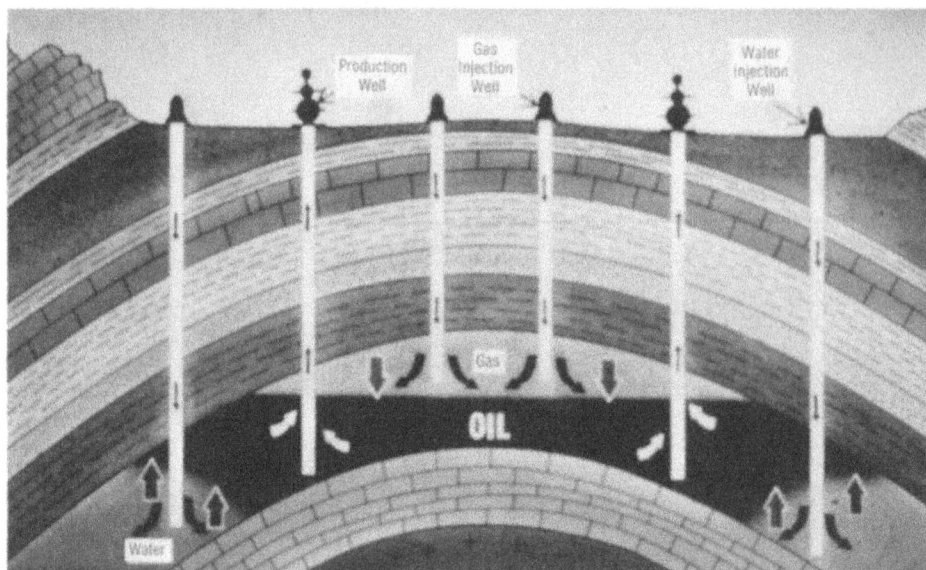

Fig. 1.5—Combined gas and water injection programs.

rapidly as fluids are withdrawn from an undersaturated reservoir until the bubblepoint is reached. Then, solution gas drive becomes the source of energy for fluid displacement. Reservoir fluid analysis, PVT behavior, and reservoir pressure data will identify an undersaturated reservoir. These reservoirs are good candidates for water injection to maintain a high reservoir pressure and to increase oil recovery.

Gas-Cap Drive

When a reservoir has a large gas cap, as shown in Fig. 1.4, there may be a large amount of energy stored in the form of compressed gas. The gas cap expands as fluids are withdrawn from the reservoir, displacing the oil by a gas drive assisted by gravity drainage. Expansion of the gas cap is limited by the desired pressure level in the reservoir and by gas production after gas cones into production wells. Reservoirs with large gas caps generally are not considered to be good waterflood candidates. Pressure has been maintained in some of these reservoirs by injection of gas into the gas cap. Gas-cap reservoirs that have an underlying water zone may have combined gas- and water-injection programs, as depicted in Fig. 1.5.

Caution is required when a combined gas/water-injection project is considered. There is a risk that oil will be displaced into the gas-cap region and will remain trapped at the end of the flood.

Gravity Drainage

Gravity drainage may be a primary producing mechanism in thick reservoirs that have good vertical communication or in steeply dipping reservoirs. Gravity drainage is a slow process because gas must migrate up structure or to the top of the formation to fill the space formerly occupied by oil. Gas migration is fast relative to oil drainage so that oil rates are controlled by the rate of oil drainage. In reservoirs where wells are drawn down to atmospheric pressure, air may enter the reservoir at the top of the producing interval if the entire interval is open.

Gravity drainage is an important producing mechanism in several large California reservoirs. These reservoirs contain heavy oil and are not waterflood candidates.

References

1. Hayes, H.J. *et al: Enhanced Oil Recovery*, Natl. Pet. Council, Washington, D.C. (Dec. 1976) 3.
2. Dickey, P.A.: "The First Oil Well," *J. Pet. Tech.* (Jan. 1959) 14–26.
3. Carll, J.F.: "The Geology of the Oil Regions of Warren, Venango, Clarion and Butler Counties," Pennsylvania Second Geological Survey, Report III (1880) 263.
4. Fettke, C.R.: "The Bradford Oil Field, Pennsylvania and New York," *Bull. M21*, Pennsylvania Geological Survey, Fourth Series (1938) 298–301.
5. Sweeney, A.E. Jr.: "A Survey at Secondary Recovery Operations and Methods Employed in the United States," paper from the Interstate Oil Compact Commission presented at the 1957 Illinois Oil and Gas Assn. Meeting, April 11.

Chapter 2
Microscopic Efficiency of Immiscible Displacement

2.1 Introduction

The presence of immiscible fluids (oil/water, oil/gas, water/gas, or oil/water and gas) in reservoir rocks alters the capacity of a rock to transmit fluids. In this chapter we examine fundamental concepts that are used to characterize the distribution and flow of immiscible fluids in porous rocks.

Two or more phases are considered immiscible at a specified temperature and pressure if a visible interface forms after the phases have been mixed vigorously with sufficient time for phase equilibrium to occur. Water and oil are immiscible under nearly all reservoir and surface conditions because the mutual solubilities of oil in water and water in oil are small. This broad definition of immiscibility includes cases in which a crude oil and a gas are in equilibrium. In this instance, the crude oil contains all of the components of the gas in proportions determined by equilibrium distribution coefficients. These systems are sometimes described as partially miscible because substantial amounts of gas may be dissolved in the oil depending on temperature, pressure, and composition.

2.2 Fundamental Principles Governing Fluid and Rock Interactions

2.2.1 Interfacial Tension (IFT)

The interface between two phases is a region of limited solubility, which is, at most, a few molecules thick. It may be visualized as a phase boundary that occurs because the attractive forces between molecules in the same phase are much larger than those that exist between molecules in different phases.

The IFT is a fundamental thermodynamic property of an interface. It is defined as the energy required to increase the area of the interface by one unit. Fig. 2.1 shows a ring tensiometer used to determine the IFT at an oil/water interface. The ring positioned at the interface is raised by applying a force. As the ring is pulled through the interface, the interfacial area increases, as depicted in Fig. 2.2. The force at the snapoff point divided by the circumference of the ring (corrected for a geometric factor) is the IFT. Other methods of measuring IFT's include sessile drop,[1] pendant drop,[2] and spinning drop[3] techniques.

Typical values of IFT's for crude-oil/water systems are presented in Table 2.1.[4] IFT's between oil and water are usually 10 to 30 dynes/cm [10 to 30 mN/m] at 77°F [25°C]. The IFT between a liquid and its vapor (or air) is referred to as surface tension. Correlations of surface tension of paraffin hydrocarbons with molecular weight and temperature are presented in Fig. 2.3.[5]

IFT is a measure of miscibility; the lower the IFT, the closer two phases approach miscibility. For example, as the critical point is approached, the properties of the liquid phase become indistinguishable from those of the vapor phase. Consequently, the IFT becomes zero at the critical point. Fig. 2.4 shows the reduction in IFT with pressure for the methane/pentane system at 100°F [38°C] as the critical pressure of 2,420 psia [16 685 kPa] is approached.[6] Low IFT's are also observed in systems containing alcohols and surfactants. Values of 10^{-5} dynes/cm [10^{-5} mN/m] have been reported for some formulations.[7] IFT's less than 10^{-3} dynes/cm [10^{-3} mN/m] are referred to as ultralow tensions.

2.2.2 Wettability

Interaction between the surface of the reservoir rock and the fluid phases confined in the pore space influences fluid distribution in rocks as well as flow properties. When two immiscible phases are placed in contact with a solid surface, one of the phases is usually attracted to the surface more strongly than the other phase. This phase is identified as the wetting phase while the other phase is the nonwetting phase.

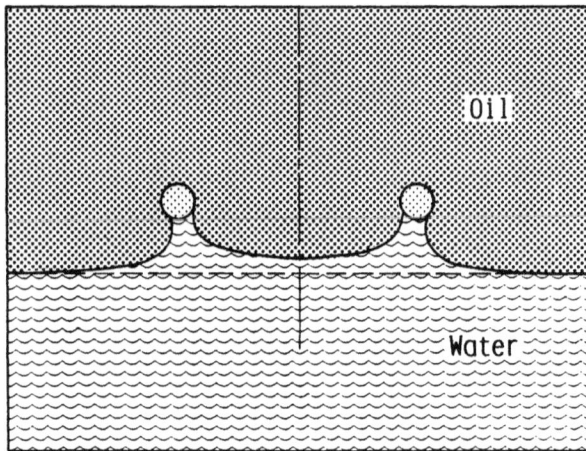

Fig. 2.1—Distension of interfacial film during IFT measurement.

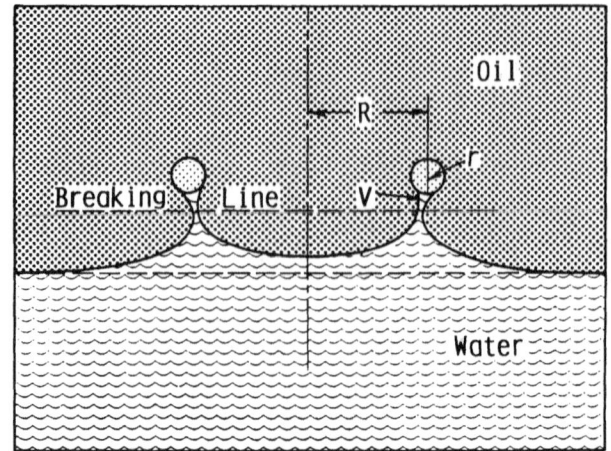

Fig. 2.2—Condition of interfacial film at breaking point.

TABLE 2.1—IFT'S FOR CRUDE OILS[4]

Field	Oil Formation	State	Oil Viscosity (cp)	IFT (dynes/cm)
West Delta	Offshore	Louisiana	30.4	17.9
Cayuga	Woodbine	Texas	82.9	17.9
Fairport	Lansing	Kansas	5.3	20.8
Bayou	Choctaw	Louisiana	16.1	15.6
Chase-Silica	Kansas City	Kansas	6.7	19.6
Hofra	Paleocene	Libya	5.1	27.1
Black Bay	Miocene	Louisiana	70.8	17.7
Bar-Dew	Bartlesville	Oklahoma	9.0	17.7
Bar-Dew	Bartlesville	Oklahoma	6.8	21.4
Eugene Island	Offshore	Louisiana	7.4	16.2
Cambridge	Second Berea	Ohio	15.3	14.7
Grand Isle	Offshore	Louisiana	10.3	16.1
Bastian Bay	Uvigerina	Louisiana	112.2	24.8
Oklahoma City	Wilcox	Oklahoma	6.7	20.1
Glenpool	Glen	Oklahoma	5.1	24.7
Cumberland	McLish	Oklahoma	5.8	18.5
Allen District	Allen	Oklahoma	22.0	25.9
Squirrel	Squirrel	Oklahoma	33.0	22.3
Berclair	Vicksburg	Texas	44.5	10.3
Greenwood-Waskom	Wacatoch	Louisiana	5.9	11.9
Ship Shoal	Miocene	Louisiana	22.2	17.3
Gilliland	—	Oklahoma	12.8	17.8
Clear Creek	Upper Bearhead	Louisiana	2.4	17.3
Ray	Arbuckle	Kansas	21.9	25.3
Wheeler	Ellenburger	Texas	4.5	18.2
Rio Bravo	Rio Bravo	California	3.8	17.8
Tatums	Tatums	Oklahoma	133.7	28.8
Saturday Island	Miocene	Louisiana	22.4	31.5
North Shongaloo-Red	Tokio	Louisiana	5.2	17.7
Elk Hills	Shallow Zone	California	99.2	12.6
Eugene Island	Miocene	Louisiana	27.7	15.3
Fairport	Reagan	Kansas	31.8	23.4
Long Beach	Alamitos	California	114.0	20.5
Colgrade	Wilcox	Louisiana	360.0	19.0
Spivey Grabs	Mississippi	Kansas	26.4	24.5
Elk Hills	Shallow Zone	California	213.0	14.2
Trix-Liz	Woodbine A	Texas	693.8	10.6
St. Teresa	Cypress	Illinois	121.7	21.6
Bradford	Devonian	Pennsylvania	2.8	9.9
Huntington Beach	South Main Area	California	86.2	16.4
Bartlesville	Bartlesville	Oklahoma	180.0	13.0
Rhodes Pool	Mississippi Chat	Kansas	43.4	30.5
Toborg	—	Texas	153.6	18.0

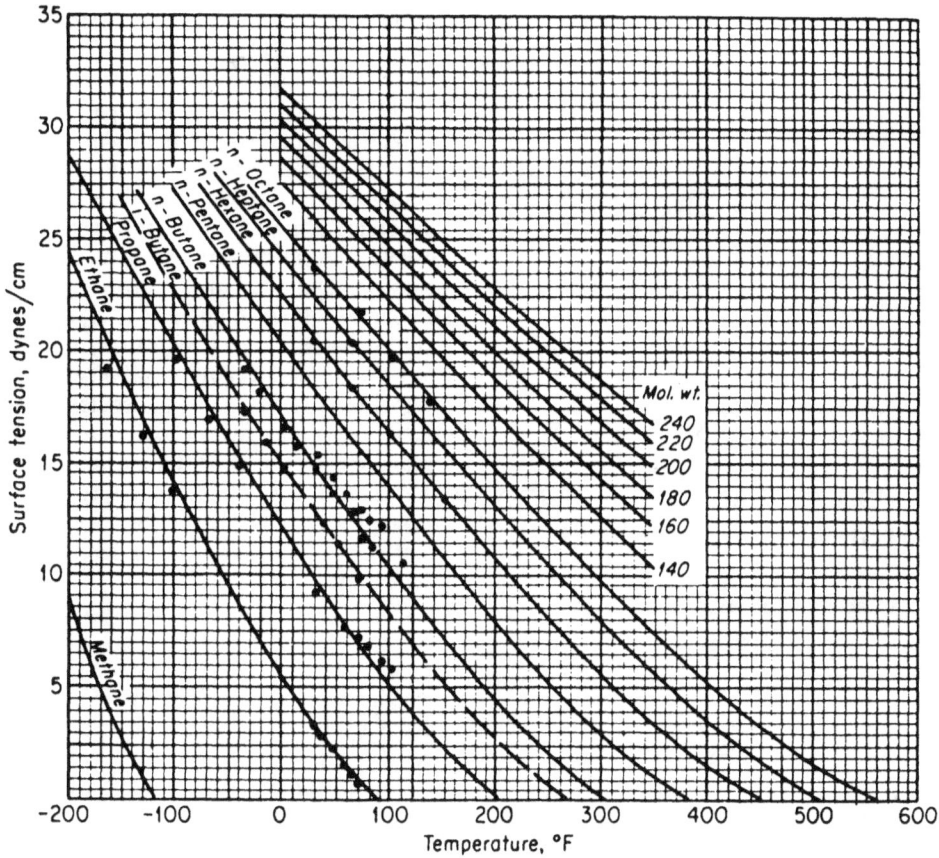

Fig. 2.3—Surface tension of paraffin hydrocarbons.[23]

Wettability is explained quantitatively by examining the force balances between two immiscible fluids at the contact line between the two fluids (water and oil) and the solid shown in Fig. 2.5.[8] The water phase spreads out over the surface in preference to oil. The forces that are present at the contact line are σ_{os}, the IFT between the solid and the oil phase; σ_{ws}, the IFT between the solid and the water phase; and σ_{ow}, the IFT between the oil and water phases. The contact angle, θ, is measured through the water phase to σ_{ow}, the tangent to the interface at the contact line. At equilibrium, the sum of the forces acting along the contact line must be zero. Eq. 2.1 describes the Young equation,[9] representing the force balance in the direction parallel to the rock surface:

$$\sigma_{os} - \sigma_{ws} = \sigma_{ow} \cos \theta. \quad \ldots \ldots \ldots \ldots \ldots (2.1)$$

Neither σ_{os} or σ_{ws} has ever been measured. Consequently, θ is the principal measure of wettability for a smooth, homogeneous surface. Examples of water-wet and oil-wet systems are shown in Fig. 2.6.[10] Water-wet systems are those with contact angles near zero. Oil-wet systems have contact angles approaching 180°. Systems with contact angles near 90° are referred to as having intermediate wettability.

Wettability is a complex function of fluid and solid properties. Wide variations in contact angle can be obtained by appropriate selection of fluid pairs or the solid. Fig. 2.7 illustrates four systems[11] that exhibit water-wet, intermediate, and oil-wet behavior on a silica surface. Fig. 2.7 also shows the contact angles for the same fluid pairs

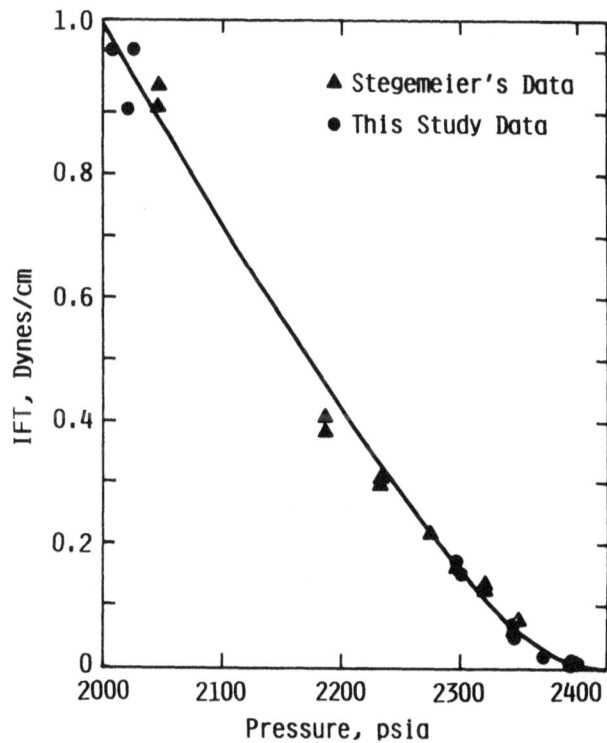

Fig. 2.4—IFT, methane/n-pentane systems at 100°F.[6]

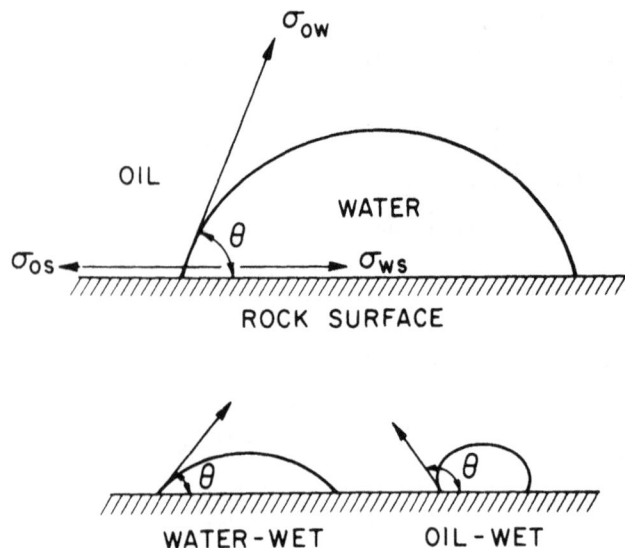

Fig. 2.5—Wettability of oil/water/solid system.[8]

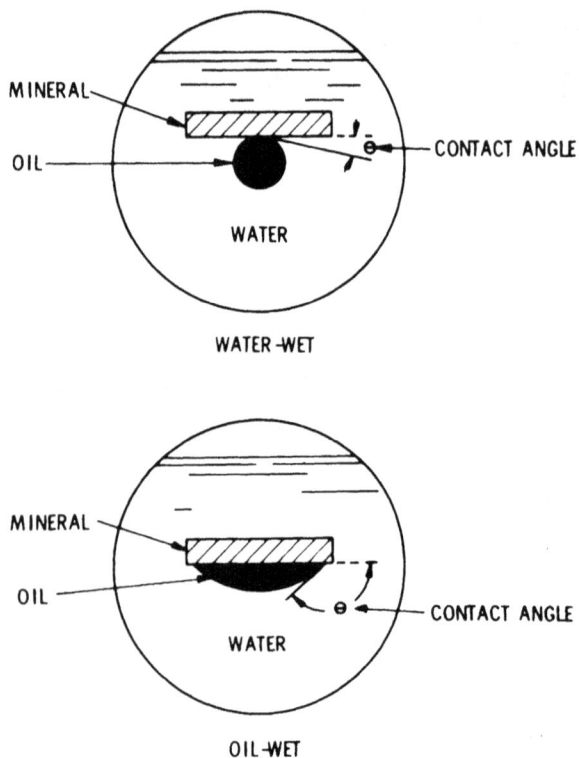

Fig. 2.6—Contact angles measured through the aqueous phase.[10]

thenic acid/water), which was water-wet on silica, became intermediate wetting on calcite. These examples demonstrate that wettability is related to rock mineralogy as well as the properties of the fluid pairs.

Why do some fluid pairs exhibit oil-wet behavior while other fluid pairs are water wetting? Wetting indicates a stronger attraction of the solid surface for one phase than the other. It is well known that water preferentially wets calcite and silica surfaces in the presence of "pure" paraffin hydrocarbons. However, addition of small amounts of polar compounds (i.e., hydrocarbons containing nitrogen, oxygen, or sulfur) or film-forming compounds to a paraffin hydrocarbon can change wettability. Adsorption of the polar molecules on the solid surface leaves the surface hydrophobic, i.e., dislikes water. Deposition of a hydrocarbon film on the solid surface produces the same effect. Thus wettability of reservoir rocks is strongly affected by the presence or absence of molecules in the crude oil that absorb or deposit on the mineral surfaces in the rock.

Crude oils are complex hydrocarbon mixtures and contain many compounds that may alter wettability.[12] It is usually not possible to determine from chemical analysis whether a crude oil will exhibit water- or oil-wetting tendencies when in contact with a specific surface. As discussed in Sec. 2.3.3, interpretation of indirect measurements and observations are used to infer wetting preference.

Some crude oils will change a water-wet surface to an oil-wet surface on contact.[13] Crude oils from the Ventura D-7[14] and East Texas[15] fields deposit oil films on solid surfaces under certain conditions. Data from geological and core analyses also provide insight into the wetting preference. Microscopic examination of rock grains sometimes reveals hydrocarbon constituents that adhere to sand grains.[16] Asphaltic constituents that are not removed by normal core-cleaning procedures (Ref. 17, Page 13) have been found on rocks like the Tensleep and Bradford sandstones. Some compounds are adsorbed strongly on solid surfaces. Porphyrins (organic metallic complexes) as well as high-molecular-weight paraffinic and aromatic compounds were detected in extracts obtained from prolonged Soxhlet extraction of reservoir cores with toluene.[18] The reservoir cores remained strongly oil wet until heated in an oxidizing atmosphere, presumedly removing the adsorbed hydrocarbons by oxidation to CO_2 and water. These examples show that (1) some crude oils contain compounds that alter wettability of surfaces in reservoir rocks and (2) some reservoir rocks contain hydrocarbons that are so strongly attached to mineral surfaces that an oil-wet preference exists on those surfaces.

Fluid properties are not the only factors influencing wettability of reservoir rocks. Surfaces encountered in reservoirs are often chemically heterogeneous. Fig. 2.8 is a photograph taken through the scanning electron microscope (SEM) of a sand grain from the North Burbank reservoir rock, which is partially coated with clay.[19] The strongly oil-wet character of the North Burbank reservoir rock is attributed to a coating of chamosite clay covering about 70% of the quartz surface.

A combination of crude-oil characteristics and mineralogy of the surface may produce different wettability conditions. The mineralogy in Fig. 2.9a leads to isolated

on a calcite surface. Hydrocarbons in Fig. 2.7 represent constituents known to be present in crude oils. The composition of the hydrocarbon phase and its interaction with the silica surface cause wettability to change. Iso-octane/water (Fluid Pair A) has the same wetting behavior for both silica and calcite surfaces. Fluid Pair B exhibited intermediate wettability on silica but water wettability on calcite. Fluid Pair C (isoquinoline/water) reversed wettability on the calcite surface while Fluid Pair D (naph-

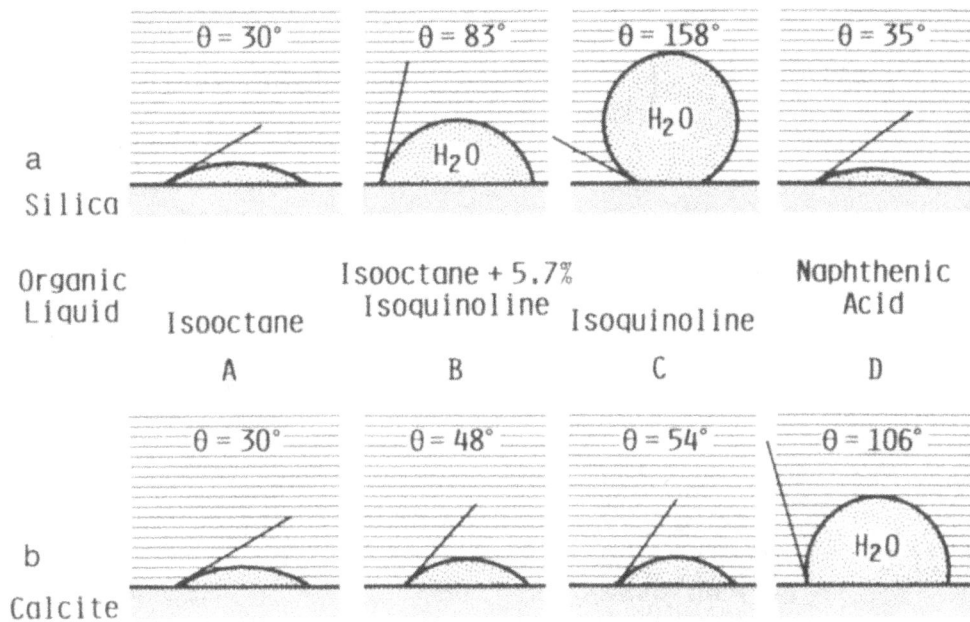

Fig. 2.7—Interfacial contact angles: (a) silica surface and (b) calcite surface.[23]

oil-wettted surfaces, sometimes referred to as a "dalmation" wetting pattern. A mixed-wettability model, depicted in Fig. 2.9b, considers some portions of the reservoir rock to the oil wet while other portions are water wet.[15] Oil films are continuous and provide a path for oil to flow at relatively low oil saturations.

2.2.3 Capillary Pressure

The concept of capillary pressure as a characteristic of a porous rock evolved from the representation of capillary phenomena in capillary tubes. An oil/water interface or an air/water interface in a large tube is flat because the wetting forces at the walls of the tube are distributed over a large perimeter and do not penetrate into the interior to any extent. Therefore, the pressures of the fluids at the interface are equal. Pores in reservoir rocks are analogous to capillary tubes in that the diameters are small. When diameters are small, surface forces induced by preferential wetting of the solid by one of the fluids extend over the entire interface, causing measureable pressure differences between the two fluid phases across the interface. As an example, consider the oil/water interface in the horizontal glass capillary tube in Fig. 2.10, which is at static equilibrium. Water strongly wets the glass surface with a contact angle near zero. If sensitive pressure gauges were attached to each end of the capillary tube to measure the water-phase pressure and the oil-phase pressure, we would observe that the oil-phase pressure is always larger than the water-phase pressure, regardless of the length of the tube. Water can be displaced from the capillary tube by injecting oil into the tube. Oil will be displaced spontaneously from the tube if the pressure of the oil phase is reduced, even though the pressure in the water phase is less than the pressure in the oil phase.

The phenomena described in the preceding paragraph are predictable from the analysis of the forces at the contact line between the interface and the solid surface. Com-

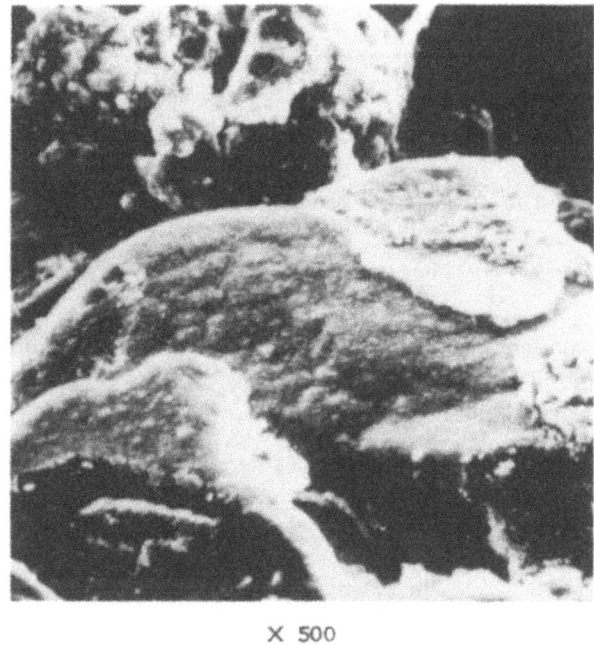

X 500

Fig. 2.8—Photomicrographs of Burbank sand grain showing chamosite coating.[19]

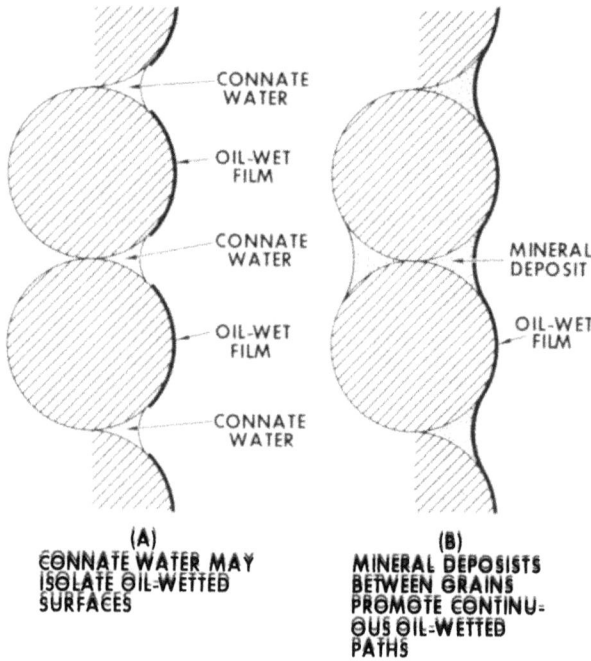

Fig. 2.9—Relationship of mineralogy to wetting conditions: (a) dalmation wetting and (b) mixed wettability.[15]

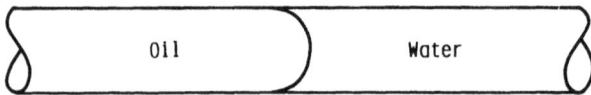

Fig. 2.10—The interface between the oil phase and the water phase in a horizontal water-wet capillary tube.

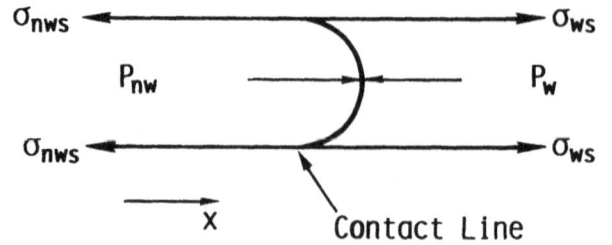

Fig. 2.11—Force balance in the x direction across a fluid interface at static equilibrium.

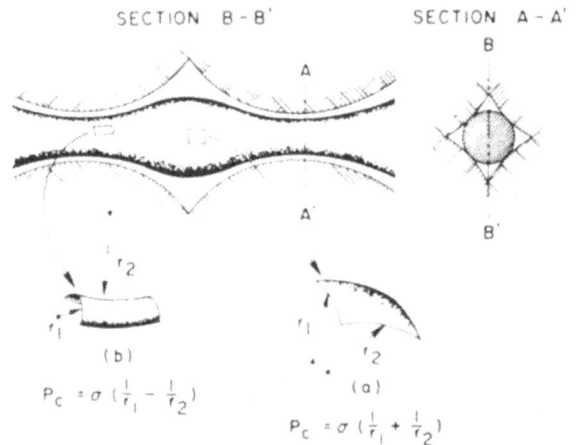

Fig. 2.12—Two types of interfacial curvature.[21]

ponents of the forces acting in the horizontal direction are illustrated in Fig. 2.11.

At equilibrium, $\Sigma F_x = 0$,

$$p_{nw}(\pi r^2) + \sigma_{ws}(2\pi r) - p_w(\pi r^2) - \sigma_{nws}(2\pi r) = 0, \quad (2.2)$$

or

$$p_{nw} - p_w = \frac{2(\sigma_{nws} - \sigma_{ws})}{r}. \quad \ldots \ldots \ldots \ldots (2.3)$$

However, from Eq. 2.1, $\sigma_{nws} - \sigma_{ws} = \sigma_{nww} \cos \theta$ so that

$$p_{nw} - p_w = \frac{2\sigma_{nww} \cos \theta}{r}. \quad \ldots \ldots \ldots \ldots (2.4)$$

The pressure difference in Eq. 2.4 is defined as P_c, the capillary pressure. By petroleum engineering convention, the capillary pressure is $p_o - p_w$ for oil/water systems. Thus P_c is negative for an oil-wet surface.

Eq. 2.4 is a special case of a generalized expression for the pressure difference across an interface that was developed by LaPlace.[20] Eq. 2.5 is the generalized relationship

$$p_o - p_w = \sigma_{ow}\left(\frac{1}{r_1^*} + \frac{1}{r_2^*}\right), \quad \ldots \ldots \ldots \ldots (2.5)$$

where r_1^* and r_2^* are the radii of curvature for the interface. If the radii of curvature are on the same side of the interface, as in the capillary tube, both have positive values. Otherwise, the smaller radius is positive and the larger radius is negative.[21] r_1^* and r_2^* are equal to $r/\cos \theta$ for a uniform, smooth capillary tube of radius r.[22]

Fluid interfaces in porous media are far too complex to apply Eq. 2.4 quantitatively. Even in a simple representation of fluid distribution, many interfaces occur along the pore, as shown in Fig. 2.12, rather than across the pore as might be envisioned from Fig. 2.10.

In Fig. 2.12, different types of curvature are depicted in a pore that contains a continuous interface. At equilibrium, the capillary pressure across the interfaces not forced against the pore wall is the same at every location on the interface. The mean radius of curvature is constant although the surface changes shape. Equal capillary pressures do not represent unique surface shapes. Thus the capillary pressure is a measure of the curvature between the two fluid phases in the porous media.

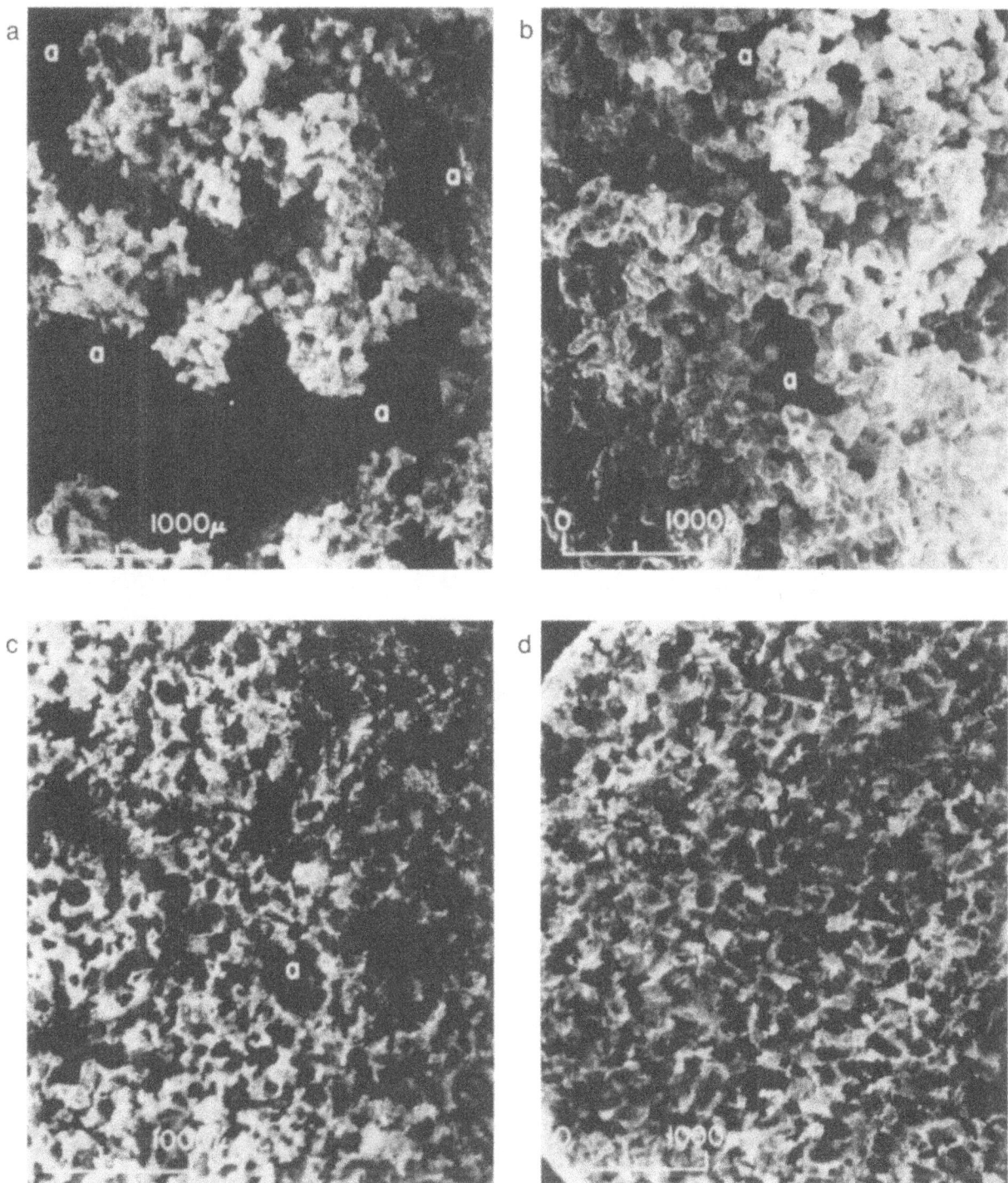

Fig. 2.13—Wood's metal acting as the nonwetting phase in Berea sandstone (low magnification): (a) 22% saturation, (b) 36% saturation, (c) 52% saturation, and (d) 73% saturation. [22]

2.3 Methods of Inferring Fluid Distributions in Porous Media

Fluid distribution is the relative location of two or more phases in the pore space. The distribution of immiscible phases in the pore space of a reservoir rock is of interest in oil recovery processes because it influences both the rates of flow of each phase and the oil recovery efficien-

cy. In this section we examine some methods that are used to infer fluid distributions in porous rocks. The term "infer" is appropriate because direct measurement of fluid distribution is not feasible. In most instances, it is necessary to interpret capillary pressure or wettability data to infer fluid distributions. However, fluid saturation, the volume fraction of a pore occupied by a phase, can be be measured directly.

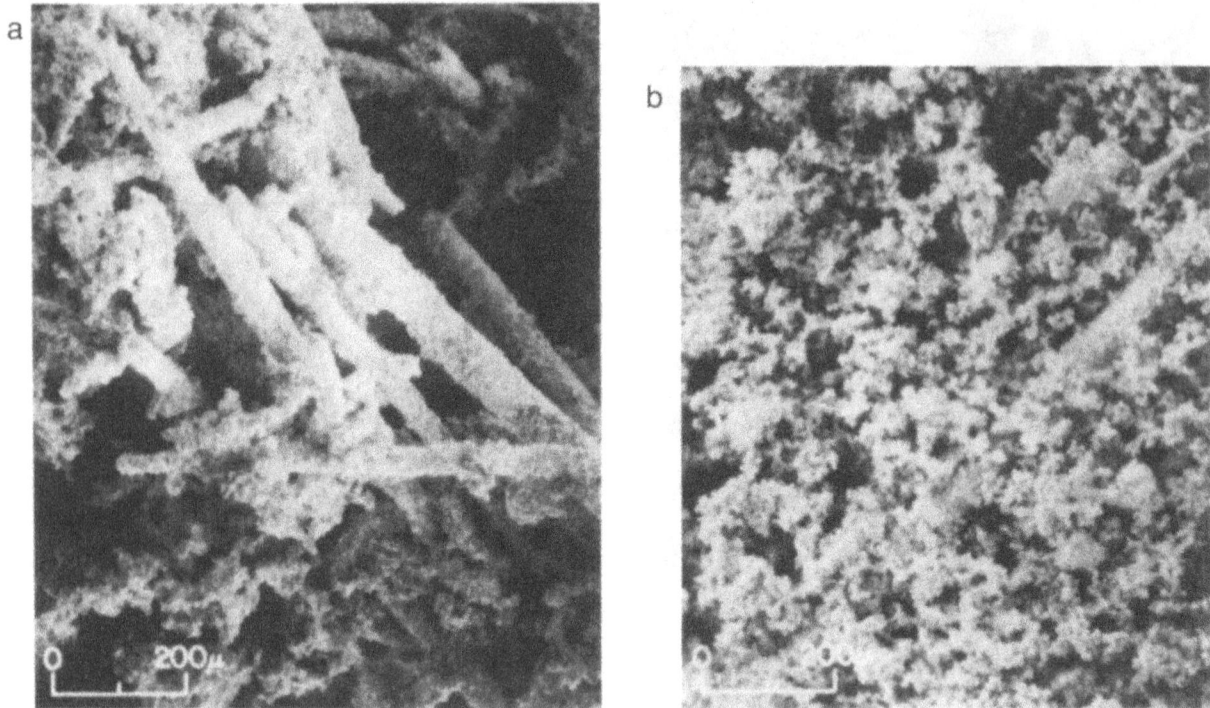

Fig. 2.14—Wood's metal pore casts at two saturations in San Andres dolomite: (a) 12% saturation and (b) 87% saturation.[22]

2.3.1 Replicas of Fluid Distributions

The nearest approximation to direct measurement of fluid distributions combines SEM with replicas of the fluid phase at specified saturations.[22] This technique provides valuable insight into fluid distributions under strong wetting conditions. The examples presented in this section are based on Wood's metal replicas of fluid distributions in sandstone and carbonate rocks. Molten Wood's metal acts as the nonwetting phase while the remaining gas is a strong wetting phase. Replicas of fluid distribution can be obtained by solidifying the Wood's metal at the desired saturation and dissolving the rock from a fresh surface with hydrofluoric or hydrochloric acid. SEM photomicrographs are taken of the Wood's metal replicas.

Figs. 2.13a through 2.13d show SEM photomicrographs of Wood's metal acting as the nonwetting phase in Berea sandstone at saturations of 22, 36, 52, and 73%. Locations marked "a" represent portions of the rock (both grain and pores) where Wood's metal did not penetrate at the particular saturation. At a saturation of 73%, the Wood's metal appears to be present in all pores.

Figs. 2.14a and 2.14b are Wood's metal casts of two saturations in San Andres dolomite. Figs. 2.15a and 2.15b

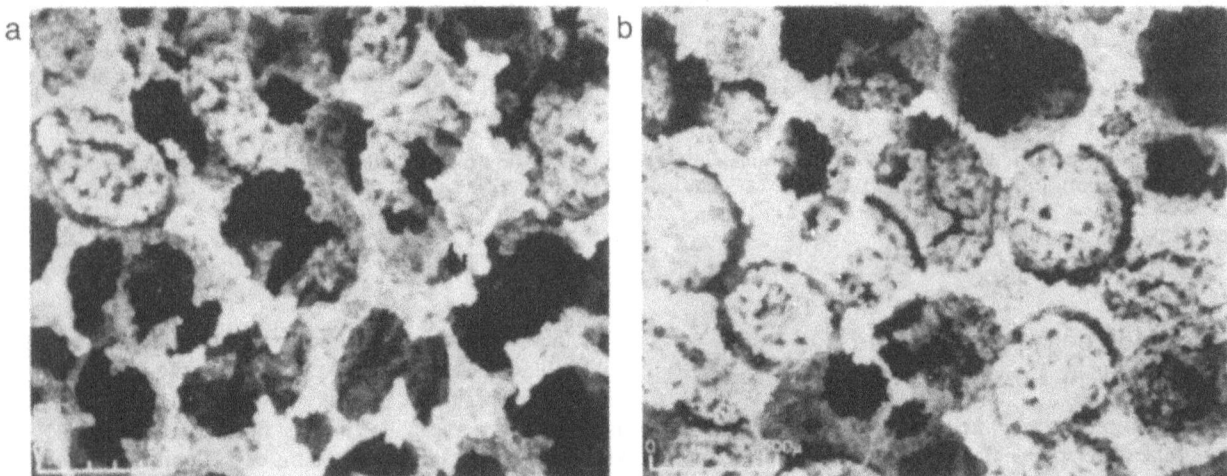

Fig. 2.15—Pore casts of Jurassic Smackover dolomite: (a) 26% saturation and (b) 79% saturation.[22]

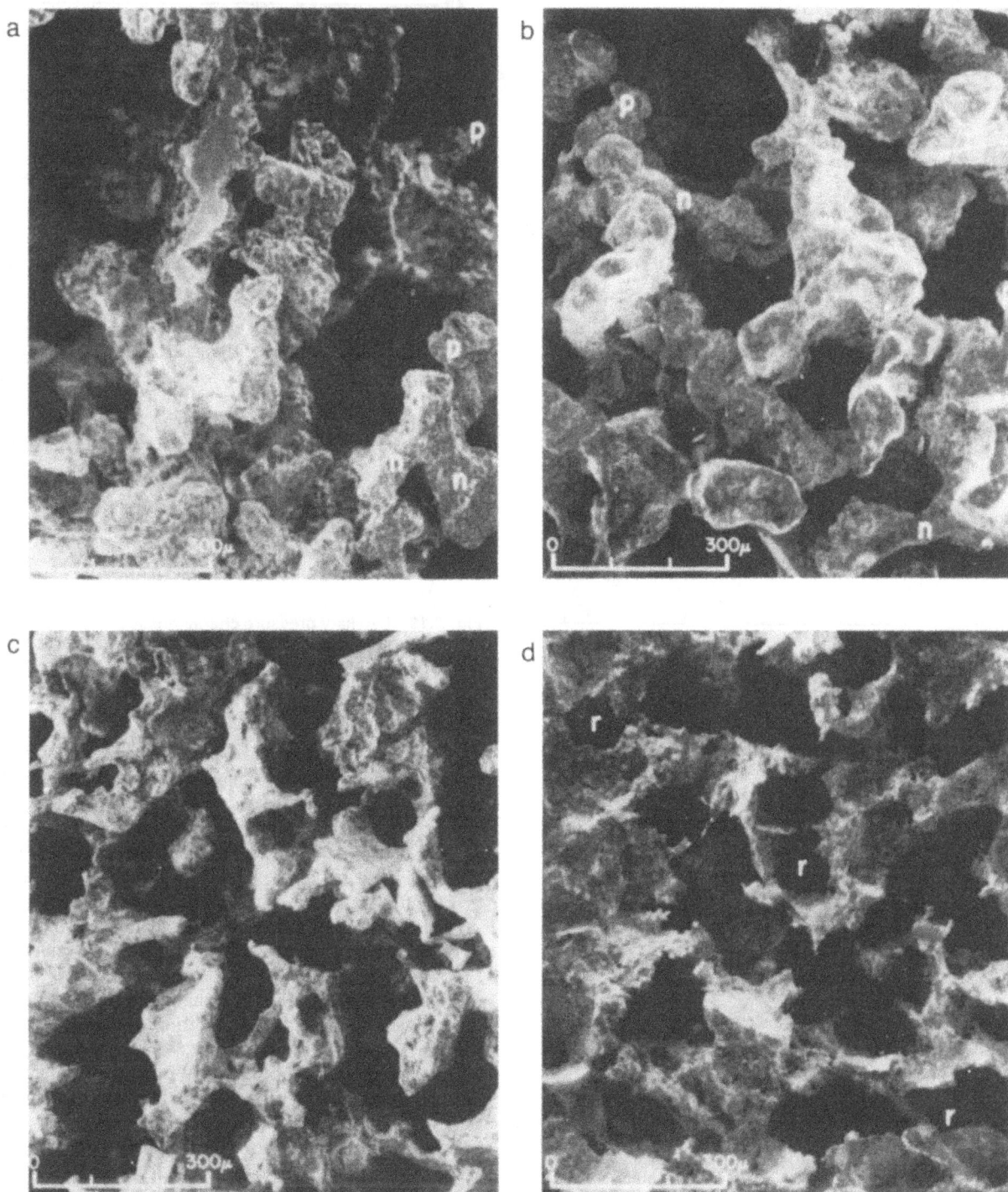

Fig. 2.16—Wood's metal acting as a nonwetting phase in Berea sandstone (high magnification): (a) 22% saturation, (b) 36% saturation, (c) 52% saturation, and (d) 73% saturation. [22]

show pore casts of Jurassic oomoldic dolomite. These photomicrographs as well as others in Ref. 22 illustrate the complex shape of the nonwetting phase at different saturations as well as the influence of pore structure (i.e., rock type) on fluid distribution in reservoir rocks.

2.3.2 Interpretation of Capillary Pressure Data

The measurement of capillary pressures in porous media has been discussed in Refs. 23 through 29 and will not

be covered here. We are concerned with the interpretation of capillary pressure data to provide information on fluid distributions.

High-magnification photomicrographs of Figs. 2.13a through 2.13d are shown in Figs. 2.16a through 2.16d. In these photographs, the Wood's metal surfaces are rounded at 22% saturation, but become increasingly angular at high saturations. This indicates that the Wood's metal penetrates progressively smaller pore spaces as the

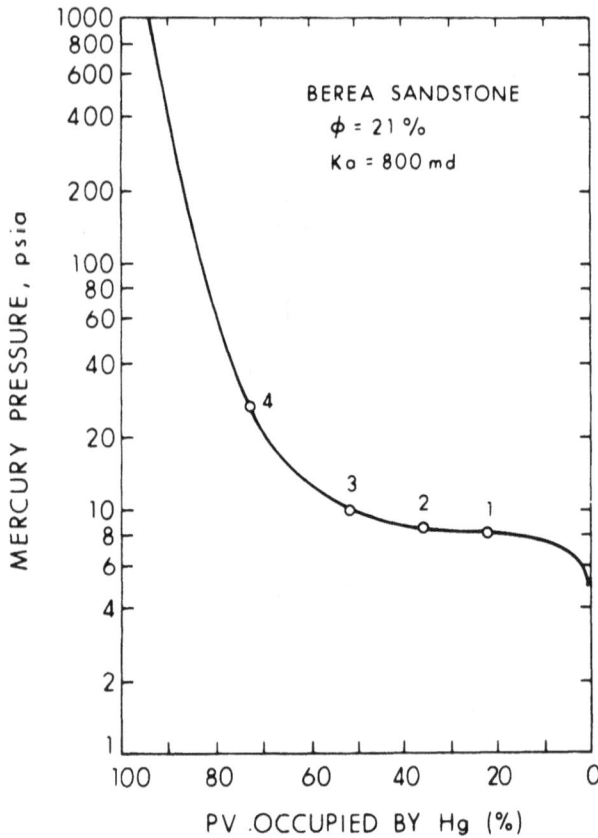

Fig. 2.17—Mercury capillary pressure curve applying to pore casts in Fig. 2.13. [22]

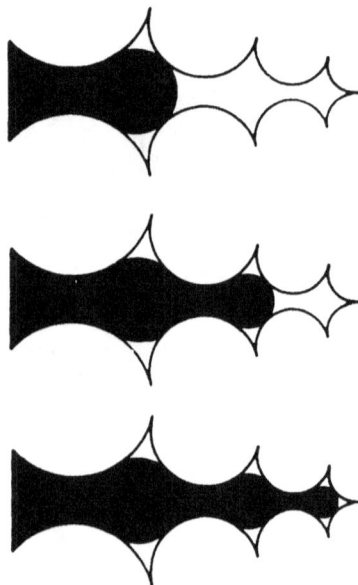

Fig. 2.18—Penetration of mercury into connected pores. [30]

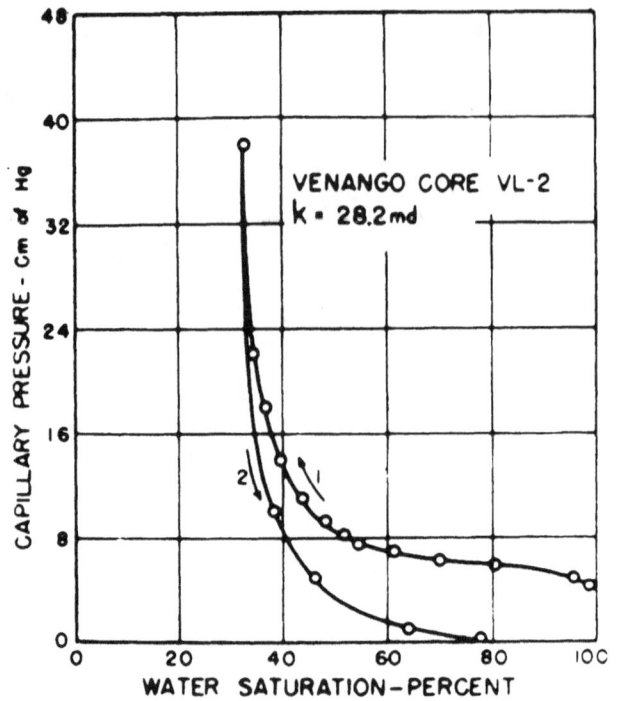

Fig. 2.19—Capillary pressure characteristics, strongly water-wet rock. Curve 1, drainage and Curve 2, imbibition. [33]

saturation increases, further confirmation that the wetting phase (in this case, air) occupies the smallest pores and crevices.

Sharpness of the interfaces also means that the radius of curvature of the interface has decreased. The capillary pressure increases as the radius of curvature decreases, as shown in Eq. 2.5. Thus, a change in capillary pressure for a known wetting condition is an indirect method of observing the change in interfacial curvature with saturation.

A continuous record of the change in the pressure of the nonwetting phase with saturation can be obtained if mercury is used as the nonwetting fluid. Since mercury has wetting properties similar to Wood's metal, the mercury distributions in Berea cores would be essentially identical to those in Figs. 2.16a through 2.16d at the same saturations. Fig. 2.17 shows how the mercury pressure increases with percentage of the PV occupied by the mercury. Points 1 through 4 correspond to the saturations in Figs. 2.16a through 2.16d.

Fig. 2.18 depicts the penetration of mercury into progressively smaller pores as the mercury saturation increases during a capillary pressure measurement. [30] By comparing the saturation history of mercury with the SEM photographs of the Wood's metal replicas, we see that the pressure-saturation history provides an indirect measure of fluid distributions as well as the size and shape of the smallest pores and crevices that are penetrated. Capillary pressure data observed with mercury may be interpreted to determine the distribution of pore entry radii. [31,32]

Capillary pressure data must be interpreted with caution because capillary pressure curves exhibit hysteresis. Saturations for a specific capillary pressure depend on the

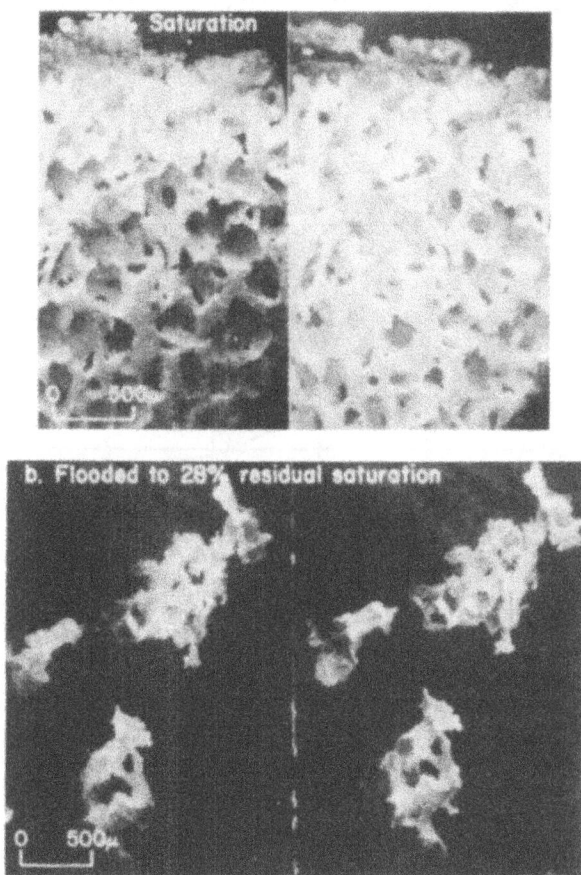

Fig. 2.20—Pore casts of Miocene sandstone (a) at initial saturation and (b) when flooded to residual Wood's metal. [22]

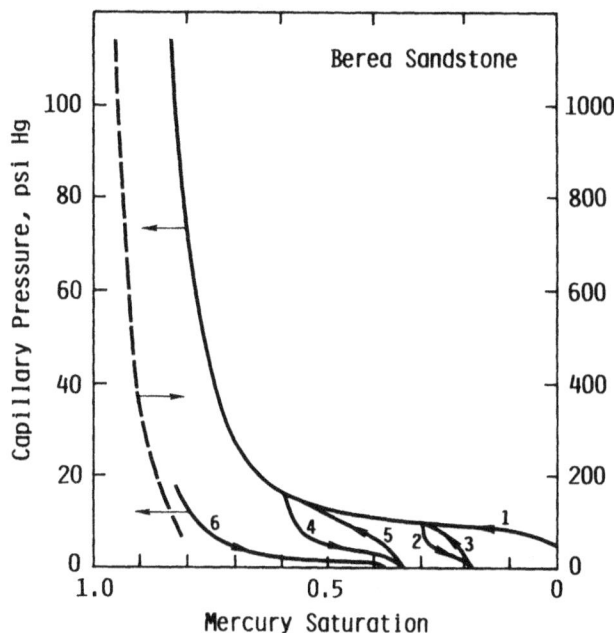

Fig. 2.21—Capillary pressure curve with hysteresis and scanning loops. [30]

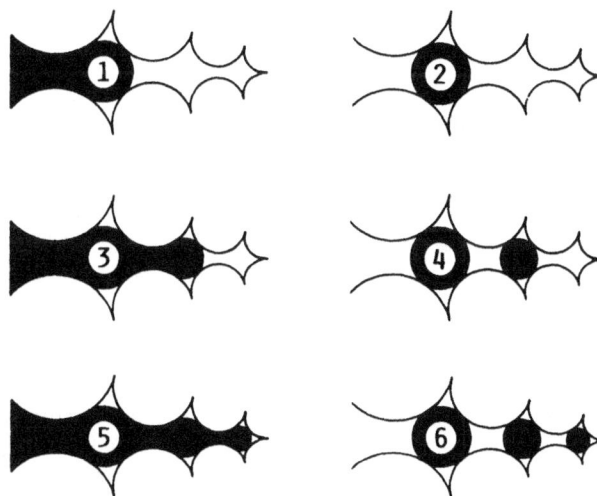

Fig. 2.22—Cross section of pore network normal to direction of flow showing fluid distributions at points on paths of Fig. 2.21. [30]

direction of saturation change. Fig. 2.19 shows capillary pressure data for a strongly water-wet rock with hysteresis. Curve 1, the drainage curve, was obtained as the water was displaced from the core by successive increases in the pressure of the nonwetting phase. Curve 2 represents the capillary pressure data when water imbibed into the core, thereby expelling the oil. Recall that, in the petroleum engineering literature, imbibition refers to the process in which the saturation of the wetting phase increases. Imbibition is often spontaneous (i.e., requires no additional applied pressure) in a strongly water-wet rock. Injection of water into a water-wet rock causes the water saturation to increase and is also called "imbibition."

Capillary pressure curves represent equilibrium states. As long as the same saturation path is followed, a unique value of the capillary pressure exists for each water saturation. It makes no difference whether the wetting-phase saturation increases spontaneously or whether the pressure on the wetting phase is increased, forcing the wetting phase into the rock. If a strongly water-wet rock saturated with 35% water, as in Fig. 2.19, is immersed in water, the oil phase will be displaced by spontaneous imbibition of water until the oil phase becomes discontinuous throughout the rock. When capillary equilibrium is reached, the water saturation will be about 78%, and the apparent capillary pressure will be zero. This may take a long time. Inside the rock, the nonwetting phase is stranded as isolated ganglia, such as those shown in Fig. 2.20. Fig. 2.20 contains stereo SEM photomicrographs

of residual Wood's metal ganglia in a Miocene sandstone. [22] Ragged edges of the ganglia are apparent, indicating small mean radii of curvature. Let us visualize the Wood's metal to be similar to the oil phase in the example discussed previously. From Eq. 2.5, the difference between wetting and nonwetting phase pressure is not zero at the ganglia interfaces. What has happened is that these ganglia no longer have a hydraulic connection to the oil phase outside the rock. At this saturation, the oil-phase pressure cannot be transmitted to the instrument used to measure oil-phase pressure. The only pressure measured is the water-phase pressure; thus the capillary pressure appears to be zero.

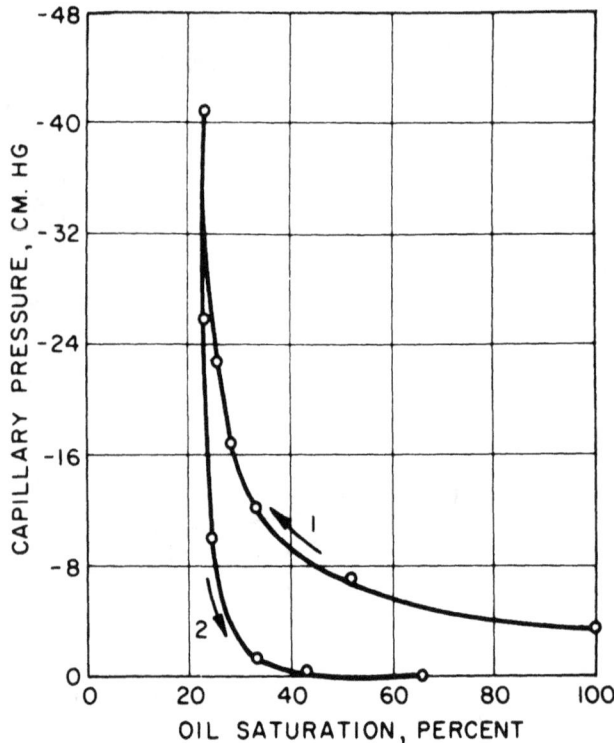

Fig. 2.23—Oil/water capillary pressure characteristics, Tensleep sandstone, oil-wet rock. Curve 1, drainage and Curve 2, imbibition.[33]

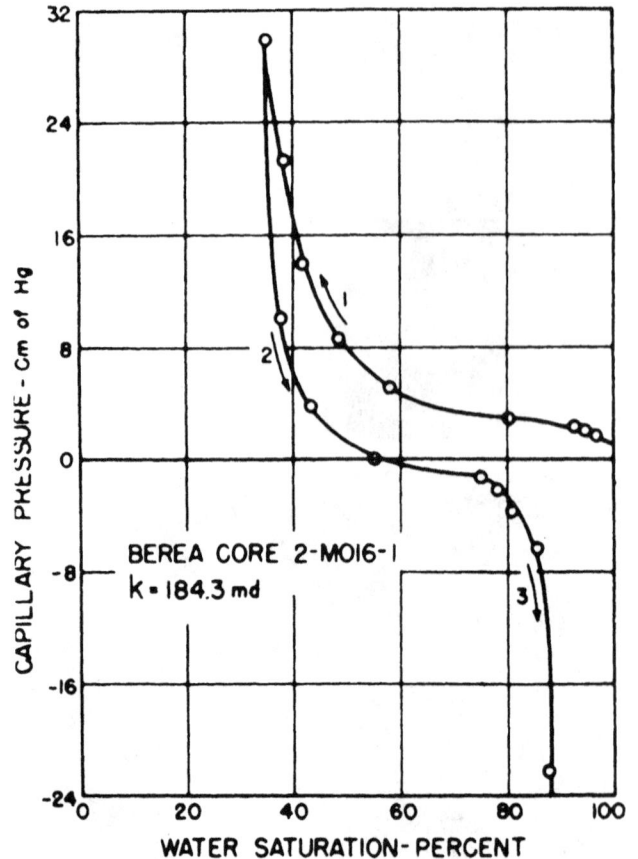

Fig. 2.24—Oil/water capillary pressure characteristics, intermediate wettability. Curve 1, drainage, Curve 2, spontaneous imbibition, and Curve 3, forced imbibition.[33]

The drainage capillary pressure curve in Fig. 2.19 exists over a much larger range of saturation change than the imbibition curve. Drainage capillary pressure data indicate different radii of curvature at the interface as well as more connectivity of the nonwetting fluid in the pore space—that is, to measure the capillary pressure at 90% saturation in Fig. 2.19, there must be at least one continuous hydraulic connection or flow path for the oil phase in the pore space.

If the oil-phase pressure is increased at some point along the imbibition curve in Fig. 2.19, the drainage capillary pressure curve will be retraced.[30] Drainage and imbibition capillary pressure curves are connected by a series of scanning loops similar to those shown in Fig. 2.21 for mercury in Berea sandstone. Loops in Fig. 2.21 were obtained following the saturation path 1-2-3-4-5-6. Scanning loops also exhibit hysteresis. Saturations depicted in Fig. 2.22 show how mercury might be trapped in pores of various sizes at the end of Paths 2, 4, and 6.

Hysteresis in capillary pressure curves is reproducible and well documented. Several authors treat this subject in more detail than required for this text, and the reader is referred to these references for further information. For our purposes, it is sufficient to understand hysteresis as a measure of shapes of fluid interfaces and the degree of interconnectedness along the drainage and imbibition paths.

Oil-wet rocks have capillary pressure behavior similar to Fig. 2.23 when water or gas is the nonwetting phase.[33] For an oil-wet rock, imbibition refers to the increase of the oil saturation, while drainage refers to increase in the water saturation. In the case of an oil-wet system, the capillary pressure is negative because of the petroleum engineering convention that defines the capillary pressure in an oil/water system to be the difference between the pressure of the oil phase and the water phase.

If the reservoir rock has intermediate wettability, a capillary pressure curve resembling Fig. 2.24 may be obtained.[33] Note that the oil phase does not become disconnected at zero capillary pressure. Fig. 2.24 shows that the water saturation can be increased to about 88% by increasing the water-phase pressure, and simultaneously displacing oil from the pore space.

Drainage capillary pressure data represent the path followed when oil migrates into a reservoir that is 100% water saturated. Therefore, drainage capillary pressure data are used to estimate connate water saturation and, consequently, the initial oil saturation. At the time of oil migration, the reservoir rock is strongly wetted by water. We know that wettability of a porous rock can be altered over long periods of time by adsorption of polar compounds from crude oil. Unless the reservoir rock remains strongly water wet, it is probable that the capillary pressure curve for the initial oil saturation is different from the curve for the reservoir rock when discovered. Thus there are uncertainties when connate water saturation is estimated from drainage capillary pressure data.

In summary, capillary pressure curves are a measure of fluid distribution, saturations, wettability, and connectedness in a porous rock. They also represent an experimental correlation of the pressure difference between two phases at equilibrium in a porous rock.

2.3.3 Determination of Wettability of Reservoir Rocks

Many methods have been investigated to measure or to infer wettability of reservoir rocks in the presence of reservoir crude oil and formation brine. These methods can be broadly divided into two categories. One category is based on experiments using uncontaminated crude oil and the dominant mineral present in the reservoir rock. The second approach uses displacement data on reservoir cores that have been obtained under conditions intended to retain original wettability. Craig (Ref. 17, Page 13) reviewed a number of these methods.

No method has been universally accepted as a unique measure of wettability. In general, the methods tend to give similar indications of wettability for rocks that are strongly oil wet or strongly water wet. Two approaches will be examined in this text: contact-angle measurement and imbibition displacement methods.

2.3.3.1 Wettability From Contact-Angle Studies.

The contact-angle method is based on the premise that wettability of reservoir rocks is governed by the presence or absence of polar or film-forming compounds in crude oil that adsorb or deposit on the mineral surface. Wettability is determined by measuring the contact angle after the uncontaminated reservoir crude oil and simulated brine are exposed to the dominant mineral present in the reservoir rock. This is usually quartz, representing the dominant mineral of sandstone reservoirs, or calcite, representing limestone reservoir rock.

Fig. 2.25a depicts one method that has been used to measure the contact angle.[10] (See also Ref. 17, Page 13.) A drop of reservoir crude oil immersed in simulated reservoir brine is confined between parallel surfaces of a crystal that is chemically homogeneous and microscopically smooth. Fig. 2.25a shows the initial condition for the experiment (Ref. 17, Page 13). During the experiment, the lower surface is moved as in Fig. 2.25b, creating an instability in interfacial forces. To re-establish equilibrium of the interface, the oil phase attempts to displace the water phase from the surface at Point A while the water phase tries to advance by displacing the oil from the surface previously covered by the oil phase at Point B. Two contact angles can be measured at the oil/brine/mineral interface on the stationary crystal. However, only the water-advancing contact angle measured at Point B correlates with other wettability indicators.[10]

The water-advancing contact angle changes with the aging of the oil/mineral/brine interface. As the water advances over a previously oil-contacted surface, the polar or film-forming compounds that adsorbed or deposited on the surface may be displaced by water or desorbed. At the same time, a rate-dependent adsorption or deposition process may occur between the solid surface and the crude oil at locations away from the oil/brine/mineral contact point. In some samples, a great length of time is required before an equilibrium contact angle is obtained, and large changes can occur between the initial contact angle and the final measurement. Fig. 2.26 shows the change in water-advancing contact angle with interface age for several oil/brine/mineral systems.

Contact angles measured for 55 oil-producing reservoirs are summarized in Table 2.2. A wettability scale may be used to classify these contact-angle data. The scale is ar-

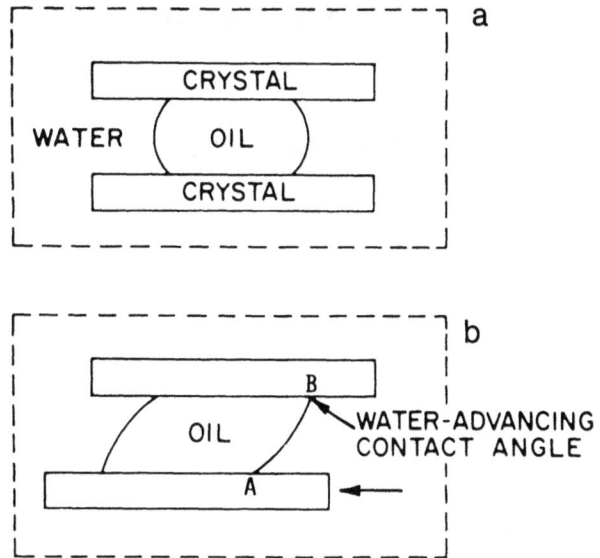

Fig. 2.25—Contact angle measurement (after Ref. 17, Fig. 2.2).

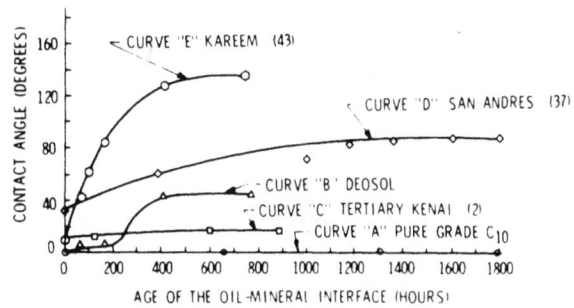

Fig. 2.26—Typical contact angle wettability situations.[10]

bitrary except at the endpoints (0 and 180°). In Table 2.2, Treiber et al. define water-wet behavior for contact angles 0 to 75°, intermediate wettability 75 to 105°, and oil wettability 105 to 180°. Fig. 2.27 shows the distribution of these data.[34] A wide distribution of wettabilities is indicated with a distinct cluster of oil-wet systems for the particular set of reservoirs investigated. The arbitrary nature of any wettability scale is evident. If another scale is defined[32] with intermediate wettability designated by contact angles between 62 and 133°, 26 of these reservoirs (47%) fall in the intermediate wettability category.

Also shown in Table 2.2 is the wettability inferred from relative permeability measurements (Sec. 2.4.3.3), on native-state cores. Of the 22 reservoirs investigated, 18 have the same wettability, as judged from fluid-flow measurements.

The contact-angle method has some limitations. Long times are required for equilibrium to be reached. Laboratory time is but a fraction of the millions of years available for equilibration in a reservoir. If adsorption occurs at a slow but undetectable rate, systems that appear water wet in a contact cell may exhibit more oil wetness in the reservoir environment. Other properties of the reservoir rock influence wettability. A contact angle measured on a smooth, homogeneous surface does not reflect surface

TABLE 2.2—SUMMARY OF RESERVOIR AND WETTABILITY INFORMATION[10]

State	Formation	Petrology*	Minerals**	Contact Angle (degrees)	Oil/Mineral Interface Age (hours)	Temperature (°F)	°API	Wettability† Contact Angle	Wettability† Relative Permeability
1. Alaska	Conglomerate	CONGL.	Q	0	800	155	36	WW	WW
2. Alaska	Tertiary Kenai Sand	SS	Q, C, Fe	15, 145, 20	900, 1,500, 900	136	38	WW	WW
3. Argentina	I and L Sands	SS	Q	20	1,300	165	—	WW	I
4. Argentina	M Sand	SS	Q	10	810	165	—	WW	WW
5. Canada	Beaverhill Lake	LS	C	140	450	225	38	OW	OW
6. Colorado	Weber	SS	Q, C	140, 145	960, 860	160	34	OW	OW
7. Iran	Burgan Sand	SS	Q	140	1,060	152	16	OW	OW
8. Louisiana	Patin Sand	SS	Q	50	1,490	170	37	WW	WW
9. Nebraska	Muddy "D"	SS	Q	10	350	200	36	WW	—
10. New Mexico	Abo Reef	DT	C	37	1,070	109	43	WW	—
11. New Mexico	Gallup	SS	Q	120	600	158	—	OW	—
12. New Mexico	Grayburg	DT	C	130	600	102	34	OW	—
13. New Mexico	San Andres	DT	C	120	1,000	100	27	OW	—
14. North Dakota	Heath	SS	C, Q	150, 100	310, 1,200	188	36	OW	—
15. North Dakota	Madison	LS	C	150	430	150	39	OW	—
16. North Dakota	Madison	LS	C	130	190	165	37	OW	—
17. North Dakota	Sanish	SS	Q	140	440	240	44	OW	—
18. Oklahoma	Deese, Eason	SS	Q	100	1,800	112	32	I	—
19. Oklahoma	Deese, Tussy	SS	Q	125	1,000	112	—	OW	—
20. Oklahoma	Humphrey	SS	Q	100	2,000	116	—	I	—
21. Oklahoma	Wilcox	SS	Q	160	750	160	36	OW	—
22. Texas (Gulf)	Glen Rose	DT	C	140	450	140	—	OW	—
23. Texas (Gulf)	Upper Frio	SS	Q	40	1,600	162	30	WW	WW
24. Texas (West)	Canyon Reef	DT	C	140	330	151	42	OW	I
25. Texas (West)	Clearfork	DT	C	125	750	109	30	OW	OW
26. Texas (West)	Clearfork	DT	C	140	950	110	40	OW	—
27. Texas (West)	Clearfork	DT	C	127	440	107	26	OW	OW
28. Texas (West)	Devonian Chert	CRT	Q	125	1,000	120	42	OW	OW
29. Texas (West)	Ellenburger	DT	C	145	450	204	50	OW	—
30. Texas (West)	5,600 ft.	DT	C	140	400	105	—	OW	—
31. Texas (West)	Grayburg	DT	C	130	580	94	36	OW	—
32. Texas (West)	Grayburg	DT	C	120	500	90	35	OW	OW
33. Texas (West)	Grayburg	DT	C	135	300	100	29	OW	OW
34. Texas (West)	Grayburg	DT	C, S	125, 45	650, 1,270	100	34	OW	OW
35. Texas (West)	Holt	DT	C	130	470	100	38	OW	—
36. Texas (West)	Lower Holt	DT	C	0	500	120	—	WW	WW
37. Texas (West)	San Andres	DT	C	88	2,400	115	30	I	OW
38. Texas (West)	San Andres	DT	C	130	600	106	33	OW	OW
39. Texas (West)	Sandy Dolomite	DT, SS	C, Q	125, 145	2,000, 1,550	90	—	OW	—
40. Texas (West)	Spraberry	SS	Q	55	1,820	108	39	WW	—
41. Texas (West)	Spraberry	SS	Q	50	1,100	115	34	WW	—
42. Texas (West)	Upper Holt	DT	C	125	300	120	—	OW	—
43. UAR	Kareem	SS	Q, S, C	160, 140, 135	1,900, 1,300, 740	174	30	OW	OW
44. Utah	Weber	SS	Q	130	500	139	32	OW	—
45. Wyoming	Dakota	SS	Q	0	810	124	33	WW	—
46. Wyoming	Madison	LS	C	140	330	100	33	OW	—
47. Wyoming	Minnelusa	SS	Q	130	630	134	27	OW	—
48. Wyoming	Minnelusa	SS	Q	135	600	170	18	OW	—
49. Wyoming	Muddy	SS	Q	0	870	130	—	WW	—
50. Wyoming	Nugget	SS	Q	0	1,200	80	14	WW	OW
51. Wyoming	Phosphoria	DT	C	115	520	88	24	OW	OW
52. Wyoming	Phosphoria	DT	C	130	400	130	25	OW	—
53. Wyoming	Tensleep	SS	Q	125	890	135	25	OW	—
54. Wyoming	Tensleep	SS	Q	120	480	100	34	OW	—
55. Wyoming	Tensleep	SS	Q	70	2,255	96	24	WW	—

*SS = sandstone, CONGL. = conglomerate, LS = limestone, DT = dolomite, and CRT = chert.
**Q = quartz, C = calcite, S = selemite, and Fe = siderite.
†WW = water-wet, I = intermediate wettability, and OW = oil-wet.

roughness of the reservoir rock,[35] pore geometry, hysteresis in contact angle,[36] or heterogeneity. Finally, measurement of the contact angle requires meticulous experimental technique. All of these limitations are moderated by the fact that uncontaminated samples of reservoir crude oil can be obtained with more certainty and at less expense than reservoir cores. For some reservoirs, contact-angle measurement may be the only feasible method to obtain some indication of reservoir wettability.

2.3.3.2 Wettability From Displacement Studies. Wettability may be inferred from displacement experiments conducted on carefully preserved cores. Procedures for obtaining and preserving native-state cores are discussed by Craig (Ref. 17, Page 22) and will not be considered further in this text. Laboratory tests are usually conduct-

ed with refined oils and simulated formation brine. In these tests, it is assumed that wettability established by the components adsorbed on the rock surface is not altered during the tests. In this section we examine imbibition/displacement tests for determining wettability. Wettability inferred from relative permeability tests is discussed in Sec. 2.4.3.3.

A simple wettability test devised by Amott[37] is based on a comparison of the amounts of oil and water that imbibe into a core to the maximum amounts of oil or water that can be forced into a core under a simulated displacement of one phase by the other. The term "imbibition" as used in this section refers to the spontaneous displacement of a portion of one phase by another. For example, water will imbibe into (spontaneously displace oil from) a strongly water-wet rock when a core saturated with oil is immersed in water. In a porous rock that is strongly oil wet, oil will imbibe, spontaneously displacing water from a water-saturated core. Imbibition occurs because of the strong preference of the rock surface for the wetting phase.

Imbibition is unsteady and some period of time is required for equilibrium to be attained.[38] In strongly wet systems, the saturation of the displaced phase remaining after imbibition of the wetting phase may be nearly equal to the residual saturation of that phase in a coreflood experiment. Residual saturations are discussed in more detail in Sec. 2.5. Rocks that exhibit intermediate wettability would be expected to imbibe small amounts of oil and/or water. In these systems, there may be no clearly defined wetting phase. Thus use of the term imbibition to define an increase in wetting-phase saturation should be used with care for intermediate-wettability systems.

The Amott[37] wettability test is conducted on a reservoir core that has been flushed with kerosene (filtered through silica gel to remove polar compounds) and water to remove most of the crude oil and formation water. Then the core is centrifuged under water to displace the kerosene saturation to residual. The test consists of four displacements. The sequence of displacements follows.

Amott Wettability Test. Initial condition of core: fully saturated with water and kerosene at residual kerosene saturation.

Test 1: determine volume of water that is spontaneously displaced from the core when it is immersed in kerosene for 20 hours.

Test 2: determine additional volume of water that is displaced from the core when the core from Test 1 is centrifuged under kerosene until water production ceases.

Test 3: immerse core from Test 2 under water for 20 hours and determine the volume of oil spontaneously displaced by water.

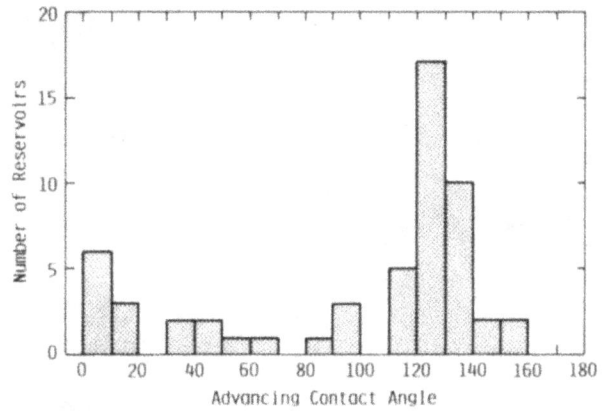

Fig. 2.27—Wettability distribution given by contact angle measurements on smooth mineral surfaces for fluids from 55 reservoirs.[34]

Test 4: centrifuge core from Test 3 under water and determine the volume of oil displaced when oil saturation is reduced to a residual level.

Two wettability indexes (A_w and A_o) can be defined from the data and are presented as follows:

Water wetting preference: A_w = volume of water spontaneously imbibed (Test 3) divided by volume of water at residual oil saturation (Test 4).

Oil wetting preference: A_o = volume of oil spontaneously imbibed (Test 1) divided by volume of oil at residual water saturation (Test 2).

Each index measures the fraction of the displaceable volume that can be displaced spontaneously in 20 hours. Thus $A_w = 1$ and $A_o = 0$ for a strongly water-wet rock while $A_w = 0$ and $A_o = 1$ for a strongly oil-wet rock. A modified method using reservoir crude oil and formation

TABLE 2.3—WETTABILITY TEST RESULTS IN FRESH CORES[37]

Core Description	Displacement by Water Ratio, A_w	Displacement by Oil Ratio, A_o	Relative Displacement Index
Fired Ohio sandstone, Core 1	0.88	0.00	0.88
Fired Ohio sandstone, Core 2	0.93	0.00	0.93
Silicon-treated Ohio sandstone, Core 1	0.00	0.84	−0.84
Silicon-treated Ohio sandstone, Core 2	0.00	0.65	−0.65
Well 1, Oil Zone A, San Joaquin Valley, CA, Core 1	0.12	0.00	0.12
Well 1, Oil Zone A, San Joaquin Valley, CA, Core 2	0.12	0.00	0.12
Well 2, Oil Zone A, San Joaquin Valley, CA, Core 1	0.17	0.00	0.17
Well 2, Oil Zone A, San Joaquin Valley, CA, Core 2	0.14	0.00	0.14
Well 3, Oil Zone A, San Joaquin Valley, CA, Core 1	0.11	0.00	0.11
Well 3, Oil Zone A, San Joaquin Valley, CA, Core 2	0.12	0.00	0.12
Well 4, gray sand below Oil Zone A, San Joaquin Valley, CA, Core 1	0.90	0.00	0.90
Well 4, gray sand below Oil Zone A, San Joaquin Valley, CA, Core 2	1.00	0.00	1.00
Oil Zone B, Sterling County, TX, Core 1	0.94	0.00	0.94
Oil Zone B, Sterling County, TX, Core 2	1.00	0.00	1.00
Bradford Third Zone, Bradford field, PA, Core 1	0.00	0.79	0.79
Bradford Third Zone, Bradford field, PA, Core 2	0.00	0.77	0.77
Bradford Third Zone, Bradford field, PA, Core 3	0.00	0.82	0.82
Bradford Third Zone, Bradford field, PA, Core 4	0.00	0.80	0.80
Oil Zone C, Ochiltreo County, TX, Core 1	0.42	0.00	0.42
Oil Zone C, Ochiltreo County, TX, Core 2	0.36	0.00	0.36
Oil Zone D, Chaves County, NM, Core 1	0.10	0.00	0.10
Oil Zone E, Alta., Canada, Core 1	0.35	0.00	0.35
Oil Zone E, Alta., Canada, Core 2	0.26	0.00	0.26
Oil Zone E, Alta., Canada, Core 3	0.24	0.00	0.24
Oil Zone E, Alta., Canada, Core 4	0.25	0.00	0.25

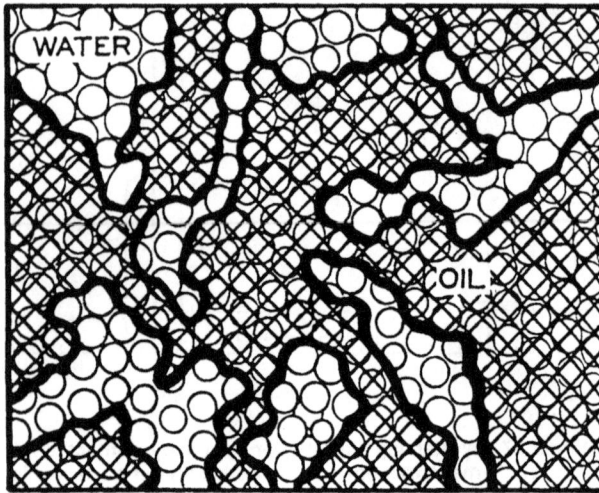

Fig. 2.28—Boundaries of flow channels for simultaneous flow of oil and water. [34]

water is described by Boneau and Clampitt. [39] A relative displacement index I_{rd}, defined by Eq. 2.6, may be introduced to express wettability with a single numerical value [39]:

$$I_{rd} = A_w - A_o. \quad \dots \dots \dots \dots \dots \dots (2.6)$$

According to this index, intermediate wettability would be indicated by I_{rd} near zero. Wettability test results for several fresh cores are shown in Table 2.3. [37]

The Amott method is relatively quick, with analysis times of a few days compared to months for contact angle measurement. The USBM method [4] characterizes wettability in terms of an index derived from capillary pressure measurements in a centrifugal field. Both are reproducible and provide some measure of the effect of pore geometry on wettability. On the other hand, there is the inherent uncertainty of whether or not the wettability of the core measured in the laboratory environment is identical to that which would be present when the rock was in the reservoir. There are substantial concerns about both contamination of the core with drilling fluids as well as change in wettability from exposure to air or during core-cleaning processes. Ample evidence of potential difficulties has been presented in other studies. [37-47] (See also Ref. 17, Page 14.)

The results obtained from the contact angle method [10] have important implications for wettability tests conducted by displacement procedures. If desorption of components that alter wettability occurs at the same relative rate as adsorption, refined oils will give correct indications of wettability in core tests because there is not enough time for desorption of these materials to occur. Two other observations follow. First, wettability tests on preserved reservoir cores should be conducted with minimal storage time since contamination by coring fluids is likely. Storage of carefully preserved cores for several months may alter wettability. Second, displacement tests conducted on outcrop core samples will not give valid indication of crude-oil wettability unless adsorption of polar or film-forming constituents occurs rapidly. The data in Fig. 2.26 imply long equilibration times for some oils and, hence, incorrect wettability evaluation for short tests on outcrop samples.

2.4 Principles of Multiphase Flow in Porous Media

2.4.1 Multiphase Fluid-Flow Concepts

Displacement of oil from a porous material by water or gas involves the simultaneous flow of two or more immiscible phases. In this section we describe multiphase fluid flows in porous media to form a basis for mathematical models developed in Sec. 2.4.2.

Most of our understanding of multiphase flow in porous materials is the result of several studies in the late 1940's and early 1950's. [48-55] These studies were conducted to identify fluid-flow patterns during multiphase flow through porous media. Multiphase flows were observed microscopically in flow cells containing glass beads, or thin sections of reservoir rock. These studies were essentially 2D because the top and bottom of the flow cells were one to three grain diameters apart.

2.4.1.1 Channel Flow. In the microscopic experiments, oil and water flowed simultaneously at constant overall rates. Each phase was observed to flow through its own network of interconnected, tortuous channels as might be visualized from Figs. 2.13 through 2.16. Boundaries of the channels varied from solid/liquid interfaces to liquid/liquid interfaces as depicted in Fig. 2.28. [48] Flows within a channel followed streamlines that conformed to the shape of the channel. Eddy (circulating) flow patterns were not observed. The width of a flow channel varied from less than one grain diameter to several grain diameters. As the saturation of one phase increased, the widths of the channels containing that phase increased. Channel flows were observed at all velocities anticipated in reservoirs.

These characteristics of two-phase flow and fluid distributions in porous media were confirmed in a unique series of experiments in unconsolidated sandpacks. [54] The displacement of oil by water from a water-wet, unconsolidated sandpack was simulated by use of colored plastic as the nonwetting phase and molten Wood's metal as the wetting phase. Fluid distributions were captured by solidifying the Wood's metal and plastic in the core at selected saturations and intervals in the displacement process. Then, motion pictures of the core were made as thin sections were continuously removed with a saw. The motion pictures revealed meandering channels for both phases, analogous to the observations in glass bead packs. Some channels were continuous throughout the porous media. Other channels were dead-end branches connected to continuous channels.

In summary, multiphase flow of water, oil, and gas at or well above velocities attainable in reservoirs occurs through a series of continuous but tortuous channels. Each phase flows through its own channels. As a consequence of this model, there must be one or more continuous flow paths in the pore space from one end of the porous media to another for a phase to flow through a reservoir material. Extensive experiments on all types of reservoir rocks have shown that, for a specified rock and fluid phase, minimum saturations are required for the wetting phase and the nonwetting phase to flow in a two-phase system under an applied pressure gradient. In the case of an oil/water system, the minimum water saturation is designated the interstitial water saturation or S_{iw}. The saturation where the flow of oil ceases is termed the residual oil saturation or S_{or}. These values are functions of wettability, IFT, and rock type and must be determined by experimental measurement. Further discussion of residual oil saturations is presented in Sec. 2.5.

2.4.1.2 Slug Flow or Dispersed Phase Flow.
A second flow mechanism occurs at velocities that are higher than those found in petroleum reservoirs. The rupture of stable oil and water flow channels was observed at the velocities of about 100 ft/D [30.5 m/d] in one study.[52] Water and oil flowed heterogeneously through the porous media as globules. In another study with a different flow cell, the flow of one phase occurred in separate, isolated slugs or globules when the total velocity exceeded 1,000 ft/D [305 m/d].[52] Oil was the dispersed phase in water-wet systems while water became dispersed in oil-wet systems. I refer to this flow mechanism as slug flow or dispersed-phase flow.

When flow is dispersed, one phase becomes discontinuous and is transported through the porous media by the continuous phase if the viscous forces exerted by the continuous phase are large. This mechanism is not important in conventional multiphase flow processes in petroleum reservoirs where velocities are on the order of 1 ft/D [0.004 mm/s]. IFT's are large (30 dynes/cm [30 mN/m]) in these flows. In some EOR processes, IFT's are reduced to levels of 10^{-5} dynes/cm [10^{-5} mN/m], a factor of 10^6 lower than exists in water or gas displacement processes. If there is an equivalence between viscous and capillary forces as indicated by correlations in Sec. 2.6, the mechanism of slug flow may be important at reservoir velocities when interfacial forces are reduced.

2.4.2 Fluid Permeability in Multiphase Flows
The permeabilities of two (or more) phases flowing through a porous rock can be computed from experimental data by assuming Darcy's law applies to each phase. When this is done in two-phase systems, the permeability of a phase varies with saturation, ranging from zero at the minimum saturation of the phase to a maximum when the second phase is immobile. Permeability-saturation correlations obtained from experimental data form the basis for reservoir engineering calculations involving two- (or three-) phase fluid flow.

The development of permeability-saturation correlations is reviewed in this section. The effects of hysteresis, wettability, and fluid properties are included.

2.4.2.1 Extension of Darcy's Law to Multiphase Flow.
Mathematical representation of multiphase flow in porous media when channel flow exists is based on the hypothesis that Darcy's law applies to each phase. Eqs. 2.7 through 2.9 describe 1D flow of two or three phases in the x direction. In Eq. 2.7, u_{ox} is the Darcy (superficial) velocity for the oil phase:

$$u_{ox} = -\frac{k_{ox}\rho_o}{\mu_o}\frac{\partial\Phi_o}{\partial x}, \dots\dots\dots\dots\dots(2.7)$$

$$u_{wx} = -\frac{k_{wx}\rho_w}{\mu_w}\frac{\partial\Phi_w}{\partial x}, \dots\dots\dots\dots\dots(2.8)$$

and

$$u_{gx} = -\frac{k_{gx}\rho_g}{\mu_g}\frac{\partial\Phi_g}{\partial x}, \dots\dots\dots\dots\dots(2.9)$$

where k_{ox}, k_{wx}, and k_{gx} are the effective permeabilities of the oil, water, and gas phases in the x direction. The potentials for each phase (Φ_o, Φ_w, and Φ_g) are defined as in Eq. 2.10 for the oil phase, i.e.,

$$\Phi_o = g(Z-Z_d) + \int_{p_{od}}^{p_o}\frac{dp_o}{\rho_o}, \dots\dots\dots\dots\dots(2.10)$$

where Z is the elevation above a horizontal datum, Z_d is the elevation of the reference datum, and p_{od} is the oil-phase pressure at the reference datum. When x is in the horizontal plane,

$$\frac{\partial\Phi_o}{\partial x} = \frac{1}{\rho_o}\frac{\partial p_o}{\partial x}, \dots\dots\dots\dots\dots(2.11)$$

and Eq. 2.7 becomes

$$u_{ox} = -\frac{k_{ox}}{\mu_o}\frac{\partial p_o}{\partial x}. \dots\dots\dots\dots\dots(2.12)$$

Similarly,

$$u_{wx} = -\frac{k_{wx}}{\mu_w}\frac{\partial p_w}{\partial x}, \dots\dots\dots\dots\dots(2.13)$$

and

$$u_{gx} = -\frac{k_{gx}}{\mu_g}\frac{\partial p_g}{\partial x}. \dots\dots\dots\dots\dots(2.14)$$

In Eqs. 2.7 through 2.14 the phase pressures are related through capillary pressure curves. The effective permeability of a phase is a function of the saturation, varying from zero when the phase is immobile at residual saturation to a maximum when the other phase or phases are at their residual saturations.

2.4.2.2 Relative Permeability.
Effective permeability data are generally presented as relative permeability data. The relative permeability is defined as the ratio of the effective permeability of a phase to a base permeability. Three different base permeabilities are used (Ref. 17, Page 20): (1) the absolute air permeability, (2) the absolute water permeability, and (3) the effective permeability to oil at residual wetting-phase saturation. For most cases, the three base permeabilities do not have the same value. It is necessary to know which base permeability was used for a particular set of relative permeability data.

2.4.2.3 Hysteresis of Relative Permeability Curves.
The relative permeability of a phase usually depends on the path that was followed to reach the saturation. This path dependency, termed hysteresis, is analogous to the variation of capillary pressure with the direction of saturation change as discussed in Sec. 2.3.2.

Relative permeability curves for two typical paths are shown in Figs. 2.29A and 2.29B. The primary drainage path shown in Fig. 2.29A represents the variation of rela-

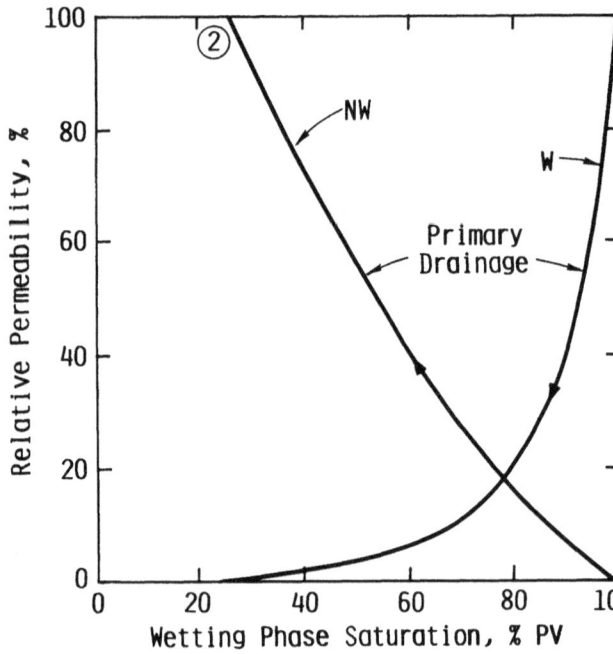

Fig. 2.29A—Primary drainage relative permeability curves for a strongly water-wet rock (after Ref. 17, Fig. 2.15).

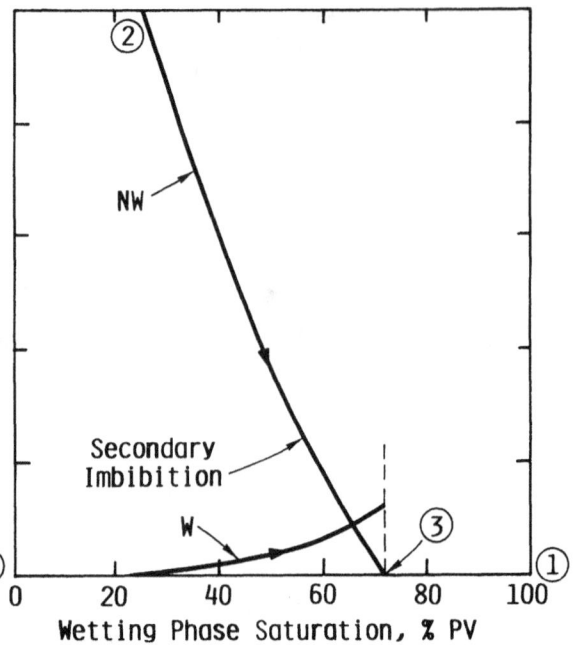

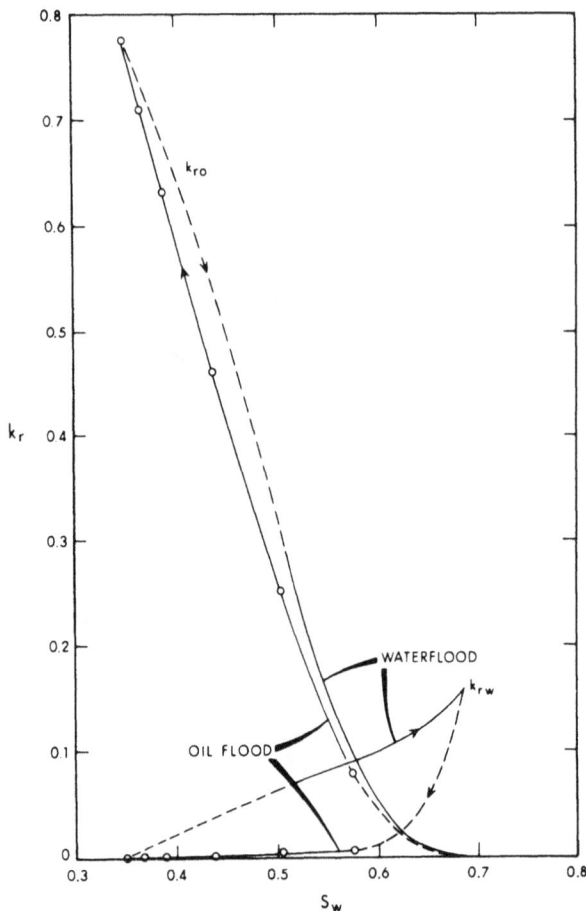

Fig. 2.29B—Secondary-imbibition relative permeability curves for a strongly water-wet rock (after Ref. 17, Fig. 2.15).

Fig. 2.30—Hysteresis of waterflood and oil-flood relative permeability curves.[61]

tive permeability with saturation when the initial saturation of the wetting phase is 1.0 and the nonwetting phase saturation is increased gradually. Arrows indicate the direction of saturation change. The curve labeled NW corresponds to the nonwetting phase. The base permeability is the permeability to the nonwetting phase at interstitial wetting-phase saturation. In Fig. 2.29A the interstitial wetting-phase saturation is about 25%. The primary drainage path is analogous to the path followed when gas displaces oil in a solution-drive reservoir or an oilflood of a water-wet reservoir that is 100% saturated with water.

Secondary imbibition curves shown in Fig. 2.29B begin at the interstitial saturation of the wetting phase and follow a path where the saturation of the wetting phase increases. This is the path that would be followed during the displacement of oil by water in a strongly water-wet rock. The relative permeability of the wetting phase reaches its maximum value at the residual saturation (Point 3) of the nonwetting phase, as noted by the dashed line in Fig. 2.29B.

The paths followed to generate the relative permeability data are 1 to 2 and 2 to 3 as noted in Figs. 2.29A and 2.29B. The relative permeability of the nonwetting phase is quite dependent on the saturation path. However, the relative permeability of a strongly wetting phase may be independent of the path over the same saturation range as can be seen by comparing Figs. 2.29A and 2.29B. Several examples of this behavior have been presented.[56-60]

In most oil production operations, oil displacement occurs along one of the two paths. Although hysteresis of relative permeability curves has been recognized, applications have been limited to making sure that the proper relative permeability curve was used in reservoir engineering calculations. For example, it is possible to generate relative permeability data corresponding to Path 1 to

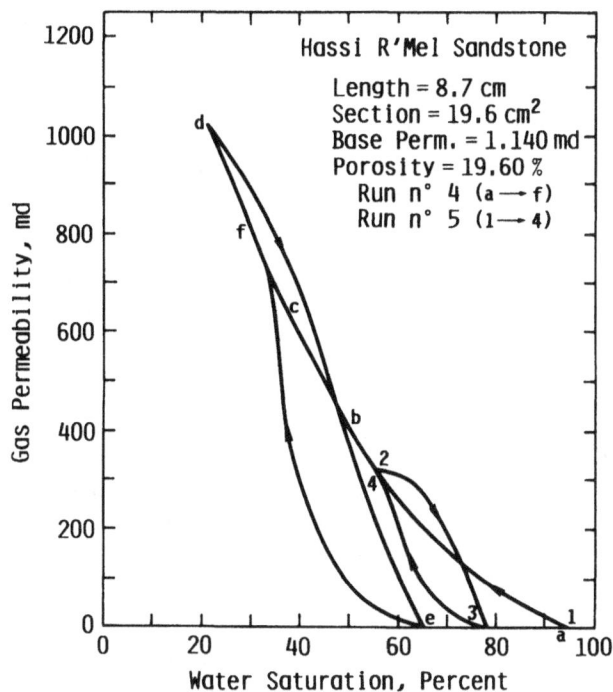

Fig. 2.31—Cyclic permeability, Hassi R'Mel core.[62]

TABLE 2.4—EFFECTS OF PRESSURE GRADIENT ON RELATIVE PERMEABILITY MEASUREMENT[69]

Oil/Water Flow Ratio	k_o/k_w	Average Brine Saturation (%)	Pressure Gradient (psi/in.)	kPa/m
infinite	infinite	14.2	2.04	554
100:1	150.6	36.9	4.71	1,278
100:1	149.4	37.2	9.16	2,486
100:1	149.4	37.3	1.27	345
100:1	152.3	37.7	0.678	184
10:1	15.76	47.3	14.34	3,893
10:1	15.36	45.9	4.93	1,338
10:1	15.04	46.3	0.994	270
1:1	1.488	53.5	1.24	337
1:1	1.510	52.9	3.05	828
1:10	0.1507	56.0	16.47	4,471
1:10	0.1507	55.6	8.14	2,210
1:10	0.1537	55.1	2.43	660
0		57.7	15.91	4,319

2 to 3 for water and oil in a strongly water-wet rock. Because a waterflood of this rock would follow Path 2 to 3, data for Path 1 to 2 would not be needed and not obtained.

Reservoirs that are candidates for EOR processes usually have saturations that are less than the initial oil saturation. Thus, the initial condition of the reservoir will be on one of the displacement paths. If the EOR process mobilizes oil, creating an oil bank or region of increased oil saturation, the saturation path must change from imbibition to drainage or vice versa. When the direction of saturation change is altered, it is necessary to obtain relative permeability curves that represent both paths. Fig. 2.30 illustrates hysteresis in oil and water relative permeability curves for a waterflood and an oilflood in a water-wet core, with arrows indicating the direction of saturation change.[61] The oilflood simulates the path that would be followed when the oil saturation increased. The base permeability in Fig. 2.30 is the absolute permeability to water. Relative permeabilities for the nonwetting oil phase are quite close for water saturations between 0.50 and 0.70. However, hysteresis was observed in the relative permeability of the water that was the wetting phase.

Hysteresis is observed in other fluid systems. Fig. 2.31 illustrates the history dependence of the gas permeability of the Hassi R'Mel sandstone for a series of cyclic paths.[62] Path 2 to 3 to 4 is a cyclic hysteresis loop that begins and ends at the same point on the primary drainage curve. Cycle d to e to f consists of a secondary imbibition path (d to e) and secondary drainage path (e to f). Point e is the irreducible or residual gas saturation. Point f is also on the primary drainage curve.

Conceptually, the hysteresis of relative permeability curves is closely related to hysteresis of capillary pressure curves. Fluid saturations do not occupy the same pore space and have the same interfacial curvature when the

nonwetting-phase saturation increases as when the wetting phase increases. Specific causes of this difference have not been well defined. Various investigators attribute hysteresis to different mechanisms. McCaffery and Bennion[60] found no hysteresis of secondary drainage and imbibition relative permeability curves for a dodecane/nitrogen system in a sintered Teflon core. Teflon is uniformly wetted by dodecane in the presence of nitrogen and the contact angle measured through dodecane is 20°. Hysteresis was observed for other fluid pairs when the contact angle measured through the wetting phase was equal to or exceeded 73°, suggesting that contact angle is an important factor. Other possible contributors to hysteresis of relative permeability curves include surface roughness and hysteresis of the contact angle.[35] Raimondi and Torcaso[63] attributed reduction of imbibition permeabilities in water-wet systems to gradual trapping of the residual oil saturation. Larson[64] developed a model of two-phase flow based on the percolation theory that qualitatively produces hysteresis in relative permeability curves.

2.4.2.4 Validity of Darcy's Law for Multiphase Flow.
A rigorous proof that verifies the extension of Darcy's law to multiphase flow in porous media has not been developed. Verification of this extension is based on experimental data from numerous multiphase-flow studies in porous media. To show that Darcy's law was valid for phase velocities expected in oil and gas reservoirs it was necessary to show that the effective permeability of a phase is independent of the pressure gradient when the measurement is made on the same saturation path.

Methods used to measure relative permeability are described in Refs. 65-71. Some methods are discussed in Chap. 3 in examples illustrating fluid-flow concepts. All measurement techniques used in laboratory determination of relative permeability must be conducted under conditions that overcome capillary end effects.[67,68] Capillary end effects are caused by the discontinuity in capillary pressures in the space between the end of the core and the end cap of the core holder and are discussed in Chap. 3. The data presented in the remainder of this section were obtained under conditions in which capillary end effects were considered negligible.

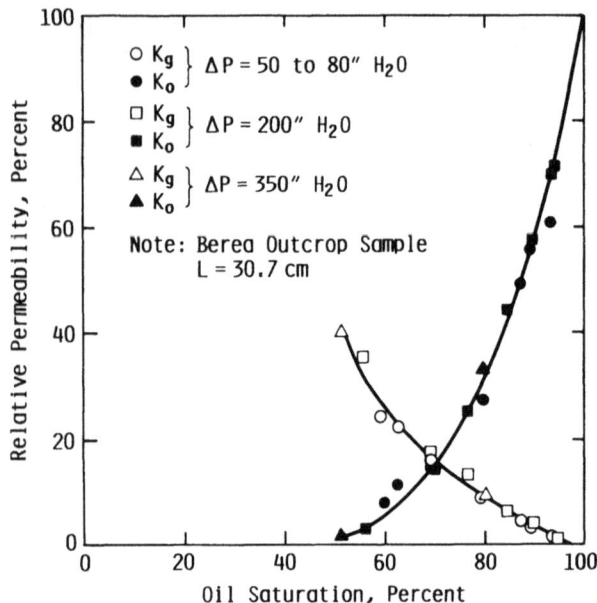

Fig. 2.32—Effect of flow rate on relative permeability curves for drainage conditions.[68]

Experimental verification of Darcy's law for multiphase flow in porous rocks was given in a series of papers presented in the 1950's. Water/oil relative permeability data on the secondary imbibition path were obtained by Geffen *et al.*[69] at different pressure gradients using a modified Pennsylvania State U. method. This method is illustrated in Example 3.5. Their data at several brine saturations are presented in Table 2.4. The ratio of oil to water permeability in Table 2.4 is not dependent on pressure gradient in the range investigated.

Gas/oil drainage relative permeability data on a Berea sandstone core at pressure gradients varying from 8.14 to 191.6 psi/ft [184 to 4471 kPa/m] using the steady-state method are shown on Fig. 2.32.[68] There is no effect of pressure gradient on drainage relative permeability curves for gas displacing oil. Refs. 67, 68, and 70 also contain experimental data that show that the relative permeability at a given saturation is not dependent on pressure gradient.

The experimental data collected over many years of laboratory measurements indicate that the relative permeability of a phase in a particular porous rock depends on fluid properties and fluid/rock interactions but not on the pressure gradient. The effect of fluid properties and fluid/rock interactions is considered in the next section.

2.4.3 Correlation of Relative Permeability Data for Porous Rocks

Relative permeability is an empirical method of correlating multiphase fluid flow in porous media with the saturation of the phases and the path followed in reaching a particular saturation. Values measured for a rock with the same fluid pair are reproducible on the same saturation path. Data obtained from steady-state measurements have been used successfully in unsteady-state models to compute transient multiphase flow. Investigators report good agreement between relative permeabilities obtained from steady-state and unsteady-state methods[65,68] for

homogeneous core plugs. In this section, the effects of the porous rock, wettability, viscosity ratio, and IFT on relative permeability curves are examined.

2.4.3.1 Correlation With Rock Properties. Many studies have been conducted in an attempt to correlate relative permeability to the properties of the porous rock for the same reservoir fluids. Rock type, pore structure (including surface roughness), and mineralogy are different in every reservoir. Furthermore, some of these properties vary within a reservoir, as discussed in Chap. 7.

Rock properties that control pore geometry may affect relative permeability curves (Ref. 17, Page 20). Microscopic examination of rock samples can provide insight into differences in relative permeability curves between samples taken from the same reservoir as well as those taken from different reservoirs. Figs. 2.33 and 2.34 show photomicrographs and relative permeability curves for two different sandstones.[72] Absolute permeabilities are 1,314 and 20 md, respectively. The photomicrograph in Fig. 2.33 indicates that this sandstone has large, well-connected pores. Relative permeability curves in Fig. 2.33 are characterized by (1) a low interstitial water saturation ($S_{iw} = 0.18$), (2) a value of k_{ro} at S_{iw} approximately equal to the absolute air permeability, and (3) the presence of two-phase flow over a wide range of saturations ($0.18 \leq S_w \leq 0.70$).

Interpretation of the photomicrograph of the second sandstone in Fig. 2.34 suggests the pores are small and well connected. The value of S_{iw} is about 0.33 while k_{ro} at S_{iw} is about 0.75. Two-phase flow extends from $0.33 \leq S_w \leq 0.74$.

The two sets of relative permeability curves can be compared in terms of pore sizes if there are no other significant geological differences between the rocks and if both have the same wetting preference, in this case strongly water wet. Under these conditions, rocks with large pores will have a low value of S_{iw} because nearly all pores are accessible to both phases and a relatively small amount of water is required to wet the surface area. In the rock with small pores, the interstitial water saturation is higher for two reasons. The surface area wetted by the water phase is larger. Also, the distribution of pore sizes probably contains small pores that can be filled only with water and thus become dead-end pores as far as the oil phase is concerned. Relative permeability to oil at S_{iw} is reduced because small pores filled with water block the flow of oil.

A similar situation exists at the residual oil saturation, S_{or}. Oil is trapped in the largest pores in both sandstones, forcing water to flow through the smaller pores. Thus k_{rw} at S_{or} is less in the sandstone containing small pores than in the sandstone with large pores. Small pores appear to control the magnitude of k_{rw} at S_{or}. These examples show how interpretation of photomicrographs, coupled with relative permeability curves, can give qualitative explanations for differences in relative permeability curves.

There has been limited success in developing quantitative methods to correlate relative permeability data with rock properties. Correlation with absolute permeability is possible for some rocks. One difficulty with correlations based on rock properties is that the porous rock is so complex that properties such as porosity and permeability are not adequate descriptors of the rock on a

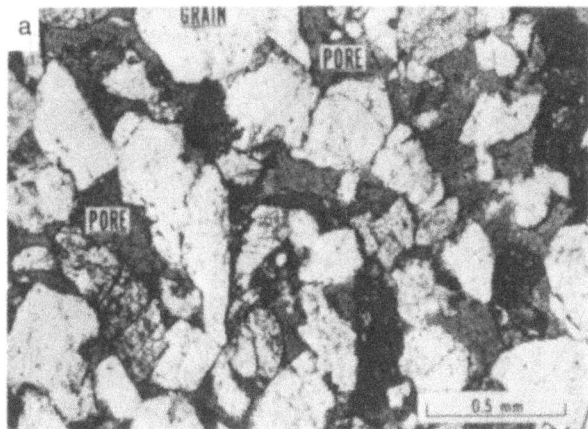

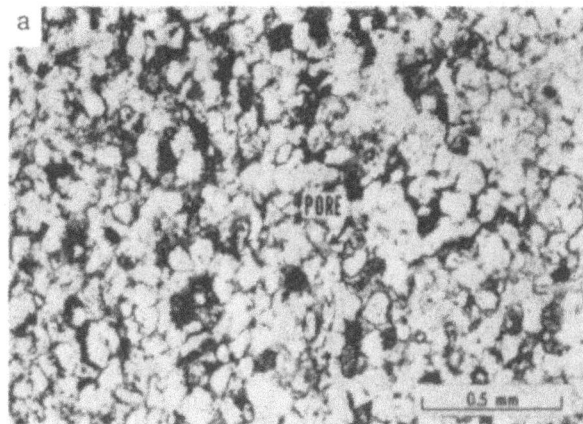

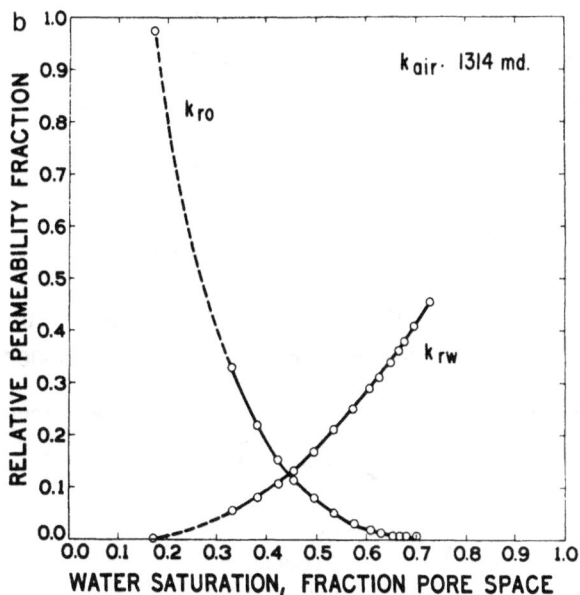

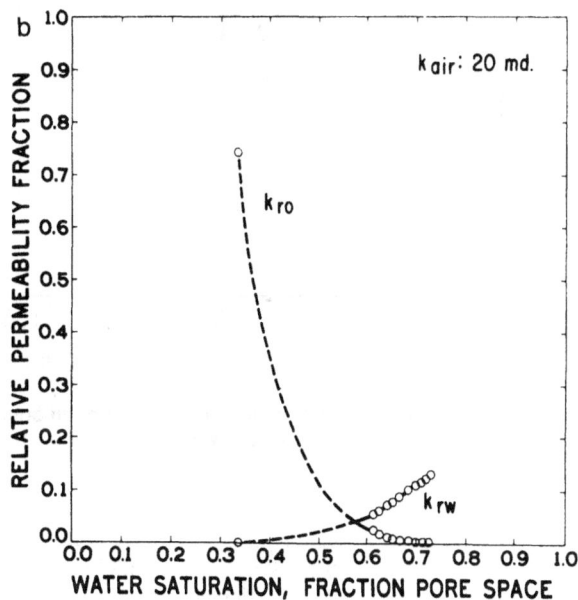

Fig. 2.33—Sandstone containing large well-connected pores: (a) photomicrograph and (b) water/oil relative permeability curve.[72]

Fig. 2.34—Sandstone containing small well-connected pores: (a) photomicrograph and (b) water/oil relative permeability curve.[72]

microscopic scale. Consequently, various averaging methods are used to obtain a relative permeability curve that approximates overall reservoir behavior (Ref. 17, Page 25).

2.4.3.2 Wettability. Wettability is the most important factor influencing relative permeability curves. It is also the most difficult variable to reproduce in field and laboratory environments. As discussed in Sec. 2.3.3, no absolute standard exists for measuring wettability in porous materials. Studies of relative permeability have been conducted on the same porous media using two approaches to alter wettability. In one approach, wettability is varied by selection of fluid pairs that exhibit different contact angles against the dominant mineral or material present in the surface of the porous material. The second approach is based on altering the wettability of the surface by adsorption of a silicon polymer. The same fluids are used in developing relative permeability curves. Different degrees of wettability are obtained by varying the concentration of polymer in the treating solution.

Figs. 2.35 and 2.36 show imbibition relative permeabilities for oil and water in a fired Torpedo sandstone core.[58] Firing stabilized the clay minerals, creating a uniform rock surface that was considered essentially pure quartz. The contact angle was varied by changing the amount of an oil-soluble surfactant dissolved in the oil. In Fig. 2.35, water saturations increase for both sets of relative permeability curves as indicated by the arrows. Relative permeability data are plotted on a semi-logarithmic scale to display smaller values accurately. Wetting conditions are expressed in terms of the contact angle measured through the water phase. Base permeabilities for Figs. 2.35 and 2.36 were the permeabilities to oil at connate water saturation.

Relative permeabilities to oil are significantly higher for slightly water-wet conditions (47°). In contrast, relative permeabilities to water are significantly higher for strongly oil-wet conditions (180°). Other relative permeability curves corresponding to contact angles of 0, 90, and 138° shown in Fig. 2.36 reveal the same trend relating wettability to relative permeability. At fixed water

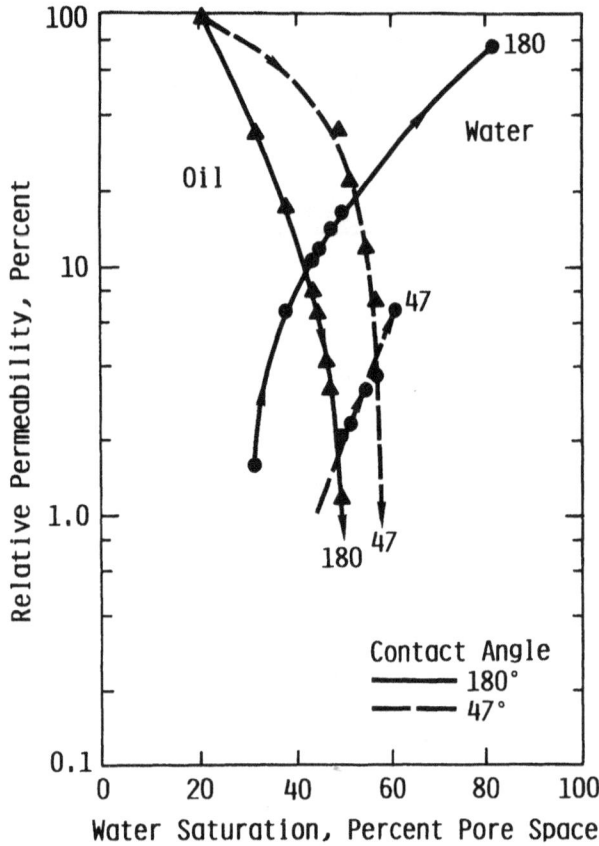

Fig. 2.35—Relative permeabilities for two wetting conditions, Torpedo sandstone as water saturation increases.[58]

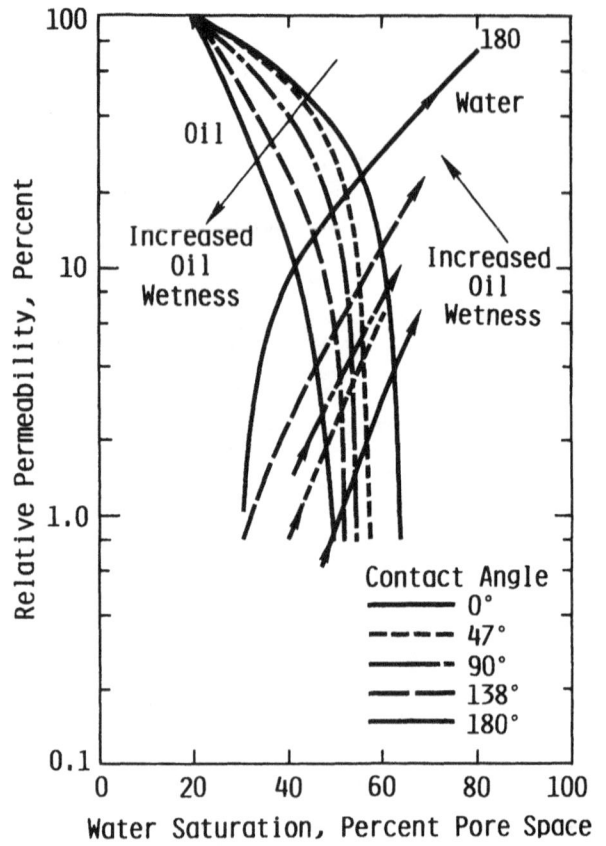

Fig. 2.36—Relative permeabilities for a range of wetting conditions (indicated by contact angle) of Torpedo sandstone as water saturation increases.[58]

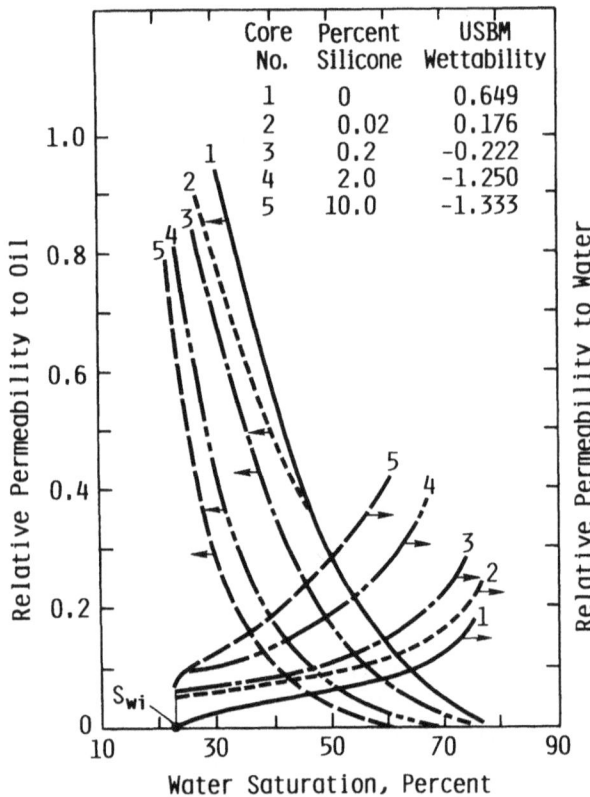

Fig. 2.37—Oil and water relative permeabilities of Squirrel sandstone cores for water-wet and oil-wet conditions.[59]

saturation, oil permeability decreases and water permeability increases as the contact angle increases. The progressive change of relative permeability curves with saturation can be seen by following the intersection of relative permeability curves ($k_{ro} = k_{rw}$) as the contact angle varies. Two trends are apparent. The intersection occurs at a lower water saturation and at larger values of the relative permeability as water wetness decreases (i.e., contact angle increases).

Fig. 2.37 shows oil and water relative permeability curves obtained with sandstone cores from the Squirrel formation that had been treated with solutions containing different concentrations of GE Dri-Film 144,[59] a silicon polymer. Different degrees of wetting were obtained by increasing the concentrations of silicon polymer in the treating solution. Wettability is expressed in terms of a wettability index devised by USBM.[4] This index ranges from 1.0 for a highly water-wet rock to -1.5 for a strongly oil-wet rock. Oil from a Squirrel reservoir was used in the tests. All relative permeability data were obtained by the unsteady-state method[65] and represent the saturation path where water saturation increases from S_{iw} to $1 - S_{or}$.

The trend of the relative permeability curves shifting toward lower water saturation is identical with the trends in Figs. 2.35 and 2.36—i.e., when water is the displacing phase in a uniformly wetted porous material, the relative permeability of water increases and the relative permeability of oil decreases as oil wetness increases. This is a general property of uniformly wetted porous materials.

Uniform Wetting—Nonuniform Wetting. A consistent trend in relative permeability data with wettability indicators (i.e., contact angle, silicon concentration) was shown in the previous section for porous media considered to have uniform wettability. However, there is growing evidence that many reservoirs do not have uniform wettability. Two examples of nonuniform wetting are presented in this section.

A mixed-wettability model was postulated to explain unusual values of oil permeability observed at low oil saturation in East Texas cores.[15] This model, depicted in Fig. 2.9, considers some portions of the rock surface to be oil wet while others are water wet. Oil permeability at low saturations is obtained by a series of connected oil-wet films that are present over some portions of the rock surface. A procedure was developed to treat outcrop and reservoir rock samples to attain mixed-wettability behavior in laboratory waterfloods.

Silicon polymers such as GE Dri-Film are used to decrease water wettability of rocks. Relative permeability data for Berea core material before and after treatment with dri-film are shown in Fig. 2.38.* Both sets of curves were obtained with the unsteady-state method, displacing oil with brine. Thus the data represent paths with increasing brine saturations. Water relative permeability increased as water wettability decreased, as expected from relative permeability trends in uniformly wetted porous media. However, the relative permeability to oil increased after dri-filming, which does not follow the behavior expected for uniformly wetted materials.

The same trends are present in an independent set of experiments[59] in Berea sandstone cores treated with various concentrations of silicon polymer. The ratio of oil permeability to water permeability was inferred from waterflood performance. At fixed water saturation, oil permeability increased more than water relative permeability as water wetness decreased. Changes with decreasing water wetness for both sets of data are opposite from those shown in Figs. 2.35 through 2.37, where oil permeability decreased and water permeability increased.

Use of silicon polymers to alter wettability of pore surfaces may not create uniform wetting on a microscopic level. Perhaps the chemical does not deposit or adsorb uniformly on mineral surfaces.

A correlation of wettability indicators with relative permeability curves is shown in Figs. 2.35 through 2.37 for uniformly wetted porous media. Other investigators have found the same relationship for uniform mixtures of porous materials that have heterogeneous wettability,[73,74] such as mixtures of Teflon particles (strongly oil wet) and glass beads (strongly water wet). This trend suggests that it would be possible to infer wettability from relative permeability curves. However, there are uncertainties in this method. To infer wettability, a reference curve should be available for a strong wetting condition, a curve that is difficult to obtain. Trends in Figs. 2.35 through 2.37 apply to uniformly wetted porous material. Systems that exhibit nonuniform wettability could not be analyzed by this procedure.

Guidelines for evaluating wetting conditions from relative permeability data have been proposed.[10] (See also Ref. 17, Page 20.) For example, the relative permeability of water in consolidated, water-wet porous media is

*Personal communication with R.L. Clampitt, Phillips Petroleum Co., 1978.

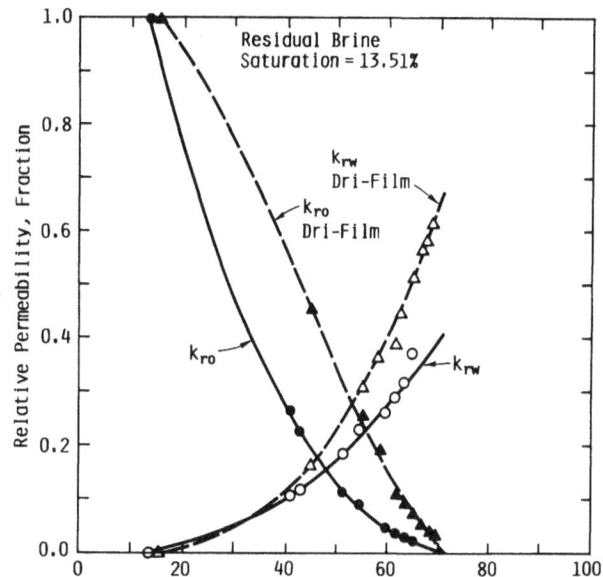

Fig. 2.38—Relative permeability curves for Berea sandstone before and after dri-film treatment.*

normally 15% or less at residual oil saturation. In contrast, oil-wet formations normally have relative water permeabilities of 50% or higher at residual oil saturation. These and similar flow properties were used to infer the wettability for the reservoirs in Table 2.2 listed under the relative permeability heading.[10] There is good agreement between wettability from contact angle measurements and wettability inferred from relative permeability curves.

Summary

In conclusion of Sec. 2.4.3.2, I want to re-emphasize the following concepts. Wettability controls the location and shape of oil and water relative permeability curves on saturation plots. Consistent trends of relative permeability curves with wettability are observed in systems believed to have uniform wettability. Porous rocks exhibiting nonuniform wetting properties, as inferred from relative permeability curves, are found in reservoirs. Nonuniform wetting conditions may be created in laboratory cores. These possibilities introduce uncertainty in the use of relative permeability curves to infer wettability of a reservoir.

2.4.3.3 IFT. Almost all of the oil produced to date is the result of displacement processes in which IFT's between water and oil, water and gas, or oil and gas range from 10 to 40 dynes/cm [10 to 40 mN/m]. Changes of IFT over this interval have little effect on relative permeability curves as long as wettability remains the same. Substantial reductions of IFT from these ranges produce small changes in relative permeabilities. For example, reduction of IFT from 35.0 to 5.0 dynes/cm [35.0 to 5.0 mN/m] increased the permeabilities of both oil and water phases by 20 to 30%.[75] The reduced IFT was obtained by using amyl alcohol for the hydrocarbon phase. Wettability was considered constant.

Some EOR processes use low or ultralow IFT's to displace oil. As IFT's decrease, the relative permeability curves should change. Less energy is required to maintain interfaces between immiscible fluids in porous media. Interfacial effects are also related to wettability and

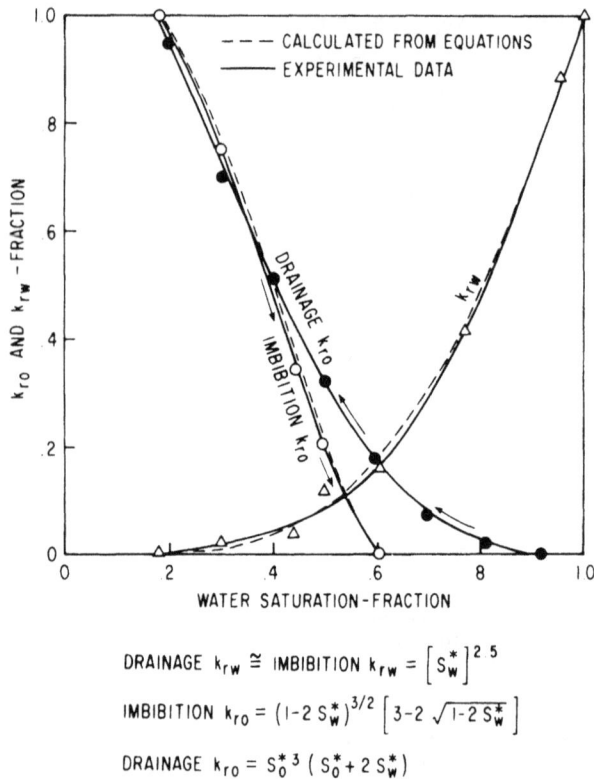

$$\text{DRAINAGE } k_{rw} \cong \text{IMBIBITION } k_{rw} = \left[S_w^* \right]^{2.5}$$

$$\text{IMBIBITION } k_{ro} = (1 - 2 S_w^*)^{3/2} \left[3 - 2 \sqrt{1 - 2 S_w^*} \right]$$

$$\text{DRAINAGE } k_{ro} = S_o^{*3} (S_o^* + 2 S_w^*)$$

Fig. 2.39—Steady-state relative permeability data for water-flood and oil floods.[77]

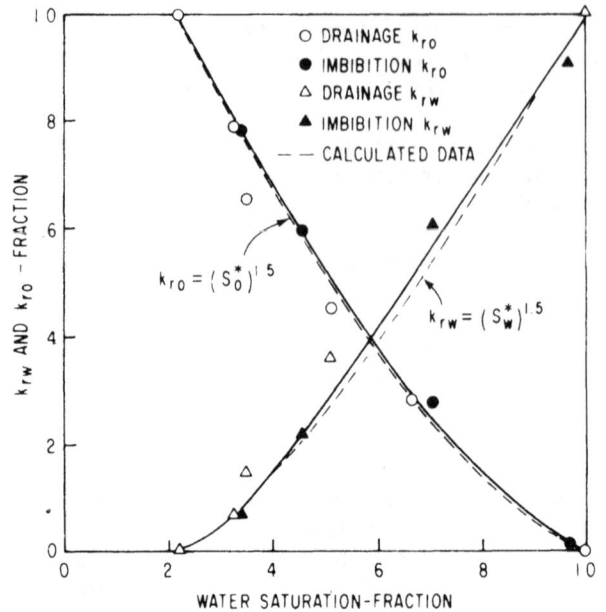

Fig. 2.40—Steady-state relative permeability data—low IFT.[77]

pore geometry through Eq. 2.5. The net effect of reduced IFT should be to increase the relative permeability of each phase. Theoretically, the relative permeabilities of two phases should approach the saturation of the phases as miscibility nears. Under these conditions, both phases are assumed to be flowing with neither phase trapped or isolated in dead-end pores.

There are limited experimental data that show the effect of IFT on relative permeability curves. Results of two investigations are included in this section to provide background material.

The effects of IFT on oil displacement were determined in Teflon cores[76] using fluid pairs selected to obtain the same contact angles. The viscosity ratio (viscosity displaced/viscosity displacing) was constant. IFT varied from 0.5 to 40.0 dynes/cm [0.5 to 40.0 mN/m] while wettability, inferred from contact angle measurements, varied over two narrow ranges, 40 to 45° and 139 to 145°. Displacement data were obtained for two cases: (1) wetting fluid displacing the nonwetting fluid and (2) nonwetting fluid displacing the wetting fluid. Relative permeability curves were not presented for most of the displacement experiments. However, the effect of IFT on the ratio of relative permeabilities can be inferred from the displacement data.

In both sets of experiments, the ratio of displaced phase permeability to displacing phase permeability increased as IFT was reduced. Thus the permeability of the displaced phase was affected to a larger extent by IFT than the displacing phase permeability. Small changes were observed when the nonwetting fluid was displaced. These changes occurred at high wetting-phase saturations near the end of the relative permeability curves. Large changes

in relative permeability ratios were observed when the wetting fluid was displaced by the nonwetting fluid. Thus the effects of IFT variations from 40 to 0.5 dynes/cm [40 to 0.5 mN/m] depends on whether the saturation is on the imbibition or drainage path.

Experimental data[77] showing the change in relative permeability curves with IFT for a system that contained a surfactant have been obtained. Fig. 2.39 shows the steady-state imbibition and drainage relative permeability curves for Loma Novia crude oil and brine in a Berea core. Fig. 2.40 shows the steady-state drainage and imbibition data for a crude-oil surfactant system that had IFT's of 0.02 to 0.0002 dynes/cm [0.02 to 0.0002 mN/m]. Relative permeabilities of both phases increased as expected, and hysteresis of relative permeability curves was not observed. It is not known if wettability of the core was changed by contact with surfactant.

2.4.3.4 Viscosity Ratio. Perhaps no technical problem has provoked as much continuing discussion as the effect of viscosity ratio on the flow of immiscible phases in porous media. Many engineers consider the relative permeability curves to be independent of viscosities of the displaced and displacing fluids. Others have reported irregular flow patterns in visual cells as well as variations in relative permeability data that have been attributed to viscosity ratio effects. This dependence has been discussed in terms of viscous fingering,[78] microscopic rock heterogeneities (Ref. 17, Page 34), relative importance of viscous forces to capillary forces in a displacement process,[79] a length effect observed on short cores,[80] and inadequate time for fluids to attain capillary equilibrium in laboratory relative permeability tests.* No single explanation appears to account for all variations.

There are no extensive studies that demonstrate the absence of a viscosity effect for all rock and fluid pairs. What is available is a collection of experiments that show mixed

*Personal communication with R.J. Blackwell, Exxon Production Research Co., 1978.

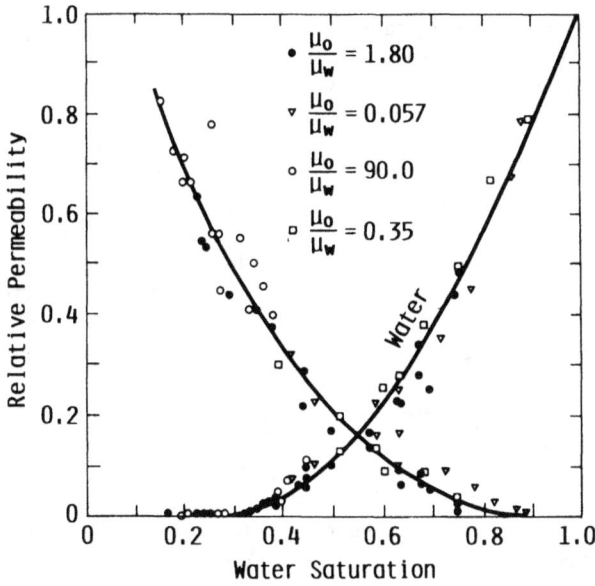

Fig. 2.41—The effect of the viscosity ratio (oil to water) on the relative permeabilities in a 100- to 200-mesh sand (Ref. 75, after Ref. 23).

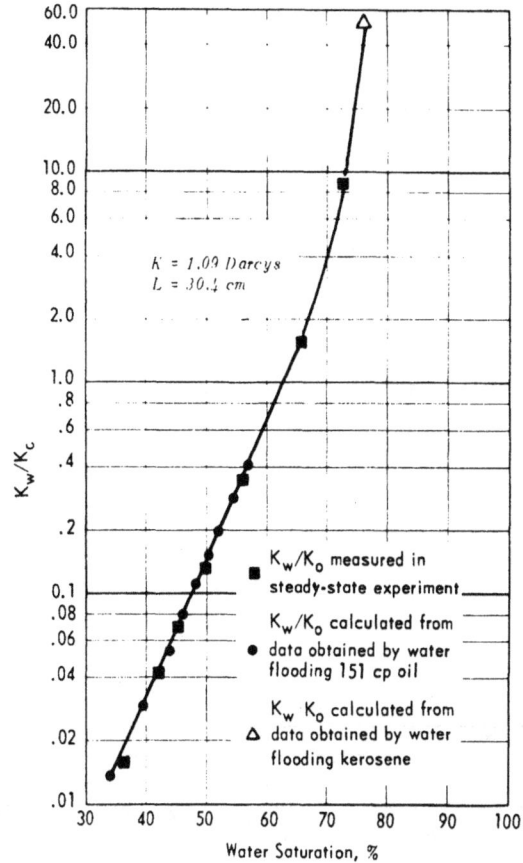

Fig. 2.42—Comparison of k_w/k_o saturation relationship measured in steady-state flow with values calculated from flooding performance. [81]

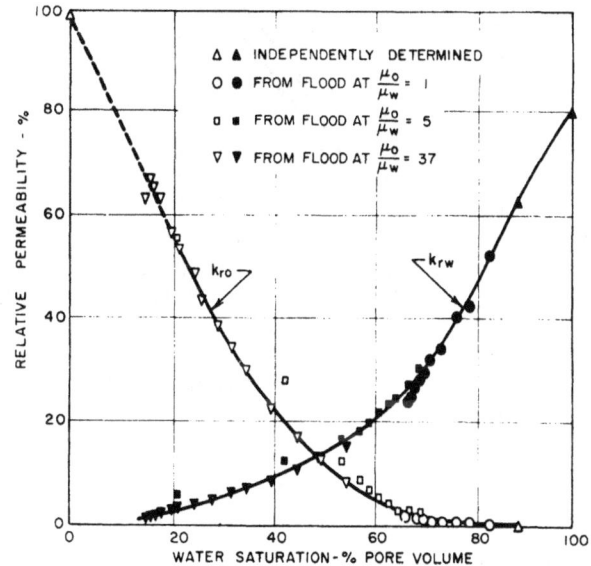

Fig. 2.43—Water/oil permeabilities from waterflood susceptability test data. Dri-filmed glass spheres, linear cell permeability = 11.5 darcies. [65]

trends. I will review some of these data to illustrate what is known as well as the areas where uncertainty still exists.

If the relative permeability curves for a particular rock are functions of saturation, wettability, and IFT, certain statements can be made about data derived with different fluid pairs. For example, we should expect the drainage relative permeability curve for a gas/water system in a strongly water-wet rock to be identical to the drainage curve for an oil/water system in the same rock. Water and oil relative permeabilities for the secondary imbibition path should be the same whether obtained with an oil viscosity of 100 cp [100 mPa·s] or with water thickened by addition of sucrose. Small differences might be expected because of slight variations in wettability. However, as long as wettability is not altered significantly, the choice of fluids would be determined by laboratory conditions, particularly the time required to obtain the experimental data. Some of the experimental data[60,65,70,75,81,82] supporting the independence of relative permeability from viscosity ratio are presented in the following paragraphs.

Relative permeabilities obtained in an unconsolidated-sandpack containing 100- to 200-mesh sand are presented in Fig. 2.41 for ratios of oil to water viscosity that varied from 0.057 to 90.[75] The experimental data have some scatter but no systematic deviations of relative permeability with viscosity. Relative permeability data from a 1.09-darcy sandpack are presented in Fig. 2.42. Steady-state k_w/k_o data for a kerosene/water system are compared to k_w/k_o data determined from analysis of waterflood performance when 1 cp [1 mPa·s] water displaced 151 cp [151 mPa·s] oil. The data in Fig. 2.42 show that relative permeability is independent of viscosity ratio for the unconsolidated sandpack for viscosity ratios of 1.8 to 151.

Fig. 2.43 shows relative permeability curves for oil/water viscosity ratios of 1:1, 5:1, and 37:1 in an 11.5-darcy packed bed of dri-filmed glass spheres.[65] The dri-film treatment decreases water wetness, but no measure of wettability was provided. The data in Fig. 2.43 show

little variation of water-phase relative permeability over the saturation range where comparable data exist. However, oil relative permeability data at a viscosity ratio of five appear to differ significantly from those obtained at a ratio of 37 for water saturations less than about 55%.

TORPEDO SANDSTONE
$k_o = 345$ md

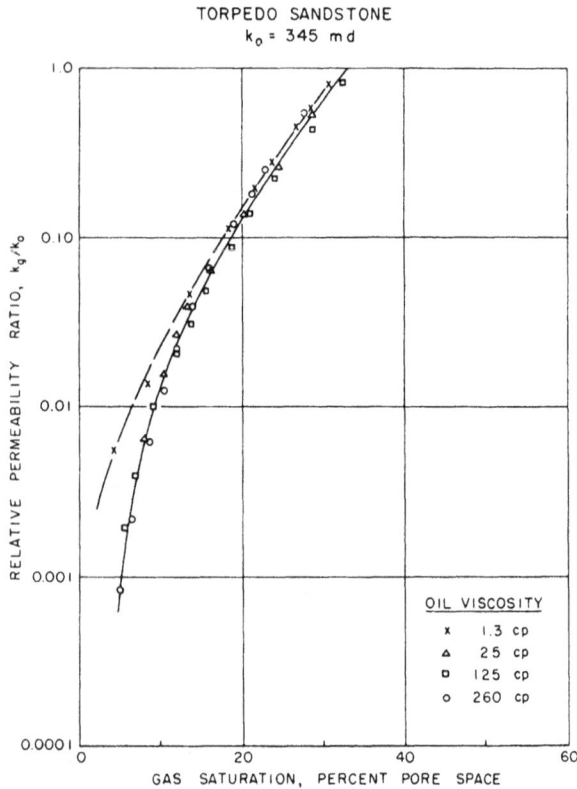

Fig. 2.44—Effect of oil viscosity on k_g/k_o obtained by displacement with gas.[70]

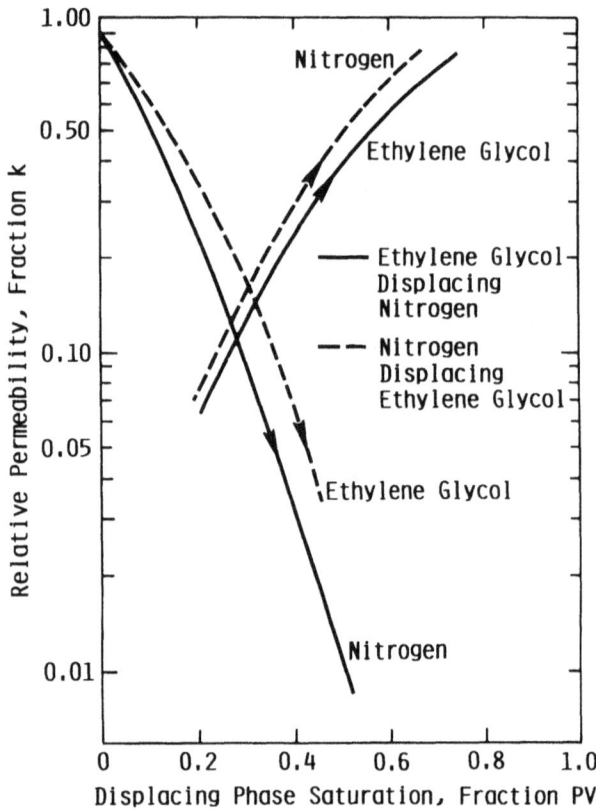

Fig. 2.45—Comparison of relative permeabilities for ethylene glycol displacing nitrogen and nitrogen displacing ethylene glycol in Core 1.[60]

Large viscosity ratios are encountered in the drainage process where one of the flowing phases is a gas. Relative permeabilities obtained along the primary drainage path are generally found with a viscous oil as the second flowing phase. Fig. 2.44 shows the ratio of k_g/k_o for oil viscosities ranging from 1.3 to 260 cp [1.3 to 260 mPa·s]. There is no effect of oil viscosity and viscosity ratio for oils between 25 and 260 cp [25 and 260 mPa·s]. The viscosity ratio varied from 1,400 to 14,600 in these experiments. The difference between the data for the 1.3-cp [1.3-mPa·s] oil and other data was attributed to the absence of a stablized zone in the displacement process. See Sec. 3.12.1 for further discussion.

The effect of viscosity ratio on steady-state relative permeabilities of fluid pairs in Teflon cores was investigated over a wide range of wetting characteristics.[60] Similar relative permeability curves were obtained on the drainage path for dodecane/water and strongly wetting liquid (dodecane or heptane)/nitrogen. Contact angles were identical for these fluid pairs. Imbibition relative permeability curves and secondary drainage curves were nearly identical.

The results of these studies indicate that the relative permeability curves obtained along the same saturation path are independent of the viscosity ratio of the fluids used to obtain the data for strongly wetted systems. However, the permeabilities of the porous media were about 1 darcy or larger, with one exception.[70] Experimental data that extend these comparisons to porous media with lower permeabilities and different mineralogy of pore structure are scarce.

Significant viscosity effects have been proposed in the literature.[60,78,79,83-85] Three investigations appear to be free from experimental uncertainties. All studies were done using Teflon porous media.

Some effects of viscosity ratio were observed for an ethylene-glycol/nitrogen system. The contact angle on Teflon was 90°[60]; thus neither phase was preferentially wetting. Drainage relative permeability curves for zero initial displacing-phase saturation are shown on Fig. 2.45. Relative permeability curves for increasing glycol saturations were as much as 60% higher than those measured for increasing nitrogen saturations. The viscosity ratio for the latter case was about 1,030.

A series of displacement experiments was run[75] in Teflon cores where viscosity ratios (viscosity displaced/viscosity displacing) of 1, 12, and 20 were used. In one set of experiments, a strongly wetting fluid displaced a nonwetting fluid. In another, the nonwetting fluid displaced a strongly wetting fluid. Relative permeability ratios inferred from the displacement experiments show substantial effects of viscosity ratio when the nonwetting fluid displaced the wetting fluid. This change in displacement behavior was attributed to viscous fingering caused by uneven fluid movement across the displacement front.

Relative permeability data obtained in sintered Teflon cores at three viscosity ratios are presented in Fig. 2.46.[79] Wettability, inferred from contact angle measurements, was approximately the same for the three fluid pairs.

The parameter R is the ratio of the viscosity of the wetting fluid to the viscosity of the nonwetting fluid. These relative permeability curves were derived from displacement experiments in which the wetting fluid was displaced

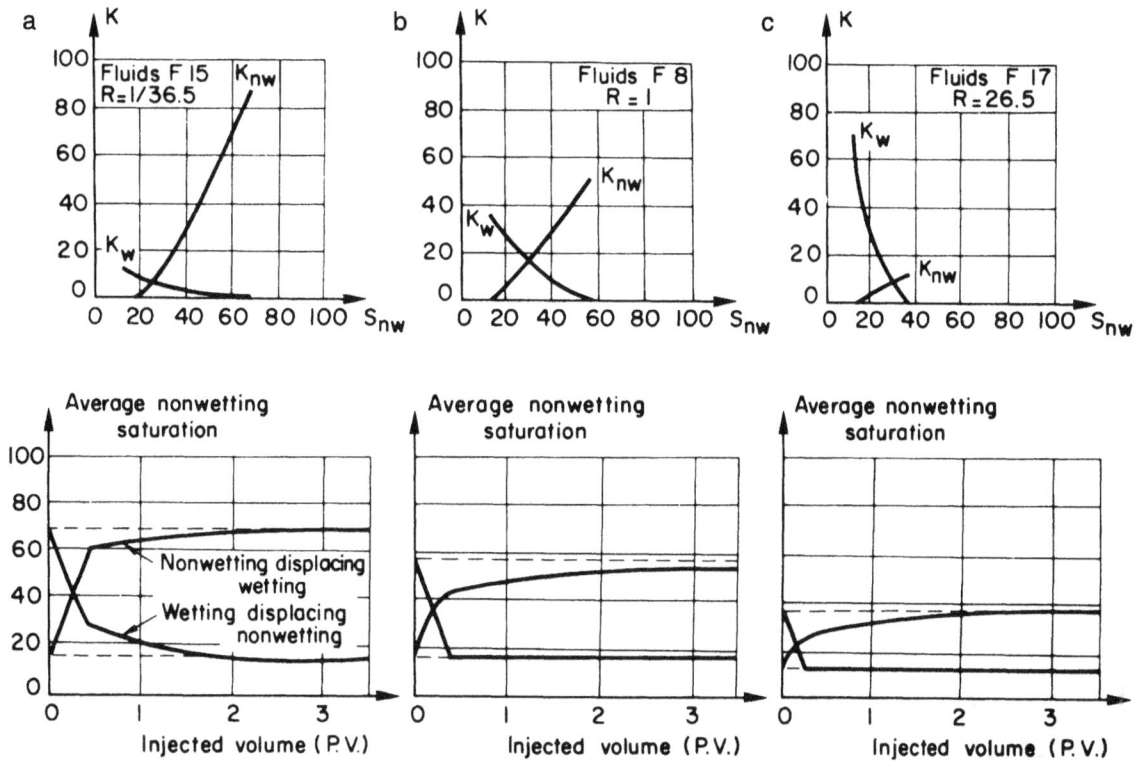

Fig. 2.46—Effect of viscosity ratio (R = viscosity wetting phase/viscosity nonwetting phase) on relative permeability curves: (a) $R = 0.027$, (b) $R = 1$, and (c) $R = 26.5$.[79]

by the nonwetting fluid. Fig. 2.46a represents the relative permeabilities when $R = 0.027$. Fig. 2.46b shows relative permeability curves when the viscosities are equal ($R = 1$). Fig. 2.46c shows relative permeability curves when the viscosity ratio was 26.5. Effects of viscosity ratio on relative permeability curves as well as residual saturations are evident from Figs. 2.46a through 2.46c. The endpoint relative permeability of the wetting phase increases as the viscosity ratio increases, while the endpoint relative permeability of the nonwetting phase decreases. Data for the displacement of the nonwetting fluid by the wetting fluid for the same range of viscosity ratios indicated little effect of viscosity ratio on displacement performance, and, thus, on relative permeability curves.

Summary

There is little effect of viscosity ratio on relative permeability curves when a wetting fluid displaces a nonwetting fluid in strongly wetted systems. Limited data for systems of intermediate wettability indicate large effects. Changes of relative permeability curves with viscosity have been observed for viscosity ratios between 1 and 26.5 for the displacement of a wetting fluid by a nonwetting fluid. This effect of viscosity ratio on relative permeability appears to be confined to viscosity ratios less than about 100.

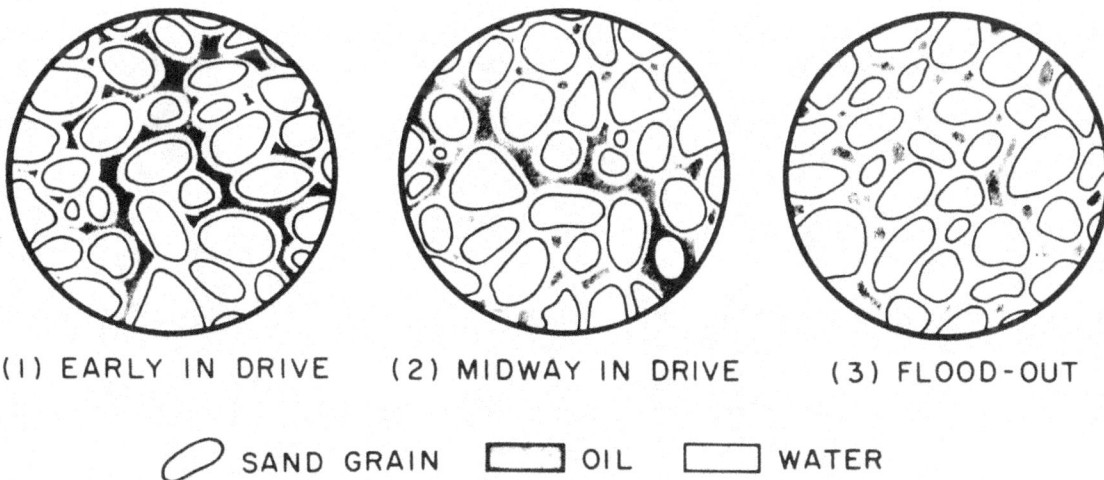

(1) EARLY IN DRIVE   (2) MIDWAY IN DRIVE   (3) FLOOD-OUT

SAND GRAIN   OIL   WATER

Fig. 2.47—Fluid distribution during waterflood of water-wet rock (Ref. 17, Fig. 2.5, after Ref. 54).

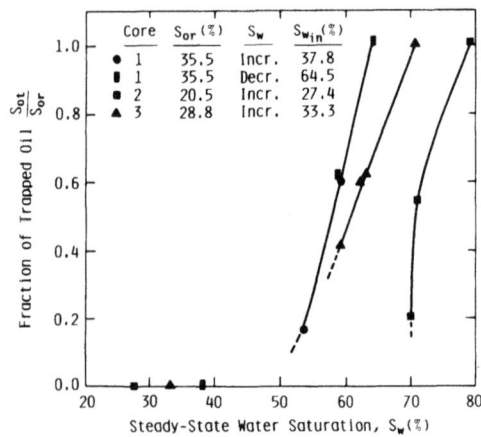

Fig. 2.48—Oil trapped on imbibition as a function of water saturation.[63]

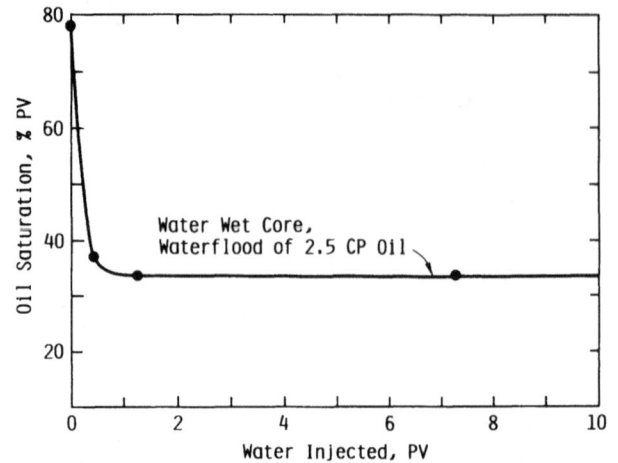

Fig. 2.49—Waterflood behavior for a water-wet core.[15]

2.5 Residual Oil Saturation

The oil saturation that remains trapped in a reservoir rock after a displacement process is dependent on many variables. These include wettability, pore size distribution, microscopic heterogeneity of the rock, and properties of the displacing fluid.

We begin this section by examining the characteristics of water-wet systems in which oil has been displaced by water to a residual saturation. It is assumed that the displacement process occurs without bypassing, which has been attributed to viscous fingering or rock heterogeneities. The value of the residual oil saturation is important for two reasons. First, it establishes the maximum efficiency for the displacement of oil by water on a microscopic level. Secondly, it is the initial saturation for EOR processes in regions of a reservoir previously swept by a waterflood.

2.5.1 Trapping of Residual Oil

The residual oil saturation in a strongly water-wet rock exists as disconnected ganglia depicted in Fig. 2.20. The trapping process has been revealed qualitatively from photographs of fluid distributions in a simulated water drive described in Sec. 2.4.1. Drawings depicting fluid distri-

butions are shown in Fig. 2.47. Some trapped oil is shown in the transition region (2) that separates the portion of the reservoir not contacted by the injected water from that portion where the remaining oil is completely trapped (3).

Raimondi and Torcaso[63] studied the trapping of residual oil by determining the fraction of the oil saturation that was hydraulically connected on the imbibition path. The oil phase is connected throughout the entire porous media over some saturation interval. As trapping occurs, part of the oil is no longer hydraulically connected to the flowing phase. This is illustrated in Fig. 2.48 where the fraction of S_{or} that has become disconnected is plotted against the average water saturation. The trapping process occurs over a narrow range of water saturations, suggesting that there is a distribution of the sizes of ganglia. A model of the trapping process based on a normal distribution of ganglion sizes has been proposed.[86]

Displacement performance on short cores also shows the characteristic behavior of a water-wet core. Fig. 2.49 shows how the oil saturation changes with the quantity of water injected.[15] The residual oil saturation of 0.335 was reached after about 1.3 PV's of water were injected. Thus the trapping process was abrupt as depicted in Figs. 2.47 and 2.48.

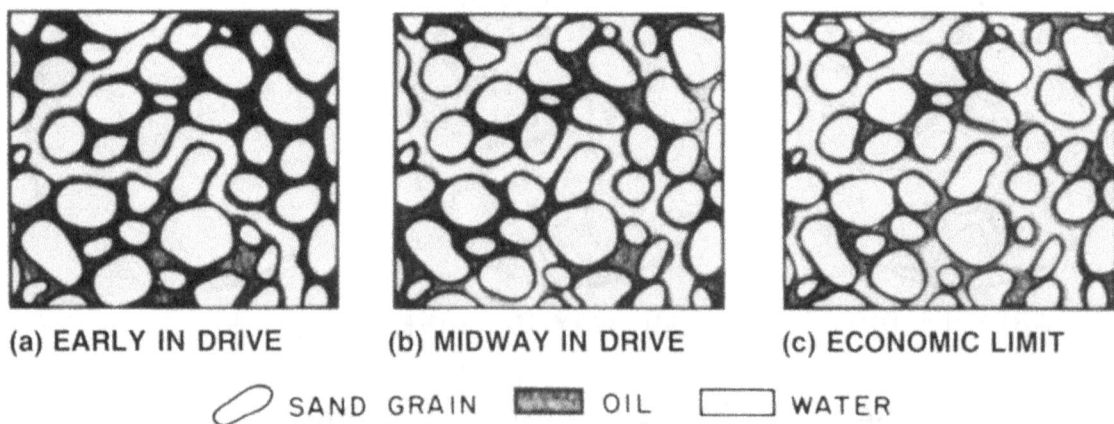

(a) EARLY IN DRIVE   (b) MIDWAY IN DRIVE   (c) ECONOMIC LIMIT

◯ SAND GRAIN   ▨ OIL   ▢ WATER

Fig. 2.50—Fluid distribution during waterflood of an oil-wet rock (Ref. 17, Fig 2.6, Ref. 54).

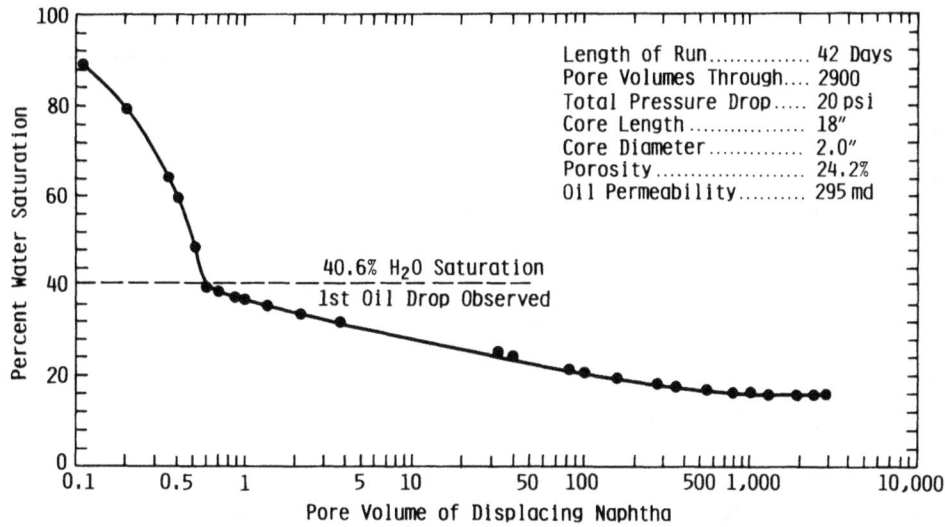

Fig. 2.51—Interstitial water saturation in a long core as a function of PV's of displacing naphtha.[87]

In strongly oil-wet porous media, the oil covers all surfaces at interstitial water saturation. Interstitial water may exist as isolated ganglia analogous to the Wood's metal replicas in Fig. 2.20, when the initial saturation is high (20 to 30%), or as spherical droplets that could occur if connate water saturation is low (10%). When oil is displaced by water, flow channels develop, as depicted in Fig. 2.50.

The trapping process in uniformly oil-wet rock differs from the process in uniformly water-wet rock. An oil film surrounds the sand grains and is connected to smaller flow channels. Oil flow persists at diminishing rates until the smallest oil channels can no longer transmit fluid under the prevailing pressure gradient. Craig described Fig. 2.50c as the saturation at the economic limit because an immobile saturation had not been reached.

Trapping of residual oil in an oil-wet rock is exactly analogous to the trapping of interstitial water during an oilflood of a strongly water-wet rock. Oilfloods are commonly done in laboratory work to reach interstitial water saturation. Fig. 2.51 presents data from a typical oilflood. Note that the water continues to be displaced after several hundred PV's of oil were injected.[87] In this case an immobile saturation had not been reached. The difference in displacement behavior between waterflood in Figs. 2.47 and 2.51 is directly related to wettability. The time required to reach a specific interstitial water saturation can be (and usually is) reduced when a viscous oil is used in the oilflood rather than naphtha.

Rocks that exhibit nonuniform wetting properties have not been studied extensively. The work of Salathiel[15] clearly demonstrates the possibility of residual oil satu-

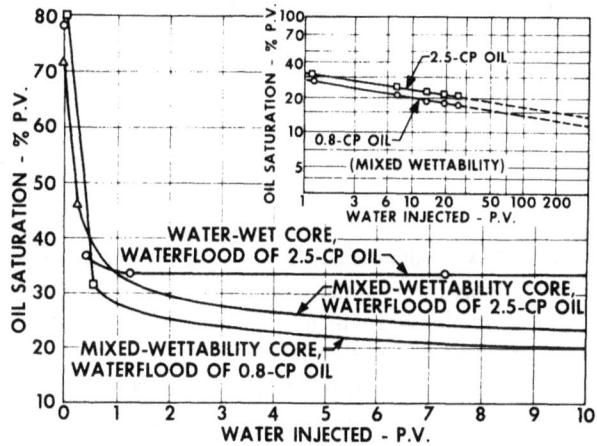

Fig. 2.52—Comparison of waterflood behavior for mixed wettability and water-wet cores (insert shows extension of mixed-wettability flooding data).[15]

rations in mixed wettability conditions that are lower than expected for strongly water-wet rock. Fig. 2.52 is a comparison of waterflood behavior for mixed wettability and water-wet cores.[15] Oil saturation in the mixed-wettability cores continued to decrease with cumulative water injected and had not reached "residual" saturations after 25 PV's of water were injected.

Table 2.5 contains a comparison of residual saturations for waterfloods on cores that had been treated to produce mixed wettability. Residual oil saturations under mixed

TABLE 2.5—RESIDUAL SATURATION AFTER 25 PV WATERFLOODING[15]

Rock Sample	Permeability (md)	Porosity (%)	S_w at Time of "Contact" (% PV)	S_o After 25 PV Waterflooding Water Wet	S_o After 25 PV Waterflooding Mixed Wettability
Boise (sandstone)	1,094	29.3	13.5	33.5	20.5
Upper Austin (sandstone)	595	28.0	20.0	30.0	22.9
Woodbine Outcrop (sandstone)	690	33.0	17.0	27.3	30.7
Upper Noodle (limestone)	620	21.2	18.9	40.5	28.1
Lissie (sandstone)	536	21.9	7.2	42.5	29.1

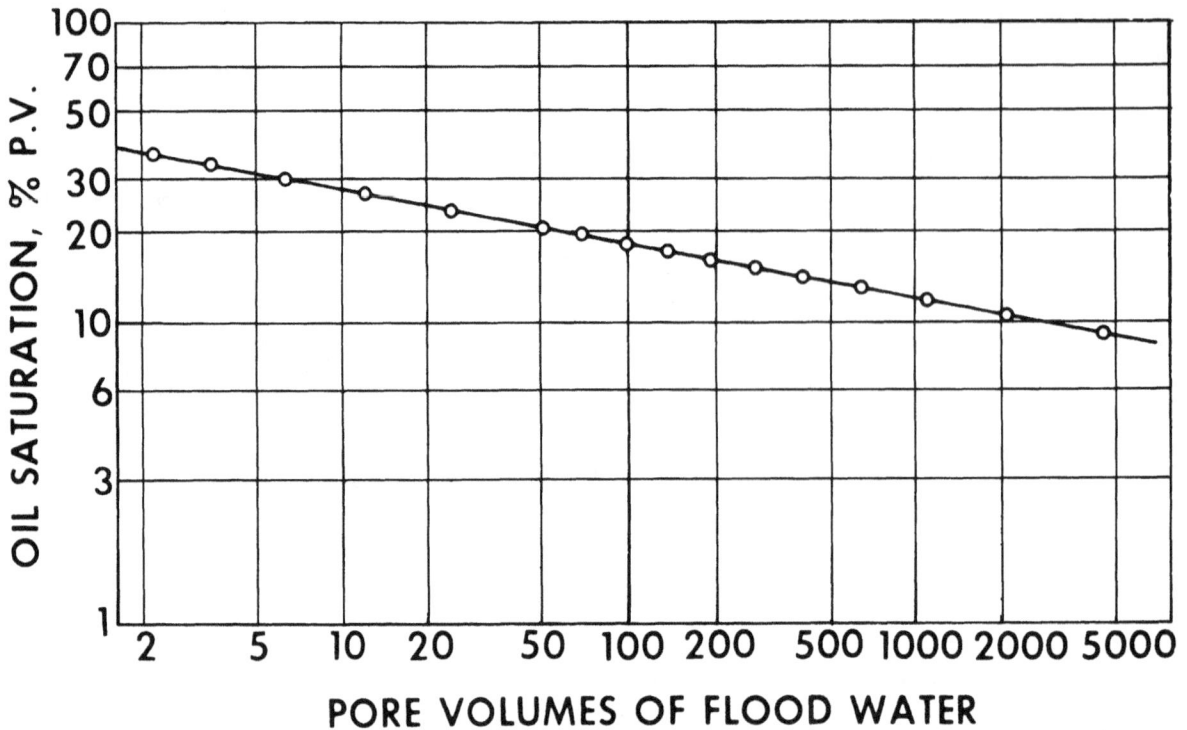

Fig. 2.53—Extended waterflood data on a preserved East Texas field core.[15]

wettability were significantly less than residual saturations when the cores were water wet. These low residual saturations were obtained at high ratios of water/oil after many PV's of water had been injected.[15]

The extent of the mixed wettability condition in oil reservoirs is not known. There are excellent data from the Woodbine reservoir, East Texas field, that support the concept of mixed wettability. Extended-waterflooding restored-state cores leave residual saturations less than 10% PV as indicated by Fig. 2.53.

These laboratory results are supported by pressure cores recovered at depths of 20 ft [6.1 m] or more below the present water/oil contact that were found to contain less than 10% PV of oil.

2.5.1.1 Microscopic Displacement Efficiency. It is useful to measure the effectiveness of the displacement process. The microscopic displacement efficiency of a waterflood, E_D, is defined as follows: E_D = stock-tank oil recovered per unit PV contacted by water divided by stock-tank oil in place at beginning of waterflood per unit contacted by water. E_D may also be expressed in terms of saturation changes by Eq. 2.15:

$$E_D = \frac{\dfrac{\overline{S_{o1}}}{B_{o1}} - \dfrac{\overline{S_o}}{B_o}}{\dfrac{\overline{S_{o1}}}{B_{o1}}}, \quad \dotsello \text{(2.15)}$$

where

$\overline{S_{o1}}$ = volumetric average oil saturation at the beginning of the waterflood, where the average pressure is $\overline{p}_1$, fraction,

$\overline{S_o}$ = volumetric average oil saturation at a particular point during the waterflood,

B_{o1} = oil FVF at pressure $\overline{p}_1$, bbl/STB [m^3/stock-tank m^3], and

B_o = oil FVF at a particular point during the waterflood, bbl/STB [m^3/stock-tank m^3].

When the oil saturation in the PV swept by water is reduced to the residual saturation (S_{or}),

$$E_D = 1 - \left(\frac{S_{or}}{\overline{\overline{S_{o1}}}}\right)\left(\frac{B_{o1}}{B_o}\right), \quad \dotsello \text{(2.16)}$$

which becomes

$$E_D = 1 - \frac{S_{or}}{\overline{\overline{S_{o1}}}} \quad \dotsello \text{(2.17)}$$

when the FVF's for oil are equal. Eq. 2.16 or 2.17 represents the maximum displacement efficiency.

Example 2.1. A reservoir is to be waterflooded immediately after drilling and completion. Waterflood tests done under simulated reservoir conditions indicate that the residual oil saturation wil be 0.28 in a core where the in-

**TABLE 2.6—EFFECT OF VISCOUS AND CAPILLARY FORCES
PRESENT AT TRAPPING ON RESIDUAL OIL SATURATION FOR
DISPLACEMENT OF OIL BY WATER IN WATER-WET CORES[93]**

	Residual Oil Saturation Fraction PV		
Core Material	Torpedo	Elgin	Berea
Base Case			
Flood rate = 2 ft/D [0.007 mm/s] μ_o/μ_w = 1.0 IFT 30 dynes/cm [30 mN/m]	0.416	0.482	0.495
Change of Viscous Forces			
Flood rate = 200 ft/D [0.007 mm/s] Displacing phase viscosity*	0.338	0.323	0.395
μ_o/μ_w = 0.055, flood rate 2 ft/D [0.007 mm/s]	0.193	0.275	0.315
Change of Capillary Forces			
IFT* = 1.5 mN/m [1.5 dynes/cm]	0.285	0.275	0.315

*Base case for other parameters.

itial oil saturation was 0.67. Estimate the displacement efficiency of the waterflood if the oil FVF is 1.45 and does not change during the waterflood. There is no initial gas saturation.

Solution. Because FVF's do not change,

$$E_D = 1 - \frac{0.28}{0.67} = 0.58.$$

Therefore, the displacement efficiency is 58% in those regions of the reservoir that have been flooded to residual oil saturation.

The oil displaced by a waterflood of a reservoir in which V_{pw} has been swept to an average oil saturation of S_{or} is given by Eq. 2.18:

$$N_{pw} = E_D V_{pw} \frac{\overline{S_{ol}}}{B_{ol}}, \quad \ldots\ldots\ldots\ldots\ldots\ldots (2.18)$$

where

N_{pw} = oil displaced by water, STB [stock-tank m³], and

V_{pw} = PV that has been swept by water to a volumetric average saturation of S_{or}.

Because the average oil saturation in the swept region decreases with PV's of water injected, E_D also varies with PV's of water injected.

2.5.1.2 Effect of Viscous and Capillary Forces on Residual Oil Saturations. In the early and mid-1950's, the effect of rate on oil recovery by waterflooding was investigated intensively.[88-93] (See also Ref. 17, Pages 39–43.) There were debates on how results from short laboratory cores could be scaled to reservoir conditions.[91,92] One part of the problem involved the relative importance of viscous forces to capillary forces on the residual oil saturation.

Table 2.6 illustrates the effects of changing viscous and capillary forces on the residual oil saturation in water-wet cores.[93] Viscous forces were changed by increasing

TABLE 2.7—COMPARISON OF RESIDUAL OIL SATURATIONS IN LONG AND SHORT WATER-WET CORES[93]

Run No.	Core Material and Sample No.	Air Permeability (md)	Viscosity[g] Ratio (μ_o/μ_w)	IFT (dyne/cm)	Flood Rate[f] (ft/D)	Residual Oil Saturation, % PV		
						Long Core	Core Plug[e]	Difference
1	Torpedo 1	400	1.3	30	1	42.9[a]	46.4	3.5
2	Torpedo 1	400	1.3	30	30	42.0[a]	44.0	2.0
3	Torpedo 2	400	1.3	30	1	44.6[a]	44.8	0.2
4	Torpedo 2	400	1.3	30	30	42.3[a]	43.6	1.3
5	Torpedo 3	400	1.3	30	2	43.1[b]	41.5	1.6
6	Torpedo 1	400	1.0	0.5	2	39.8[a]	38.6	1.2
7	Torpedo 1	400	1.0	0.5	20	28.3[a]	29.0	0.7
8	Bandera	14	1.3	30	2	39.6[c]	42.4	2.8
9	Woodbine	2,500	1.3	30	60	37.2[c]	34.8	2.4
10	Berea	350	1.3	30	180	38.5[d]	38.2	0.3
11	Berea	350	1.0	0.5	5	25.5[d]	29.5	4.0

Average deviation = 1.7% PV

[a] 3-ft × 2-in. diameter.
[b] 9-ft × 2-in. diameter.
[c] 1-ft × 2-in. diameter.
[d] 8-ft × 2-in. diameter.
[e] 1-in. × ⅞-in. diameter.
[f] Linear rate of advance of flood front.
[g] The viscosity of each phase is approximately 1 cp in this series of experiments.

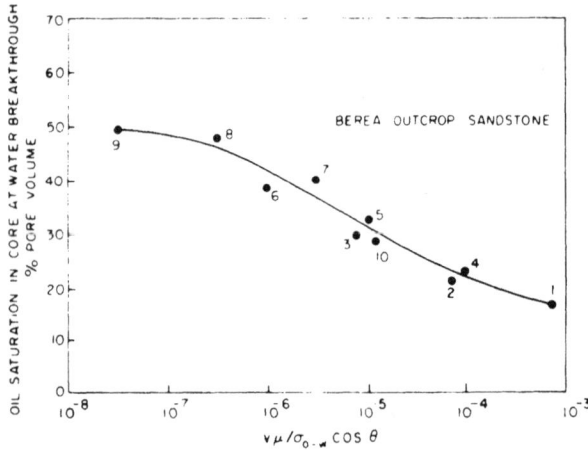

Fig. 2.54—Correlation of N_{ca}/cos θ with oil saturation in core at breakthrough before period (Ref. 94 after Ref. 93). (Note: Moore and Slobod used μ in centipoise. Abrams changed their scale to correspond to μ in poises.)

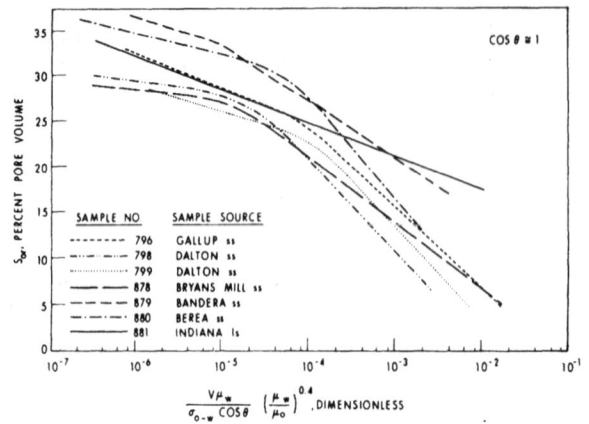

Fig. 2.56—Effect of fluid viscosity, IFT, and flow velocity on S_{or} for various rock samples.[94]

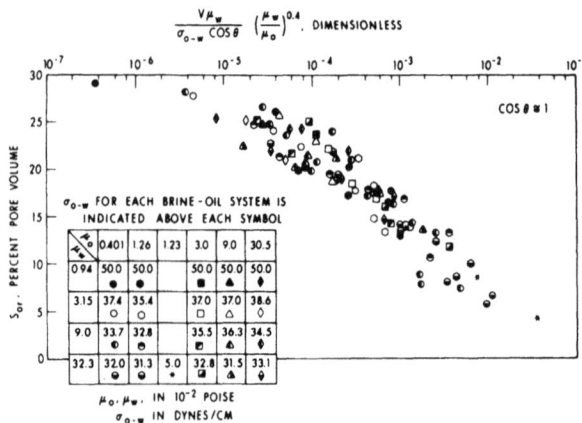

Fig. 2.55—Effect of fluid viscosity, IFT, and flow velocity on S_{or} for Sample 799.[94]

Residual oil saturations obtained from short core plugs (1 in. [2.54 cm]) should be the same as saturations from long cores (1 to 10 ft [0.305 to 3.05 m]) as long as properties of the rock and fluids are identical. Table 2.7 contains a comparison of results for waterfloods on short and long cores.[93] The residual saturations agree within experimental error.

2.5.2 Correlation of Residual Oil Saturation With Capillary and Viscous Forces

The dependence of residual oil saturation on the capillary and viscous forces present at the time of trapping was demonstrated by Moore and Slobod[93] and verified by the extensive experiments of Abrams[94] for water-wet porous media. Using concepts of dimensional analysis and scaling, Moore and Slobod[93] proposed that the residual oil saturation should be a function of a dimensionless group representing the ratio of viscous forces to capillary forces. Eq. 2.19 defines this group, where

the velocity and/or viscosity of the displacing phase. Capillary forces were altered by adding tertiary butyl alcohol to fluids to reduce IFT. The connate and injected fluids were identical. For example, the viscosity ratio (μ_o/μ_w) was changed by increasing the viscosity of both connate and injected fluid. Thus viscosities and IFT's were those existing at the instant that trapping of the residual phase occurred. The results in Table 2.6 were considered representative of several thousand flooding tests on core plugs as well as many long core displacements.

The significance of these data is the experimental verification of the dependence of residual oil saturation on the relative magnitude of capillary and viscous forces that are present at the time of trapping for the displacement of oil by water from a water-wet rock. These results were confirmed in an extensive study by Abrams.[94] The development of a theory to correlate these observations as well as those of other investigators is presented in Sec. 2.5.3.

In water-wet cores, the fact that residual oil exists as disconnected ganglia has further implications. The length of the ganglia, on the order of 1 to 10 pore diameters,[95] is small compared to cores used in laboratory tests.

$$\frac{\text{viscous forces}}{\text{capillary forces}} = \frac{v\mu_w}{\sigma_{ow}\cos\theta}, \dots\dots\dots(2.19)$$

- v = interstitial velocity, which is u/ϕ, ft/D [m/s],
- μ_w = viscosity of connate and displacing fluid, cp [Pa·s],
- σ_{ow} = IFT between oil and displacing fluid, dynes/cm [mN/m], and
- θ = contact angle.

Many authors proposed similar groups to relate the magnitude of viscous forces to capillary forces. Melrose and Brandner[95] introduced the term capillary number (N_{ca}) defined by Eq. 2.20.

$$N_{ca} = \frac{v\mu_w}{\sigma_{ow}}. \dots\dots\dots\dots\dots(2.20)$$

This terminology will be used in all correlations in this text.

Fig. 2.54 illustrates the correlation of oil saturation at water breakthrough for one set of the data from Moore and Slobod[93] with $N_{ca}/\cos\theta$. The data in Fig. 2.54 represent the oil saturation in the cores when water was first detected in the effluent. Since only a small amount of oil was produced after breakthrough of water, the results are representative of the trapping process. The correlation shows a decrease in oil saturation at the time of trapping as the ratio of viscous to capillary force increases.

The generality of this correlation was demonstrated by Abrams.[94] Abrams investigated the influence of IFT, fluid viscosity, and flow velocity on S_{or} for six different sandstones and a limestone. All rock samples were treated to make them strongly water wet. Abrams correlated the residual oil saturation using a modified capillary number. The velocity, v, in Eq. 2.20 was replaced by $v/(S_{oi}-S_{or})$ for floods conducted at constant injection rate. Scatter was reduced by adding a term to represent viscosity effects. The modified capillary number is approximated by Eq. 2.21 for waterfloods at constant injection rate:

$$N_{cam}=\frac{v\mu_w}{(S_{oi}-S_{or})\sigma_{ow}\cos\theta}\left(\frac{\mu_w}{\mu_o}\right)^{0.4}. \quad\ldots\ldots (2.21)$$

Fig. 2.55 shows the correlation of experimental data for a 140-md Dalton sandstone using the viscosity term. Fig. 2.56 shows the correlation of data from the seven rock samples. Normal waterfloods are represented by values of the modified capillary number $(<10^{-6})$.

All sandstone correlations have a characteristic behavior. At low values of the modified capillary number $(<10^{-6})$, the residual oil saturation varies only slightly. This is the region where capillary forces dominate the displacement process. Each sandstone has a distinct break that represents the transition from a displacement process dominated by capillary forces to one dominated by a competition between viscous forces and capillary forces. The transition occurs for values of N_{cam} in the range of 10^{-4} to 10^{-5}. Correlations are different for each rock type, as might be anticipated from rock properties.

The correlation of residual oil saturation with the ratio of viscous to capillary forces has important applications for water-wet porous rocks. First, it is possible to demonstrate the independence of residual oil saturation from flood velocity at rates expected in reservoirs. Second, the determination of residual oil saturation after a waterflood as well as ultimate recovery has been done by conducting flood tests or flood pot tests[96] on small cores at high rates (Ref. 17, Page 22). High rates are used to overcome capillary end effects (see Sec. 3.2.5). The correlations show that residual oil saturations can be reduced below "normal waterflood residual" if the laboratory coreflood is conducted at a large value of N_{cam}.

Example 2.2. The residual oil saturation of a strongly water-wetted sandstone core was determined to be 0.258 in a routine laboratory waterflood. Initial oil saturation was 0.68. The core was waterflooded by injecting water at a constant pressure drop. At the end of the waterflood, the flood rate was 0.0069 cu in./min [0.113 cm³/s]. Porosity of the core was 0.187 and the diameter was 1.106 in. [2.81 cm]. IFT between the oil and the water is 53

dynes/cm [53 mN/m] and the viscosity of the oil is 6 cp [6 mPa·s]. What is the probability that the residual oil saturation is representative of the value expected in a field flood?

Solution. The residual oil saturation will be independent of viscous forces when $N_{cam}\leq10^{-6}$.

Because the sandstone is strongly water-wet, $\cos\theta=1.0$.

$$v=\frac{q}{A\phi}=\frac{0.113\text{ cm}^3/s}{\frac{\pi}{4}(2.81\text{ cm})^2(0.187)}$$

$$=0.0182\text{ cm/s}$$

$$=1.82\times10^{-4}\text{ m/s};$$

$$N_{cam}=\left[\frac{v\mu_w}{(S_{oi}-S_{or})\sigma_{ow}}\right]\left(\frac{\mu_w}{\mu_o}\right)^{0.4}$$

$$=\left[\frac{(1.82\times10^{-4}\text{ m/s})(0.001\text{ Pa}\cdot s)}{(0.422)0.053\text{ Pa}}\right]$$

$$\times\left(\frac{0.001\text{ Pa}\cdot s}{0.006\text{ Pa}\cdot s}\right)^{0.4}$$

$$=(8.13\times10^{-6})\,(0.4884)$$

$$=3.97\times10^{-6}.$$

From Fig. 2.56 this value of the modified capillary number is in the area where viscous forces have a small effect on residual saturations in most sandstone rocks. However, a reduction of 1 to 2% saturation could be attributed to viscous forces generated by high flow rates if the core is similar to Sample No. 799 in Fig. 2.55.

2.5.2.1 The Effect of Trapped Gas on Residual Oil Saturation.

Waterfloods in solution gas-drive reservoirs usually begin after reservoir pressure has declined and GOR's have become excessive. At this point, there is an appreciable free-gas saturation in the pore space. When water is injected into a porous medium containing oil, water, and gas, residual saturations of both oil and gas may remain. The effects of a gas saturation on residual oil saturation following water injection will be considered in this section. The mechanics of the displacement process will be discussed in Chap. 3.

The injection of water into a solution gas-drive reservoir usually occurs at rates that cause repressurization of the reservoir. If pressures are high enough, the gas that has been trapped by the displacement process will dissolve in the oil with no effect on subsequent residual oil saturations. Craig (Ref. 17, Page 40) has outlined a procedure to estimate the pressure required to dissolve trapped gas.

The presence of a trapped gas saturation at the time residual oil is trapped by water has a substantial effect on the residual oil saturation in preferentially water-wet

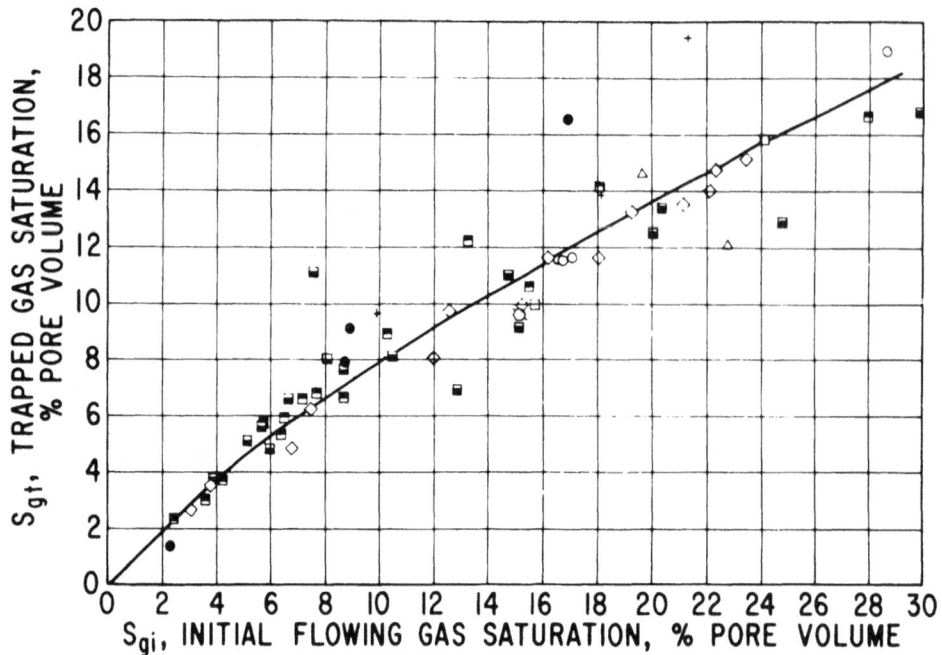

Fig. 2.57— Correlation of initial and trapped gas saturation for preferentially water-wet and oil-wet rocks (Ref. 17, Fig. 3.21).

rock.[96,97] Craig (Ref. 17, Page 41) summarized results of several investigators in Fig. 2.57, which is a correlation of the reduction in residual oil saturation with the initial gas saturation. In consolidated, water-wet rock, a trapped gas saturation reduces the residual oil saturation. The amount of reduction may be estimated from Fig. 2.58 if experimental data are not available for the specific reservoir rock. The data base for the correlation in Figs. 2.57 and 2.58 is limited to sandstones and synthetic media (alundum). Richardson and Perkins[89] show much smaller effects of a gas saturation on residual oil saturations in unconsolidated sand. A change of 2% saturation was detected between waterfloods conducted with no initial gas saturation and those with an initial gas saturation of 25%.

Oil-wet rocks[96] have not been studied extensively. The available data from dri-filmed alundum cores indicate no effect of trapped gas on residual oil saturations.

2.5.3 Conceptual Models—Trapping of Residual Oil

Residual oil is trapped in water-wet rocks as ganglia that may be confined to a single pore or multiple pores. Trapping occurs when the forces exerted on the oil by the flowing water cannot overcome the capillary forces generated by the preferential wettability of the rock by the water.

Basic features of the trapping process can be illustrated when the displacement of oil by water in the pore doublet model in Fig. 2.59 is considered. Extensive dis-

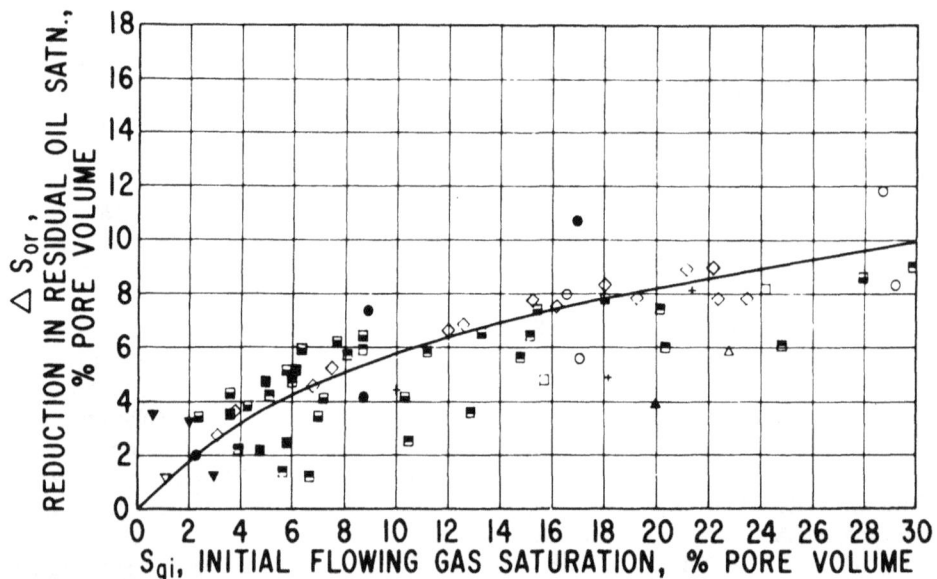

Fig. 2.58—Effect of initial gas saturation on waterflood oil recovery, preferentially water-wet rocks (Ref. 17, Fig. 3.22).

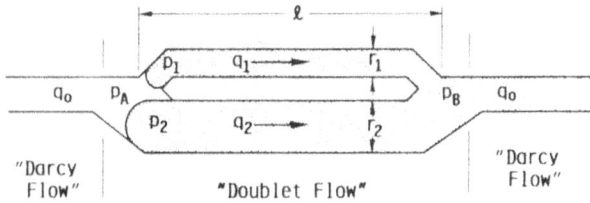

Fig. 2.59—Pore model for displacement at back of oil filament.[98]

TABLE 2.8—COMPARISON OF VISCOUS AND CAPILLARY PRESSURE DROP IN CIRCULAR WATER-WET PORES, PORE VELOCITY AT 1 ft/D [3.53 μm/s]

Pore Radius, r (μm)	Viscous Pressure Drop, $8\,\mu Lv/r^2$ (Pa)	Capillary Pressure Drop, ΔP_c (Pa)	Total Pressure Drop, $p_A - p_B$ (Pa)
2.5	2.26	24,000	$-23,998$
5	0.56	12,000	$-12,000$
10	0.141	6,000	$-6,000$
25	0.023	2,400	$-2,400$
50	0.0056	1,200	$-1,200$
100	0.0014	600	-600

cussions of the pore doublet model can be found in Refs. 98 through 101.

In Fig. 2.59, water displaces oil from two pores that have radii r_1 and r_2, respectively. The two pores are connected at Points A and B to form a pore doublet. The pressures at Points A and B remain constant until trapping occurs. For the purposes of this example, the viscosities and densities of the oil and water phases are equal. Pore 1 is assumed to be smaller than Pore 2. Oil is trapped if displacement proceeds faster in one pore than the other *and* if there is insufficient pressure difference between Points A and B to displace the isolated oil drop from the pore that has the slower displacement rate.

The trapping process in the pore doublet model can be simulated by estimating the velocity of the water in each pore from elementary models of fluid flow and capillary forces. If the densities of both phases are constant, the flow of each phase will be steady, and flow rate can be computed from Poiseuille's equation for laminar flow in a circular tube.[102] When v_1 is the velocity in Pore 1, the pressure drop caused by viscous forces between the flowing fluid and the pore walls is given by Eq. 2.22:

$$\Delta p_1 = \frac{8\mu L_1 v_1}{r_1^{\,2}}, \qquad \dots \dots \dots \dots (2.22)$$

where L_1 is the length of the pore filled with the particular phase.

Because the pores are preferentially wetted by water, there is a difference in pressure between the water and the oil across the oil/water interface as discussed in Sec. 2.2.3. The pressure in the oil phase is higher than the pressure in the water phase as given by Eq. 2.23:

$$\Delta P_c = p_{oi} - p_{wi} = \frac{2\sigma \cos \theta}{r}. \qquad \dots \dots \dots (2.23)$$

If we consider the pressure distribution between Points A and B after water enters Pore 1, we note that

$$p_A - p_B = p_A - p_{wi} + p_{wi} - p_{oi} + p_{oi} - p_B, \quad (2.24)$$

where

$p_A - p_{wi}$ = pressure drop caused by viscous forces in water phase,

$p_{wi} - p_{oi}$ = pressure change across interface caused by capillary force, and

$p_{oi} - p_B$ = pressure drop in oil phase caused by viscous forces.

Substitution of Eqs. 2.22 and 2.23 into Eq. 2.24 for Pore 1 yields Eq. 2.25:

$$p_A - p_B = \frac{8\mu_w L_w v_1}{r_1^{\,2}} - \frac{2\sigma \cos \theta}{r_1} + \frac{8\mu_o L_o v_1}{r_1^{\,2}}.$$
$$\dots \dots \dots \dots \dots \dots (2.25)$$

Because

$$\mu_o = \mu_w = \mu \text{ and } L \simeq L_w + L_o,$$

$$\Delta p_{AB} = \frac{8\mu L v_1}{r_1^{\,2}} - \Delta P_{c1}. \qquad \dots \dots \dots (2.26)$$

where

$\dfrac{8\mu L v_1}{r_1^{\,2}}$ = viscous pressure drop and

ΔP_{c1} = capillary pressure.

It is helpful to examine some numerical values of the two terms on the right side of Eq. 2.26. Consider the displacement of oil by water in a single pore of radius r at a velocity of 1 ft/D [3.53×10^{-6} m/s]. The length of the pore is 500 μm, the viscosity is 1 cp [1 mPa·s], and the IFT is 30 dynes/cm [30 mN/m]. The contact angle, θ, is zero. Table 2.8 shows numerical values of $p_A - p_B$ for different pore radii.

It can be seen from Table 2.8 that $p_A - p_B$ is always negative for the displacement of a nonwetting fluid at rates anticipated in reservoir rocks, as noted by Stegemeier[98] because capillary forces dominate viscous forces in strongly wetted systems.

In the pore doublet model, the same pressure difference ($p_A - p_B$) is present across both pores until water breaks through in one of the pores. Trapping occurs in the pore that has the smallest velocity. For Pore 2, the velocity is obtained from Eq. 2.27.

$$p_A - p_B = \frac{8\mu L v_2}{r_2^{\,2}} - \Delta p_{c2}, \qquad \dots \dots \dots (2.27)$$

where $\Delta p_{c2} = \dfrac{2\sigma \cos \theta}{r_2}$.

TABLE 2.9—PRESSURE DROP REQUIRED TO OBTAIN A
SPECIFIED VELOCITY IN PORE 1 OR 2

Pore 1		Pore 2	
$v_1 = 0$	$\Delta p_{AB} = -\Delta P_{c_1}$	$v_2 = 0$	$\Delta p_{AB} = -\Delta P_{c_2}$
$v_1 > 0$	$\Delta p_{AB} > -\Delta P_{c_1}$	$v_2 > 0$	$\Delta p_{AB} > -\Delta P_{c_2}$
$v_1 < 0$	$\Delta p_{AB} < -\Delta P_{c_1}$	$v_2 < 0$	$\Delta p_{AB} < -\Delta P_{c_2}$

TABLE 2.10—DISPLACEMENT VELOCITY IN SMALL PORE
($r_1 = 2.5$ μm) WHEN DISPLACEMENT VELOCITY IN
LARGE PORE v_2 IS 1 ft/D [3.53 μm/s]

r_2/r_1	Δp_{AB} Pa	v_1 (ft/D)	[m/s]
2	− 12,000	5,326	0.0188
4	− 6,000	7,960	0.0281
10	− 2,400	9,575	0.0338
20	− 1,200	10,085	0.0356
40	− 600	10,368	0.0366

Pressure drops corresponding to zero, positive, and negative velocities in each pore are given in Table 2.9.

Velocities v_1 and v_2 must be positive for displacement to occur simultaneously in both pores. This can happen only when $\Delta p_{AB} > -\Delta P_{c1}$ and $\Delta p_{AB} > -\Delta P_{c2}$. Because $r_2 > r_1$, $\Delta P_{c2} < \Delta P_{c1}$. Simultaneous displacement occurs when $\Delta p_{AB} > -\Delta P_{c2}$. An interesting result from this analysis is illustrated in Table 2.10, in which the displacement velocity in Pore 1 was computed at different ratios of r_2/r_1 when the displacement velocity in Pore 2 was 1 ft/D [3.53×10^{-6} m/s]. The radius of the small pore is 2.5 μm. Other parameters are the same as those used to prepare Table 2.8.

Displacement velocities in Pore 1 in Table 2.10 are 5,000 to 10,000 times larger than the velocity in Pore 2. Thus oil is always displaced from the smaller Pore 1 first when the pores are strongly water wet.

At the moment that all oil in Pore 1 has been displaced, the pressure at B drops and p_A becomes larger than p_B. The oil in Pore 2 may be displaced by water if water does not cut off the oil phase in Pore 2 at B. When cutoff occurs, the oil phase is isolated as a globule. If the velocity remains constant through Pore 1, the pressure difference $p_A - p_B$ caused by frictional losses in that pore is now available to force the isolated oil globule from Pore 2, as depicted in Fig. 2.60.

The pressure difference across a motionless globule caused by capillary forces is

$$p_{w1} - p_{w2} = (p_{w1} - p_{o1}) + (p_{o2} - p_{w2}), \quad \ldots \ldots (2.28)$$

because

$$p_{o1} = p_{o2}.$$

Substituting from Eq. 2.23,

$$\Delta p_w = p_{w1} - p_{w2} = -\frac{2\sigma}{r_A^*} + \frac{2\sigma}{r_B^*}, \quad \ldots \ldots \ldots (2.29)$$

where r_A^* and r_B^* are the radii of curvature at the left and right interfaces. From Eq. 2.29 we see that the isolated globule can be displaced from Pore 2 if $\Delta p_{AB} > \Delta p_w$.

Thus, isolation of the oil in the larger pore leads to trapping when capillary forces holding the globule in Pore 2 are equal to or greater than the viscous forces caused by flow through pores that have been displaced. This may occur by change in the pore radius at B or by a difference in contact angle.

In Sec. 2.3.3, the contact angle was introduced as an indicator of wettability. Two contact angles were described in Fig. 2.25. The advancing contact angle is observed when the water phase advances over a surface not previously covered by water. The receding contact angle is measured as the water phase is displaced from a surface. These contact angles are nearly identical for clean, homogeneous, smooth surfaces.[35]

Hysteresis of the contact angle has been discussed for years[35] in relation to the Jamin effect. Hysteresis can occur when smooth surfaces are contaminated or when there is roughness. Advancing (θ_A) and receding (θ_R) contact angles are shown on Fig. 2.61, assuming $p_{w1} > p_{w2}$. The radius of curvature of the interface that has a contact angle θ in a cylindrical pore of radius r_2 is given by Eq. 2.30.

$$r = \frac{r_2}{\cos \theta}. \quad \ldots \ldots \ldots (2.30)$$

If the globule is at static equilibrium but near the onset of motion, the pressure drop across the globule in Fig. 2.61 is described by Eq. 2.31.

$$p_{w1} - p_{w2} = \frac{2\sigma}{r_2}(\cos \theta_R - \cos \theta_A). \quad \ldots \ldots (2.31)$$

Because $\theta_R < \theta_A$ in this illustration, $\cos \theta_A < \cos \theta_R$. Eq. 2.31 represents the minimum pressure required to displace the globule through Pore 2 when hysteresis of the contact angle is present.

The approximate Δp_{AB} available to displace the isolated globule is the value corresponding to the flow of the water through the smaller pore at reservoir velocities.

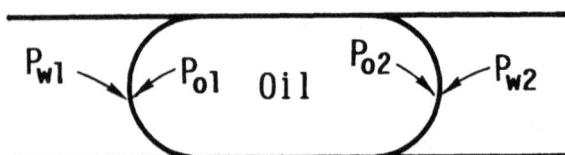

Fig. 2.60—Isolated oil globule in Pore 2 after water breakthrough in Pore 1—equal contact angles.

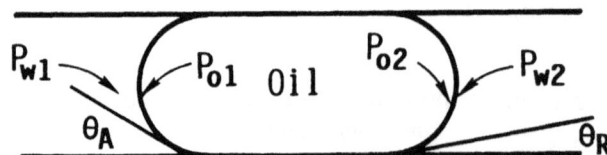

Fig. 2.61—Advancing and receding contact angles for displacement of Pore 2 after water breakthrough in Pore 1.

Some values of Δp_{AB} caused by friction losses are presented in Col. 2 of Table 2.8 for a reservoir velocity of 3.53 μm/s. From these values, it is possible to estimate the change in radius of curvature that is required to hold a globule motionless when Δp_{AB} is known for Pore 1.

Example 2.3. Water flowing in Pore 1 of the pore doublet model has reached Point B, isolating the oil in Pore 2. Radius of Pore 1 is 2.5 μm, and the velocity is 3.53 μm/s. Find the radius of curvature at the downstream end of the drop that would hold the drop motionless if the radius of Pore 2 is 10 μm.

Solution. The pressure drop caused by viscous flow in Pore 1 was computed from Eq. 2.22 in an earlier example and is given in Col. 2 of Table 2.8. From Table 2.8,

$$\Delta p_{AB} = 2.26 \text{ Pa}.$$

The capillary pressure at the upstream interface of Pore 2 was computed from Eq. 2.23 ($\Delta P_c = 6000$ Pa) and appears in Col. 3 of Table 2.8.
Substituting in Eq. 2.29,

$$2.26 = -6,000 + 2\sigma/r_B^*$$

$$1/r_B^* = 6002.26/2\sigma$$

$$= \frac{6002.26 \text{ Pa}}{2(0.03 \text{ Pa}\cdot\text{s})}$$

Note that Pa = N/m^2 and $\sigma = 0.03$ Pa·s = 100 037.7 m^{-1} and $r_B^* = 9.996$ μm.

A decrease in the radius of curvature by 0.004 μm on the downstream side would be sufficient to trap the isolated oil in the 10-μm pore.

The isolated oil could also be trapped in a uniform pore by hysteresis in contact angles. If the receding contact angle (θ_R) is 0°, the advancing contact angle (θ_A) for the motionless globule can be computed from Eq. 2.31.

$$\frac{2\sigma}{r_2}(\cos\theta_R - \cos\theta_A) = 2.26,$$

$$\frac{(2)(0.03 \text{ N/m})}{10 \ \mu\text{m}}(\cos\theta_R - \cos\theta_A) = 2.26,$$

and

$$\cos\theta_R - \cos\theta_A = 3.767 \times 10^{-4}.$$

Because $\theta_R = 0$°,

$$\cos\theta_A = 0.99962,$$

and

$$\theta_A = 1.57°.$$

According to this example, either a change in contact angle caused by hysteresis or a small change in pore radius, r_2, would be sufficient to trap the oil globule.

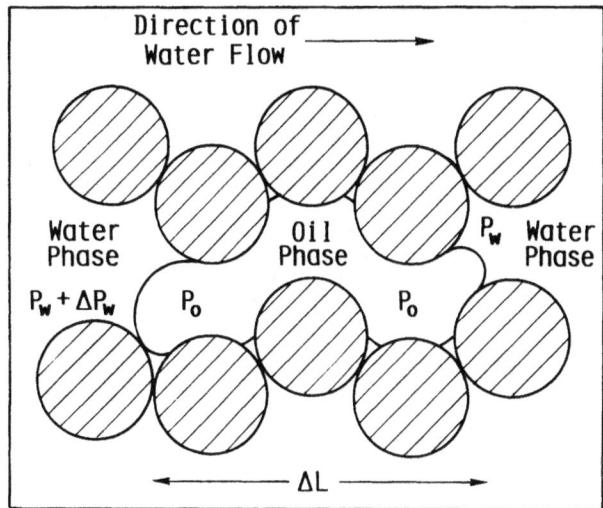

Fig. 2.62—Configuration of trapped oil ganglion.[95]

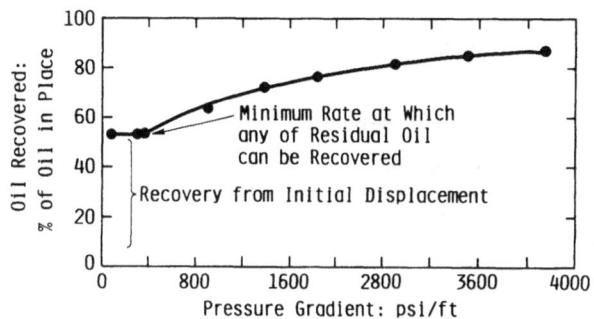

Fig. 2.63—Pressure gradient required to recover the disconnected residual oil.[88]

The pore doublet model is not intended to be an exact representation of the trapping process in porous media. Tortuosity and connectedness of the pore space are not adequately represented. The pore doublet model does illustrate the essential features of the trapping process. The examples show the dominance of capillary forces over viscous forces in the trapping process as well as the influence of capillary forces on the movement of fluids through small pores. Second-order effects, such as entrance losses and transient fluid flow, have been neglected. Other illustrations of trapping in the pore doublet model are presented by Stegemeier.[98]

2.6 Mobilization of Residual Oil

Residual oil is confined in porous media by interfacial forces that exist between the oil and water acting in the pores. Fig. 2.62 illustrates a trapped oil ganglion extending over several pore diameters.[95] Conceptually, one would expect to be able to displace the oil ganglion by increasing the viscous forces that tend to push the ganglion out or by decreasing the interfacial forces that hold the ganglion in place. Several studies provide experimental verification of this expectation.[88,93-95,103,104]

The effect of viscous forces (pressure gradient) on residual oil saturations is shown in Fig. 2.63 for a series of experiments in a Berea sandstone core.[88] The Berea core was saturated with water and flooded to an intersti-

TABLE 2.11—EFFECTS OF VISCOUS AND CAPILLARY FORCES ON THE MOBILIZATION OF RESIDUAL OIL [103]

		Saturating Fluid			Critical, i.e., Oil or Gas First Observed					Conditions at Maximum Rate		
Run No.	Core No.	Non-wetting	Wetting (connate water)	Displacing Fluid	Observed (psi)	(psi/ft)	IFT (dyne/cm)	(psi/ft) (dyne/cm)	Rate (cm²/sec × 10³)	Maximum Δp (psi) Observed	Maximum (psi/ft) (dyne/cm)	Additional Recovery, % S_{or}
1	B-17	air	brine	brine	18.47	112.5	59	1.91	128	20.7	2.15	32.8
2	B-TC	air	brine	1% x – 100	6.1	65.3	30.8	2.10	16.7	10.03	3.5	36.1
3	B-18	oil	brine	brine	13.38	87.4	35.9	2.43	28.5	52.5	9.6	15.0
4	B-19	oil	brine	brine	10.98	72.6	35.9	2.04	34.2	96.0	17.7	29.4
5	B-19	oil	brine	0.1% x – 100	1.77	11.7	4.18	2.80	2.2	22.3	41.0	47.7
6	B-19	oil	aqueous phase	aqueous phase	2.00	13.2	5.25	2.52	1.4	35.3	44.5	57.2
Tests With Initial Rate Greater Than Critical												
7	B-19	oil	brine	0.1% x – 100	3.96	26.2	4.18	6.03	3.1	10.5	21.6	55.2
8	B-19	oil phase (20% isopropanol in total mixture)	aqueous phase	aqueous alcohol phase	4.11	27.2	5.25	5.17	3.27	49	62	50.5
Surfactant and Oil Equilibrated Prior to Experiment												
	B-20	oil phase	0.1% to 100	0.1% to 100	2.35	15.6	3.47	4.5	0.264		382	98

TABLE 2.12—MOBILIZATION OF TRAPPED OIL IN ROCK FROM SANDSTONE RESERVOIRS [104]

Core No.	Permeability to Air (md)	% of Residual Oil Recovered at Maximum $\Delta p/L\sigma$	Maximum $\Delta p/L\sigma$ $\left(\dfrac{\text{lbf/sq in./ft}}{\text{dyne/cm}}\right)$	Critical Value of $\Delta p/L\sigma$ $\left(\dfrac{\text{lbf/sq in./ft}}{\text{dyne/cm}}\right)$
2-269	40.8	17.8	52.3	13.60
2-205	52.1	9.3	52.8	20.10
D-8-9	95.5	4.8	82.6	23.80
J-13	189.0	7.9	61.3	16.30
S-12	242.0	15.7	61.9	11.40
ON-2	320.0	28.2	86.7	4.03
T-4	381.0	22.4	56.2	5.12
Ber-1	346.0	25.1	46.2	6.13
D-1-8	503.0	17.1	26.3	1.71
SP-6	533.0	24.8	15.2	1.90
JK-1	585.0	11.7	20.4	3.81
SP-5	613.0	27.2	30.9	1.12
DOT	635.0	29.8	21.5	2.01
21X	752.0	26.4	30.9	1.66
80	1,450.0	34.1	24.2	0.73
MG-6	2,190.0	48.2	12.2	0.3

tial water saturation of 29% by injection of oil. Then, it was flooded with brine until no oil was produced. As shown in Fig. 2.63, the residual oil saturation was independent of pressure gradient up to a value of about 370 psi/ft [8370 kPa/m]. At this point, the interstitial velocity of the water in the core was 350 ft/D [1.2 mm/s].

As the pressure gradient increased about 370 psi/ft [8370 kPa/m], more oil was displaced from the core. A residual oil saturation of about 11% remained in the core when the experiment was terminated at about 3,750 psi/ft [84.8 MPa/m].

Reservoir pressure gradients are on the order of a few psi/ft. Thus all water rates and pressure gradients in these experiments were larger than any values expected in reservoirs. The example shows that there was a minimum pressure gradient required to mobilize residual oil in this core and that some residual oil held strongly in place by capillary forces can be displaced if the pressure gradient or viscous forces exerted by the flowing water are large enough.

A corresponding relationship between oil recovery and IFT was developed in the mid to late 1960's. Wagner and Leach[6] demonstrated that the residual nonwetting oil phase could be displaced by a nonwetting hydrocarbon vapor phase at reservoir pressure gradients when the IFT was less than 0.7 dynes/cm [0.7 mN/m]. Taber[103] and Taber et al.[104] conducted a systematic investigation of the effects of viscous forces ($\Delta p/L$) and capillary forces (σ_w) on Berea sandstone cores to determine the onset of residual oil mobilization as well as the relationship between residual oil saturation and ($\Delta p/L$). Several different fluid pairs were investigated with IFT's ranging from 3.5 to 59 dynes/cm [3.5 to 59 mN/m]. In all cases, initial oil production was obtained when a "critical" value of $\Delta p/L\sigma$ was attained, irrespective of whether the critical value was obtained by increasing $\Delta p/L$ or decreasing σ. The "critical" values of $\Delta p/L$ ranged from 1.91 to 2.80 for Berea cores as shown in Table 2.11. Also shown in Table 2.11 is the additional oil recovery from successive increases in $\Delta p/L\sigma$ to the maximum value investigated.

The generality of a critical $\Delta p/L\sigma$ is shown in Table 2.12 where displacement data are reported for 16 cores. Most of these cores were sandstones from oil-producing reservoirs.

2.6.1 Capillary Number Correlations

The correlation of residual oil saturation with $\Delta p/L\sigma$ is consistent with the concept of the capillary number. These two concepts can be related through Darcy's law,[95] as in Eq. 2.32.

$$N_{ca}^* = \frac{u\mu_w}{\sigma_{ow}} = \left(\frac{kk_{rw}}{\mu_w}\frac{\Delta p}{L}\right)\frac{\mu_w}{\sigma_{ow}} = \frac{kk_{rw}}{\sigma_{ow}}\frac{\Delta p}{L}, \quad ..(2.32)$$

where $N_{ca}^* = \phi N_{ca}$.

Taber's "critical" $\Delta p/L\sigma$ translates to an equivalent $(N_{ca})_{\text{crit}}$ for mobilization of residual oil. Fig. 2.64 shows the correlation between residual oil saturation on Taber's Runs 6 and 9 with capillary number.[30]

The shape of the correlation is similar to capillary number correlations shown in Fig. 2.56 for the trapping of connected oil. For $N_{ca}^* < 10^{-5}$ there is no mobilization

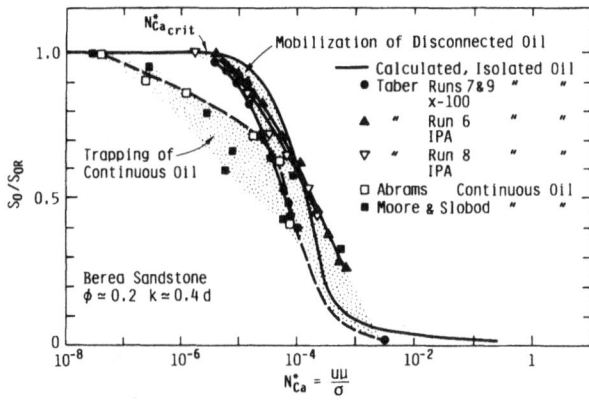

Fig. 2.64—Correlation of residual oil saturation with capillary number at trapping and mobilization. Reprinted with permission from Stegemeier, G.L.: "Mechanisms of Entrapment and Mobilization of Oil in Porous Media," *Improved Oil Recovery by Surfactant and Polymer Flooding,* D.O. Shah and R.S. Schecter (eds.), Academic Press, New York City (1977) 55–59.

TABLE 2.13—EFFECT OF VISCOUS AND CAPILLARY FORCES ON MOBILIZATION OF TRAPPED OIL[93]

Core Material	Residual Oil Saturation, Fraction PV	
	Torpedo	Berea
Base Case		
Flood rate = 2 ft/D [7.06 μm/s] $\mu_o/\mu_w = 1.0$ IFT = 30 dynes/cm [30 mN/m]	0.416	0.495
Change of Viscous Forces After Trapping		
Increase velocity to 200 ft/D [0.007 mm/s] after trapping	0.381	0.426
Change viscosity ratio $(\mu_o/\mu_w) = 0.055$ after trapping	0.41	0.488
Change of Interfacial Forces		
Decrease IFT to 1.5 mN/m [1.5 dyne/cm] after trapping	0.41	0.480

of residual oil. As $N_{ca}^* > 10^{-5}$, the fraction of the residual oil mobilized increases sharply with increasing capillary number.

Experimental data indicate another interesting difference between the trapping and mobilization processes. At the same oil saturation, trapping occurs at a smaller value of the capillary number than mobilization. This is illustrated in Fig. 2.64, which has different correlations for trapping of connected oil and mobilization of disconnected oil when $10^{-7} < N_{ca}^* < 10^{-5}$. Once trapping occurs, the residual oil resists mobilization even though it is flooded at capillary numbers larger than those present at trapping. Table 2.13 illustrates this effect for two types of porous rock from the data of Moore and Slobod.[93]

2.6.2 Models for Mobilization of Residual Oil

Several models have been developed to simulate the conditions necessary to mobilize residual oil. Because the pore space is very complex in rocks and representation is not exact, certain parameters in the models are selected to fit the model to the experimental data.

A model of the mobilization process based on the critical pressure gradient required to move a trapped ganglion is shown in Fig. 2.62.[95] Eq. 2.33 defines the critical capillary number developed by Melrose and Brandner[95] for the case when $\theta = 0$,

$$(N_{ca})_{crit} = \frac{k_{rw}J(S_w)^2 G}{4N\bar{r}_{dr}}, \qquad (2.33)$$

where

$$G = L\frac{\bar{r}_{imb}}{\bar{r}_{dr}}, \qquad (2.34)$$

and

$$J(S_w) = \frac{P_c}{\sigma}\sqrt{\frac{k}{\phi}}. \qquad (2.35)$$

In Eq. 2.33, N represents the length of the trapped ganglion in particle diameters and $J(S_w)$ is the Leverett number obtained from drainage capillary pressure data. The

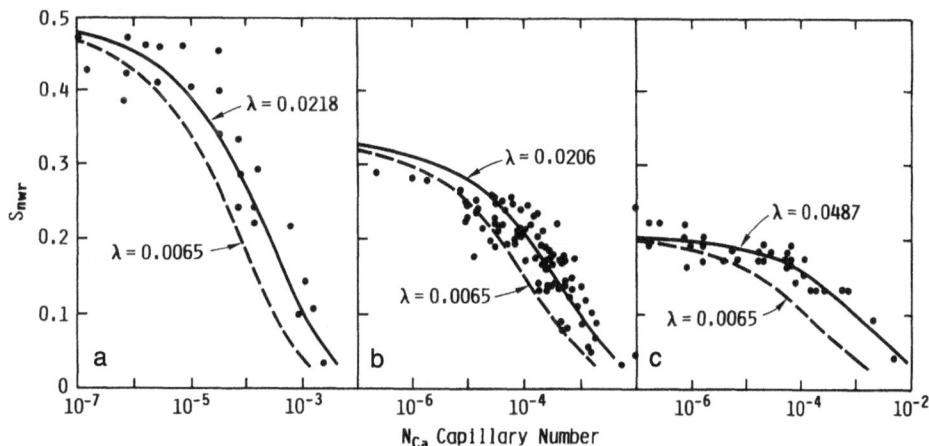

Fig. 2.65—Residual nonwetting phase saturation as a function of capillary number, experiment and theory; (a) Foster and McMillin's data, $\sigma = 2$; (b) Abram's data, $\sigma = 3$; and (c) Lefebvre du Prey's data, $\sigma = 5$. Reprinted by permission from *Nature,* Vol. 268, No. 5619, Copyright ©1977, Macmillan Journals Ltd.[105]

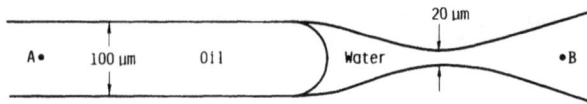

Fig. 2.66—Oil drop in pore throat for Problem 2.5.

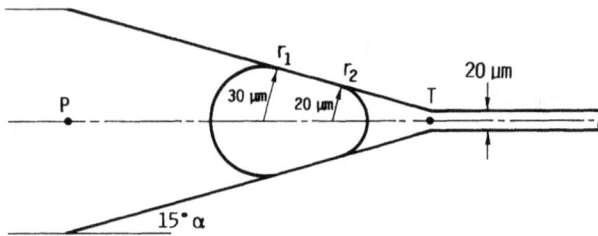

Fig. 2.67—Oil drop trapped in a conical pore, Problem 2.6.

terms $\bar{r}_{imb}$ and $\bar{r}_{dr}$ represent the mean radii of curvature of the drainage and imbibition interfaces at the front and back of the trapped ganglion when the contact angle is zero.

Capillary numbers computed from Eq. 2.33 were reported to be in the range observed experimentally for oil mobilization. Another feature of Eq. 2.33 is that large ganglia are mobilized at smaller critical capillary numbers. This result has been verified[86] in scaled experiments involving single ganglia in a packed bed of spheres.

A similar[33] model incorporates different radii of curvature at the front and back of the ganglion. Terms representing the ratio of pore neck radius to pore body radius and a geometric correction factor to account for contact angles of an interface passing through a toroidal opening are introduced. The solid lines on Fig. 2.64 show the residual oil saturation as a function of capillary number computed from this model for Taber's data in Berea sandstone. The agreement is excellent considering the complexity of pore space.

In another approach, Larson et al.[105] applied percolation theory to simulate trapping of a residual phase in porous media. They were able to fit several sets of residual-oil-saturation/capillary-number correlations with two parameters λ and σ. The parameter λ is a measure of interfacial curvature and the topology of the pore surface, while σ describes the connectedness of the pore space. Computed S_{or}/capillary correlations are shown in Fig. 2.65.

All models of the trapping and mobilization must be considered correlative at this stage of their development. That is, matches between experimental data can be obtained by appropriate choice of model parameters. The prediction of trapping and mobilization for a specific reservoir rock requires an independent method of deriving model parameters from measurable properties of the porous media and fluids.

Problems

2.1. A horizontal capillary tube (radius 10 μm) contains an oil/water interface. The pressure in the oil phase is 14.5 psi [100 kPa]. Determine the pressure in the water phase if the contact angle is 30° and the IFT is 30 dynes/cm [30 mN/m].

2.2. A horizontal capillary tube (radius 1 μm) filled with water connects two large vessels. One vessel contains air at a pressure of 14.5 psi [100 kPa] while the other contains water at a pressure of 14.5 psi [100 kPa]. The capillary tube is strongly water wet ($\theta=0$). IFT between air and water is 72 dynes/cm [72 mN/m]. How much must the pressure in the vessel containing air be increased to displace the water from the capillary tube?

Note: The SI unit for pressure is the pascal (Pa).
   1 Pa=1 N/m^2.
   1 atm=101,325 Pa or 101.325 kPa.

2.3. A capillary tube with radius 20 μm and 3.93 in. [100 mm] in length is filled with water. At time=0, one end of the capillary tube is inserted vertically in a beaker containing oil ($\gamma_o=0.798$) and kept 0.79 in. [20 mm] below the surface of the oil. Then, the other end is exposed to atmospheric pressure. What is the equilibrium distribution of oil and water in the capillary tube if the surface of the tube is (a) strongly oil wet ($\theta=180°$), (b) intermediate wetted ($\theta=90°$), and (c) moderately water wet ($\theta=45°$). IFT between the oil and water is 25 dynes/cm [25 mN/m] and the IFT between air and water is 72 dynes/cm [72 mN/m].

2.4. An oil drop is trapped by water in a cylindrical capillary tube that has an inside radius of 5 μm. The water phase is at atmospheric pressure 14.696 psi [101.325 kPa], and exhibits an IFT of 35 dynes/cm [35 mN/m] against the oil phase. The capillary tube is strongly oil wet (contact angle=180°). Viscosity of the oil is 0.7 cp [0.7 mPa·s] and the viscosity of the water is 1 cp [1.0 mPa·s].

a. Sketch the drop as it would be seen inside the tube when viewed from the side.

b. Compute the pressure in the oil phase in pascals.

2.5. A large oil drop is displaced through a smooth circular pore by water. The pore shown in Fig. 2.66 has a diameter of 100 μm. Near the end of the pore is a throat that has a diameter of 20 μm.

a. Assuming the throat is smooth and the entire pore is water wet with a contact angle of 30°, compute the pressure drop (in pascals) between Points A and B required to displace the drop through the throat.

b. Compute the pressure drop (in pascals) between Points A and B required to displace the oil through the throat if the entire pore is strongly oil wet.

Properties of the oil and water are as follows:

$\rho_w = 1.0$ g/cm^3
$\rho_o = 0.7$ g/cm^3
$\sigma_{ow} = 20$ mN/m
$\mu_o = 6$ mPa·s, and
$\mu_w = 1$ mPa·s.

2.6. An oil bubble is lodged in the conical pore shown in Fig. 2.67. The pore is strongly water wet (contact angle =0°), and there is no fluid flow through the pore. Radii of curvature of the drop are 20 and 30 μm, respectively, as noted on Fig. 2.67. The pore walls have a uniform slope of 15° from Point P to T. Pore throat radius is 10 μm.

Dullien[106] shows that the radii of curvature for a cone are given by Eqs. 2.36 and 2.37 where r_{1b} and r_{2b} are the radii of the bases of cones that pass through Points 1 and 2.

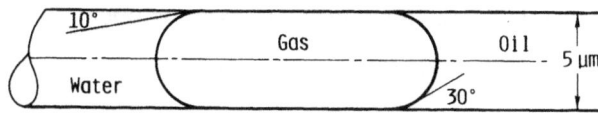

Fig. 2.68—Interfaces between fluids in a circular pore.

$$r_1^* = \frac{r_{1b}}{\cos(\theta - \alpha)} \quad \ldots\ldots\ldots\ldots\ldots\ldots (2.36)$$

and

$$r_2 = \frac{r_{2b}}{\cos(\theta + \alpha)}, \quad \ldots\ldots\ldots\ldots\ldots\ldots (2.37)$$

where θ = contact angle and α = slope of wall.

a. Determine the pressure in the oil bubble and the pressure at Point P when the pressure at Point T is 1 atm [101.325 kPa].

b. What is the minimum pressure that can be applied at Point P that will push the drop through the pore throat? Hint: The volume of the drop is constant but radii of curvature change as the drop approaches the pore throat. The volume of the drop may be estimated with consideration of the drop to be two hemispheres of radii r_1^* and r_2^* separated by the frustrum of a cone with bases r_1^* and r_2^*. It will be helpful to note that the radius of the base (r) is related to the distance from the pore throat (x by Eq. 2.38):

$$\frac{x}{r - 10} = \tan(90 - \alpha). \quad \ldots\ldots\ldots\ldots\ldots (2.38)$$

c. Suppose the IFT between the oil phase and the water phase is 0.1 dynes/cm [0.1 mN/m] at the rear of the drop. What effect would this change have on the results of (a) and (b) if the IFT on the pore neck side (T) remained unchanged?

d. Consider the available pressure difference between Points P and T to be 0.03 psi [0.207 kPa]. What IFT is required to displace the drop through the pore throat?

2.7. The entrance pressure or threshold pressure is the capillary pressure that must be exerted to force the non-wetting phase into a capillary tube or porous rock that is completely saturated with the wetting phase. An oil-wet pore (contact angle 145°) with radius 2 μm is filled with oil. It is desired to displace the oil from the pore with brine that has an IFT of 18 dynes/cm [18 mN/m] in contact with oil. What is the entrance pressure required for the water to enter the pore?

2.8. Fig. 2.68 shows a gas bubble confined in a 5.0-μm-diameter capillary tube that is wet by water in the presence of gas and by oil in the presence of gas. The drop is motionless. The contact angles measured through the wetting phases were 10 and 30°, respectively. Determine the pressure in the water and gas phases if the pressure in the oil phase is 14.5 psi [100 kPa]. IFT's are 72 dynes/cm [72 mN/m] at the water/gas interface and 31 dynes/cm [31 mN/m] at the gas/oil interface.

2.9. A reservoir that is 100 ft [30.5 m] thick has an underlying water zone. Core analyses indicate that the reservoir may be represented as a three-layer system. The

TABLE 2.14—OIL/WATER CAPILLARY PRESSURE DATA DRAINAGE PATH*

k = 900 md		k = 50 md		k = 10 md	
S_w (%)	P_c (psi)	S_w (%)	P_c (psi)	S_w (%)	P_c (psi)
100	0.78	100	1.27	100	1.82
40	1.06	70.0	1.43	90	1.82
38.5	1.15	67.7	1.62	87.7	1.85
36.9	1.25	66.2	1.73	86.2	1.87
35.4	1.27	64.6	1.85	84.2	1.92
33.8	1.32	63.1	1.96	83.1	1.98
32.3	1.43	61.5	2.22	81.5	2.08
30.8	1.62	60.0	2.54	80.0	2.26
28.5	1.89	58.5	2.88	78.5	2.54
26.9	2.17	58.1	3.00	76.9	2.88
25.4	2.45	56.9	3.35	76.2	3.00
23.8	2.91	55.4	3.92	75.4	3.23
23.5	3.0	53.8	4.62	73.8	3.83
22.7	3.46	52.3	5.65	72.3	4.62
20.8	4.06	51.8	6.00	70.0	6.00
19.2	6.0	50.8	6.92	67.7	9.00
17.7	9.0	50.0	7.85	65.8	12.0
16.9	12.0	49.4	9.00	64.2	15.0
16.8	15.0	47.7	12.0		
		46.6	15.0		

*From Ref. 23, Fig. 3–26, after Ref. 107.

upper layer is 20 ft [6.1 m] thick and has a permeability of 900 md. The middle layer is 50 ft [15.2 m] thick and has a permeability of 50 md. Permeability of the lower layer is 10 md. The layers are hydraulically connected, and the free water level is at the bottom of the lower layer.

Estimate the saturation distribution in the reservoir from the capillary pressure data given in Table 2.14. The density of the oil and water are 0.8708 and 1.0 g/cm³, respectively. The reservoir is in capillary (vertical) equilibrium and, thus, there is no fluid flow in the vertical direction. Show that the capillary pressure at elevation Z above a horizontal datum is given by Eq. 2.39:

$$P_c = P_{ci} - g(\rho_o - \rho_w)(Z - Z_i), \quad \ldots\ldots\ldots\ldots (2.39)$$

where

Z_i = elevation of $S_w = 1.0$ above the datum, ft [m], and

P_{ci} = capillary pressure corresponding at $S_w = 1.0$.

Plot the saturation as a function of the distance above the free-water level. Identify the oil/water transition zone. Determine average water and oil saturations in each zone. Would the saturation distribution change if the 900- and 10-md zones were interchanged?

Notes: (a) At the interface between two layers, the oil cap and water phases are continuous such that the oil pressures in each layer are identical, and the water pressures are identical. (b) The oil/water transition zone includes intervals where the water saturation is not equal to the connate water saturation. (c) It is convenient to choose the free water level as the datum.

2.10. What is the significance of a zero value of the capillary pressure in

a. a strongly water-wet rock?

b. a porous rock with intermediate wettability?

2.11. The porous media shown in Fig. 2.69 consists of two horizontal layers that are hydraulically connected and in capillary equilibrium. Properties of each layer are

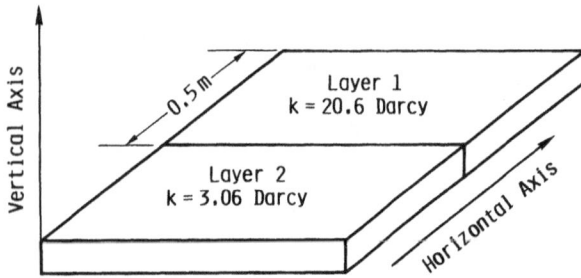

Fig. 2.69—Two horizontal layers at capillary equilibrium.

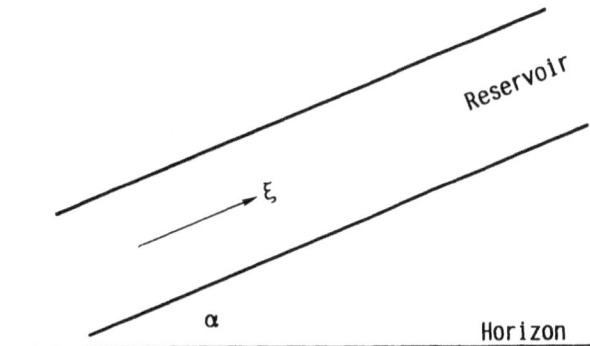

Fig. 2.70—Orientation of reservoir inclined at an angle α with respect to the horizon.

TABLE 2.15—PROPERTIES OF POROUS ROCK

Layer	Permeability (darcies)	Porosity	Width (m)	Thickness (m)
1*	20.6	0.35	0.5	0.0254
2	3.06	0.25	0.5	0.0254

*A sample taken from the center of Layer 1 had a water saturation of 0.4.

The volume of water displaced spontaneously from the cores when immersed in kerosene for 20 hours $= 0.03$ cu in. [0.5 cm^3].

An additional volume of water expelled from the core when centrifuged under kerosene $= 0.119$ cu in. [1.95 cm^3].

The volume of oil spontaneously displaced by water $= 0.006$ cu in. [1.0 cm^3].

An additional volume of kerosene displaced by water when centrifuged $= 0.109$ cu in. [1.8 cm^3].

The PV of the core was 0.21 cu in. [3.4 cm^3].

Final water saturation determined by Soxhlet extraction was 0.80.

Compute the relative displacement index. Are the data internally consistent?

2.16. A wettability method based on centrifuging core samples is described by Donaldson *et al.*[4] In this method, a cylindrical core sample 0.75 in. [1.9 cm] in diameter by 0.9 in. [2.3 cm] long is used. Suppose the wettability tests were conducted on cubic core samples that were $0.8 \times 0.8 \times 0.8$ in. [$2 \times 2 \times 2$ cm]. Would you expect to obtain the same wettability index as would be found with cylindrical samples?

2.17. Describe the nature of multiphase flows in porous media (a) when viscous forces are low and (b) when viscous forces are large.

2.18. Fig. 2.70 shows a reservoir that is inclined at an angle α with respect to the horizon. Write the expression that describes the flow of oil in the direction ξ. Define all variables.

2.19. Laboratory cores are often reduced to interstitial water saturation by oil flooding a core that is 100% saturated with water. Sketch the relative permeability curves for the oil and water phases assuming base permeability is the permeability to oil at interstitial water saturation and the rock is strongly water wet.

2.20. (a) Sketch the oil and water relative permeability curves for the displacement of oil by water from a strongly water-wet reservoir. The waterflood begins at interstitial saturation. Label key points of these curves.

(b) On the same plot, sketch (using dashed curves) the oil and water relative permeability curves for a waterflood of a uniformly oil-wet reservoir.

2.21. Discuss the effects of viscosity ratio and IFT on relative permeability curves in uniformly water-wet porous rocks.

2.22. Craig (Ref. 17, Page 25) observes that relative permeability curves for oil-wet rocks are independent of interstitial water saturation as long as the interstitial water

given in Table 2.15. The relationship between capillary pressure and water saturation is given by Eqs. 2.40 and 2.41 where z is a correlating parameter. Units of capillary pressure are kPa and the permeability is expressed in m^2.

$$P_c = \frac{10^2 z}{\sqrt{\dfrac{k}{3.02 \times 10^{-12}}}}, \quad \dots\dots\dots\dots\dots (2.40)$$

and

$$S_w = \frac{0.85}{1 + 1.845z} - \frac{2z(z - 0.19)}{1 + 144z^4}. \quad \dots\dots\dots (2.41)$$

(a) Determine the water saturation distribution in each layer, neglecting the effect of gravity forces. (b) Determine the volumetric average water saturation for Layers 1 and 2 combined.

2.12. Fig. 2.17 contains a drainage capillary curve for Berea sandstone that has two points marked 4 and 1. Explain why the capillary pressure is higher at Point 4 than at Point 1.

2.13. Define wettability and discuss the causes of wettability differences in oil reservoirs.

2.14. Describe the concept of mixed wettability and indicate how this wettability model differs from a system that exhibits uniform wettability.

2.15. The following data were reported from a series of experiments to determine wettability of a rock by the Amott method.[37] The core was prepared for the test as described in Sec. 2.3.3.2.

saturation is less than 0.20. What does this observation mean in terms of location of interstitial water in strongly oil-wet rocks?

2.23. The mechanism of slug flow was discussed in Sec. 2.4.1.2. Estimate the range of capillary numbers that characterize the onset of slug flow.

2.24. Compare the microscopic displacement efficiency of a waterflood of a strongly water-wet rock with the microscopic displacement efficiency of a waterflood in a strongly oil-wet rock.

2.25. Fig. 2.51 shows the change in interstitial water saturation as a function of PV of naphtha injected for an oilflood of a water-wet rock.

a. Define a microscopic displacement efficiency E_D for this process assuming $B_o = B_w = 1.00$.

b. Plot E_D vs. PV of naphtha injected.

c. Compare (b) with E_D for the waterflood of a water-wet core in Fig. 2.49, by plotting these data as a dashed curve on the graph prepared in (b).

2.26. When a waterflood is conducted in a reservoir that was depleted by solution gas drive, there is an initial gas saturation. Discuss the effect of an initial gas saturation on the recovery of oil by waterflooding.

2.27. Describe the trapping process during a waterflood of an oil-wet reservoir.

2.28. High-molecular-weight polymers in dilute concentrations can increase the viscosity of water significantly. For example, a few hundred ppm of certain polymers will raise the viscosity of water from 1 to 100 cp [1 to 100 mPa·s]. You are assigned to investigate the possibility of installing polymer flood in a strongly water-wet reservoir. What would the effect of a polymer flood be on the residual oil saturation if (a) the polymer were injected at the beginning of the waterflood or (b) the polymer were injected near the end of the waterflood.

Properties of the reservoir rock and fluids are given in Table 2.16. The average interstitial velocity in the reservoir is estimated to be 2 ft/D [7.06 μm/s].

2.29. The data in Table 2.17 were reported as representative of the waterflood recovery from tests on three core plugs that were flooded to 99.9% water cut.

Are these data acceptable for estimates of waterflood residual saturations?

TABLE 2.16—PROPERTIES OF RESERVOIR ROCK AND FLUIDS—PROBLEM 2.28

Porosity	0.25
IFT, mN/m [dyne/cm]	15 [15]
Viscosity of oil, cp [mPa·s]	6 [6]
Permeability of water, md	50

2.30. Waterflood susceptibility tests (flood pot tests) are commonly performed on core plugs cut from fresh core. In these tests, the core is flooded to a high WOR and saturations are determined. Typical results are shown in Table 2.18.

The waterflood was conducted with a pressure drop of 40 psi [276 kPa] across a 2-in. [5-cm] core. The reservoir rock is believed to be strongly water wet. Crude-oil gravity is 36°API [0.8 g/cm³].

When a gas saturation is present, there is always a possibility that residual oil saturations will be affected. The vacant space is either solution gas (assume methane) or air. If the vacant space is solution gas, some can redissolve in the oil. (Maximum solubility is 20 scf/STB [0.57 std m³/stock-tank m³] at an average pressure of 20 psi [138 kPa].) However, if the gas is air, there will be no dissolution at the pressures anticipated.

Examine the postwaterflood saturations and determine the effects, if any, of the vacant space on the residual oil saturation. For the purpose of this problem, assume that the average pressure in the core is 20 psi [138 kPa] throughout the displacement. The average pressure in the core rises to this value before the outlet end of the core is opened to atmospheric pressure.

Is the residual saturation correct? If not, what is your estimate of the correct value?

2.31 Richardson and Perkins[89] found residual oil saturations of 0.20 in a sandpack when flooded at 423.6 μm/s. The sand was considered to be strongly water wet. The oil was kerosene and the IFT between the kerosene and the flood water was 37 dynes/cm [37 mN/m]. Viscosity of the kerosene was 1.8 cp [1.8 mPa·s] while the viscosity of the water was 0.89 cp [0.89 mPa·s].

a. What is the possibility that the residual oil saturation was affected by the velocity of the waterflood?

b. Estimate the flood velocity required to reduce the residual oil saturation to 0.15.

c. Check your predictions with the data in Ref. 89.

2.32. In laboratory experiments, outcrop core material, such as Berea, Boise, and Torpedo sandstones, are pre-

TABLE 2.17—WATERFLOOD SUSCEPTIBILITY TEST RESULTS

Sample Number	Permeability to Water at Residual Saturation, k_w (darcies)	Porosity, ϕ (%)	Initial Oil Saturation (% PV)	Residual Oil Saturation (% PV)	Flooding* Pressure (kPa)
1	0.013	21.2	68.1	26.5	1,034
2	0.004	18.7	66.8	25.8	2,827
3	0.0014	17.4	67.5	32.0	7,446

*Across 2 in. [5 cm].

TABLE 2.18—SAMPLE B 102

	Before Waterflood	After Waterflood
Oil saturation	0.41	0.21
Water saturation	0.38	0.62
Vacant	0.21	0.17
	1.00	1.00

Porosity = 0.18

TABLE 2.19—DISPLACEMENT DATA[76]

Wetting Displacing Nonwetting

Permeability of Teflon core, darcy	0.430
Porosity	0.285
Initial saturation of wetting phase	0.2

IFT (mN/m)	Viscosity of Displacing Liquid (mPa·s)	Viscosity of Displaced Liquid (mPa·s)	Saturation of Displaced Phase
40.0	0.30	0.953	0.33
4.8	*	*	0.31
1.1	2.36	1.05	0.29
0.5	2.36	0.953	0.26

Nonwetting Displacing Wetting

Permeability of Teflon core, darcy	0.400
Porosity	0.280
Initial saturation of nonwetting phase	0.3

IFT (mN/m)	Viscosity of Displacing Liquid (mPa·s)	Viscosity of Displaced Liquid (mPa·s)	Saturation of Displaced Phase
40.0	0.953	2.36	0.36
4.8	*	*	0.32
2.1	2.88	1.188	0.28
1.1	0.953	2.05	0.274
0.5	0.953	2.05	0.263

*Unavailable data points.

TABLE 2.20—PROPERTIES OF NEW OIL RECOVERY CHEMICAL

pH	9.4
Viscosity, cp [mPa·s]	40 [40]
Density, lbm/cu ft [g/cm^3]	68.6 [1.1]
IFT, dyne/cm [mN/m]	0.1 [0.1]

TABLE 2.21—RESERVOIR ROCK AND FLUID PROPERTIES FOR PROBLEM 2.35

Reservoir rock	sandstone
Area, acre [m^2]	100 [404 687]
Thickness (average), ft [m]	16.4 [5]
Porosity	0.2
Permeability, md	30
Interstitial water saturation	0.40
Residual oil saturation	0.27
Water viscosity at reservoir temperature, cp [mPa·s]	1 [1]
Oil viscosity at reservoir temperature, cp [mPa·s]	6 [6]
IFT between water and oil, dyne/cm [mN/m]	23 [23]

pared for waterflood tests by the following procedure. A dry core is placed in a core holder, evacuated, and saturated with brine. Then, oil is injected to reduce the water content to an interstitial saturation. When a low oil viscosity is used (1 cp [1 mPa·s]), water saturations on the order of 0.30 are observed in Berea cores.

In some tests, it is necessary to have water saturations less than 0.30. It has been suggested that lower water saturations might be obtained by using a mineral oil with a viscosity of 150 cp [150 mPa·s] for the oilflood. The viscous oil would be miscibly displaced from the core after the oilflood by flooding with several PV's of the test oil. Is this suggestion likely to work? Why or why not?

2.33. Data reported by Mungan[76] for residual saturations of the displacing phase when wetting and nonwetting fluids were the displacing phases are summarized in Table 2.19. All floods were conducted in Teflon cores 12 in. [30.48 cm] in length, at a Darcy velocity of 105.9 μm/s.

a. Develop a capillary number correlation for each set of data.

b. Compare the data of Mungan with the correlations in Fig. 4 of Ref. 79.

c. Why are different correlations obtained when the wetting and nonwetting phases are residual saturations in the same porous media?

2.34. The model in Section 2.5.3 illustrates the trapping process in a pore doublet model when oil and water have equal viscosities. Some reservoirs contain oils with viscosities from 200 to 1,000 cp [200 to 1000 mPa·s]. What conclusions can be drawn from Example 2.2 when the oil viscosity is in this range? The viscosity of the water is 1 cp [1 mPa·s] and IFT is 30 dynes/cm [30 mN/m].

2.35. You are in a meeting with a representative of a chemical company that has developed a new chemical for recovery of oil in a field that has been waterflooded. Properties of aqueous solutions of the chemical are shown in Table 2.20.

The chemical salesperson describes a successful application of the chemical in an Oklahoma field where the operator added a "few barrels" of the chemical to injected water and saw the oil cut increase in several producing wells. Then, the salesperson proposes a field test in a field under your supervision that is near its economic limit (i.e., watered out and at residual oil saturation). Properties of this field are summarized in Table 2.21. The reservoir is considered strongly water wet.

Evaluate the potential of this chemical to mobilize residual oil in your reservoir, assuming it could be injected at the same rates you are currently using in the waterflood.

2.36. Fig. 2.63 contains a set of data illustrating the effect of pressure gradient on the recovery of discontinuous residual oil. Selected data are summarized in Table 2.22.[88]

Compute the capillary number for each data point using Eq. 2.32. For the purpose of this problem, the IFT is 30 dynes/cm [30 mN/m], the viscosity of water is 1 cp [1 mPa·s], and the absolute permeability of the rock is 156 md. Porosity is 0.20.

TABLE 2.22—EFFECT OF PRESSURE GRADIENT ON RESIDUAL OIL SATURATION AFTER TRAPPING[88]

S_o PV	Pressure Gradient (psi/ft)
0.355	70
0.355	370
0.284	600
0.213	1,200
0.142	2,400
0.114	3,200

Relative permeability data for water are given by the following equations where $S_{wi}=0.29$:

$$S_w^* = \frac{S_w - S_{wi}}{1 - S_{wi}};$$

$$k_{rw} = S_w^{*\,2.5}.$$

Conversion factors:
$(21.02)(1\ \text{psi/ft}) = \text{kPa/m};$
$\text{Pa} = \text{N/m}^2.$

The capillary number is dimensionless.
Compare the ratio of S_o/S_{or} at your values of the capillary number with the correlations of Fig. 2.64.

Nomenclature

A = cross-sectional area perpendicular to flow, sq ft [m²]
A_o = oil-wetting preference
A_w = water-wetting preference
B_o = oil FVF, STB/bbl [stock-tank m³/m³]
B_{o1} = oil FVF at Pressure 1, STB/bbl [stock-tank m³/m³]
E_D = microscopic displacement efficiency
F_x = force in x direction, lbf [N]
g = acceleration of gravity, ft/sec² [cm/s²]
G = correlating parameter of Melrose and Brandner
I_{rd} = relative displacement index
$J(S_w)$ = Leverett number, dimensionless
k = permeability, darcy
k_b = base permeability for relative permeability curves, darcy
k_g = permeability to gas, darcy
k_o = permeability to oil, darcy
k_{rg} = relative permeability to gas, fraction
k_{ro} = relative permeability to oil, fraction
k_{rw} = relative permeability to water, fraction
k_w = permeability to water, darcy
L = length, ft [m]
N = length of trapped ganglion in particle diameters, dimensionless
N_{ca} = capillary number based on interstitial velocity, dimensionless
N_{ca}^* = capillary number based on darcy velocity, dimensionless

N_{cam} = modified capillary number, dimensionless
N_{pw} = oil displaced by waterflood, STB [stock-tank m³]
p = pressure, psi [Pa]
p_A = pressure at Point A, psi [Pa]
p_B = pressure at Point B, psi [Pa]
p_g = gas-phase pressure, psi [Pa]
p_{nw} = pressure of nonwetting phase
p_o = oil-phase pressure, psi [Pa]
p_{od} = oil-phase pressure at horizontal datum with elevation Z_d, psi [kPa]
p_{oi} = oil-phase pressure on oil side of an oil/water-drop interface, psi [Pa]
p_w = pressure of wetting phase or water, psi [Pa]
p_{wi} = pressure of water phase on water side of an oil/water-drop interface, psi [Pa]
$\overline{p_1}$ = average reservoir pressure at beginning of a waterflood, psi [Pa]
P_c = capillary pressure, psi [Pa]
q = volumetric flow rate, B/D [m³/s]
r = radius, ft [m]
r^* = radius of curvature, ft [m]
r_1^* = principal radius of curvature for an interface, ft [m]
r_2^* = principal radius of curvature for an interface, ft [m]
$\overline{r}_{dr}$ = mean radius of curvature of drainage interface for $\theta=0$, ft [m]
$\overline{r}_{imb}$ = mean radius of curvature of imbibition interface for $\theta=0$, ft [m]
S_{iw} = interstitial water saturation where the water phase is immobile under the applied pressure gradient, fraction
S_o = oil saturation, fraction
$\overline{S}_o$ = volumetric average oil saturation at a particular point during the waterflood, fraction
S_{oi} = initial oil saturation, fraction
S_{or} = residual oil saturation, fraction
S_{ot} = oil saturation not hydraulically connected, fraction
$\overline{S}_{o1}$ = volumetric average oil saturation at beginning of waterflood where the pressure is $\overline{p_1}$, fraction
S_w = water saturation, fraction
S_{wi} = initial water saturation, fraction
u = darcy velocity, ft/D [m/d]
u_{gx} = darcy velocity of gas phase in x direction, ft/D [m/d]
u_{ox} = darcy velocity of oil phase in x direction, ft/D [m/d]
u_{wx} = darcy velocity of water phase in x direction, ft/D [m/d]
v = velocity in a pore or interstitial velocity, ft/D [m/d]
V_{pw} = PV swept
x = distance in the x direction of a coordinate system, ft [m]

y = distance in the y direction of coordinate system, ft [m]

z = distance in the z direction of a coordinate system, ft [m]

Z = elevation relative to a horizontal datum at Z_d, ft [m]

Z_d = elevation of a horizontal datum, ft [m]

Z_i = elevation of saturation $S_w = 1.0$ relative to datum Z_d, ft [m]

α = angle of reservoir with respect to horizon

γ_o = specific gravity of oil

θ = contact angle measured through water phase, degrees

θ_A = advancing contact angle, degrees

θ_R = receding contact angle, degrees

λ = correlating parameter of Larson *et al.*

μ = viscosity, cp [mPa·s]

μ_g = gas viscosity, cp [mPa·s]

μ_o = oil viscosity, cp [mPa·s]

μ_w = water viscosity, cp [mPa·s]

ρ = density, lbm/cu ft [g/cm^3]

ρ_g = gas density, lbm/cu ft [g/cm^3]

ρ_o = oil density, lbm/cu ft [g/cm^3]

ρ_w = water density, lbm/cu ft [g/cm^3]

σ = IFT, dynes/cm [mN/m] (correlating parameter of Larson *et al.*)

σ_{nws} = surface tension between nonwetting and solid phases, dynes/cm [mN/m]

σ_{nww} = IFT between nonwetting and wetting phases, dynes/cm [mN/m]

σ_{os} = surface tension between oil and solid phases, dynes/cm [mN/m]

σ_{ow} = IFT between oil and water phases, dynes/cm [mN/m]

σ_{ws} = surface tension between water and solid phases, dynes/cm [mN/m]

ϕ = porosity, fraction

Φ = potential, sq ft/D^2 [m^2/s^2]

Φ_g = gas-phase potential, sq ft/D^2 [m^2/s^2]

Φ_o = oil-phase potential, sq ft/D^2 [m^2/s^2]

Φ_w = water-phase potential, sq ft/D^2 [m^2/s^2]

Subscripts

A = advancing or location A

B = location B

g = gas phase

o = oil phase

R = receding

s = solid phase

w = water phase

x = x direction

y = y direction

z = z direction

1 = location 1

2 = location 2

References

1. Healy, R.N. and Reed, R.L.: ''Physicochemical Aspects of Microemulsion Flooding,'' *Soc. Pet. Eng. J.* (Oct. 1974) 491–501; *Trans.*, AIME, **257**.

2. Hough, E.W., Rzasa, M.J., and Wood, B.B.: ''Interfacial Tensions at Reservoir Pressures and Temperatures, Apparatus and the Water-Methane System,'' *Trans.*, AIME (1951) **192**, 57–60.

3. Cayias, J.L., Schechter, R.S., and Wade, W.H.: ''The Measurement of Low Interfacial Tension Via the Spinning Drop Technique,'' *Adsorption at Interfaces*, K.L. Mittal (ed.), Symposium Series, ACS, Washington, D.C. (1975) **8**.

4. Donaldson, E.C., Thomas, R.D., and Lorenz, P.B.: ''Wettability Determination and Its Effect on Recovery Efficiency,'' *Soc. Pet. Eng. J.* (March 1969) 13–20.

5. Katz, D.L. and Saltman, W.: ''Surface Tension of Hydrocarbons,'' *Ind. and Eng. Chem.* (Jan. 1939) **31**.

6. Wagner, O.R. and Leach, R.O.: ''Improving Oil Displacement Efficiency by Wettability Adjustment,'' *J. Pet. Tech.* (April 1959) 65–72; *Trans.*, AIME, **216**.

7. Foster, W.R.: ''A Low-Tension Waterflooding Process,'' *J. Pet. Tech.* (Feb. 1973) 205–10; *Trans.*, AIME, **255**.

8. Raza, S.H., Treiber, L.E., and Archer, D.L.: ''Wettability of Reservoir Rocks and Its Evaluation,'' *Prod. Monthly* (April 1968) **33**, No. 4, 2–7.

9. Adamson, A.W.: *Physical Chemistry of Surfaces*, second edition, Interscience Publishers, New York City (1967) 353.

10. Treiber, L.E., Archer, D.L., and Owens, W.W.: ''A Laboratory Evaluation of the Wettability of Fifty Oil-Producing Reservoirs,'' *Soc. Pet. Eng. J.* (Dec. 1972) 531–40; *Trans.*, AIME, **253**.

11. Benner, F.C. and Bartell, F.E.: ''The Effect of Polar Impurities Upon Capillary and Surface Phenomena in Petroleum Production,'' *Drill. and Prod. Prac.*, API (1941) 341–48.

12. Denekas, M.O., Mattax, C.C., and Davis, G.T.: ''Effect of Crude Oil Components on Rock Wettability,'' *J. Pet. Tech.* (Nov. 1959) 330–33; *Trans.*, AIME, **255**.

13. Katz, D.L.: ''Possibilities of Secondary Recovery for the Oklahoma City Wilcox Sand,'' *Trans.*, AIME (1942) **146**, 28–53.

14. Reisberg, J. and Doscher, T.M.: ''Interfacial Phenomena in Crude-Oil-Water Systems,'' *Prod. Monthly* (Nov. 1956) 43–50.

15. Salathiel, R.A.: ''Oil Recovery by Surface Film Drainage in Mixed-Wettability Rocks,'' *J. Pet. Tech.* (Oct. 1973) 1216–24; *Trans.*, AIME, **255**.

16. Nutting, P.G.: ''Some Physical and Chemical Properties of Reservoir Rocks Bearing on the Accumulation and Discharge of Oil,'' *Problems of Petroleum Geology*, AAPG (1934) 825–32.

17. Craig, F.F. Jr.: *The Reservoir Engineering Aspects of Waterflooding*, Monograph Series, SPE, Richardson, TX (1971) **3**.

18. Jennings, H.Y. Jr.: ''Surface Properties of Natural and Synthetic Porous Media,'' *Prod. Monthly* (March 1957) 20–24.

19. Trantham, J.C. and Clampitt, R.L.: ''Determination of Oil Saturation After Waterflooding in an Oil-Wet Reservoir—The North Burbank Unit Track 97 Project,'' *J. Pet. Tech.* (May 1977) 491–500.

20. Hiemenz, P.C.: *Principles of Colloid and Surface Chemistry*, Marcel Dekker Inc., New York City (1977) 235.

21. Pickerell, J.J., Swanson, B.F., and Hickman, W.B.: ''Application of Air-Mercury and Oil-Air Capillary Pressure Data in the Study of Pore Structure and Fluid Distribution,'' *Soc. Pet. Eng. J.* (March 1966) 55–61; *Trans.*, AIME, **237**.

22. Swanson, B.F.: ''Visualizing Pores and Nonwetting Phase in Porous Rock,'' *J. Pet. Tech.* (Jan. 1979) 10–18.

23. Amyx, J.W., Bass, D.M. Jr., and Whiting, R.L.: *Petroleum Reservoir Engineering*, McGraw Hill Book Co., New York City (1960) 142–50.

24. Bruce, W.A. and Welge, H.J.: ''The Restored State Method for Determination of Oil in Place and Connate Water,'' *Drill. and Prod. Prac.*, API (1947) 166–74.

25. Purcell, W.R.: ''Capillary Pressures—Their Measurement Using Mercury and the Calculation of Permeability Therefrom,'' *Trans.*, AIME (1949) **186**, 39–48.

26. Slobod, R.L., Chambers, A., and Prehn, W.L. Jr.: ''Use of Centrifuge for Determining Connate Water, Residual Oil, and Capillary Pressure Curves of Small Core Samples,'' *Trans.*, AIME (1951) **192**, 127–34.

27. Brown, H.W.: ''Capillary Pressure Investigations,'' *Trans.*, AIME (1951) **192**, 67–74.

28. Hassler, G.L. and Brunner, E.: ''Measurement of Capillary Pressures in Small Core Samples,'' *Trans.*, AIME (1945) **160**, 114–23.

29. Szabo, M.T.: ''New Methods for Measuring Imbibition Capillary Pressure and Electrical Resistivity Curves by Centrifuge,'' *Soc. Pet. Eng. J.* (June 1974) 243–52.

30. Stegemeier, G.L.: "Mechanisms of Entrapment and Mobilization of Oil in Porous Media," *Improved Oil Recovery by Surfactant and Polymer Flooding*, D.O. Shah and R.S. Schechter (eds.), Academic Press, New York City (1977) 55-91.
31. Burdine, N.T., Gournay, L.S., and Reicharty, P.O.: "Pore Size Distribution of Petroleum Reservoir Rocks," *Trans.*, AIME (1950) **189**, 195-204.
32. Ritter, H.L. and Drake, L.C.: "Pore Size Distribution in Porous Materials," *Ind. Eng. Chem.* (Dec. 1945) **17**, 782-86.
33. Killins, C.R., Nielson, R.F., and Calhoun, J.C. Jr.: "Capillary Desaturation and Imbibition in Rocks," *Prod. Monthly* (Feb. 1953) **18**, No. 2, 30-39.
34. Morrow, N.R.: "Capillary Pressure Correlations For Uniformly Wetted Porous Media," *J. Cdn. Pet. Tech.* (Oct.-Dec. 1976) 49-69.
35. Morrow, N.R.: "The Effects of Surface Roughness Contact Angle With Special Reference to Petroleum Recovery," *J. Cnd. Pet. Tech.* (Oct.-Dec. 1975) 42-53.
36. Morrow, N.R. and McCaffery, F.G.: "Displacement Studies in Uniformly Wetted Porous Media," *Proc.*, Soc. of Chemical Industry Intl. Symposium on Wetting, Loughborough, England (Sept. 1976) 289-319.
37. Amott, E.: "Observations Relating to the Wettability of Porous Rock," *Trans.*, AIME (1959) **216**, 156-92.
38. Blair, P.M.: "Calculation of Oil Displacement by Counter-Current Water Imbibition," *Soc. Pet. Eng. J.* (Sept. 1964) 195-202; *Trans.*, AIME, **231**.
39. Boneau, D.F. and Clampitt, R.L.: "A Surfactant System for the Oil-Wet Sandstone of the North Burbank Unit," *J. Pet. Tech.* (May 1977) 501-06.
40. Grist, D.M., Langley, G.O., and Neustadter, E.L.: "The Dependence of Water Permeability on Core Cleaning Methods in the Case of Some Sandstone Samples," *J. Cdn. Pet. Tech.* (April-June 1975) 48-52.
41. Kennedy, H.T., Van Meter, O.E., and Jones, R.G.: "Saturation Determination of Rotary Cores," *Pet. Eng.* (Jan. 1954) **26**, No. 1, 1352-62.
42. Jenks, L.H. *et al.*: "Fluid Flow Within a Porous Medium Near a Diamond Core Bit," *J. Cdn. Pet. Tech.* (Oct.-Dec. 1968) 172-80.
43. Jenks, L.H. *et al.*: "Coring for Reservoir Connate Water Saturations," *J. Pet. Tech.* (Aug. 1969) 932.
44. Willmon, G.J.: "A Study of Displacement Efficiency in the Redwater Field," *J. Pet. Tech.* (April 1967) 449-56.
45. Bobeck, J.E., Mattax, C.C., and Denekas, M.O.: "Reservoir Rock Wettability—Its Significance and Evaluation," *J. Pet. Tech.* (July 1958) 155-60; *Trans.*, AIME, **213**.
46. Richardson, J.G., Perkins, F.M. Jr., and Osoba, J.S.: "Differences in Behavior of Fresh and Aged East Texas Woodbine Cores," *J. Pet. Tech.* (June 1955) 86-91; *Trans.*, AIME, **204**.
47. Mungan, N.: "Certain Wettability Effects in Laboratory Waterfloods," *J. Pet. Tech.* (Feb. 1966) 247-52; *Trans.*, AIME, **237**.
48. Chatenever, A. and Calhoun, J.C. Jr.: "Visual Examinations of Fluid Behavior in Porous Media," *Trans.*, AIME (1952) **195**, 149-56.
49. Chatenever, A.: "Flow of Oil and Water in Porous Media," *Oil and Gas J.* (May 26, 1952) 174-75.
50. Wilson, D.A., Calhoun, J.C. Jr., and Chatenever, A.: "Fluid Saturations in a Porous Medium," *Oil and Gas J.* (May 26, 1952) 175-79, 204-05.
51. Moore, T.F. and Blum, H.A.: "Wettability in Surface-Active Agent Waterflooding," *Oil and Gas J.* (Dec. 8, 1952) 108.
52. Chatenever, A.: "Microscopic Behavior of Fluids in Porous Systems," Final Report API Project 47b (April 1957).
53. Kimbler, O.K. and Caudle, B.H.: "New Technique for Study of Fluid Flow and Phase Distribution in Porous Media," *Oil and Gas J.* (Dec. 16, 1957) 85.
54. Wilson, D.A.: "Fluid Distributions on Porous Systems," a motion picture produced by Amoco Production Co., Chicago, IL (1952).
55. Chatenever, A., Indra, M.K., and Kyte, J.R: "Microscopic Observations of Solution Gas Drive Behavior," *J. Pet. Tech.* (June 1959) 13-15.
56. Schneider, F.N. and Owens, W.W: "Sandstone and Carbonate Two- and Three-Phase Relative Permeability Characteristics," *Soc. Pet. Eng. J.* (March 1970) 75-84; *Trans.*, AIME, **249**.
57. Leverett, M.C. and Lewis, W.B.: "Steady Flow of Gas/Oil/Water Mixtures Through Unconsolidated Sand," *Trans.*, AIME (1941) **142**, 107-16.
58. Owens, W.W. and Archer, D.L.: "The Effect of Rock Wettability on Oil-Water Relative Permeability Relationships," *J. Pet. Tech.* (July 1971) 873-78; *Trans.*, AIME, **251**.
59. Donaldson, E.C. and Crocker, M.E: "Review of Petroleum Oil Saturation and Its Determination," Report BERC/RI-77/15, U.S. DOE, Bartlesville Energy Research Center (Dec. 1977).
60. McCaffery, F.G. and Bennion, D.W.: "The Effect of Wettability on Two-Phase Relative Permeabilities," *J. Cdn. Pet. Tech.* (Oct.-Dec. 1974) 42-53.
61. Jones, S.C. and Roszelle, W.O.: "Graphical Techniques for Determining Relative Permeability From Displacement Experiments," *J. Pet. Tech.* (May 1978) 807-17.
62. Colona, J., Brissand, F., and Millet, J.L.: "Evolution of Capillarity and Relative Permeability Hysteresis," *Soc. Pet. Eng. J.* (Feb. 1972) 28-38; *Trans.*, AIME, **253**.
63. Raimondi, P. and Torcaso, M.A.: "Distribution of the Oil Phase Obtained Upon Imbibition of Water," *Soc. Pet. Eng. J.* (March 1964) 49-55; *Trans.*, AIME, **231**.
64. Larson, R.G.: MS thesis, Dept. of Chemical Engineering and Materials Science, U. of Minnesota, Minneapolis (1977).
65. Johnson, E.F., Bossler, D.P., and Naumann, V.O.: "Calculation of Relative Permeability from Displacement Experiments," *J. Pet. Tech.* (Jan. 1959) 61-63; *Trans.*, AIME, **216**.
66. Loomis, A.G. and Crowell, D.C.: "Relative Permeability Studies: Gas-Oil and Water-Oil Systems," *Bull.* 599, U.S. Bureau of Mines, Washington (1962).
67. Osoba, J.S. *et al.*: "Laboratory Measurements of Relative Permeability," *Trans.*, AIME (1951) **192**, 47-56.
68. Richardson, J.G. *et al.*: "Laboratory Determination of Relative Permeability," *Trans.*, AIME (1952) **195**, 187-96.
69. Geffen, T.M. *et al.*: "Experimental Investigation of Factors Affecting Laboratory Relative Permeability Measurements," *Trans.*, AIME (1951) **192**, 99-110.
70. Owens, W.W., Parrish, D.R., and Lamoreaux, W.E.: "An Evaluation of a Gas Drive Method for Determining Relative Permeability Relationships," *J. Pet. Tech.* (Dec. 1956) 275-80; *Trans.*, AIME, **207**.
71. Welge, H.J.: "A Simplified Method for Computing Oil Recovery by Gas or Water Drive," *Trans.*, AIME (1952) **195**, 91-98.
72. Morgan, J.T. and Gordon, D.T.: "Influence of Pore Geometry on Water-Oil Relative Permeability," *J. Pet. Tech.* (Oct. 1970) 1199-1208.
73. Fatt, I. and Klikoff, W.A.: "Effect of Fractional Wettability on Multiphase Flow Through Porous Media," *J. Pet. Tech.* (Oct. 1959) 71-76; *Trans.*, AIME, **213**.
74. Singhai, A.K., Mukherjee, D.P., and Somerton, W.H.: "Effect of Heterogeneous Wettability on Flow of Fluids Through Porous Media," *J. Cdn. Pet. Tech.* (July-Sept. 1976) 63-70.
75. Leverett, M.C.: "Flow of Oil-Water Mixtures Through Unconsolidated Sands," *Trans.*, AIME (1939) **132**, 149-71.
76. Mungan, N.: "Interfacial Effects in Immiscible Liquid-Liquid Displacement in Porous Media," *Soc. Pet. Eng. J.* (Sept. 1966) 247-53; *Trans.*, AIME, **237**.
77. Talash, A.W.: "Experimental and Calculated Relative Permeability Data for Systems Containing Tension Additives," paper SPE 5810 presented at the 1976 SPE Improved Oil Recovery Symposium, Tulsa, March 22-24.
78. van Meurs, P. and van der Poel, C.: "A Theoretical Description of Water Drive Processes Involving Viscous Fingering," *Trans.*, AIME (1958) **213**, 103-112.
79. Lefebvre du Prey, E.J.: "Factors Affecting Relative Permeabilities of a Consolidated Porous Medium," *Soc. Pet. Eng. J.* (Feb. 1973) 39-47.
80. Rathmell, J.J., Braun, P.H., and Perkins, T.K.: "Reservoir Waterflood Residual Oil Saturation From Laboratory Tests," *J. Pet. Tech.* (Feb. 1973) 175-85; *Trans.*, AIME, **255**.
81. Richardson, J.G.: "The Calculation of Waterflood Recovery From Steady State Relative Permeability Data," *J. Pet. Tech.* (May 1957) 64-66; *Trans.*, AIME, **210**.
82. Donaldson, E.C., Lorenz, P.B., and Thomas, R.D.: "The Effects of Viscosity and Wettability on Oil and Water Relative Permeabilities," paper SPE 1562 presented at the 1966 SPE Annual Meeting, Dallas, Oct. 2-5.
83. Odeh, A.S.: "Effect of Viscosity Ratio on Relative Permeability," *J. Pet. Tech.* (Dec. 1959) 346-53; *Trans.*, AIME, **216**.
84. Baker, P.E.: "Discussion of Effect of Viscosity Ratio on Relative Permeability," *J. Pet. Tech.* (Nov. 1960) 65-66; *Trans.*, AIME, **219**.

85. Downie, J. and Crane, F.E.: "Effect of Viscosity on Relative Permeability," *Soc. Pet. Eng. J.* (June 1961) 59–60; *Trans.,* AIME, **222**.

86. Ng, K.M., Davis, H.T., and Scriven, L.E.: "Visualization of Blob Mechanics in Flow Through Porous Media," *Chem. Engr. Sci.* (1978) **33**, No. 8, 1009–17.

87. Slobod, R.L.: "Attainment of Connate Water in Long Cores by Dynamic Displacement," *Trans.,* AIME (1950) **189**, 359–63.

88. Jordan, J.K., McCardell, W.M., and Hocott, C.R.: "Effect of Rate on Oil Recovery by Waterflooding," report for Humble Oil and Refining Co., Houston (1956).

89. Richardson, J.G. and Perkins, F.M. Jr.: "A Laboratory Investigation on the Effect of Rate on Recovery of Oil by Waterflooding," *J. Pet. Tech.* (April 1957) 114–21; *Trans.,* AIME, **210**.

90. Perkins, F.M. Jr.: "An Investigation of the Role of Capillary Forces in Laboratory Waterfloods," *J. Pet. Tech.* (Nov. 1957) 49–51; *Trans.,* AIME, **210**.

91. Rapoport, L.A. and Leas, W.J.: "Properties of Linear Waterfloods," *J. Pet. Tech.* (May 1953) 139–48; *Trans.,* AIME, **198**.

92. Kyte, J.R. and Rapoport, L.A.: "Linear Waterflood Behavior and End Effects in Water-Wet Porous Media," *J. Pet. Tech.* (Oct. 1958) 47–50; *Trans.,* AIME, **213**.

93. Moore, T.F. and Slobod, R.L.: "The Effect of Viscosity and Capillarity on the Displacement of Oil by Water," *Prod. Monthly* (Aug. 1956) 20–30.

94. Abrams, A.: "The Influence of Fluid Viscosity, Interfacial Tension, and Flow Velocity on Residual Oil Saturation Left by Waterflood," *Soc. Pet. Eng. J.* (Oct. 1975) 437–47.

95. Melrose, J.C. and Brandner, C.F.: "Role of Capillary Forces in Determining Microscopic Displacement Efficiency for Oil Recovery by Waterflooding," *J. Cdn. Pet. Tech.* (Oct.–Dec. 1974) 54–62.

96. Kyte, J.R. *et al.*: "Mechanism of Waterflooding in the Presence of Free Gas," *J. Pet. Tech.* (Sept. 1956) 215–21; *Trans.,* AIME, **207**.

97. Holmgren, C.R. and Morse, R.A.: "Effect of Free Gas Saturation on Oil Recovery by Waterflooding," *Trans.,* AIME (1951) **192**, 135–40.

98. Stegemeier, G.L.: "Relationship of Trapped Oil Saturation to Petrophysical Properties of Porous Media," paper SPE 4754 presented at the 1974 SPE Improved Oil Recovery Symposium, Tulsa, April 22–24.

99. Rose, W. and Witherspoon, P.A.: "Trapping Oil in a Pore Doublet," *Prod. Monthly* (Dec. 1956) 32–38.

100. Slobod, R.L.: "Comments on Trapping Oil in a Pore Doublet," *Prod. Monthly* (Jan. 1957) 17.

101. Rose, W. and Cleary, J.: "Further Indications of Pore Doublet Theory," *Prod. Monthly* (Jan. 1958) 20–25.

102. Bird, R.B., Stewart, W.E., and Lightfoot, E.N.: *Transport Phenomena,* John Wiley and Sons Inc., New York City (1960) 46.

103. Taber, J.J.: "Dynamic and Static Forces Required to Remove a Discontinuous Oil Phase from Porous Media Containing Both Oil and Water," *Soc. Pet. Eng. J.* (March 1969) 3–12.

104. Taber, J.J., Kirby, J.C., and Schroeder, F.U.: "Studies on the Displacement of Residual Oil: Viscosity and Permeability Effects," Symposium Series, AIChE, New York City (1973) **69**, No. 127, 53–56.

105. Larson, R.G., Scriven, L.E., and Davis, H.T.: "Percolation Theory of Residual Phase in Porous Media," *Nature* (1977) **268**, No. 5619, 409–13.

106. Dullien, F.A.L.: "Effects of Pore Structure on Capillary and Flow Phenomena in Sandstones," *J. Cdn. Pet. Tech.* (July–Sept. 1975) 48–55.

107. Wright, H.T. Jr. and Wooddy, L.D. Jr.: "Formation Evaluation at the Borregas and Seeligson Field, Brooks and Jim Wells County, Texas" paper presented at the 1985 AIME Symposium on Formation Evaluation, Calgary, Sept. 29–Oct. 2.

SI Metric Conversion Factors

°API	$141.5/(131.5 + °API)$		$= \text{g/cm}^3$
cp	$\times\ 1.0^*$	E−03	$= \text{Pa·s}$
dyne	$\times\ 1.0^*$	E−02	$= \text{mN}$
ft	$\times\ 3.048^*$	E−01	$= \text{m}$
°F	$(°F − 32)/1.8$		$= °C$
in.	$\times\ 2.54^*$	E+00	$= \text{cm}$
lbm	$\times\ 4.535\ 924$	E−01	$= \text{kg}$
psi	$\times\ 6.894\ 757$	E+00	$= \text{kPa}$
sq in.	$\times\ 6.451\ 6^*$	E+00	$= \text{cm}^2$

*Conversion factor is exact.

Chapter 3
Macroscopic Displacement Efficiency of a Linear Waterflood

3.1 Introduction

Macroscopic efficiency is the term used to describe the displacement efficiency of a waterflood in a specified volume of reservoir rock. Oil displacement processes nearly always vary with time. Thus macroscopic displacement efficiency also changes with time. The approach followed in this chapter uses partial differential equations to represent the conservation of mass and Darcy's law for multiphase fluid flow. Solutions of these equations for specified reservoir geometries yield displacement rate/time estimates. In some cases, partial mathematical solutions can be obtained with a desk calculator and graph paper. Correlations developed from scaled laboratory experiments are another form of solutions to these equations. Large problems in heterogeneous reservoirs may be solvable only with the use of numerical simulators.

Chapter 3 introduces the equations describing multiphase flow in porous media. Solutions of these equations are developed that describe the macroscopic displacement efficiency in linear (1D) waterfloods.

3.2 Development of Equations Describing Multiphase Flow in Porous Media

The flow of fluids through porous media is described by the continuity equation, which is the partial differential equation describing the law of conservation of mass at every point in the porous medium. A derivation of this relationship is included in the next section.

3.2.1 Continuity Equation for Porous Media With Fluid Flow

Consider a small element of a porous medium shown in Fig. 3.1 that has dimensions Δx, Δy, and Δz. For the purposes of this example, consider the flow of two fluids—oil and water. Darcy flow is assumed.

Now a material balance will be written for each phase as it flows through the differential element. The positive directions are indicated by the arrows in Fig. 3.1. The procedure will be illustrated for the oil phase.

The law of conservation of mass, written in terms of rates, is described as follows.

$$
\begin{pmatrix} \text{Mass of oil entering} \\ \text{the differential} \\ \text{element in the} \\ \text{time increment } \Delta t \end{pmatrix} - \begin{pmatrix} \text{Mass of oil leaving} \\ \text{the differential} \\ \text{element in the} \\ \text{time increment } \Delta t \end{pmatrix}
$$

$$
= \begin{pmatrix} \text{Mass of oil that} \\ \text{accumulates within} \\ \text{the differential} \\ \text{element in the} \\ \text{time increment } \Delta t \end{pmatrix}. \quad \dots\dots\dots\dots (3.1)
$$

Referring to Fig. 3.1, the mass of oil entering $\Delta x \Delta y \Delta z$ in the time increment Δt is defined as

$$
[(\rho_o u_{ox})\big|_x \Delta y \Delta z \Delta t] + [(\rho_o u_{oy})\big|_y \Delta x \Delta z \Delta t]
$$

$$
+ [(\rho_o u_{oz})\big|_z \Delta x \Delta y \Delta t], \quad \dots\dots\dots\dots\dots (3.2)
$$

where the first quantity is the mass of oil flowing through Plane x in Δt, the second quantity is the mass of oil flowing through Plane y in Δt, and the third quantity is the mass of oil flowing through Plane z in Δt.

The mass of oil leaving $\Delta x \Delta y \Delta z$ in the time increment Δt is

$$
(\rho_o u_{ox})\big|_{x+\Delta x} \Delta y \Delta z \Delta t + (\rho_o u_{oy})\big|_{y+\Delta y} \Delta x \Delta z \Delta t
$$

$$
+ (\rho_o u_{oz})\big|_{z+\Delta z} \Delta x \Delta y \Delta t. \quad \dots\dots\dots\dots\dots (3.3)
$$

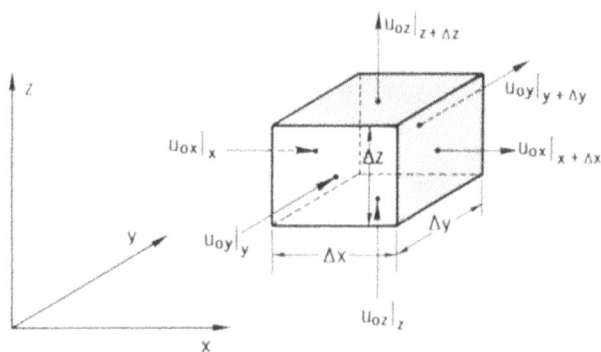

Fig. 3.1—Differential element of porous rock.

Oil accumulates within the differential element by change of oil saturation, variation of density with pressure and temperature, and the change in the porosity of the differential element caused by large changes in net confining pressure.

The mass of oil that accumulates in $\Delta x \Delta y \Delta z$ during the time increment Δt is

$$(\rho_o S_o \phi \Delta x \Delta y \Delta z)\big|_{t+\Delta t} - (\rho_o S_o \phi \Delta x \Delta y \Delta z)\big|_t,$$

$$\cdots\cdots\cdots\cdots\cdots\cdots (3.4)$$

where $\phi \Delta x \Delta y \Delta z$ is the incremental PV.

Substituting Eqs. 3.2 through 3.4 into Eq. 3.1 and rearranging them gives Eq. 3.5 for the oil phase.

$$-[(\rho_o u_{ox})\big|_{x+\Delta x}\Delta y \Delta z \Delta t - (\rho_o u_{ox})\big|_x \Delta y \Delta z \Delta t]$$

$$-[(\rho_o u_{oy})\big|_{y+\Delta y}\Delta x \Delta z \Delta t - (\rho_o u_{oy})\big|_y \Delta x \Delta z \Delta t]$$

$$-[(\rho_o u_{oz})\big|_{z+\Delta z}\Delta x \Delta y \Delta t - (\rho_o u_{oz})\big|_z \Delta x \Delta y \Delta t]$$

$$=(\rho_o S_o \phi)\big|_{t+\Delta t}\Delta x \Delta y \Delta z - (\rho_o S_o \phi t)\big|_t \Delta x \Delta y \Delta z.$$

$$\cdots\cdots\cdots\cdots\cdots\cdots (3.5)$$

Eq. 3.5 can be expressed as a partial differential equation by recalling the definition of a derivative from differential calculus.

$$\frac{\partial g}{\partial x}\bigg|_{y,z,t} = \lim_{\Delta x \to 0} \frac{g(x+\Delta x,y,z,t) - g(x,y,z,t)}{\Delta x}. \quad \cdots (3.6)$$

Dividing both sides of Eq. 3.5 by $\Delta x \Delta y \Delta z \Delta t$, we obtain

$$-\frac{(\rho_o u_{ox})\big|_{x+\Delta x} - (\rho_o u_{ox})\big|_x}{\Delta x}$$

$$-\frac{(\rho_o u_{oy})\big|_{y+\Delta y} - (\rho_o u_{oy})\big|_y}{\Delta y}$$

$$-\frac{(\rho_o u_{oz})\big|_{z+\Delta z} - (\rho_o u_{oz})\big|_z}{\Delta z}$$

$$=\frac{(\rho_o S_o \phi)\big|_{t+\Delta t} - (\rho_o S_o \phi)\big|_t}{\Delta t}. \quad \cdots\cdots (3.7)$$

The limit of Eq. 3.7 as Δx, Δy, Δz, and Δt tend to zero is Eq. 3.8, the continuity equation for the oil phase. The limit $\Delta x \Delta y \Delta z \to 0$ means an infinitesimally small portion of a porous medium through which fluids flow. Clearly, if $\Delta x \Delta y \Delta z$ went to zero, Point x,y,z could be a molecule of the rock, oil, or water.

$$-\frac{\partial}{\partial x}(\rho_o u_{ox}) - \frac{\partial}{\partial y}(\rho_o u_{oy}) - \frac{\partial}{\partial z}(\rho_o u_{oz})$$

$$=\frac{\partial}{\partial t}(\rho_o S_o \phi). \quad \cdots\cdots\cdots (3.8)$$

Similarly, for the water phase,

$$-\frac{\partial}{\partial x}(\rho_w u_{wx}) - \frac{\partial}{\partial y}(\rho_w u_{wy}) - \frac{\partial}{\partial z}(\rho_w u_{wz})$$

$$=\frac{\partial}{\partial t}(\rho_w S_w \phi). \quad \cdots\cdots\cdots (3.9)$$

These equations assume that there is no dissolution of oil in the water phase. We frequently refer to this assumption by stating that there is no mass transfer between the phases.

Some rocks are compressible.[1] In a rigorous development of the continuity equation for porous media, the rock also satisfies the continuity equation relative to the fixed x,y,z coordinate system. That is,

$$-\frac{\partial}{\partial x}(\rho_f u_{fx}) - \frac{\partial}{\partial y}(\rho_f u_{fy}) - \frac{\partial}{\partial z}(\rho_f u_{fz})$$

$$=\frac{\partial}{\partial t}[\rho_f(1-\phi)]. \quad \cdots\cdots\cdots (3.10)$$

Therefore, Eqs. 3.8 through 3.10 represent the law of conservation of mass applied to an infinitesimal volume of porous medium through which fluids are flowing. Fortunately, the velocity of reservoir rock is negligible and Eq. 3.10 can be approximated by Eq. 3.11 for most petroleum engineering applications.

$$\frac{\partial}{\partial t}[\rho_f(1-\phi)]=0. \quad \dots\dots\dots\dots\dots(3.11)$$

3.2.2 Flow Equations for Each Phase
The continuity equations, while conceptually correct, are expressed in velocities that cannot be measured. To develop a representation that can be used for engineering work, we apply Darcy's law to each phase. Thus substituting Eq. 2.7 into Eq. 3.8 gives

$$\frac{\partial}{\partial x}\left(\frac{\rho_o{}^2 k_{ox}}{\mu_o}\frac{\partial \Phi_o}{\partial x}\right)+\frac{\partial}{\partial y}\left(\frac{\rho_o{}^2 k_{oy}}{\mu_o}\frac{\partial \Phi_o}{\partial y}\right)$$

$$+\frac{\partial}{\partial z}\left(\frac{\rho_o{}^2 k_{oz}}{\mu_o}\frac{\partial \Phi_o}{\partial z}\right)=\frac{\partial}{\partial t}(\rho_o S_o\phi). \quad \dots\dots(3.12)$$

Recall that Φ_o, the oil phase potential, was defined by Eq. 2.10. Applying the Leibnitz rule[2] for differentiation of an integral, we obtain

$$\frac{\partial \Phi_o}{\partial x}=\frac{\partial}{\partial x}\left[g(Z-Z_d)+\int_{P_{od}}^{p_o}\frac{dp_o}{\rho_o}\right]=g\frac{\partial Z}{\partial x}+\frac{1}{\rho_o}\frac{\partial p_o}{\partial x}.$$

$$\dots\dots\dots\dots\dots\dots\dots(3.13)$$

Analogous equations hold for the partial differentiation of Φ_o with respect to y and z. The elevation Z is measured in the vertical direction from a reference datum. When x and y are in the horizontal plane,

$$\frac{\partial Z}{\partial x}=0 \text{ and } \frac{\partial Z}{\partial y}=0.$$

The coordinate axis z is in the same plane as Z. Thus

$$\frac{\partial Z}{\partial z}=+1 \text{ if } z \text{ is directed upward, and}$$

$$\frac{\partial Z}{\partial z}=-1 \text{ if } z \text{ is directed downward.}$$

In Fig. 3.1, z is directed upward and is collinear with Z. Eq. 3.12 becomes

$$\frac{\partial}{\partial x}\left(\frac{\rho_o k_{ox}}{\mu_o}\frac{\partial p_o}{\partial x}\right)+\frac{\partial}{\partial y}\left(\frac{\rho_o k_{oy}}{\mu_o}\frac{\partial p_o}{\partial y}\right)$$

$$+\left[\frac{\partial}{\partial z}\frac{\rho_o k_{oz}}{\mu_o}\left(\frac{\partial p_o}{\partial z}+\rho_o g\right)\right]=\frac{\partial}{\partial t}(\rho_o S_o\phi).$$

$$\dots\dots\dots\dots\dots\dots\dots(3.14)$$

Eq. 3.14 is simply an expression for the law of conservation of mass in terms of parameters, which in principle can be measured.

Similarly, for the water phase

$$\frac{\partial}{\partial x}\left(\frac{\rho_w k_{wz}}{\mu_w}\frac{\partial p_w}{\partial x}\right)+\frac{\partial}{\partial y}\left(\frac{\rho_w k_{wy}}{\mu_w}\frac{\partial p_w}{\partial y}\right)$$

$$+\frac{\partial}{\partial z}\left(\frac{\rho_w k_{wz}}{\mu_w}\frac{\partial p_w}{\partial z}+\rho_w g\right)=\frac{\partial}{\partial t}(\rho_w S_w\phi).$$

$$\dots\dots\dots\dots\dots\dots\dots(3.15)$$

These equations, when solved for a particular geometry, can give pressure and saturation distributions as well as phase velocities at any point x,y,z in the porous media.

In the next section the solution of Eqs. 3.14 and 3.15 for various geometries is considered, including the assumptions that are necessary to obtain the solutions.

3.3 Steady-State Solutions to Fluid-Flow Equations in Linear Systems
The equations developed in Sec. 3.2 are general expressions for the conservation of mass during the flow of oil and water through a porous rock in three dimensions. We will solve these equations in order of increasing complexity, beginning with the problem of steady flow of two phases in a linear porous rock. Steady flow of two phases is of interest primarily for interpretation of laboratory experiments to determine relative permeabilities.

3.3.1 Steady Linear Flow
Flow is considered steady when there are no changes with time. For the purposes of this section, we assume flow is in the x direction and flow is one-dimensional (1D) in the horizontal plane. Eqs. 3.16 and 3.17 represent the steady flow of oil and water phases in the x direction.

$$\frac{\partial}{\partial x}\left(\frac{\rho_o k_{ox}}{\mu_o}\frac{\partial p_o}{\partial x}\right)=0 \quad \dots\dots\dots\dots(3.16)$$

and

$$\frac{\partial}{\partial x}\left(\frac{\rho_w k_{wx}}{\mu_w}\frac{\partial p_w}{\partial x}\right)=0. \quad \dots\dots\dots\dots(3.17)$$

These equations can be solved to obtain the oil- and water-phase pressure distributions if relative permeability curves are available.

It is possible to simplify Eqs. 3.16 and 3.17 further by making assumptions about the fluid properties. The densities and viscosities of water and oil are functions of pressure and temperature. Temperature is constant while pressure gradients are usually small.

If both oil and water densities and viscosities are considered constants in Eqs. 3.16 and 3.17, they may be removed as in Eqs. 3.18 and 3.19. The term "incompressible" is used to describe the assumption of constant densities.

$$\frac{d}{dx}\left(k_o\frac{dp_o}{dx}\right)=0 \quad \dots\dots\dots\dots\dots(3.18)$$

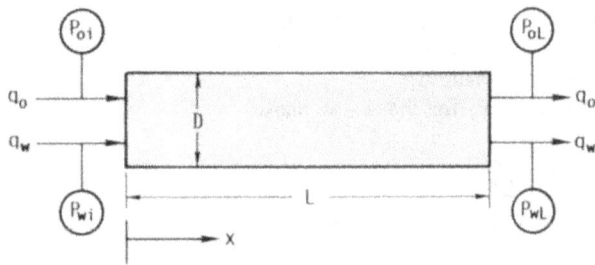

Fig. 3.2—Schematic of steady two-phase flow in a porous rock.

and

$$\frac{d}{dx}\left(k_w \frac{dp_w}{dx}\right)=0. \qquad (3.19)$$

Oil- and water-phase pressures are assumed to be correlated through the capillary pressure curve for the specific saturation path.

Let us consider the application of equations to the flow of two phases in porous media.

Example 3.1. Water and oil are pumped through a porous rock at constant rate as depicted in Fig. 3.2.

The rock is homogeneous with porosity ϕ, diameter d, and length L. Volumetric flow rates q_o and q_w are measured. Oil- and water-phase pressures are measured at each end of the core. Saturation of the core is determined gravimetrically. It is desired to determine the permeabilities of each phase.

We solve Eqs. 3.18 and 3.19 subject to the following boundary conditions.

Oil Phase
At $x=0$,

$$q_o = -\frac{k_o A}{\mu_o}\frac{dp_o}{dx} \qquad (3.20)$$

and

$$p_o = p_{oi}; \qquad (3.21)$$

at $x=L$,

$$p_o = p_{oL}, \qquad (3.22)$$

where A = cross-sectional area.

Water Phase
At $x=0$,

$$q_w = -\frac{k_w A}{\mu_w}\frac{dp_w}{dx} \qquad (3.23)$$

and

$$p_o = p_{wi}; \qquad (3.24a)$$

at $x=L$,

$$p_w = p_{wL}. \qquad (3.24b)$$

The oil-phase equation will be solved to illustrate the procedure. Integrating Eq. 3.18 with respect to x, we obtain

$$k_o \frac{dp_o}{dx} = C_o. \qquad (3.25)$$

The constant of integration, C_o, can be evaluated by applying the boundary condition at $x=0$, represented by Eq. 3.20.

$$C_o = -\frac{q_o \mu_o}{A}. \qquad (3.26)$$

If the laboratory experiment were performed in a manner that S_w was uniform in the core ($0<x<L$), k_o would be independent of x and Eq. 3.25 could be integrated directly as in Eq. 3.27. Oil permeability is computed from Eq. 3.28.

$$\int_{p_{oi}}^{P_{oL}} dp_o = \frac{C_o}{k_o}\int_0^L dx \qquad (3.27)$$

and

$$k_o = \frac{q_o \mu_o L}{A(p_{oi}-p_{oL})}. \qquad (3.28)$$

Define the base permeability for relative permeability curves, k_b, as the permeability to oil at interstitial water saturation, S_{iw}:

$$k_{ro} = \frac{k_o}{k_b}. \qquad (3.29)$$

Expressions for the water-phase permeability and relative permeability were obtained in a similar manner and are given by Eqs. 3.30 and 3.31:

$$k_w = \frac{q_w \mu_w L}{A(p_{wi}-p_{wL})} \qquad (3.30)$$

and

$$k_{rw} = \frac{k_w}{k_b}. \qquad (3.31)$$

Example 3.2 assumes the laboratory experiment was conducted under conditions where phase pressures were measured at each end of the core, and the oil and water saturations were uniform throughout the core; the situation is examined now where these assumptions are not valid.

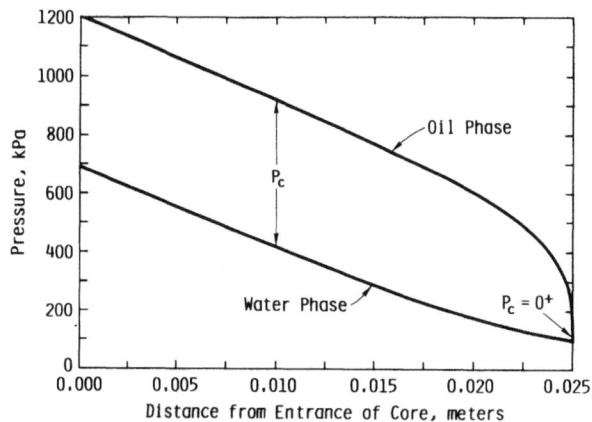

Fig. 3.3—Capillary pressure discontinuity during steady two-phase flow in laboratory cores.

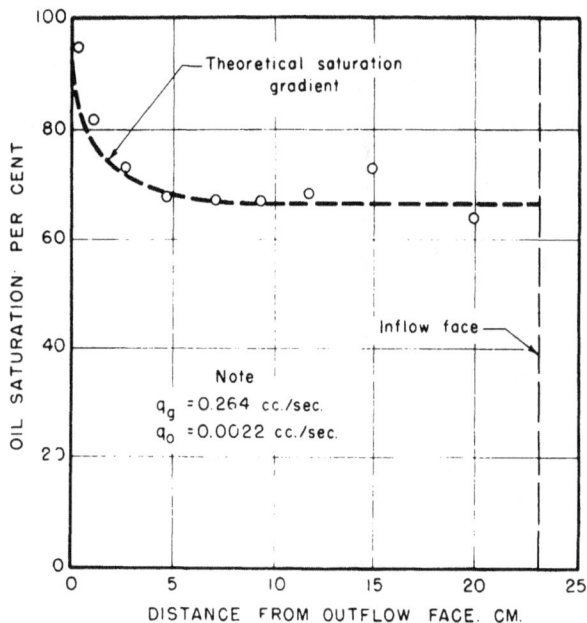

Fig. 3.4—Experimental and computed saturation profiles demonstrating capillary end effect in steady-state flow of oil and gas in a laboratory model.[3]

3.3.2 Capillary End Effect

A phenomenon known as the capillary end effect occurs under certain conditions in laboratory experiments involving the steady flow of two immiscible phases. Oil and water phases are in capillary equilibrium throughout the porous rock. The difference between the oil- and water-phase pressures is given by the capillary pressure curve corresponding to the saturation path (drainage or imbibition) and the water saturation. At the end of the core, the two phases enter a common header where the pressures of both phases are essentially equal. Thus the capillary pressure must undergo a rapid change from a finite value in the core to a value close to zero outside the core. Fig. 3.3 shows this rapid change in oil- and water-phase pressures near the end of a core. We assume that oil and water enter the core through separate distribution channels in the inlet header. Capillary entrance effects are neglected. Inlet oil and water pressures are determined by the saturation corresponding to the ratio of oil and water injection rates.

The abrupt change in capillary pressure caused by the boundary of the core has another effect. For the system to remain at capillary equilibrium, the saturation of the wetting phase must increase to the value corresponding to "zero" capillary pressure. Thus the end effect causes a saturation gradient near the end of the core. Fig. 3.4 shows such a saturation gradient determined experimentally during the steady flow of oil and gas through a Berea sandstone core.[3] Oil is the wetting phase.

The change in wetting-phase saturation can be predicted from the solutions of Eqs. 3.16 and 3.17. In Example 3.2, we will assume that the capillary pressure at the end of the core is zero and examine the effect of this boundary condition on the saturation distribution in the core during the steady flow of two phases. Other parameters are identical to those used in Example 3.1.

Example 3.2. This example is a continuation of Example 3.1 with a different set of boundary conditions.

The boundary conditions for both water and oil phases at $x=L$ are given by Eqs. 3.32 and 3.33:

$$P_c = 0 \quad \dots\dots\dots\dots\dots\dots\dots\dots\dots\dots (3.32)$$

and

$$p_{wL} = p_{oL} = p_L. \quad \dots\dots\dots\dots\dots\dots\dots (3.33)$$

From the capillary pressure curves (i.e., Fig. 2.19), there must be a saturation change in the core from S_{wi} at the inlet of the core where $P_c = p_{oi} - p_{wi}$ to S_{wL} at the exit of the core corresponding to $P_c = 0$. In actuality, P_c could not be zero in a strongly water-wet core on the imbibition path because there would be no permeability to oil at zero capillary pressure. However, the results would not change appreciably if a small but arbitrary value of P_c was assumed to exist at $x=L$.

A consequence of a saturation gradient in the core is that we can no longer consider k_o to be constant from $0 < x < L$ as it is in Example 3.1. Eqs. 3.18 and 3.19 can be rearranged and solved for the saturation distribution if relative permeability and capillary pressure curves are known along the same saturation path. The solution will be described in the following paragraphs.[3]

Rearranging Eqs. 3.18 and 3.19, we obtain Eqs. 3.34 and 3.35.

$$\frac{dp_o}{dx} = -\frac{q_o \mu_o}{A k_o} \quad \dots\dots\dots\dots\dots\dots (3.34)$$

and

$$\frac{dp_w}{dx} = -\frac{q_w \mu_w}{A k_w}. \quad \dots\dots\dots\dots\dots\dots (3.35)$$

Recall the definition of the capillary pressure:

$$\frac{dP_c}{dx} = \frac{dp_o}{dx} - \frac{dp_w}{dx} = \frac{q_w \mu_w}{A k_w} - \frac{q_o \mu_o}{A k_o}. \quad \dots\dots (3.36)$$

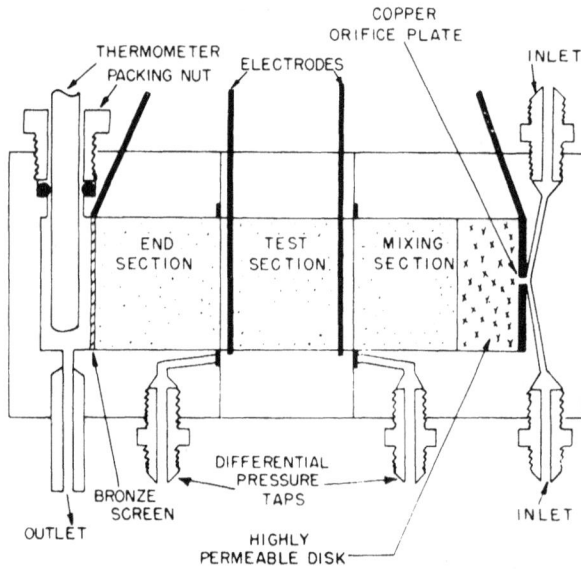

Fig. 3.5—Three-section core assembly—modified PSU method for determination of relative permeability.[4]

Because the capillary pressure curve is a function of the water saturation, the chain rule of differentiation can be used to replace dP_c/dx—that is,

$$\left(\frac{dP_c}{dx}\right) = \left(\frac{\partial P_c}{\partial S_w}\right)\left(\frac{dS_w}{dx}\right). \quad \dots\dots\dots\dots (3.37)$$

Substituting Eq. 3.37 into Eq. 3.36, we obtain

$$\left(\frac{\partial P_c}{\partial S_w}\right)\left(\frac{dS_w}{dx}\right) = \frac{q_w\mu_w}{Ak_w} - \frac{q_o\mu_o}{Ak_o}. \quad \dots\dots (3.38)$$

Eq. 3.38 can be rearranged as Eq. 3.39 and solved by graphical or numerical integration.

$$\int_{S_w}^{S_{wL}} \frac{\left(\frac{\partial P_c}{\partial S_w}\right)dS_w}{\frac{q_w\mu_w}{Ak_w} - \frac{q_o\mu_o}{Ak_o}} = \int_x^L dx. \quad \dots\dots\dots (3.39)$$

The integration is done between x and $x=L$ where the boundary condition $P_c=0$ is applied at $S_w=S_{wL}$. From a practical point of view, Eq. 3.39 is rewritten as an integral where distance is measured from the outlet end of the core—that is, if $x_1 = L-x$, Eq. 3.39 can be rewritten as Eq. 3.40 with x_1 the distance from the end of the core. Thus

$$\int_{S_w}^{S_{wL}} \frac{\left(\frac{\partial P_c}{\partial S_w}\right)dS_w}{\frac{q_w\mu_w}{Ak_w} - \frac{q_o\mu_o}{Ak_o}} = -\int_0^{L-x} dx_1. \quad \dots\dots (3.40)$$

The dashed curve in Fig. 3.4 is the predicted saturation gradient obtained by solution of Eq. 3.40 when oil and gas were the flowing phase.

Other computations supported by experimental data[3] showed that the saturation gradient could be driven into a small region near the end of the core by making the two fluids flow at higher rates. This practice is used to reduce end effects in laboratory cores.

End effects encountered during the steady flow of two phases through a porous rock may be reduced by using "boundary effect absorbers" placed on both ends of the core sample. Fig. 3.5 illustrates a three-section core assembly[4] where the test core is placed between mixing and end sections to prevent end effects from causing saturation changes in the test section. This approach is referred to as the modified Pennsylvania State U. (PSU) method[4] for determination of relative permeabilities.

In the modified PSU method, the test core is in hydraulic contact (and capillary equilibrium) with porous media of similar properties at the end sections. Operating conditions are chosen to confine the saturation changes caused by end effects to the two end sections. The test core is assumed to have uniform saturation that is determined gravimetrically or by electrical resistance measurements as illustrated in Fig. 3.5. Individual phase pressures are not measured. Usually the nonwetting phase pressure is measured. In interpreting the data, it is reasoned that errors introduced by neglecting the difference in phase pressures are small because the change in total pressure over the length of the test section is much larger than the change in the capillary pressures.

Solutions to Eqs. 3.18 and 3.19 for oil- and water-phase permeabilities are given by Eqs. 3.41 and 3.42.

$$k_o = \frac{q_o\mu_o L}{A\Delta p} \quad \dots\dots\dots\dots\dots\dots\dots\dots\dots (3.41)$$

and

$$k_w = \frac{q_w\mu_w L}{A\Delta p}, \quad \dots\dots\dots\dots\dots\dots\dots\dots (3.42)$$

where Δp is the pressure drop across the test section measured through the nonwetting phase.

In Example 3.3, the effect of assuming equal oil- and water-phase pressures in the determination of phase permeabilities using the PSU method is examined.

Example 3.3. Fig. 3.5 shows a modified PSU apparatus used to obtain data for calculation of permeabilities of oil and water during the steady flow of two phases. We assume that the differential pressure taps measure the nonwetting-phase pressure. The pressure drop across the test section is used to compute the permeabilities of both phases as in Eqs. 3.41 and 3.42 for the case of an oil/water system. In this example, we consider the error in computed permeabilities caused by the use of Eqs. 3.41 and 3.42.

The pressure drop measured between the differential pressure taps is $p_{o1} - p_{o2}$. If the rock is strongly water-wet, the oil permeability will be correct. The true water permeability, k_w*, would be computed from Eq. 3.43,

$$k_w* = \frac{q_w \mu_w L}{A(p_{w1} - p_{w2})}, \quad \dots\dots\dots\dots\dots (3.43)$$

but the water-phase pressures are not known.

The capillary pressure curve is a correlation of oil-phase pressure to water-phase pressure. P_c will be positive when the rock is strongly water-wet and when

$$p_{w1} - p_{w2} = (p_{o1} - p_{o2}) - (P_{c1} - P_{c2}).$$

Thus

$$k_w* = \frac{k_w(p_{o1} - p_{o2})}{(p_{o1} - p_{o2}) - (P_{c1} - P_{c2})} \quad \dots\dots\dots (3.44)$$

and

$$k_w* = k_w\left(\frac{1}{1 - \dfrac{P_{c1} - P_{c2}}{p_{o1} - p_{o2}}}\right). \quad \dots\dots\dots (3.45)$$

Capillary pressure is a function of water saturation. If there is no saturation gradient in the core, $P_{c1} = P_{c2}$ and $k_{rw}* = k_w$. The same result is obtained if $(p_{o1} - p_{o2}) \gg (P_{c1} - P_{c2})$, as is usually assumed in the PSU method.

Consider the effect of small changes in the water saturation in the test section. Saturation gradients on the order of 2% saturation are not unusual over test sections 1.2 to 3 in. [3.05 to 7.62 cm] long. Table 3.1 contains imbibition capillary pressure data at several water saturations.[5] Also included in Table 3.1 are estimates of the oil-phase pressure difference $(p_{o1} - p_{o2})$, which would be needed if $k_w/k_w* = 0.98$ and 0.80.

For this porous medium, the capillary pressure change over 2% saturation ranges from 30.3 psi [208.9 kPa] at $S_w = 0.373$ to 0.12 psi [0.83 kPa] at $S_w = 0.782$. Thus at $S_w = 0.373$ the pressure drop in the oil phase would have to be more than 1,515 psi [10 466 kPa] in the test section for the computed water-phase permeability to be within 98% of the true permeability. In contrast, a pressure change in the oil phase of 1.45 psi [9.99 kPa] at an average water saturation of 0.782 would be required to obtain the water-phase permeability that would be within 98% of the true value.

This example shows how capillary pressure differences affect computed values of wetting-phase (water) permeabilities when nonwetting-phase pressures are measured. Maintenance of large pressure changes in the oil phase by operation at high flow rates can reduce these errors. Substantial errors in wetting-phase permeabilities may occur at a low wetting-phase saturation because small changes in saturation produce large changes in capillary pressure. In some cases, pressure drops required to reduce the effect of a small saturation change may not be attainable in the experimental apparatus.

Although all examples in this section were based on oil and water phases, the equations for oil/gas and gas/water

TABLE 3.1—ESTIMATED PRESSURE REQUIRED TO OVERCOME CAPILLARY PRESSURE GRADIENTS BY THE PSU METHOD

S_w (PV)	ΔS_w (PV)	P_c (psi)	ΔP_c (psi)	Δp_o for $k_w/k_w* = 0.98$ (psi)	Δp_o for $k_w/k_w* = 0.80$ (psi)
0.363	0.009	58.8	20.6	1,029.5	103.0
0.372	0.012	38.2	9.7	485.0	48.5
0.384	0.013	28.5	5.9	294.5	29.5
0.397	0.013	22.6	5.0	249.5	25.0
0.410	0.018	17.7	4.7	235.0	23.5
0.418	0.012	13.0	2.7	133.0	13.3
0.440	0.013	10.3	1.9	95.0	9.5
0.453	0.017	8.4	1.6	81.5	8.2
0.470	0.022	6.8	1.5	73.0	7.3
0.492	0.021	5.3	1.3	66.0	6.6
0.513	0.022	4.0	1.1	53.5	5.4
0.535	0.034	2.0	0.78	38.9	3.9
0.569	0.030	2.1	0.78	39.2	3.9
0.599	0.043	1.4	0.54	26.9	2.7
0.642	0.031	0.81	0.29	14.5	1.5
0.673	0.064	0.52	0.38	18.9	1.9
0.737	0.034	0.15	0.12	5.8	0.6
0.771	0.022	0.03	0.03	1.5	0.15
0.793					

flow have the same form when appropriate adjustments are made for fluid properties. A gas is compressible, so it is not possible to remove the density from the equivalent form of Eqs. 3.16 and 3.17. Under some conditions, the density is represented by the ideal gas law. Steady flow of two phases in porous media is seldom encountered outside the laboratory. The concepts illustrated in this section are applicable to the analysis of unsteady-state flows in subsequent sections.

3.4 The Frontal Advance Equation for Unsteady 1D Displacement

The displacement of one fluid by another fluid is an unsteady-state process because the saturations of the fluids change with time. This causes changes in relative permeabilities and either pressure or phase velocities. Fig. 3.6 shows four representative stages of a linear waterflood at interstitial water saturation.

Initial water and oil saturations are uniform, as shown in Fig. 3.6a. Injection of water at flow rate q_t causes oil to be displaced from the reservoir. A sharp water saturation gradient develops, as in Fig. 3.6b. Water and oil flow simultaneously in the region behind the saturation change. There is no flow of water ahead of the saturation change because the permeability to water is essentially zero. Eventually, water arrives at the end of the reservoir, as shown in Fig. 3.6c. This point is called "breakthrough." After breakthrough, the fraction of water in the effluent increases as the remaining oil is displaced. Fig. 3.6d depicts the water saturation in a linear system late in the displacement.

Two methods to predict displacement performance will be presented. The first method is the Buckley-Leverett, or frontal advance, model, which can be solved easily with graphical techniques. The second method is the generalized treatment of two-phase flow leading to a set of partial differential equations that can be solved on a digital computer with numerical techniques.

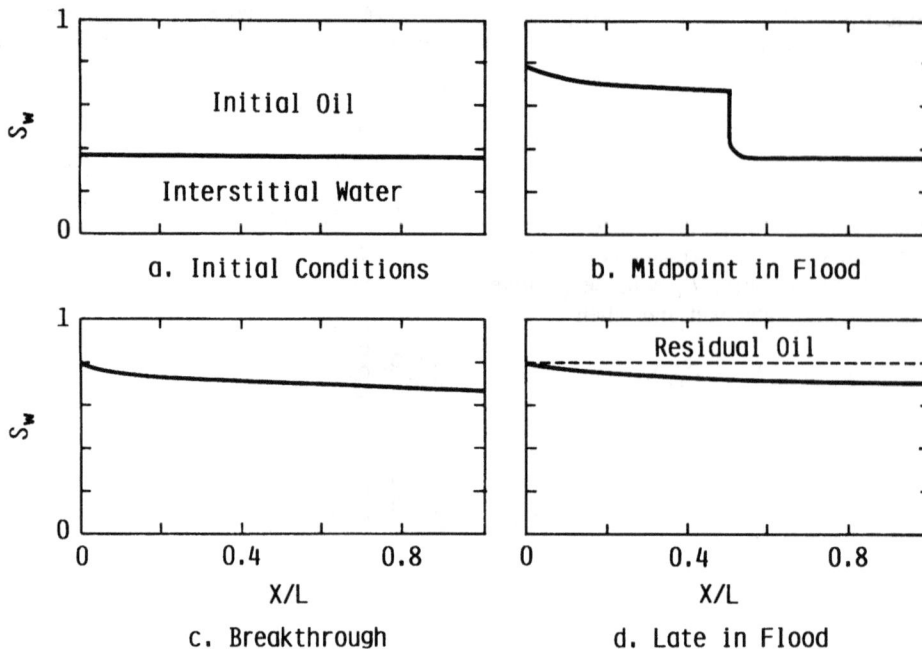

Fig. 3.6—Saturation distributions during different stages of a waterflood.

3.4.1 Buckley-Leverett Model

The Buckley-Leverett model[6] was developed by application of the law of conservation of mass to the flow of two fluids (oil and water) in one direction (x). When oil is displaced by water from a linear system, Eqs. 3.8 and 3.9 become Eqs. 3.46 and 3.47. The subscript x will be dropped because all discussion applies to the x direction.

$$-\frac{\partial}{\partial x}(\rho_o u_{ox}) = \frac{\partial}{\partial t}(\rho_o S_o \phi) \quad \dots \dots \dots \dots (3.46)$$

and

$$-\frac{\partial}{\partial x}(\rho_w u_{wx}) = \frac{\partial}{\partial t}(\rho_w S_w \phi). \quad \dots \dots \dots \dots (3.47)$$

Eqs. 3.46 and 3.47 may also be written in terms of volumetric flow rates q_o and q_w by multiplying both sides of the equations by the cross-sectional area available for flow (A). Thus

$$-\frac{\partial}{\partial x}(\rho_o q_o) = A\frac{\partial}{\partial t}(\rho_o S_o \phi) \quad \dots \dots \dots \dots (3.48)$$

and

$$-\frac{\partial}{\partial x}(\rho_w q_w) = A\frac{\partial}{\partial t}(\rho_w S_w \phi). \quad \dots \dots \dots (3.49)$$

In the Buckley-Leverett model, water and oil are considered incompressible and thus ρ_o and ρ_w are constant. Porosity is also constant, so Eqs. 3.48 and 3.49 become

$$-\frac{\partial q_o}{\partial x} = A\phi\frac{\partial S_o}{\partial t} \quad \dots \dots \dots \dots \dots (3.50)$$

and

$$-\frac{\partial q_w}{\partial x} = A\phi\frac{\partial S_w}{\partial t}. \quad \dots \dots \dots \dots \dots (3.51)$$

The sum of Eqs. 3.50 and 3.51 is Eq. 3.52:

$$-\frac{\partial(q_o + q_w)}{\partial x} = A\phi\frac{\partial}{\partial t}(S_o + S_w). \quad \dots \dots \dots (3.52)$$

Because $S_w + S_o = 1.0$,

$$\frac{\partial(q_o + q_w)}{\partial x} = 0, \quad \dots \dots \dots \dots \dots \dots (3.53)$$

or $q_o + q_w = q_t$ = constant.

Saturations q_o and q_w vary with distance x. However, because oil and water are assumed to be incompressible, the total volumetric flow rate at any time t is constant for every position x in the linear system. Eq. 3.51 is used to continue the development of the Buckley-Leverett model. The fractional flow of a phase, f, is defined as the volume fraction of the phase that is flowing at x, t.

For oil and water phases,

$$f_o = \frac{q_o}{q_t} = \frac{q_o}{q_w + q_o} \quad \dots \dots \dots \dots \dots (3.54)$$

and

$$f_w = \frac{q_w}{q_t} = \frac{q_w}{q_w + q_o}. \quad \dots \dots \dots \dots \dots (3.55)$$

Because the fractional flow is a volume balance,

$$f_o + f_w = 1.0. \quad \dots \dots \dots \dots \dots \dots (3.56)$$

Substituting Eq. 3.55 into Eq. 3.51, we obtain

$$-\frac{\partial f_w}{\partial x}=\frac{\phi A}{q_t}\frac{\partial S_w}{\partial t}. \quad \dots\dots\dots\dots\dots (3.57)$$

To develop a solution for Eq. 3.57, it is necessary to obtain an equivalent form of Eq. 3.57, which involves one dependent variable (i.e., either f_w or S_w). In the Buckley-Leverett model, an expression for $\partial S_w/\partial t$ is obtained following the chain rule of differentiation.

The derivation begins by observing that the water saturation in the porous rock is a function of two independent variables, x and t. Thus we can write

$$S_w=S_w(x,t) \quad \dots\dots\dots\dots\dots\dots\dots (3.58)$$

or

$$dS_w=\left(\frac{\partial S_w}{\partial x}\right)_t dx+\left(\frac{\partial S_w}{\partial t}\right)_x dt. \quad \dots\dots\dots (3.59)$$

If there is interest in what happens to a particular saturation, S_w, it is possible to set $dS_w=0$ in Eq. 3.59 and after some rearrangement to obtain Eq. 3.60:

$$\left(\frac{dx}{dt}\right)_{S_w}=-\frac{\left(\dfrac{\partial S_w}{\partial t}\right)_x}{\left(\dfrac{\partial S_w}{\partial x}\right)_t}. \quad \dots\dots\dots\dots (3.60)$$

The term $(dx/dt)_{S_w}$ is the velocity at which the saturation, S_w, moves through the porous medium. Later, Eq. 3.60 will be used to eliminate $\partial S_w/\partial t$ from Eq. 3.57.

When the fractional flow of water is assumed to be only a function of water saturation, Eq. 3.61 can be derived by application of the chain rule: if $f_w=f_w(S_w)$, then

$$\left(\frac{\partial f_w}{dx}\right)_t=\left(\frac{\partial f_w}{\partial S_w}\right)_t\left(\frac{\partial S_w}{\partial x}\right)_t. \quad \dots\dots\dots\dots (3.61)$$

Substituting Eqs. 3.60 and 3.61 into Eq. 3.57 yields

$$-\left(\frac{\partial f_w}{dS_w}\right)_t\left(\frac{\partial S_w}{\partial x}\right)_t=-\frac{\phi A}{q_t}\left(\frac{\partial S_w}{\partial x}\right)_t\left(\frac{dx}{dt}\right)_{S_w}$$

$$\dots\dots\dots\dots\dots\dots\dots (3.62)$$

or

$$\left(\frac{dx}{dt}\right)_{S_w}=\frac{q_t}{\phi A}\left(\frac{\partial f_w}{\partial S_w}\right)_t. \quad \dots\dots\dots\dots (3.63)$$

Eq. 3.63 is the Buckley-Leverett equation (also called the frontal advance equation), which states that in a linear displacement process, each water saturation moves through the porous rock at a velocity that can be computed from the derivative of the fractional flow with respect to water saturation. Three assumptions were made in developing Eq. 3.63: (1) incompressible flow, (2) that the fractional flow of water is a function only of the water

saturation, and (3) no mass transfer between phases. In the next section we develop an expression for the fractional flow of water.

3.4.2 Fractional Flow Equation

An expression for the fractional flow of water will be developed from Eqs. 3.8 and 3.9. For the purpose of this section, the coordinate axis for the x direction will be assumed to be at an angle of α degrees above the horizontal plane.

From the definition of fractional flow,

$$q_w=f_w q_t \quad \dots\dots\dots\dots\dots\dots\dots\dots\dots (3.64)$$

and

$$q_o=(1-f_w)q_t. \quad \dots\dots\dots\dots\dots\dots\dots (3.65)$$

Substituting these equations into Eqs. 3.8 and 3.9,

$$(1-f_w)q_t=-\frac{k_o A}{\mu_o}\left(\frac{\partial p_o}{\partial x}+\rho_o g \sin \alpha\right) \quad \dots\dots(3.66)$$

and

$$f_w q_t=-\frac{k_w A}{\mu_w}\left(\frac{\partial p_w}{\partial x}+\rho_w g \sin \alpha\right). \quad \dots\dots\dots(3.67)$$

Rearranging Eqs. 3.66 and 3.67,

$$-(1-f_w)\frac{q_t}{A}\frac{\mu_o}{k_o}=\frac{\partial p_o}{\partial x}+\rho_o g \sin \alpha \quad \dots\dots\dots(3.68)$$

and

$$-f_w\frac{q_t}{A}\frac{\mu_w}{k_w}=\frac{\partial p_w}{\partial x}+\rho_w g \sin \alpha. \quad \dots\dots\dots\dots(3.69)$$

The derivatives of oil- and water-phase pressures can be represented in terms of the capillary pressure by subtracting Eq. 3.69 from Eq. 3.68. Noting that

$$\frac{\partial P_c}{\partial x}=\frac{\partial p_o}{\partial x}-\frac{\partial p_w}{\partial x}, \quad \dots\dots\dots\dots\dots\dots (3.70)$$

we obtain

$$-\frac{q_t}{A}\frac{\mu_o}{k_o}+\frac{q_t}{A}f_w\left(\frac{\mu_o}{k_o}+\frac{\mu_w}{k_w}\right)$$

$$=\frac{\partial P_c}{\partial x}+(\rho_o-\rho_w)g \sin \alpha. \quad \dots\dots\dots\dots(3.71)$$

Solving Eq. 3.71 for f_w,

$$f_w = \frac{\dfrac{\mu_o}{k_o}}{\dfrac{\mu_o}{k_o}+\dfrac{\mu_w}{k_w}}$$

$$+ \frac{\dfrac{A}{q_t}\left[\dfrac{\partial P_c}{\partial x}+(\rho_o-\rho_w)g\sin\alpha\right]}{\dfrac{\mu_o}{k_o}+\dfrac{\mu_w}{k_w}}, \quad \dots\dots (3.72)$$

or

$$f_w = \frac{1}{1+\left(\dfrac{k_o}{k_w}\right)\left(\dfrac{\mu_w}{\mu_o}\right)}$$

$$+ \frac{\dfrac{k_oA}{\mu_oq_t}\left[\dfrac{\partial P_c}{\partial x}+(\rho_o-\rho_w)g\sin\alpha\right]}{1+\left(\dfrac{k_o}{k_w}\right)\left(\dfrac{\mu_w}{\mu_o}\right)}. \quad \dots (3.73)$$

In oilfield units (barrels, days, darcies, centipoise, feet, pounds per square inch, grams per cubic centimeter), Eq. 3.73 becomes

$$f_w = \frac{1}{1+\left(\dfrac{k_o}{k_w}\right)\left(\dfrac{\mu_w}{\mu_o}\right)}$$

$$+ \frac{\dfrac{1.127k_oA}{\mu_oq_t}\left[\dfrac{\partial P_c}{\partial x}+0.433(\rho_o-\rho_w)g\sin\alpha\right]}{1+\left(\dfrac{k_o}{k_w}\right)\left(\dfrac{\mu_w}{\mu_o}\right)}.$$

$$\dots\dots\dots\dots\dots (3.74)$$

To derive Eqs. 3.72 and 3.73, we assumed that Darcy's law describing the flow of water and oil and the difference between oil- and water-phase pressures is represented by the capillary pressure curve corresponding to the saturation path. Gravity forces are considered only by including the component of the velocity that acts in the direction of flow (i.e., x).

3.4.3 Development of the Frontal Advance Solution

Solution of the frontal advance equation (Eq. 3.63) for specified boundary conditions forms the basis for prediction of immiscible displacement in a linear system. Eq. 3.63 states that a particular water saturation propagates through a porous rock at a constant velocity. This velocity, $(dx/dt)_{S_w}$, is determined uniquely by the water saturation through the fractional flow equation.

Consider the porous rock that is saturated initially with oil and water and is at interstitial water saturation, S_{iw}. At $t=0+$, water is injected into the rock at a constant rate, q_t. As time progresses, a water saturation profile develops in the porous rock that varies from S_{iw}, as long as no water has reached the end of the core, to $1-S_{or}$ at $x=0$. A sketch of the water saturation at time t is shown in Fig. 3.6b.

The location x_{Sw1} of any saturation, S_{w1}, can be obtained by integrating Eq. 3.63 with respect to time as illustrated in Eqs. 3.75 and 3.76.

$$\int_0^{x_{S_w}}dx_{S_w}=\frac{q_t}{A\phi}\int_0^t\left(\frac{\partial f_w}{\partial S_w}\right)_t dt. \quad \dots\dots (3.75)$$

When $\partial f_w/\partial S_w$ is only a function of S_w, Eq. 3.75 can be integrated directly to obtain Eq. 3.76.

$$x_{Sw}=\frac{q_t t}{A\phi}\left(\frac{\partial f_w}{\partial S_w}\right)_{Sw}. \quad \dots\dots (3.76)$$

Thus if $\partial f_w/\partial S_w$ could be determined accurately from a plot of f_w (computed from Eq. 3.73) vs. S_w, the location of all saturations could be determined as long as the distance x_{Sw} is less than or equal to the length of the porous medium. There are some limitations to Eq. 3.76 that will be discussed in Sec. 3.12.

In principle, the derivative of the fractional flow with respect to S_w can be obtained by differentiating Eq. 3.73, where the permeabilities of oil and water phases are clearly functions of water saturation. The term that is difficult to evaluate is $\partial P_c/\partial x$. When $\partial P_c/\partial x=0$ or when q_t is large, f_w can be computed directly from the relative permeability data. However, these assumptions are valid for only a portion of the range of water saturations.

The capillary pressure for a specific saturation path is a function of water saturation (discussed in Chap. 2). The capillary pressure derivative can be written also as Eq. 3.77,

$$\frac{\partial P_c}{\partial x}=\left(\frac{\partial P_c}{\partial S_w}\right)\left(\frac{\partial S_w}{\partial x}\right), \quad \dots\dots (3.77)$$

which shows that $\partial P_c/\partial x$ will be small when $\partial P_c/\partial S_w$ and/or $\partial S_w/\partial x$ are small. In a linear displacement process, this situation will occur at moderate to high water saturations. At low water saturations, $\partial P_c/\partial S_w$ is large, as seen in Fig. 3.6b. Steep saturation gradients, $\partial S_w/\partial x>>0$, must be present near the front of the invading water region. Thus there is a range of water saturations where f_w and $\partial f_w/\partial S_w$ cannot be computed from Eq. 3.73 because $\partial S_w/\partial x$ is not available unless numerical solutions are used.

Next, we consider how the water saturation range where $\partial P_c/\partial x=0$ is found so that f_w can be computed from Eqs. 3.78 or 3.79:

$$f_w = \frac{1}{1+\left(\dfrac{k_o}{k_w}\right)\left(\dfrac{\mu_w}{\mu_o}\right)}+\frac{\dfrac{k_oA}{\mu_oq_t}(\rho_o-\rho_w)g\sin\alpha}{1+\left(\dfrac{k_o}{k_w}\right)\left(\dfrac{\mu_w}{\mu_o}\right)}.$$

$$\dots\dots\dots\dots\dots (3.78)$$

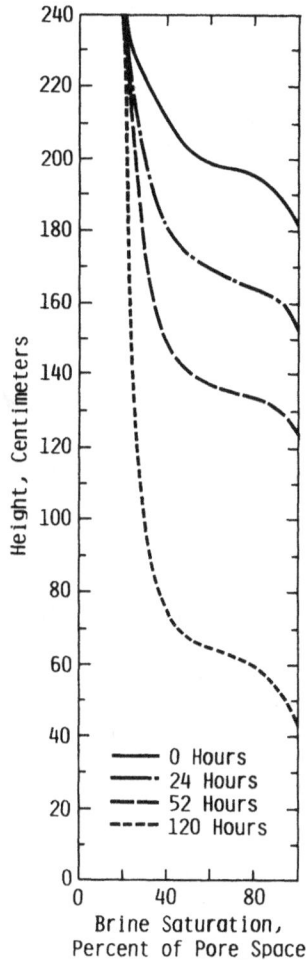

Fig. 3.7—Formation of stabilized zone during brine displacement by gas in a vertical sandpack.[7]

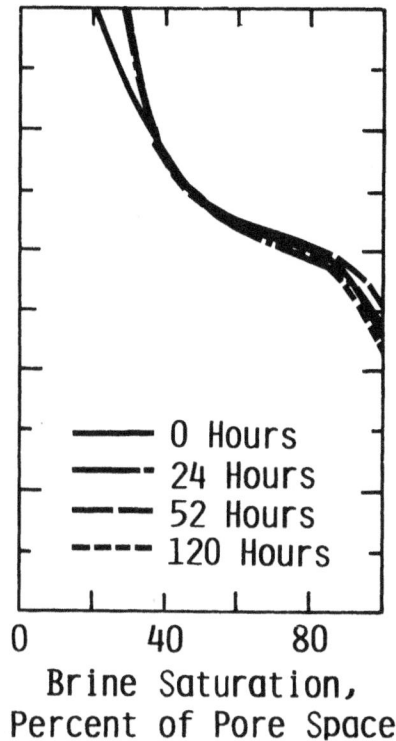

Fig. 3.8—Stabilized zones in displacement Curves A, B, C, and D.[7]

When x is in the horizontal plane, $\alpha=0$, and there is no gravity term,

$$f_w = \frac{1}{1+\left(\dfrac{k_o}{k_w}\right)\left(\dfrac{\mu_w}{\mu_o}\right)}. \qquad \ldots\ldots\ldots\ldots (3.79)$$

The rationale for selection of the saturation range comes from the intuition of Buckley and Leverett,[6] the experimental observations of Terwilliger et al.,[7] and the mathematical construction of von Neumann presented by Welge.[8] We will begin by reviewing the experimental observations of Terwilliger et al.[7] to develop insight into the displacement process.

Saturation distributions in a vertical tube packed with sand were measured during the displacement of brine by gas. A constant displacement rate was obtained by withdrawing brine from the bottom of the sandpack. Fig. 3.7 illustrates experimentally determined brine saturation profiles at four different times that were characteristic of certain operating conditions. The saturation profiles were divided into two parts. The portion of the water-saturation/distance profile from $S_w=1.0$ to $S_w=0.6$ had the same shape as can be seen by overlaying Curves A, B, C, and D as in Fig. 3.8.

Because all curves have the same shape, each saturation within this zone moved at the same velocity. The term "stabilized zone" refers to the region at the leading edge of the displacement process where saturations moved at *equal velocities*. Brine saturations less than 0.4 moved at various velocities. The term "nonstabilized zone" was introduced to describe this region. Velocities of saturations (Eq. 3.76) appeared to be constant in all zones at different times, verifying an important conclusion previously discussed in the development of the frontal advance equation.

A typical fractional curve computed from Eq. 3.79 for an oil/water system is shown in Fig. 3.9. Also shown in Fig. 3.9 is a tangent to the fractional flow curve that originates from the initial water saturation. The point of tangency defines the "breakthrough" or "flood front" saturation, S_{wf}. This saturation is equivalent to the saturation obtained by Buckley and Leverett by intuitive arguments. Subsequently it was recognized[7] that this tangent intersected the fractional flow curve at the saturation common to both stabilized and nonstabilized zones. Saturations greater than S_{wf} satisfy the fractional flow equations given by Eq. 3.79, while fractional flows for saturations less than S_{wf} do not.

The shape of the saturation profile between the initial water saturation and the breakthrough saturation cannot be predicted from the frontal advance solution. An approximation originating from the Buckley-Leverett solution is to consider the saturation change to be a step increase from the interstitial water saturation, S_{iw}, to the flood front saturation, S_{wf}. This is often called a "shock" because all saturations less than S_{wf} move at the velocity of the flood front. Saturations greater than S_{wf} move

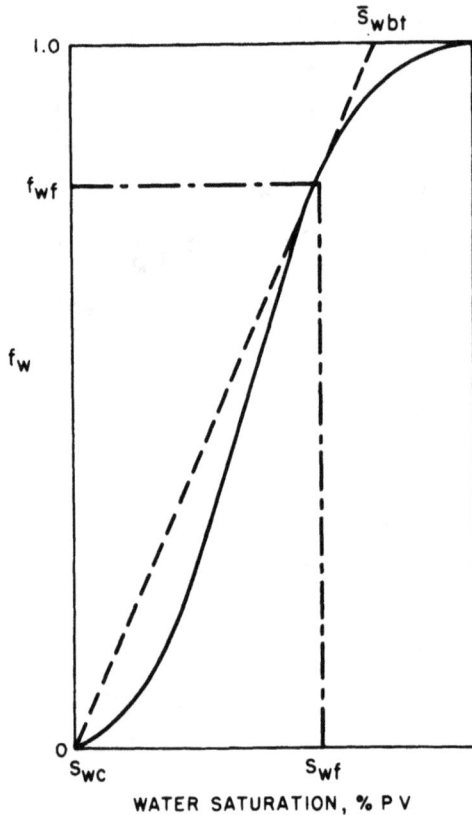

Fig. 3.9—Determination of flood front saturation (after Ref. 9, Fig. 3.10).

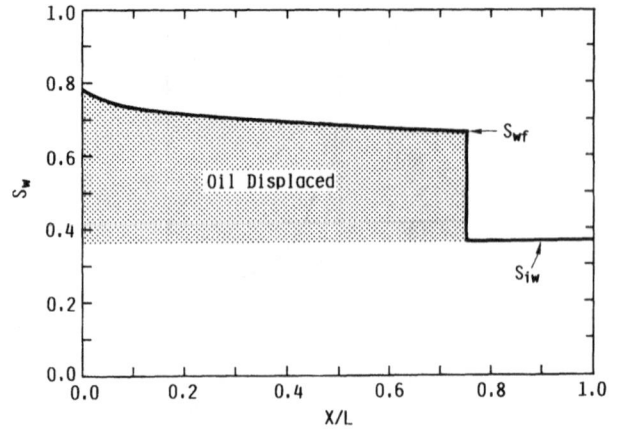

Fig. 3.10—Saturation profile computed from the Buckley-Leverett solution.

at velocities computed from Eq. 3.63 by evaluating the derivative of the fractional flow curve at S_w. At time t, the location x of saturation S_w may be computed for $S_{iw} \leqslant S_w \leqslant 1 - S_{or}$ with Eq. 3.63 to obtain

$$x_{Sw} = \frac{q_t t}{A\phi} \left(\frac{\partial f_w}{\partial S_w} \right)_{S_{wf}}$$

for $S_{iw} < S_w \leqslant S_{wf}$, and

$$x_{Sw} = \frac{q_t t}{A\phi} \left(\frac{\partial f_w}{\partial S_w} \right)_{S_w}$$

for $S_{wf} \leqslant S_w \leqslant 1 - S_{or}$.

Fig. 3.10 shows the saturation distribution obtained from the Buckley-Leverett solution.

The Buckley-Leverett solution describes immiscible displacement in linear systems when stabilized flow occurs. Fortunately, the width of the stabilized zone is relatively narrow in many situations. Errors made by neglecting this zone in linear displacement calculations are not usually large. To use the frontal advance solution for analysis and prediction of oil displacement in linear systems, it is necessary to know when the displacement is stabilized. For the purpose of this section, we will assume that stabilized flow occurs. Criteria for determining whether a displacement is stabilized are presented in Sec. 3.12.1. Stabilized flow is believed to occur in many reservoir displacement processes. Laboratory displacements may be designed to attain stabilized flow.

3.5 Estimating the Displacement Performance for a Linear Waterflood at Constant Injection Rate

In simulating the displacement performance, we are interested in estimating the volume of oil displaced at any time, the rate of oil production, and the volume of water that must be handled per volume of oil once water production begins. Expressions that can be used to calculate these quantities are developed and illustrated in this section.

Oil Displaced. Until water arrives at the end of a system, oil will be produced at the same rate as water is injected for an incompressible system where the interstitial water was assumed to be immobile. When water breakthrough occurs, a water saturation gradient exists from the inlet to the end of the system. The volume of water in the system between $x = x_1$ and $x = x_2$ can be obtained by integrating Eq. 3.80.

$$V_w = \int_{x_1}^{x_2} S_w A\phi dx, \dots\dots\dots\dots\dots\dots\dots (3.80)$$

where V_w is the volume of water in the porous rock between x_1 and x_2.

The volume of oil displaced from the region is

$$V_o = V_w - A\phi(x_2 - x_1)S_{iw}, \dots\dots\dots\dots (3.81)$$

where V_o is the volume of oil displaced from the interval $x_1 \leqslant x \leqslant x_2$.

Welge[8] and Craig (Ref. 9, Page 108) developed solutions to Eq. 3.80. The following development parallels their solutions.

Let $\overline{S_w}$ represent the volumetric average water saturation for the Region $x_1 \leqslant x \leqslant x_2$. Then

$$\overline{S_w} = \frac{\displaystyle\int_{x_1}^{x_2} S_w A\phi dx}{\displaystyle\int_{x_1}^{x_2} A\phi dx}. \dots\dots\dots\dots\dots\dots (3.82)$$

For constant values of ϕ and A, Eq. 3.82 reduces to

$$\overline{S_w} = \frac{\int_{x_1}^{x_2} S_w dx}{x_2 - x_1}. \quad \dots \dots \dots \dots \dots (3.83)$$

The integrand in Eq. 3.83 can be evaluated by use of Eq. 3.76. The derivative of the product xS_w is expressed in Eq. 3.84.

$$d(xS_w) = S_w dx + x dS_w. \quad \dots \dots \dots \dots \dots (3.84)$$

The integrand $S_w dx = d(xS_w) - x dS_w$.

Substitution into Eq. 3.81 with corresponding changes of integration limits yields Eq. 3.85.

$$\overline{S_w} = \frac{1}{x_2 - x_1} \int_1^2 [d(xS_w) - x dS_w], \quad \dots \dots \dots (3.85)$$

$$\overline{S_w} = \frac{1}{x_2 - x_1} \int_{x_1 S_{w1}}^{x_2 S_{w2}} d(xS_w) - \frac{1}{x_2 - x_1} \int_1^2 x dS_w,$$

$$\dots \dots \dots \dots \dots \dots (3.86)$$

and

$$\overline{S_w} = \frac{x_2 S_{w2} - x_1 S_{w1}}{x_2 - x_1} - \frac{1}{x_2 - x_1} \int_1^2 x dS_w. \quad \dots \dots (3.87)$$

Now consider the remaining integral in Eq. 3.87. From Eq. 3.86,

$$\int_1^2 x dS_w = \int_1^2 \frac{q_t t}{A\phi} \left(\frac{\partial f_w}{\partial S_w}\right)_{S_w} dS_w, \quad \dots \dots \dots (3.88)$$

$$\int_1^2 x dS_w = \frac{q_t t}{A\phi} \int_1^2 \left(\frac{\partial f_w}{\partial S_w}\right)_{S_w} dS_w, \quad \dots \dots \dots (3.89)$$

and

$$\int_1^2 x dS_w = \frac{q_t t}{A\phi} \int_1^2 df_w. \quad \dots \dots \dots \dots \dots (3.90)$$

Therefore,

$$\int_1^2 x dS_w = \frac{q_t t}{A\phi} (f_{w2} - f_{w1}). \quad \dots \dots \dots \dots (3.91)$$

Thus the expression for the average water saturation for the interval $x_1 \leqslant x \leqslant x_2$ is given by Eq. 3.92.

$$\overline{S_w} = \frac{x_2 S_{w2} - x_1 S_{w1}}{x_2 - x_1} - \left(\frac{q_t t}{A\phi}\right) \frac{(f_{w2} - f_{w1})}{(x_2 - x_1)}. \quad . (3.92)$$

When $x_1 = 0$ and sufficient time has passed for water to have arrived at the end of the core ($x_2 = L$), the average water saturation in the core is

$$\overline{S_w} = S_{w2} - \frac{q_t t}{A\phi L} (f_{w2} - f_{w1}). \quad \dots \dots \dots \dots (3.93)$$

Usually, $f_{w1} = 1.0$ at $x = 0$ and Eq. 3.93 becomes

$$\overline{S_w} = S_{w2} + \frac{q_t t}{A\phi L} (1 - f_{w2}). \quad \dots \dots \dots \dots (3.94)$$

We note that $q_t t$ represents the total volume of water injected (W_i), while $A\phi L$ is the PV of the porous rock, V_p. We define Q_i by Eq. 3.95 as the number of PV's of water injected.

$$Q_i = \frac{W_i}{A\phi L}. \quad \dots \dots \dots \dots \dots \dots \dots (3.95)$$

For constant injection rate,

$$Q_i = \frac{q_t t}{A\phi L}, \quad \dots \dots \dots \dots \dots \dots \dots (3.96)$$

and Eq. 3.94 becomes

$$\overline{S_w} = S_{w2} + Q_i (1 - f_{w2}). \quad \dots \dots \dots \dots \dots (3.97)$$

Because the displaced hydrocarbon saturation is $\overline{S_w} - S_{iw}$, the cumulative oil displaced, N_p, is given by Eq. 3.98.

$$N_p = V_p (\overline{S_w} - S_{iw}), \quad \dots \dots \dots \dots \dots (3.98)$$

where the FVF was assumed to be 1.0.

One further simplification is possible. At the end of the system ($x = L$) the water saturation is S_{w2} after water arrives. From Eq. 3.76, we have

$$x_{Sw2} = L = \frac{q_t t}{A\phi} \left(\frac{\partial f_w}{\partial S_w}\right)_{S_{w2}}$$

or

$$Q_1 = \frac{1}{\left(\dfrac{\partial f_w}{\partial S_w}\right)_{S_{w2}}}, \quad \dots \dots \dots \dots (3.99)$$

so that

$$\overline{S_w} = S_{w2} + \frac{(1 - f_{w2})}{f'_{Sw2}}, \quad \dots \dots \dots \dots (3.100)$$

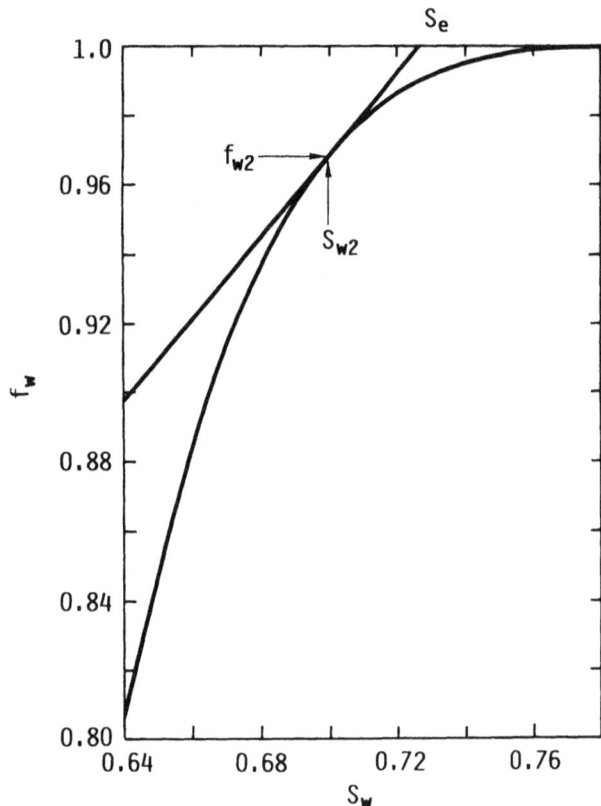

Fig. 3.11—Extrapolation of tangent to fractional flow curve to determine $\overline{S}_w$.

where

$$f'_{Sw2} = \left(\frac{\partial f_w}{\partial S_w}\right)_{S_{w2}} . \quad \dots\dots\dots\dots\dots\dots (3.101)$$

Fig. 3.11 shows a tangent drawn to the fractional flow curve at a saturation $S_{w2} \geqslant S_{wf}$. The tangent intersects the $f_w = 1.0$ line at S_t. We now show that S_e is $\overline{S}_{w2}$. From Fig. 3.11,

$$\left(\frac{\partial f_w}{\partial S_w}\right)_{S_{w2}} = \frac{1-f_w}{S_e - S_{w2}} . \quad \dots\dots\dots\dots\dots (3.102)$$

From Eq. 3.100,

$$\left(\frac{\partial f_w}{\partial S_w}\right)_{S_{w2}} = \frac{1-f_{w2}}{\overline{S}_w - S_{w2}} . \quad \dots\dots\dots\dots\dots (3.103)$$

Comparison of Eqs. 3.102 and 3.103 shows that $S_e = \overline{S}_w$ and the average saturation after breakthrough can be obtained by finding the intersection of the tangent to the $f_w - S_w$ curve with $f_w = 1.0$.

Production Rates. The fractional flow of water is determined from the frontal advance solution for every value of S_{w2}. Thus

$$q_{w2} = \frac{f_{w2}q_t}{B_w} , \quad \dots\dots\dots\dots\dots\dots (3.104)$$

$$q_{o2} = \frac{f_{o2}q_t}{B_o} , \quad \dots\dots\dots\dots\dots\dots (3.105)$$

and

$$q_{o2} = \frac{(1-f_{w2})q_t}{B_o} . \quad \dots\dots\dots\dots\dots (3.106)$$

WOR. The WOR is a measure of the efficiency of the displacement at a point in the process. In production operations, it represents the volume of water that must be handled to produce a unit volume of oil. Eq. 3.107 defines the WOR for a linear system.

$$F_{wo} = \left(\frac{f_{w2}}{f_{o2}}\right)\left(\frac{B_o}{B_w}\right) . \quad \dots\dots\dots\dots (3.107)$$

Time Required for Displacement. Because the injection rate does not vary with time, the value of the time corresponding to Q_i PV's of fluid injected is obtained from Eq. 3.108:

$$t = \frac{Q_i}{q_t/A\phi L} . \quad \dots\dots\dots\dots\dots\dots (3.108)$$

The developments presented in the previous sections are the basis of a procedure to predict displacement performance. Example 3.4 illustrates the use of the frontal advance solution to estimate waterflood performance in a linear reservoir.

Example 3.4. A waterflood is under consideration for a narrow "shoestring" reservoir that is 300 ft [91.44 m] wide, 20 ft [6.1 m] thick, and 1,000 ft [305 m] long. The reservoir is horizontal and has a porosity of 0.15 and an initial water saturation of 0.363, which is considered immobile. It is proposed to drill a row of injection wells at one end of the reservoir and flood the reservoir by injecting water at a rate of 338 B/D [53.7 m³/d]. Viscosities of the oil and water are 2.0 and 1.0 cp [0.002 and 0.001 Pa·s], respectively. Relative permeability data corresponding to the displacement of oil by water are given by Eqs. 3.109 and 3.110.[5] The residual oil saturation is 0.205. Base permeability is the absolute permeability to oil at interstitial water saturation, which is assumed to be equal to the absolute permeability. Oil and water FVF's are 1.0:

$$k_{ro} = (1-S_{wD})^{2.56} \quad \dots\dots\dots\dots\dots (3.109)$$

TABLE 3.2—RELATIVE PERMEABILITIES AND FRACTIONAL FLOWS, VISCOSITY RATIO = 2.0

Index	S_w	k_{rw}	k_{ro}	f_w
1	0.363	0.000	1.000	0.000
2	0.380	0.000	0.902	0.000
3	0.400	0.000	0.795	0.000
4	0.420	0.000	0.696	0.001
5	0.440	0.001	0.605	0.004
6	0.460	0.003	0.522	0.011
7	0.480	0.006	0.445	0.026
8	0.500	0.011	0.377	0.055
9	0.520	0.018	0.315	0.103
10	0.540	0.028	0.260	0.179
11	0.560	0.042	0.210	0.285
12	0.580	0.060	0.168	0.418
13	0.600	0.084	0.131	0.562
14	0.620	0.113	0.099	0.696
15	0.640	0.149	0.073	0.805
16	0.660	0.194	0.051	0.884
17	0.680	0.247	0.034	0.936
18	0.700	0.310	0.021	0.968
19	0.720	0.384	0.011	0.985
20	0.740	0.470	0.005	0.995
21	0.760	0.570	0.002	0.999
22	0.795	0.780	0.000	1.000

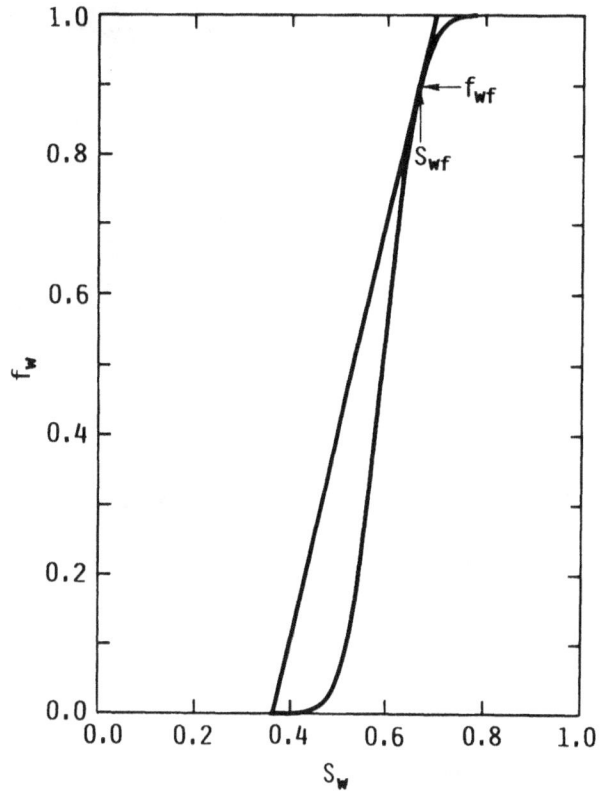

Fig. 3.12—Fractional flow curve with tangent drawn to find S_{wf}.

and

$$k_{rw} = 0.78 S_{wD}^{3.72}, \quad \dots \dots \dots (3.110)$$

where

$$S_{wD} = \frac{(S_w - S_{iw})}{(1 - S_{or} - S_{iw})}, \quad \dots \dots \dots (3.111)$$

which represents normalized water saturation.

Estimates of the oil displacement rate and cumulative oil displaced are required as functions of injection time.

Solution.

From Eq. 3.79,

$$f_w = \frac{1}{1 + \frac{k_{ro}}{k_{rw}} \frac{\mu_w}{\mu_o}}.$$

Values of f_w computed at water saturation increments of 0.02 are presented in Table 3.2.

The graph of f_w vs. water saturation is presented in Fig. 3.12. Fig. 3.13 is a plot of f_w vs. S_w on an expanded scale to enable accurate construction of tangents after breakthrough.

A tangent drawn to Fig. 3.12 from $S_{iw} = 0.363$ intersects the fractional flow curve at $S_w = 0.665$. Thus the stabilized zone includes all water saturations from $S_w = 0.363$ to 0.65. Sometimes, it is difficult to determine the exact point where the tangent to the fractional flow curve intersects the curve. This occurs when the fractional flow curve does not change rapidly with saturation, as is the case in Fig. 3.12. The value of the breakthrough saturation, S_{wf}, obtained by estimating the point of tangency can be checked by applying an overall material balance.

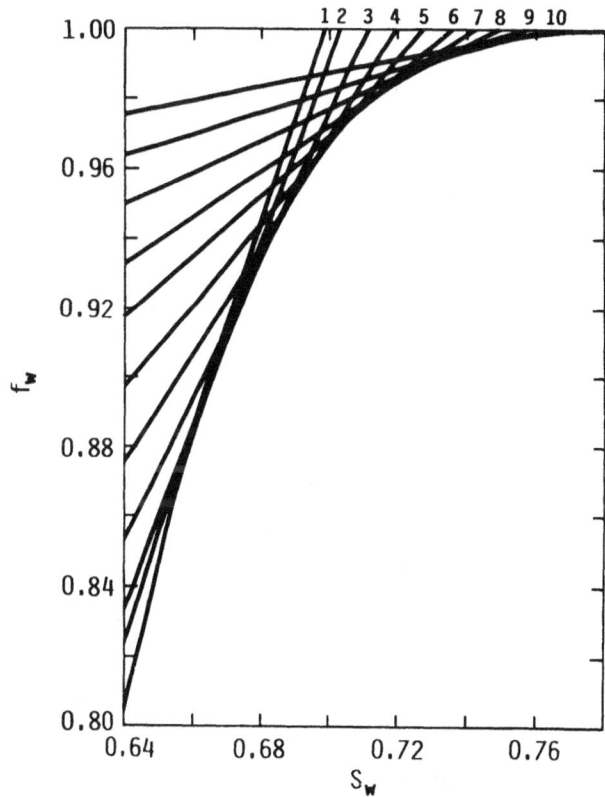

Fig. 3.13—Expanded fractional flow curve for computation of performance after breakthrough.

**TABLE 3.3—CALCULATED RESULTS FROM FRONTAL ADVANCE SOLUTION,
CONSTANT WATER INJECTION RATE OF 338 B/D**

Number	S_{w2}	$\overline{S_{w2}}$	f_{w2}	Q_i (fraction PV)	Time (days)	N_p (STB)	q_o (B/D)	WOR (bbl/STB)
0	0.363		0.000	0.173	82.0	27,729	338.0	0.0
1	0.665	0.700	0.899	0.337	159.8	54,016	34.1	8.9
2	0.670	0.703	0.913	0.379	179.7	54,497	29.4	10.5
3	0.680	0.713	0.936	0.516	244.7	56,100	21.6	14.0
4	0.690	0.721	0.953	0.660	313.0	57,392	15.9	20.3
5	0.700	0.730	0.968	0.938	444.8	58,825	10.8	30.3
6	0.710	0.736	0.977	1.130	535.9	59,786	7.8	42.5
7	0.720	0.741	0.984	1.313	622.6	60,972	5.4	61.5
8	0.730	0.750	0.990	2.000	948.4	62,030	3.4	99.0
9	0.740	0.758	0.995	3.600	1,707.0	63,312	1.7	199.0
10	0.750	0.766	0.997	5.333	2,529.0	64,595	1.0	322.3

The cumulative oil recovered at breakthrough is computed from Eq. 3.98 with $\overline{S_w} = \overline{S_{wf}}$. When the connate water is immobile, the cumulative oil recovered at breakthrough is $q_t t$. Recall that in Eq. 3.96,

$$Q_i = \frac{q_t t}{A\phi L}.$$

At breakthrough,

$$Q_{ibt} = \frac{q_t t_{bt}}{A\phi L} \quad \dots\dots\dots\dots\dots\dots\dots\dots (3.112)$$

and

$$Q_{ibt} = (\overline{S_{wf}} - S_{iw}). \quad \dots\dots\dots\dots\dots\dots (3.113)$$

Because

$$Q_{ibt} = \frac{1}{\left(\dfrac{\partial f_w}{\partial S_w}\right)_{S_{wf}}},$$

then

$$Q_{ibt} = \frac{(\overline{S_{wf}} - S_{wf})}{1 - f_{Swf}}. \quad \dots\dots\dots\dots\dots (3.114)$$

If the tangent construction is correct, the values of Q_{ibt} from Eqs. 3.113 and 3.114 will be the same. Otherwise, a trial-and-error procedure can be used to find the $\overline{S_{wf}}$ that satisfies both equations.

In this example, the average saturation at breakthrough is estimated to be about 0.70 from Fig. 3.12. Thus, from Eq. 3.113,

$$Q_{ibt} = 0.337.$$

The breakthrough saturation appears to be 0.665 from Fig. 3.13, and $f_{Swf} = 0.899$. Substituting into Eq. 3.114,

$$Q_{ibt} = \frac{(0.70 - 0.665)}{(1 - 0.899)} = 0.347.$$

Although closer agreement is possible, additional calculations are not justified because of the difficulty in reading values accurately from the fractional flow curves. A value of 0.337 will be used in this example.

Oil recovery at breakthrough may be computed using the average water saturation.

$$N_p = V_p(\overline{S_{wf}} - S_{iw}),$$

where

$$V_p = A\phi L$$

$$= \frac{(300\ \text{ft})(20\ \text{ft})(1,000\ \text{ft})(0.15)}{5.615\ \text{cu ft/bbl}}$$

$$= 160,285\ \text{bbl}$$

and

$$N_p = (160,285)(0.70 - 0.363)$$

$$= 54,016\ \text{bbl}.$$

Time to reach breakthrough is given by

$$t = \frac{Q_{ibt} V_p}{q_t},$$

where

$$t = 474.2 Q_i$$
$$= (474.2)(0.337)$$
$$= 159.8\ \text{days}.$$

The WOR is computed from Eq. 3.107 where WOR is equal to the volume water per volume oil produced:

$$F_{wo} = \frac{f_{w2}}{f_{o2}}$$

$$= \frac{0.899}{1 - 0.899}$$

$$= 8.9.$$

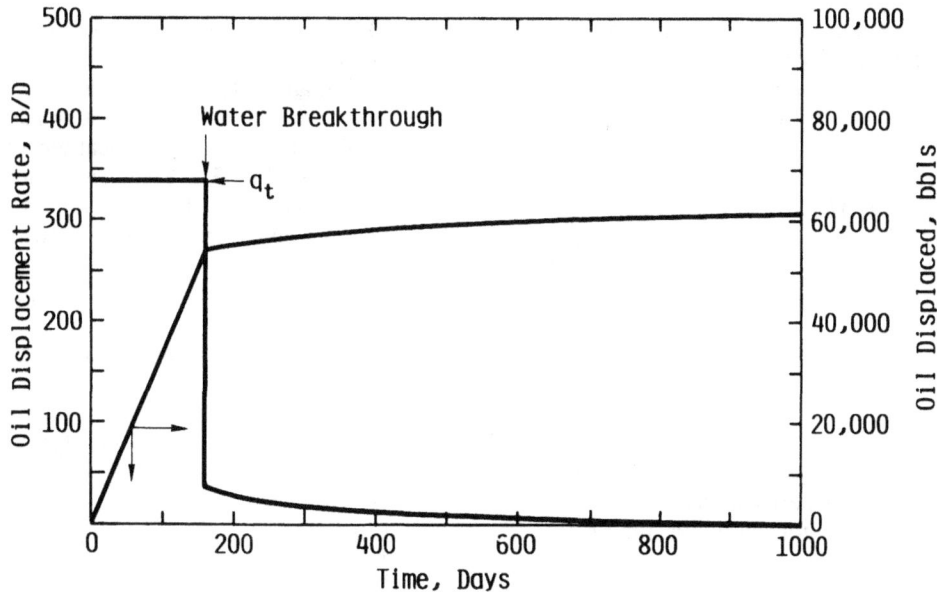

Fig. 3.14—Computed displacement performance from frontal advance solution.

Performance after breakthrough was found by selecting nine values of S_{w2} listed in Table 3.3 and determining f_{w2} and f'_{Sw2} for each S_{w2} by drawing tangents to the expanded fractional flow curve in Fig. 3.13. Results of these calculations are summarized in Table 3.3.

Fig. 3.14 shows the variation of oil production rate and cumulative oil production with time. The sharp drop in oil rate that occurs at 150.7 days is caused by the breakthrough of water at the end of the reservoir. The frontal advance solution instantaneously drops the fractional flow of oil from 1.0 to $1.0 - f_{Swf}$ at this point. In this example, about 54,016 bbl [8588 m³] of oil is recovered in 150.7 days before breakthrough.

3.6 Linear Waterflood at Constant Pressure Drop

Economic considerations usually lead to waterflooding a reservoir at the maximum rate possible. In many reservoirs, this rate is determined by the pressure drop that can be maintained across the system. Pressure in the injection well may be limited by the operator's desire to operate below the fracture pressure of the formation. At the producing well, control of gas and sand or other operating problems may also lead to maintenance of a specified backpressure. Thus operation at constant pressure change between injection and production wells is encountered frequently in field applications.

When the pressure change across a linear system remains constant, the injection rate usually varies with time. The frontal advance solution for constant pressure drop can be obtained by reviewing the development of the frontal advance solution from Eq. 3.63, the frontal advance equation:

$$\left(\frac{dx}{dt}\right)_{S_w} = \frac{q_t}{A\phi}\left(\frac{\partial f_w}{\partial S_w}\right)_t.$$

When this equation was derived, no assumption was made regarding the variation of q_t with time. However, because fluids are assumed to be incompressible, a change in flow rate at one end of a linear system is propagated instantaneously to the other end. This means the total flow rate q_t may vary with time but is constant for every location x at the same time t.

In Sec. 3.4, the location of saturation S_w, x_{Sw}, at any time was found by integrating Eq. 3.63 as in Eq. 3.115.

$$\int_0^{x_{Sw}} dx_{Sw} = \int_0^t \left(\frac{q_t}{A\phi}\right)\left(\frac{\partial f_w}{\partial S_w}\right)_{S_w} dt. \quad \ldots \ldots (3.115)$$

When q_t is a function of time, integration of Eq. 3.115 leads to Eq. 3.116.

$$x_{Sw} = \frac{1}{A\phi}\left(\frac{\partial f_w}{\partial S_w}\right)_{S_w} \int_0^t q_t dt. \quad \ldots \ldots \ldots (3.116)$$

Define Q_i as follows:

$$Q_i = \frac{\int_0^t q_t dt}{V_p} = \frac{W_i}{V_p}. \quad \ldots \ldots \ldots \ldots (3.117)$$

Then

$$x_{Sw} = LQ_i f'_{Sw}. \quad \ldots \ldots \ldots \ldots (3.118)$$

Eq. 3.118 is identical to Eq. 3.76 derived for constant injection rate when Eq. 3.76 is written in terms of Q_i. Therefore, fractional flows, oil recovery, and WOR for a linear displacement process depend only on Q_i and the fractional flow curve. It makes no difference whether the rate is constant or varies arbitrarily with time.

When pressure change is constant in a linear displacement process, the total flow rate at time t is expressed by Eq. 3.119 or 3.120, assuming the pressure gradients in oil and water phases are equal.

$$q_t = -\left(\frac{k_{ro}}{\mu_o} + \frac{k_{rw}}{\mu_w}\right)k_b A \frac{dp}{dx}, \quad \dots \dots (3.119)$$

or

$$q_t = -\lambda_r k_b A \frac{dp}{dx}, \quad \dots \dots (3.120)$$

where λ_r is the total relative mobility and k_b is the base permeability for relative permeability data.

Because q_t is constant for every x at time t,

$$q_t \int_0^L \frac{dx}{\lambda_r} = -k_b A \int_{p_i}^{p_p} dp. \quad \dots \dots (3.121)$$

Integrating and rearranging this equation obtains

$$q_t = \frac{k_b A(p_i - p_p)}{\int_0^L \lambda_r^{-1} dx}, \quad \dots \dots (3.122)$$

where

$$\lambda_r^{-1} = \frac{1}{\lambda_{ro} + \lambda_{rw}} \quad \dots \dots (3.123)$$

is the apparent or effective viscosity of the oil and water phases at saturation S_w.

Eq. 3.122 must be solved by numerical techniques for most problems. In the next section, we introduce a solution technique based on the average apparent viscosity.[10]

Concept of the Average Apparent Viscosity. The average apparent viscosity for a linear displacement process over the distance L is defined by Eq. 3.124:

$$\overline{\lambda^{-1}} = \frac{\int_0^L \lambda_r^{-1} dx}{\int_0^L dx}. \quad \dots \dots (3.124)$$

If the integral in Eq. 3.124 can be evaluated, the total flow rate q_t can be computed from Eq. 3.125:

$$q_t = \frac{k_b A(p_i - p_o)}{\overline{\lambda^{-1}} L}. \quad \dots \dots (3.125)$$

In oilfield units (darcies, centipoise, feet, pounds per square inch, B/D), Eq. 3.125 is

$$q_t = \frac{1.127 k_b A(p_i - p_p)}{\overline{\lambda^{-1}} L}. \quad \dots \dots (3.126)$$

Eq. 3.118 relates the saturation distribution to f'_{Sw} at a given value of Q_i. The saturation distribution can be computed at all x, $0 \le x \le L$, for $S_{w2} \ge S_{wf}$. Therefore, $\overline{\lambda^{-1}}$ could be determined by evaluating the integral in Eq. 3.124 with numerical or graphical techniques.

Inspection of Eq. 3.118 suggests another approach. When Q_i is fixed, x_{Sw} is a function only of f'_{Sw}. The variables x_{Sw} and x are identical. Thus

$$dx = LQ_i df'_{Sw}. \quad \dots \dots (3.127)$$

Changing variables and limits of integration, we obtain

$$\frac{\int_0^L \lambda_r^{-1} dx}{\int_0^L dx} = \frac{\int_{f'_{Swo}}^{f'_{Sw2}} \lambda_r^{-1}(LQ_i df'_{Sw})}{\int_{f'_{Swo}}^{f'_{Sw2}} LQ_i df'_{Sw}} \quad \dots (3.128)$$

or

$$\frac{\int_0^L \lambda_r^{-1} dx}{\int_0^L dx} = \frac{\int_o^{f'_{Sw2}} \lambda_r^{-1} df'}{f'_{Sw2}}, \quad \dots \dots (3.129)$$

where f'_{Sw2} is the derivative of the fractional flow curve at $x = L$, $S_w = S_{w2}$, and f'_{Sw0} is the derivative of f_w at $x = 0$. If $S_w = 1 - S_{or}$ at $x = 0$, then $f'_{Sw0} = 0$. Also after breakthrough $\overline{\lambda^{-1}} = \overline{\lambda_2^{-1}}$. Thus

$$\overline{\lambda_2^{-1}} = \frac{\int_0^{f'_{Sw2}} \lambda_r^{-1} df'_{Sw}}{f'_{Sw2}}. \quad \dots \dots (3.130)$$

Evaluation of $\overline{\lambda_2^{-1}}$ for all saturations $S_{wf} \le S_w \le 1 - S_{or}$ permits calculation of q_t from Eq. 3.125 at and after breakthrough. For the interval before breakthrough, the integral in Eq. 3.124 is expressed as the sum of two integrals.

$$\int_0^L \lambda_r^{-1} dx = \int_0^{x_{Swf}} \lambda_r^{-1} dx + \int_{x_{Swf}}^L \lambda_r^{-1} dx, \quad \dots \dots (3.131)$$

$$\int_0^{x_{Swf}} \lambda_r^{-1} dx = x_{Swf} \overline{\lambda^{-1}}_{Swf}, \quad \dots \dots (3.132)$$

when $x_{Swf} \le L$, $S_{w2} = S_{iw}$, $\lambda^{-1}_{rw} = 0$, and $\lambda_r^{-1} = \lambda^{-1}_{ro}$; therefore,

$$\overline{\lambda^{-1}} = \frac{x_{Swf}}{L} \overline{\lambda^{-1}}_{Swf} + \left(1 - \frac{x_{Swf}}{L}\right)\lambda^{-1}_{ro}. \quad \dots \dots (3.133)$$

Because $x_{Sw} = LQ_i f'_{Swf}$,

$$\overline{\lambda^{-1}} = \lambda_{ro}^{-1} + (\overline{\lambda^{-1}}_{Swf} - \lambda^{-1}_{ro}) Q_i f'_{Swf}. \quad \ldots \ldots (3.134)$$

Thus the average apparent viscosity before breakthrough varies linearly with PV's injected.

Estimation of the q_t/Time Relationship. The last step in the analysis of a linear waterflood is to relate the flow rate q_t to time. Eq. 3.125, combined with Eqs. 3.130 and 3.133, provides a direct relationship between q_t and Q_i—that is, a value of q_t may be computed for every value of Q_i. Because

$$Q_i = \frac{\int_0^t q_t dt}{V_p}, \quad \ldots \ldots \ldots \ldots \ldots \ldots (3.135)$$

a method is needed to extract values of time from pairs of q_t and Q_i values. One method is illustrated in this section. Let t^n and t^{n+1} represent two successive times where $t^{n+1} > t^n$. Eq. 3.135 may be written for each time as

$$Q_i^n = \frac{\int_0^{t^n} q_t dt}{V_p} \quad \ldots \ldots \ldots \ldots \ldots (3.136)$$

and

$$Q_i^{n+1} = \frac{\int_0^{t^{n+1}} q_t dt}{V_p}. \quad \ldots \ldots \ldots \ldots (3.137)$$

Subtracting Eq. 3.136 from Eq. 3.137 gives

$$Q_i^{n+1} - Q_i^n = \frac{\int_0^{t^{n+1}} q_t dt - \int_0^{t^n} q_t dt}{V_p} \quad \ldots (3.138)$$

and

$$-Q_i^n = \frac{1}{V_p} \int_{t^n}^{t^{n+1}} q_t dt. \quad \ldots \ldots \ldots \ldots (3.139)$$

Assuming that q_t in Eq. 3.139 can be approximated by $(q_t^n + q_t^{n+1})/2$, we obtain

$$t^{n+1} = t^n + \frac{2(Q_i^{n+1} - Q_i^n)V_p}{(q_t^{n+1} + q_t^n)}. \quad \ldots \ldots \ldots (3.140)$$

When $n=0$, then $t^n = 0$ and $Q_i^n = 0$. For this case, t^1 is estimated with Eq. 3.141.

$$t^1 = \frac{2Q_i^1 V_p}{q_t^1 + q_t^o}. \quad \ldots \ldots \ldots \ldots \ldots (3.141)$$

The value of q_t^o is computed from Eq. 3.142 because when fluids are incompressible the applied pressure drop is propagated instantaneously through the reservoir, causing the initial flow rate to be approximated by the flow of oil at interstitial water saturation (S_{iw}).

$$q_t^o = \frac{1.127 k_b A (p_i - p_p)}{\lambda_r^{-1} L}. \quad \ldots \ldots \ldots \ldots (3.142)$$

Example 3.5 illustrates the computation of waterflood performance in a linear reservoir at constant pressure drop.

Example 3.5. The reservoir in Example 3.4 is to be waterflooded by maintaining a pressure drop of 500 psi [3448 kPa] between the injection wells and production wells. All other parameters are the same as those used in Example 3.4. Permeability to oil at interstitial water saturation (base permeability) is 200 md.

The method described in this section was used to solve the problem with a computer program. Eqs. 3.143 through 3.147 represent the functional forms used in the calculations. Appendix A contains programs that compute k_{ro}, k_{rw}, f_w, f'_{Sw}, and S_{wf} from these functions.

$$k_{ro} = \alpha_1 (1 - S_{wD})^m, \quad \ldots \ldots \ldots \ldots (3.143)$$

$$k_{rw} = \alpha_2 S_{wD}, \quad \ldots \ldots \ldots \ldots \ldots (3.144)$$

$$S_{wD} = \frac{S_w - S_{iw}}{1 - S_{or} - S_{iw}}, \quad \ldots \ldots \ldots \ldots (3.145)$$

$$f_w = \frac{S_{wD}^n}{S_{wD}^n + A(1 - S_{wD})^m}, \quad \ldots \ldots \ldots (3.146)$$

and

$$\frac{\partial f_w}{\partial S_w} =$$

$$\frac{AB + [n S_{wD}^{n-1}(1 - S_{wD})^m + m S_{wD}^n (1 - S_{wD})^{m-1}]}{[S_{wD}^n + A(1 - S_{wD})^m]^2},$$

$$\ldots \ldots \ldots \ldots \ldots \ldots (3.147)$$

where

$$A = \frac{\alpha_1 \mu_w}{\alpha_2 \mu_o}$$

and

$$B = \frac{1}{1 - S_{or} - S_{iw}}.$$

**TABLE 3.4—COMPUTED PERFORMANCE AT LINEAR WATERFLOOD AT
CONSTANT PRESSURE DROP OF 500 psi**

Time (days)	q_o (B/D)	q_w (B/D)	q_t B/D	WOR (bbl/STB)	Q_i (PV)	N_p (PV)	N_p (bbl)	$\bar{\lambda^{-1}}$ (cp)
0.0	338.16	0.0	338.16	0.0	0.0	0.0	0.0	2.000
16.4	318.05	0.0	318.05	0.0	0.034	0.034	5,383.6	2.126
33.8	300.21	0.0	300.21	0.0	0.067	0.067	10,767.2	2.253
52.2	284.25	0.0	284.25	0.0	0.101	0.101	16,150.8	2.379
71.7	269.91	0.0	269.91	0.0	0.134	0.134	21,534.5	2.506
92.1	256.95	0.0	256.95	0.0	0.168	0.168	26,918.1	2.632
113.6	245.17	0.0	245.17	0.0	0.202	0.202	32,301.7	2.759
136.0	234.43	0.0	234.43	0.0	0.235	0.235	37,685.3	2.885
159.5	224.59	0.0	224.59	0.0	0.269	0.269	43,068.9	3.011
183.9	215.54	0.0	215.54	0.0	0.302	0.302	48,452.5	3.138
209.4	21.60	185.59	207.19	8.59	0.336	0.336	53,836.1	3.264
224.8	20.43	190.99	211.43	9.35	0.356	0.338	54,159.0	3.199
241.1	19.31	196.48	215.79	10.17	0.378	0.340	54,483.8	3.134
258.5	18.23	202.06	220.29	11.08	0.401	0.342	54,810.4	3.070
277.1	17.19	207.72	224.90	12.09	0.427	0.344	55,138.4	3.007
296.8	16.18	213.46	229.65	13.19	0.455	0.346	55,467.7	2.945
317.9	15.22	219.29	234.51	14.41	0.486	0.348	55,797.9	2.884
340.3	14.29	225.21	239.50	15.76	0.519	0.350	56,158.9	2.824
364.3	13.40	231.20	244.60	17.26	0.555	0.352	56,460.4	2.765
389.9	12.54	237.27	249.82	18.92	0.594	0.354	56,792.3	2.707
417.2	11.72	243.42	255.15	20.76	0.638	0.356	57,124.3	2.651
446.6	10.94	249.65	260.59	22.82	0.685	0.358	57,456.2	2.595
478.0	10.19	255.96	266.14	25.12	0.736	0.361	57,788.0	2.541
511.7	9.47	262.33	271.80	27.69	0.793	0.363	58,119.4	2.488
548.0	8.79	268.78	277.57	30.58	0.855	0.365	58,450.4	2.437
587.1	8.14	275.30	283.44	33.83	0.924	0.367	58,780.7	2.386
629.2	7.52	281.88	289.40	37.49	0.999	0.369	59,110.2	2.337
674.7	6.93	288.53	295.46	41.63	1.082	0.371	59,438.8	2.289
724.0	6.37	295.25	301.62	46.33	1.174	0.373	59,766.5	2.242
777.5	5.85	302.02	307.87	51.67	1.275	0.375	60,093.0	2.197
835.7	5.35	308.86	314.20	57.78	1.388	0.377	60,418.2	2.153
899.2	4.87	315.75	320.62	64.77	1.514	0.379	60,742.1	2.109
968.6	4.43	322.69	327.12	72.82	1.654	0.381	61,064.7	2.067
1,044.7	4.01	329.69	333.70	82.12	1.811	0.383	61,385.6	2.027
1,128.5	3.62	336.74	340.36	92.92	1.987	0.385	61,705.0	1.987
1,220.9	3.26	343.83	347.09	105.52	2.186	0.387	62,022.7	1.949

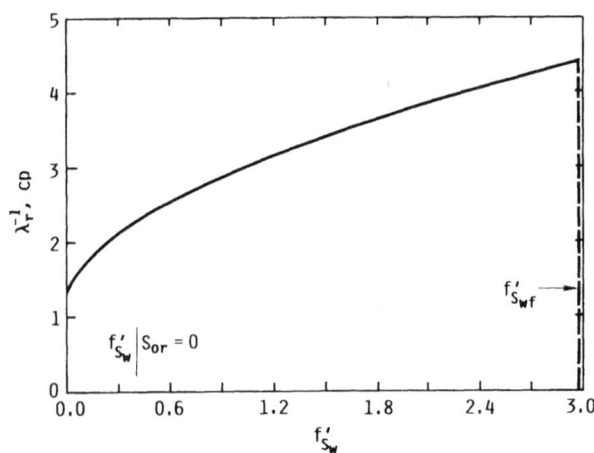

Fig. 3.15—Apparent viscosity as a function of f'_{Sw} for Example 3.5.

In Example 3.4,
$\alpha_1 = 1.0$,
$\alpha_2 = 0.78$,
$\mu_o = 2.0$ cp [2.0 mPa·s],
$S_{iw} = 0.363$,
$m = 2.56$,
$n = 3.72$,
$\mu_w = 1.0$ cp [1 mPa·s], and
$S_{or} = 0.205$.

In computing the displacement performance before breakthrough, we subdivided Q_{ibt} into 10 increments. This is equivalent to advancing the flood front through the reservoir in increments of 0.1 L. Performance after breakthrough was computed by subdividing the saturation interval $(1-S_{or})$ to S_{wf} into 50 increments.

Table 3.4 contains computed results to a WOR of 106. Flow rates and times were computed from Eqs. 3.126, 3.140, and 3.141 after the average apparent viscosities were determined.

Calculation of Average Apparent Viscosities. Evaluation of the average apparent viscosity for the interval $S_{wf} \leqslant S_w \leqslant 1-S_{or}$ was accomplished by numerical integration of Eq. 3.124. Values of λ_r^{-1} were computed for each of the 51 values of S_w. Fig. 3.15 shows the relationship between λ_r^{-1} and f'_{Sw}. Average apparent viscosi-

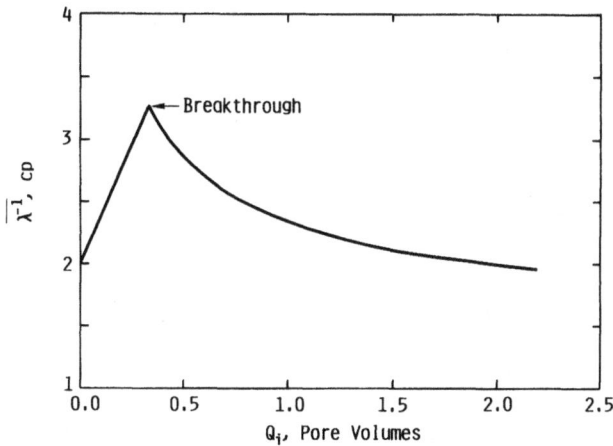

Fig. 3.16—Variation of the average apparent viscosity with PV's injected.

TABLE 3.5—RESULTS FROM COMPUTATION OF AVERAGE APPARENT VISCOSITY

S_{w2} (fraction)	f'_{Sw}	$\overline{\lambda^{-1}}_{Sw}$	Q_i (PV)
0.7950	0.	1.282	∞
0.7924	0.001	1.297	732.583
0.7898	0.004	1.317	240.704
0.7871	0.008	1.336	123.870
0.7845	0.013	1.356	76.592
0.7819	0.019	1.377	52.368
0.7793	0.026	1.398	38.154
0.7766	0.034	1.420	29.043
0.7740	0.044	1.442	22.827
0.7714	0.054	1.465	18.385
0.7688	0.066	1.489	15.096
0.7662	0.079	1.514	12.589
0.7635	0.094	1.539	10.634
0.7609	0.110	1.565	9.080
0.7583	0.128	1.592	7.824
0.7557	0.147	1.620	6.795
0.7530	0.168	1.648	5.943
0.7504	0.191	1.678	5.228
0.7478	0.216	1.708	4.624
0.7452	0.243	1.740	4.109
0.7425	0.273	1.772	3.667
0.7399	0.304	1.805	3.285
0.7373	0.339	1.839	2.954
0.7347	0.375	1.875	2.664
0.7321	0.415	1.911	2.410
0.7294	0.458	1.949	2.186
0.7268	0.503	1.987	1.987
0.7242	0.552	2.027	1.811
0.7216	0.604	2.067	1.654
0.7189	0.660	2.109	1.514
0.7163	0.720	2.153	1.388
0.7137	0.784	2.197	1.275
0.7111	0.852	2.242	1.174
0.7085	0.924	2.289	1.082
0.7058	1.001	2.337	0.999
0.7032	1.083	2.386	0.924
0.7006	1.169	2.437	0.855
0.6980	1.261	2.488	0.793
0.6953	1.358	2.541	0.736
0.6927	1.460	2.595	0.685
0.6901	1.568	2.651	0.638
0.6875	1.682	2.707	0.594
0.6848	1.802	2.765	0.555
0.6822	1.927	2.824	0.519
0.6796	2.059	2.884	0.486
0.6770	2.197	2.945	0.455
0.6744	2.341	3.007	0.427
0.6717	2.491	3.070	0.401
0.6691	2.647	3.134	0.378
0.6665	2.809	3.199	0.356
0.6639	2.977	3.264	0.336

ties were computed from the data in Fig. 3.15 with the trapezoidal rule. Table 3.5 presents values of the average apparent viscosity in the order computed. Also included in Table 3.5 are values of Q_i that correspond to f'_{Sw2}. Note that $\overline{\lambda^{-1}}_{Sw2}$ is a unique function of Q_i for a particular set of fluid and rock properties.

Before breakthrough, $\overline{\lambda^{-1}}$ is a linear function of Q_i for $0 \leqslant Q_i \leqslant Q_{ibt}$. From Eq. 3.134,

$$\overline{\lambda^{-1}} = (\overline{\lambda^{-1}}_{Swf} - \lambda_{ro}^{-1})Q_i f'_{Swf} + \lambda_{ro}^{-1},$$

where

$$\lambda_{ro}^{-1} = \frac{\mu_o}{(k_{ro})_{S_{iw}}} = 2.0 \text{ cp}.$$

From Table 3.4, $S_{wf} = 0.6639$, where

$$\overline{\lambda^{-1}} = 3.264 \text{ cp},$$
$$f'_{Swf} = 2.978, \text{ and}$$
$$Q_{ibt} = 0.336.$$

Therefore,

$$\overline{\lambda^{-1}} = (3.264 - 2.0)Q_i(2.978) + 2.0 = 3.764Q_i + 2.0$$

for $Q_i \leqslant 0.336$.

Fig. 3.16 is a plot of the average apparent viscosity vs. PV's injected for the linear waterflood in Example 3.5. Note that the average apparent viscosity increases linearly until breakthrough and declines thereafter.

The injection rate, q_t, is computed from Eq. 3.126 for each value of Q_i. For example, at $Q_i = 0.401$, $\overline{\lambda^{-1}} = 3.070$ cp:

$$q_t = \frac{(1.127)(0.200)(20 \text{ ft})(300 \text{ ft})(500 \text{ psi})}{(3.070 \text{ cp})(1,000 \text{ ft})}$$

$$= 220.3 \text{ B/D}.$$

Time from the beginning of injection is calculated using Eqs. 3.140 and 3.141. Starting from $t=0$, we determine the time corresponding to the first value of Q_i (0.0336 or 0.034).

$$q_t^0 = 338.16 \text{ B/D at } Q_i^0 = 0 \ (t^0 = 0),$$

$$q_t^1 = 318.05 \text{ B/D at } Q_i^1 = 0.0336,$$

$$t^{n+1} = t^n + \frac{2(Q_i^{n+1} - Q_i^n)V_p}{(q_t^{n+1} + q_t^n)},$$

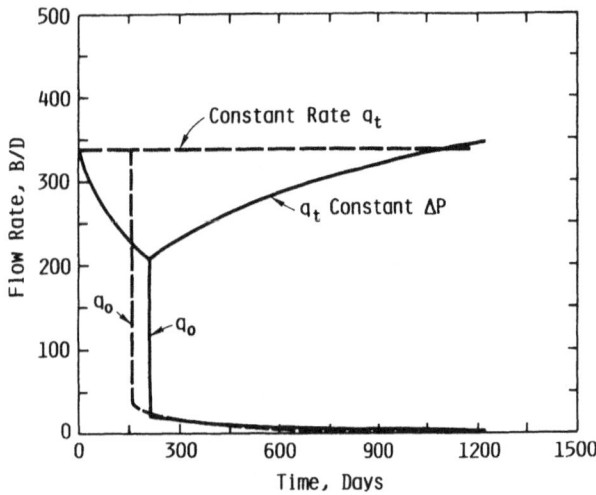

Fig. 3.17—Comparison of total production rates and oil production rates for constant-rate and constant-pressure waterfloods.

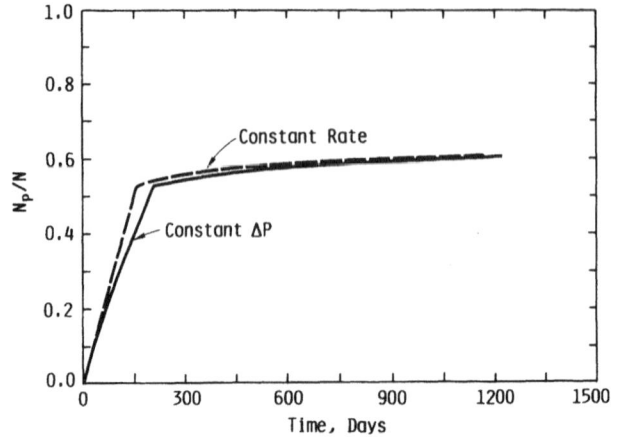

Fig. 3.18—Comparison of oil recovery as a function of time for constant-rate and constant-pressure-drop displacements.

and

$$t^1 = 0 + \frac{(2)(0.0336-0)(160,285)}{(338.16+318.05)} = 16.4 \text{ days}.$$

Calculation of q_t and Time. Total injection rates for the constant-pressure flood decline with time until shortly after breakthrough because the average apparent viscosity behind the flood front is greater than the viscosity of the oil bank ahead of the front. After breakthrough, total production rate (and thus the injection rate) rises as the average apparent viscosity decreases with increased oil displacement. Oil production rates for the two cases of constant rate and constant pressure changes are compared in Fig. 3.17. Oil production rates follow the trends of total injection rate.

The sharp decline in oil production at breakthrough is caused by the high displacement efficiency for the particular set of fluid and rock properties used in the simulation. About 78% of the recoverable oil is displaced at breakthrough. Cumulative oil production curves are shown in Fig. 3.18. A longer time interval is required to waterflood the reservoir under constant-pressure operation. WOR's were not plotted because WOR's for both types of floods are the same at equal values of Q_i.

3.7 Equivalence of Constant-Rate and Constant-Pressure Solutions of the Frontal Advance Equation

The frontal advance solution for a linear waterflood conducted at constant rate may be used to predict performance of a constant-pressure flood if the pressure has been determined for the constant-rate solution. This can be seen by comparing the solutions for the two cases. Eq. 3.148 shows that the location of the saturation S_w in a linear displacement is dependent only on Q_i, the number of PV's injected.

$$x_{S_w} = LQ_i f'_{S_w}. \quad\dotfill (3.148)$$

Thus the saturation profiles are identical at the same value of Q_i whether the displacement process occurs at constant-rate or at constant-pressure difference.

Similarly, at any time t, the total rate and total pressure changes are related by Eq. 3.149:

$$q_t = \frac{k_b A(p_i - p_p)}{\displaystyle\int_0^L \frac{dx}{\lambda_{ro} + \lambda_{rw}}}. \quad\dotfill (3.149)$$

Introducing the definition of $\overline{\lambda^{-1}}$, we obtain

$$q_t = \frac{k_b A(p_i - p_L)}{\overline{\lambda^{-1}} L}. \quad\dotfill (3.150)$$

Rearranging Eq. 3.150 yields

$$\overline{\lambda^{-1}} = \frac{k_b A(p_i - p_p)}{q_t L}. \quad\dotfill (3.151)$$

Because the average apparent viscosity $\overline{\lambda^{-1}}$ is a function of only Q_i, the following relationship must hold at equal values of Q_i.

$$\overline{\lambda^{-1}} =$$

$$\underbrace{\frac{k_b A(p_i - p_p)}{q_t L}}_{\text{Constant Rate}} = \overbrace{\frac{k_b A(p_i - p_p)}{q_t L}}^{\text{Constant } \Delta p} = \overbrace{\frac{k_b A(p_i - p_p)}{q_t L}}^{\text{Variable } \Delta p}$$
$$\underbrace{\phantom{\frac{k_b A(p_i - p_p)}{q_t L}}}_{\text{Variable Rate}}$$

$$\dotfill (3.152)$$

Therefore, a plot of $\overline{\lambda^{-1}}$ vs. Q_i defines a unique relationship between flow rate and pressure drop for the frontal advance solution of a linear waterflood.

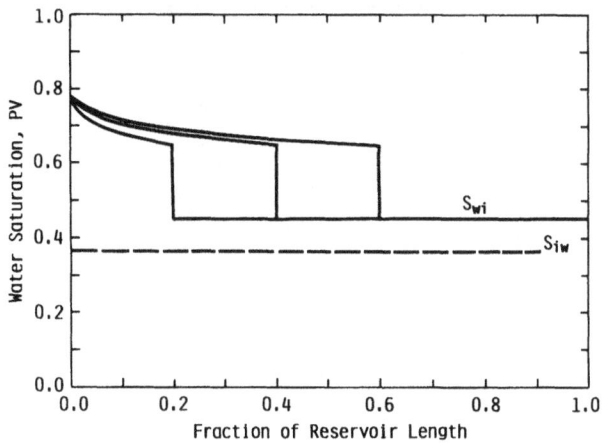

Fig. 3.19—Saturation distribution during waterflooding with mobile initial water saturation with increasing injected water volumes.

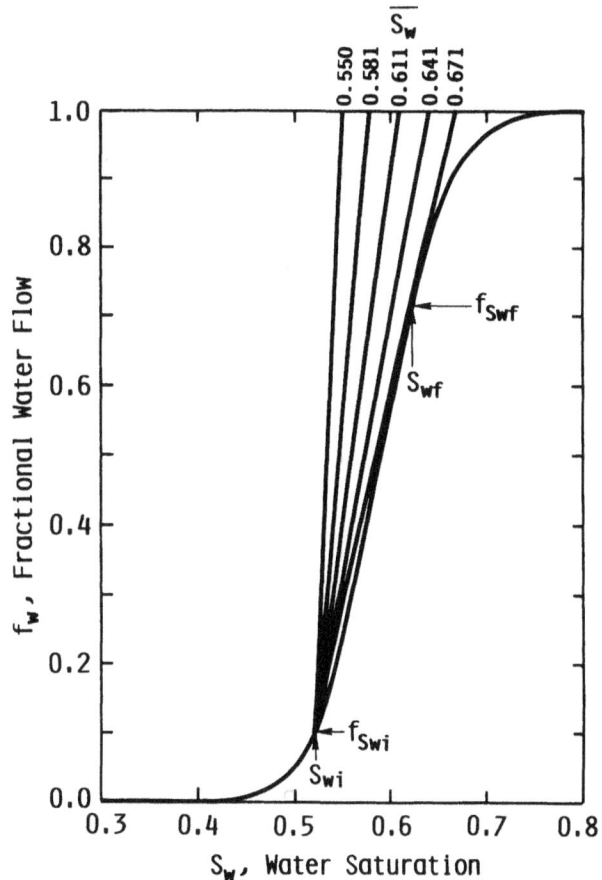

Fig. 3.20—Construction of tangent to determine breakthrough saturation when initial water saturation is mobile.

Example 3.6 illustrates the use of this relationship to compute the pressure drop during the waterflood conducted at a constant rate in Example 3.4. Base permeability is assumed to be 0.1 darcy.

Example 3.6. When the injection rate is constant in a waterflood, the pressure drop usually changes. Find the pressure drop when $Q_i = 0.378$. Rearranging Eq. 3.152 yields

$$p_i - p_p = \frac{\overline{\lambda^{-1}} q_t L}{1.127 k_b A}.$$

Referring to Table 3.4 or Table 3.5 at $Q_i = 0.378$, we find $\overline{\lambda^{-1}} = 3.134$ cp [3.134 mPa·s]. Solving for Δp in Eq. 3.152, we obtain

$$p_i - p_p =$$

$$\frac{(3.134 \text{ cp})(338 \text{ B/D})(1,000 \text{ ft})}{1.127 \text{ B/D} \left[\dfrac{(\text{cp})(\text{ft})}{(\text{darcy})(\text{sq ft})(\text{psi})} \right] (0.100 \text{ darcy})(20 \text{ ft})(300 \text{ ft})}$$

$$= 1,564 \text{ psi}.$$

3.8 Linear Waterflood With a Mobile Initial Water Saturation

The frontal advance solution developed in Sec. 3.4 assumes the interstitial water is immobile. No water production occurs until the flood front arrives at the end of the system. In this section it is shown how the frontal advance solution is modified to include the effects of a mobile initial water saturation.

Fig. 3.19 depicts a linear system that has a uniform water saturation that is mobile. Visualize this system as if water and oil were injected and produced at a constant ratio corresponding to the steady flow of oil and water with $f_{wi} = f_w(S_{wi})$ on the appropriate saturation path. A waterflood of this system would be represented by a change in f_w at $x = 0$ from $f_w(S_{wi})$ to $f_w = 1.0$ for time

greater than 0. Fig. 3.19 shows the saturation distribution at a point in the process assuming the displacement is stabilized—i.e., a flood front forms.

Oil and water production rates are given by Eqs. 3.153 and 3.154 until the flood front saturation arrives:

$$q_w = f_{wi} q_t \quad \dots \dots \dots \dots \dots \dots \dots \dots \dots \dots (3.153)$$

and

$$q_o = (1 - f_{wi}) q_t. \quad \dots \dots \dots \dots \dots \dots \dots \dots (3.154)$$

The flood front saturation is found with a modification of the tangent construction procedure described in Sec. 3.4.3 developed by Craig (Ref. 9, Page 108).

The derivation is based on an overall material balance on the water phase. Let f_{wi} be the fractional flow of water corresponding to the initial water saturation. Then, at any time in the displacement before arrival of the water at the end of the system,

$$\int_0^t q_t dt = \int_0^t f_{wi} q_t dt + \int_0^L (S_w - S_{wi}) \phi A dx, \quad \dots (3.155)$$

where the first quantity is the cumulative injected water, the second quantity is cumulative water produced, and the

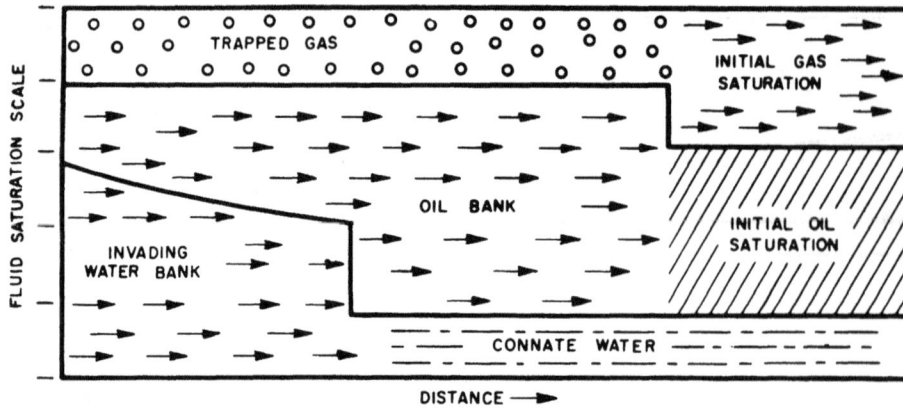

Fig. 3.21—Saturation profile during a waterflood in a depleted reservoir when a trapped gas saturation exists.[11]

third quantity is water accumulated by change in water saturation. Dividing by the PV of the system, ϕAL,

$$Q_i = f_{wi} Q_i + \frac{1}{\phi AL} \int_0^L (S_w - S_{wi}) A\phi dx. \quad \ldots \ldots (3.156)$$

Using the definition of $\overline{S_w}$ from Eq. 3.82,

$$\overline{S_w} = \frac{\displaystyle\int_{x_1}^{x_2} S_w A\phi dx}{\displaystyle\int_{x_1}^{x_2} A\phi dx},$$

and integrating between $x_1 = 0$ and $x_2 = L$, we obtain from Eqs. 3.156 and 3.82

$$Q_i = f_{wi} Q_i + \overline{S_w} - S_{wi}. \quad \ldots \ldots \ldots \ldots (3.157)$$

Rearranging Eq. 3.157 yields Eq. 3.158:

$$\frac{1}{Q_i} = \frac{1 - f_{wi}}{\overline{S_w} - S_{wi}}, \quad \ldots \ldots \ldots \ldots \ldots (3.158)$$

which states that the average water saturation for the region $0 < x < L$ with $S_{w2} = S_{wi}$ is found by drawing a straight line from the fractional flow curve at (f_{wi}, S_{wi}) with slope $1/Q_i$ to $f_w = 1.0$. Fig. 3.20 illustrates several lines drawn from f_{wi}, S_{wi} to the axis at $f_w = 0.1$. The fractional flow curve is from Example 3.4. In Fig. 3.20, $f_{wi} = 0.1$ when $S_{wi} = 0.52$. Also shown on Fig. 3.20 is the value of $\overline{S_w}$ for each Q_i. Note that Eq. 3.158 is consistent with the derivations of Sec. 3.5 when the interstitial water is immobile ($f_{wi} = 0$ and $\overline{S_w} = S_{wi}$). In this case, $Q_i = \overline{S_w} - S_{wi}$.

In Sec. 3.5 an expression for the average water saturation was derived from the frontal advance equation. Eq. 3.97 is valid when the water saturation at $x_2 = L$ is S_{w2}, where $S_{w2} > S_{wi}$. The saturation S_{w2} must satisfy the material balance given by Eq. 3.158 when the flood front

arrives at the end of the system. Rearranging Eq. 3.97 yields

$$\frac{1}{Q_{i2}} = \frac{1 - f_{w2}}{\overline{S_w} - S_{w2}}. \quad \ldots \ldots \ldots \ldots \ldots (3.159)$$

Comparison of Eqs. 3.158 and 3.159 shows that Point f_{w2}, S_{w2} must be the intersection of the tangent drawn from f_{wi}, S_{wi} to f_w as shown in Fig. 3.20. The saturation S_{w2} is the only saturation on the computed fractional flow curve that satisfies the material balance at the instant the water saturation changes from S_{wi} to S_{w2}. Thus S_{w2} at the point of tangency is the breakthrough saturation when the initial water saturation is mobile. Performance after breakthrough is computed in the same manner described in Sec. 3.5.

Minimum Conditions Required To Establish a Flood Front. A flood front is characterized by a saturation discontinuity between S_{wi} and S_{wf} at breakthrough. An oil bank forms when the velocity of the flood front saturation exceeds or is equal to the velocity of the initial saturation. Velocities of flood front saturation and the initial saturation are given by Eqs. 3.160 and 3.161:

$$v_{Swf} = \frac{q_t}{A\phi} \left(\frac{\partial f_w}{\partial S_w} \right)_{S_{wf}} \quad \ldots \ldots \ldots \ldots (3.160)$$

and

$$v_{Swi} = \frac{q_t}{A\phi} f_{wi}. \quad \ldots \ldots \ldots \ldots \ldots (3.161)$$

The flood front contains all velocities $v_{Swi} < v_{Sw} < v_{Swf}$. Therefore, a flood front can form when

$$f_{wi} < \left(\frac{\partial f_w}{\partial S_w} \right)_{S_{wf}}. \quad \ldots \ldots \ldots \ldots (3.162)$$

In principle, the flood front saturation is found from the intersection of the tangent to the fractional flow curve drawn from f_{wi}, S_{wi}. Thus a saturation discontinuity will

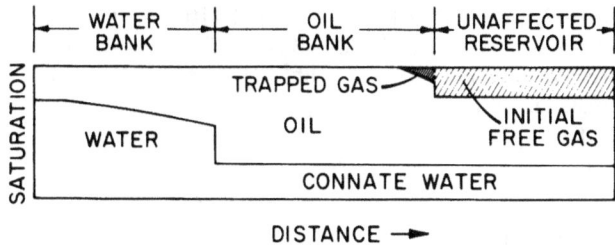

Fig. 3.22—Saturation profile during waterflood of a depleted reservoir when trapped gas is redissolved (Ref. 9, Page 40).

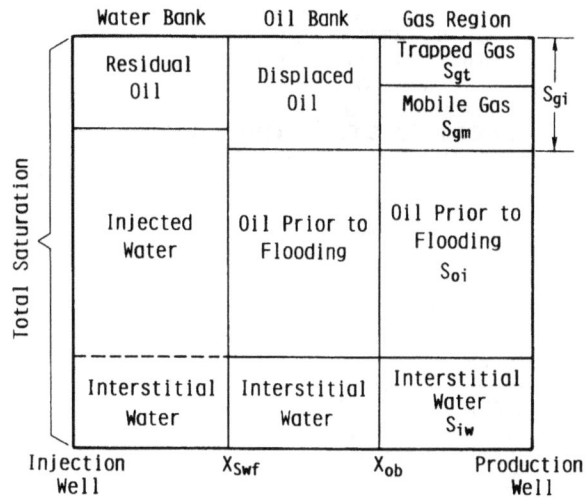

Fig. 3.23—Fluid saturation regions between the producing and injecting wells under immiscible fluid injection (after Prats et al.)[12]

exist as long as a tangent can be drawn to the curve. When a tangent cannot be drawn to the curve, the entire displacement behaves as if performance was computed after breakthrough.

Application of this approach to predict flood front saturation is limited by the extent to which capillary pressure gradients affect the true fractional flow curve. It is not known whether the stabilized-zone concept for mobile connate water saturation is valid for all connate water saturations.

3.9 Linear Waterflood Performance in a Depleted Reservoir

Most reservoirs that are candidates for waterflooding have been produced under primary operation for some period of time. When the principal producing mechanism is dissolved gas drive, a mobile gas saturation exists at the beginning of the waterflood. In this section we develop methods of estimating displacement performance when there is an initial gas saturation and the connate water saturation is considered immobile.

Injection of water into a depleted reservoir causes the displacement of both oil and gas. Although three phases are present, frequently it is possible to represent the displacement process as a series of connected regions where only two phases are flowing. Fig. 3.21 depicts one model of the displacement process in which mobile gas is displaced by the oil bank created by the injected water.[11] The oil bank displaces the gas immiscibly, leaving a residual gas saturation at the leading edge of the oil bank. Because the mobility ratio for the displacement of gas by oil is quite favorable ($\ll 0.1$), the average oil saturation in the oil bank is the breakthrough oil saturation or $1 - S_{iw} - S_{gr}$.

3.9.1 Displacement Performance When Trapped Gas Dissolves in Oil

Economic considerations usually dictate waterflooding at the highest feasible injection rates. Reservoir pressure increases, leading to the possibility that some or all of the gas trapped by the oil bank will be redissolved. Craig (Ref. 9, Page 40) presents a method for estimating the pressure at which all trapped gas will go back into solution. Relatively small increases in pressure frequently are required to redissolve the trapped gas. Thus, in the design of most waterfloods, the assumption of zero residual gas saturation is a good starting point.

Displacement performance in the absence of a trapped gas saturation is estimated as follows. It is assumed that

the oil bank is formed quickly so that all fluid flow involves only two phases. Fig. 3.22 shows the saturation distribution corresponding to this waterflood model. Oil displaced by the injected water displaces the mobile gas (S_{gm}), leaving a trapped gas saturation (S_{gt}) at the leading edge of the oil bank. Trapped gas is dissolved "instantaneously" at the leading edge of the bank with displaced oil filling the pore space equivalent to the volume occupied by the trapped gas. There is little oil production until the oil bank reaches the production well because gas mobility is much greater than the oil mobility under the same pressure gradient. The time required for the oil bank to reach the production well is termed the fill-up time. During this time interval, a volume of gas equivalent to $V_p * S_{gm}$ at reservoir conditions will be displaced. In the waterflood model, the frontal advance solution describes the displacement performance in the water bank. In effect, this region is a normal waterflood with initial oil saturation at $1 - S_{wi}$.

Fig. 3.23 is a model depicting fluid saturations during the waterflood.[12] The location of the leading edge of the oil bank is found by an overall material balance on the oil phase—that is,

$$A\phi L S_{oi} = \int_0^t q_t f_{oi} \, dt + (1 - \overline{S_{wf}}) A\phi x_{Swf}$$

$$+ A\phi (x_{ob} - x_{Swf}) S_{ob} + A\phi (L - x_{ob}) S_{oi}. \quad \ldots \ldots (3.163)$$

Solving for x_{ob}, we obtain

$$\frac{x_{ob}}{L} = \frac{f_{oi} Q_i + (1 - \overline{S_{wf}} - S_{ob}) \dfrac{x_{Swf}}{L}}{S_{oi} - S_{ob}}. \quad \ldots \ldots (3.164)$$

The flood front position (x_{Swf}) is obtained from the material balance on water.

$$(\overline{S_{wf}} - S_{wr}) A\phi x_{Swf} = \int_0^t q_t \, dt. \quad \ldots \ldots \ldots (3.165)$$

TABLE 3.6—
DISPLACEMENT PERFORMANCE FOR
DEPLETED RESERVOIR, CONSTANT
INJECTION RATE 338 B/D,
INITIAL GAS SATURATION = 0.15,
MOBILE GAS SATURATION = 0.10, AND
TRAPPED GAS REDISSOLVES IN OIL.

Q_i (PV)	N_p (bbl)	q_o (B/D)	Time (days)
0.000	0	0.0	0.0
0.050	0	0.0	23.7
0.100	0	0.0	47.4
0.149	0	0.0	70.7
0.150	0	338.0	71.0*
0.200	8,014	338.0	94.8
0.250	16,029	338.0	118.6
0.300	24,043	338.0	142.3
0.336	29,813	338.0	159.3
0.337	29,973	35.0	159.8**
0.379	30,454	29.4	179.7
0.516	32,051	21.6	244.7
0.660	33,339	15.8	313.0
0.938	34,782	10.8	444.8
1.130	35,744	7.8	535.9
1.313	36,545	5.5	622.6
2.000	37,988	3.4	948.4

*Oil bank arrival.
**Water breakthrough.

Recalling the definition of Q_i, we can determine that

$$\frac{x_{Swf}}{L} = \frac{Q_i}{\overline{S_{wf}} - S_{iw}}. \qquad \ldots \ldots \ldots \ldots \ldots \ldots (3.166)$$

Substitution of Eq. 3.166 into Eq. 3.164, noting that $S_{ob} = 1 - S_{iw}$, gives x_{ob}/L:

$$\frac{x_{ob}}{L} = \frac{Q_i}{S_{ob} - S_{oi}}(1 - f_{oi}). \qquad \ldots \ldots \ldots \ldots (3.167)$$

Because $S_{ob} = S_{oi} + S_{gi}$, when the residual gas saturation redissolves in the oil,

$$\frac{x_{ob}}{L} = \frac{(1 - f_{oi})Q_i}{S_{gi}}. \qquad \ldots \ldots \ldots \ldots \ldots \ldots (3.168)$$

Example 3.7 illustrates the effect of an initial gas saturation on the performance of a linear waterflood.

Example 3.7. This example is an extension of Example 3.4 when the initial gas saturation is 0.15 and the residual gas saturation is 0.05. Interstitial water saturation is 0.363 as in Example 3.4, but the initial oil saturation is now 0.487. The displacement of gas by oil follows the imbibition path because oil is the wetting phase for the displacement of gas in the presence of immobile interstitial water.

First, we compute the oil in place and the fill-up time. The volume of the reservoir in Example 3.4 is 160,285 bbl [25 483 m^3]. Because of depletion drive, the initial oil in place is 78,059 bbl [12 410 m^3]. Oil left as residual saturation at the end of a complete waterflood

(WOR = infinite) is 160,285 bbl [12 410 m^3] multiplied by 0.205, which equals 32,858 bbl [5224 m^3]. Maximum recovery would be 45,201 bbl [7186 m^3] if flooded to residual saturation.

During fill-up, $f_{oi} = 0$ at most gas saturations near S_{gi}. Thus

$$\frac{x_{ob}}{L} = \frac{Q_i}{S_{gi}}. \qquad \ldots \ldots \ldots \ldots \ldots \ldots (3.169)$$

Displacement of the oil bank is the same as in Example 3.5 except oil production begins when $Q_i = S_{gi}$. The time required to inject this volume of water is the fill-up time, t_f—that is,

$$Q_{if} = \frac{\displaystyle\int_0^{t_f} q_t \, dt}{A\phi L}, \qquad \ldots \ldots \ldots \ldots \ldots (3.170)$$

$$Q_{if} = S_{gi},$$

or

$$W_{if} = S_{gi} A\phi L, \qquad \ldots \ldots \ldots \ldots \ldots \ldots (3.171)$$

where W_{if} is the volume of water injected to fill-up.

From Eq. 3.171, W_{if} equals $(0.15)(160,285) = 24,043$ bbl.

The mobile gas saturation is 0.10, so the reservoir volume of the displaced gas is 16,029 bbl [2548 m^3]. The remaining 8,014 bbl [1274 m^3] of the water injected during fill-up displaced 8,014 bbl [1274 m^3] of oil into pore space initially occupied by the trapped gas that redissolved into the oil.

In Example 3.4, the injection rate was constant at 338.0 B/D [53.7 m^3/d]. With the same rate, the fill-up time is computed from W_{if}:

$$t_f = \frac{24,043 \text{ bbl}}{338.0 \text{ B/D}} = 71.1 \text{ days}.$$

Therefore, from $t = 0$ to $t = t_f$, only gas would be produced at the end of the linear system. At fill-up, the flood front saturation is computed from Eq. 3.166. Recalling from Example 3.4, $S_{wf} = 0.70$,

$$\frac{x_{Swf}}{L} = \frac{0.15}{0.70 - 0.363}$$

$$= 0.445.$$

The region from $0.445 \leq x/L < 1.0$ now has an oil bank with an oil saturation of $1 - S_{iw} = 0.637$. By material balance, the volume of oil that is displaced when water breaks through at the end of the system is given by Eq. 3.172.

$$N_{pbt} = \left(1 - \frac{x_{Swf}}{L}\right)(\overline{S_{wf}}\text{fill-up} - S_{iw})V_p. \quad \ldots (3.172)$$

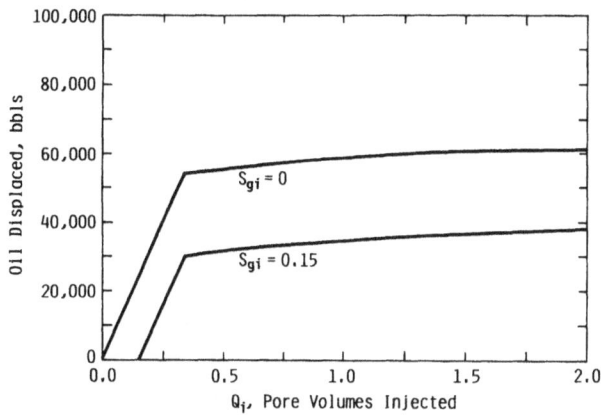

Fig. 3.24—Effect of an initial gas saturation on oil production from a linear waterflood.

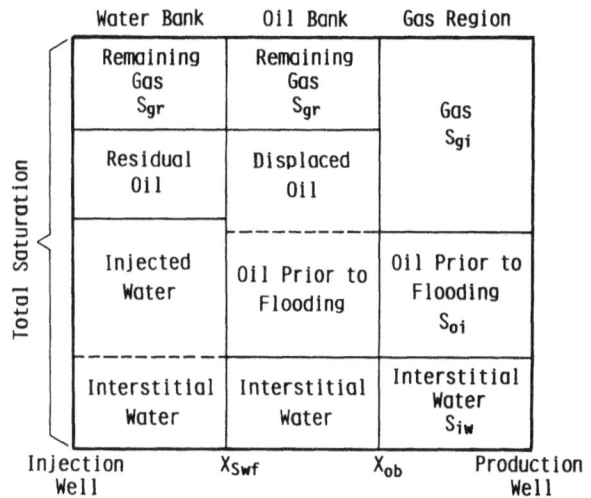

Fig. 3.25—Fluid saturation regions between the producing and injecting wells under immiscible fluid injection (after Prats et al.[12])

Thus

$$N_{pbt} = (0.555)(0.70 - 0.363)(160,285 \text{ bbl})$$

$$= 29,979 \text{ bbl}.$$

The remaining 15,222 bbl [2420 m³] of the displaceable oil (based on S_{or}) is recovered after breakthrough. Production performance after breakthrough is identical to the estimated values in Example 3.4 with one exception. Oil recovery after breakthrough is computed from Eq. 3.173:

$$N_p = N_{pbt} + (\overline{S_w} - \overline{S_{wf}})V_p. \quad \ldots \ldots \ldots \ldots (3.173)$$

Computed results are summarized in Table 3.6.

The effect of an initial gas saturation on the displacement performance of a linear waterflood is shown in Fig. 3.24. The upper curve on Fig. 3.24 represents data from Table 3.3 with $S_{gi} = 0$. The lower curve is from Table 3.6. The principal effect of an initial gas saturation in this model is to delay arrival of the oil bank by 0.15 PV. Gas produced at a volumetric rate (at reservoir conditions) is 0.1/0.15 of the injection rate during the fill-up time. Of course, overall recovery is 0.15 PV less.

3.9.2 Waterflood When Trapped Gas is Present

Reservoir pressures may not be high enough to cause trapped gas to dissolve in the oil. Displacement performance can be simulated with the frontal advance solution and material balances if the trapped gas is incompressible. Otherwise, numerical techniques are required to account for changes in saturation and fluid flows as the volume occupied by the trapped gas changes with pressure during the displacement.

The approach introduced in the previous section is used to develop expressions for the displacement performance. An oil bank is assumed to form rapidly and to displace the gas, leaving a residual gas saturation S_{gr}. The residual gas saturation is constant throughout the remainder of the flood. Saturation distributions in the oil and water banks for the displacement model are shown in Fig. 3.25.[12] A displacement process where there is no flow of the displaced phase after arrival of the flood front (in this case the oilflood front) is often described as "piston-like" displacement.

Location of the oil bank is found by writing a material balance around the oil. This material balance is represented by Eq. 3.174.

$$\frac{x_{ob}}{L} = \frac{Q_i(1 - f_{oi})}{S_{gi} - S_{gr}}. \quad \ldots \ldots \ldots \ldots \ldots (3.174)$$

Oil saturation in the oil bank is $1 - S_{iw} - S_{gr}$. By analogy, the material balance on the water phase gives Eq. 3.166.

The displacement of the oil bank by the injected water in the presence of trapped gas may be visualized as a waterflood in a system that has its PV reduced by S_{gr}—that is, there is little effect of the trapped gas on displacement performance. The principal exception is the reduction of residual oil saturation in the presence of trapped gas as discussed in Chap. 2. This shifts the endpoint of the relative permeability curves toward lower oil saturation. Otherwise, relative permeability curves and displacement performance are not influenced by trapped gas.

Adaptation of the frontal advance solution for trapped gas is straightforward. We define the effective PV, water, and oil saturations by Eqs. 3.175 through 3.177.

$$V_{pe} = V_p(1 - S_{gr}), \quad \ldots \ldots \ldots \ldots \ldots (3.175)$$

$$S_{we} = \frac{S_w}{1 - S_{gr}}, \quad \ldots \ldots \ldots \ldots \ldots (3.176)$$

and

$$S_{oe} = \frac{S_o}{1 - S_{gr}}. \quad \ldots \ldots \ldots \ldots \ldots (3.177)$$

Relative permeability curves are expressed in terms of gas-free saturations with residual oil saturation adjusted to include effects of the trapped gas saturation.

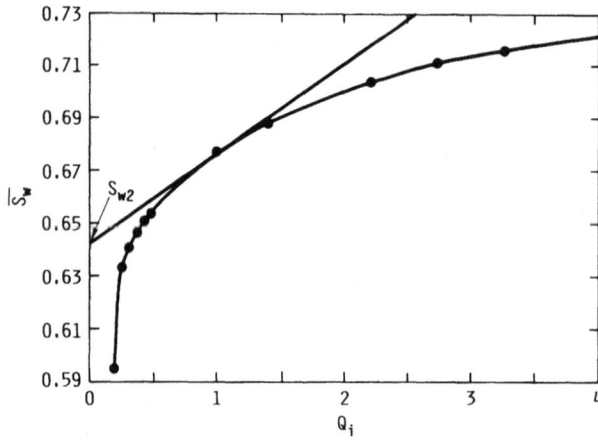

Fig. 3.26—Determination of water saturation (S_{w2}) and permeability ratio from displacement data.

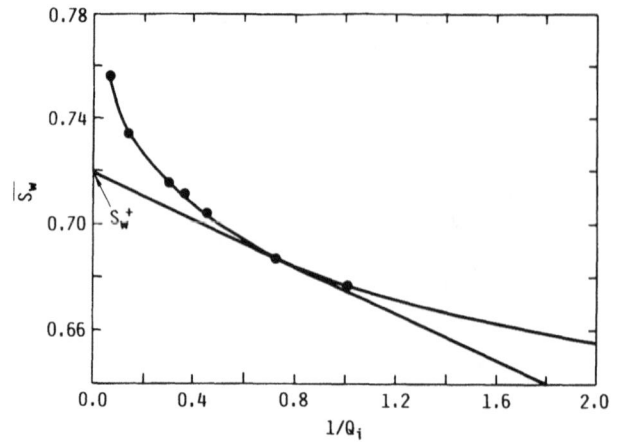

Fig. 3.27—Estimation of S_{w2} by method of Jones and Roszelle at large values of Q_i.

The frontal advance solution is obtained for the effective system. Thus

$$\frac{x_{Swfe}}{L} = Q_{ie} f'_{Swfe} \quad \dots\dots\dots\dots\dots (3.178)$$

and

$$\overline{S_{wfe}} = S_{wfe} + Q_{ibte}(1 - f_{Swfe}). \quad \dots\dots\dots (3.179)$$

For the total system,

$$Q_i = \frac{Q_{ie}}{1 - S_{gr}} \quad \dots\dots\dots\dots\dots\dots (3.180)$$

and

$$\overline{S_{wf}} = \overline{S_{wfe}}(1 - S_{gr}). \quad \dots\dots\dots\dots (3.181)$$

The displacement performance for the entire flood is found by computing the oil bank position from Eq. 3.174 at selected values of Q_{ie} obtained from the frontal advance solution. After breakthrough of the water, performance is obtained from the frontal advance solution based on the effective system.

Fill-up time is computed from Eqs. 3.182 and 3.183.

$$Q_{if} = S_{gi} - S_{gr} \quad \dots\dots\dots\dots\dots\dots (3.182)$$

and

$$Q_{if} = \frac{\int_0^{t_f} q_t \, dt}{V_p}. \quad \dots\dots\dots\dots\dots (3.183)$$

For constant-rate injection,

$$t_f = \frac{V_p(S_{gi} - S_{gr})}{q_t}. \quad \dots\dots\dots\dots (3.184)$$

In developing the displacement models for cases of zero or finite residual gas saturation, we assumed oil and water banks formed quickly after the beginning of injection. The formation of the water and oil banks involves flow of three phases. Accurate simulation of this process requires simultaneous solution of three partial differential equations representing the flows of oil, water, and gas in the waterflood. Techniques are available that can solve these equations.[13] However, the model presented in this section preserves the essential features of the displacement process.

3.10 Determination of Relative Permeability Curves From Analysis of Linear Displacement Data

An important application of the frontal advance solution is the calculation of relative permeability ratios and individual relative permeability curves from linear displacement data obtained in the laboratory on reservoir core samples. The data must be obtained under conditions where stabilized flow is attained and capillary end effects are minimized. These problems are discussed in Sec. 3.12.

3.10.1 Relative Permeability Ratios

The determination of relative permeability ratios from linear displacement data is relatively straightforward. In a linear displacement experiment, the fractional flow of water is known as a function of cumulative water injected or PV's injected. The average water saturation is computed from the material balance on the water phase. Initial water saturation, usually immobile, is determined from the material balance before the displacement process. With these data, the relative permeability ratios can be computed.

The water saturation at the end of the core is not known but can be computed from Eq. 3.93—that is,

$$S_{w2} = \overline{S_w} + Q_i(f_{w2} - f_{w1}). \quad \dots\dots\dots (3.185)$$

f_{w1} will be 1.0 for most cases so that

$$S_{w2} = \overline{S_w} - Q_i f_{o2}. \quad \dots\dots\dots\dots (3.186)$$

**TABLE 3.7—DISPLACEMENT DATA FOR DETERMINATION
OF RELATIVE PERMEABILITY RATIOS**

Cut	Oil Production (% PV)	Water Production (% PV)	N_p (% PV)	Q_i (fraction PV)	$1/Q_i$	S_w (fraction)
1	19.88	0.00	19.88	0.1988	5.030	0.5954
2	3.80	2.02	23.68	0.2570	3.891	0.6334
3	0.74	3.96	24.42	0.3140	3.185	0.6408
4	0.54	5.19	24.96	0.3713	2.693	0.6462
5	0.43	5.50	25.39	0.4306	2.322	0.6505
6	0.31	5.62	25.70	0.4899	2.041	0.6536
7	2.33	48.60	28.03	0.9992	1.001	0.6769
8	1.08	39.49	29.11	1.4049	0.712	0.6877
9	1.59	79.25	30.70	2.2133	0.452	0.7036
10	0.77	52.05	31.47	2.7415	0.365	0.7113
11	0.43	51.43	31.90	3.2601	0.307	0.7156
12	1.86	374.68	33.76	7.0255	0.142	0.7342
13	2.21	732.59	35.76	14.3735	0.070	0.7563

Total PV, cm³	2.58
Initial water saturation	0.396
Viscosity water/viscosity oil	0.529
Injection rate, cm³/hr	40

Values of k_o/k_w corresponding to S_{w2} are computed from f_{w2} by rearranging Eq. 3.79.

$$\left(\frac{k_o}{k_w}\right)_{S_{w2}} = \frac{\mu_o}{\mu_w}\left(\frac{1}{f_{w2}} - 1\right).$$

In the displacement experiment, oil and water production are collected over some increment of time. There is no measuring instrument that determines f_{o2} continuously. The average fractional flow of oil for a given time increment is computed with Eq. 3.187:

$$f_{o2} = \lim_{\Delta W_i \to 0} \frac{N_{p|W_{i+\Delta W_i}} - N_{p|W_i}}{\Delta W_i} = \frac{dN_p}{dW_i}, \quad \ldots (3.187)$$

where dN_p is the cumulative oil produced during the time increment Δt and ΔW_i is the water injected during the same time period. Eq. 3.187 can be expressed in terms of the average water saturation and PV's injected as in Eq. 3.188:

$$dN_p = A\phi L d\overline{S_w}$$

and

$$dW_i = A\phi L dQ_i.$$

Therefore,

$$f_{o2} = \frac{d\overline{S_w}}{dQ_i}. \quad \ldots\ldots\ldots\ldots\ldots\ldots\ldots (3.188)$$

Thus the fractional flow of oil can be obtained by differentiating a plot of the average water saturation vs. cumulative PV's injected.

**TABLE 3.8—CALCULATION OF RELATIVE
PERMEABILITY RATIOS**

Q_i (PV)	$\overline{S_w}$ (PV)	S_w^+ (PV)	S_{w2} (PV)	f_{o2}	f_{w2}	k_w/k_o
0.1988	0.5984			1.0	0	0
0.2570	0.6334		0.574	0.231	0.769	1.76
0.3140	0.6408		0.608	0.104	0.896	3.56
0.3713	0.6462		0.615	0.084	0.916	5.77
0.4306	0.6505		0.624	0.062	0.938	8.07
0.4899	0.6536		0.628	0.0527	0.948	9.59
0.9992	0.6769		0.643	0.0339	0.966	15.06
1.4049	0.6877		0.656	0.023	0.977	22.55
1.4049	0.6877	0.719	0.654	0.024	0.976	21.50
2.2133	0.7036		0.670	0.015	0.985	34.30
2.2133	0.7036	0.737	0.672	0.014	0.986	36.5
2.7415	0.7113	0.741	0.682	0.011	0.989	48.3
3.2601	0.7156	0.744	0.687	0.009	0.991	60.3
7.0255	0.7342	0.758	0.710	0.003	0.997	155.6
14.3735	0.7563	0.787	0.726	0.002	0.998	247.1

Fig. 3.26 is a plot of $\overline{S_w}$ vs. Q_i for displacement data of Example 3.7. A tangent is drawn to the curve at $Q_i = 1.0$. Let S_w^* equal the value of S_w when $Q_i = 0$. Then

$$\frac{d\overline{S_w}}{dQ_i} = \frac{\overline{S_w} - S_w^*}{Q_i - 0}. \quad \ldots\ldots\ldots\ldots\ldots\ldots (3.189)$$

The intersection of the tangent with the ordinate ($Q_i = 0$) can be shown to be S_{w2}.[10]

Substitution of Eq. 3.188 into Eq. 3.185 yields

$$S_{w2} = \overline{S_w} - Q_i \frac{d\overline{S_w}}{dQ_i}. \quad \ldots\ldots\ldots\ldots\ldots (3.190)$$

From Eq. 3.190,

$$S_{w2} = \overline{S_w} - Q_i\left(\frac{\overline{S_w} - S_w^*}{Q_i}\right). \quad \ldots\ldots\ldots\ldots (3.191)$$

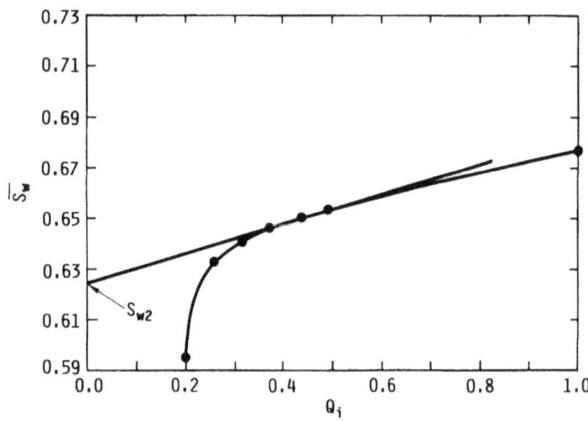

Fig. 3.28—Expanded plot of $\overline{S}_w$ vs. Q_i to obtain slopes for computation of S_{w2} and relative permeability ratios.

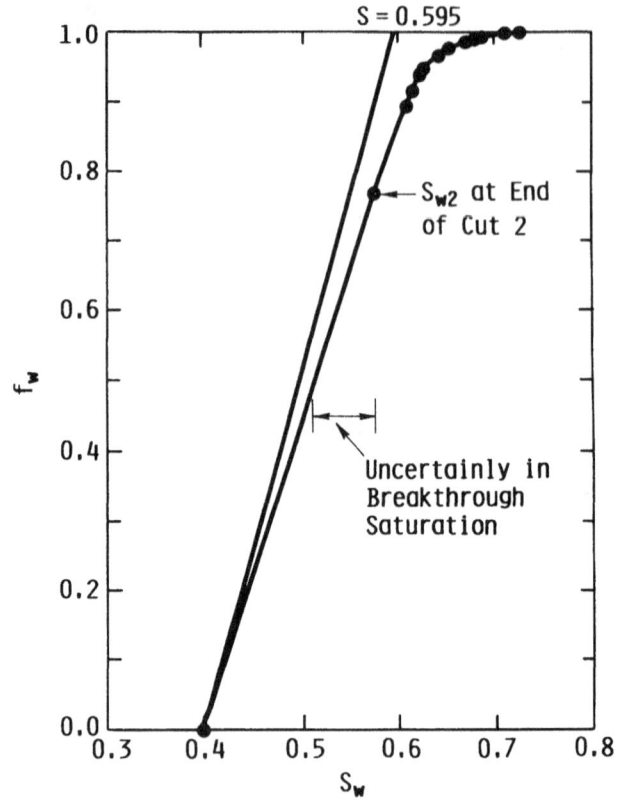

Fig. 3.29—Fractional flow curve obtained from experimental data.

Thus

$$S_{w2} = S_w^*. \qquad\qquad (3.192)$$

At low oil cuts, the slope $\overline{dS_w}/dQ$ is small and long extensions of tangents amplify errors. Jones and Roszelle[10] developed a graphical method that reduces these errors. They plot the average water saturation, $\overline{S}_w$, vs. the reciprocal of the PV's injected ($1/Q_i$), as shown in Fig. 3.27. The tangent to the S_w curve intersects the ordinate ($1/Q_i = 0$) at $S_w{}^+$. Outflow end saturations are computed from Eq. 3.193:

$$S_{w2} = 2\overline{S}_w - S_w{}^+. \qquad\qquad (3.193)$$

Computation of S_{w2} from experimental data requires differentiation of the plot of $\overline{S}_w$ vs. Q_i. When this graph is smooth, the derivatives are easy to compute. Sometimes there are discontinuities in the experimental data. In these cases, it is necessary to smooth the data before differentiation. Data may be smoothed by hand with the appropriate curve or with numerical techniques.

The last part of the analysis is to determine the breakthrough saturation S_{wf}. This can be done by plotting f_{w2} vs. S_{w2}. A tangent from S_{iw} to f_w intersects the fractional flow curve at the breakthrough saturation as discussed in Sec. 3.5. Example 3.8 illustrates the computation of the relative permeability ratio from laboratory displacement data.

Example 3.8. Displacement data[14] for Core 4 from the Dominguez field, California, are shown in Table 3.7.

Cumulative PV's of water injected (Q_i) and average water saturations ($\overline{S}_w$) are also included in Table 3.7.

Graphs of average water saturation vs. Q_i and $1/Q_i$ for these data are shown in Figs. 3.26 and 3.27. Values of S_{w2} or $S_w{}^+$ determined from tangent construction are presented in Table 3.8.

Fig. 3.28 is a plot of $\overline{S}_w$ vs. Q_i on an expanded scale. This procedure is necessary to obtain accurate slopes when large changes of average saturation occur over small ranges of Q_i. The last columns of Table 3.8 contain f_{o2}, f_{w2}, and k_w/k_o computed from values of S_{w2} and Q_i.

At $Q_i = 1.4049$ and 2.2133, S_{w2} was determined by both extrapolation methods illustrated in Figs. 3.26 and 3.27. The agreement between values calculated by the two methods is within the precision of the data.

Fig. 3.29 is the fractional-flow/saturation relationship for this example. The breakthrough saturation cannot be determined accurately because the first value of S_{w2} corresponds to f_{w2} at the end of the second cut. However, the breakthrough saturation must be between 0.51, the minimum saturation for formation of an oil bank, and S_{w2} at the end of Cut 2. The range of uncertainty is indicated in Fig. 3.29. Computed relative permeability ratios are presented in Fig. 3.30.

3.10.2 Permeabilities of Each Phase

Individual relative permeabilities may be calculated from linear displacement data when the pressure differences across the core are measured during displacement at constant rate. The method was developed by Johnson et al.[15] and is referred to as the unsteady-state method. Jones and Roszelle[10] developed a graphical technique that is equivalent to the equations of Johnson et al. The graphical technique is easier to use and has the capability of more accurate determination of relative permeabilities from the experimental data.

The graphical technique for constant-rate displacement will be illustrated in this section. Application to a constant-pressure-drop displacement is discussed in Ref. 10.

In Sec. 3.5 we observed that when a displacement is conducted at constant injection rate, the pressure across

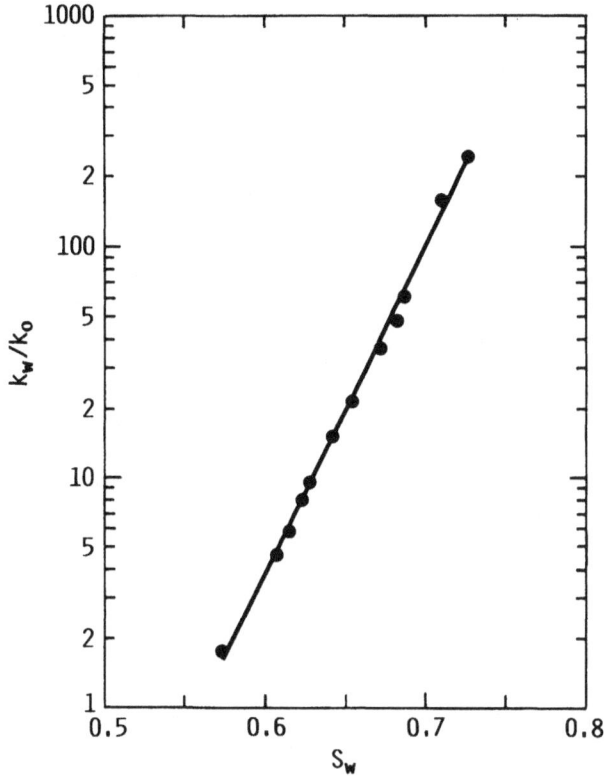

Fig. 3.30—Computed relative permeability ratio vs. water saturation.

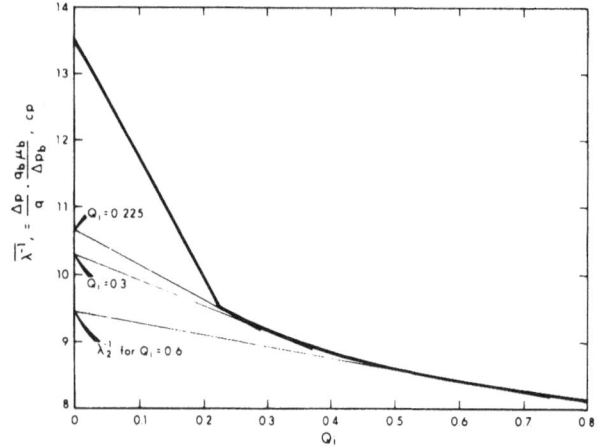

Fig. 3.31—Construction for determining outlet-end effective viscosity from average values. [10]

the core changes with time and, consequently, the volume of water injected. At any instant of time the total pressure drop is related to fluid and rock properties by Eq. 3.194.

$$\Delta p = -\int_0^L \frac{\frac{q_t}{A}dx}{k_b\left(\frac{k_{ro}}{\mu_o}+\frac{k_{rw}}{\mu_w}\right)}, \quad \ldots\ldots\ldots(3.194)$$

where k_b is the base permeability for the relative permeability data. The base permeability is determined for single phase flow through the core. Eq. 3.195 gives $k_b A$ in terms of experimental measurements.

$$k_b A = \frac{q_b \mu_b L}{-\Delta p_b}. \quad \ldots\ldots\ldots\ldots\ldots(3.195)$$

In Eq. 3.195, Δp_b is the pressure drop measured over the core when a single fluid phase with viscosity μ_b flows at rate q_b.

Substitution of Eq. 3.195 into Eq. 3.194 gives

$$\Delta p = \frac{q_t \Delta p_b}{q_b \mu_b L}\int_0^L \frac{dx}{\frac{k_{ro}}{\mu_o}+\frac{k_{rw}}{\mu_w}}. \quad \ldots\ldots\ldots(3.196)$$

Define the effective or apparent viscosity with Eq. 3.197 and the average apparent viscosity with Eq. 3.198:

$$\lambda_r^{-1} = \left(\frac{k_{ro}}{\mu_o}+\frac{k_{rw}}{\mu_w}\right)^{-1} \quad \ldots\ldots\ldots(3.197)$$

and

$$\overline{\lambda^{-1}} = \frac{\int_0^x \overline{\lambda_r^{-1}}\,dx}{\int_0^x dx}. \quad \ldots\ldots\ldots(3.198)$$

At the outlet end of the core (Point 2), $x=L$. Thus

$$\overline{\lambda^{-1}} = \overline{\lambda_2^{-1}} = \frac{q_b \mu_b \Delta p}{q_t \Delta p_b}. \quad \ldots\ldots\ldots(3.199)$$

The average apparent viscosity is computed from experimental measurements with Eq. 3.199.

Computation of individual relative permeabilities requires values of the apparent viscosity at a point where the saturation is known. The point where most information is known is at the outlet of the core where f_{o2} and S_{w2} are computed from linear displacement data with Eqs. 3.200 and 3.201:

$$k_{ro} = \frac{\mu_o f_{o2}}{\lambda_2^{-1}} \quad \ldots\ldots\ldots\ldots(3.200)$$

and

$$k_{rw} = \frac{\mu_w f_{w2}}{\lambda_2^{-1}}. \quad \ldots\ldots\ldots\ldots(3.201)$$

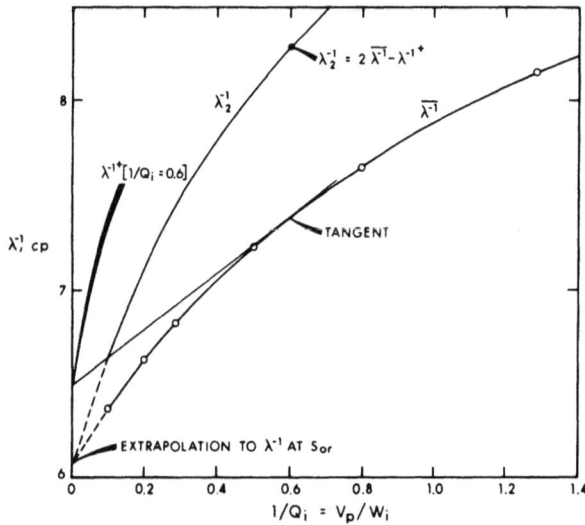

Fig. 3.32—Construction for determining point effective viscosity at large throughput paths. [10]

these conditions a plot of $\overline{\lambda^{-1}}$ vs. $1/Q_i$ is presented as in Fig. 3.32. A tangent to the curve at a specific value of $1/Q_i$ intersects the ordinate at λ^{-1+}. The apparent viscosities are computed from Eq. 3.203:

$$\lambda_2^{-1} = \overline{2\lambda^{-1}} - \lambda^{-1+}. \dots\dots\dots\dots\dots (3.203)$$

As discussed earlier, experimental data must be smooth before they can be differentiated. It may be necessary to smooth data either graphically or numerically before values of λ_2^{-1}, k_{rw}, and k_{ro} can be extracted from the experimental data.

The graphical method also may be used to analyze displacement data when the pressure drop is held constant and the rate varies with time. An example illustrating the technique is included in Ref. 10. Example 3.9 illustrates the use of the Jones and Roszelle method [10] to obtain relative-permeability/saturation curves for a constant-rate displacement.

Example 3.9. Table 3.9 contains data obtained during the displacement of oil by water at constant rate. [10] Relative permeabilities and corresponding saturations are to be determined from the data. The base condition for relative permeability values is the permeability to water at 100% water saturation.

First, it is necessary to express the displacement in terms of Q_i, $\overline{S_w}$, and $\overline{\lambda^{-1}}$. Computed values are given in Table 3.9. Saturations at the outlet end are determined following the procedures described in this section. Figs. 3.26 and 3.27 illustrate the construction of tangents for determining saturations at the outlet end. Calculated values of S_w are given in Table 3.9.

Apparent viscosities at the outlet end are determined by preparing plots of λ^{-1} vs. Q_i or $1/Q_i$. Figs. 3.31 and 3.32 are the plots for this example showing the construction of tangents to determine λ_2^{-1} or λ^{-1+}. Intercepts determined from the graph are included in Table 3.10. The remainder of the calculation is straightforward. Computed relative permeabilities and saturations are shown in Fig. 3.33. The curves begin at $S_w = 0.35$ where the

Only the average apparent viscosity is determined from experimental data. Thus it is necessary to develop a relationship between $\overline{\lambda_2^{-1}}$ and λ_2^{-1} to compute values of λ_2^{-1}. Jones and Roszelle [10] developed a procedure for this purpose. Eq. 3.202 was derived with linear scaling principles. A system is linearly scalable if the saturation is a function only of Distance x and the cumulative water injected. [16] This condition is met by the frontal advance solution after breakthrough.

$$\lambda_2^{-1} = \overline{\lambda^{-1}} - Q_i \frac{d\overline{\lambda^{-1}}}{dQ_i}. \dots\dots\dots\dots (3.202)$$

Fig. 3.31 shows the construction of tangents to the graph of $\overline{\lambda^{-1}}$ vs. Q_i. The intercept on the ordinate at $Q_i = 0$ is λ_2^{-1}, the apparent viscosity at the outlet end of the core. When Q_i is large, the slope of the graph becomes small and extrapolation errors are possible. For

TABLE 3.9—DATA FOR CALCULATION OF OIL/WATER RELATIVE PERMEABILITIES FROM A CONSTANT-RATE WATERFLOOD DISPLACEMENT [10]

W_i (mL)	N_p (mL)	$-\Delta p$ (psi)	Q_i $= W_i/V_p$ (PV)	$\overline{S_w}$ $= S_{iw} + N_p/V_p$ (fraction)	$\overline{\lambda^{-1}}$ $= \Delta p \mu_b q_b/(\Delta p_b q)$ (cp)
0.00	0.00	138.6	0.000	0.350	13.50
3.11	3.11	120.4	0.100	0.450	11.73
7.00*	7.00	97.5	0.225	0.575	9.50
11.20	7.84	91.9	0.360	0.602	8.95
16.28	8.43	87.9	0.523	0.621	8.56
24.27	8.93	83.7	0.780	0.637	8.15
39.20	9.30	78.5	1.260	0.649	7.65
62.30	9.65	74.2	2.001	0.660	7.23
108.90	9.96	70.0	3.500	0.670	6.82
155.60	10.11	68.1	5.000	0.675	6.63
311.30	10.30	65.4	10.000	0.681	6.37

*Point of water breakthrough.

$V_p = 31.13$ mL, $L = 5.002$ in., $d = 1.500$ in., $\Delta p_b/q_b = 0.1245$ psi/[mL/hr] at $S_w = 1.000$, $\mu_b = \mu_w = 0.970$ cp, $k = 35.4$ md, $\mu_o = 10.45$ cp, $\phi = 0.215$, $S_{iw} = 0.350$, and $q = 80.0$ mL/hr.

TABLE 3.10—CALCULATED VALUES FOR OIL AND WATER RELATIVE PERMEABILITIES [10]

Q_i (PV)	$\dfrac{1}{Q_i}$	Average $\bar{S}_w$ (fraction)	Intercept S_{w2} (fraction)	Average λ^{-1} (cp)	Intercept λ_2^{-1} (cp)	$\dfrac{f_{o2}}{=(\bar{S}_w - S_{w2})/Q_i}$	f_{w2} $= 1 - f_{o2}$	k_{rw} $= f_w \mu_w / \lambda_2^{-1}$	k_{ro} $= f_o \mu_o / \lambda_2^{-1}$
0.000	∞	0.350	0.350	13.50	13.50	1.000	0.000	0.000	0.774
0.100	10.000	0.450	0.350	11.73	13.50	1.000	0.000	0.000	0.774
0.225	4.444	0.575	0.511	9.50	10.66	0.291	0.709	0.065	0.285
0.300	3.333	0.593	0.534	9.16	10.30	0.197	0.803	0.076	0.200
0.600	1.667	0.627	0.580	8.42	9.45	0.078	0.922	0.095	0.087
1.000	1.000	0.643	0.617	7.88	8.95	0.026	0.974	0.106	0.030
2.500	0.400	0.664	0.646	7.05	7.79	0.007	0.993	0.124	0.010
5.000	0.200	0.675	0.664	6.63	7.11	0.002	0.998	0.136	0.003
10.000	0.100	0.681	0.676	6.37	6.69	0.001	0.999	0.145	0.002
∞	0.000	0.687	0.687	6.07	6.07	0.000	1.000	0.160	0.000

value of $k_{rw} = 0$ and $k_{ro} = 0.74$. The dashed line represents the region of the relative permeability curve where the frontal advance solution does not apply. The first available k_{rw} and k_{ro} values are those corresponding to $S_{w2} = 0.511$.

3.11 Factors That Control Displacement Efficiency of a Linear Waterflood

3.11.1 Effects of Reservoir Rock and Fluid Properties on the Efficiency of Linear Displacement

The frontal advance solution may be used to explore the effects of reservoir rock and fluid parameters on the rate of oil displacement.

Wettability. In Chap. 2, we discussed relative permeability curves and the variables that determine the shape of the relative permeability curves. Wettability was shown to be a principal factor controlling relative permeabilities. This property of the reservoir rock/fluids is fixed for the purposes of waterflooding a given reservoir. The effects of wettability changes will be examined in this section to help develop some understanding of fractional flow curves and how they change for different wetting conditions.

Fig. 2.35 contains relative permeability curves corresponding to two wetting conditions in Torpedo sandstone cores for the displacement of oil by water. [17] Values of the relative permeabilities were interpolated from Fig. 2.35 and were used to compute fractional flow curves for water displacing oil with a viscosity ratio $\mu_o / \mu_w = 25.0$. These curves are shown in Fig. 3.34.

Fractional flow curves are quite different, as might be expected from examination of Eq. 3.79:

$$f_w = \cfrac{1}{1 + \left(\cfrac{k_o}{k_w}\right)\left(\cfrac{\mu_w}{\mu_o}\right)}.$$

In uniformly wetted materials, a decrease in water wetness is reflected by an increase in the water permeability and a decrease in oil permeability. The fractional flow curve for the oil-wet case is much steeper than the water-wet curve. This leads to significant differences in displacement rates, as shown in Fig. 3.35. The initial water saturation in both examples was 0.20.

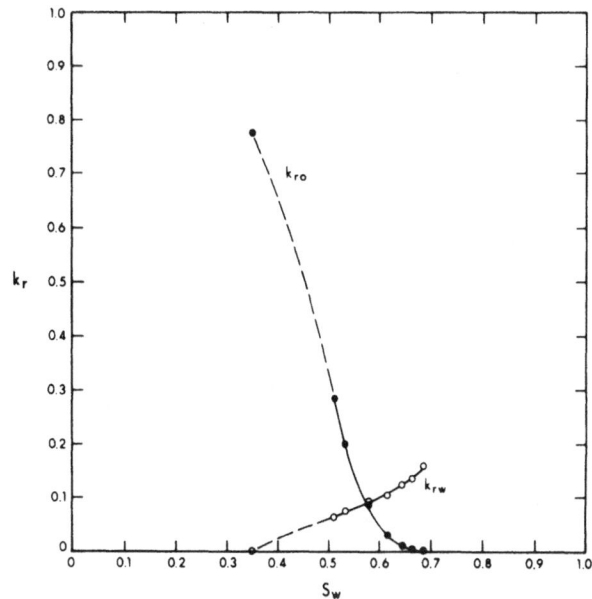

Fig. 3.33—Relative permeability curves from example calculation. [10]

The displacement of oil by water from a strongly oil-wet rock is less efficient than from a strongly water-wet rock. In the oil-wet rock, oil displacement continues after many PV's have been injected, while in the water-wet rock, the process is completed for all practical purposes within a few PV's. In the example (presented in Fig. 3.35), the oil saturation for the water-wet core is about 8.9% less than the oil saturation for the oil-wet rock after 2 PV's of injection. In general, residual oil saturations in uniformly wetted porous media are higher when the rock is strongly oil-wet. This means that displacement will occur at a slower rate and probably at lower ultimate recovery in a strongly oil-wet reservoir.

The wettability of a rock and the relative permeability curves representing flow of water and oil are fixed for all practical purposes as far as waterflooding is concerned. One might ask why we should be concerned about wettability and the shape of relative permeability curves. From a practical viewpoint, there are cases where no more data are available than the endpoints of the relative permeability curves (i.e., oil permeability at interstitial water saturation, water permeability at residual oil saturation,

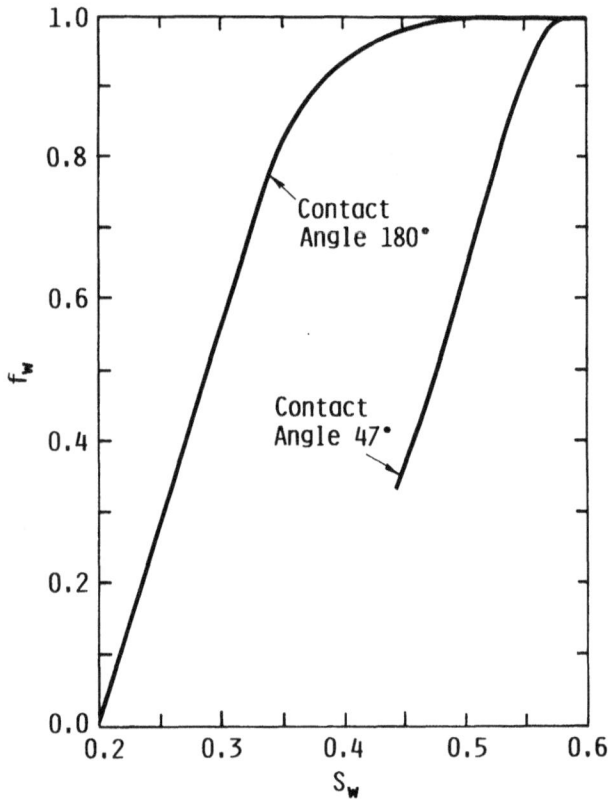

Fig. 3.34—Fractional flow curves for waterfloods of water- and oil-wet rocks at an oil/water viscosity ratio of 25.

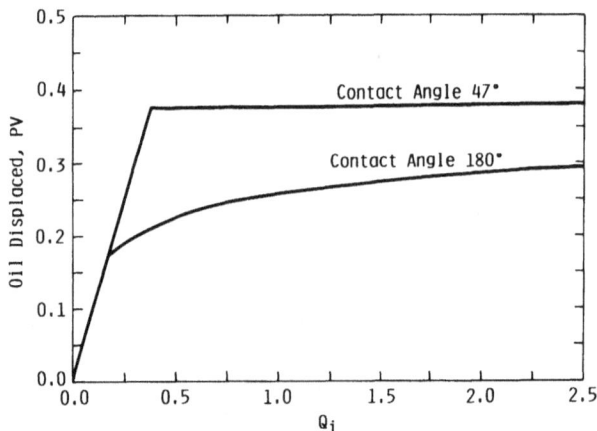

Fig. 3.35—Effect of wettability on oil displacement by water injection.

and some indication of water or oil wetness). Approximate relative permeability curves may have to be developed from this information using the knowledge of the shapes of the permeability curves. In the reservoir environment, an actual waterflood is so complex that many factors contribute to the observed production response. Some of these factors will be discussed later in Chaps. 6 and 7.

Analysis of waterflood performance is frequently a process in which the cause of a particular response is

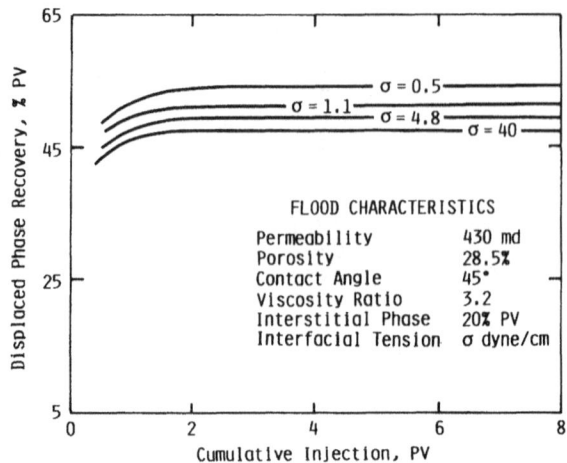

Fig. 3.36—Effect of IFT on displacement of a nonwetting by a wetting liquid.[19]

determined by process of elimination; i.e., potential causes are examined in terms of the contribution to the observed response. An example of this type of analysis was presented by Salathiel.[18] Portions of this example were presented in Secs. 2.2 and 2.3. Some feeling for shapes of relative permeability curves and their effects on displacement performance is useful in sorting out causes, even if done subjectively.

Interfacial Tension (IFT). When IFT decreases, the relative permeability curves for both phases increase, as shown in Fig. 2.38. The effects of decreasing the IFT from 40 to 0.5 dynes/cm [40 to 0.5 mN/m] on displacement efficiency are shown in Figs. 3.36 and 3.37. Fig. 3.36 illustrates recovery efficiency for the displacement of a nonwetting phase by a wetting phase.[19] Fig. 3.37 shows data for the displacement of the wetting phase by the nonwetting phase. The viscosity ratio (viscosity displaced/viscosity displacing) was 3.2 for all the data, and the wettability, as indicated by the contact angle, was constant in each set of experiments. Decreasing the IFT causes the relative permeability of the displaced phase to increase more than that of the displacing phase. The displaced phase is recovered with less cumulative fluid injected as the IFT decreases. The effect of wettability on displacement efficiency is illustrated also. The porous media in Figs. 3.36 and 3.37 were made from compacted teflon powder.

Viscosity Ratio. Viscosity ratio is the remaining parameter that influences displacement efficiency when gravity effects are absent. Fig. 3.38 compares the waterflood behavior for water displacing oils with viscosities of 151 and 1.8 cp [151 and 1.8 mPa·s][20] from an unconsolidated sandpack. The curves in Fig. 3.38 represent the calculated displacement performance from the frontal advance solution using relative permeability data from Fig. 2.42. Good agreement was obtained.

The effect of increased viscosity ratios can be illustrated by examining how the fractional-flow curves change with viscosity ratio. From Eq. 3.79, the fractional flow of water becomes progressively larger as the viscosity of oil increases relative to water. This produces a steeper fractional-flow curve at low saturations and a long tail

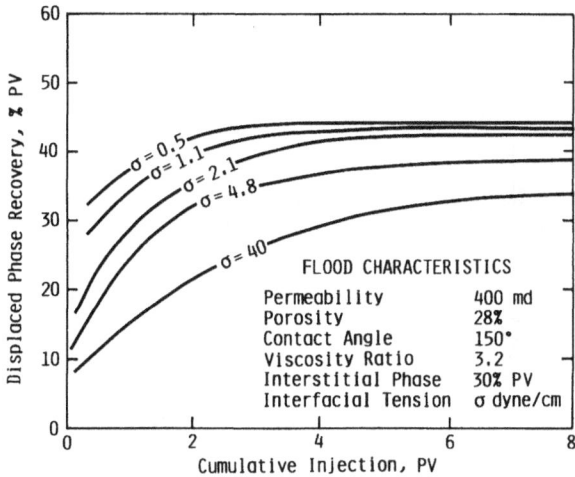

Fig. 3.37—Effect of IFT on displacement of a wetting by a nonwetting liquid.[19]

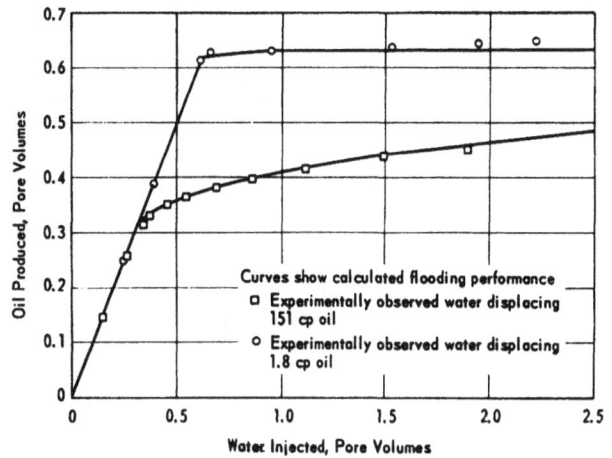

Fig. 3.38—Comparison of waterflooding response with results calculated from the frontal advance equation.[20]

at high saturations. Fig. 3.39 shows fractional-flow curves for oil/water viscosity ratios of 1, 5, 10, 50, 100, 150, 200, and 500 for the displacement of oil by water. The relative permeability data for unconsolidated sand in Fig. 2.40 were used to prepare these curves.

The displacement PV's-injected curve for larger oil/water viscosity ratios shows some of the trends previously discussed in connection with displacement from an oil-wet core. Comparing Figs. 3.34 and 3.39, it can be seen that breakthrough of water occurs early because the fractional flow curve is steeper at lower saturations. Oil production is prolonged over many PV's of water injection.

Ultimate recovery is independent of viscosity ratio because relative permeability curves are independent of the viscosity ratio when the wetting fluid displaces a nonwetting fluid. Because hundreds of PV's of water may be required to reach this point for large viscosity ratios, the equivalence of ultimate recovery and thus endpoint permeabilities is not usually observed. The economic limit for this type of waterflood will be reached well before the last oil is displaced.

Gravity Forces. Some reservoirs are inclined at an angle with the horizontal. When water displaces oil linearly in an inclined reservoir, it is necessary to include the contribution of gravity forces to the fractional-flow curve. Recall Eq. 3.78, which expressed in oilfield units is

$$f_w = \cfrac{1}{1+\left(\cfrac{k_o}{k_w}\right)\left(\cfrac{\mu_w}{\mu_o}\right)}$$

$$+\cfrac{\cfrac{0.433k_oA}{\mu_oq_t}(\rho_o-\rho_w)g\sin\alpha}{1+\left(\cfrac{k_o}{k_w}\right)\left(\cfrac{\mu_w}{\mu_o}\right)}.$$

Because the oil density is less than the water density, gravity forces reduce the fractional flow of water when water is moving updip. Water injected at the crest of a structure would move faster under the influence of gravity forces than at the same total rate in a horizontal reservoir.

3.11.2 Mobility Ratio—Immiscible Displacement Processes

Displacement efficiency of a linear process varies with fluid and rock properties as illustrated in Sec. 3.11.1. Because of this, it is difficult to compare performance when these parameters change. Engineers sought a single parameter that would be an accurate indicator of displacement performance. The concept of the mobility ratio evolved from these efforts to correlate displacement data for linear as well as areal systems.[21]

When a waterflood is assumed to be piston-like, oil flows at interstitial water saturation ahead of the sharp front while water flows at residual oil saturation behind the front. The mobility ratio for a very sharp front displacement is given by Eq. 3.204:

$$M=\frac{\lambda_D}{\lambda_d}, \quad\dotfill(3.204)$$

where λ_D is the mobility of the displacing phase behind the front and λ_d is the mobility of the displaced phase ahead of the front.

For waterflood under piston-like displacement,

$$\lambda_D=\lambda_w=\left(\frac{k_w}{\mu_w}\right)_{S_{or}} \quad\dotfill(3.205)$$

and

$$\lambda_d=\lambda_o=\left(\frac{k_o}{\mu_o}\right)_{S_{iw}}, \quad\dotfill(3.206)$$

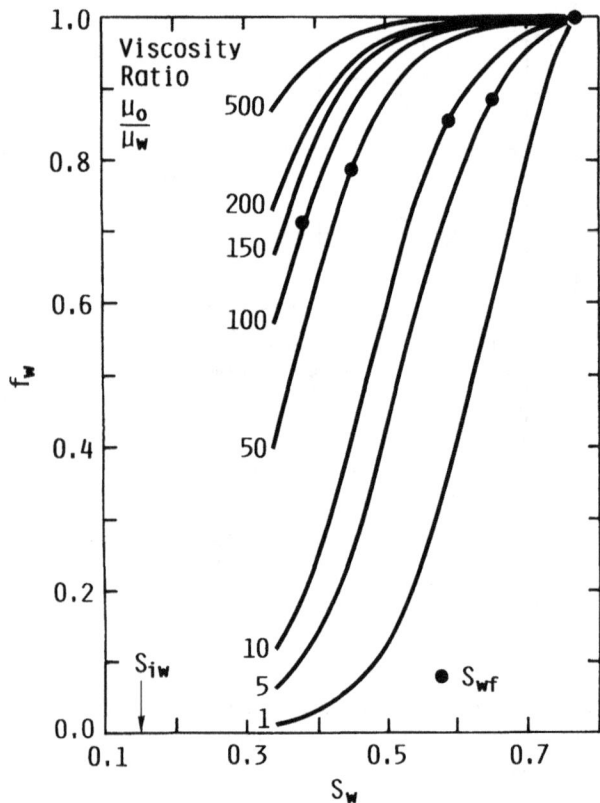

Fig. 3.39—Effect of viscosity ratio on fractional flow.

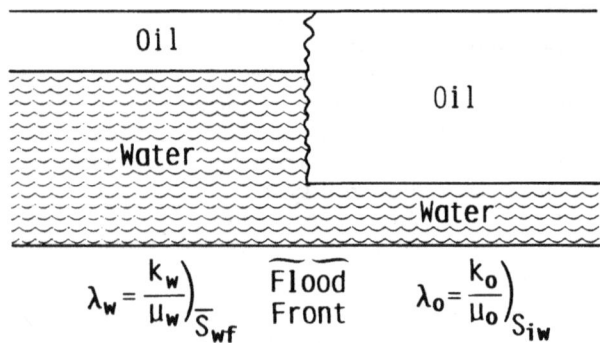

Fig. 3.40—Evaluation of relative permeabilities for computation of M_S.

and the mobility ratio is given by Eq. 3.207:

$$M = \frac{\left(\dfrac{k_w}{\mu_w}\right)_{S_{or}}}{\left(\dfrac{k_o}{\mu_o}\right)_{S_{iw}}} = \frac{\left(\dfrac{k_{rw}}{\mu_w}\right)_{S_{or}}}{\left(\dfrac{k_{ro}}{\mu_o}\right)_{S_{iw}}} . \quad \ldots \ldots \ldots (3.207)$$

The permeabilities for water and oil are the endpoints of permeability/water-saturation curves for a waterflood and are often available from flood-pot tests on reservoir cores.

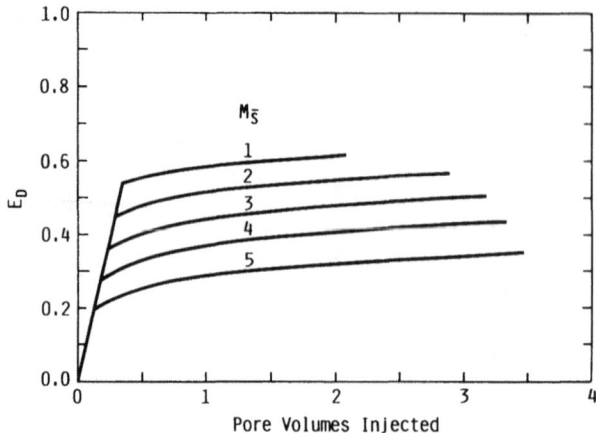

Fig. 3.41—Effect of mobility ratio (M_S) on displacement efficiency of a linear waterflood.

Most waterfloods do not have piston-like displacement. The mobility ratio commonly associated with waterflooding is designated also with the symbol M in most waterflooding literature. In this text the symbol $M_{\bar{S}}$ is adopted to conform as closely as possible to SPE reservoir symbols. $M_{\bar{S}}$ is defined by Eq. 3.208:

$$M_{\bar{S}} = \frac{\left(\dfrac{k_{rw}}{\mu_w}\right)_{\overline{S_{wf}}}}{\left(\dfrac{k_{ro}}{\mu_o}\right)_{S_{iw}}} . \quad \ldots \ldots \ldots \ldots \ldots (3.208)$$

In Eq. 3.208, the apparent mobility of the water phase k_{rw}/μ_w is computed by evaluating the relative permeability to water at $\overline{S_{wf}}$, the average water saturation in a linear waterflood at breakthrough.

Fig. 3.40 illustrates the locations where relative permeabilities are determined to compute the mobility ratio for a waterflood, $M_{\bar{S}}$.

The mobility ratio, $M_{\bar{S}}$, may be visualized as a measure of the relative rate of oil movement ahead of the front to water movement behind the front, assuming pressure gradients in both phases are equal. A mobility ratio of 1.0 indicates that oil and water are moving at the same relative rate. When $M_{\bar{S}} < 1.0$, water moves slower than the oil, leading to high water saturations at breakthrough as well as high displacement efficiency. Values of $M_{\bar{S}} > 1.0$ indicate that the water behind the front is moving faster than the oil ahead of the front. Displacement efficiency at water breakthrough is reduced and the remaining oil is recovered after a large volume of water has been injected.

A series of displacement calculations is presented in Figs. 3.41 and 3.42 to illustrate the effect of mobility ratio on displacement efficiencies of a linear waterflood. Relative permeability data from Example 3.4 and the oil viscosities in Table 3.11 were used to obtain displacement performance. Calculations show that displacement efficiency decreases as the mobility ratio increases. Thus, at the same displacement efficiency, large volumes of fluid must be injected and higher WOR's are produced when

**TABLE 3.11—PARAMETERS FOR
MOBILITY RATIO EFFECTS**

$M_{\bar{S}}$	μ_o (cp)
1.0	1.599
2.0	7.941
3.0	30.965
4.0	124.803
5.0	630.739

the mobility ratio increases. We also observe from Table 3.11 that relative permeabilities have a strong impact on the mobility ratio because the mobility ratio is significantly lower than the viscosity ratio.

The mobility ratio is defined for immiscible displacement processes (i.e., water displacing oil, gas displacing oil, or gas displacing water) by Eq. 3.209:

$$M_{\bar{S}_i} = \frac{(\lambda_i)\,\overline{S_i}}{(\lambda_d)\,\overline{S_d}}, \quad\dots\dots\dots\dots\dots\dots(3.209)$$

where λ_i is the mobility of injected fluid evaluated at $\overline{S_i}$, the average saturation of the injected fluid at breakthrough, and λ_d is the mobility of displaced fluid evaluated at $\overline{S_d}$, the saturation of the displaced fluid ahead of the front. Mobility ratios for water displacing oil vary from 0.02 to 10 while mobility ratios as large as 1,000 may be computed for gas displacing oil. For miscible displacement, the mobility ratio becomes the ratio of viscosities as indicated by Eq. 3.210. This mobility ratio is called the total mobility ratio, M_t:

$$M_t = \frac{\mu_d}{\mu_D}. \quad\dots\dots\dots\dots\dots\dots\dots(3.210)$$

M_t differs from $M_{\bar{S}}$ in that a single phase flows behind the displacing front. An equivalent total mobility ratio may be defined for a linear waterflood that incorporates two-phase flow behind the flood front. Eq. 3.211 defines the total mobility of a linear waterflood when the initial water saturation is immobile:

$$M_t = \frac{\lambda_{ro}^{-1}}{\overline{\lambda}^{-1}_{Swf}}, \quad\dots\dots\dots\dots\dots\dots(3.211)$$

where $\overline{\lambda}^{-1}_{Swf}$ is the average apparent viscosity at S_{wf}.

3.12 Limitations of the Frontal Advance Solution

The frontal advance solution developed and applied in the previous sections has limitations that should be acknowledged. Some come from assumptions used in developing the model while others relate to the conditions encountered in reservoir rock. In developing the frontal advance equation, the two immiscible fluids were considered incompressible. Flow in one direction was assumed. The rock was homogeneous and isotropic and the initial fluid saturations were uniform throughout the region simu-

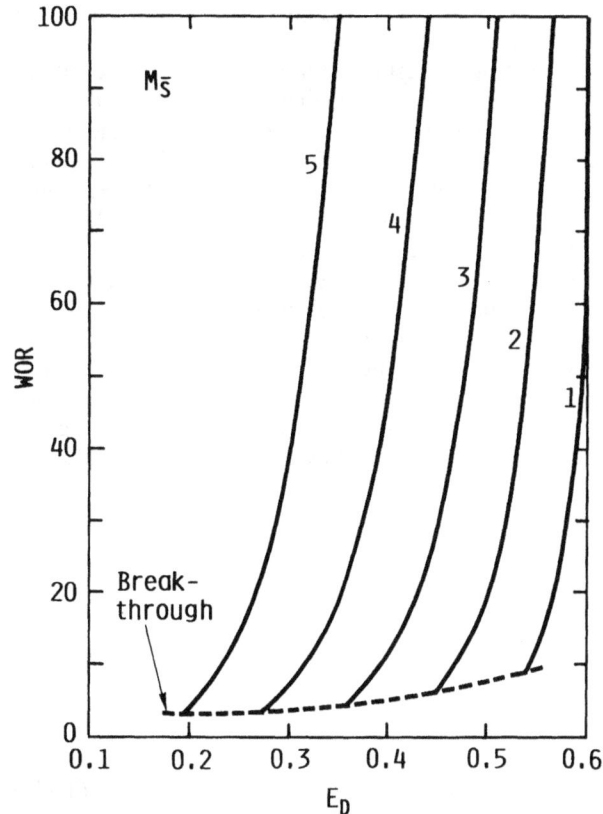

Fig. 3.42—Variation of WOR with displacement efficiency for several mobility ratios.

lated. Applications to problems where these assumptions are not approximated closely will lead to uncertainty in the differences between computed and observed displacement performance. In these cases, numerical models that do not include the assumptions may be required to describe the waterflood. There are few guidelines other than experience to assist the engineer in judging how closely the model assumptions must approach the physical situation. Some insight may be developed from technical papers that compare the Buckley-Leverett solution with results from numerical models. Sec. 3.13 contains an introduction to the capabilities of 1D numerical simulators.

In Secs. 3.12.1 and 3.12.2, we consider two limitations that may be present in linear displacements at certain flow rates.

3.12.1 Stabilization of a Linear Displacement Process

The frontal advance solution developed in Sec. 3.4 applies to a stabilized displacement process. When a stabilized zone forms, the breakthrough saturation and the recovery-at-PV-injected curve (after breakthrough) should be independent of the length of the rock and the rate of injection. Rate and length effects were observed in many early laboratory experiments. Also, interpretation of displacement data was complicated by the presence of capillary end effects in some systems. Displacement behavior became independent of injection rate and length for systems with long lengths and high injection rates. An empirical correlation based on dimensional analysis was developed by Rapoport and Leas[22] to judge when a flood was stabilized. Displacement data for linear floods were independent of length, rate of injection, and viscosity of the injected water when the numerical value of the scaling coefficient, $Lu_t\mu_w$, exceeded a critical value. Critical values of scaling coefficients obtained in laboratory experiments range from 0.4 to 5.0 cm$^2 \cdot$kPa$\cdot$s/min [0.835×10^{-9} to 5.85×10^{-9} N], or in oilfield units, from 0.62 to 7.75 sq ft-cp/D.

Rate and length effects occur when capillary forces become important in the displacement process. The effect of capillary forces can be examined by converting the frontal advance equation to a dimensionless form. Referring to Eqs. 3.59, 3.64, and 3.76, the frontal advance equation written in terms of saturation changes is given by Eq. 3.212, where gravity terms have been neglected:

$$\phi\frac{\partial S_w}{\partial t}+\frac{q_t}{A}\frac{\partial}{\partial x}\left[\frac{1}{1+\left(\dfrac{k_o}{k_w}\right)\left(\dfrac{\mu_w}{\mu_o}\right)}\right.$$

$$+\left.\frac{\dfrac{k_o}{\mu_o u_t}\dfrac{\partial P_c}{\partial x}}{1+\left(\dfrac{k_o}{k_w}\right)\left(\dfrac{\mu_w}{\mu_o}\right)}\right]=0. \quad\dots\dots\dots\dots (3.212)$$

If

$$f_w=\frac{1}{1+\left(\dfrac{k_o}{k_w}\right)\left(\dfrac{\mu_w}{\mu_o}\right)},$$

then Eq. 3.212 may be written as

$$\frac{\partial S_w}{\partial t}+\frac{q_t}{A\phi}\left(\frac{\partial f_w}{\partial S_w}\right)\left(\frac{\partial S_w}{\partial x}\right)$$

$$-\frac{\partial}{\partial x}\left[\frac{k_o f_w}{\mu_o u_t}\frac{\partial P_c}{\partial x}\right]=0. \quad\dots\dots\dots\dots (3.213)$$

Define dimensionless distance, x_D, and time, t_D, variable as in Eq. 3.214 and 3.215:

$$x_D=\frac{x}{L} \quad\dots\dots\dots\dots\dots\dots\dots\dots (3.214)$$

and

$$t_D=\frac{tu_t}{L\phi}. \quad\dots\dots\dots\dots\dots\dots\dots (3.215)$$

After some rearrangement, Eq. 3.213 becomes

$$\frac{\partial S_w}{\partial t_D}+\frac{df_w}{dS_w}\frac{\partial S_w}{\partial x_D}$$

$$-\frac{k_o}{\left(\dfrac{\mu_o}{\mu_w}\right)(Lu_t\mu_w)}\frac{\partial}{\partial x_D}\left[k_o f_w\left(\frac{dP_c}{dS_w}\right)\left(\frac{\partial S_w}{\partial x_D}\right)\right]$$

$$=0. \quad\dots\dots\dots\dots\dots\dots\dots\dots (3.216)$$

The scaling coefficient $Lu_t\mu_w$ evolves from the dimensional analysis.

Capillary effects in Eq. 3.216 are represented by the last term. This term will decrease as the scaling coefficient $Lu_t\mu_w$ increases. As noted earlier, critical values of the scaling coefficient are determined from experimental data.

Example 3.10 illustrates the use of the scaling coefficient to determine the stability of a linear waterflood in a reservoir.

Example 3.10. The reservoir in Example 3.4 is 1,000 ft [305 m] long, and it was flooded at an average frontal velocity (u_t/ϕ) of 1 ft/D [0.004 mm/s]. The porosity of the reservoir was 0.19. Water viscosity is 0.7 cp [0.7 kPa$\cdot$s].

Estimate the scaling coefficient for the displacement process and determine whether the displacement was stabilized.

Solution.

The value of the scaling coefficient must be determined. It is convenient to use oilfield units.

The value of u_t is 0.19 ft/D. Thus

$$Lu_t\mu_w=(1,000\text{ ft})(0.19\text{ ft/D})(0.7\text{ cp})$$

$$=133\text{ sq ft-cp/D}.$$

This value is an order of magnitude above the values observed in laboratory experiments for formation of a stabilized zone. Stabilized flow occurs under essentially all field conditions where displacement is linear and gravity effects are negligible.

An important use of the frontal advance solution is to compute relative permeability ratios and relative permeabilities from linear displacement tests in laboratory cores. For the frontal advance equation to be applicable to laboratory cores, it is necessary to estimate the operating conditions for stabilized flooding as well as conditions required to overcome capillary end effects.

The displacement rate required to obtain stabilized flow in a laboratory experiment is estimated in Example 3.11.

Example 3.11. It is desired to conduct a laboratory waterflood under stabilized conditions in a core 1 in. [2.54 cm] in diameter and 1.97 in. [5 cm] long. The porosity of the core is 0.15.

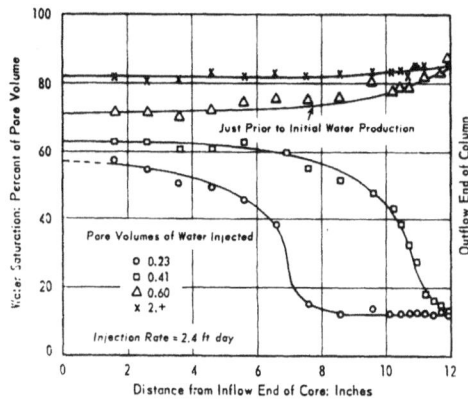

Fig. 3.43—Saturation profiles at a low flood rate. [23]

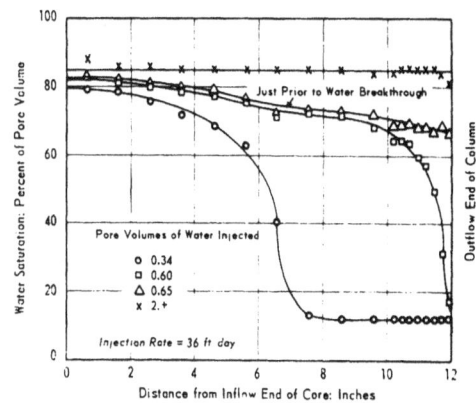

Fig. 3.44—Saturation profiles at a high flood rate. [23]

Estimate the volumetric injection rate in cubic meters per second if the critical scaling coefficient is 5.85×10^{-9} N. Viscosity of the water is 1.0 cp [1.0 kPa·s].

$$L u_t \mu_w = (0.05 \text{ m})(u_t)(0.001 \text{ Pa·s}) \frac{\text{N/m}^2}{\text{Pa}}$$

$$= 5 \times 10^{-5} u_t.$$

If the critical scaling coefficient is 5.85×10^{-9} N,

$$u_t = \frac{5.85 \times 10^{-9} \text{ N}}{5 \times 10^{-5}}$$

$$= 1.17 \times 10^{-4} \text{ m/s},$$

and

$$q = u_t A$$

$$= (1.17 \times 10^{-4} \text{ m/s}) \left(\frac{\pi}{4} \right) (0.0254 \text{ m})^2$$

$$= 5.93 \times 10^{-8} \text{ m}^3/\text{s}.$$

3.12.2 Capillary End Effects During Displacement in Linear Systems

Capillary end effects analogous to those discussed in Sec. 3.3.2 may lead to incorrect interpretation of experimental data from laboratory corefloods. In this section, we investigate the outlet end effect and criteria for minimizing this effect in displacement experiments. Figs. 3.43 and 3.44 show water saturation profiles at two different flow rates in a water-wet sandpack. [23] Steep saturation gradients characteristic of the displacement process are shown before arrival of the saturation front. Of particular importance is the saturation distribution just before water breakthrough at the outlet end of the core.

In Fig. 3.43, water accumulated at the outlet end during the displacement process. Water saturations are larger at the end than in the interior of the core. At the higher frontal velocity in Fig. 3.44, water breakthrough occurs

at about 0.65 PV injected. There is no accumulation of water at the end of the core and the water saturation profile declines from inlet to outlet as expected.

The end effect shown in Fig. 3.43 is caused by the discontinuity in the pressures of the oil and water phases at the end of the core. This effect can be explained by considering the difference in oil- and water-phase pressures at two stages in a waterflood of a strongly water-wet rock. [24] Before the arrival of the water at the end of core, the oil phase is continuous from the end of the core to the end plate of the core holder. The water phase is immobile and its pressure is determined by the capillary pressure curve for the imbibition path. For a water-wet rock, the water-phase pressure is less than the oil-phase pressure. At low injection rates such as in Fig. 3.43, the oil-phase pressure in the space between the core and the end plate exceeds the water-phase pressure. No flow of water can occur as long as $p_w < p_o$. As a result, water arriving at the end of the core accumulates until the water-phase pressure in the core exceeds the oil-phase pressure in the end space of the core holder.

Accumulation of water at the outlet end of the core eventually leads to sufficient water-phase pressure for flow; this happens as follows. Increased water saturation decreases the oil-phase permeability. Because oil is flowing at a constant rate, the oil-phase pressure must rise. Water-phase pressure rises because of the capillary-pressure/saturation relationship. The end effect is enhanced and the capillary pressure curve decreases sharply as the water saturation increases (see Fig. 2.19).

The effect of capillary pressure gradients is minimized (but not eliminated) in the displacement shown in Fig. 3.44 because the pressure gradients in both oil and water phases were large in comparison to the capillary pressure gradient. The water phase has sufficient pressure to flow from the core on arrival at the outlet end and does not accumulate as was observed at lower rates.

The outlet end effect creates the potential for misinterpretation of laboratory data from water-wet cores. Because water accumulates at the end of the core, breakthrough of water is delayed and a displacement test may appear to be stabilized when it is not. [24] The accumulation of water at the end of the core also causes difficulties in computing relative permeabilities from experimental data. Both pressure drop and fractional flow data

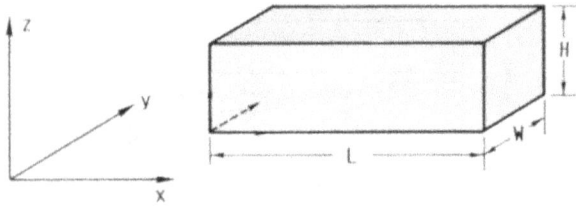

Fig. 3.45—Mathematical model of a reservoir.

may reflect capillary end effects. Batycky *et al.*,[25] however, show how relative permeability data may be derived from data influenced by capillary effects with a numerical simulator.

Rule-of-thumb criteria based on laboratory experience are used to select operating conditions that minimize capillary end effects. The end effect is considered small in water-wet cores when flow is stabilized.[24] Thus the scaling parameters for stabilized flow that were described in Sec. 3.12.1 may be used for this purpose. Other criteria include performing the waterflood at an overall pressure drop of 50 psi [345 kPa][24] or maintaining frontal velocities (u_t/ϕ) near 30 to 50 ft/D [9.1 to 15.2 m/d].

3.13 Solution of Linear Displacement Equations With Numerical Models

The frontal advance solution has unparalleled ease of solution, particularly for the engineer who wants quick estimates that can be obtained with a few hours of calculations at the desk or with a few seconds on a small computer. The development of high-speed digital computers created a quantum jump in the capability of engineers to solve the partial differential equations describing multiphase flow. Numerical methods that permit reasonably complete solution of linear flow equations, including the effects of capillary forces, have been developed. We will illustrate the capability of numerical techniques by reviewing some solutions that have been developed. No attempt will be made to show how numerical solutions are obtained. This is beyond the scope of this text and is covered elsewhere.

3.13.1 Linear Waterflood Behavior Including Capillary Pressure Effects

The equations describing the flow of water and oil phases in a linear system were developed in Sec. 3.2.2. These relationships will be applied in the following example.

Consider the displacement of oil by water from a linear, horizontal reservoir.[26] The reservoir is homogeneous and isotropic—that is, the permeability does not vary with position (x) or direction of the coordinate axis as discussed in Chap. 2. Initial water saturation is uniform throughout the reservoir and is immobile. The reservoir contains only oil and water. Properties of the oil and water, relative permeability, and capillary pressure curves are available for the saturation path. Incompressible flow and constant viscosities are assumed. The injection rate is constant.

Fig. 3.45 depicts the mathematical model of the reservoir. The reservoir is defined by the limits $0 \leqslant y \leqslant W$,

$0 \leqslant x \leqslant L$, and $0 \leqslant z \leqslant H$. Because flow is linear, no changes occur with the y and z coordinates. Eq. 3.217 describes the change in water saturations with respect to position at any point in the model. The oil saturation equation is not included because $S_o = 1 - S_w$. These partial differential equations are nonlinear because the capillary pressure gradient and relative permeability curves are functions of the water saturation:

$$\frac{\partial}{\partial x}\left[\frac{k_w k_o \dfrac{dP_c}{dS_w}}{(k_w \mu_o + k_o \mu_w)}\frac{\partial S_w}{\partial x}\right]$$

$$+ \frac{d}{dS_w}\left(\frac{u_t}{1+\dfrac{k_o \mu_w}{k_w \mu_o}}\right)\frac{\partial S_w}{\partial x} = \phi\frac{\partial S_w}{\partial t}, \quad \ldots (3.217)$$

where

$$k_w = k_w(S_w), \quad \ldots\ldots\ldots\ldots\ldots\ldots\ldots\ldots(3.218)$$

$$k_o = k_o(S_w), \quad \ldots\ldots\ldots\ldots\ldots\ldots\ldots\ldots(3.219)$$

$$P_c = P_c(S_w), \quad \ldots\ldots\ldots\ldots\ldots\ldots\ldots\ldots(3.220)$$

and

$$S_w = S_{iw} \quad \ldots\ldots\ldots\ldots\ldots\ldots\ldots\ldots(3.221)$$

at $t=0$, $0 \leqslant x \leqslant L$.

Because Eq. 3.217 is second order in x, two boundary conditions are needed where S_w or a derivative of S_w is known as a function of time.

One boundary condition is developed at the inlet of the reservoir where water enters at a constant injection rate. At $x=0$,

$$q_w = q_t = -\left(\frac{k_w}{\mu_w}\right)A\left(\frac{\partial p_w}{\partial x}\right). \quad \ldots\ldots\ldots(3.222)$$

Because there is no flow of oil at the inlet face, it is assumed that

$$\frac{\partial p_o}{\partial x} = 0.$$

Thus at $x=0$,

$$\frac{\partial P_c}{\partial x} = \frac{q_t \mu_w}{Ak_w}. \quad \ldots\ldots\ldots\ldots\ldots\ldots(3.223)$$

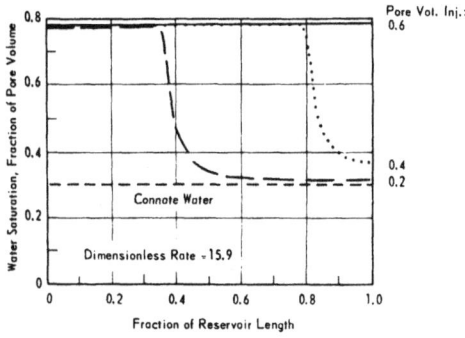

Fig. 3.46—Computed saturation distributions at several points in time.

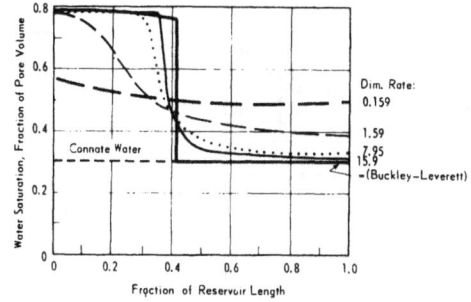

Fig. 3.47—Saturation distributions for four cases at 0.2 PV of water injected. [27]

Rewriting in terms of the saturation gradient at $x=0$,

$$\frac{\partial P_c}{\partial x} = \left(\frac{dP_c}{dS_w}\right)\left(\frac{\partial S_w}{\partial x}\right) \quad \dots\dots\dots\dots\dots (3.224)$$

and

$$\frac{\partial S_w}{\partial x} = \frac{\left(\dfrac{q_t \mu_w}{A k_w}\right)}{\left(\dfrac{dP_c}{dS_w}\right)} \cdot \quad \dots\dots\dots\dots\dots (3.225)$$

The other boundary condition is developed at $x=L$. Several possible boundary conditions have been used. For example, Douglas et al. [26] used observations from laboratory experiments to devise a suitable boundary condition. They assumed no water would be produced until the water saturation at the end of the reservoir reached a specified value, S_{wres}. Thereafter, the water saturation was held fixed for the remainder of the flood—that is,

$$q_w = 0, \quad S_w < S_{wres} \quad \dots\dots\dots\dots\dots\dots (3.226)$$

and S_w was held constant at S_{wres},

$$S_w > S_{wres}. \quad \dots\dots\dots\dots\dots\dots\dots\dots (3.227)$$

Fig. 3.46 shows computed saturations according to the boundary conditions of Eqs. 3.225 through 3.227. The saturation profile is sharp with rounding at the "foot" of the profile caused by capillary forces. Fig. 3.47 shows the computed saturation distributions for four dimensionless rates along with the profile computed from the Buckley-Leverett solution. Recall that this solution considers the capillary pressure gradient to be zero. Dimensionless rates of 0.159 and 1.59 are so low that a stabilized zone does not form. Capillary forces dominate the displacement process. As the dimensionless rate increases, the saturation profiles approach the frontal advance solution because capillary effects are confined to a smaller

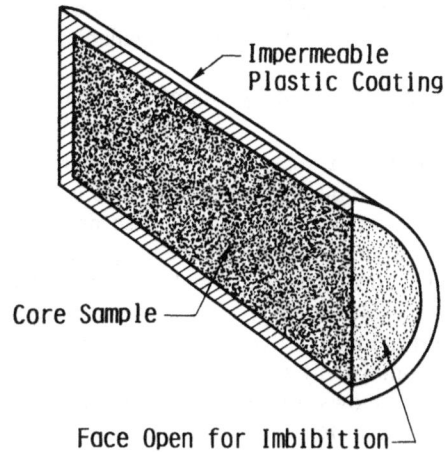

Fig. 3.48—Linear imbibition test sample. [29]

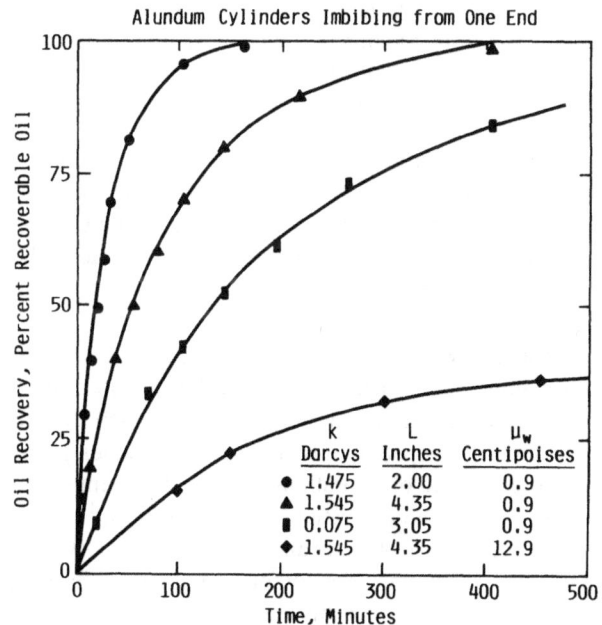

Fig. 3.49—Linear imbibition test data. [29]

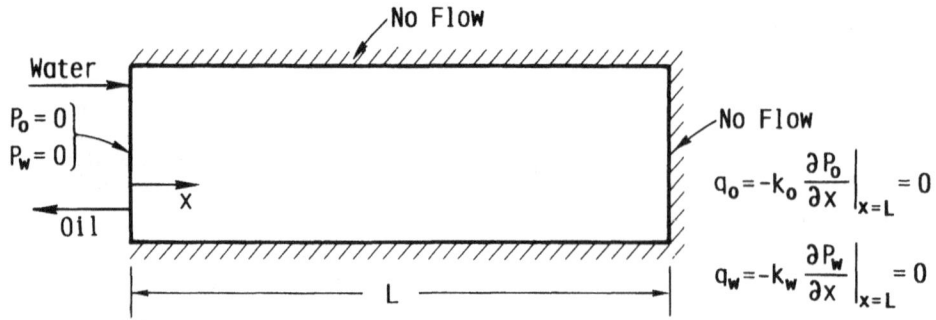

Fig. 3.50—Mathematical model of linear imbibition process showing boundary conditions.

part of the reservoir. The Buckley-Leverett solution is a limiting case that is valid at large dimensionless rates. Other solutions of this problem are available.[27,28]

3.13.2 Oil Displacement by Countercurrent Imbibition

In strongly water-wet rocks, oil displacement may occur by spontaneous imbibition whenever there is sufficient difference between oil- and water-phase pressure gradients. We will examine an example of spontaneous imbibition with countercurrent flow of water and oil to illustrate the displacement process as well as the capability of numerical methods to simulate a displacement process that is dominated by capillary forces.

Fig. 3.48[29] shows a core sample encased in plastic with one end open for imbibition. The core is strongly water-wet and is saturated with oil. At $t=0$, the open face is exposed to water. Water imbibes into the core because of the difference in pressure gradients of the water and oil phases caused by capillary forces. Oil flows countercurrent to the flow of water. Fig. 3.49 shows experimental imbibition data[29] for different rock and fluid properties. In the short core (0.167 ft [0.051 m]), displacement was complete within 200 minutes, with oil saturation reduced to residual values. Note that if the core had intermediate wettability or was strongly oil-wet, no water would imbibe.

A displacement process that is dominated by capillary forces such as those illustrated in Figs. 3.48 and 3.49 cannot be represented by the frontal advance solution. Good simulations are possible with numerical techniques. The model developed by Blair[30] to represent the process will be presented in this section.

Although water and oil flow countercurrent to each other, flow is still linear and the partial differential equations developed in Sec. 3.2.2 apply. In this model, the solution is developed by solving for oil- and water-phase pressures instead of saturations. The development of the model follows the derivation of Douglas *et al.*[13]

The linear flows of water and oil are described by Eqs. 3.228 and 3.229 when gravity forces are negligible (i.e., $\alpha=0$) and the fluid densities are constant:

$$\frac{\partial}{\partial x}\left(\frac{k_o \partial p_o}{\mu_o \partial x}\right) = \phi \frac{\partial S_o}{\partial t} \quad\dots\dots\dots\dots\dots (3.228)$$

and

$$\frac{\partial}{\partial x}\left(\frac{k_w \partial p_w}{\mu_w \partial x}\right) = \phi \frac{\partial S_w}{\partial t}. \quad\dots\dots\dots\dots (3.229)$$

The saturation derivatives may be expressed in terms of Eq. 3.230, recalling that $S_o = 1 - S_w$. Thus

$$\frac{\partial S_w}{\partial t} = \left(\frac{\partial S_w}{\partial P_c}\right)\left(\frac{\partial P_c}{\partial t}\right) = S'\left(\frac{\partial p_o}{\partial t} - \frac{\partial p_w}{\partial t}\right)$$

$$\dots\dots\dots\dots\dots\dots\dots (3.230)$$

and

$$\frac{\partial S_o}{\partial t} = -\frac{\partial S_w}{\partial t}. \quad\dots\dots\dots\dots\dots\dots (3.231)$$

Substituting into Eqs. 3.228 and 3.229, we have

$$\frac{\partial}{\partial x}\left(\frac{k_o}{\mu_o}\frac{\partial p_w}{\partial x}\right) = -\phi S'\left(\frac{\partial p_o}{\partial t} - \frac{\partial p_w}{\partial t}\right) \quad\dots (3.232)$$

and

$$\frac{\partial}{\partial x}\left(\frac{k_w}{\mu_w}\frac{\partial p_w}{\partial x}\right) = +\phi S'\left(\frac{\partial p_o}{\partial t} - \frac{\partial p_w}{\partial t}\right). \quad\dots (3.233)$$

Eqs. 3.232 and 3.233 form a set of partial differential equations in p_o and p_w, which must be solved simultaneously subject to the boundary conditions for a particular problem. Because oil and water pressures appear as second derivatives, a total of four boundary conditions must be found, two for each phase.

Fig. 3.50 illustrates the model of linear imbibition, corresponding to the core in Fig. 3.48. Flow occurs only at the inflow face. Thus a no-flow boundary condition applies for both oil and water phases at the end of the core.

At $x=L$, $q_o=0$ and $q_w=0$, which leads to the following equations:

$$\frac{\partial p_o}{\partial x} = 0 \quad\dots\dots\dots\dots\dots\dots\dots\dots (3.234)$$

and

$$\frac{\partial p_w}{\partial x} = 0. \quad \dots\dots\dots\dots\dots\dots\dots\dots\dots\dots (3.235)$$

The boundary condition at the $x=0$ is difficult to formulate for this problem because there is a discontinuity in phase pressure between the inflow face and the end of the core. Blair[30] approximated this boundary condition by requiring that the oil- and water-phase pressures be equal for $t \geq 0$ at $x=0$.

Fig. 3.51 illustrates computed pressure and saturation profiles after 6.6 hours of simulated imbibition.[30] A steep pressure gradient is shown in the water phase, causing water to imbibe toward the end of the reservoir. The water-phase pressure in the core is negative, as expected when the water pressure is zero at the core inlet. The gradient in the oil phase is in the opposite direction. The oil-pressure curve is not as steep as the water-pressure curve because the oil permeability is higher than the water permeability. Fig. 3.52 compares results of simulated imbibition tests for five different oil viscosities.[30]

The examples presented in this section show the capability of numerical techniques to solve multiphase fluid flow problems in 1D. This can be done when the rock and fluid properties are known. Several cases can be examined with relatively minor changes in input data. The cost of a numerical solution is the investment in programs and the expense of running the simulations. Numerical solutions will be discussed later in connection with multidimensional examples.

Problems

3.1. A well is producing oil and water at a WOR of 1:1 (i.e., 1 bbl [0.16 m^3] of water per barrel of oil). The water cut has increased gradually over a long period of

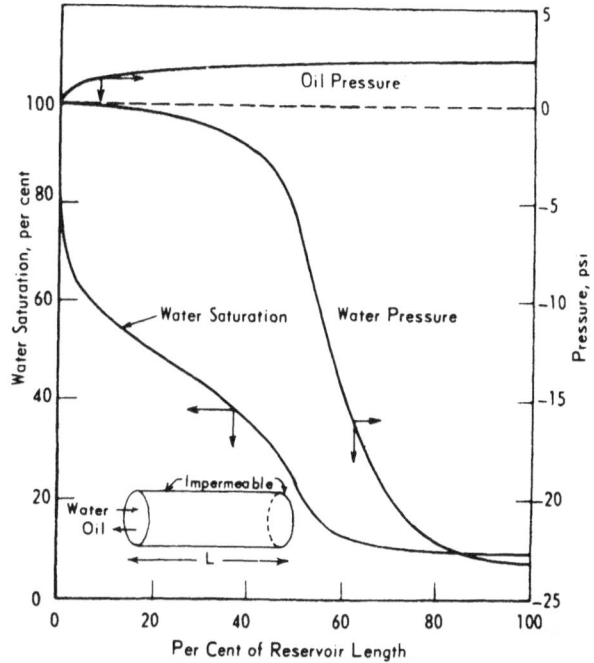

Fig. 3.51—Pressure and saturation profiles at 6.6 hours (linear imbibition, waterfront at $L/2$), $\mu_o = 5$ cp, $\mu_w = 1$ cp, $S_{lw} = 9.2\%$, $\varphi = 32.1\%$, $l = 50$ cm, $A = 7.92$ cm^2, elapsed time = 6.6 hours, in-place oil produced = 22%, and $k = 200$ md.[30]

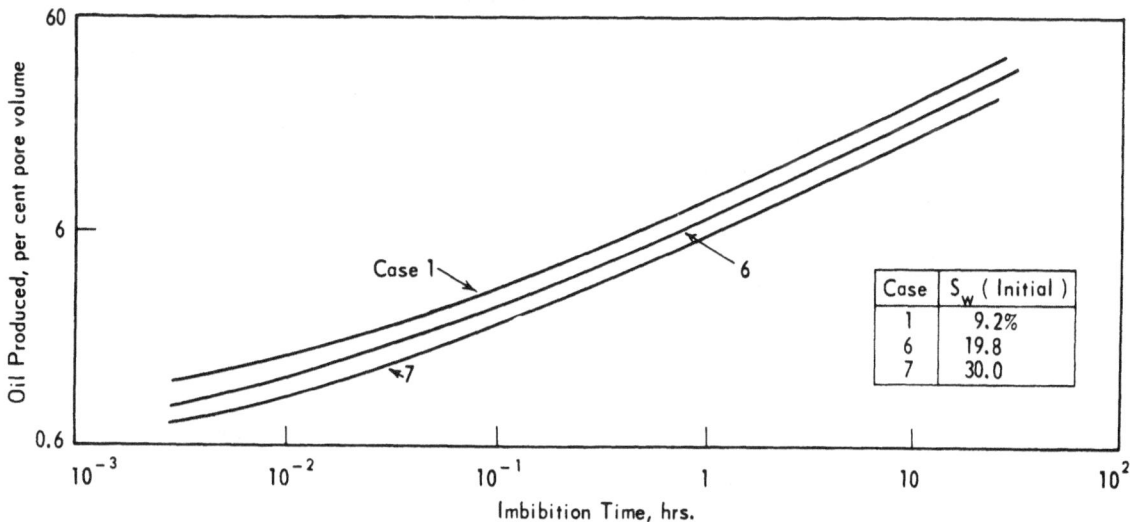

Fig. 3.52—Oil produced by linear countercurrent imbibition. Comparison for three different water saturations.[30]

TABLE 3.12—TEST DATA

L = length of core, cm	7.50
A = cross-sectional area of core, cm^2	7.89
S_{iw} = interstitial water saturation, %	30
ϕ = porosity, %	23.4
μ_w = viscosity of water at mean test conditions, cp	0.766
μ_o = viscosity of oil at mean test conditions, cp	1.20
W_d = dry weight of core, g	178.8830
W_w = water-saturated weight of core, g	192.6580
T_t = test temperature, °C	32
ρ_w = density of water at mean test conditions, g/cm^3	0.9945
ρ_o = density of oil at mean test conditions, g/cm^3	0.7494
V_p = pore volume, cm^3	13.85

Steady-State Data

Run	Time Interval for Rate Measurements (seconds)	Volume of Fluid Collected During Time Interval (cm^3)		Pressure Drop Across Core (atm)	Weight of Saturated Core at End of Run, W_c (g)
		Water	Oil		
1	311.6	0.70	9.32	0.983	190.5363
2	331.1	1.93	8.04	1.020	190.8359
3	349.3	3.50	6.53	1.170	191.0360
4	343.2	5.78	4.22	1.240	191.2567
5	336.5	7.90	2.10	1.340	191.4252

time. Laboratory studies have resulted in correlation of relative permeability data in terms of Eq. 3.111 for strongly water-wet rock.

Oil Relative Permeability

imbibition:

$$k_{ro} = 0.75(1 - S_{wD})^2. \quad\ldots\ldots\ldots\ldots\ldots\ldots (3.236)$$

drainage:

$$k_{ro} = 0.75(1 - S_{wD})^3. \quad\ldots\ldots\ldots\ldots\ldots\ldots (3.237)$$

Water Relative Permeability

drainage and imbibition:

$$k_{rw} = 0.1 S_{wD}^2, \quad\ldots\ldots\ldots\ldots\ldots\ldots\ldots (3.238)$$

where S_{wD} is given by Eq. 3.111.

Estimate the water saturation in the region around the well if the water viscosity is 0.5 cp [0.5 mPa·s] and the oil viscosity is 5.0 cp [5.0 mPa·s]. The interstitial water saturation is 0.30 and the residual oil saturation is 0.25.

3.2. An oil well produces from a formation that is 10 ft [3.05 m] thick with permeability and porosity of 20.1 md and 19.6%, respectively. Viscosity of the crude oil is 6.3 cp [6.3 mPa·s] at the reservoir temperature of 75°F [24°C]. Average water saturation is 40% and there is no free or dissolved gas saturation. The reservoir pressure at a distance of 660 ft [183 m] from the well is 1,200 psi [8274 kPa]. Estimate the steady-state production rates of water and oil (in barrels per day) assuming the wellbore radius is 6 in. [15.24 cm] and the pressure in the well is maintained at 200 psi [1.38 mPa]. Use the relative permeability data given in Fig. 2.41.

3.3. The data in Table 3.12 were obtained to determine the oil and water permeabilities of a core by the steady-state method. The pressure is measured in the water phase.

Assume that pressure gradients in oil and water phases are equal. Prepare a plot of k_o and k_w vs. water saturation. There is no initial gas saturation.

3.4. In Problem 3.3, permeabilities were computed for the oil and water phases at five saturations. Additional data taken on the core indicate

$$k_o = 43.7 \text{ md at } S_{iw} = 0.30$$

and

$$k_w = 18.5 \text{ md at } S_{or} = 0.27.$$

Absolute permeability of the core to liquid is 60 md.

a. Prepare a plot of relative permeability vs. water saturation using the absolute liquid permeability as the base permeability.

b. Define a dimensionless water saturation with Eq. 3.111.

Plot the relative permeability from Part a vs. the dimensionless water saturation.

c. Correlations in the form of Eqs. 3.143 and 3.144, respectively,

$$k_{ro} = \alpha_1 (1 - S_{wD})^m$$

and

$$k_{rw} = \alpha_2 S_{wD}^{\,n},$$

are used to represent relative permeability data where α_1, α_2, m, and n are constants determined by fitting the data to the correlation. Using the data computed from Problem 3.3 and S_{wD} from Part b, determine the coefficients that yield the best fit using least-squares linear regression.

3.5. The relative permeability data in Table 3.13 were obtained from a core sample taken from the Waltersburg

TABLE 3.13—RELATIVE PERMEABILITY MEASUREMENTS ON RUDOLPH 5, LEVEL 2,293 FT, AS MEASURED BY ELF-AQUITAINE, AUG. 1979[31]

S_w (fraction)	k_{rw}	k_{ro}
0.16	0.0	1.0
0.20	0.004	0.870
0.25	0.008	0.722
0.30	0.013	0.572
0.35	0.017	0.413
0.40	0.022	0.250
0.45	0.033	0.178
0.50	0.044	0.105
0.55	0.055	0.055
0.60	0.073	0.015
0.65	0.092	0.006
0.70	0.110	0.0

TABLE 3.14—COMPUTED SATURATION DISTRIBUTION FOR CAPILLARY END EFFECT

Distance From Core Inlet (m)	Water Saturation (fraction PV)
0.0249996	0.8423601
0.0249988	0.8385902
0.0249972	0.8348202
0.0249946	0.8310502
0.0249906	0.8272803
0.0249851	0.8235103
0.0249777	0.8197404
0.0249681	0.8159705
0.0249561	0.8122005
0.0249413	0.8084305
0.0249235	0.8046606
0.0249023	0.8008906
0.0248775	0.7971207
0.0248487	0.7933507
0.0248156	0.7895808
0.0247778	0.7858108
0.0247351	0.7820409
0.0246872	0.7782709
0.0246335	0.7745010
0.0245739	0.7707310
0.0245079	0.7669611
0.0244351	0.7631911
0.0243551	0.7594212
0.0242674	0.7556512
0.0241716	0.7518813
0.0240671	0.7481113
0.0239533	0.7443414
0.0238297	0.7405714
0.0236955	0.7368015
0.0235498	0.7330315
0.0233919	0.7292616
0.0232205	0.7254916
0.0230345	0.7217217
0.0228324	0.7179517
0.0226126	0.7141818
0.0223729	0.7104118
0.0221109	0.7066419
0.0218235	0.7028719
0.0215068	0.6991020
0.0211559	0.6953320
0.0207644	0.6915621
0.0203239	0.6877921
0.0198225	0.6840222
0.0192436	0.6802522
0.0185619	0.6764823
0.0177373	0.6727123
0.0166992	0.6689424
0.0153046	0.6651724
0.0148257	0.6641724
0.0142858	0.6631724
0.0136675	0.6621724
0.0129448	0.6611724
0.0120758	0.6601724
0.0109875	0.6591724
0.0095337	0.6581724
0.0073490	0.6571724
0.0029715	0.6561724
0.0	0.6551724

formation in the Storms pool, White County, IL.[31] The data were taken under conditions where the water saturation increased during the tests.

Correlation of relative permeability data with water saturations is often done by fitting empirical functions to the experimental data. The following functional form is found frequently in petroleum engineering literature:

$$\frac{k_{ro}}{k_{rw}} = ae^{bS_w}, \dots\dots\dots\dots\dots\dots\dots(3.239)$$

where a and b are constants.

a. Evaluate this function as a possible correlation for the Storms data. (It is helpful to plot k_{ro}/k_{rw} vs. S_w on semilog paper to determine intervals where the above equation fits the data.)

b. Compare computed and experimental data.

c. Evaluate the functional forms used in Problem 3.4 to correlate the experimental data and compare experimental data with computed values.

3.6. In the determination of relative permeability by the steady-state method, a saturation gradient develops at the end of the core because of the discontinuity in capillary pressure. This discontinuity exists because fluids flow from a region where there is high capillary pressure to the open region in the header where the capillary pressure is zero.

In Sec. 3.3.2 a method is described to compute the saturation distribution in a core when there is a capillary end effect at the outlet of the core during the steady flow of two phases. The solution is based on numerical integration of Eq. 3.40.

A computer program that solves Eq. 3.40 was used to generate the water saturation as a function of distance from the outlet end of the core. The saturation distribution corresponding to the data and the relative-permeability/capillary-pressure relationships given by Eqs. 3.240 through 3.243 is presented in Table 3.14.

Write a computer program that does the following.

1. Compute the average water saturation in the core and compare it with the water saturation in the region of the core where capillary end effects are not important.

2. Compute the pressure (in pascals) in the water and oil phases as a function of position in the core. The water-phase pressure at the core outlet is 1 atm [101.325 kPa]. In solving for the pressure distributions, begin at the core outlet and integrate toward the inlet. Solve for the water-phase pressure distribution and compute oil-phase pressure from the capillary pressure correlation. Plot oil- and water-phase pressures as a function of distance from the inlet of the core.

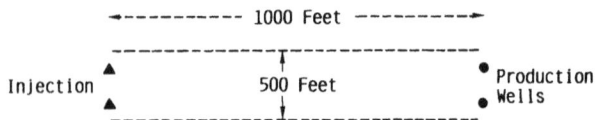

Fig. 3.53—Flooding plan for a narrow reservoir.

TABLE 3.15—ROCK AND FLUID PROPERTIES—PROBLEM 3.6

Diameter of core, m	0.0508
Length of core, m	0.0250
Absolute permeability, μm^2	0.330
Porosity, %	20.9
Oil viscosity, mPa·s	2.0
Water viscosity, mPa·s	1.0
Oil rate, m^3/s	1×10^{-6}
Water rate, m^3/s	1×10^{-6}

3. Compute the relative permeabilities to oil and water for the entire core using the pressure data obtained in Step 2. Compare these values to k_{ro} and k_{rw} in the region unaffected by end effects.

Table 3.15 gives properties of the rock and the fluids. There are no capillary entrance effects—that is, oil- and water-phase pressures at the entrance correspond to the capillary pressure at the existing water saturation.

The relative permeability and capillary pressure relationships are given by the following equations[23]:

$$P_c = \frac{10^3 z}{\sqrt{\dfrac{k_b}{3.02 \times 10^{-12}}}}, \qquad \ldots \ldots \ldots \ldots (3.240)$$

where the units of P_c are kilopascals. The variable z is a correlating parameter.

$$S_w = \frac{0.85}{1+1.845z} - \frac{2z(z-0.19)}{1+144z^4}, \qquad \ldots \ldots \ldots (3.241)$$

$$k_{rw} = \frac{0.6}{(1+3.75z)(1+207z^2)}, \qquad \ldots \ldots \ldots (3.242)$$

and

$$k_{ro} = \left(\frac{z^2}{0.2+z^2}\right)\left[\frac{0.0167z}{0.0135+(z-0.2765)^2}+0.753\right].$$

$$\ldots \ldots \ldots \ldots \ldots \ldots \ldots \ldots \ldots \ldots \ldots (3.243)$$

3.7. A long, narrow reservoir will be waterflooded with the flooding pattern shown in Fig. 3.53. The reservoir is 500 ft [152.4 m] wide, 20 ft [6.096 m] thick, and 1,000 ft [304.8 m] long. Properties of the reservoir rock and fluids are given in Table 3.16. Relative permeability data from Table 3.2 should be used.

Estimate the linear waterflood performance of this reservoir when the water injection rate for *each* well is 5 B/D/ft [2.6 $m^3/d/m$] net sand thickness.

a. Prepare a fractional flow curve and determine the breakthrough saturation.

b. Prepare a graph of water saturation vs. distance at the instant of time when the flood front (S_{wf}) is 500 ft [152.4 m] from the injection wells.

c. Determine the oil production rate as a function of PV's injected to a WOR of 100:1. Plot the oil rate (barrels per day) vs. PV's injected.

d. Determine the cumulative oil displaced (N_p in barrels) as a function of PV's injected (Q_i) to a WOR of 100:1. Plot N_p vs. Q_i, with N_p as the ordinate.

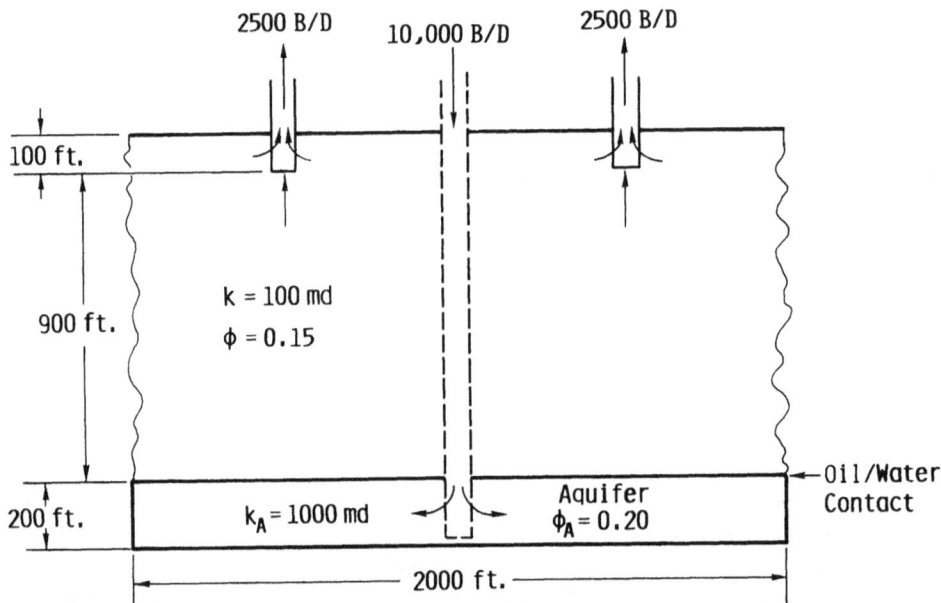

Fig. 3.54—Cross section of reef reservoir showing reservoir and aquifer intervals.

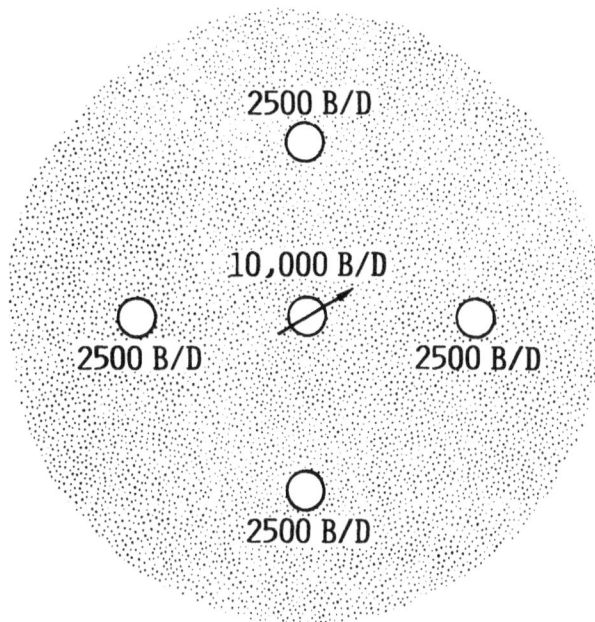

Fig. 3.55—Well layout—Problem 3.8.

TABLE 3.16—ROCK AND FLUID PROPERTIES—PROBLEM 3.7

Porosity	0.206
Permeability, md	23
S_{oi}	0.637
Water, cp	1.0
Oil, cp	10.0

TABLE 3.17—PROPERTIES OF RESERVOIR ROCK AND FLUIDS—PROBLEM 3.8

Thickness, ft	1,000
Porosity	0.15
Permeability to oil at interstitial water saturation, md	100
Interstitial water saturation	0.30
Residual oil saturation	0.35
Oil viscosity, cp	10
Water viscosity, cp	1
Oil density, lbm/cu ft	49.9
Water density, lbm/cu ft	62.4
Initial reservoir pressure (p_i), psia	5,000
Bubblepoint pressure (p_{bp}), psia	2,695
Current pressure, psia	4,000
FVF at p_i, bbl/STB	1.355
FVF at p_{bp}, bbl/STB	1.391

TABLE 3.18—RELATIVE PERMEABILITY DATA—PROBLEM 3.8

S_w (fraction)	k_{rw}	k_{ro}	$\dfrac{k_{ro}}{k_{rw}}\dfrac{\mu_w}{\mu_o}$
0.300	0.0	1.00	∞
0.335	0.001	0.729	72.90
0.370	0.004	0.512	12.80
0.405	0.009	0.343	3.811
0.440	0.016	0.216	1.350
0.475	0.025	0.125	0.500
0.510	0.036	0.064	0.178
0.545	0.049	0.027	0.055
0.580	0.064	0.008	0.013
0.615	0.081	0.001	0.001
0.650	0.100	0.000	0.000

e. Plot the WOR vs. cumulative oil produced on semilog paper. Use the semilog scale for WOR on the ordinate.

3.8. Oil is sometimes found in carbonate structures called reefs. These structures are limited in areal extent but may be several hundred feet thick. Fig. 3.54 is a cross section through a reef-type reservoir that is approximately 2,000 ft [609.6 m] in diameter and 1,000 ft [304.8 m] thick. Four production wells are completed in the upper part of the reservoir, as shown in Figs. 3.54 and 3.55. The reservoir is underlain by an aquifer that has high permeability (1,000 md) but also has limited areal extent. Because of this, there will be limited water influx as the reservoir pressure declines. Table 3.17 contains properties of the reservoir oil and rock.

Although the reservoir has produced by fluid and rock expansion since discovery, the oil is highly undersaturated. It will be necessary to waterflood the reservoir as soon as possible to maintain the reservoir pressure at desired levels. One flooding plan involves injection of water into the aquifer through the proposed well that would be drilled to the aquifer in the center of the structure, as depicted by the dashed lines in Fig. 3.54. This would create a bottomwater drive because the aquifer has high vertical and horizontal permeability.

You are requested to estimate the waterflood performance of the bottom waterflood when the water injection rate is maintained at 10,000 B/D [1589.87 m^3/d] throughout the life of the flood. Production rates of 2,500 B/D [397.47 m^3/d] would be controlled at each production well so that the flood can be treated as a linear displacement process.

Determine the following.

a. The time required for breakthrough of water into the production well.

b. The cumulative waterflood recovery in stock-tank barrels at breakthrough, assuming the average reservoir pressure is 4,000 psia [27.58 mPa].

Values of k_{ro}, k_{rw}, and $(k_{ro}/k_{rw})(\mu_w \mu_o)$ are given at selected water saturations in Table 3.18. Base permeability for the relative permeability curve is the permeability to oil at interstitial water saturation, S_{iw}.

3.9 The gas cap reservoir shown in Fig. 3.56 will be produced by allowing the gas cap to expand and to displace the oil. A constant pressure will be maintained in the expanding gas cap by injecting gas in the well that penetrates the top of the reservoir. The gas injection rate will be equivalent to 6,000 B/D [953.9 m^3/d]. Producing wells in the reservoir are perforated at the bottom of the interval to minimize gas coning. Properties of the gas and reservoir oil are summarized in Table 3.19. Relative permeability data for the drainage cycle are given in Fig. 3.57.

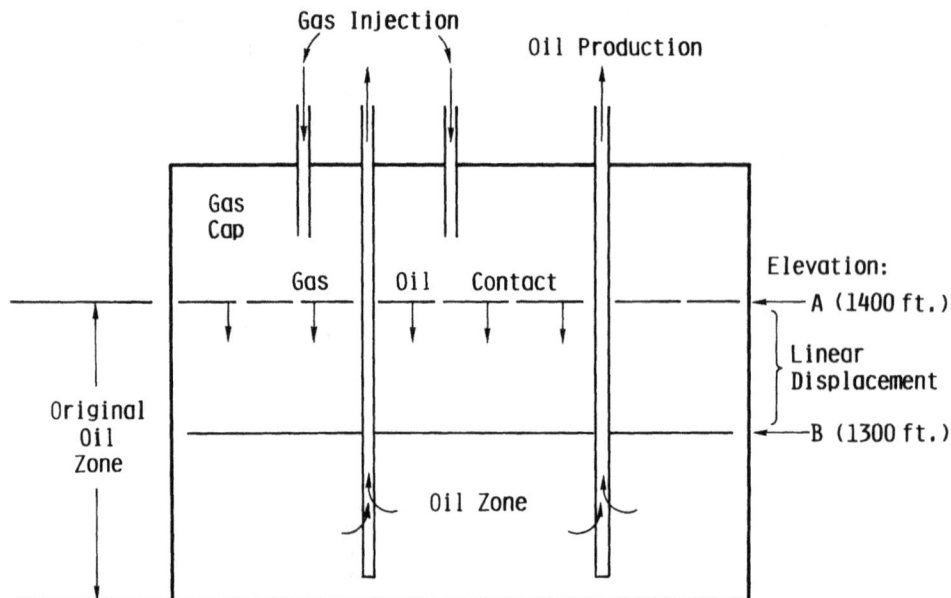

Fig. 3.56—Gas-cap reservoir with gas injection.

TABLE 3.19—PROPERTIES OF RESERVOIR AND RESERVOIR FLUIDS—PROBLEM 3.9

Absolute permeability, md	1,000
Porosity	0.30
Interstitial water saturation	0.15
Oil saturation in oil zone	0.85
Initial gas saturation in oil zone	0
PV between elevations A and B, bbl	50×10^6
Areal extent of gas/oil contact, acres	215

Properties of reservoir fluids
(at reservoir temperature and pressure):

	Gas	Oil	Water
Density, g/cm^3	0.0556	0.8859	1.0
Viscosity, cp	0.015	0.8	1.0

It is desired to estimate the oil recovery as a function of PV's of gas injected (measured at reservoir temperature and pressure) for the region between Elevations A and B in Fig. 3.56. You may assume that linear displacement occurs in this interval. The initial water saturation is immobile. None of the injected gas dissolves in the reservoir oil—i.e., the oil is saturated at existing reservoir conditions. Neglect the compressibility of the gas in displacement calculations.

Estimate the volume of oil displaced from the Interval A to B as a function of PV's of gas injected. The calculation procedure should consider that gas flows from $f_g = 0$ to 0.99.

3.10. Problem 3.10 illustrates the computation of waterflood performance in a linear system when the pressure difference between the inlet and outlet of the system is constant. You will use all parameters from Problem 3.7 except q_t. In this problem, the pressure difference is fixed at 2,000 psi [13.79 mPa].

When the pressure difference is fixed, the total flow rate varies with time. The method based on evaluation

of the average apparent viscosity will be used to compute q_t at particular values of Q_i. Time corresponding to Q_i will be computed using the method described in Sec. 3.6.

Write a computer program that will produce the following output.

a. The waterflood performance of the linear system from the beginning of the flood to WOR $\leq$ 100. Include in your calculations the oil rate, water rate, total injection rate, WOR, Q_i, N_p(PV), N_p, and $\overline{\lambda^{-1}}$.

b. A table that contains $(Q_i, \overline{\lambda^{-1}_{S_w}})$ pairs from the beginning of the flood ($Q_i = 0$) to the end of the flood (WOR $\leq$ 100). The interval before breakthrough should be subdivided into 10 increments.

c. Find the elapsed time (in days) corresponding to every value of Q_i used in your computations. Production rates should be reported in barrels per day.

d. Print your output in tabular form in the following order: time, q_o, q_w, q_t, WOR, Q_i, N_p(PV), N_p, and $\overline{\lambda^{-1}}$ in the same format as Table 3.4.

e. Plot the total injection rate vs. time.

f. Compute the mobility ratio, $M_{\bar{s}}$, for the waterflood as in Sec. 3.11.2.

g. The last part of this problem is based on the equivalence between constant-rate and constant-pressure solutions of the frontal advance equation. Using the results obtained in Part d, determine the pressure difference in pounds per square inch at WOR = 16.0 when the total injection rate is constant at 200 B/D [31.79 m^3/d] as in Problem 3.7.

Note: Three computer subprograms SWF, PROP, and APVIS, are included in Appendix A to compute certain parameters used in the solution of this problem. SWF computes breakthrough saturation and f'_{Swf} from generalized relative permeability relationships. PROP computes k_{ro}, k_{rw}, f_w, and f'_w for specified values of S_w when oil and water viscosities are known. APVIS computes the average apparent viscosity for a linear system by methods described in Sec. 3.6.

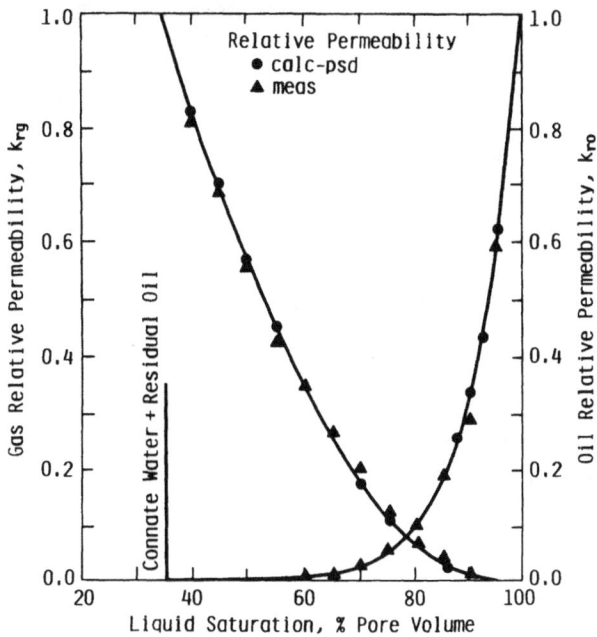

Fig. 3.57—Relative permeability data for Problem 3.9.[32]

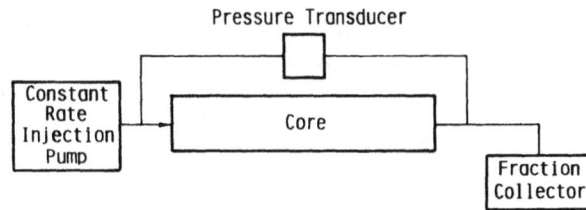

Fig. 3.58—Apparatus for waterflood experiments at constant injection rate in Problem 3.11.

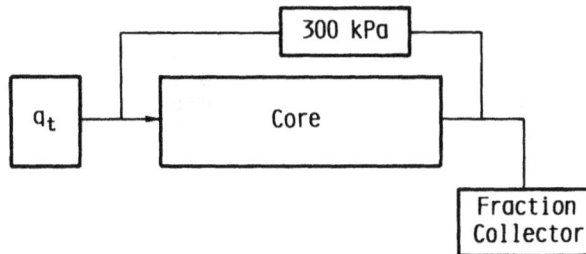

Fig. 3.59—Apparatus for waterflood experiments at constant pressure drop in Problem 3.12.

3.11. It is desired to develop a waterflood experiment in the petroleum engineering laboratory. As the engineer in charge of displacement experiments, you have been assigned the responsibility of designing a linear displacement experiment. Data from this experiment will be used to compute relative permeability ratio and relative permeabilities for the particular fluid/rock system being investigated.

Fig. 3.58 shows the principal items that must be specified.

Your task is to select *one* pump and *one* pressure transducer from Table 3.20 that meets the following experimental conditions.

a. Instantaneous WOR of 100 must be reached in no more than 2 hours.

b. Pressure transducers must have the lowest range possible to obtain the highest precision.

The following parameters are representative of the core material that will be used in the experiment.

$$L = 0.984 \text{ ft } [0.30 \text{ m}],$$
$$d = 0.164 \text{ ft } [0.05 \text{ m}],$$
$$\phi = 0.2,$$
$$k_b = k_{o|S_{iw}} = 0.150 \text{ darcy, permeability to oil at } S_{iw},$$
$$S_{oi} = 0.75,$$
$$S_{or} = 0.25,$$
$$\mu_o = 2.5 \text{ cp } [2.5 \text{ mPa·s}], \text{ and}$$
$$\mu_w = 1.0 \text{ cp } [1.0 \text{ mPa·s}].$$

The core is saturated with oil and water at the beginning of the displacement. The initial water saturation is 0.25.

Relative permeability curves are represented by the following relationships:

$$k_{ro} = (1 - S_{wD})^2 \quad \dots\dots\dots\dots\dots\dots\dots (3.244)$$

TABLE 3.20—DISPLACEMENT PUMPS AND PRESSURE TRANSDUCERS AVAILABLE—PROBLEM 3.11

Pump Number	Rate (mL/hr)	Transducer Number	Pressure Range (kPa)
P-A	6	T-1	0 to 7.0
P-B	12	T-2	0 to 14.0
P-C	24	T-3	0 to 34.0
P-D	48	T-4	0 to 68.0
P-E	96	T-5	0 to 170.0
P-F	120	T-6	0 to 340.0
P-G	200	T-7	0 to 700.0
P-H	300	T-8	0 to 1700.0
P-I	400	T-9	0 to 3400.0
P-J	500		

and

$$k_{rw} = 0.15 S_{wD}^3, \quad \dots\dots\dots\dots\dots\dots\dots (3.245)$$

where S_{wD} is given by Eq. 3.111.

Table 3.21 contains apparent viscosity data that are representative of the rock-fluid pairs that will be used in the laboratory experiment.

3.12. Fig. 3.59 shows the displacement apparatus described in Problem 3.11 designed so that the pressure drop was constant at 43.51 psi [300 kPa].

Using the data in Problem 3.11, compute the injection rate into the core (in cubic centimeters per hour) when the WOR is 25.

3.13. The reservoir shown in Fig. 3.60 is being waterflooded using an edgewater drive. Properties of the reservoir rock and fluids are given in Table 3.22.

TABLE 3.21—APPARENT VISCOSITY DATA FOR LINEAR WATERFLOOD	
Q_i (PV)	$\overline{\lambda^{-1}}$ (cp)
0.	2.50
0.046	3.07
0.093	3.64
0.139	4.20
0.186	4.77
0.232	5.34
0.279	5.91
0.325	6.47
0.372	7.04
0.418	7.61
0.465	8.18
0.478	8.14
0.491	8.11
0.505	8.08
0.520	8.05
0.535	8.01
0.552	7.98
0.569	7.95
0.587	7.92
0.606	7.88
0.627	7.85
0.648	7.82
0.671	7.79
0.695	7.75
0.720	7.72
0.747	7.69
0.776	7.66
0.807	7.62
0.839	7.59
0.874	7.56
0.912	7.53
0.952	7.50
0.995	7.47
1.042	7.43
1.093	7.40
1.147	7.37
1.207	7.34
1.272	7.31
1.343	7.28
1.421	7.25
1.507	7.22
1.603	7.19
1.709	7.16
1.828	7.13
1.963	7.10
2.115	7.07
2.291	7.04
2.493	7.02
2.730	6.99
3.010	6.96
3.346	6.93
3.759	6.90
4.275	6.88
4.939	6.85
5.826	6.82
7.068	6.80
8.934	6.77
12.045	6.74
18.271	6.72
36.995	6.69
∞	6.67

TABLE 3.22—PROPERTIES OF ROCK AND FLUIDS	
Initial oil saturation	0.75
Residual oil saturation	0.30
Oil viscosity, cp	10.0
Oil density, g/cm^3	0.9
Water viscosity, cp	1.0
Water density, g/cm^3	1.0
Reservoir thickness, ft	20
Reservoir width, ft	1,000
Reservoir length, ft	2,500
Porosity	0.25
Base permeability, md	500
Oil FVF, bbl/STB	1.23

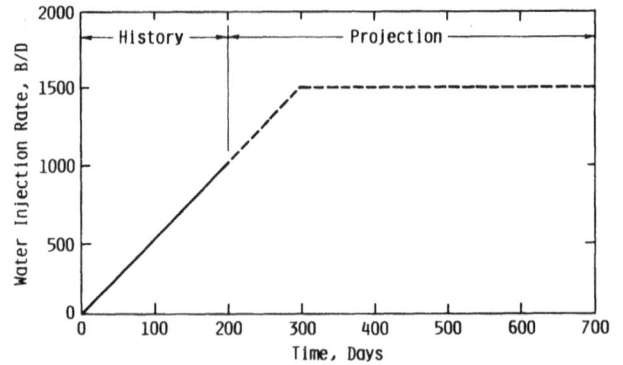

Fig. 3.61—Injection-rate history and projection for Problem 3.13.

Because of limited supply water, the injection rate has varied linearly with time for the first 300 days of the flood as depicted in Fig. 3.61 and is expected to remain constant at 1,500 B/D [239 m^3/d].

Relative permeability curves are represented by the following relationships:

$$k_{ro} = 0.8(1 - S_{wD})^{2.06} \quad \ldots\ldots\ldots\ldots\ldots\ldots (3.246)$$

and

$$k_{rw} = 0.2 S_{wD}^{2.33}, \quad \ldots\ldots\ldots\ldots\ldots\ldots\ldots (3.247)$$

where S_{wD} is given by Eq. 3.111.

Table 3.23 contains apparent viscosity data when the waterflood was designed under the assumption that the flood would operate at constant pressure drop. Estimate the pressure drop between injection and production wells after 1 year of water injection.

3.14. Suppose the reservoir in Problem 3.13 contained an initial water saturation of 0.45 when the waterflood began. All other parameters remain the same. Determine the cumulative volume of water injected when the WOR reaches 1.0 in the producing wells.

3.15. A reservoir is to be waterflooded using edgewater drive as shown in Fig. 3.62. Properties of the reservoir rock and fluids are given in Table 3.24. The reservoir is located at a depth of 4,000 ft [1219 m] and is assumed to be homogeneous.

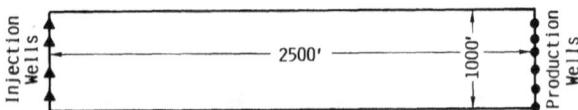

Fig. 3.60—Waterflood of a reservoir with an edgewater drive.

TABLE 3.23—APPARENT VISCOSITY DATA FOR LINEAR WATERFLOOD—PROBLEM 3.13

$\mu_o = 10$ cp and $\mu_w = 1$ cp

Q_i (PV)	$\overline{\lambda}^{-1}$ (cp)	Q_i (PV)	$\overline{\lambda}^{-1}$ (cp)
0.0	12.50	0.734	8.04
0.016	12.40	0.780	7.91
0.032	12.30	0.831	7.78
0.048	12.20	0.885	7.65
0.065	12.10	0.944	7.52
0.081	12.00	1.009	7.40
0.097	11.90	1.080	7.28
0.113	11.80	1.157	7.16
0.129	11.70	1.241	7.05
0.145	11.60	1.334	6.94
0.162	11.50	1.437	6.83
0.178	11.40	1.550	6.72
0.194	11.29	1.676	6.62
0.210	11.19	1.816	6.52
0.226	11.09	1.972	6.43
0.242	10.99	2.148	6.33
0.259	10.89	2.346	6.24
0.275	10.79	2.570	6.15
0.291	10.69	2.827	6.07
0.307	10.59	3.121	5.98
0.323	10.49	3.463	5.90
0.337	10.33	3.862	5.82
0.352	10.17	4.335	5.75
0.368	10.00	4.901	5.67
0.386	9.84	5.589	5.60
0.404	9.68	6.441	5.53
0.424	9.52	7.522	5.46
0.446	9.36	8.930	5.40
0.469	9.21	10.835	5.34
0.494	9.05	13.543	5.27
0.521	8.90	17.675	5.22
0.550	8.75	24.698	5.16
0.582	8.60	39.095	5.10
0.616	8.46	83.938	5.05
0.652	8.32	∞	5.00
0.692	8.18		

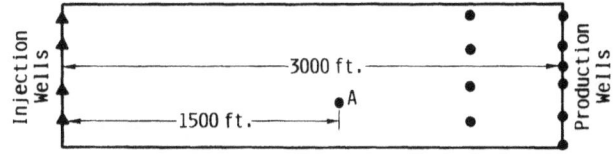

Fig. 3.62—Edgewater drive in the reservoir in Problem 3.15. Observation well is located at Point A.

TABLE 3.24—PROPERTIES OF ROCK AND FLUIDS—PROBLEM 3.15

Initial oil saturation	0.73
Residual oil saturation	0.35
Oil viscosity, cp	5.0
Oil density, g/cm^3	0.9
Water viscosity, cp	0.7
Water density, g/cm^3	1.0
Reservoir thickness, ft	25
Reservoir width, ft	2,000
Reservoir length, ft	3,000
Porosity	0.35
Base permeability (k_{ro} at S_{iw}), md	250
Oil FVF, bbl/STB	1.1

TABLE 3.25—APPARENT VISCOSITY DATA FOR LINEAR WATERFLOOD—PROBLEM 3.15

Q_i (PV)	$\overline{\lambda}^{-1}$ (cp)	Q_i (PV)	$\overline{\lambda}^{-1}$ (cp)
0.0	5.00	0.611	8.43
0.017	5.22	0.638	8.37
0.034	5.45	0.668	8.31
0.050	5.67	0.701	8.24
0.067	5.90	0.735	8.18
0.084	6.12	0.772	8.12
0.101	6.35	0.812	8.06
0.117	6.57	0.856	8.00
0.134	6.80	0.903	7.94
0.151	7.02	0.954	7.88
0.168	7.25	1.010	7.83
0.185	7.47	1.071	7.77
0.201	7.70	1.138	7.71
0.218	7.92	1.212	7.66
0.235	8.15	1.294	7.60
0.252	8.37	1.385	7.55
0.269	8.60	1.486	7.50
0.285	8.82	1.600	7.44
0.302	9.05	1.729	7.39
0.319	9.27	1.876	7.34
0.336	9.50	2.044	7.29
0.346	9.43	2.239	7.24
0.358	9.36	2.467	7.19
0.370	9.29	2.738	7.14
0.383	9.22	3.064	7.10
0.396	9.16	3.464	7.05
0.410	9.19	3.965	7.00
0.426	9.02	4.610	6.96
0.441	8.95	5.473	6.91
0.458	8.89	6.683	6.87
0.476	8.82	8.500	6.83
0.495	8.76	11.534	6.79
0.516	8.69	17.606	6.75
0.537	8.62	35.836	6.71
0.560	8.56	∞	6.67
0.584	8.50		

You are responsible for the design of the waterflood including selection of water injection equipment. Water injection pumps are designed to operate at a constant rate. It is desired to operate the waterflood so that fracturing of the formation is avoided. The fracture gradient in the area is 0.8 psi/ft [18.1 kPa/m]. Thus the maximum bottomhole pressure (BHP) in the injection wells cannot exceed 3,200 psi [22 063 kPa]. Production wells are on pump with an average pressure of 100 psi [690 kPa] maintained at the midpoint of the productive interval.

Relative permeability correlations are given by Eqs. 3.111, 3.244, and 3.245. Apparent viscosity data corresponding to these correlations are given in Table 3.25.

Table 3.26 contains a series of pumps that are available for this waterflood.

a. Select the pump with the highest injection rate that could be used within the fracturing pressure constraint if the economic limit of the flood occurs at WOR=50. Show your reasoning for selecting the pump.

b. Estimate the time (days) required for breakthrough when the waterflood is operated at the constant injection rate determined in Part a.

c. What is the wellhead pressure (in pounds per square inch) of the injection wells when breakthrough occurs in the producing wells?

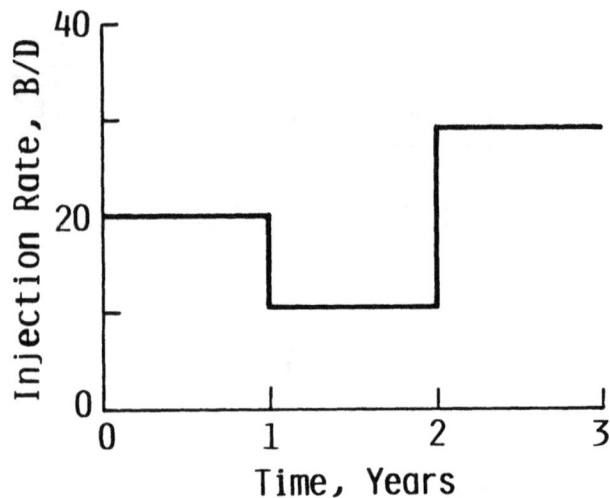

Fig. 3.63—Injection-rate history for Problem 3.16.

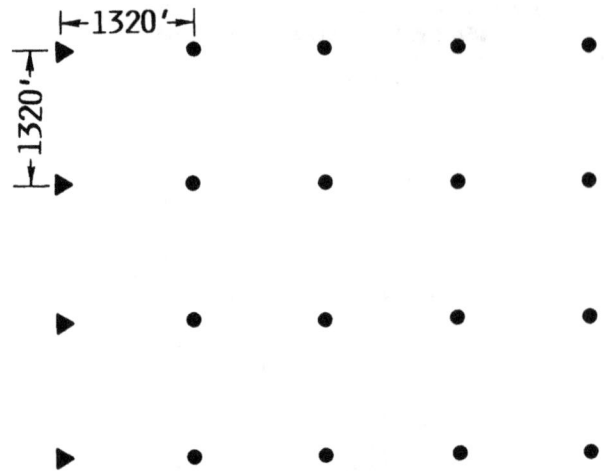

Fig. 3.64—Well arrangement for evaluation of a proposed edgewater flood in Problem 3.17.

TABLE 3.26—PUMP SELECTION GUIDE: TRIPLEX PUMPS AT 300 REV/MIN—PROBLEM 3.15

Plunger Diameter (in.)	Injection Rate (B/D)
¾	185
⅞	250
1	329
1⅜	621
1½	737
1⅝	864
1¾	1,005
2	1,313
2¼	1,659
2½	2,023
2¾	2,469
3	2,743

TABLE 3.27—RESERVOIR AND FLUID PROPERTIES—PROBLEM 3.15

Reservoir Properties	
Length, ft	220
Width, ft	200
Thickness, ft	17.7
Porosity	0.18
Absolute permeability, md	100
Initial oil saturation	0.76
Residual oil saturation	0.30

Fluid Properties	
Oil viscosity, cp	19
Water viscosity, cp	1
Oil FVF, bbl/STB	1.04

d. An abandoned well located at Point A in Fig. 3.62 will be used to follow the progress of the waterflood. At what time (days) in the flood would you expect to see a water cut of 10% in fluid samples withdrawn from this well?

e. Compute the total oil recovery (in stock-tank barrels) when the WOR is 50.

3.16. A waterflood in a linear system has been under way for 3 years. During this period, the injection rate varied as shown in Fig. 3.63.

Reservoir rock and fluid properties are summarized in Table 3.27. Relative permeability data for this reservoir are presented in Eqs. 3.248 and 3.249. Average apparent viscosities at selected values of Q_i are tabulated in Table 3.28.

$$k_{ro} = 0.4(1 - S_{wD})^2 \quad\dots\dots\dots\dots\dots\dots (3.248)$$

and

$$k_{rw} = 0.3 S_{wD}^{2.5}. \quad\dots\dots\dots\dots\dots\dots (3.249)$$

1. At what time should water production begin at the end of the reservoir? Express your answer in days from the beginning of injection.

2. What is the pressure drop (in pounds per square inch) across the linear system after 3 years of injection?

3. The flood will be terminated when the WOR (at stock-tank conditions) is 100:1.

a. Calculate the oil recovery (in barrels) when floodout occurs.

b. Assume the injection rate remains constant at 30 B/D [5 m³/d] for $t \geq 3$ years. Determine the flood life in days.

3.17. The reservoir shown in Fig. 3.64 has been depleted by solution gas drive from an initial pressure of 1,710 psia [11 790 kPa] to a current pressure of 424.7 psia [2928.2 kPa]. The reservoir was developed on 40-acre [161 875-m²] spacing and covers an area of about 800 acres [3 237 498 m²]. Properties of the reservoir and the reservoir oil are given in Table 3.29. It is planned to waterflood the reservoir. One option is to convert edge wells to water injection as indicated in Fig. 3.64. Oil would be displaced to the interior wells where it would be produced. Each row of interior wells would be converted to injection as they water out.

You are requested to estimate the performance of the first line of production wells by assuming that the displacement process can be approximated by a linear water-

TABLE 3.28—SELECTED VALUES OF AVERAGE APPARENT VISCOSITY FOR LINEAR WATERFLOOD—PROBLEM 3.16

Q_i (PV)	$\bar{\lambda}^{-1}$ (cp)	Q_i (PV)	$\bar{\lambda}^{-1}$ (cp)
0.0	47.50	0.909	9.63
0.011	46.12	1.005	9.21
0.023	44.74	1.111	8.82
0.034	43.35	1.228	8.44
0.045	41.97	1.359	8.09
0.057	40.59	1.503	7.76
0.068	39.21	1.664	7.46
0.079	37.83	1.844	7.17
0.091	36.44	2.043	6.89
0.102	35.06	2.266	6.63
0.113	33.68	2.515	6.39
0.125	32.30	2.793	6.17
0.136	30.92	3.106	5.95
0.147	29.53	3.458	5.75
0.159	28.15	3.854	5.56
0.170	26.77	4.302	5.38
0.181	25.39	4.812	5.21
0.193	24.01	5.392	5.05
0.204	22.62	6.057	4.90
0.215	21.24	6.823	4.76
0.227	19.86	7.710	4.63
0.245	18.95	8.745	4.50
0.266	18.06	9.964	4.38
0.290	17.20	11.412	4.27
0.317	16.38	13.154	4.17
0.347	15.59	15.278	4.07
0.380	14.83	17.913	3.97
0.417	14.11	21.251	3.88
0.459	13.43	25.596	3.80
0.505	12.79	31.453	3.72
0.556	12.18	39.731	3.65
0.613	11.61	52.248	3.58
0.676	11.07	73.249	3.51
0.746	10.56	115.462	3.45
0.823	10.08	242.541	3.39
		∞	3.33

TABLE 3.29—RESERVOIR AND FLUID DATA—PROBLEM 3.17

Area, acres	800
Average permeability, md	10.0
Porosity, %	12
Net pay thickness, ft	60
Bottomhole injection pressure limit, psig (SP* = 1,200 psig)	3,170
Reservoir pressure at start of waterflood, psig	400
Initial gas saturation, %	18
Initial water saturation, %	13
Residual oil saturation	0.30
Oil viscosity at 400 psig, cp	2.27
Water viscosity, cp	0.65
Oil gravity, °API	32
Initial oil FVF at 1,710 psig	1.228
Oil FVF at 400 psig	1.113
PV, bbl [m³]	$4,469 \times 10^4$
Original oil in place at bubblepoint, STB	$3,166 \times 10^4$
Estimated ultimate primary recovery at 100 psig abandonment pressure (5,700 bbl/acre or 14.4% of original oil in place), STB	$4,559 \times 10^3$

*Surface pressure.

TABLE 3.30—RESERVOIR AND FLUID PROPERTIES*—PROBLEM 3.19

Thickness, ft	20
Porosity	0.25
Absolute permeability, md	100
Initial water saturation	0.25
Residual oil saturation	0.28
Oil viscosity, cp	3.5
Water viscosity, cp	0.9
Oil FVF	1.10

$$k_{rw} = 0.276\, S_{wD}^{1.46}$$

$$k_{ro} = 0.335\, (1 - S_{wD})^{1.98}$$

*Base permeability for relative permeability curves is the absolute permeability.

flood operating at a constant pressure drop between the row of injection wells and the first row of production wells. BHP in the injection wells will be limited to 3,170 psig [21 856 kPa]. Fluid levels in production wells will be maintained so that the average sandface pressure is 500 psig [3447 kPa]. Assume trapped gas goes back into solution. Relative permeability curves are represented by the following correlations.

$$k_{ro} = (1 - S_{wD})^{1.82}, \quad \dots\dots\dots\dots\dots\dots (3.250)$$

$$k_{rw} = 0.295 S_{wD}^{4.59}, \quad \dots\dots\dots\dots\dots\dots (3.251)$$

and S_{wD} is given by Eq. 3.111.

a. Determine the oil displacement rate as a function of time to a WOR of 25 at the first row of producing wells.

b. What is the volume of oil (in stock-tank barrels) displaced when the first line of production wells is producing at a WOR of 25?

c. How does the water injection rate change with time? What water injection capacity would be needed for the four wells?

d. Prepare an explanation of why you would or would not recommend further consideration of this flooding plan.

3.18. Fluid flow in the region around an injection well that has not been fractured is essentially radial and thus 1D. The continuity equations for the radial flow of oil and water are given by Eqs. 3.252 and 3.253.

$$-\frac{1}{r}\frac{\partial}{\partial r}(r\rho_o u_o) = \frac{\partial}{\partial t}(\rho_o S_o \phi) \quad \dots\dots\dots\dots (3.252)$$

and

$$-\frac{1}{r}\frac{\partial}{\partial r}(r\rho_w u_w) = \frac{\partial}{\partial r}(\rho_w S_w \phi). \quad \dots\dots\dots (3.253)$$

a. Assuming that oil and water are incompressible, show that the frontal advance equation in radial coordinates is given by Eq. 3.254:

$$\frac{d(r^2)}{dt}\bigg|_{S_w} = \frac{q_t}{\pi h \phi}\left(\frac{\partial f_w}{\partial S_w}\right). \quad \dots\dots\dots\dots (3.254)$$

TABLE 3.31—DATA FOR MP-4, SAMPLE 21[33]

$S_{or} = 0.262$, $S_{iw} = 0.422$, $\phi = 0.235$,
$k_{abs} = 142$ md, and k_o at $S_{iw} = 99$ md.

Drainage Relative Permeability

S_w	k_{rw}	k_{ro}
0.738	0.626	0.0
0.613	0.0094	0.320
0.590	0.0045	0.432
0.577	0.003	0.486
0.567	0.0021	0.535
0.559	0.0015	0.574
0.541	0.00037	0.623
0.523	0.00045	0.750
0.502	0.00022	0.825
0.483	0.00011	0.898
0.467	0.000067	0.898
0.451	0.000038	0.959
0.440	0.000019	0.991
0.422	0.0	1.000

TABLE 3.32—CHESNEY MP-4[33]

	Sample 14	Sample 3
k_a, md	499	32
k_o at S_{iw}, md	390	30
k_w at S_{or}, md	133	14
S_{iw}	24.8	27.2
S_{or}	23.3	26.2
ϕ	28.1	13.5

Imbibition Relative Permeability Data*

Sample 14			Sample 3		
S_w	k_{rw}	k_{ro}	S_w	k_{rw}	k_{ro}
35.7	0.036	0.536	30.3	0.009	0.722
41.0	0.061	0.390	34.2	0.022	0.485
46.6	0.083	0.260	38.1	0.036	0.325
57.1	0.104	0.163	42.6	0.057	0.193
55.3	0.121	0.103	46.3	0.079	0.128
59.2	0.146	0.061	49.2	0.104	0.088
62.9	0.172	0.034	52.2	0.132	0.056
66.5	0.202	0.018	55.6	0.167	0.035
69.5	0.232	0.009	58.6	0.209	0.021
72.1	0.260	0.005	62.1	0.262	0.012
73.8	0.280	0.003	65.3	0.314	0.0065
76.7	0.318	—	68.5	0.369	0.0035
			70.7	0.409	0.0017
			73.8	0.469	0.0017

*Base permeability is permeability to oil at connate water saturation.

b. Develop a procedure for predicting waterflood performance during the time fluid flow is radial when the injection rate is constant.

3.19. Water has been injected at a rate of 2 B/D/ft [1 m³/d/m] net sand into a well that has an effective wellbore radius (r_w) of 0.5 ft [0.15 m]. Properties of the reservoir and fluids are given in Table 3.30.
a. Determine the water saturation distribution as a function of distance from the wellbore after 1, 50, and 100 days of injection, assuming radial flow.
b. Plot water saturation from Part a vs. r/r_w.
c. How could the results of Parts a and b be used to compute the injection well pressure if the reservoir pressure of radius r_e is constant at some value p_e?

3.20. A reservoir core is to be prepared for waterflooding tests. Properties of the core are given in Table 3.31. The core is water-wet and is at residual oil saturation. [32] It is desired to oilflood the core to interstitial water saturation, S_{iw}. Using the relative permeability data from Table 3.31, determine the number of PV's of oil to be injected to reach a water saturation of 0.43 when the viscosity of the oil is 10 cp [10 mPa·s]. The viscosity of the water is 1.0 cp [1.0 kPa·s].

3.21. Relative permeability may vary with rock permeability (as discussed in Sec. 2.4.3.1) even though core samples were taken from the same formation just a few feet apart. Imbibition relative permeability data for two such rock samples are given in Table 3.32. Determine the effects of differences in relative permeability curves for a waterflood in each zone by computing the displacement efficiency of each waterflood to a WOR of 80:1. Compare displacement performance by plotting microscopic displacement efficiency E_D vs. PV's injected. The viscosities of the oil and water are 5.2 and 1.0 cp [5.2 and 1.0 mPa·s], respectively.

3.22. Honarpour *et al.* [34] developed empirical correlations for estimating two-phase relative permeability in consolidated rock. For sandstones, the following correlations were obtained as a function of wettability. Water-wet:

$$k_{rw} = 0.035388 \frac{(S_w - S_{iw})}{(1 - S_{iw} - S_{orw})} - 0.010874$$

$$\times \left(\frac{S_w - S_{orw}}{1 - S_{iw} - S_{orw}} \right)^{2.9} + 0.56556(S_w)^{3.6}$$

$$\times (S_w - S_{iw}). \quad \dots\dots\dots\dots\dots (3.255)$$

Oil-wet and intermediate:

$$k_{rw} = 1.5814 \left(\frac{S_w - S_{iw}}{1 - S_{iw}} \right)^{1.91}$$

$$-0.58617 \left(\frac{S_w - S_{orw}}{1 - S_{iw} - S_{orw}} \right) (S_w - S_{iw})$$

$$-1.2484\phi (1 - S_{iw})(S_w - S_{iw}). \quad \dots\dots\dots (3.256)$$

All wetting conditions:

$$k_{ro} = 0.76067 \left[\frac{\left(\frac{S_o}{1 - S_{iw}} \right) - S_{orw}}{1 - S_{orw}} \right]^{1.8}$$

$$\times \left(\frac{S_o - S_{orw}}{1 - S_{iw} - S_{orw}} \right)^{2.0} + 2.631\phi (1 - S_{orw})$$

$$\times (S_o - S_{orw}), \quad \dots\dots\dots\dots\dots (3.257)$$

TABLE 3.33—WATERFLOOD DATA*

W_i (cm^3)	N_p (cm^3)	Δp (psi)
0.00	0.00	16.31
1.50	1.50	15.93
3.02**	3.02	15.55
3.87	3.18	14.03
6.00	3.48	11.94
7.69	3.65	11.18
9.09	3.77	10.71
13.81	4.10	9.60
20.00	4.42	8.66
32.20	4.80	7.42
48.40	5.02	6.48
100.00	5.25	5.22
200.00	5.36	4.51
400.00	5.42	4.06

*From S.C. Jones, 1985.
**Water breakthrough.

TABLE 3.34—EXPERIMENTAL DATA FOR PROBLEM 3.23*

q, cm^3/min	3.00
μ_w, cp	0.811
μ_o, cp	5.52
Temperature, °F	85
Core plug length, in.	3.031
Core plug diameter, in.	0.997
PV, cm^3	9.67
Air permeability, md	647
Brine permeability, md	571
S_{oi}	0.724
Upstream pressure, psia	1,000
Confining pressure, psia	3,000
$\Delta p/q$ at 100% brine saturation, psi/(cm^3/min)	0.532
$\Delta p/q$ at initial oil saturation, psi/(cm^3/min)	5.437

TABLE 3.35—EXPERIMENTAL DATA FOR PROBLEM 3.24*

Length, in.	2.992
Diameter, in.	0.998
PV, cm^3	9.880
k_w, md	409
k_a, md	471
μ_w, cp	0.830
μ_o, cp	11.2
q at constant rate, cm^3/min	1.0

where S_{orw} is the residual oil saturation to water, fraction, and S_{iw} is the interstitial water saturation, fraction.

a. Compare these relative permeability correlations with the data in Table 3.31.

b. Rework Problem 3.21 using the correlations that give the best fit to the data. Note that Appendix A contains a generalized root-finding program that determines S_{wf} from relative permeability correlations.

3.23. A constant-rate waterflood in a sucrosic dolomite core plug was performed to determine oil/water relative permeability curves. Cumulative water injection, oil production, and overall pressure-drop data are given in Table 3.33. These data were taken from a much larger data set and were slightly smoothed. Other pertinent data are given in Table 3.34.

With the Jones-Roszelle graphical technique, find the k_{ro} and k_{rw} curves relative to the absolute brine permeability. Indicate how to convert the relative permeabilities so that they are relative to the effective oil permeability before the waterflood started.

3.24. A sucrosic dolomite core plug was oilflooded to obtain secondary-drainage oil/water relative permeability curves. See Tables 3.34 and 3.35 for the available data. Find the fractional flow of brine, k_{rw}, and k_{ro} as functions of the outflow-face water saturation (S_{w2}). Base permeability is the absolute permeability to brine. The water saturation at the end of the oilflood was found to be 0.325 by Dean Stark Soxhlet extraction and is assumed to be the interstitial saturation, S_{iw}.

3.25. Johnson, Bossler, and Naumann (JBN)[15] developed the first method of determining oil and water relative permeabilities from the analysis of waterflood performance. The JBN method produces relative permeability curves equivalent to the Jones and Roszelle method, but the base permeability is the permeability to oil at connate water saturation.

Because the JBN method is based on the Buckley-Leverett solution, relative permeabilities can be obtained for saturations greater than or equal to S_{wf}. In the JBN

TABLE 3.36—OILFLOOD PERFORMANCE DATA*

Volume Oil Injected (cm^3)	Volume Water Produced (cm^3)	Overall Pressure Drop (psi)
0.000	0.000	2.73
2.000	2.000	12.34
3.920**	3.920	21.57
4.000	3.925	21.54
5.000	3.987	21.14
6.000	4.029	20.87
7.000	4.058	20.69
8.000	4.075	20.50
9.000	4.083	20.57
10.000	4.085	20.55
14.000	4.087	20.54
50.000	4.087	20.54

*From S.C. Jones, 1985.
**Oil breakthrough.

method, the fractional flow of oil at the outlet face of the core is given by Eq. 3.188:

$$f_{o2} = \frac{\overline{dS_w}}{dQ_i}.$$

The water saturation at the outlet of the core (S_{w2}) is obtained from Eq. 3.186:

$$S_{w2} = \overline{S_w} - Q_i f_{o2}.$$

Individual phase permeabilities corresponding to S_{w2} are computed from the overall pressure-drop data that are measured from the beginning of the waterflood as a function of Q_i. Let

$$I_r = (q/A\Delta p)/(q_s/A\Delta p_s), \quad \ldots\ldots\ldots\ldots(3.258)$$

where

I_r = relative injectivity,
q = volumetric displacement rate, cm^3/hour,
q_s = volumetric displacement rate at beginning of waterflood,
Δp = overall pressure drop, psi, and
Δp_s = overall pressure drop at beginning of waterflood, psi.

Johnson *et al.* show that the oil-phase relative permeability is related to the process variables by Eq. 3.259:

$$\frac{f_{o2}}{k_{ro}} = \frac{d\left(\dfrac{1}{Q_i I_r}\right)}{d\left(\dfrac{1}{Q_i}\right)}. \quad \ldots\ldots\ldots\ldots(3.259)$$

Values of k_{ro} at S_{w2} can be determined from the experimental data by differentiating the graph of $1/Q_i I_r$ vs. $1/Q_i$ graphically or numerically. The relative permeability of the water phase is computed from Eq. 3.260:

$$k_{rw} = \left(\frac{1-f_{o2}}{f_{o2}}\right)\frac{\mu_w}{\mu_o}k_{ro}. \quad \ldots\ldots\ldots\ldots(3.260)$$

Determine the relative permeability curves for water and oil phases with the JBN method for the displacement data in Problem 3.23.

3.26. Solve Problem 3.24 with the JBN method described in Problem 3.25. What adjustments must be made in the method to handle calculation of secondary-drainage relative permeability curves?

Nomenclature

A = cross-sectional area available for flow, sq ft [m^2]
B = constant defined by $1/(1-S_{or}-S_{iw})$
B_o = oil FVF, bbl/STB [m^3/stock-tank m^3]
B_w = water FVF, bbl/STB [m^3/stock-tank m^3]
d = diameter, ft [m]
E_D = microscopic displacement efficiency
f = volume fraction of flowing phase
f_o = fractional flow of oil
f_{o2} = fractional flow of oil at outlet of linear system
f_{Swf} = fractional flow of water at breakthrough
f_w = fractional flow of water
f_{wi} = fractional flow of water at initial water saturation, S_{wi}
f_{w1} = fractional flow of water at position x_1
f_{w2} = fractional flow of water at position x_2
f'_2 = derivative of f_w with respect to S_w at x_2
f'_{Sw} = derivative of f_w with respect to S_w

f'_{Swr} = derivative of f_w with respect to S_w at S_{iw}
F_{wo} = water/oil ratio, bbl/STB [m^3/stock-tank m^3]
g = gravity constant
k_b = base permeability for relative permeability data, darcies
k_o = permeability to oil, darcies
k_{ox} = permeability to oil in the x direction, darcies
k_{oy} = permeability to oil in the y direction, darcies
k_{oz} = permeability to oil in the z direction, darcies
k_{ro} = relative permeability to oil
k_{rw} = relative permeability to water
k_w = permeability to water, darcies
k_w^* = permeability to water when water-phase pressure gradient is known in steady-state method, darcies
k_{wx} = permeability to water in the x direction, darcies
k_{wy} = permeability to water in the y direction, darcies
k_{wz} = permeability to water in the z direction, darcies
L = length, ft [m]
M = mobility ratio based on endpoints
$M_{\bar{S}}$ = mobility ratio based on average saturation behind the flood front
M_t = total mobility ratio
N_p = cumulative oil production, STB [stock-tank m^3]
N_{pbt} = cumulative oil production at breakthrough, STB [stock-tank m^3]
p_i = initial pressure, psi [kPa]
p_L = pressure at outlet of linear core, psi [kPa]
p_o = pressure at inlet of linear core, psi [kPa]
p_{od} = oil-phase pressure at reference horizontal datum, psi [kPa]
p_{oi} = initial oil-phase pressure, psi [kPa]
P_{oL} = oil-phase pressure at L, psi [kPa]
p_{o1} = oil-phase pressure at x_1, psi [kPa]
p_{o2} = oil-phase pressure at x_2, psi [kPa]
p_p = pressure at producing well, psi [kPa]
p_w = water-phase pressure, psi [kPa]
p_{wi} = water-phase pressure at inlet of linear core, psi [kPa]
p_{wL} = water-phase pressure at L, psi [kPa]
p_{w1} = water-phase pressure at x_1, psi [kPa]
p_{w2} = water-phase pressure at x_2, psi [kPa]
P_c = capillary pressure, psi [kPa]
P_{c1} = capillary pressure at x_1, psi [kPa]
P_{c2} = capillary pressure at x_2, psi [kPa]
q_b = base injection rate, B/D [m^3/d]
q_o = oil production rate, B/D [m^3/d]
q_t = total injection or production rate, B/D [m^3/d]
q_t^n = total injection rate at timestep n, B/D [m^3/d]

q_w = water production rate, B/D [m^3/d]

Q_i = cumulative PV's injected

Q_{ibt} = cumulative PV's injected at breakthrough

Q_{ibte} = effective PV's of water injected at breakthrough

Q_{ie} = effective PV's of water injected

Q_{if} = cumulative PV's of water injected at fill-up

S' = $\partial S_w / \partial P_c$

S_e = extrapolated water saturation (Fig. 3.11), fraction

S_{gi} = initial gas saturation, fraction

S_{gm} = mobile gas saturation, fraction

S_{gr} = residual gas saturation, fraction

S_{gt} = trapped gas saturation, fraction

S_{iw} = interstitial water saturation where water phase is immobile under applied pressure gradient, fraction

S_o = oil saturation, fraction

S_{ob} = oil saturation in oil bank, fraction

S_{oe} = effective oil saturation, fraction

S_{oi} = initial oil saturation, fraction

S_{or} = residual oil saturation, fraction

S_w = water saturation, fraction

$\overline{S_w}$ = average water saturation in linear system, fraction

S_{wD} = dimensionless water saturation

S_{we} = effective water saturation when trapped gas is present, fraction

S_{wf} = water saturation at breakthrough, fraction

$\overline{S_{wfe}}$ = effective average water saturation at breakthrough, fraction

S_{wi} = initial water saturation, fraction

S_{wL} = water saturation at $x=L$

S_{wres} = water saturation at end of linear core when water begins to flow in the presence of capillary end effects during linear displacement, fraction

S_{w1} = water saturation at x_1, fraction

S_{w2} = water saturation at x_2, fraction

S_w^* = extrapolated water saturation, Jones and Roszelle method (Fig. 3.27), fraction

S_w^+ = extrapolated water saturation, Jones and Roszelle method (Fig. 3.27), fraction

t = time, days

t_{bt} = time at breakthrough, days

t_D = dimensionless time

t_f = fill-up time, days

t^n = time corresponding to timestep n, days

u_{fx} = rock velocity in x direction, ft/D [m/d]

u_{fy} = rock velocity in y direction, ft/D [m/d]

u_{fz} = rock velocity in z direction, ft/D [m/d]

u_{ox} = oil velocity in the x direction, ft/D [m/d]

u_{oy} = oil velocity in the y direction, ft/D [m/d]

u_{oz} = oil velocity in the z direction, ft/D [m/d]

u_{Swf} = velocity of flood front saturation, ft/D [m/d]

u_t = total Darcy velocity in linear displacement, ft/D [m/d]

u_{wx} = water velocity in the x direction, ft/D [m/d]

u_{wy} = water velocity in the y direction, ft/D [m/d]

u_{wz} = water velocity in the z direction, ft/D [m/d]

V_o = oil volume, bbl [m^3]

V_p = pore volume

V_{pe} = effective PV when there is trapped gas, bbl [m^3]

V_w = water volume, bbl [m^3]

W_i = cumulative water injected, bbl [m^3]

W_{if} = cumulative water injected at fill-up, bbl [m^3]

W_{in} = cumulative water injected at timestep n, bbl [m^3]

x = position in x coordinate system, ft [m]

x_D = dimensionless distance in the x direction

x_{ob} = location of oil bank, ft [m]

x_{Sw} = location of water saturation on the x axis, ft [m]

x_{Swf} = location of the flood front saturation on the x axis, ft [m]

x_{Swfe} = location of the effective flood front saturation on the x axis, ft [m]

x_1 = position of Point 1 on x axis, ft [m]

x_2 = position of Point 2 on x axis, ft [m]

y = position on the y coordinate system, ft [m]

z = position on the z coordinate system, ft [m]

Z = elevation with respect to the horizontal

Z_d = elevation of datum, ft [m]

λ_d = mobility of displaced phase, darcy/cp [darcy/Pa·s]

λ_D = mobility of displacing phase, darcy/cp [darcy/Pa·s]

λ_i = mobility of injected phase, darcy/cp [darcy/Pa·s]

λ_o = mobility of oil phase, darcy/cp [darcy/Pa·s]

λ_r = total relative mobility at S_w, cp^{-1} [Pa·s^{-1}]

λ_r^{-1} = reciprocal total relative mobility or apparent viscosity, cp [Pa·s]

λ_{ro}^{-1} = apparent viscosity of oil phase, cp [Pa·s]

λ_w = mobility of water phase, darcy/cp [darcy/Pa·s]

λ^{-1+} = extrapolated apparent viscosity, Jones and Roszelle method, cp [Pa·s]

λ_2^{-1} = apparent viscosity when water saturation S_{w2} is at the end of the core, cp [Pa·s]

$\overline{\lambda^{-1}}$ = average apparent viscosity for a linear system, cp [Pa·s]

$\overline{\lambda_2^{-1}}$ = average apparent viscosity when $S_w = S_{w2}$ at the end of the core, cp [Pa·s]

$\overline{\lambda_{Swf}^{-1}}$ = average apparent viscosity when flood front saturation reaches end of core, cp [Pa·s]

μ_d = viscosity of displaced phase, cp [Pa·s]

μ_D = viscosity of displacing phase, cp [Pa·s]

μ_o = oil-phase viscosity, cp [Pa·s]

μ_w = water-phase viscosity, cp [Pa·s]

ρ_f = rock density, lbm/cu ft [kg/m^3]

ρ_o = oil density, lbm/cu ft [kg/m^3]

ρ_w = water density, lbm/cu ft [kg/m^3]

ϕ = porosity, fraction

Φ_o = oil-phase potential, sq ft/sec^2 [m^2/s^2]

References

1. Amyx, J.W., Bass, D.M. Jr., and Whiting, R.L.: *Petroleum Reservoir Engineering,* McGraw-Hill Book Co. Inc., New York City (1960) 57.
2. Bird, R.B., Stewart, W.E., and Lightfoot, E.N.: *Transport Phenomena,* John Wiley and Sons, New York City (1960) 68.
3. Richardson, J.G. *et al.*: "Laboratory Determination of Relative Permeability," *Trans.*, AIME (1952) **195**, 187–96.
4. Geffen, T.M. *et al.*: "Experimental Investigation of Factors Affecting Laboratory Relative Permeability Measurements," *Trans.*, AIME (1951) **192**, 99–110.
5. McEwen, C.R.: "A Numerical Solution of the Linear Displacement Equation With Capillary Pressure," *J. Pet. Tech.* (Aug. 1959) 45–48; *Trans.*, AIME (1959) **216**.
6. Buckley, S.E. and Leverett, M.C.: "Mechanism of Fluid Displacement in Sands," *Trans.*, AIME (1942) **146**, 107–16.
7. Terwilliger, P.L. *et al.*: "An Experimental and Theoretical Investigation of Gravity Drainage Performance," *Trans.*, AIME (1951) **192**, 285–96.
8. Welge, H.J.: "A Simplified Method for Computing Oil Recoveries by Gas or Water Drive," *Trans.*, AIME (1952) **195**, 91–98.
9. Craig, F.F. Jr.: *Reservoir Engineering Aspects of Waterflooding,* Monograph Series, SPE, Richardson, TX (1971) **3**, 108.
10. Jones, S.C. and Roszelle, W.O.: "Graphical Techniques for Determining Relative Permeability From Displacement Experiments," *J. Pet. Tech.* (May 1978) 807–17; *Trans.*, AIME, **265**.
11. Kyte, J.R. *et al.*: "Mechanism of Waterflooding in the Presence of Free Gas," *J. Pet. Tech.* (Sept. 1956) 215–21; *Trans.*, AIME, **207**.
12. Prats, M. *et al.*: "Prediction of Injection Rate and Production History for Multifluid Five-Spot Floods," *J. Pet. Tech.* (May 1959) 95–105; *Trans.*, AIME, **216**.
13. Douglas, J. Jr., Peaceman, D.W., and Rachford, H.H. Jr.: "A Method for Calculating Multidimensional Immiscible Displacement," *Trans.*, AIME (1959) **216**, 297–306.
14. Loomis, A.G. and Crowell, D.C.: "Relative Permeability Studies: Gas-Oil and Water-Oil Systems," *Bull.*, USBM **559** (1962) 27.
15. Johnson, E.F., Bossler, D.P., and Naumann, V.O.: "Calculation of Relative Permeability from Displacement Experiments," *J. Pet. Tech.* (Jan. 1959) 61–63; *Trans.*, AIME, **216**.
16. Parsons, R.W. and Jones, S.C.: "Linear Scaling in Slug-Type Processes—Application to Micellar Flooding," *Soc. Pet. Eng. J.* (Feb. 1977) 11–26.
17. Owens, W.W. and Archer, D.L.: "The Effect of Rock Wettability on Oil/Water Relative Permeability Relationships," *J. Pet. Tech.* (July 1971) 873–78; *Trans.*, AIME, **251**.
18. Salathiel, R.A.: "Oil Recovery by Surface Film Drainage in Mixed-Wettability Rocks," *J. Pet. Tech.* (Oct. 1973) 1216–24; *Trans.*, AIME, **255**.
19. Mungan, N.: "Interfacial Effect in Immiscible Liquid-Liquid Displacement in Porous Media," *Soc. Pet. Eng. J.* (Sept. 1966) 247–53; *Trans.*, AIME, **237**.
20. Richardson, J.G.: "The Calculation of Waterflood Recovery From Steady-State Relative Permeability Data," *J. Pet. Tech.* (May 1957) 64–66; *Trans.*, AIME, **210**.
21. Craig, F.F. Jr., Geffen, T.M., and Morse, R.A.: "Oil Recovery Performance of Pattern Gas or Water Injection Operations From Model Tests," *J. Pet. Tech.* (Jan. 1955) 7–15; *Trans.*, AIME, **204**.
22. Rapoport, L.A. and Leas, W.J.: "Properties of Linear Waterfloods," *J. Pet. Tech.* (May 1953) 139–48; *Trans.*, AIME, **198**.
23. Perkins, F.M. Jr.: "An Investigation of the Role of Capillary Forces in Laboratory Waterfloods," *J. Pet. Tech.* (Nov. 1957) 49–51; *Trans.*, AIME, **210**.
24. Kyte, J.R. and Rapoport, L.A.: "Linear Waterflood Behavior and End Effects in Water-Wet Porous Media," *J. Pet. Tech.* (Oct. 1958) 47–50; *Trans.*, AIME, **213**.
25. Batycky, J.R. *et al.*: "Interpreting Relative Permeability and Wettability From Unsteady-State Displacement Measurements," *Soc. Pet. Eng. J.* (June 1981) 296–308.
26. Douglas, J. Jr., Blair, P.M., and Wagner, R.J.: "Calculation of Linear Waterflood Behavior Including the Effects of Capillary Pressure," *Trans.*, AIME (1958) **213**, 96–102.
27. Fayers, F.J. and Sheldon, J.W.: "The Effect of Capillary Pressure and Gravity on Two-Phase Fluid Flow in a Porous Medium," *Trans.*, AIME (1959) **216**, 147–55.
28. Hovanessian, S.A. and Fayers, F.J.: "Linear Waterflood With Gravity and Capillary Effects," *Soc. Pet. Eng. J.* (March 1961) 32–36; *Trans.*, AIME, **222**.
29. Mattax, C.C. and Kyte, J.R.: "Imbibition Oil Recovery From Fractured, Water-Drive Reservoir," *Soc. Pet. Eng. J.* (June 1962) 177–84; *Trans.*, AIME, **225**.
30. Blair, P.M.: "Calculation of Oil Displacement by Counter-Current Water Imbibition," *Soc. Pet. Eng. J.* (Sept. 1964) 195–202; *Trans.*, AIME, **231**.
31. "Enhanced Oil Recovery by Improved Waterflooding," D.M. Boghassian (ed.), *Bull.* DOE/ET/12065-26, Natl. Technical Information Service, Springfield, VA (Aug. 1980).
32. Burdine, N.T., Gournay, L.S., and Reichertz, P.P.: "Pore Size Distribution of Petroleum Reservoir Rocks," *Trans.*, AIME, **189**, 195–204.
33. "El Dorado Micellar-Polymer Demonstration Project," First Annual Report, G. Rosenwald (ed.), BERC/TPR-75/1, Natl. Technical Information Service, Springfield, VA (Oct. 1975).
34. Honarpour, M. *et al.*: "Empirical Equations for Estimating Two-Phase Relative Permeability in Consolidated Rock," *J. Pet. Tech.* (Dec. 1982) 2905–08.

SI Metric Conversion Factors

°API 141.5/(131.5 + °API)		= g/cm^3
atm × 1.013 250*	E+05	= Pa
bbl × 1.589 873	E−01	= m^3
cp × 1.0*	E−03	= Pa·s
ft × 3.048*	E−01	= m
°F (°F−32)/1.8		= °C
in. × 2.54*	E+00	= cm
psi × 6.894 757	E+00	= kPa
sq in. × 6.451 6*	E+00	= cm^2

*Conversion factor is exact.

Chapter 4
Immiscible Displacement in Two Dimensions—Areal

4.1 Introduction

Fundamental concepts of immiscible displacement processes for linear systems were presented in Chap. 3. Oil displacement in homogeneous reservoirs that have strong natural water drives may be approximated by linear models introduced in Chap. 3. Two-dimensional (2D) displacement is observed in linear systems if permeability varies with position in the reservoir, if gravity segregation of injected and displaced fluids occurs, and/or when capillary forces are large relative to viscous forces. Reservoirs with limited sources of energy require fluid injection through one or more injection wells to displace oil to production wells. If the reservoir is homogeneous and isotropic, the flood front may be vertical so that the displacement process occurs primarily in two dimensions across the area of the pattern or reservoir. Otherwise, the displacement will be three-dimensional (3D).

Chap. 4 introduces methods to estimate displacement performance in two dimensions. We will study areal (2D) displacement in reservoirs that are homogeneous and isotropic. Methods of estimating displacement efficiency for these systems are presented in order of increasing complexity.

4.2 Fluid Flow Equations in 2D— Areal Displacement

Two-dimensional flows are encountered in all displacement processes where fluids are injected into a reservoir. The movement of fluids is controlled by the arrangement of injection and production wells as well as reservoir heterogeneities. Reservoir heterogeneities, particularly in the vertical direction, and gravity segregation cause flows to vary in three dimensions. Discussion of 3D displacement will appear in Sec. 5.8. For the purposes of this section, we will consider reservoirs in the horizontal plane that are homogeneous and have uniform thickness.

Reservoir displacement processes are frequently conducted in patterns where a specific configuration of injection wells and production wells is repeated across the field. Fig. 4.1 illustrates several flooding patterns used commonly in waterflooding.[1] Other reservoirs may be flooded from injection wells located in the center of a field or from the edge of a field as depicted in Fig. 4.2.[2]

Prediction of displacement performance has been an evolutionary process. Earlier methods yielded quick estimates of displacement performance. More sophisticated mathematical models require some knowledge of computer skills. In this section, methods of estimating performance are presented in order of increasing complexity.

All models have one factor in common: they represent approximate solutions to the 2D fluid flow equations given by Eqs. 4.1 and 4.2.

$$\frac{\partial}{\partial x}\left(\frac{k_o \rho_o}{\mu_o}\frac{\partial p_o}{\partial x}\right) + \frac{\partial}{\partial y}\left(\frac{k_o \rho_o}{\mu_o}\frac{\partial p_o}{\partial y}\right)$$

$$= \frac{\partial}{\partial t}\left(\rho_o S_o \phi\right) \quad \dots \dots \dots \dots \dots \dots \dots (4.1)$$

and

$$\frac{\partial}{\partial x}\left(\frac{k_w \rho_w}{\mu_w}\frac{\partial p_w}{\partial x}\right) + \frac{\partial}{\partial y}\left(\frac{k_w p_w}{\mu_w}\frac{\partial p_w}{\partial y}\right)$$

$$= \frac{\partial}{\partial t}\left(\rho_w S_w \phi\right), \quad \dots \dots \dots \dots \dots \dots (4.2)$$

where

$$p_o - p_w = P_c(S_w), \quad \dots \dots \dots \dots \dots \dots \dots (4.3)$$

$$k_o = k_o(S_w), \quad \dots \dots \dots \dots \dots \dots \dots \dots (4.4)$$

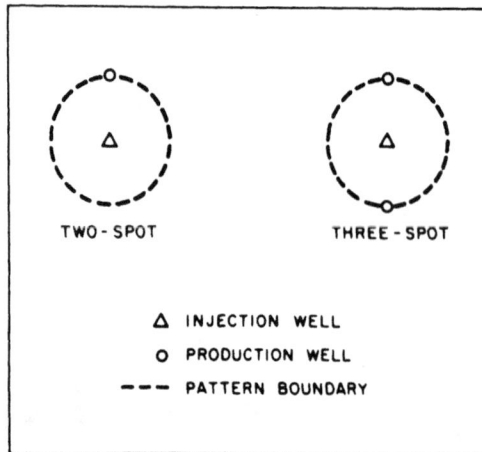

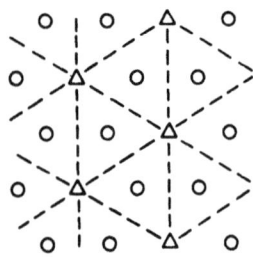
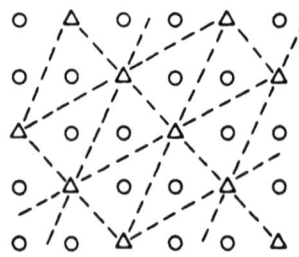
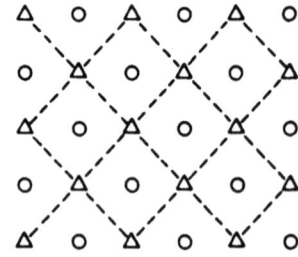

REGULAR FOUR-SPOT   SKEWED FOUR-SPOT   FIVE-SPOT

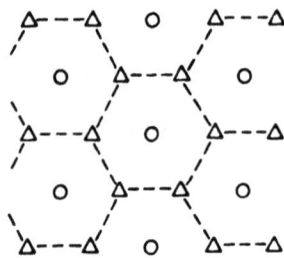
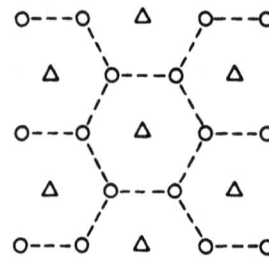

SEVEN-SPOT   INVERTED SEVEN-SPOT

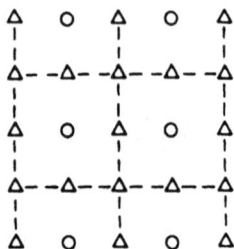
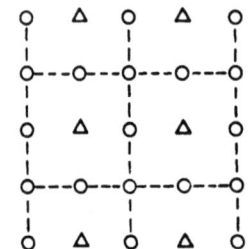
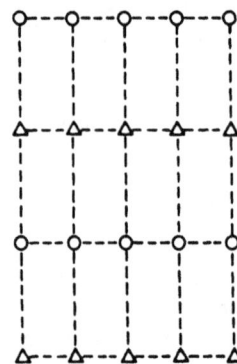
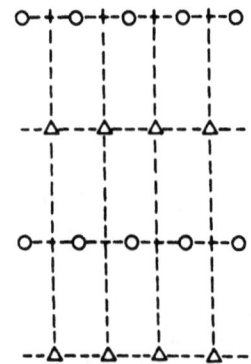

NORMAL NINE-SPOT   INVERTED NINE-SPOT   DIRECT LINE DRIVE   STAGGERED LINE DRIVE

Fig. 4.1—Flooding patterns.[1]

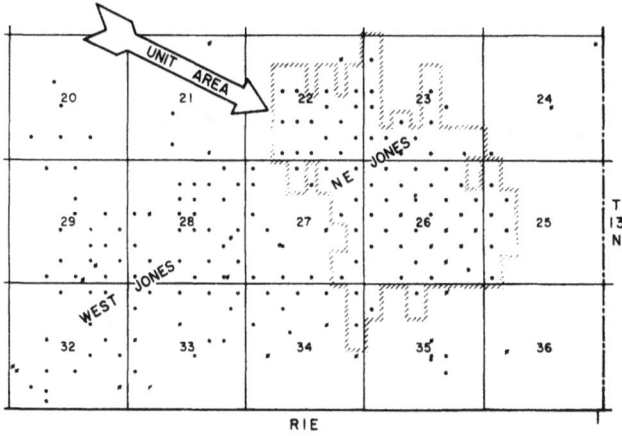

Fig. 4.2—Well arrangement with center and edge injection wells in northeast Jones Cleveland sand unit.[2]

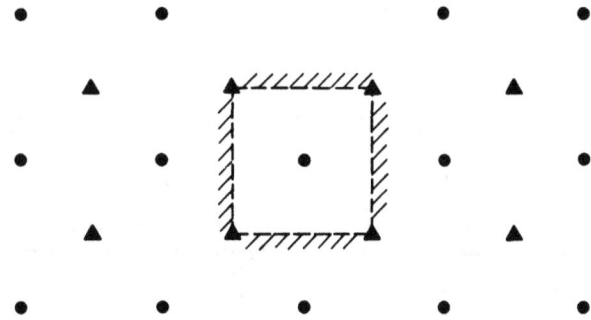

Fig. 4.3—Five-spot pattern in a fully developed field with no-flow boundaries.

and

$$k_w = k_w(S_w). \quad \dots\dots\dots\dots\dots\dots\dots\dots\dots (4.5)$$

In Sections 4.3 through 4.5, we will examine solutions to Eqs. 4.1 and 4.2 for the geometry of the five-spot pattern.

4.3 Displacement in a Five-Spot Pattern

A regular five-spot pattern in Fig. 4.1 consists of a production well surrounded by four injection wells. An inverted pattern has an injection well surrounded by four production wells. When either pattern is repeated in a large reservoir, the ratio of injection wells to production wells approaches 1.0. Consequently, models of displacement performances refer to an ideal five-spot pattern, as in Fig. 4.3, where there are four injection wells and one production well. It is assumed that the injection rates are equal to the production rates. Thus flow is symmetric around each injection well with 0.25 of the injection rate from each well confined to the pattern. The dashed lines connecting injection wells in Fig. 4.3 represent the boundaries of the pattern area. In a homogeneous reservoir where injection and production rates are equal, the dashed boundaries also represent no-flow boundaries. Therefore, analysis of a five-spot pattern in a reservoir can be simplified by examining the behavior of a single five-spot pattern.

Further simplifications are possible. Each quadrant is symmetric, as shown in Fig. 4.4. The cross-hatched areas are no-flow boundaries for the quadrant, and a line drawn from injection well to production well subdivides a quadrant into two symmetrical parts. This line also is a no-flow boundary. In some cases, the displacement performance of one-eighth of a five-spot pattern is used to estimate the behavior of the full pattern.

Typical displacement performances of five-spot floods are shown in Fig. 4.5 for four different oil/water viscosity ratios. These data were obtained in laboratory sandpacks simulating a quadrant of a five-spot pattern.[3]

4.4 Correlations Developed From Scaled Laboratory Models—The CGM Model

Correlations of displacement performance have been developed from experiments with scaled laboratory models.

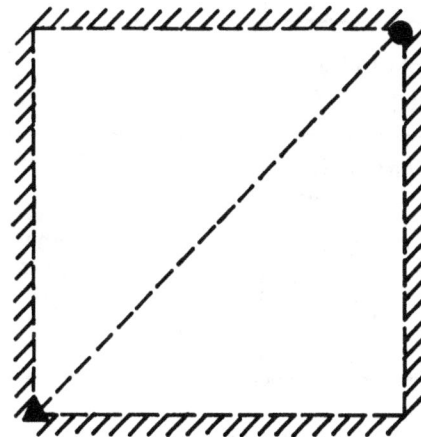

Fig. 4.4—Quadrant of a five-spot in a homogeneous reservoir indicating the plane of symmetry.

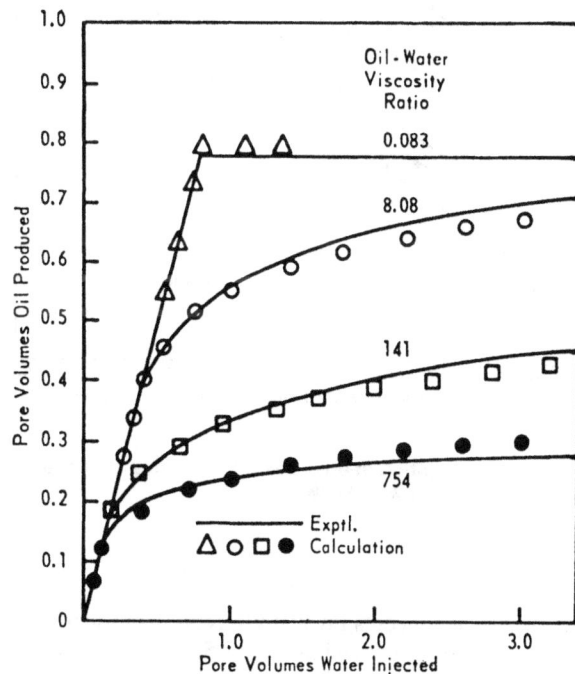

Fig. 4.5—Oil recovery from waterflooding five-spot models.[3]

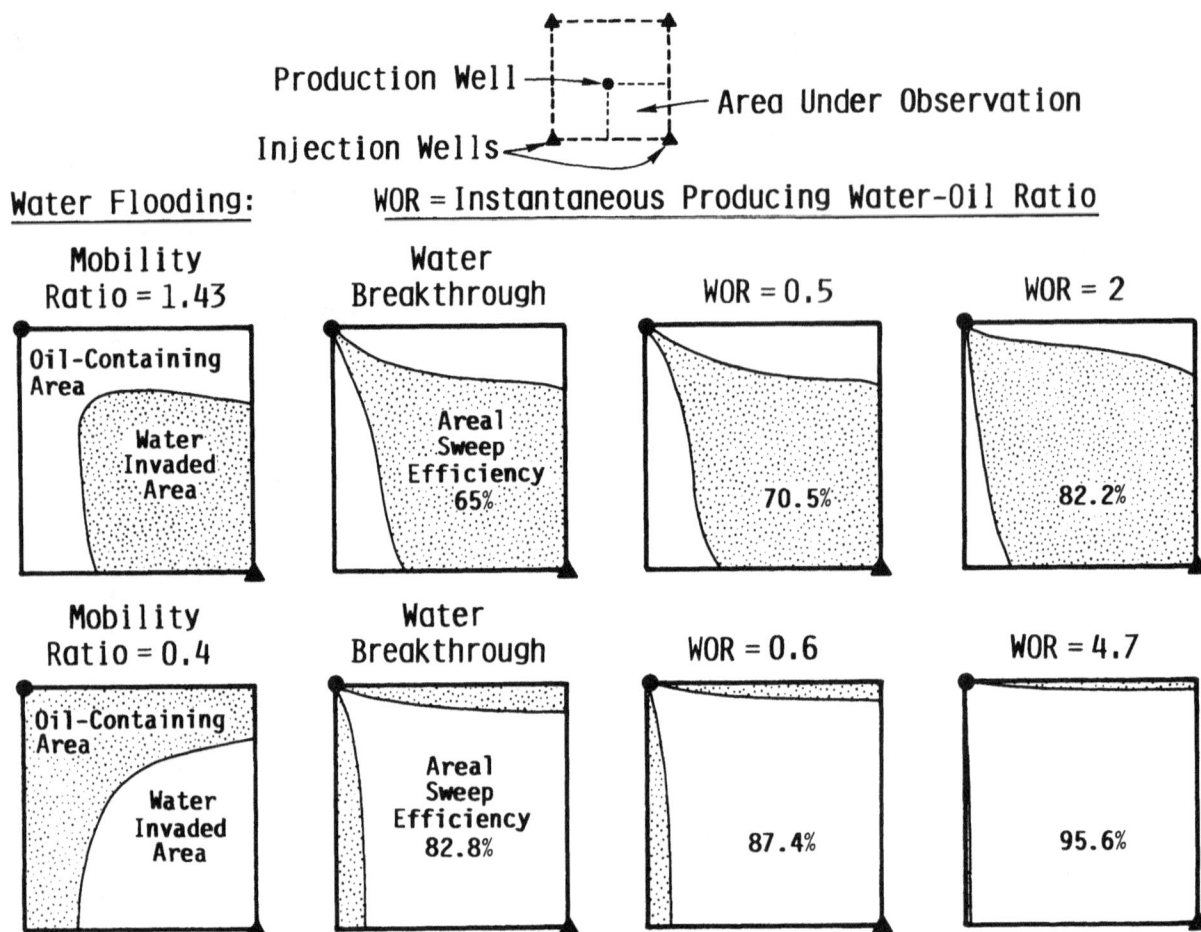

Fig. 4.6—X-ray shadowgraphs of flood progress in scaled five-spot patterns.[11]

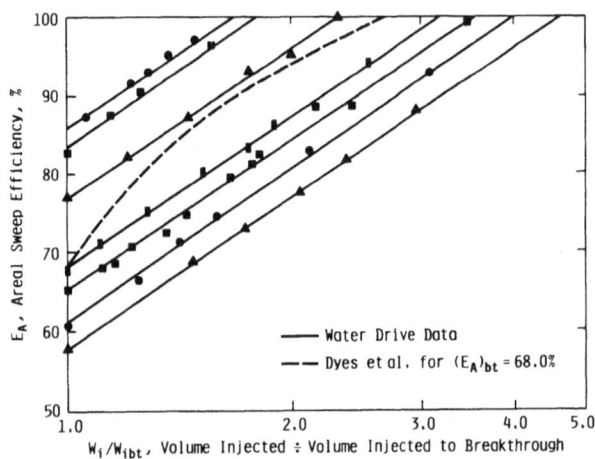

Fig. 4.7—Correlation of areal sweep efficiency after break-through with W_i/W_{ibt}.

Fundamental concepts involved in conducting scaled experiments are introduced in Sec. 5.5. Nearly all laboratory experiments used miscible fluids.[4-10] Application to immiscible displacement processes is based on the premise that displacement performance is determined primarily by pattern geometry and viscous forces. Capillary forces are not considered to be important in the gross movements of fluids in a pattern.[4] Correlations based on miscible displacements approximate results obtained from immiscible displacement if recovery in the displaced region is complete when the saturation front arrives. Otherwise, a saturation gradient exists in the swept region with continued displacement of oil.[11] The correlations derived from miscible displacement data are generally applicable to miscible displacement and are useful only for immiscible displacement processes for a narrow range of conditions. Therefore, these correlations are considered in Chap. 6 with applications limited to problems where the mobility ratio is small or when an estimate of displacement performance is required in a relatively short period of time.

Craig-Geffen-Morse (CGM) Correlation. Craig *et al.*[11] obtained experimental data in horizontal laboratory models representing a quadrant of a five-spot. Immiscible fluids were used with gravity and capillary forces scaled. Areal sweep efficiencies were determined from X-ray shadowgraphs taken during various stages of the displacement. Fig. 4.6 illustrates shadowgraphs of two model floods for two mobility ratios.

Experimental data for a variety of oil and aqueous systems were correlated empirically. Fig. 4.7 shows the correlation of areal sweep efficiencies at and after break-

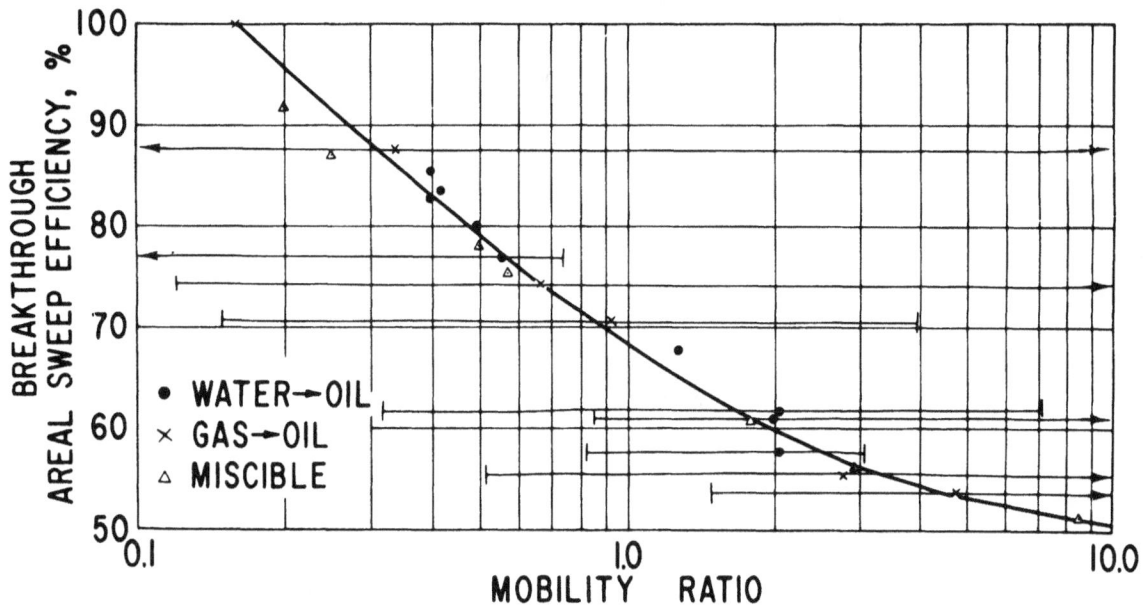

Fig. 4.8—Correlation of areal sweep efficiency at breakthrough with mobility ratio for miscible and immiscible displacement in five-spot pattern floods (Ref. 1, Fig. 4.1).

through with the ratio of volume injected to volume injected at breakthrough. The data in Fig. 4.7 are also represented by Eq. 4.6 or 4.7.

$$E_A = E_{Abt} + 0.633 \log \frac{W_i}{W_{ibt}} \quad \ldots \ldots \ldots \ldots (4.6)$$

or

$$E_A = E_{Abt} + 0.274 \ln \frac{W_i}{W_{ibt}}, \quad \ldots \ldots \ldots \ldots (4.7)$$

where E_{Abt} is the areal sweep efficiency at breakthrough of the displacing fluid, and E_A is the fraction of the area that has been swept to an average water saturation of $\overline{S_{wf}}$.

Areal sweep efficiencies at breakthrough from model results were correlated with mobility ratio ($M_{\bar{S}}$) or M_t as in Fig. 4.8. The mobility ratio was chosen arbitrarily so that areal sweep efficiencies from immiscible displacement tests were in agreement with the correlation for miscible displacement developed by Dyes *et al.*[5] at breakthrough. It was found that the mobility ratios could be computed from relative permeability data when the water permeability was evaluated at the average water saturation behind the front as determined from the frontal advance solution. That is, when the mobility ratio for the waterflood was $M_{\bar{S}}$.

The correlation of E_{Abt} with $M_{\bar{S}}$ can also be represented by Eq. 4.8.[*]

$$E_{Abt} = 0.54602036 + \frac{0.03170817}{M_{\bar{S}}} + \frac{0.30222997}{e^{M_{\bar{S}}}}$$

$$- 0.00509693 \ M_{\bar{S}}. \quad \ldots \ldots \ldots \ldots (4.8)$$

[*]Negahban, Shahin: personal communication, U. of Kansas, Lawrence, KS (1982).

Eq. 4.8, which is valid for $0.16 \leq M_{\bar{S}} \leq 10$, was obtained by fitting Fig. E.9 of Ref. 1.

Prediction of Displacement Performance. Estimation of waterflood performance with the CGM method correlation can be done in a few hours at a desk with a pocket calculator and graph paper. The model is adapted easily to computer solution. (A detailed example of the procedure is described in Appendix E of Ref. 1.) The development of the model to emphasize fundamental properties of immiscible displacement processes in five-spot patterns will be reviewed.

The development in the following section considers the displacement of oil by water in a five-spot pattern with no initial gas saturation. Areal sweep efficiency and oil recovery efficiency at breakthrough are readily determined from the correlations in Eq. 4.7 and Fig. 4.9. At breakthrough, oil recovery is given by Eq. 4.9 when $B_o = 1.0$.

$$N_{pbt} = E_{Abt} \left(\overline{S_{wf}} - S_{iw} \right) V_p, \quad \ldots \ldots \ldots \ldots (4.9)$$

where $\overline{S_{wf}}$ is the average displacing-phase saturation at breakthrough in a linear flood as computed from frontal advance solution.

Production after breakthrough is estimated from Eq. 4.10.

$$N_p = E_A (\overline{S_{w5}} - S_{iw}) V_p, \quad \ldots \ldots \ldots \ldots (4.10)$$

where $\overline{S_{w5}}$ is the average water saturation in a region swept by the injected fluid. The key to this model is the assumption made to evaluate $\overline{S_{w5}}$. A new variable, Q_i^*, is defined to represent the number of water-contacted PV's in the five-spot pattern—that is, Q_i^* equals the volume of water injected divided by the volume of the five-spot contacted by the injected water.

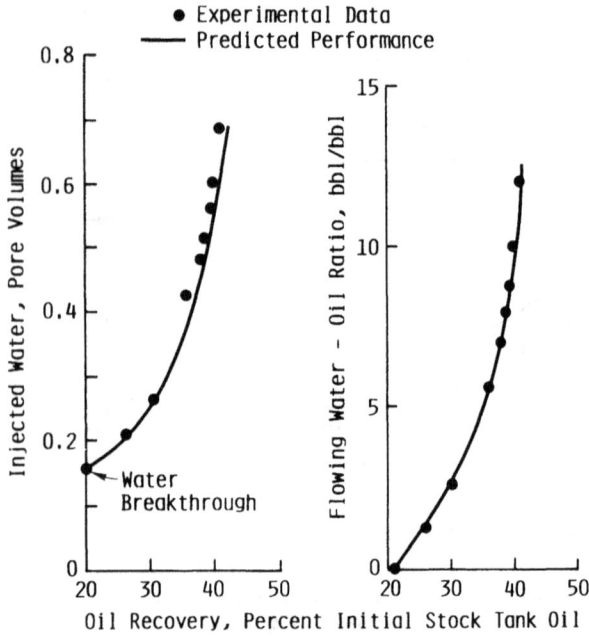

Fig. 4.9—Performance of five-spot waterflood predicted with the CGM model. [11]

For example, at breakthrough

$$Q^*_{ibt} = \frac{W_{ibt}}{E_{Abt} V_p} = \overline{S_{wf}} - S_{iw}.$$

At breakthrough

$$f'_{Sw5} = \frac{1}{Q^*_i}.$$

Thus when $Q_{i2} = Q^*_i$,

$$f'_{Sw2} = f'_{Sw5} \quad \dots\dots\dots\dots\dots\dots\dots\dots (4.11)$$

and

$$\overline{S_{w5}} = S_{w2} + f_{o2} Q^*_i. \quad \dots\dots\dots\dots\dots\dots (4.12)$$

Values of f_{o2} and S_{w2} are obtained from the frontal advance solution at

$$f'_{Sw2} = \frac{1}{Q^*_i}. \quad \dots\dots\dots\dots\dots\dots\dots\dots (4.13)$$

Evaluation of Q^*_i is outlined in the next section.

Evaluation of Q^*_i. At breakthrough, the volume of the five-spot contacted by the injected water is $E_{Abt} V_p$. Because

$$W_{ibt} = E_{Abt} V_p (\overline{S_{wf}} - S_{iw}),$$

$$Q^*_{ibt} = \frac{W_{ibt}}{E_{Abt} V_p}, \quad \dots\dots\dots\dots\dots\dots\dots (4.14)$$

$$Q^*_{ibt} = \overline{S_{wf}} - S_{iw}. \quad \dots\dots\dots\dots\dots\dots (4.15)$$

It is useful to illustrate the difference between Q^*_i and Q_i. For the five-spot pattern,

$$Q_i = \frac{W_i}{V_p}.$$

Thus

$$Q_{ibt} = E_{Abt} (\overline{S_{wf}} - S_{iw}).$$

After breakthrough, the change in the number of water-contacted PV's (dQ^*_i) with injected fluid (dW_i) is defined by Eq. 4.16:

$$dQ^*_i = \frac{dW_i}{E_A V_p}. \quad \dots\dots\dots\dots\dots\dots\dots\dots (4.16)$$

When Eq. 4.16 is integrated from the breakthrough ($W_i/W_{ibt} = 1.0$) to W_i/W_{ibt}, Eq. 4.17 is obtained.

$$\frac{Q^*_i}{Q^*_{ibt}} = 1 + E_{Abt} \int_1^{W_i/W_{ibt}} \frac{d\left(\dfrac{W_i}{W_{ibt}}\right)}{E_A}. \quad \dots\dots (4.17)$$

Eq. 4.17 applies until $E_A = 1.0$. The value of W_i required to sweep the pattern completely, termed W_{i100}, is computed from Eq. 4.18.

$$W_{i100} = W_{ibt} e^{(1 - E_{Abt})/0.274}. \quad \dots\dots\dots\dots (4.18)$$

Let Q^*_{i100} represent the value of Q^*_i at $E_A = 1.0$. Then Q^*_i for $W_i > W_{i100}$ is given by Eq. 4.19.

$$Q^*_i = Q^*_{i100} + \frac{W_i - W_{i100}}{V_p}. \quad \dots\dots\dots\dots (4.19)$$

Numerical values of Q^*_i/Q^*_{ibt} may be obtained by substituting Eq. 4.7 for E_A into Eq. 4.17 and performing the integration. The resulting expression is Eq. 4.20.

$$\frac{Q^*_i}{Q^*_{ibt}} = 1 + a_1 e^{-a_1} [Ei(a_2) - Ei(a_1)], \quad \dots\dots (4.20)$$

where

$$a_1 = 3.65 \, E_{Abt},$$
$$a_2 = a_1 + \ln(W_i/W_{ibt}), \text{ and}$$
$$W_{ibt} \le W_i \le W_{i100},$$

and $Ei(x)$ is the Ei function. Eq. 4.21 is an infinite-series representation for $Ei(x)$ that is accurate to 10^{-8}. [12]

$$Ei(x) = 0.57721557 + \ln x + \sum_{n=1}^{\infty} x^n/(nn!). \quad \dots (4.21)$$

Eq. 4.21 is evaluated easily on many pocket calculators or by a short computer program. Thus it is possible to compute Q_i^*/Q_{ibt}^* for any E_{Abt} and W_i for which the correlation is valid. A program named CGM is included in Appendix A. The program computes the values of Q_i^*/Q_{ibt}^* at a specific value of W_i when W_{ibt} and E_{Abt} are given.

Craig[1] gives Q_i^*/Q_{ibt}^* for a wide range of W_i/W_{ibt} with E_{Abt} ranging from 50 to 99% in Table E.9 of his monograph.

Estimation of WOR. The variation of WOR after breakthrough is estimated by separating the displaced area into two distinct regions. A newly swept region is defined as the region just swept by the injected fluid. The previously swept region contains all reservoir volume where the water saturation is greater than S_{wf}. Oil is displaced in the region behind the front where $S_w > S_{wf}$. Performance in this region is estimated by assuming that all produced water comes from the previously swept zone while oil is produced from both newly swept and previously swept regions. At the end of the previously swept region where the production well is located, the fractional flow of oil is obtained from fractional flow data with Eq. 4.11. The WOR for $E_A > 1.0$ is f_{w2}/f_{o2}. As E_A increases from E_{Abt} to 1.0, the amount of oil produced by the expanding newly swept region must be included to determine the WOR at the producing well.

An expression for the WOR in the producing well ($E < 1.0$) may be derived from the volumes of oil and water produced when an incremental volume of water is injected. Let $dW_i =$ incremental volume of injected water. Then

$$dW_i = dW_p + dN_{ps} + dN_{pu}, \quad \dots\dots\dots\dots (4.22)$$

where

dW_p = incremental volume of water produced,
dN_{ps} = incremental oil produced from previously swept zone, and
dN_{pu} = incremental oil produced from newly swept zone.

According to the model assumptions,

$$f_{o2} = \frac{dN_{ps}}{dW_p + dN_{ps}}. \quad \dots\dots\dots\dots (4.23)$$

Thus

$$dW_p = (1.0 - f_{o2})(dW_i - dN_{pu}) \quad \dots\dots\dots (4.24)$$

and

$$dN_p = f_{o2}(dW_i - dN_{pu}) + dN_{pu}. \quad \dots\dots\dots (4.25)$$

Because

$$F_{wo} = dW_p/dN_p, \quad \dots\dots\dots\dots (4.26)$$

then

$$F_{wo} = \frac{(1.0 - f_{o2})(1.0 - dN_{pu}/dW_i)}{f_{o2}(1.0 - dN_{pu}/dW_i) + dN_{pu}/dW_i}. \quad \dots (4.27)$$

The newly swept region is considered to be at the breakthrough saturation, S_{wf}, computed from the frontal advance solution. Thus the amount of oil displaced from an incremental increase in swept area (dE_A) is

$$dN_{pu} = (S_{wf} - S_{iw})V_p dE_A.$$

At breakthrough

$$W_{ibt} = (\overline{S_{wf}} - S_{iw})E_{Abt}V_p,$$

so that

$$dN_{pu} = dE_A \left(\frac{W_{ibt}}{E_{Abt}}\right) \left(\frac{S_{wf} - S_{iw}}{\overline{S_{wf}} - S_{iw}}\right). \quad \dots\dots\dots (4.28)$$

The volume of oil displaced per unit of fluid injected (or produced) is given by Eq. 4.29.

$$\frac{dN_{pu}}{dW_i} = \left(\frac{dE_A}{dW_i}\right) \left(\frac{W_{ibt}}{E_{Abt}}\right) \left(\frac{S_{wf} - S_{iw}}{\overline{S_{wf}} - S_{iw}}\right). \quad \dots (4.29)$$

Examination of Eq. 4.29 shows that dE_A/dW_i is the derivative of Eq. 4.7. From Eq. 4.7,

$$\frac{dE_A}{dW_i} = \frac{0.274}{W_i}.$$

Therefore, in the newly swept area, the incremental volume of oil displacement per unit of fluid injection is

$$\frac{dN_{pu}}{dW_i} = \frac{0.274 W_{ibt}}{W_i} \left(\frac{1}{E_{Abt}}\right) \left(\frac{S_{wf} - S_{iw}}{\overline{S_{wf}} - S_{iw}}\right).$$

$$\dots\dots\dots\dots\dots\dots\dots\dots\dots\dots\dots (4.30)$$

The volume of oil produced from the expanding front can be computed directly from Eq. 4.30 by integrating from breakthrough to W_i. Eq. 4.31 gives the volume of oil that is displaced by the expanding flood front when the water saturation is changed from S_{iw} to S_{wf} as the front arrives.

$$\int_{N_{pbt}}^{N_{pu}} dN_{pu} = 0.274 \frac{W_{ibt}}{E_{Abt}} \left(\frac{S_{wf} - S_{iw}}{\overline{S_{wf}} - S_{iw}}\right) \int_{W_{ibt}}^{W_i} \frac{dW_i}{W_i}$$

and

$$N_{pu} = N_{pbt} + 0.274 \frac{W_{ibt}}{E_{Abt}} \left(\frac{S_{wf} - S_{iw}}{\overline{S_{wf}} - S_{iw}}\right) \ln\left(\frac{W_i}{W_{ibt}}\right).$$

$$\dots\dots\dots\dots\dots\dots\dots\dots (4.31)$$

Remember, when $E_A = 1.0$, $W_i = W_{i100}$ and no further oil can be displaced by the expanding flood front.

Fig. 4.9 compares experimental data and the computed performance for a five-spot waterflood on a scaled laboratory model.[11]

Example 4.1 illustrates the application of the CGM model for a waterflood where the initial gas saturation is zero. An example of the CGM model, including the effects of an initial gas saturation, is presented in Appendix E of Ref. 1.

Example 4.1. A reservoir is to be waterflooded with a five-spot pattern. Properties of the rock and fluids are the same as in Examples 3.4 and 3.5. For the purpose of this example we will consider the pattern area to be 10 acres [40 469 m^2]. FVF's for oil and water are assumed to be 1.0.

The calculation procedure can be subdivided into a series of computational steps.

1. Determine the mobility ratio and E_{Abt}. The average saturation behind the front at breakthrough, $\overline{S}_{wf}$, was determined in Example 3.4 as 0.70. Thus $\overline{S}_{wD}$ can be computed from Eq. 3.111:

$$\overline{S}_{wD} = \frac{\overline{S}_{wf} - 0.363}{0.432}$$

$$= 0.78.$$

From Eq. 3.110,

$$k_{rw} = 0.78 S_{wD}^{3.72}$$

$$= 0.31.$$

Thus

$$M_{\overline{s}} = \frac{(0.31)/(1.0)}{(1.0)/(2.0)}$$

$$= 0.62.$$

The value of E_{Abt} is about 0.76 at $M_{\overline{s}} = 0.62$ on Fig. 4.8.

2. Compute V_p, W_{ibt}, W_{i100}, and Q^*_{ibt}. Pore volume (V_p):

$$V_P = (10 \text{ acres})(43,560 \text{ sq ft/acre})(20.0 \text{ ft})$$

$$\times (1.0 \text{ bbl}/5.615 \text{ cu ft})(0.15)$$

$$= 232,734 \text{ bbl}.$$

Water injected at breakthrough (W_{ibt}):

$$W_{ibt} = E_A(\overline{S}_{wf} - S_{iw})V_p$$

$$= 0.76(0.699 - 0.363)(232,734)$$

$$= 59,431 \text{ bbl}.$$

Water-contacted PV's at breakthrough (Q^*_{ibt}):

$$Q^*_{ibt} = \overline{S}_{wf} - S_{iw}$$

$$= 0.699 - 0.36$$

$$= 0.336.$$

3. Estimate the maximum value of W_i when the WOR reaches some desired level. $(W_i)_{max}$ can be estimated from the frontal advance solution by assuming that $Q_{i5} = Q_{i2}$ at some WOR—e.g., 200. Then $f_{w2} = 0.995$, $S_{w2} = 0.74$, and $f'_{Sw2} = 0.306$ from Table 3.5. Thus

$$(Q_i)_{max} = 1.0/0.306$$

$$= 3.27,$$

and $(W_i)_{max}$ is about 761,039 bbl [120 996 m^3], which we will round off to 760,000 bbl [120 830 m^3].

4. The remaining computations were performed with a short computer program. The sequence of operations is (1) the interval from W_{ibt} to W_{i100} was subdivided into nine increments so that 10 values of W_i would be available from W_{ibt} to W_{i100}; (2) the interval from W_{i100} to $(W_i)_{max}$ was subdivided also into 19 increments to provide more detail after 100% of the area was swept; the number of increments in (1) and (2) is arbitrary; (3) values of Q_i^* were computed for $W_{ibt} < W_i < W_{i100}$ using Eq. 4.20; (4) values of Q_i^* were computed using Eq. 4.19 for $W_{i100} < W_i < W_{imax}$; and (5) the areal sweep efficiency was evaluated for $W_{ibt} < W_i < W_{i100}$.

Computed results are shown in Cols. 1 through 4 of Table 4.1. The difference between W_{ibt} in Table 4.1 and the value computed in this example is caused by rounding off.

5. Displacement performance was computed from the values of Q_i^* determined for each W_i: (1) because $f'_{Sw} = 1.0/Q_i^*$, a value of S_{w2} was obtained by linear interpolation from Table 3.5 for each Q_i^*; (2) values of f_{w2}, f_{o2}, $\overline{S}_{w2}$, and N_p are computed for each W_i; (3) the WOR is computed with Eq. 4.27 until $E_A = 1.0$; thereafter, $F_{wo} = f_{w2}/f_{o2}$; and (4) PV's of fluid injected (Q_i) were computed for each W_i as W_i/V_p. Results of these computations are summarized in Cols. 4 through 11 of Table 4.1.

Estimating Injection Rates. The CGM model estimates waterflood performance in a five-spot in terms of cumulative volume of water injected. When the injection rate is constant and known, oil displacement rates can be determined as a function of time. Otherwise, the CGM method does not provide information on the timing of production or the rates at which fluids can be injected or produced.

Many waterfloods operate under conditions where the pressure drop is known. When the mobility ratio is unity, the injection rate for a five-spot pattern can be calculated from Eq. 4.32:

$$i = \frac{3.541 \, kh\Delta p}{\mu \left(\ln \dfrac{d}{r_w} - 0.619 \right)}, \quad \dots \dots \dots \dots (4.32)$$

where

d = distance from the injection well to the production well, ft [m], and

r_w = effective wellbore radius, ft.

Units of Eq. 4.32 are barrels per day, darcies, feet, pounds per square inch, and centipoise. The mobility ratio is usually not 1.0 in a waterflood. Consider the pressure drop between injection well and production well to be constant. When the injected fluid is less mobile than the displaced fluids, the injection rate will decline as the displacement

TABLE 4.1—DISPLACEMENT PERFORMANCE OF A FIVE-SPOT PATTERN, CGM MODEL

W_i (bbl)	$\dfrac{W_i}{W_{ibt}}$	E_A (fraction)	$Q_i^{\bullet}$	S_{w2} (fraction)	f_{o2}	$\overline{S_{w5}}$ (fraction)	N_p (PV)	Q_i (PV)	N_p (bbl)	F_{wo} (bbl/bbl)
59,409	1.00	0.76	0.336	0.664	0.104	0.699	0.255	0.255	59,409	1.5
68,658	1.16	0.80	0.387	0.670	0.087	0.704	0.272	0.295	63,413	1.9
77,906	1.31	0.83	0.435	0.675	0.075	0.708	0.288	0.335	66,917	2.3
87,155	1.47	0.87	0.482	0.679	0.065	0.711	0.301	0.374	70,038	2.7
96,403	1.62	0.89	0.527	0.683	0.058	0.714	0.313	0.414	72,858	3.1
105,652	1.78	0.92	0.571	0.686	0.053	0.716	0.324	0.454	75,429	3.4
114,900	1.93	0.94	0.614	0.689	0.048	0.718	0.334	0.494	77,791	3.8
124,149	2.09	0.96	0.656	0.691	0.044	0.720	0.344	0.533	79,977	4.2
133,397	2.25	0.98	0.697	0.693	0.041	0.722	0.352	0.573	82,011	4.6
142,645	2.40	1.00	0.737	0.695	0.038	0.724	0.361	0.613	83,912	25.1
175,138	2.95	1.00	0.876	0.701	0.031	0.728	0.365	0.753	85,029	31.6
207,630	3.49	1.00	1.016	0.706	0.025	0.732	0.369	0.892	85,935	38.3
240,123	4.04	1.00	1.156	0.711	0.022	0.736	0.373	1.032	86,695	45.4
272,615	4.59	1.00	1.295	0.714	0.019	0.738	0.375	1.171	87,345	52.7
305,107	5.14	1.00	1.435	0.717	0.016	0.741	0.378	1.311	87,914	60.3
337,600	5.68	1.00	1.574	0.720	0.014	0.743	0.380	1.451	88,413	68.2
370,092	6.23	1.00	1.714	0.723	0.013	0.745	0.382	1.590	88,857	76.3
402,584	6.78	1.00	1.854	0.725	0.012	0.747	0.384	1.730	89,254	84.7
435,077	7.32	1.00	1.993	0.727	0.011	0.748	0.385	1.869	89,611	93.3
467,569	7.87	1.00	2.133	0.729	0.010	0.749	0.386	2.009	89,945	102.1
500,061	8.42	1.00	2.273	0.731	0.009	0.751	0.388	2.149	90,249	111.1
532,554	8.96	1.00	2.412	0.732	0.008	0.752	0.389	2.288	90,521	120.4
565,046	9.51	1.00	2.552	0.734	0.008	0.753	0.390	2.428	90,785	129.9
597,538	10.06	1.00	2.691	0.735	0.007	0.754	0.391	2.567	91,020	139.6
630,031	10.60	1.00	2.831	0.736	0.007	0.755	0.392	2.707	91,248	149.5
662,523	11.15	1.00	2.971	0.737	0.006	0.756	0.393	2.847	91,451	159.6
695,015	11.70	1.00	3.110	0.739	0.006	0.757	0.394	2.986	91,654	169.8
727,508	12.25	1.00	3.250	0.740	0.006	0.758	0.395	3.126	91,833	180.4
760,000	12.79	1.00	3.389	0.741	0.005	0.758	0.395	3.266	92,011	191.0

continues. Similarly, injection rates will increase with the fraction of the pattern displaced when the injected fluid is more mobile than the displaced fluid.

Caudle and Witte[6] used these concepts to develop a correlation between injection rate, mobility ratio, and areal sweep efficiency from data obtained for miscible displacement in a scaled five-spot pattern. They correlated their data in terms of the ratio of the injection rate of the displacement to the injection rate at unit mobility ratio. This term, defined by Eq. 4.33, is called the conductance ratio.

$$\gamma = \frac{i}{i_b}. \qquad \qquad (4.33)$$

Fig. 4.10 shows the correlation between the conductance ratio, area swept (E_A), and mobility ratio (M).[6]

The conductance, defined by Eq. 4.33, is the ratio of the injection rate during the miscible displacement to the injection rate computed from Eq. 4.32 assuming the displaced phase is flowing in the pattern as a single phase under the same pressure drop. For a waterflood, i_b, is the injection rate of oil at S_{iw}. Craig[1] used Fig. 4.10 to estimate water injection rates by evaluating the mobility ratio at $M_{\bar{S}}$ before breakthrough. After breakthrough the mobility ratio of the five-spot was computed from Eq. 4.34:

$$M_{\overline{Sw5}} = \frac{\left(\dfrac{k_{rw}}{\mu_w}\right)_{\overline{S_{w5}}}}{\left(\dfrac{k_{ro}}{\mu_o}\right)_{S_{iw}}}, \qquad \qquad (4.34)$$

where $\overline{S_{w5}}$ is the average water saturation in the swept region computed from Eq. 4.12.

Example 4.2 shows how the injection rate can be estimated for the five-spot pattern in Example 4.1. The time corresponding to this injection rate is determined by the method described in Sec. 3.6 for the performance of a linear waterflood at constant pressure drop. Injection rates may also be estimated by using approximation models developed in Chap. 6.

Example 4.2. Estimate the injection rate for the waterflood in Example 4.1 at 40,000 and 114,900 bbl [6360 and 18 268 m³] of water injected with Craig's adaptation of the Caudle and Witte correlation. The pressure drop is 500 psi [3.4 MPa] between the injection well and production well. Permeability of the oil at interstitial water saturation is 100 md. An effective wellbore radius of 0.5 ft [0.15 m] is assumed.

Solution: we begin by computing the base flow rate for a five-spot pattern assuming oil is flowing under a pressure drop of 500 psi [3.4 MPa] between the injection well and the production well. In a 10-acre [40 469-m²] five-spot pattern, $d = 660/\sqrt{2} = 466.7$ ft [142 m].

From Eq. 4.32,

$$i_b = \frac{(3.541)(0.1 \text{ darcy})(20 \text{ ft})(500 \text{ psi})}{(2.0 \text{ cp})\left[\ln\left(\dfrac{466.7}{0.5}\right) - 0.619\right]}$$

$$= 284.7 \text{ B/D}.$$

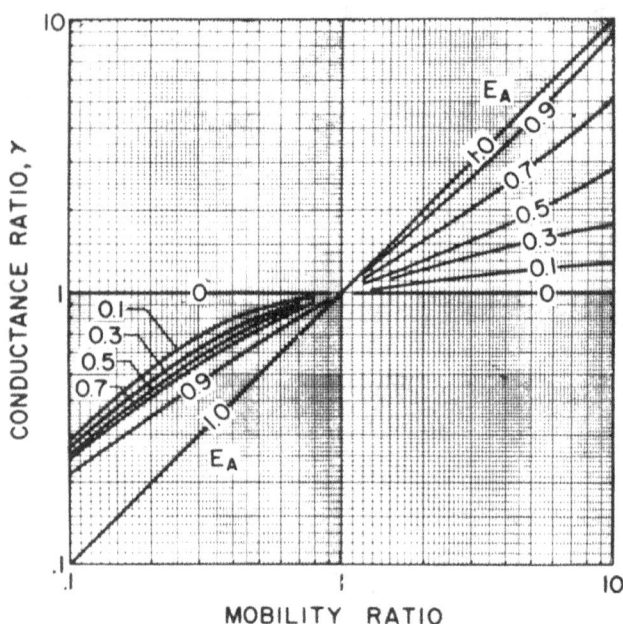

Fig. 4.10—Correlation of conductance ratio for five-spot patterns—miscible displacement (after Caudle and Witte[6]).

From Table 4.1 we note that breakthrough occurs at 59,409 bbl [9445 m³], so $W_i = 40,000$ bbl [6360 m³] occurs before breakthrough. When $E_A < E_{Abt}$, the mobility of the swept region is $M_{\bar{S}} = 0.62$.

$$E_A = \frac{W_i}{V_p(\overline{S_{wf}} - S_{iw})}$$

$$= \frac{40,000}{(232,734)(0.699 - 0.363)}$$

$$= 0.51.$$

The conductivity ratio from Fig. 4.10 for $M_{\bar{S}} = 0.62$ and $E_A = 0.51$ is 0.84. Because $i = \gamma i_b$, $i = 239.1$ B/D [38 m³/d].

At $W_i = 114,900$, breakthrough has occurred and $E_A = 0.94$. From Table 4.1, $\overline{S_{w5}} = 0.718$. Following Craig's adaptation,

$$M_{\overline{Sw5}} = \frac{\left(\frac{k_{rw}}{\mu_w}\right)_{\overline{S_{w5}}}}{\left(\frac{k_{ro}}{\mu_o}\right)_{S_{iw}}},$$

$$k_{rw} = 0.78 S_{wD}^{3.72},$$

$$S_{wD} = \frac{0.718 - 0.363}{1 - 0.205 - 0.363},$$

$$= 0.822,$$

$$(k_{rw})_{\overline{Sw5}} = 0.376, \text{ and}$$

$$M_{\overline{Sw5}} = \frac{0.376/1.0}{1.0/2.0}$$

$$= 0.752.$$

From Fig. 4.10 at $M_{\overline{Sw5}} = 0.75$ and $E_A = 0.94$, we find $\gamma = 0.80$. The injection rate is 227.7 B/D [36.2 m³/d] when $W_i = 114,900$ bbl [18 268 m³].

A final point to remember is that when $E_A = 1.0$, $\gamma = M_{\overline{Sw5}} = 1.56$ and the injection rate becomes 443.4 B/D [71 m³/d]. This is the maximum injection rate possible under the conditions assumed for this example.

4.5 Streamtube Models

A waterflood may be developed with injection and production wells located in some nonuniform arrangements. When this happens, two methods are available to simulate displacement performance. One method is based on numerical simulation of the fluid flow equations in two dimensions. An introduction to numerical simulation of 2D displacement processes is included in Sec. 4.7.

The second method that will be developed in this text is referred to as the streamtube or stream-channel approach.[13-27] The advantage of this method, originated by Higgins and Leighton[13] in 1962,[16-29] is that approximate solutions of displacement problems can often be obtained without resorting to mathematically complex and expensive numerical simulators.

Streamlines are paths followed by fluid particles as they traverse from an injection well to a production well. The region bounded by two streamlines is called a "streamtube" or "stream channel" because there is no fluid flow across streamlines. Streamlines for the steady flow of a single phase (unit mobility ratio) have been determined for regular displacement patterns. Streamlines for arbitrary arrangements of injection and production wells may be determined by superposition[28-30] or numerical solution.[31]

Streamtube models assume that immiscible displacement processes follow the same streamlines as those determined for the steady flow of a homogeneous fluid in the porous media. The region to be simulated is divided into streamtubes based on streamlines determined for the flow of a single-phase fluid. A linear displacement model is used to simulate immiscible displacement within the streamtube. The performance of the region is simulated by combination of the performances of the streamtubes making up the region at the same points in time. Computation of displacement performance requires numerous repetitive calculations. Consequently, it is well suited for solution on a digital computer.

Streamtube models developed for a five-spot pattern will be presented to illustrate the general procedure. Fig. 4.11 shows streamlines in a quadrant of a five-spot pattern. In steady-state flow, all streamlines begin at an injection well and terminate at a production well. The area between two streamlines is a streamtube, as indicated by the shaded area between Streamlines 2 and 3 on Fig. 4.11. Streamtubes have different volumes. Furthermore, the area available for fluid flow in any channel changes as fluids travel from the injection well to the production well. Fig. 4.12 depicts a streamtube or stream channel between the injection and production wells. The streamline drawn through the center of the streamtube is the path of the coordinate ξ that will be used to locate the position of a fluid element in a streamtube. We will refer to the coordinate ξ as the streamtube coordinate.[31]

Displacement of oil by water is simulated by solving the frontal advance equation in streamtube coordinates.

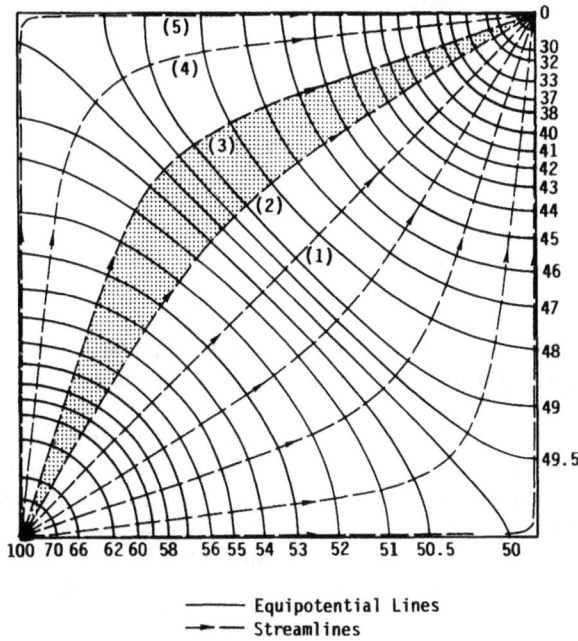

Fig. 4.11—Streamlines and equipotential lines for flow of a single phase in a quadrant of a five-spot.[31]

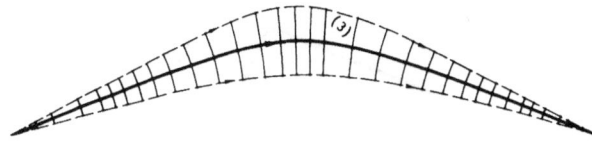

Fig. 4.12—Streamtube bounded by Streamlines 2 and 3.

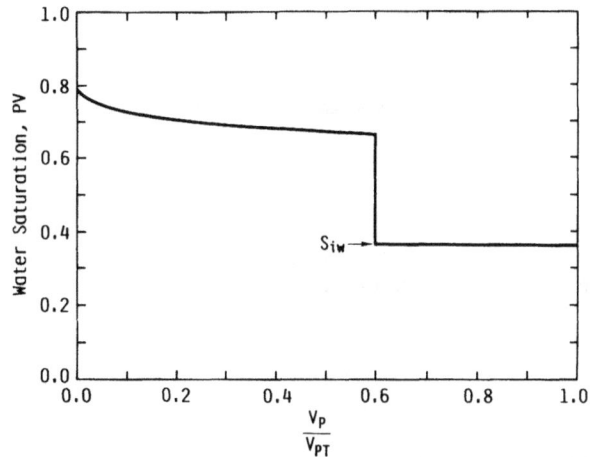

Fig. 4.13—Saturation distribution on a streamtube at $Q_i = 0.202$ for Example 4.2.

Eq. 3.63 is the frontal advance equation that was developed in Sec. 3.4:

$$\left(\frac{dx}{dt}\right)_{S_w} = \frac{q_t}{A\phi}\left(\frac{\partial f_w}{\partial S_w}\right)_{S_w}.$$

The frontal advance equation written in terms of the streamtube coordinate ξ is expressed as Eq. 4.35.

$$\left(\frac{d\xi}{dt}\right)_{S_w} = \frac{q_t}{A(\xi)\phi(\xi)}\left(\frac{\partial f_w}{\partial S_w}\right)_{S_w}. \quad\ldots\ldots\ldots (4.35)$$

In deriving Eq. 4.35 for the case of variable area, $A(\xi)$, we assume that flow perpendicular to the coordinate ξ occurs instantaneously.

Eq. 4.35 may be integrated with respect to ξ and t to find the location of saturation S_w at any time t.

$$\int_{\xi_I}^{\xi} A(\xi)\phi(\xi)d\xi = \int_0^t q_t f'_{Sw}dt. \quad\ldots\ldots\ldots (4.36)$$

The integral on the left side of Eq. 2.36 is $V_p(\xi)$, the PV that has saturations ranging from $1-S_{or}$ at $\xi=\xi_I$ to S_w at ξ—that is,

$$V_p(\xi) = \int_{\xi_I}^{\xi} A(\xi)\phi(\xi)d\xi. \quad\ldots\ldots\ldots (4.37)$$

It follows that

$$V_{pT} = \int_{\xi_I}^{\xi_p} A(\xi)\phi(\xi)d\xi.$$

Recalling the definition of Q_i from Eq. 3.117,

$$V_{pSw} = V_{pT}Q_i f'_{Sw}, \quad\ldots\ldots\ldots (4.38)$$

where Q_i is the number of PV's of fluid injected into the streamtube and V_{pT} is the PV of the streamtube.

There is a direct analogy between the frontal advance solution for linear displacement and displacement in a streamtube. This can be seen by comparing saturation profiles at the same cumulative PV's injected. When Q_i is fixed, Eq. 4.38 can be used to find water saturation that exists at each location $V_p (0 \le V_p \le V_{pT})$. Because

$$f'_{Sw} = \left(\frac{V_p}{V_{pT}}\right)\left(\frac{1}{Q_i}\right), \quad\ldots\ldots\ldots (4.39)$$

we can find the value of S_w corresponding to f'_{Sw} and generate the saturation profile in terms of the displaced volume ratio V_p/V_{pT} for a streamtube. Fig. 4.13 shows the saturation profile in a streamtube when $Q_i = 0.30$ for the parameters of Example 3.4. The same procedure was used in Chap. 3 to obtain the saturation profile at the specified value of Q_i. In linear displacement,

$$f'_{Sw} = \left(\frac{x_{Sw}}{L}\right)\left(\frac{1}{Q_i}\right). \quad\ldots\ldots\ldots (4.40)$$

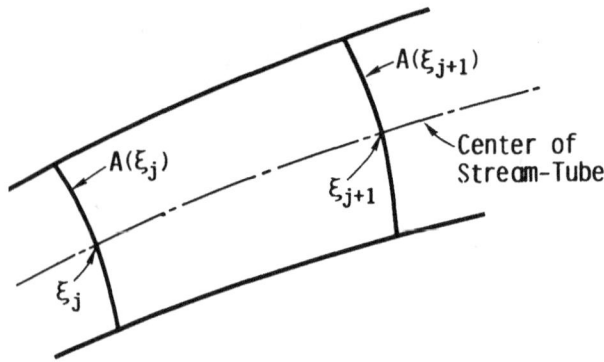

Fig. 4.14—Section of streamtube.

**TABLE 4.2—CHANNEL AREAS FOR
FIVE-SPOT PATTERN[14]**

Stream Channel	A^*/A_T	A^*
1	0.2214	706
2	0.2453	782
3	0.2880	918
4	0.2453	782

A_T (one-eighth pattern) = 3,188.

Comparison of Eq. 4.39 and 4.40 shows that saturation profiles at a specified value of Q_i are identical when expressed in terms of V_p/V_{pT} because

$$\frac{x_{Sw}}{L} = \frac{x_{Sw}\phi A}{L\phi A} = \frac{V_p}{V_{pT}} \quad \dots\dots\dots\dots\dots (4.41)$$

for a linear system.

The analogy between saturation profiles for linear and streamtube displacement models has other implications. It can be shown that the average water saturation in the streamtube is obtained by the Welge method introduced in Sec. 3.5. Once the volume of a streamtube is known, the displacement performance can be computed as a function of Q_i. For example, the cumulative oil displacement from a streamtube i is computed from Eq. 4.42 when there is no initial gas saturation.

$$N_{pi} = (\overline{S_{w2}} - S_{iw})V_{pTi}, \quad \dots\dots\dots\dots\dots (4.42)$$

where $\overline{S_{w2}}$ is the average water saturation in the streamtube corresponding to Q_i.

The PV of a streamtube, V_{pT}, is defined by Eq. 4.43.

$$V_{pT} = A^*\phi h, \quad \dots\dots\dots\dots\dots\dots (4.43)$$

where

A^* = surface area enclosed by the pair of streamlines that define the streamtube,

ϕ = porosity of streamtube (assumed constant), and

h = thickness of streamtube (usually unit thickness).

The value of A^* may be determined by graphical or by numerical techniques.[17] Table 4.2 has the ratio of A^*/A_T from Ref. 14 for Streamtubes 1 through 4 of Fig. 4.12, where A_T is the surface area of one-eighth of the five-spot.

Displacement performance is computed as if the streamtube were linear. The average water saturation in the streamtube is obtained by the same procedures used for linear displacement presented in Sec. 3.5.

Therefore, once the volume of a streamtube is known, the displacement performance can be computed as a function of Q_i.

Displacement performance of a pattern is estimated by combining the results from individual streamtubes at the same points in time. The solution of the frontal advance equation in streamtube coordinates yields displacement performance in terms of Q_i for each streamtube. Thus it is necessary to develop a method to compute the relationship between Q_i and time. One method is presented in the following section.

The total flow rate at any point in a streamtube is expressed by Eq. 4.44.

$$q_t = -(\lambda_{ro} + \lambda_{rw})k_b A(\xi)h\frac{dp}{d\xi}. \quad \dots\dots\dots (4.44)$$

When the pressure drop between injection and production wells is constant, the total flow rate in a given stream channel is obtained by integrating Eq. 4.44 from $\xi=\xi_I$ to $\xi=\xi_p$ to obtain Eq. 4.45.

$$q_t = \frac{k_b(p_I - p_p)h}{\displaystyle\int_{\xi_I}^{\xi_p} \frac{\lambda_r^{-1}d\xi}{A(\xi)}}. \quad \dots\dots\dots\dots\dots (4.45)$$

The integral in Eq. 4.45 may be evaluated numerically or graphically when λ_r^{-1} and $A(\xi)$ are known for $\xi_I \le \xi \le \xi_p$. Fig. 4.14 shows a section of a streamtube between stream coordinates ξ_j and ξ_{j+1}. Values of $A(\xi_j)$ and $A(\xi_{j+1})$ correspond to the lengths of curved isopotential lines that pass through the points ξ_j and ξ_{j+1}. (As indicated earlier, these lengths may be determined by graphical or by numerical techniques.)

The PV for the interval $\xi_I \le \xi$ is computed with Eq. 4.46.

$$V_{p\xi} = \int_{\xi_I}^{\xi} \phi A(\xi)d\xi. \quad \dots\dots\dots\dots\dots (4.46)$$

Integration of the $A(\xi) - \xi$ data for each streamtube produces a correlation between $V_{p\xi}$ and ξ. Rearranging the frontal advance solution in streamtube coordinates, we obtain

$$f'_{Sw} = V_p/(V_{pT}Q_i) \quad \dots\dots\dots\dots\dots (4.47)$$

for $S_w \ge S_{wf}$. Eq. 4.47 defines f'_{sw} for $S_w \ge S_{wf}$ and therefore λ_r^{-1} for all $V_{p\xi}$. Before breakthrough,

$\lambda_r^{-1} = \lambda_{ro}^{-1}$ for $V_{pS_{wf}} < V_{p\xi} \leq V_{pT}$. After breakthrough, values of $f'_{Sw} \leq f'_{Sw2}$ are located in streamtube PV by Eq. 4.48.

$$V_p = V_{pT} Q_{i2} f'_{Sw}. \qquad \ldots \ldots \ldots \ldots \ldots \ldots (4.48)$$

Solution of Eq. 4.45 for a streamtube produces pairs of q_t and Q_i for the complete displacement of the streamtube. The relationship between q_t, Q_i, and time is determined following the procedure described in Sec. 3.6 for a linear waterflood at constant pressure drop. Displacement performance of the entire area is obtained by adding the results from streamtubes at the same times.

Higgins and Leighton Method. In the streamtube model developed by Higgins and Leighton,[32] a streamtube is divided into n cells that have equal volume. Displacement is simulated by advancing the flood front saturation one cell for each production step until breakthrough occurs in the cell n. Because each streamtube is divided into n cells, the integral in Eq. 4.45 becomes

$$\int_{\xi_I}^{\xi_P} \frac{d\xi}{A(\xi)(\lambda_{ro} + \lambda_{rw})} = \int_{\xi_I}^{\xi_1} \frac{d\xi}{A(\xi)(\lambda_{ro} + \lambda_{rw})} +$$

$$\ldots + \int_{\xi_{n-1}}^{\xi_n} \frac{d\xi}{A(\xi)(\lambda_{ro} + \lambda_{rw})}. \qquad \ldots \ldots \ldots \ldots (4.49)$$

Higgins and Leighton approximate the integrals by defining cell averages as in Eqs. 4.51 through 4.53. Let

L_j = average length of Cell j,
A_j = average cross-sectional area of Cell j per unit thickness,
$\overline{\lambda_{ro}}$ = average relative mobility of the oil phase in Cell j, and
$\overline{\lambda_{rw}}$ = average apparent mobility of the water phase in Cell j.

Then

$$\int_{\xi_{j-1}}^{\xi_j} \frac{d\xi}{A(\xi)(\lambda_{ro} + \lambda_{rw})} = \frac{L_j}{A_j} \left(\frac{1}{\overline{\lambda_{ro}} + \overline{\lambda_{rw}}} \right)_j. \quad \ldots (4.50)$$

The ratio L_j/A_j is called the shape factor and is designated G_j in the remaining expressions. Eq. 4.45 becomes

$$\int_{\xi_I}^{\xi_P} \frac{d\xi}{A(\xi)(\lambda_{ro} + \lambda_{rw})} = \sum_{j=1}^{n} \frac{G_j}{(\overline{\lambda_{ro}} + \overline{\lambda_{rw}})_j}. \quad \ldots (4.51)$$

Values of G_j for stream channels needed to simulate five-spot, seven-spot, and linedrive patterns are presented in Ref. 14 and Appendix B. Refs. 17 through 24 provide background material on the evaluation of G_j. All values of G_j are based on streamlines obtained for unit mobility ratio flows in homogeneous, isotropic porous materials. Parsons[30] shows how streamtube models may be applied to anisotropic reservoirs.

Average apparent mobilities in Eqs. 4.50 and 4.51 are defined by Eqs. 4.52 and 4.53. For the oil phase,

$$\overline{\lambda_{roj}} = \frac{\int_{f'_{j-1}}^{f'_j} df'}{\int_{f'_{j-1}}^{f'_j} \lambda_{ro}^{-1} df'}. \qquad \ldots \ldots \ldots \ldots \ldots (4.52)$$

For the water phase,

$$\overline{\lambda_{rwj}} = \frac{\int_{f'_{j-1}}^{f'_j} df'}{\int_{f'_{j-1}}^{f'_j} \lambda_{rw}^{-1} df'}, \qquad \ldots \ldots \ldots \ldots (4.53)$$

where $f'_j = f'_{Swj}$.

Eqs. 4.52 and 4.53 define the average apparent mobility in the same manner as in Sec. 3.6 for linear systems. It should be pointed out that use of Eqs. 4.52 and 4.53 to obtain the average apparent mobilities for a streamtube is an approximation while the corresponding expression for linear displacement is an exact representation for the frontal advance solution.

Displacement computations in the Higgins-Leighton model proceed by advancing the breakthrough saturation from the end of one cell to the next until breakthrough occurs at the end of Cell n. Thus, when Cell k has been displaced to S_{wf}, the values of f'_{Swj} at the end of Cells j, $j \leq k$, are given by Eqs. 4.54 through 4.58.

$$f'_{Swj} = (V_{pj}/V_{pT})(1.0/Q_{ik}), \qquad \ldots \ldots \ldots (4.54)$$

where

$$Q_{ik} = (1/f'_{Swf})(V_{pk}/V_{pT}). \qquad \ldots \ldots \ldots (4.55)$$

Because cells have equal volume,

$$\Delta V_p = \frac{V_{pT}}{n}. \qquad \ldots \ldots \ldots \ldots \ldots \ldots (4.56)$$

Thus for $k \leq n$,

$$Q_{ik} = \frac{k}{(n)(f'_{Swf})}, \qquad \ldots \ldots \ldots \ldots \ldots (4.57)$$

and for $j \leq k$,

$$f'_{Swj} = (j/n)(1.0/Q_{ik}). \qquad \ldots \ldots \ldots \ldots (4.58)$$

After breakthrough, the value of f'_{Sw2} or Q_i may be fixed at the end of Cell n. In the Higgins-Leighton computer program,[18] f'_{Sw2} is selected. Then

$$Q_{i2} = 1/f'_{Sw2} \qquad \ldots \ldots \ldots \ldots \ldots (4.59)$$

and

$$f'_{Swj} = (j/n)f'_{Sw2}. \qquad \ldots \ldots \ldots \ldots (4.60)$$

Eqs. 4.58 and 4.60 give the values of f'_j at the end of each Cell j that are needed to compute $\overline{\lambda_{roj}}$ and $\overline{\lambda_{rwj}}$ in Eqs. 4.52 and 4.53. Before breakthrough at the end of the last cell, $q_o = q_t$, and q_t is computed from Eq. 4.61.

$$q_t = \frac{k_b(p_I - p_p)h}{\sum_{j=1}^{k} \frac{G_j}{(\overline{\lambda_{ro}} + \overline{\lambda_{rw}})_j} + \sum_{j=k+1}^{n} \frac{G_j}{\overline{\lambda_{roi}}}}. \quad \ldots (4.61)$$

TABLE 4.3—STREAMTUBE PARAMETERS FOR FIG. 4.15

Cell Number	$\xi(J)$ (ft)	$A(J)$ (sq ft/ft)	$V(J)$ (cu ft/ft)	$V(J)/V_T$	$G(J)$
0	0.50	0.12	0.0	0.0	
1	52.78	12.90	340.31	0.025	19.0666
2	74.64	18.24	680.62	0.050	1.4182
3	91.41	22.34	1,020.93	0.075	0.8296
4	105.55	25.79	1,361.24	0.100	0.5886
5	118.01	28.84	1,701.56	0.125	0.4566
6	129.27	31.59	2,041.87	0.150	0.3730
7	139.63	34.12	2,382.18	0.175	0.3154
8	149.27	36.48	2,722.49	0.200	0.2732
9	158.33	38.69	3,062.80	0.225	0.2410
10	166.89	40.78	3,403.11	0.250	0.2156
11	175.04	42.77	3,743.42	0.275	0.1950
12	182.82	44.68	4,083.73	0.300	0.1780
13	190.29	46.50	4,424.04	0.325	0.1638
14	197.47	48.25	4,764.35	0.350	0.1516
15	204.40	49.95	5,104.67	0.375	0.1412
16	211.10	51.59	5,444.98	0.400	0.1321
17	217.60	53.17	5,785.29	0.425	0.1240
18	223.91	54.72	6,125.60	0.450	0.1170
19	230.04	56.21	6,465.91	0.475	0.1106
20	236.02	57.68	6,806.22	0.500	0.1050
21	242.00	56.21	7,146.53	0.525	0.1050
22	248.13	54.72	7,486.84	0.550	0.1106
23	254.44	53.17	7,827.15	0.575	0.1170
24	260.94	51.59	8,167.46	0.600	0.1240
25	267.64	49.95	8,507.78	0.625	0.1321
26	274.57	48.25	8,848.09	0.650	0.1412
27	281.75	46.50	9,188.40	0.675	0.1516
28	289.22	44.68	9,528.71	0.700	0.1638
29	297.00	42.77	9,869.02	0.725	0.1780
30	305.15	40.78	10,209.33	0.750	0.1950
31	313.71	38.69	10,549.64	0.775	0.2156
32	322.77	36.48	10,889.95	0.800	0.2410
33	332.41	34.12	11,230.26	0.825	0.2732
34	342.77	31.59	11,570.57	0.850	0.3154
35	354.03	28.84	11,910.88	0.875	0.3730
36	366.49	25.79	12,251.20	0.900	0.4566
37	380.63	22.34	12,591.51	0.925	0.5886
38	397.40	18.24	12,931.82	0.950	0.8296
39	419.26	12.90	13,272.13	0.975	1.4182
40	471.54	0.12	13,612.44	1.000	19.0666

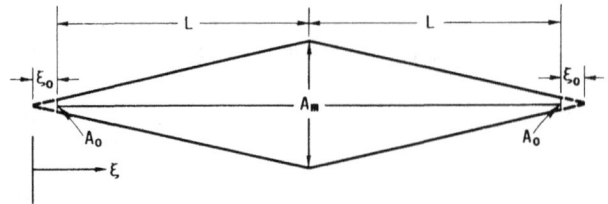

Fig. 4.15—Streamtube constructed for Example 4.3.

and

$$V(\xi) = \left[\frac{A_o + A(\xi)}{2} \right] (\xi - \xi_o), \quad \ldots \ldots \ldots (4.64)$$

and for $L < \xi < 2L - \xi_o$,

$$A(\xi) = (2L - \xi) \left(\frac{A_m}{L} \right) \quad \ldots \ldots \ldots \ldots (4.65)$$

and

$$V(\xi) = \left(\frac{A_o + A_m}{2} \right) (L - \xi_o) + \left(\frac{A_m + A(\xi)}{2} \right) (\xi - L)$$

$$\ldots \ldots \ldots \ldots \ldots (4.66)$$

where
$V_p(\xi) = \phi V(\xi)$,
ξ_o = distance from origin to beginning of streamtube, ft [m],
A_o = cross-sectional area per unit thickness at ξ_o,
L = distance from origin to middle of streamtube, and
A_m = cross-sectional area per unit thickness at middle of streamtube ($\xi = L$).

In Eqs. 4.63 and 4.65, the cross-sectional area is a linear function of streamtube coordinate ξ. Thus the shape factor for the increment ξ_{j-1} to ξ_j can be computed analytically and is given by Eqs. 4.67 through 4.69.

$$G_j = \int_{\xi_{j-1}}^{\xi_i} \frac{d\xi}{A(\xi)} \quad \ldots \ldots \ldots \ldots (4.67)$$

$$= \frac{L}{A_m} \ln \left(\frac{\xi_j}{\xi_{j-1}} \right) \text{ for } \xi_j \le L \quad \ldots \ldots (4.68)$$

$$= \frac{L}{A_m} \ln \left(\frac{2L - \xi_j - 1}{2L - \xi_j} \right) \text{ for } L \le \xi_j \le 2L - \xi_o.$$

$$\ldots \ldots \ldots \ldots \ldots (4.69)$$

After breakthrough

$$q_t = \frac{k_b (p_i - p_p) h}{\sum\limits_{j=1}^{n} \frac{G_j}{(\bar{\lambda}_{ro} + \bar{\lambda}_{rw})_j}} \quad \ldots \ldots \ldots \ldots (4.62)$$

Prediction of waterflood performance of a single streamtube is illustrated in Example 4.3.

Example 4.3. This example illustrates the computation of waterflood performance in a streamtube that has a simple geometric shape. [33] The streamtube in Fig. 4.15 consists of two isosceles trapezoids of equal volume that are joined at the base. Unit thickness is assumed so that $A(\xi)$ is the cross-sectional area perpendicular to the direction of flow per unit thickness. Distance ξ is measured from the origin. The streamtube is defined when values of A_m, L, and ξ are specified. Expressions for streamtube parameters $A(\xi)$ and $V_p(\xi)$ developed with trigonometry are given as Eqs. 4.63 through 4.66. For $\xi_o < \xi < L$,

$$A(\xi) = \xi \left(\frac{A_m}{L} \right) \quad \ldots \ldots \ldots \ldots (4.63)$$

TABLE 4.4—WATERFLOOD PERFORMANCE OF A STREAMTUBE, CONSTANT PRESSURE DROP = 500 psia

Production Step	Time (days)	Oil Rate (B/D)	Water Rate (B/D)	Total Rate (B/D)	F_{wo} (bbl/bbl)	Q_i (PV)	N_p (PV)	Production Step	Time (days)	Oil Rate (B/D)	Water Rate (B/D)	Total Rate (B/D)	F_{wo} (bbl/bbl)	Q_i (PV)	N_p (PV)
0	0.0	22.4	0.0	22.4	0.0	0.0	0.0	46	159.25	1.2	15.0	16.2	12.6	0.440	0.345
1	3.02	18.0	0.0	18.0	0.0	0.008	0.008	47	168.15	1.1	15.3	16.5	13.4	0.460	0.346
2	6.27	19.6	0.0	19.6	0.0	0.017	0.017	48	176.93	1.1	15.6	16.7	14.2	0.480	0.348
3	9.33	20.3	0.0	20.3	0.0	0.025	0.025	49	185.60	1.1	15.8	16.9	15.0	0.500	0.349
4	12.29	20.8	0.0	20.8	0.0	0.034	0.034	50	194.17	1.0	16.0	17.1	15.8	0.520	0.350
5	15.20	21.2	0.0	21.2	0.0	0.042	0.042	51	202.64	1.0	16.3	17.3	16.6	0.540	0.351
6	18.06	21.5	0.0	21.5	0.0	0.050	0.050	52	211.02	0.9	16.5	17.4	17.5	0.560	0.353
7	20.89	21.7	0.0	21.7	0.0	0.059	0.059	53	219.31	0.9	16.7	17.6	18.3	0.580	0.354
8	23.70	21.9	0.0	21.9	0.0	0.067	0.067	54	227.52	0.9	16.9	17.8	19.1	0.600	0.355
9	26.48	22.0	0.0	22.0	0.0	0.076	0.076	55	235.65	0.9	17.1	18.0	20.0	0.620	0.356
10	29.25	22.2	0.0	22.2	0.0	0.084	0.084	56	243.71	0.8	17.3	18.1	20.9	0.640	0.357
11	32.00	22.3	0.0	22.3	0.0	0.092	0.092	57	251.71	0.8	17.5	18.3	21.7	0.660	0.357
12	34.73	22.4	0.0	22.4	0.0	0.101	0.101	58	259.63	0.8	17.6	18.4	22.6	0.680	0.358
13	37.46	22.5	0.0	22.5	0.0	0.109	0.109	59	267.49	0.8	17.8	18.6	23.5	0.700	0.359
14	40.17	22.5	0.0	22.5	0.0	0.118	0.118	60	275.29	0.7	18.0	18.7	24.4	0.720	0.360
15	42.87	22.6	0.0	22.6	0.0	0.126	0.126	61	283.04	0.7	18.1	18.8	25.3	0.740	0.361
16	45.57	22.7	0.0	22.7	0.0	0.134	0.134	62	290.72	0.7	18.3	19.0	26.2	0.760	0.361
17	48.26	22.8	0.0	22.8	0.0	0.143	0.143	63	298.36	0.7	18.4	19.1	27.1	0.780	0.362
18	50.94	22.8	0.0	22.8	0.0	0.151	0.151	64	305.94	0.7	18.6	19.2	28.0	0.800	0.363
19	53.61	22.9	0.0	22.9	0.0	0.160	0.160	65	313.47	0.6	18.7	19.4	28.9	0.820	0.364
20	56.28	22.9	0.0	22.9	0.0	0.168	0.168	66	320.96	0.6	18.9	19.5	29.9	0.840	0.364
21	58.94	23.0	0.0	23.0	0.0	0.176	0.176	67	328.40	0.6	19.0	19.6	30.8	0.860	0.365
22	61.60	23.0	0.0	23.0	0.0	0.185	0.185	68	335.80	0.6	19.1	19.7	31.7	0.880	0.365
23	64.25	23.0	0.0	23.0	0.0	0.193	0.193	69	343.15	0.6	19.3	19.9	32.7	0.900	0.366
24	66.90	23.1	0.0	23.1	0.0	0.202	0.202	70	350.46	0.6	19.3	19.9	33.7	0.920	0.367
25	69.55	23.1	0.0	23.1	0.0	0.210	0.210	71	357.74	0.6	19.5	20.0	34.6	0.940	0.367
26	72.19	23.1	0.0	23.1	0.0	0.218	0.218	72	364.98	0.6	19.6	20.2	35.6	0.960	0.368
27	74.84	23.1	0.0	23.1	0.0	0.227	0.227	73	372.18	0.5	19.7	20.3	36.6	0.980	0.368
28	77.49	23.1	0.0	23.1	0.0	0.235	0.235	74	379.34	0.5	19.8	20.3	37.5	1.000	0.369
29	80.14	23.0	0.0	23.0	0.0	0.244	0.244	75	386.48	0.5	19.9	20.4	38.5	1.020	0.369
30	82.79	23.0	0.0	23.0	0.0	0.252	0.252	76	393.57	0.5	20.1	20.6	39.5	1.040	0.370
31	85.45	23.0	0.0	23.0	0.0	0.260	0.260	77	400.63	0.5	20.2	20.7	40.5	1.060	0.370
32	88.11	22.9	0.0	22.9	0.0	0.269	0.269	78	407.66	0.5	20.2	20.7	41.5	1.080	0.371
33	90.79	22.8	0.0	22.8	0.0	0.277	0.277	79	414.67	0.5	20.3	20.8	42.5	1.100	0.371
34	93.47	22.7	0.0	22.7	0.0	0.285	0.285	80	421.64	0.5	20.4	20.9	43.6	1.120	0.372
35	96.17	22.6	0.0	22.6	0.0	0.294	0.294	81	428.58	0.5	20.6	21.0	44.6	1.140	0.372
36	98.88	22.4	0.0	22.4	0.0	0.302	0.302	82	435.49	0.5	20.6	21.1	45.6	1.160	0.373
37	101.62	22.2	0.0	22.2	0.0	0.311	0.311	83	442.38	0.4	20.7	21.1	46.6	1.180	0.373
38	104.40	21.8	0.0	21.8	0.0	0.319	0.319	84	449.25	0.4	20.8	21.2	47.7	1.200	0.373
39	107.24	21.2	0.0	21.2	0.0	0.327	0.327	85	456.08	0.4	20.9	21.3	48.7	1.220	0.374
40	110.63	14.9	0.0	14.9	0.0	0.336	0.336	86	462.88	0.4	21.0	21.4	49.8	1.240	0.374
41	112.64	1.5	13.4	14.9	8.7	0.340	0.336	87	469.67	0.4	21.1	21.5	50.8	1.260	0.375
42	122.28	1.5	13.8	15.2	9.5	0.360	0.338	88	476.43	0.4	21.1	21.5	51.9	1.280	0.375
43	131.76	1.4	14.1	15.5	10.3	0.380	0.340	89	483.17	0.4	21.2	21.6	53.0	1.300	0.375
44	141.07	1.3	14.4	15.7	11.0	0.400	0.342	90	489.89	0.4	21.3	21.7	54.0	1.320	0.376
45	150.23	1.3	14.8	16.0	11.8	0.420	0.343								

Table 4.3 contains streamtube parameters computed from Eqs. 4.63 through 4.69 by subdividing the streamtube into 40 increments of equal volume. The values of ξ_o, A_m, and L are:

$$\xi_o = 0.5 \text{ ft } [0.15 \text{ m}],$$
$$A_m = 57.675 \text{ sq ft/ft } [17.58 \text{ m}^2/\text{m}], \text{ and}$$
$$L = 236.02 \text{ ft } [71.94 \text{ m}].$$

The total streamtube volume per unit thickness is 13,612.44 cu ft/ft [385.45 m³/m].

Displacement performance will be calculated for a reservoir that has the same properties used in Examples 3.4 and 3.5. Relative permeability and fractional flow data are taken from Example 3.4. Porosity is 0.15 and thickness is 20.0 ft [6.1 m]. The PV of the streamtube (V_{pT}) is 7,273 bbl [1156 m³].

In Example 3.4, the flood-front saturation was 0.664 and $f'_{Swf}=2.978$. Thus breakthrough at the end of the streamtube occurs when $V_{Sw2}=V_{pS_{wf}}=V_{pT}$. After breakthrough, the saturations $0.664<S_{w2}<0.795$ arrive at the end of the streamtube. Values of f'_{Sw2}, S_{w2}, and

Q_i are determined when S_{w2} is specified in the linear waterflood. Displacement performance for the streamtube is given in Table 4.4. Note that S_{w2}, S_w, f_{w2}, F_{wo}, Q_i, and oil recovery (expressed as PV) are identical to the values presented in Table 3.3 for a linear waterflood. Total volume of oil displaced in the streamtube is less than the linear model because the PV is less. The principal difference between the linear and streamtube models is in the computation of flow rates.

Computation of the flow rate corresponding to a pressure difference of 500 psia [3.45 MPa] will be illustrated for the value of Q (0.66) when $S_{w2}=0.69$. Although the computations are not difficult, they are too tedious to do by hand, so it is convenient to do them on the computer. Numerical results presented in this example were obtained with a small computer program.

The total flow rate in the streamtube is computed with Eq. 4.45.

$$q_t = \frac{k_b(p_I - p_p)h}{\displaystyle\int_{\xi_I}^{\xi_p} \frac{\lambda_r^{-1} d\xi}{A(\xi)}}.$$

**TABLE 4.5—SATURATION PROFILE AND AVERAGE VISCOSITIES
COMPUTATION OF FLOW RATE IN STREAMTUBE AT $Q_i = 0.66^*$**

Cell Number	$\Sigma(J)$ (ft)	$\dfrac{V_p(J)}{V_{pT}}$	f'_{Sw}	S_w (fraction)	$\overline{\lambda_r^{-1}}$ (cp)	$G(J)$
1	52.78	0.025	0.038	0.776	1.428	19.0666
2	74.64	0.050	0.076	0.767	1.586	1.4182
3	91.41	0.075	0.114	0.760	1.698	0.8296
4	105.55	0.100	0.152	0.775	1.791	0.5886
5	118.01	0.125	0.189	0.751	1.875	0.4566
6	129.27	0.150	0.227	0.747	1.949	0.3730
7	139.63	0.175	0.265	0.743	2.019	0.3154
8	149.27	0.200	0.303	0.740	2.085	0.2732
9	158.33	0.225	0.341	0.737	2.145	0.2410
10	166.89	0.250	0.379	0.734	2.204	0.2156
11	175.04	0.275	0.417	0.732	2.260	0.1950
12	182.82	0.300	0.455	0.730	2.312	0.1780
13	190.29	0.325	0.492	0.727	2.363	0.1638
14	197.47	0.350	0.530	0.725	2.413	0.1516
15	204.40	0.375	0.568	0.723	2.462	0.1412
16	211.10	0.400	0.606	0.721	2.510	0.1321
17	217.60	0.425	0.644	0.720	2.551	0.1240
18	223.91	0.450	0.682	0.718	2.597	0.1170
19	230.04	0.475	0.720	0.716	2.644	0.1106
20	236.02	0.500	0.758	0.715	2.679	0.1050
21	242.00	0.525	0.795	0.713	2.725	0.1050
22	248.13	0.550	0.833	0.712	2.762	0.1106
23	254.44	0.575	0.871	0.710	2.803	0.1170
24	260.94	0.600	0.909	0.709	2.842	0.1240
25	267.64	0.625	0.947	0.708	2.878	0.1321
26	274.57	0.650	0.985	0.706	2.918	0.1412
27	281.75	0.675	1.023	0.705	2.953	0.1516
28	289.22	0.700	1.061	0.704	2.989	0.1638
29	297.00	0.725	1.098	0.703	3.026	0.1780
30	305.15	0.750	1.136	0.702	3.058	0.1950
31	313.71	0.775	1.174	0.700	3.099	0.2156
32	322.77	0.800	1.212	0.699	3.123	0.2410
33	332.41	0.825	1.250	0.698	3.165	0.2732
34	342.77	0.850	1.288	0.697	3.192	0.3154
35	354.03	0.875	1.326	0.696	3.227	0.3730
36	366.49	0.900	1.364	0.695	3.264	0.4566
37	380.63	0.925	1.402	0.694	3.285	0.5886
38	397.40	0.950	1.439	0.693	3.325	0.8296
39	419.26	0.975	1.477	0.692	3.355	1.4182
40	471.54	1.000	1.515	0.691	3.381	19.0666

*Total resistance $(\Sigma G_j \overline{\lambda_{rj}^{-1}}) = 123.23$ cp.

Every Q_i in Table 4.4 determines a saturation distribution that is known at each value of ξ_j in the streamtube. Thus λ_r^{-1} is also known from Eq. 3.123, recalling that $\mu_o = 2$ cp [2 mPa·s] and $\mu_w = 1.0$ cp [1.0 mPa·s]. For this example, we approximate Eq. 4.50 as

$$\int_{\xi_{j-1}}^{\xi_j} \frac{d\xi}{A(\xi)(\lambda_{ro} + \lambda_{rw})} = G_j \overline{\lambda_{rj}^{-1}}, \qquad \ldots \ldots \ldots (4.70)$$

where

$$\overline{\lambda_{rj}^{-1}} = \frac{\displaystyle\int_{f'_{j-1}}^{f'_j} \lambda_r^{-1} df'}{f'_j - f'_{j-1}}. \qquad \ldots \ldots \ldots \ldots \ldots (4.71)$$

Referring to Sec. 3.6, it can be shown that

$$\overline{\lambda_{rj}^{-1}} = \frac{f'_j \overline{\lambda_{Swj}^{-1}} - f'_{j-1} \overline{\lambda_{Swj-1}^{-1}}}{f'_j - f'_{j-1}}, \qquad \ldots \ldots \ldots (4.72)$$

where $\overline{\lambda_{Swj}^{-1}}$ is the average apparent viscosity computed in Sec. 3.6 for a linear system. Values of $\overline{\lambda_{Sw}^{-1}}$ are given in Table 3.4. With these expressions, the integral in Eq. 4.45 becomes

$$\int_{\xi_I}^{\xi_P} \frac{\lambda_r^{-1} d\xi}{A(\xi)} = \sum_{j=1}^{n} G_j \overline{\lambda_{rj}^{-1}}. \qquad \ldots \ldots \ldots \ldots (4.73)$$

When $Q_i = 0.66$ $(S_{w2} = 0.69)$, the saturation distribution computed from Eq. 4.48 is given in Table 4.5. This distribution was found by computing the values of f'_{Swj} from Eq. 4.60 and then by using linear interpolation to find the corresponding values of S_{wj} from Table 4.4. The value of

$$\sum_{j=1}^{n} G_j \lambda_{rj}^{-1}$$

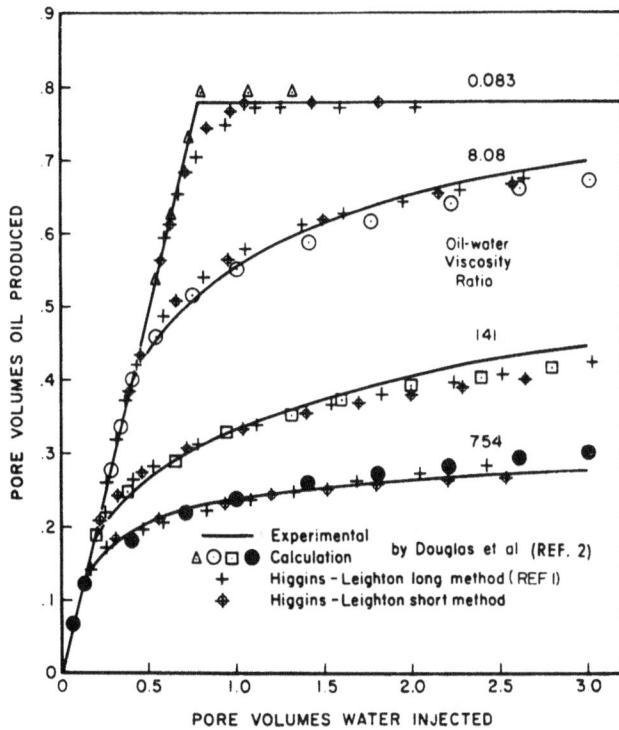

Fig. 4.16—Comparison of experimental and computed water-flood performance for five-spot waterflood in laboratory models. [32]

TABLE 4.6—SHAPE FACTORS AND
STREAMTUBE PARAMETERS FOR A FIVE-SPOT
PATTERN [13,18]

Cell Number	Five-Spot Channel Number			
	1	2	3	4
1	17.372	17.880	19.620	36.582
2	1.506	1.558	1.696	2.920
3	0.886	0.990	0.950	1.848
4	0.631	0.639	0.670	1.197
5	0.495	0.495	0.536	0.928
6	0.397	0.397	0.440	0.768
7	0.325	0.331	0.363	0.643
8	0.286	0.282	0.306	0.535
9	0.256	0.245	0.262	0.451
10	0.231	0.220	0.229	0.378
11	0.211	0.200	0.203	0.310
12	0.196	0.185	0.182	0.253
13	0.183	0.172	0.166	0.205
14	0.173	0.162	0.152	0.167
15	0.166	0.155	0.139	0.139
16	0.161	0.150	0.128	0.121
17	0.158	0.147	0.118	0.108
18	0.155	0.144	0.110	0.096
19	0.152	0.139	0.105	0.088
20	0.150	0.124	0.101	0.082
21	0.150	0.124	0.101	0.082
22	0.152	0.139	0.105	0.088
23	0.155	0.144	0.110	0.096
24	0.158	0.147	0.118	0.108
25	0.161	0.150	0.128	0.121
26	0.166	0.155	0.139	0.139
27	0.173	0.162	0.152	0.167
28	0.183	0.172	0.166	0.205
29	0.196	0.185	0.182	0.253
30	0.211	0.200	0.203	0.310
31	0.231	0.220	0.229	0.378
32	0.256	0.245	0.262	0.451
33	0.286	0.282	0.306	0.535
34	0.325	0.331	0.363	0.643
35	0.397	0.397	0.440	0.768
36	0.495	0.495	0.536	0.928
37	0.631	0.639	0.670	1.197
38	0.886	0.900	0.950	1.848
39	1.506	1.558	1.696	2.920
40	17.372	17.880	19.620	36.582
Volume fraction	0.2237	0.2478	0.2909	0.2376
Angle between streamlines, degrees	13.13	12.94	12.13	6.80

$d/r_w = 486.22$

from the last two columns of Table 4.5 is 123.23. Substituting into Eq. 4.45 we have

$$q_t = \frac{(1.127)(0.200 \text{ darcy})(500.0 \text{ psi})(20.04 \text{ ft})}{123.23 \text{ cp}}$$

$$= 18.29 \text{ B/D},$$

where the constant 1.127 accounts for unit conversion.

Total flow rates computed at each stage of the displacement are summarized in Table 4.4. Waterflood performance was simulated at 40 values of Q_i to breakthrough and 40 values of Q_i after breakthrough. Time in Table 4.4 was computed by the procedure described in Sec. 3.6.

The streamtube in Fig. 4.15 could be one of several streamtubes representing an area to be flooded. Composite performance of a collection of streamtubes originating and terminating at the same injection and production wells is obtained by combining results from different streamtubes at equal values of injection time. A computer program that does waterflood calculations based on the Higgins-Leighton model is available in Ref. 18. The program is written in FORTRAN and includes treatment of mobile gas saturation in displacement calculations as discussed in Sec. 3.9. The program handles multiple layers as discussed in Chap. 5.

The Higgins-Leighton model was used to simulate the performance of the laboratory waterfloods in a five-spot that were shown in Fig. 4.16. The agreement is excellent and indicated that streamlines and streamtubes obtained for unit mobility ratio were reasonable approximations for those that exist at other mobility ratios. Further in-

vestigation by Martin and Wegner [31] has shown that mobility ratio has little effect on streamtube results for many problems of practical interest. Fixed streamtube results were within 10% of variable streamtube results with the exception of direct linedrive patterns for mobility ratios of 0.1 and 1.0. Martin et al. [34] discuss difficulties in predicting waterflood behavior in isolated five-spot patterns. Other examples illustrating the use of streamtube models are presented in Chap. 6.

Simulation of pattern floods with the Higgins-Leighton model requires shape factors for the streamtubes that represent symmetric elements of the pattern. For the five-spot pattern, there are four streamtubes in the one-eighth-pattern area, which is the symmetric element. Shape factors for each of these streamtubes are presented in Table 4.6. Also included in Table 4.6 are the angles between streamlines at the injection and production wells, volume fraction of the one-eighth-pattern occupied by each streamtube, and the ratio of wellbore radius to distance between the injection and production well used to compute first and last shape factors for each streamtube. The latter values are needed in the calculations.

The Higgins-Leighton model combines results from individual streamtubes at the same point in time to simulate pattern performance. We will discuss the methods used to combine results for displacement of several streamtubes at constant pressure drop in Chap. 5 in connection with simulation of waterflood performance in layered systems. It can be shown that all shape factors except the first and last are independent of the actual size of the pattern. The first and last shape factors in most floods are in the region where the flow is radial. Under these conditions, the first or last shape factor is derived following the procedure in Eqs. 4.74 through 4.78.

Flow in the first and the last cells of a streamtube is assumed to be radial. For a radial segment bounded by r_w and r_1, G_1 is defined by Eq. 4.74.

$$G_1 = \int_{r_w}^{r_1} \frac{dr}{2\pi r \Delta \theta} \quad \ldots\ldots\ldots\ldots\ldots\ldots (4.74)$$

$$= \frac{1}{2\pi\Delta\theta} \ln \frac{r_1}{r_w}, \quad \ldots\ldots\ldots\ldots\ldots\ldots (4.75)$$

where

r_1 = radius of the end of the first cell,
r_w = wellbore radius, and
$\Delta\theta$ = angle between streamlines, which define the streamtube radians.

Eq. 4.75 may also be written with the use of dimensionless distances by introducing d, the distance between the injection and production wells as in Eq. 4.76.

$$G_1 = \frac{1}{2\pi\Delta\theta} \ln \left(\frac{r_1/d}{r_w/d} \right). \quad \ldots\ldots\ldots\ldots\ldots (4.76)$$

A similar expression can be written for the shape factor for the last cell of the streamtube.

Because G_1 and G_n are computed with a specific wellbore radius, it is necessary to adjust these shape factors to the effective wellbore radius of the pattern that is to be simulated. Eqs. 4.78 and 4.79 show how the adjusted value G_1^* is obtained from the angle between streamlines (θ), the original wellbore radius used to derive the shape factor (r_w), and the effective wellbore used to derive the shape factor (r_w) and the effective wellbore radius (r_w^*) for the problem to be simulated.

Let r_w^*, r_1^*, and d^* represent the parameters for the problem to be simulated. Because of the geometric similarity,

$$\frac{r_1^*}{d^*} = \frac{r}{d}. \quad \ldots\ldots\ldots\ldots\ldots\ldots (4.77)$$

The reader must show that

$$G_n^* = G_1 + \frac{1}{2\pi\Delta\theta} \left[\ln \frac{r_w/d}{(r_w^*/d^*)} \right] \quad \ldots\ldots\ldots\ldots (4.78)$$

for Cell 1, and

$$G_n^* = G_n + \frac{1}{2\pi\Delta\theta} \left[\ln \frac{(r_w/d)}{(r_w^*/d^*)} \right] \quad \ldots\ldots\ldots\ldots (4.79)$$

for Cell n. In Eqs. 4.78 and 4.79 the radii of the injection well (Cell 1) and production well (Cell n) are assumed to be equal.

This calculation is done as part of the Higgins-Leighton program described in Ref. 18 for the five-spot pattern. Program revision is necessary for other patterns. Shape factors and other parameters for linedrive, staggered linedrive, and seven-spot patterns are included in Appendix B.

Example 4.4 illustrates the use of the Higgins-Leighton program (described in Ref. 18) to estimate the waterflood performance in a 40-acre [161 875-m^2] five-spot pattern.

Example 4.4. A reservoir is to be waterflooded on 40-acre [161 875-m^2] spacing with five-spot patterns. Thickness of the reservoir is 25 ft [7.6 m] and the average porosity is 18. There is no initial gas saturation. Other properties of the rocks and fluids are given below. The flood is operated under a constant pressure drop of 750 psi. The effective wellbore radius is 1.5 ft [0.46 m]. Relative permeability data are correlated in terms of the dimensionless water saturation, S_{wD}, as

$$k_{ro} = 0.402(1 - S_{wD})^{2.06}$$

and

$$k_{rw} = 0.248 \, S_{wD}^{2.33},$$

where

$$S_{wD} = \frac{S_w - S_{iw}}{1 - S_{or} - S_{iw}}.$$

Rock properties are $k_b = 50$ md, $S_{or} = 0.3$, and $S_{iw} = 0.3$. Fluid properties are $\mu_o = 1.62$ cp [1.62 mPa·s] and $\mu_w = 1.0$ cp [1.0 mPa·s].

Results computed with the Higgins-Leighton program are presented in Table 4.7.

4.6 Comparison of CGM and Streamtube Models for a Five-Spot Pattern

The areal displacement models presented in this section were selected because (1) they produce adequate simulations of immiscible displacement in two dimensions, and (2) the mathematical and/or computational skills are within the limits established for this text. Streamline models

TABLE 4.7—SUMMARY OF PREDICTED WATERFLOOD PERFORMANCE BY
THE HIGGINS-LEIGHTON STREAMTUBE MODEL IN A HOMOGENEOUS
RESERVOIR

Time (days)	q_o (B/D)	q_w (B/D)	q_t (B/D)	F_{wo}	N_p (PV)	S_g (PV)	Q_i (PV)
91.3	179.7	0.0	179.7	0.00	0.016	0.000	0.016
182.5	183.6	0.0	183.6	0.00	0.027	0.000	0.027
273.8	186.4	0.0	186.4	0.00	0.039	0.000	0.039
365.0	187.3	0.0	187.3	0.00	0.051	0.000	0.051
456.3	187.6	0.0	187.6	0.00	0.063	0.000	0.063
547.5	187.6	0.0	187.6	0.00	0.075	0.000	0.075
638.8	187.5	0.0	187.5	0.00	0.087	0.000	0.087
730.0	187.3	0.0	187.3	0.00	0.099	0.000	0.099
1,095.0	185.9	0.0	185.3	0.00	0.146	0.000	0.146
1,460.0	183.5	0.0	183.5	0.00	0.193	0.000	0.193
1,825.0	177.6	0.0	177.6	0.00	0.239	0.000	0.239
2,190.0	121.3	25.7	147.0	0.21	0.277	0.000	0.281
2,555.0	80.3	52.2	132.5	0.65	0.299	0.000	0.316
2,920.0	37.0	76.9	113.9	2.08	0.315	0.000	0.348
3,285.0	35.6	79.1	114.6	2.22	0.325	0.000	0.378
3,650.0	33.8	81.0	114.8	2.40	0.334	0.000	0.408
4,015.0	28.9	82.8	111.7	2.86	0.342	0.000	0.438
4,380.0	8.1	97.2	105.3	12.02	0.345	0.000	0.466
4,745.0	7.5	99.0	106.4	13.21	0.347	0.000	0.494
5,110.0	7.0	100.0	107.5	14.44	0.349	0.000	0.522
5,475.0	6.5	102.0	108.5	15.72	0.351	0.000	0.551
5,840.0	6.1	103.3	109.4	17.04	0.352	0.000	0.580
6,205.0	5.7	104.6	110.3	18.41	0.354	0.000	0.609
6,570.0	5.3	105.8	111.1	19.82	0.355	0.000	0.638
6,935.0	5.0	106.9	112.6	22.79	0.358	0.000	0.668
7,300.0	4.7	107.9	113.3	24.34	0.359	0.000	0.698
7,665.0	4.5	108.8	114.0	25.94	0.360	0.000	0.729
8,030.0	4.2	109.8	114.6	27.57	0.361	0.000	0.790
8,760.0	3.8	111.4	115.2	29.25	0.362	0.000	0.821

(Higgins-Leighton) require the availability of a digital computer. Although the CGM model can be used along with the set of graphs and tables provided by Craig[1] and a pocket calculator, the model is easy to code on programmable pocket calculators or personal computers.

Each model presented in Secs. 4.4 and 4.5 has the capability to simulate waterflood performance in a homogeneous five-spot pattern. They are similar in that some form of the Buckley-Leverett equation is used to compute displacement performance after breakthrough. Thus we should expect the models to produce comparable estimates of displacement efficiency for the same reservoir and rock parameters.

Waterflood performances for a series of five-spot floods were computed with the CGM model and the Higgins-Leighton model and are shown in Table 4.8. The parameters in Example 4.1 were used except for oil viscosity. Oil viscosity was varied to obtain mobility ratios ($M_{\bar{S}}$) of about 0.5, 1.0, 2.0, and 5.0. Computed results show that the two methods are in good agreement over a wide range of Q_i and mobility ratios for this set of relative permeability curves. Either model could be used to estimate displacement performance. However, the Higgins-Leighton model produces estimates of flow rates at different times when the pressure drop is fixed, while the CGM model must use correlations such as Fig. 4.10 when the injection rate is not constant.

4.7 Prediction of Areal Displacement by 2D Numerical Simulators

Displacement of oil by water in a 2D areal flood can be simulated by numerical solution of Eqs. 4.1 through 4.5.

Arrangement of production and injection wells is arbitrary. Reservoir and fluid properties are considered uniform with respect to depth but may vary with position (x, y) throughout the reservoir. Results of two simulations are presented in this section to show the potential of the method. Development of these models, which are sometimes referred to as 2D or black-oil reservoir simulators, is described in Refs. 35 through 40.

4.7.1 Simulation of Five-Spot Flood Pattern

Douglas et al.[3] solved Eqs. 4.1 through 4.5 for the experimental conditions presented in Fig. 4.15. Points plotted in Fig. 4.15 are computed values. The numerical solution also produces saturation and pressure distributions at grid nodes used in the numerical solutions. Saturation contours, obtained by connecting points of constant saturation, show the movement of the displacing front as a function of time. The effects of capillary and gravity forces were included in the Douglas et al. 2D model.

4.7.2 Flood-Front Simulation

Some displacement projects involve injection of water at the edge of a large field to maintain reservoir pressure or to supplement a natural water drive. Injection wells are not positioned so that flood performance can be analyzed as a regular pattern. There may be anisotropy in permeability and/or porosity across the field. All of these possibilities could be handled by solving Eqs. 4.1 through 4.5 in two dimensions with the Douglas et al. 2D model.

It is not always necessary to use a two-phase, 2D model such as the Douglas et al. model to study a 2D displacement process. A less complicated model, called a flood

TABLE 4.8—COMPARISON OF OIL RECOVERY PREDICTIONS FOR FIVE-SPOT PATTERNS BY THE CGM AND HIGGINS-LEIGHTON MODELS

$M_S = 0.5$			$M_S = 1.0$			$M_S = 2.0$			$M_S = 5.0$		
Q_i	CGM	HL	Q_i	CGM	HL	Q_i	CGM	HL	Q_i	CGM	HL
0.04	0.044	0.044	0.04	0.044	0.044	0.04	0.037	0.037	0.02	0.022	0.022
0.09	0.094	0.094	0.09	0.094	0.094	0.08	0.080	0.080	0.04	0.041	0.041
0.20	0.199	0.199	0.15	0.146	0.146	0.13	0.125	0.125	0.05	0.047	0.047
0.30	0.286	0.273	0.20	0.199	0.199	0.17	0.159	0.172	0.05	0.054	0.054
0.40	0.321	0.303	0.25	0.229	0.246	0.23	0.184	0.206	0.06	0.061	0.061
0.51	0.350	0.321	0.30	0.249	0.273	0.29	0.207	0.228	0.07	0.066	0.067
0.56	0.362	0.325	0.35	0.266	0.291	0.36	0.229	0.242	0.07	0.071	0.074
0.62	0.369	0.328	0.40	0.281	0.303	0.44	0.249	0.256	0.08	0.075	0.081
0.68	0.371	0.331	0.45	0.295	0.314	0.53	0.268	0.265	0.09	0.080	0.089
0.74	0.373	0.334	0.56	0.320	0.325	0.63	0.285	0.271	0.10	0.085	0.095
0.80	0.375	0.336	0.68	0.340	0.331	0.74	0.289	0.277	0.11	0.091	0.101
0.86	0.376	0.339	0.80	0.344	0.336	0.86	0.293	0.282	0.13	0.098	0.105
0.93	0.378	0.341	0.93	0.348	0.341	0.98	0.297	0.287	0.14	0.104	0.110
1.00	0.379	0.343	1.00	0.350	0.343	1.11	0.301	0.291	0.16	0.111	0.115
1.07	0.381	0.345	1.07	0.351	0.345	1.25	0.304	0.296	0.18	0.118	0.120
1.14	0.382	0.347	1.14	0.353	0.347	1.40	0.308	0.299	0.33	0.153	0.138
1.51	0.388	0.355	1.51	0.360	0.355	1.55	0.311	0.303	0.55	0.167	0.154
1.91	0.392	0.362	1.91	0.366	0.362	1.71	0.314	0.306	0.85	0.178	0.167
2.33	0.395	0.367	2.33	0.371	0.367	1.88	0.317	0.309	1.25	0.188	0.178
2.77	0.398	0.371	2.77	0.375	0.371	2.05	0.319	0.312	1.78	0.199	0.188
3.23	0.401	0.375	3.23	0.378	0.375	3.00	0.331	0.325	2.42	0.208	0.197

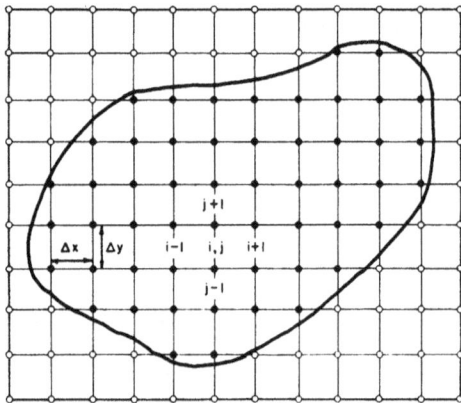

Fig. 4.17—Subdivision of a reservoir into gridblocks for numerical solution.[41]

front simulator,[41] is frequently adequate for initial work. This model estimates the position of the flood front when the displacement process is piston-like and the mobility ratio is 1.0. In piston-like displacement, the oil saturation decreases from S_{oi} to S_{or} when the flood front arrives. Fluids are considered incompressible and effects of gravity and capillary forces are neglected. When these assumptions are made, Eqs. 4.1 through 4.5 become

$$\frac{\partial}{\partial x}\left(k\frac{\partial p}{\partial x}\right) + \frac{\partial}{\partial y}\left(k\frac{\partial p}{\partial y}\right) = 0. \quad \ldots\ldots\ldots\ldots (4.80)$$

Eq. 4.80 applies to a reservoir with constant thickness h. Variable thickness can be included assuming that vertical flow occurs instantaneously as thickness changes with x and y. Eq. 4.81 can be derived in the same manner as

Eqs. 4.1 through 4.5 and 4.80 when this assumption is made.

$$\frac{\partial}{\partial x}\left(kh\frac{\partial p}{\partial x}\right) + \frac{\partial}{\partial y}\left(kh\frac{\partial p}{\partial y}\right) = 0. \quad \ldots\ldots\ldots\ldots (4.81)$$

Fig. 4.17 shows a typical grid for a 2D areal model. The numerical solution produces values of p at each grid node. These are used to compute velocities of points leaving injection wells to find the flood front. This procedure is analogous to that for streamtube models. An important use of numerical models is to simulate different flooding options. Figs. 4.18 and 4.19 illustrate flood-front patterns for a reservoir at different times for two different injection well arrangements.

Two-dimensional numerical simulations have capabilities not available in other methods. Reservoir properties may change with spatial position. Thus, although the surface well arrangement may be a five-spot or some other regular pattern, the reservoir flow pattern may be quite different. Ref. 42 illustrates the use of a flood-front model in a waterflood in the Kaybob reservoir.

4.8 Limitations of 2D Areal Simulators

Numerical simulators permit solution of the fluid-flow equations at many grid points within the reservoir boundaries. To obtain meaningful solutions it is necessary to know how reservoir and fluid properties vary with position. Relative permeability curves are needed. In most simulations of displacement processes there are properties that are not known as well as those that have large uncertainties. It is usual practice to use a period of production (several years) before simulating the desired time interval for a history match. Adjustments in reservoir properties and/or relative permeability curves are made until computed reservoir pressures are within acceptable agreement with observed pressures. Figs. 4.20 and 4.21 illustrate the permeability map for a reservoir before and

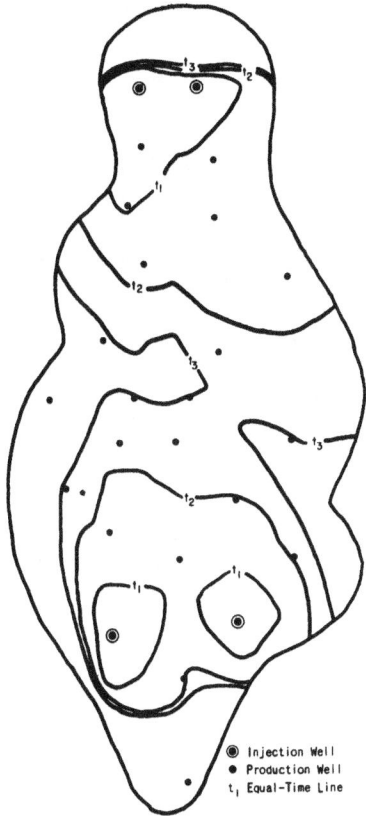

Fig. 4.18—Typical flood pattern resulting from field study with injection wells at ends of reservoir. [42]

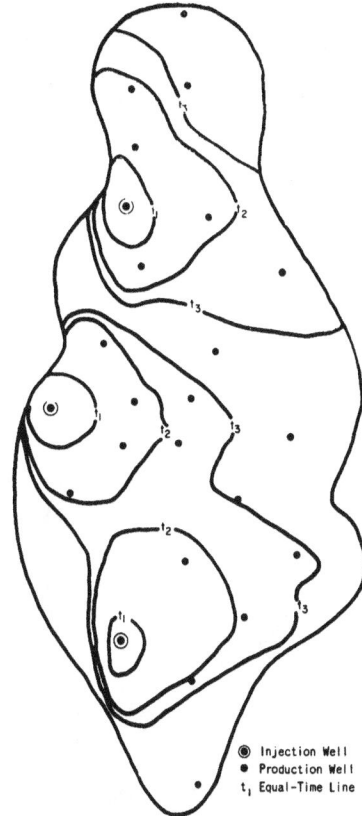

Fig. 4.19—Typical flood pattern resulting from field study edge injection. [42]

PERMEABILITIES (md) OBTAINED FROM
CORE ANALYSIS
ORIGINAL PERMEABILITIES IN OUTLINED REGIONS
WERE MULTIPLIED BY NUMBERS INDICATED TO
OBTAIN A HISTORY MATCH

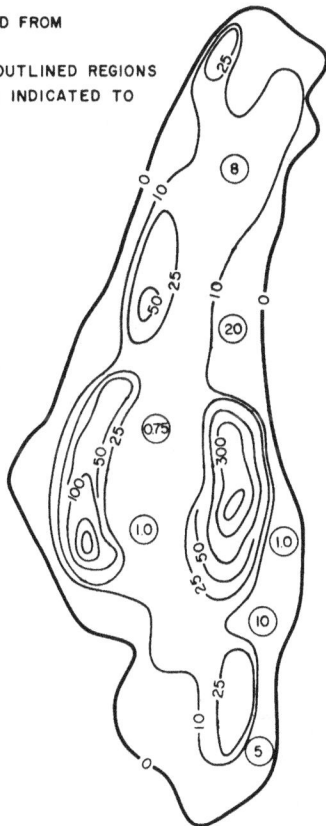

Fig. 4.20—Kaybob permeability map before history match. [42]

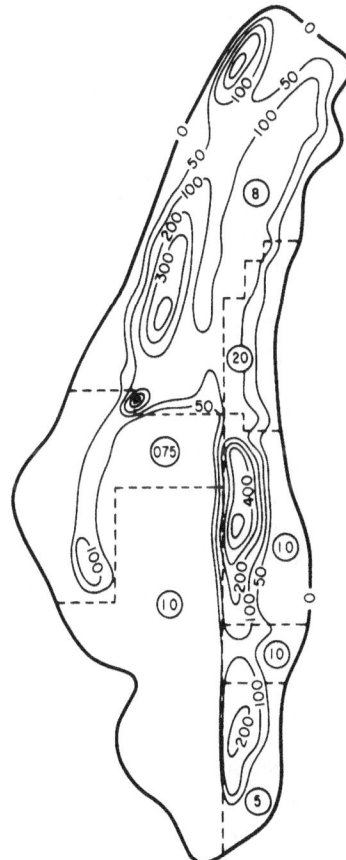

Fig. 4.21—Kaybob permeability after history match. [42]

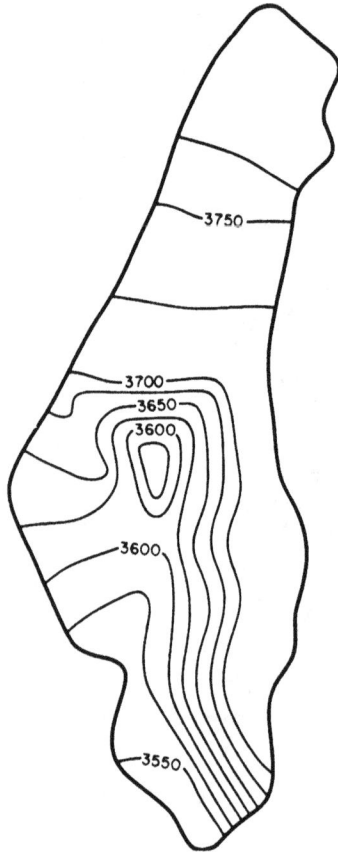

Fig. 4.22—Kaybob isobaric map, observed pressures, June 1961; hydrocarbon = PV weighted average pressure is 3,685 psig. [42]

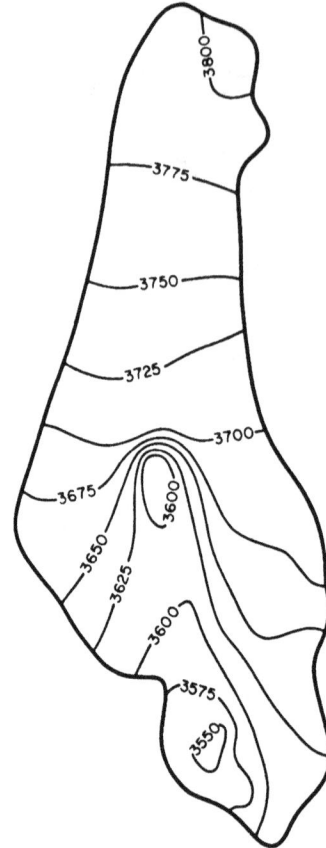

Fig. 4.23—Kaybob isobaric map, calculated pressures, June 1961; hydrocarbon = PV weighted average pressure is 3,703 psig. [42]

after a history match.[42] Figs. 4.22 and 4.23 illustrate the pressure match in June 1961. The flood-front position is shown in Fig. 4.24.

The capability of solving problems mathematically often exceeds the quantity and quality of reservoir data required by the model. Thus numerical simulators may also be viewed as a tool to help the engineer understand the behavior of a complex geological environment. Further discussion on the role of geology in the design and operation of waterfloods is included in Chap. 7.

Problems

4.1. A waterflood is being planned for a reservoir that has properties corresponding to those in Table 4.9. The waterflood will be conducted with 10-acre [404 687-m^2] five-spot patterns. You are requested to prepare estimates of the waterflood performance for a five-spot pattern to $F_{wo} = 100$ with the CGM model.

Results are to be presented in a table with the following headings: W_i (bbl), W_i/W_{ibt}, E_A, Q_i^* (PV), S_{w2}, f_{o2}, $\overline{S}_{w5}$, N_p (PV), Q_i (PV), N_p (bbl), and F_{wo}.

While the computations may be done by hand calculator, you will find the problem easier to solve by writing a small computer program. The following programs in Appendix A will be used to generate relative permeability, fractional flow data, and Q_i^*/Q_{ibt}^*.

> SWF: Computes breakthrough saturation and f'_{Swf} from generalized relative permeability relationships.

> PROP: Computes k_{ro}, k_{rw}, f_w, f'_w for specified values of S_w when oil and water viscosities are known.

> CGM: Computes the value of Q_i^*/Q_{ibt}^* at a specific value of W_i when W_{ibt} and E_{Abt} are given.

Use of the CGM model involves stepping through the calculation in increments of W_i. The computations produce values of Q_i corresponding to f_{w2}. To compute the displacement performance, it will be necessary to find the value of S_{w2} corresponding to f_{w2}. One way to do this is to prepare a table of values of $f_{w2} - S_{w2}$ from S_{wf} to $1 - S_{or}$. A specific S_{w2} can be found for a particular value of f_{w2} by use of linear interpolation. It is useful to check your program with the parameters in Example 4.1.

Note: You will find it helpful in checking the results in Example 4.1 if you subdivide the interval between W_{ibt} and W_{i100} into nine increments and the interval between W_{i100} and $(W_i)_{max}$ into 19 increments.

4.2. Evaluate the performance of a five-spot waterflood with the CGM method. Water injection rate is 10 B/D/ft [5.22 m^3/d/m] sand and is constant throughout the flood. There is no initial gas saturation. Relative permeability data from Problem 3.1 will be used. Your evaluation should include preparation of plots of oil rate (STB) vs. time, WOR vs. cumulative PV injected on a semilog scale, and recovery (percent of original oil in place) vs. cumula-

Rge. 19 W.5.M

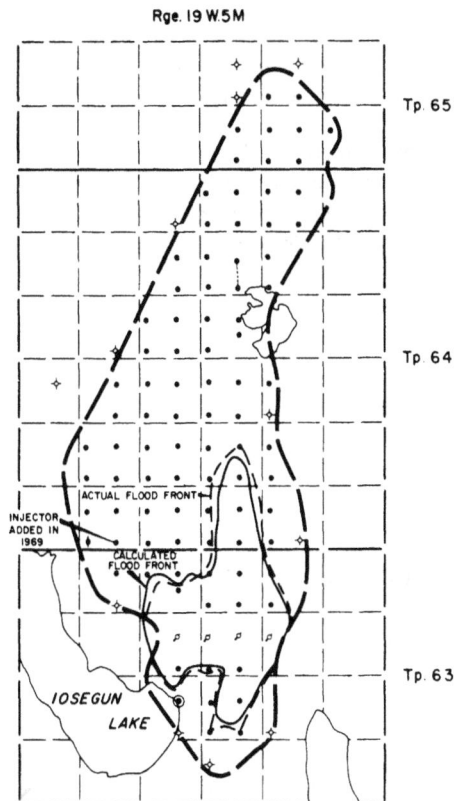

Fig. 4.24—Flood-front location, Kaybob waterflood.[42]

TABLE 4.9—RESERVOIR, FLUID AND ROCK PROPERTIES FOR PROBLEM 4.1

Porosity	0.206
Permeability, md	23
S_{oi}	0.637
Water viscosity, cp	1.0
Oil viscosity, cp	10.0
Formation thickness, ft	20
S_{or}	0.205

Relative Permeability Parameters	
α_1	1.0
m	2.56
α_2	0.78
n	3.72

TABLE 4.10—RESERVOIR, FLUID AND ROCK PROPERTIES FOR PROBLEM 4.2

Pattern area, acres	160
Porosity, %	11
Thickness, ft	5
Permeability, md	50
S_{iw}, %	35
S_{or}, %	25
B_o, bbl/STB	1.4
B_w, bbl/STB	1.0
μ_o, cp	1.22
μ_w, cp	0.9

TABLE 4.11—RESERVOIR AND FLUID PROPERTIES FOR PROBLEM 4.5

Area, acres	167
Net thickness, ft	17.5
Permeability, md	100
S_{oi}	0.80
S_{or}	0.30
Porosity	0.19
Oil gravity, °API	28.6
Oil viscosity, cp	0.7
Water viscosity, cp	1.0
B_o, bbl/STB	1.25
B_w, bbl/STB	1.00

tive PV's injected. Cutoff WOR is 50. Estimate the pressure drop between the injection and production wells as a function of time.

Other information needed for this problem is given in Table 4.10.

4.3. Suppose the initial water saturation in the reservoir described in Problem 4.2 was 40% so that water was mobile at the beginning of the waterflood. Can the CGM model be adapted to handle a mobile water saturation? Prepare a discussion with arguments supporting your case.

4.4. Rework Problem 4.2 when there is an initial gas saturation of 0.2. The relative permeability of gas is 0.001 and the viscosity of gas is 0.015 cp [0.015 mPa·s]. Assume that the waterflood is operated at a high enough pressure to redissolve the trapped gas. See Appendix E of Ref. 1 for the procedure to incorporate an initial gas saturation into the computations. This method is discussed in Chap. 6.

4.5. A waterflood is planned for the Delta field. Properties of the reservoir and the oil are summarized in Table 4.11. The field will be developed with 5-acre [20 234-m²] five-spot patterns. With the CGM model, estimate the following: (1) the areal sweep efficiency when 30,000 bbl [4770 m³] of water have been injected, (2) oil recovery at breakthrough in barrels, and (3) the injection rate schedule (barrels per day) for the field if the pressure drop between injection and production wells is maintained constant at 500 psi [3.5 MPa]. Radius of injection and production wells is 1.0 ft [0.3 m].

Relative permeability data are represented by Eqs. 4.82 and 4.83.

$$k_{ro} = 0.621(1 - S_{wD})^{1.638} \quad \dots\dots\dots\dots\dots (4.82)$$

and

$$k_{rw} = 0.74 S_{wD}^{2.146}, \quad \dots\dots\dots\dots\dots\dots (4.83)$$

with S_{wD} given by Eq. 3.111. The initial gas saturation is zero.

4.6. Problem 3.17 concerns the evaluation of an edgewater drive in a large reservoir to supplement declining reservoir energy. The reservoir shown in Fig. 3.64 could also be waterflooded with five-spot patterns by drilling injection wells in the center of each block to make 20 patterns. With the reservoir and fluid parameters from Problem 3.17, predict the waterflood performance of this field under pattern development. What are the advantages of five-spot waterflooding for this reservoir? Could other patterns be used?

TABLE 4.12—RESERVOIR AND FLUID PROPERTIES FOR PROBLEM 4.7

S_{iw}	0.2
S_{or}	0.3
Oil viscosity, cp	6.0
Water viscosity, cp	1.0
B_o, bbl/STB	1.6
B_w, bbl/STB	1.05

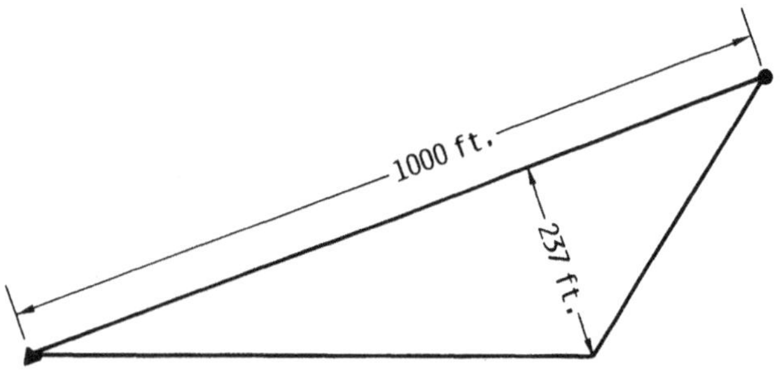

Fig. 4.25—Geometry of streamtube.

4.7. A reservoir is to be waterflooded on 5-acre [20 234-m²] spacing. The reservoir is 30 ft [9 m] thick and has a porosity of 0.20. The primary producing mechanism was solution gas drive, and a gas saturation of 0.10 is estimated to exist in the reservoir when waterflooding begins. Relative permeability data have been obtained and correlated in terms of dimensionless saturation as in Eqs. 4.84 and 4.85. Other parameters are given in Table 4.12.

$$k_{ro} = 0.75(1 - S_{wD})^2 \quad \dots \dots \dots \dots \dots (4.84)$$

and

$$k_{rw} = 0.30 S_{wD}^3, \quad \dots \dots \dots \dots \dots \dots (4.85)$$

with S_{wD} given by Eq. 3.111.

If water injection rate is 100 B/D [15.9 m³/d], determine the time in days when the first oil will be produced, assuming that the trapped gas saturation is redissolved. Estimate the waterflood performance to a WOR of 100 with the CGM model.

4.8. The method used to compute WOR in the CGM model should yield results comparable to those obtained by computing the WOR from incremental water and oil production. F_{wo} can be computed using finite-difference approximations as in Eq. 4.86. Recalling from the definition of F_{wo} in Eq. 4.26,

$$F_{wo} = \frac{dW_p}{dN_p},$$

$$F_{wo} = \frac{W_p \big|_{i+1} - W_p \big|_{i-1}}{N_p \big|_{i+1} - N_p \big|_{i-1}}, \quad \dots \dots \dots \dots (4.86)$$

where

W_p = cumulative water produced, and
N_p = cumulative oil produced, STB [stock-tank m³].

Compute the WOR between breakthrough and $E_A = 1.0$ from the computations presented in Table 4.1. Compare your values of F_{wo} with those presented in Table 4.1.

4.9. Another method of computing the WOR for the CGM model can be derived from the equations used to compute displacement performance for the region between E_{Abt} and $E_A = 1.0$. Because

$$F_{wo} = \frac{dW_p}{dN_p} \quad \dots \dots \dots \dots \dots \dots (4.87)$$

and

$$W_p = W_i - N_p,$$

then

$$\frac{dW_p}{dN_p} = \frac{V_p}{(dN_p/dQ_i)} - 1, \quad \dots \dots \dots \dots (4.88)$$

where $N_p = E_A(\overline{S_{w5}} - S_{iw})$.

Show that

$$F_{wo} = \frac{1}{f_{o2} + \dfrac{0.2749(\overline{S_{w5}} - S_{iw})}{Q_i}} - 1. \quad \dots \dots (4.89)$$

Use Eq. 4.89 to compute the WOR in Table 4.1. Compare your results with those obtained in Problem 4.8.

4.10. One advantage of streamtube models is that the total flow rate can be computed at every point in the displacement process. Demonstrate your understanding of streamtube models by calculating the initial oil rate in barrels per day from Stream Channel 2 of a direct linedrive pattern. Shape factors (G_j) determined by Higgins and Leighton for a direct linedrive pattern are given in Table B.1 of Appendix B. A pressure drop of 500 psi [3.4 MPa] is maintained between injection and production wells. The effective radius of the production well is 3.0 ft [0.9 m], while the effective radius of the injection well is 1.0 ft [0.3 m]. Well spacing is 20 acres [8.1 ha]. Permeability to oil at connate water saturation is 75 md. The oil viscosity is 2.5 cp [2.5 mPa·s].

4.11. The injection rate for a streamtube model at unit mobility can be computed from shape factors presented in Table 4.6 and Tables B.1 through B.3 of Appendix B. Compare the injection rates for five-spot, direct linedrive, staggered linedrive, and seven-spot patterns with those obtained from analytical expressions. Equations for injection rates at unit mobility are given in Table 6.1. Your computations should assume a pattern area of 10 acres [40 469 m²], fluid viscosity of 2 cp [2 mPa·s], and an average permeability of 200 md. The pressure difference between injection and production wells is 1,000 psi [6.9 MPa]. Report your computed rates in barrels per day per foot. The wellbore radius is 0.5 ft [0.15 m] for all patterns.

4.12. A streamtube for one part of an irregular waterflood pattern is shown in Fig. 4.25. Assume that the reservoir and rock properties are identical to those in Problem 4.1.

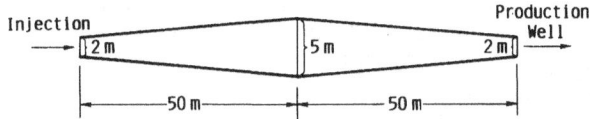

Fig. 4.26—Streamtube for Problem 4.14.

Compute the cumulative oil recovery (barrels) from this streamtube when the WOR is 50.

Compute the volume of water injected (barrels) when the WOR is 50.

Estimate the oil production rate from this streamtube at the beginning of the waterflood and the water injection rate when the entire streamtube has been reduced to residual oil saturation. The pressure drop across the streamtube is 750 psia [5 MPa]. Wellbore radius is 0.5 ft [0.15 m].

4.13. The data in Table 4.13 are streamtube parameters for Streamtube 4 in one-eighth of a five-spot. One of the streamlines of Streamtube 4 is a straight line drawn from the injection well to the production well. This streamtube contains 21.4% of the volume of one-eighth of the five-spot.

Demonstrate your understanding of streamtube modeling concepts by determining the following parameters for a waterflood of a 10-acre [40 469-m^2] five-spot pattern.

How many barrels of water have been injected into Streamtube 4 when water production begins?

What is the water saturation at $\xi = 2.822$ when the WOR is 10 at the end of the streamtube?

The reservoir is 10 ft [3 m] thick and has a porosity of 0.20. Other reservoir, fluid, and relative permeability relationships are from Problem 4.5.

4.14. A streamtube for one part of an irregular pattern is shown in Fig. 4.26.

What is the volume of water injected when the flood front reaches the midpoint of the streamtube?

Compute the cumulative oil recovery (in barrels) from this streamtube when the WOR is 20.

Compute the volume of water injected (in barrels) when the WOR is 20.

Use fluid and rock properties from Example 3.4.

4.15. A reservoir drilled on 160-acre [647 500-m^2] spacing for primary production will be waterflooded in a pattern arrangement. There are several possibilities. The reservoir could be waterflooded with a direct linedrive pattern by converting alternate rows of production wells to injectors. Another alternative is to drill infill wells to obtain 40-acre [161 857-m^2] five-spot patterns. Further infill drilling could create 20- or 10-acre [80 938- or 40 469-m^2] patterns if justified economically. The reservoir is 6,400 acres [25.9×10^6 m^2] in areal extent. Reservoir rock and fluid properties are given in Table 4.14.

With the Higgins-Leighton model, prepare waterflood performance estimates for this field assuming that all wells are drilled before the waterflood begins. Compare the development schemes in terms of rate, timing, and implied capital requirement based on wells drilled and water injection facilities required. Discuss how an operator would proceed with a project evaluation.

TABLE 4.13—STREAMTUBE PARAMETERS—STREAMTUBE 4

ξ (L)	$A(\xi)$ (L^2)	$V_p(\xi)$ (L^3)
0.0156	0.00174	0.000007
0.6048	0.1386	0.01048
1.342	0.3066	0.05155
2.080	0.4667	0.1232
2.822	0.5954	0.2226
3.5677	0.6593	0.3410
4.3145	0.6378	0.4636
5.0578	0.5354	0.5723
5.7972	0.3864	0.6555
6.5347	0.2155	0.70833
7.4040	0.0139	0.73009

Total length of streamtube coordinate = 7.4040.
Total volume of Streamtube 4 = 0.73009.
Fraction of pattern volume in Streamtube 4 = 0.214.
Length, area, and volume units are scaled—that is,
Distance ξ is a linear fraction of the actual length.

TABLE 4.14—RESERVOIR AND FLUID PROPERTIES FOR PROBLEM 4.15

Area, acres	6,400
Thickness, ft	10
Porosity	0.20
S_{oi}	0.75
S_{or}	0.25
S_{gi}	0.18
Permeability, md	75
B_o, bbl/STB	1.40
B_w, bbl/STB	1.00
Oil viscosity, cp	1.5
Water viscosity, cp	0.45
k_{rg}	0.10
r_w, ft	1.5
Δp, psi	2,000

Relative permeability relationships are given by Eqs. 4.90 and 4.91.

$$k_{ro} = 0.335(1 - S_{wD})^{2.06} \quad \ldots\ldots\ldots\ldots\ldots (4.90)$$

and

$$k_{rw} = 0.276 S_{wD}^{1.45}, \quad \ldots\ldots\ldots\ldots\ldots\ldots (4.91)$$

with S_{wD} given by Eq. 3.111.

4.16. In Appendix E of Ref. 1, a sample calculation of waterflood performance for a five-spot pattern is presented by use of the CGM model. The example calculation is for a 5-ft [1.5-m] layer that has a permeability of 31.5 md when the flood is operated under a constant pressure drop of 3,000 psi [21 MPa]. Properties of the reservoir are given in Table 4.15. Relative permeability data are given in Table 4.16. Predict the waterflood performance with the Higgins-Leighton model. You will have to adjust your results for the FVF because the Higgins-Leighton program assumes that the FVF's for oil and water are 1.0. Compare your results to those presented in Ref. 1 in Tables E.6 through E.8 and identify possible causes for differences.

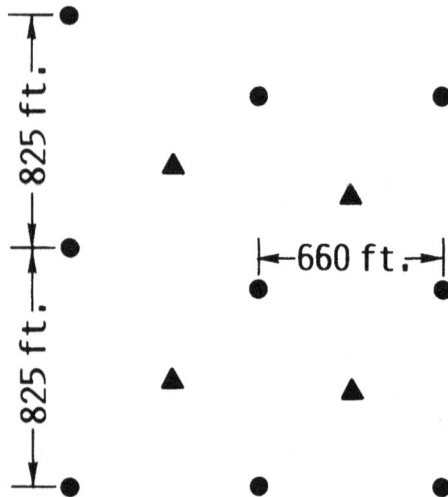

Fig. 4.27—Flood pattern with irregular well arrangement.

TABLE 4.15—PROPERTIES OF EXAMPLE RESERVOIR FOR PROBLEM 4.16

Well spacing, acres	20
Thickness, ft	50
Average permeability, md	10
Porosity, %	20
Interstitial water saturation, %PV	10
Current average gas saturation, %PV	15
Oil viscosity at current reservoir pressure, cp	1.0
Water viscosity, cp	0.5
Reservoir pressure, psi	1.000
Current oil recovery, % of initial oil in place	10.4
Oil reservoir volume factor, at original saturation pressure	1.20
Five-spots in existing wells	40
Flooding pattern	
Pattern area,* acres	40
Wellbore radius, ft	1.0

*Complete five-spot pattern contains two total wells, one injector and one producer.

$\Delta p = 3,000$ psi between injection and production wells.

4.17. Sometimes surface topology controls locations of production wells. This is not a problem during primary production. Irregular patterns (such as those depicted in Fig. 4.27), however, are formed when this type of field is waterflooded. You are requested to estimate the waterflood performance of this reservoir with the reservoir rock and fluid properties given in Problem 4.1. The waterflood will be operated at an injection rate of 200 B/D [32 m^3/d], which is split evenly between the four injection wells. What assumptions are necessary to estimate waterflood performance with the methods described in this chapter? Do you need additional information or programs? Be specific in your response.

Nomenclature

A = cross-sectional area for flow, sq ft [m^2]

A_j = cross-sectional area at Position j in a streamtube, sq ft [m^2]

A_m = area in middle of the streamtube, sq ft [m^2]

A_o = area at beginning of streamtube, sq ft [m^2]

A_T = surface area of symmetric element in a streamtube model, sq ft [m^2]

A^* = surface area enclosed by the pair of streamlines that define the streamtube, sq ft [m^2]

B_o = oil FVF, bbl/STB [m^3/stock-tank m^3]

B_w = water FVF, bbl/STB [m^3/stock-tank m^3]

d = distance between injection well and production well, ft [m]

d^* = distance between injection well and production well in a streamtube model different from the d/r_w used to derive shape factors for the streamtube, ft [m]

E_A = fraction of the area swept to flood-front saturation

E_{Abt} = areal sweep efficiency at breakthrough, fraction

f_{o2} = fractional flow of oil—q_o/q_t

f_w = fractional flow of water—q_w/q_t

f_{w2} = fractional flow of water corresponding to S_{w2}

f' = derivative of f_w with respect to S_w

f'_{Sw} = derivative of f_w with respect to S_w

f'_{Sw2} = f'_{Sw} when saturation S_{w2} is at the end of the system

TABLE 4.16—RELATIVE PERMEABILITY AND FRACTIONAL FLOW DATA FOR PROBLEM 4.16

Water Saturation, S_w (fraction)	Relative Permeability		Fractional Flow of Water, f_w
	Oil, k_{ro} (fraction)	Water, k_{rw} (fraction)	
0.10	1.000	0.000	0.0000
0.30	0.373	0.070	0.2729
0.40	0.210	0.169	0.6168
0.45	0.148	0.226	0.7533
0.50	0.100	0.300	0.8571
0.55	0.061	0.376	0.9250
0.60	0.033	0.476	0.9665
0.65	0.012	0.600	0.9901
0.70	0.000	0.740	1.0000

f'_{Sw5} = f'_{Sw} for five-spot pattern in the CGM model

F_{wo} = water/oil ratio, bbl/bbl [m^3/m^3]

g = gravitational constant, ft/sec^2 [m/s^2]

G_j = shape factor in the Higgins-Leighton model, L_j/A_j, ft^{-1} [m^{-1}]

h = thickness of reservoir, ft [m]

i = injection rate, B/D [m^3/d]

i_b = injection rate when oil flows at interstitial water saturation in a pattern flood under a specified pressure drop, B/D [m^3/d]

j = index of cell location in a streamtube

k = absolute liquid permeability, darcies

k_b = base permeability for relative permeabilities, darcies

k_o = permeability to oil phase, darcies

k_{ro} = relative permeability to oil phase, fraction

k_{rw} = relative permeability to water phase, fraction

k_w = water permeability, darcies

L = length of system, ft [m]

L_j = length of Cell j in streamtube, ft [m]

$M_{\bar{S}}$ = mobility ratio based on average water saturation behind the flood front

$M_{\overline{Sw5}}$ = mobility ratio for a five-spot based on the average water saturation in the swept area

M_t = total mobility ratio

N_p = oil produced, STB [stock-tank m^3]

N_{pbt} = oil produced to breakthrough, STB [stock-tank m^3]

N_{pi} = oil produced from Streamtube i, STB [stock-tank m^3]

N_{ps} = oil displaced from previously swept region—CGM model, STB [stock-tank m^3]

N_{pu} = oil displaced from newly swept region—CGM model, STB [stock-tank m^3]

p = pressure, psi [kPa]

p_I = pressure at injection well, psi [kPa]

p_o = oil-phase pressure, psi [kPa]

p_p = pressure at production well, psi [kPa]

p_w = water-phase pressure, psi [kPa]

Δp = $p_I - p_p$, psi [kPa]

P_c = capillary pressure $= p_o - p_w$, psi [kPa]

q_o = oil production rate, B/D [m^3/d]

q_t = total production rate, B/D [m^3/d]

q_w = water production rate, B/D [m^3/d]

Q_i = PV's of fluid injected

Q_{i2} = PV's of fluid injected when S_{w2} arrives at end of linear system

Q_i^* = number of water-contacted PV's injected

Q_{ibt}^* = number of water-contacted PV's at breakthrough

$(Q_i)_{max}$ = maximum number of water-contacted PV's

r = radius, ft [m]

r_1 = radius at Position 1, ft [m]

r_w = wellbore radius, ft [m]

r_1^* = radius at Position 1 in a pattern where d^* is the distance between injection and production wells, ft [m]

S_g = gas saturation, fraction

S_{iw} = interstitial water saturation where the water phase is immobile under the applied pressure gradient, fraction

S_o = oil saturation, fraction

S_{oi} = initial oil saturation, fraction

S_{or} = residual oil saturation, fraction

S_w = water saturation, fraction

S_{wf} = flood-front saturation

S_{w2} = water saturation at end of a linear system, fraction

$\overline{S_{wD}}$ = dimensionless average water saturation, $(\overline{S_w} - S_{iw})/(1 - S_{or} - S_{iw})$

$\overline{S_{w5}}$ = average water saturation in five-spot, fraction

V_p = pore volume, bbl [m^3]

V_{pSw} = PV displaced to saturation S_w, bbl [m^3]

V_{pSwf} = PV displaced to S_{wf}, bbl [m^3]

V_{pT} = total PV of a streamtube, bbl [m^3]

V_{pTi} = total PV of Streamtube i, bbl [m^3]

$V_{p\xi i}$ = PV of streamtube at Location ξ_i, bbl [m^3]

W_i = cumulative volume of water injected, bbl [m^3]

x, y, z = coordinates

γ = conductance ratio—i/i_b

$\Delta\theta$ = angle between adjacent streamlines

λ_r^{-1} = apparent viscosity, cp [Pa·s]

λ_{ro} = relative mobility of oil, cp^{-1} [(Pa·s)$^{-1}$]

λ_{rw} = relative mobility of water, cp^{-1} [(Pa·s)$^{-1}$]

$\overline{\lambda_r^{-1}}$ = average apparent viscosity, cp [Pa·s]

$\overline{\lambda_{ro}}$ = average relative mobility of oil, cp^{-1} [(Pa·s)$^{-1}$]

$\overline{\lambda_{rw}}$ = average relative mobility of water, cp^{-1} [(Pa·s)$^{-1}$]

μ = viscosity, cp [Pa·s]

μ_o = viscosity of oil, cp [Pa·s]

μ_w = viscosity of water, cp [Pa·s]

ξ = streamtube coordinate

ξ_I = streamtube coordinate at the injection wellbore radius, ft [m]

ξ_P = streamtube coordinate at the production wellbore radius, ft [m]

ρ_o = density of oil, lbm/cu ft [kg/m^3]

ρ_w = density of water, lbm/cu ft [kg/m^3]

ϕ = porosity, fraction

References

1. Craig, F.F. Jr.: *The Reservoir Engineering Aspects of Waterflooding*, Monograph Series, SPE, Richardson, TX (1971) **3**.
2. Wattenbarger, R.A., Howell, B.L., and Loye, P.E.: "A Successful Peripheral Water Flood in a Thin Pennsylvanian Reservoir," *J. Pet. Tech.* (Nov. 1964) 1238–42.
3. Douglas, J. Jr., Peaceman, D.W., and Rachford, H.H.: "A Method for Calculating Multi-Dimensional Immiscible Displacement," *Trans.* AIME, (1959) **216**, 297–308.

4. Slobod, R.L. and Caudle, B.H.: "X-Ray Shadowgraph Studies of Areal Sweepout Efficiencies," *Trans.*, AIME (1952) **195**, 265-70.

5. Dyes, A.B., Caudle, B.H., and Erickson, R.A.: "Oil Production After Breakthrough as Influenced by Mobility Ratio," *J. Pet. Tech.* (April 1954) 27-32; *Trans.*, AIME, **201**.

6. Caudle, B.H. and Witte, M.D.: "Production Potential Changes During Sweep-Out in a Five-Spot System," *J. Pet. Tech.* (Dec. 1959) 63-65; *Trans.*, AIME, **216**.

7. Caudle, B.H. and Loncaric, I.G.: "Oil Recovery in Five-Spot Pilot Floods," *J. Pet. Tech.* (June 1960) 132-36; *Trans.*, AIME, **219**.

8. Kimbler, O.K., Caudle, B.H., and Cooper, H.E. Jr.: "Areal Sweepout Behavior in a Nine-Spot Injection Pattern," *J. Pet. Tech.* (Feb. 1964) 199-202; *Trans.*, AIME, **231**.

9. Caudle, B.H., Hickman, B.M., and Silberberg, I.H.: "Performance of the Skewed Four-Spot Injection Pattern," *J. Pet. Tech.* (Nov. 1968) 1315-19; *Trans.*, AIME, **243**.

10. Guckert, L.G.: "Areal Sweepout Performance of Seven and Nine-Spot Flood Patterns," MS thesis, Pennsylvania State U., University Park, PA (1961).

11. Craig, F.F. Jr., Geffen, T.M., and Morse, R.A.: "Oil Recovery Performance of Pattern Gas or Water Injection Operations from Model Tests," *J. Pet. Tech.* (Jan. 1955) 7-15; *Trans.*, AIME, **204**.

12. *Handbook of Mathematical Functions*, M. Abramowitz and I.A. Stegan (eds.), Applied Mathematics Series 55, Natl. Bureau of Standards, Washington, DC (1970) 229.

13. Higgins, R.V. and Leighton, A.J.: "Computer Predictions of Water Drive of Oil and Gas Mixtures Through Irregularly Bonded Porous Media—Three-Phase Flow," *J. Pet. Tech.* (Sept. 1962) 1048-54; *Trans.*, AIME, **225**.

14. Higgins, R.V., Boley, D.W., and Leighton, A.J.: "Aids to Forecasting the Performance of Waterfloods," *J. Pet. Tech.* (Sept. 1964) 1076-82; *Trans.*, AIME, **231**.

15. Higgins, R.V. and Leighton, A.J.: "Principles of Computer Techniques for Calculating Performance of a Five-Spot Waterflood Two-Phase Flow," RI 6305, USBM (1963).

16. Higgins, R.V., Boley, D.W., and Leighton, A.J.: "Computer Techniques for Calculating Shape Factors and Channel Volumes from a Potentiometric Model for Uses in Waterflood Performance Calculations," RI 6760, USBM (1966).

17. Higgins, R.V. and Leighton, A.J.: "Improved Method for Calculating Areas and Shape Factors of Flow Rates," RI 7111, USBM (1968).

18. Leighton, A.J. and Higgins, R.V.: "Improved Method to Predict Multiphase Waterflood Performance for Constant Rates or Pressures," RI 8055, USBM (1975).

19. LeBlanc, J.L. and Caudle, B.H.: "A Streamline Model for Secondary Recovery," *Soc. Pet. Eng. J.* (March 1971) 7-12.

20. Gursa, B. and Helander, D.P.: "Shape Factor Analysis for Peripheral Waterflood Prediction by the Channel-Flow Technique," *Prod. Monthly* (April 1967) **31**, 2-8.

21. Kantar, K. and Helander, D.P.: "Graphical Technique for Interpolating Shape Factors for Peripheral Waterflood Systems," *Prod. Monthly* (June 1967) **31**, 2-4.

22. Kantar, K. and Helander, D.P.: "Simplified Shape Factor Analysis Applied to an Unconfined Peripheral Waterflooding System," *Prod. Monthly* (April 1967) **31**, 4.

23. Torres, S.R., Salazar, J.T., and Rosales, P.: "Calculation of Shape Factors in Two-Dimensional Fluid Injection Model," *Oil and Gas J.* (Sept. 28, 1970) 61-65.

24. Henley, D.H.: "Method for Studying Waterflooding Using Analog, Digital and Rock Models," *Proc.* 24th Technical Conference on Petroleum, Pennsylvania State U., University Park, PA (Oct. 23-25, 1963).

25. Higgins, R.V. and Leighton, A.J.: "Numerical Methods for Determining Streamlines and Isopressures for Use in Fluid Flow Studies," RI 7621, USBM (1972).

26. Wang, G.C.: "Streamline Model Studies of Viscous Waterfloods," PhD dissertation, U. of Texas, Austin, TX (1970).

27. James, Lin Jer-Kuan: "An Image-Well Method for Bounding Arbitrary Reservoir Shapes in the Streamline Model," PhD dissertation, U. of Texas, Austin, TX (1972).

28. Collins, R.E.: *Flow of Fluids Through Porous Materials*, Reinhold Publishing Corp., New York City (1961).

29. Hurst, W.: "Determination of Performance Curve in Five-Spot Waterflood," *Pet. Eng.* (April 1953) B-40 - B-48.

30. Parsons, R.W.: "Directional Permeability Effects in Developed and Unconfined Five-Spots," *J. Pet. Tech.* (April 1972) 487-94; *Trans.*, AIME, **253**.

31. Martin, J.C. and Wegner, R.E.: "Numerical Solution of Multiphase, Two-Dimensional Incompressible Flow Using Stream-Tube Relationships," *Soc. Pet. Eng. J.* (Oct. 1979) 313-23; *Trans.*, AIME, **267**.

32. Higgins, R.V. and Leighton, A.J.: "A Computer Method of Calculating Two-Phase Flow in Any Irregularly Bounded Porous Medium," *J. Pet. Tech.* (June 1962) 679-83; *Trans.*, AIME, **225**.

33. Patton, J.T., Coats, K.H., and Colegrove, G.T.: "Prediction of Polymer Flood Performance," *Soc. Pet. Eng. J.* (March 1971) 72-84; *Trans.*, AIME, **251**.

34. Martin, J.C., Woo, P.T., and Wegner, R.E.: "Failure of Streamtube Methods to Predict Waterflood Performance of an Isolated Inverted Five-Spot at Favorable Mobility Ratios," *J. Pet. Tech.* (Feb. 1973) 151-53.

35. Peaceman, D.W.: *Fundamentals of Numerical Reservoir Simulation*, Elsevier Scientific Publishing Co., New York City (1977).

36. Aziz, K. and Settari, A.: *Petroleum Reservoir Simulation*, Applied Science Publishers Ltd., London (1979).

37. Thomas, G.W.: *Principles of Hydrocarbon Reservoir Simulation*, second edition, IHRDC, Boston, MA (1981).

38. Critchlow, H.B.: *Modern Reservoir Engineering—A Simulation Approach*, Prentice Hall Inc., Englewood Cliffs, NJ (1977).

39. *Numerical Simulation*, Reprint Series, SPE, Richardson, TX (1973) **11**.

40. McCulloch, R.C., Longton, J.R., and Spivak, A.: "Simulation of High Relief Reservoirs, Rainbow Field, Alberta, Canada," *J. Pet. Tech.* (Nov. 1969) 1399-1408.

41. McCarty, D.G. and Barfield, E.C.: "The Use of High-Speed Computers for Predicting Flood-Out Patterns," *Trans.*, AIME (1958) **213**, 139-45.

42. Mann, L. and Johnson, G.A.: "Predicted Results of Numeric Grid Models Compared With Actual Field Performance," *J. Pet. Tech.* (Nov. 1970) 1390-98.

SI Metric Conversion Factors

acre $\times$ 4.046 873	E+03	= m^2
bbl $\times$ 1.589 873	E−01	= m^3
cp $\times$ 1.0*	E−03	= Pa·s
cu ft $\times$ 2.831 685	E−02	= m^3
ft $\times$ 3.048*	E−01	= m
psi $\times$ 6.894 757	E+00	= kPa
sq ft $\times$ 9.290 304*	E−02	= m^2

*Conversion factor is exact.

Chapter 5
Vertical Displacement in Linear and Areal Models

5.1 Introduction

Methods of predicting waterflood performance in linear and areal systems that were presented in Chaps. 3 and 4 considered the displacement process to be uniform in the vertical cross section. Thus there is no saturation change with vertical position. We know that displacement may not proceed uniformly in reservoirs for several reasons. The main reason is that oil reservoirs are usually not homogeneous and isotropic. Geologic processes lead to variations of permeabilities and porosities areally and vertically. The largest changes usually occur in permeabilities that may vary an order of magnitude or more in a vertical cross section through a reservoir. These differences in rock properties cause fluid flow rates to vary with vertical position. Even in homogeneous reservoirs, gravity and capillary forces may induce vertical flows of oil and water that influence displacement efficiency.

This chapter presents methods of accounting for the effects of vertical fluid flow on waterflood displacement efficiency. These methods are developed for linear systems to illustrate concepts. Extensions of these concepts to areal models allows approximation of performance in three dimensions. The use of numerical simulators in solving multidimensional problems is illustrated.

5.2 Two-Dimensional (2D) Displacement in Uniform Stratified Reservoirs— Layered Models

This section concerns displacement efficiency in a linear stratified reservoir. We will examine methods used to predict displacement performance for two classes of problems. The first class, called layered methods, assumes that displacement performance is controlled primarily by vertical heterogeneities. Crossflows between layers that have different permeability and/or porosity are neglected. In effect, layers are viewed as if they are separated by impermeable shale barriers.

While shale barriers do exist in some reservoirs, they may not be continuous from well to well nor impermeable. Vertical flows or crossflows of oil, water, and gas may occur because of gravity, capillary, or viscous forces that change over the vertical cross section. Models that account for fluid movement caused by these effects are presented in Secs. 5.3 through 5.5.

Fig. 5.1 shows a cross section between an injection well and a production well in a reservoir that has significant permeability change with depth. The reservoir has been subdivided into layers with properties k_j, ϕ_j, S_{wj}, and h_j. The initial water saturation is uniform in each layer.

A waterflood initiated in this reservoir may be conducted at a specified injection rate or a specified potential change between injection and production wells. It makes no difference which approach is used as far as this section is concerned. Calculations with constant potential change are less complicated and will be used to illustrate simulation of displacement performance in layered reservoirs.

Consider a displacement process in the reservoir shown in Fig. 5.1, in which the potential change between injection well and production well is constant. Changes in average density of the produced fluid caused by two-phase production also will be neglected. Thus the pressure difference across each layer is the same. The subscript j is introduced to identify the layer in the simulation. Corresponding parameters describing the layers are

t_k = cumulative injection time,

N_{pj}^k = cumulative oil displaced at t_k from Layer j,

V_{pj} = pore volume of Layer j,

q_{tj}^k = instantaneous water injection rate in Layer j at time t_k, and

f_{wj}^k = fractional flow of water leaving Layer j at time t_k.

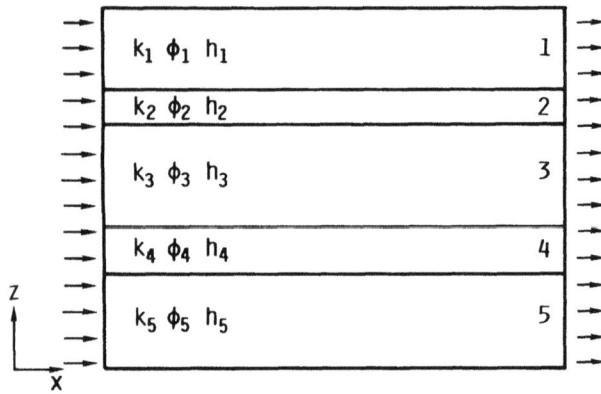

Fig. 5.1—Reservoir subdivided into five noncommunicating layers that have different properties.

Reservoir simulators produce estimates of N_{pj}^k, q_{tj}^k, and f_{wj}^k for a given layer at time t_k.

Composite performance at any time can be estimated by combination of the performance from each layer at the same cumulative injection time, t. Thus, when $t = t_1 = t_2 \ldots t_k$,

$$N_p = \sum_{j=1}^{m} N_{pj}^k, \quad\ldots\ldots\ldots\ldots\ldots\ldots (5.1)$$

$$q_t = \sum_{j=1}^{m} q_{tj}^k, \quad\ldots\ldots\ldots\ldots\ldots\ldots (5.2)$$

and

$$f_w = \frac{\sum_{j=1}^{m} f_{wj} q_{tj}}{q_t}, \quad\ldots\ldots\ldots\ldots\ldots (5.3)$$

where m is the number of layers. Other parameters—such as oil cut, oil production rate, and PV's injected—can be determined in the same manner. Because the layers perform independently, it makes no difference which model is selected to estimate performance. If sufficient data are available, simulating each individual layer and combining results may be desirable.

In many cases, the quality of reservoir data is such that a single initial water saturation and a single set of relative permeability curves are used for every layer. Displacement calculations are simplified considerably because it is necessary to compute the displacement performance for only one layer. Performance of all other layers can be computed by the procedure outlined in the following paragraphs.

For this illustration, consider linear displacement in the layers in Fig. 5.1. From the frontal advance solution, the location of any saturation $S_w \geq S_{wf}$ is given by Eq. 5.4.

$$x_{Sw} = L Q_i f'_{S_w}. \quad\ldots\ldots\ldots\ldots\ldots\ldots (5.4)$$

When a common set of relative permeability curves is used, Eq. 5.4 states that the saturation distributions in each layer are identical at the same value of Q_i. Thus, performance of all layers can be derived from one layer by development of appropriate relations between the various Q_i's.

In a layered reservoir, the layers are connected hydraulically only at injection and production wells. When the pressure difference between injection and production wells is constant, the flow rate in each layer will vary with time as the flow resistance changes because of the advancing displacement front. Estimation of waterflood performance for a constant-pressure flood was discussed in Sec. 3.6.

Prediction of displacement performance from several layers from the displacement performance of a single layer is based on computation of the time, t_j, injection rate, q_{tj}, and recovery, N_{pj}, at equal values of Q_i. The subscript j identifies the j layer. A value of $j = 1$ is assigned to the layer where Q_i, q_{ti}, N_p, etc., have been computed. When $Q_i = Q_{ij}$,

$$\frac{p_I - p_p}{\left(\int_0^L \lambda_r^{-1}\, dx\right)_1} = \frac{p_I - p_p}{\left(\int_0^L \lambda_r^{-1}\, dx\right)_j}, \quad\ldots\ldots\ldots\ldots (5.5)$$

so that

$$\frac{q_{t1}}{k_1 h_1} = \frac{q_{tj}}{k_j h_j}. \quad\ldots\ldots\ldots\ldots\ldots\ldots (5.6)$$

Therefore,

$$q_{tj} = \frac{k_j h_j}{k_1 h_1} q_{t1}. \quad\ldots\ldots\ldots\ldots\ldots\ldots (5.7)$$

The remaining step is to determine t_j for each value of t_1. From Eq. 5.6, when $Q_{i1} = Q_{ij}$,

$$\frac{\int_0^{t_1} q_{t1}\, dt}{V_{p1}} = \frac{\int_0^{t_j} q_{tj}\, dt}{V_{pj}}. \quad\ldots\ldots\ldots\ldots (5.8)$$

When the time t is estimated from the relationship between Q_i and q_t, as done in Sec. 3.6,

$$Q_{i1}^n = \frac{1}{V_{p1}} \sum_{k=1}^{n} \frac{(q_{t1}^k + q_{t1}^{k-1})}{2}(t_j^k - t_j^{k-1}) \quad\ldots\ldots (5.9)$$

or

$$Q_{ij}^n = \frac{1}{V_{pj}} \sum_{k=1}^{n} \frac{(q_{tj}^k + q_{tj}^{k-1})}{2}(t_j^k - t_j^{k-1}), \quad\ldots\ldots (5.10)$$

where the superscript k is the index on the number of values of Q_i computed for Layer 1. The substitution of Eq. 5.7 into Eq. 5.10 yields

$$Q_{ij}^n = \frac{1}{V_{pj}}\left(\frac{k_j h_j}{k_1 h_1}\right) \sum_{k=1}^{n} \left(\frac{q_{t1}^k + q_{t1}^{k-1}}{2}\right)(t_j^k - t_j^{k-1}).$$

$$\ldots\ldots\ldots\ldots\ldots\ldots\ldots\ldots (5.11)$$

TABLE 5.1—PERMEABILITY DISTRIBUTION

Layer Number	Permeability (md)
1	406.8
2	240.7
3	167.4
4	116.4
5	68.8

TABLE 5.2—WATERFLOOD PERFORMANCE OF LAYER 1 BASED ON SCALING BASE CASE RESULTS FROM TABLE 3.4

Time (days)	q_o (B/D)	q_w (B/D)	q_t (B/D)	WOR	Q_i (PV)	N_p (PV)	N_p (bbl)
0.0	137.56	0.0	137.56	0.0	0.0	0.0	0.0
8.1	129.38	0.0	129.38	0.0	0.034	0.034	1,076.7
16.6	122.13	0.0	122.13	0.0	0.067	0.067	2,153.4
25.7	115.63	0.0	115.63	0.0	0.101	0.101	3,230.2
35.3	109.80	0.0	109.80	0.0	0.134	0.134	4,306.9
45.3	104.53	0.0	104.53	0.0	0.168	0.168	5,383.6
55.9	99.74	0.0	99.74	0.0	0.202	0.202	6,460.3
66.9	95.37	0.0	95.37	0.0	0.235	0.235	7,537.1
78.4	91.36	0.0	91.36	0.0	0.269	0.269	8,613.8
90.4	87.68	0.0	87.68	0.0	0.302	0.302	9,690.5
102.9	8.79	75.50	84.28	8.59	0.336	0.336	10,767.2
110.5	8.31	77.69	86.01	9.35	0.356	0.338	10,831.8
118.5	7.86	79.93	87.78	10.18	0.378	0.340	10,896.8
127.1	7.42	82.20	89.61	11.08	0.401	0.342	10,962.1
136.2	6.99	84.50	91.49	12.08	0.427	0.344	11,027.7
145.9	6.58	86.84	93.42	13.19	0.455	0.346	11,093.5
156.3	6.19	89.21	95.40	14.41	0.486	0.348	11,159.6
167.3	5.81	91.62	97.43	15.76	0.519	0.350	11,225.8
179.1	5.45	94.05	99.50	17.25	0.555	0.352	11,292.1
191.7	5.10	96.52	101.63	18.92	0.594	0.354	11,358.5
205.1	4.77	99.02	103.80	20.77	0.638	0.356	11,424.9
219.6	4.45	101.56	106.01	22.82	0.685	0.358	11,491.2
235.0	4.15	104.12	108.27	25.12	0.736	0.361	11,557.6
251.6	3.85	106.72	110.57	27.70	0.793	0.363	11,623.9
269.4	3.58	109.34	112.92	30.58	0.855	0.365	11,690.1
288.6	3.31	111.99	115.30	33.82	0.924	0.367	11,756.1
309.3	3.06	114.67	117.73	37.48	0.999	0.369	11,822.0
331.7	2.82	117.37	120.19	41.63	1.082	0.371	11,887.8
355.9	2.59	120.11	122.70	46.35	1.174	0.373	11,953.3
382.3	2.38	122.86	125.24	51.63	1.275	0.375	12,018.6
410.9	2.18	125.64	127.82	57.73	1.388	0.377	12,083.6
422.1	1.98	128.45	130.43	64.84	1.514	0.379	12,148.4
476.2	1.80	131.27	133.07	72.84	1.654	0.381	12,212.9
513.6	1.63	134.12	135.75	82.22	1.811	0.383	12,277.1
554.8	1.47	136.99	138.46	93.02	1.987	0.385	12,341.0
600.2	1.33	139.87	141.20	105.47	2.186	0.387	12,404.5
650.6	1.19	142.77	143.96	120.20	2.410	0.389	12,467.7
706.7	1.06	145.69	146.75	137.75	2.664	0.391	12,530.6
769.4	0.94	148.63	149.57	158.16	2.954	0.393	12,593.0
839.9	0.83	151.58	152.41	183.55	3.285	0.395	12,655.0
919.4	0.72	154.54	155.27	213.43	3.667	0.397	12,716.7
1,009.8	0.63	157.52	158.15	249.81	4.109	0.399	12,777.9
1,113.2	0.55	160.51	161.05	294.45	4.624	0.400	12,838.7
1,232.4	0.47	163.50	163.97	349.50	5.228	0.402	12,899.0
1,370.8	0.40	166.51	166.91	417.66	5.943	0.404	12,958.9
1,533.2	0.34	169.52	169.86	502.07	6.795	0.406	13,018.4
1,725.1	0.28	172.54	172.82	614.70	7.824	0.408	13,077.3
1,956.6	0.23	175.56	175.79	770.66	9.080	0.410	13,135.8
2,237.7	0.19	178.59	178.78	954.37	10.634	0.412	13,193.8
2,585.3	0.15	181.62	181.77	1,240.17	12.589	0.413	13,251.3
3,023.8	0.11	184.65	184.76	1,621.11	15.096	0.415	13,308.3
3,589.9	0.09	187.68	187.77	2,196.95	18.385	0.417	13,364.8

Recall that $V_{p1} = Lbh_1\phi_1$ and $V_{pj} = Lbh_j\,\phi_j$.

Comparing Eq. 5.9 with Eq. 5.11, we see that t_j for any layer is given by Eq. 5.12.

$$t_j^n = \left(\frac{k_1}{k_j}\right)\left(\frac{\phi_j}{\phi_1}\right)t_1^n. \quad\quad\quad\quad (5.12)$$

Example 5.1 illustrates application of this approach to estimate the waterflood performance when the reservoir in Example 3.5 has five noncommunicating layers of equal thickness.

Example 5.1. Performance of a uniform, homogeneous reservoir may be used to predict the displacement efficiency of the same reservoir when it is heterogeneous. One set of relative permeability curves is assumed to represent fluid/rock interactions. Initial and final oil saturations are assumed to be independent of permeability differences.

The procedure will be illustrated for a linear waterflood in the reservoir previously discussed in Example 3.5. The reservoir described in Sec. 3.5 is subdivided into five layers of equal thickness. Permeabilities of the layers are given in Table 5.1. These permeabilities were chosen so that the average permeability weighted on thickness was 200 md with a Dykstra-Parsons permeability variation (V) of 0.5. The term V is discussed in Sec. 5.8.1. The pressure drop is constant at 500 psi [3.5 MPa]. With these choices, Example 5.1 is identical to Example 3.5 with the exception of the permeability distribution.

The computed performance of a uniform, homogeneous reservoir with a thickness of 20.0 ft [6.1 m] and a permeability of 200 md is summarized in Table 3.4. Computation of equivalent performance from the data in Table 3.4 will be illustrated for Layer 1. Then we will show how results from individual layers can be combined to obtain the composite performance of the layered reservoir.

In the following section, the subscript b is used to identify values from Table 3.4 that are referred to as the base computations. The subscript j identifies a specific layer. When $Q_{ib} = Q_{ij}$, the times required to reach a given Q_i are related by Eq. 5.13 when $\phi_b = \phi_1 = \phi_j$.

$$t_1 = \frac{k_b}{k_1}t_b. \quad\quad\quad\quad\quad (5.13)$$

Let $Q_{ib} = 0.168$. From Table 3.4, $t_b = 92.1$ days. Then

$$t_1 = \frac{(200\ md)}{(406.8\ md)}(92.1\ days)$$

$$= 45.3\ days.$$

The value of Q_{ib} is 0.168 at t_b. Because of the higher permeability in Layer 1, the same value of Q_i is reached in Layer 1 in only 45.3 days. Production rates in Layer 1 are given by Eq. 5.14.

$$q_{t1} = (k_1h_1)/(k_bh_b)q_{tb}$$

$$= \frac{(406.8\ md)(4.0\ ft)(256.95\ B/D)}{(200.0\ md)(20.0\ ft)}$$

$$= 104.53\quad B/D. \quad\quad\quad\quad (5.14)$$

Because $q_{wb} = 0$, then $q_b = q_{ob}$, and $q_{ti} = q_{oi}$. However, if $q_{wb} \neq 0$, then

$$q_{w1} = \frac{(k_1h_1)}{(k_bh_b)}q_{wb}.$$

Table 5.2 contains the performance for Layer 1 computed from the base-case results in Table 3.4.

**TABLE 5.3—SELECTED RESULTS FOR LAYERS 1 THROUGH 5
COMPUTED FROM SINGLE-LAYER PERFORMANCE OF EXAMPLE 5.3**

		Layer 1		Layer 2		Layer 3		Layer 4		Layer 5	
Q_i	N_p (bbl)	t_1 (days)	q_{o1} (B/D)	t_2 (days)	q_{o2} (B/D)	t_3 (days)	q_{o3} (B/D)	t_4 (days)	q_{o4} (B/D)	t_5 (days)	q_{o5} (B/D)
0.0	0.0	0.0	137.56	0.0	81.40	0.0	56.61	0.0	39.36	0.0	23.27
0.034	0.034	8.1	129.38	13.6	76.55	19.6	53.24	28.2	37.02	47.7	21.88
0.067	0.067	16.6	122.13	28.1	72.26	40.4	50.26	58.1	34.94	98.3	20.65
0.101	0.101	25.7	115.63	43.4	68.42	62.4	47.58	89.7	33.09	151.7	19.56
0.134	0.134	35.3	109.80	59.6	64.97	85.7	45.18	123.2	31.42	208.4	18.57
0.168	0.168	45.3	104.53	76.5	61.85	110.0	43.01	158.2	29.91	267.7	17.68
0.202	0.202	55.9	99.74	94.4	59.01	135.7	41.04	195.2	28.54	330.2	16.87
0.235	0.235	66.9	95.37	113.0	56.43	162.5	39.24	233.7	27.29	395.3	16.13
0.269	0.269	78.4	91.36	132.5	54.06	190.6	37.60	274.1	26.14	463.7	15.45
0.302	0.302	90.4	87.68	152.8	51.88	219.7	36.08	316.0	25.09	534.6	14.83
0.336	0.336	102.9	8.79	174.0	5.20	250.2	3.62	359.8	2.51	608.7	1.49
0.356	0.338	110.5	8.31	186.8	4.92	268.6	3.42	386.3	2.38	653.5	1.41
0.378	0.340	118.5	7.86	200.3	4.65	288.1	3.23	414.3	2.25	700.9	1.33
0.401	0.342	127.1	7.42	214.8	4.39	308.8	3.05	444.2	2.12	751.5	1.25
0.427	0.344	136.2	6.99	230.2	4.14	331.1	2.88	476.1	2.00	805.5	1.18
0.455	0.346	145.9	6.58	246.6	3.89	354.6	2.71	510.0	1.88	862.8	1.11
0.486	0.348	156.3	6.19	264.1	3.66	379.8	2.55	546.2	1.77	924.1	1.05
0.519	0.350	167.3	5.81	282.8	3.44	406.6	2.39	584.7	1.66	989.2	0.98
0.555	0.352	179.1	5.45	302.7	3.23	435.2	2.24	625.9	1.56	1,059.0	0.92
0.594	0.354	191.7	5.10	324.0	3.02	465.8	2.10	669.9	1.46	1,133.4	0.86
0.638	0.356	205.1	4.77	346.7	2.82	498.4	1.96	716.8	1.36	1,212.8	0.81
0.685	0.358	219.6	4.45	371.1	2.63	533.6	1.83	767.4	1.27	1,298.3	0.75
0.736	0.361	235.0	4.15	397.2	2.45	571.1	1.71	821.3	1.19	1,389.5	0.70
0.793	0.363	251.6	3.85	425.2	2.28	611.4	1.59	879.2	1.10	1,487.5	0.65
0.855	0.365	269.4	3.58	455.3	2.12	654.7	1.47	941.6	1.02	1,593.0	0.60
0.924	0.367	288.6	3.31	487.8	1.96	701.4	1.36	1,008.8	0.95	1,706.7	0.56
0.999	0.369	309.3	3.06	522.8	1.81	751.7	1.26	1,081.1	0.88	1,829.1	0.52
1.082	0.371	331.7	2.82	560.6	1.67	806.1	1.16	1,159.3	0.81	1,961.3	0.48
1.174	0.373	355.9	2.59	601.6	1.53	865.0	1.07	1,244.0	0.74	2,104.7	0.44
1.275	0.375	382.3	2.38	646.0	1.41	928.9	0.98	1,335.9	0.68	2,260.2	0.40
1.388	0.377	410.9	2.18	694.4	1.29	998.4	0.90	1,435.9	0.62	2,429.4	0.37
1.514	0.379	442.1	1.98	747.2	1.17	1,074.3	0.82	1,545.0	0.57	2,614.0	0.34
1.654	0.381	476.2	1.80	804.8	1.07	1,157.2	0.74	1,664.3	0.52	2,815.7	0.30
1.811	0.383	513.6	1.63	868.1	0.97	1,248.1	0.67	1,795.0	0.47	3,036.9	0.28
1.987	0.385	554.8	1.47	937.7	0.87	1,348.3	0.61	1,939.0	0.42	3,280.5	0.25
2.186	0.387	600.2	1.33	1,014.5	0.78	1,458.7	0.55	2,097.8	0.38	3,549.1	0.22
2.410	0.389	650.6	1.19	1,099.6	0.70	1,581.1	0.49	2,273.9	0.34	3,847.1	0.20
2.664	0.391	706.7	1.06	1,194.4	0.63	1,717.4	0.44	2,469.9	0.30	4,178.8	0.18
2.954	0.393	769.4	0.94	1,300.4	0.56	1,869.8	0.39	2,689.0	0.27	4,549.4	0.16
3.285	0.395	839.9	0.83	1,419.4	0.49	2,041.0	0.34	2,935.2	0.24	4,966.0	0.14
3.667	0.397	919.4	0.72	1,553.9	0.43	2,234.3	0.30	3,213.2	0.21	5,436.3	0.12
4.109	0.399	1,009.8	0.63	1,706.6	0.37	2,453.9	0.26	3,529.0	0.18	5,970.6	0.11
4.624	0.400	1,113.2	0.54	1,881.4	0.32	2,705.3	0.22	3,890.5	0.16	6,582.3	0.09
5.228	0.402	1,232.4	0.47	2,082.8	0.28	2,994.7	0.19	4,306.9	0.13	7,286.6	0.08
5.943	0.404	1,370.8	0.40	2,316.7	0.24	3,331.2	0.16	4,790.7	0.11	8,105.2	0.07
6.795	0.406	1,533.2	0.34	2,591.2	0.20	3,725.8	0.14	5,358.2	0.10	9,065.4	0.06
7.824	0.408	1,725.7	0.28	2,916.6	0.17	4,193.7	0.12	6,031.1	0.08	10,203.8	0.05
9.080	0.410	1,956.6	0.23	3,306.9	0.13	4,754.8	0.09	6,838.1	0.07	11,569.2	0.04
10.634	0.412	2,237.7	0.19	3,781.8	0.11	5,437.8	0.08	7,820.3	0.05	13,230.8	0.03
12.589	0.413	2,585.3	0.15	4,369.3	0.09	6,282.6	0.06	9,035.2	0.04	15,286.3	0.02
15.096	0.415	3,023.8	0.11	5,110.4	0.07	7,348.1	0.05	10,567.7	0.03	17,879.1	0.02
18.385	0.417	3,589.9	0.09	6,067.2	0.05	8,723.9	0.04	12,546.2	0.02	21,226.5	0.01

Values of t_j, q_{oj}, q_{wj}, and q_{tj} can be computed for layers for any time t_b given in Table 3.4. The result of these computations is a set of t_j, q_{oj}, q_{wj}, and q_{tj} for each Layer j, which correspond to the values of Q_{ib} and N_{pb} in Table 3.4. Selected values from these calculations are summarized in Table 5.3. By scanning Table 5.3 we can see how the displacement process changes with time in each layer. For example, breakthrough occurs in 103 days in Layer 1, compared to 608.7 days in Layer 5. Breakthrough times are summarized in Table 5.4.

The last step in the computation is to combine the performance of five separate layers at the same cumulative injection time to obtain the displacement behavior of the entire reservoir at that time. As an example, we will estimate performance of the reservoir after 365 days of injection. Linear interpolation will be used to obtain values of q_{oj}, q_{wj}, q_{tj}, Q_{ij}, and N_{pj} at $t_j = 365$ from adjacent times. Layer 1 values for adjacent times are summarized in Table 5.5.

The interpolation factor is

$$\frac{(365.0 - 355.9)}{(382.3 - 355.9)} = 0.3447.$$

Thus, to obtain q_{wj} at $t=365.0$, we assume that q_w varies linearly with time over a short time interval.

$$q_{w1} = 120.11 + 0.3447 \, (122.86 - 120.11).$$

In this example, the time intervals are close together, and interpolation results in small changes in rates.

The values presented in Table 5.6 were obtained from Table 5.3 by linear interpolation at $t=365$ days.

Rates were obtained by addition of the rates of each layer as indicated. However, Q_i and N_p must be computed with the PV in each layer.

$$Q_i = \sum_{j=1}^{m} \frac{Q_{ij} V_{pj}}{\displaystyle\sum_{j=1}^{m} V_{pj}}$$

and

$$N_p = \frac{\displaystyle\sum_{j=1}^{m} N_{pj} V_{pj}}{\displaystyle\sum_{j=1}^{m} V_{pj}},$$

Because porosities and initial water saturations were assumed to be identical in all layers for Example 5.1,

$$Q_i = \frac{\displaystyle\sum_{j=1}^{m} h_j Q_{ij}}{\displaystyle\sum_{j=1}^{m} h_j}$$

$$= (1.0/5.0)(1.209 + 0.673 + 0.468 + 0.340 + 0.220)$$

$$= 0.582$$

and

$$N_p = \frac{\displaystyle\sum_{j=1}^{m} h_j N_{pj}}{\displaystyle\sum_{j=1}^{m} h_j}$$

$$= (1.0/5.0)(0.374 + 0.358 + 0.347 + 0.336 + 0.220)$$

$$= 0.327.$$

The calculations outlined in the preceding section serve to illustrate the procedure at a single value of time. Computation of complete performance to floodout or subdivision into additional layers requires many calculations. When such information is needed, the calculation procedure should be programmed and calculations performed on a computer.

Table 5.7 contains the composite performance of the five-layer reservoir computed following the procedures outlined in this example. Values at $t=365.0$ days serve as a check on the computations.

Displacement performance for uniform and five-layer systems is compared in Fig. 5.2. Vertical heterogeneities cause a reduction of recovery at low values of Q_i.

TABLE 5.4—COMPARISON OF BREAKTHROUGH TIMES

Layer	Breakthrough Time (days)
1	103.0
2	174.0
3	250.2
4	359.8
5	608.7

TABLE 5.5—VALUES OF COMPUTED DISPLACEMENT PERFORMANCE USED TO FIND PERFORMANCE AT $t=365$ DAYS BY LINEAR INTERPOLATION

t_1 (days)	q_{o1} (B/D)	q_{w1} (B/D)	Q_{i1} (PV)	N_{p1} (PV)
355.9	2.59	120.11	1.174	0.373
382.3	2.38	122.86	1.275	0.375

TABLE 5.6—INTERPOLATED DISPLACEMENT PERFORMANCE FOR FIVE-LAYER RESERVOIR FROM SINGLE-LAYER RESULTS OF EXAMPLE 3.5 AT $t=365.0$ DAYS

Layer	q_{oj} (B/D)	q_{wj} (B/D)	q_{tj} (B/D)	Q_{ij} (PV)	N_{pj} (PV)
1	2.52	121.06	123.58	1.209	0.374
2	2.68	59.72	62.40	0.673	0.358
3	2.64	36.17	38.81	0.468	0.347
4	2.48	21.72	24.20	0.340	0.336
5	16.47	0.00	16.47	0.220	0.220
Totals	26.79	238.67	265.46		

Although the recoveries appear to be quite close as Q_i increases, different costs of production could cause termination of the flood before nearly equivalent recoveries were attained.

Oil rates for single- and five-layer simulations are shown in Fig. 5.3. Abrupt changes in oil rates (and WOR) are observed as breakthrough occurs in each layer. These changes are pronounced because much of the oil is displaced at breakthrough. Fig. 5.4 shows the oil production curve when the reservoir was subdivided into 10 layers. As the number of layers increases, more flood fronts arrive at different times, thereby smoothing the production curve. The problem of determining the number of layers for an adequate simulation is presented in Sec. 5.8.1.

5.3 Gravity Segregation and Crossflows in Linear Reservoirs

In linear reservoirs ($L>>h$), overall movement of fluids occurs in the direction of principal axis of the reservoir during displacement. The efficiency of a displacement process, however, may be influenced by fluid flow in the direction perpendicular to the main axis of flow. Mechanisms causing vertical flows or crossflows are investigated in this section.

Fig. 5.5 shows photographs of model waterfloods at two injection rates.[1] At a low injection rate (Fig. 5.5a),

TABLE 5.7—COMPOSITE PERFORMANCE OF FIVE-LAYER RESERVOIR

Time (days)	q_o (B/D)	q_w (B/D)	q_t (B/D)	WOR	Q_i (PV)	N_p (PV)	N_p (bbl)
0.0	338.19	0.0	338.19	0.0	0.0	0.0	0.0
30.4	295.35	0.0	295.35	0.0	0.060	0.060	9,563.2
60.8	266.61	0.0	266.61	0.0	0.113	0.113	18,093.6
91.3	245.48	0.0	245.48	0.0	0.161	0.161	25,871.7
121.7	156.86	80.76	237.62	0.51	0.207	0.198	31,689.0
152.1	147.98	88.24	236.22	0.60	0.252	0.227	36,334.4
182.5	96.47	140.25	236.72	1.45	0.297	0.252	40,329.6
212.9	91.92	148.85	240.77	1.62	0.342	0.270	43,198.4
243.3	88.00	156.53	244.53	1.78	0.388	0.287	45,931.7
273.8	54.19	195.67	249.86	3.61	0.435	0.298	47,840.7
304.2	52.02	203.36	255.38	3.91	0.483	0.309	49,448.0
334.6	50.06	210.42	260.48	4.20	0.532	0.318	51,002.5
365.0	26.80	238.63	265.44	8.90	0.582	0.327	52,386.4
395.4	25.67	245.38	271.04	9.56	0.633	0.332	53,191.0
425.8	24.68	251.66	276.33	10.20	0.685	0.337	53,949.9
456.3	23.76	257.53	281.29	10.84	0.737	0.341	54,688.7
486.7	22.94	263.04	285.98	11.46	0.791	0.346	55,396.7
517.1	22.19	268.23	290.42	12.09	0.846	0.350	56,085.0
547.5	21.50	273.10	294.60	12.70	0.902	0.354	56,750.3
577.9	20.87	277.71	298.58	13.31	0.958	0.358	57,391.5
608.3	20.27	282.10	302.37	13.91	1.015	0.362	58,022.2
638.8	7.13	299.25	306.37	42.00	1.073	0.363	58,248.4
669.2	6.77	303.42	310.20	44.79	1.131	0.365	58,459.3
699.6	6.45	307.45	313.89	47.68	1.190	0.366	58,661.5
730.0	6.15	311.22	317.37	50.57	1.251	0.367	58,851.2
760.4	5.88	314.88	320.76	53.59	1.311	0.368	59,034.9
790.8	5.62	318.34	323.97	56.62	1.372	0.369	59,208.4
821.3	5.38	321.69	327.07	59.77	1.434	0.370	59,376.2
851.7	5.16	324.90	330.06	62.98	1.496	0.371	59,536.2
882.1	4.95	327.96	332.91	66.22	1.559	0.372	59,689.7
912.5	4.76	330.94	335.70	69.53	1.622	0.373	59,837.3
942.9	4.58	333.78	338.36	72.91	1.686	0.374	59,979.7
973.3	4.41	336.49	340.91	76.28	1.751	0.375	60,114.8
1,003.8	4.25	339.19	343.44	79.86	1.816	0.376	60,248.4
1,034.2	4.10	341.70	345.80	83.35	1.882	0.377	60,373.3
1,064.6	3.96	344.15	348.11	86.98	1.947	0.377	60,496.0
1,095.0	3.82	346.57	350.39	90.71	2.014	0.378	60,614.7
1,125.4	3.69	348.86	352.56	94.43	2.080	0.379	60,728.5
1,155.8	3.57	351.07	354.65	98.23	2.147	0.380	60,839.2
1,186.3	3.46	353.23	356.69	102.08	2.215	0.380	60,945.5
1,216.7	3.35	355.34	358.70	106.03	2.283	0.381	61,049.7
1,247.1	3.25	357.38	360.63	110.03	2.351	0.382	61,150.4
1,277.5	3.15	359.31	362.47	113.94	2.420	0.382	61,246.2
1,307.9	3.06	361.23	364.29	118.02	2.489	0.383	61,340.9
1,338.3	2.97	363.10	366.07	122.20	2.558	0.383	61,432.9
1,368.8	2.89	364.93	367.81	126.43	2.627	0.384	61,521.9
1,399.2	2.81	366.65	369.46	130.64	2.697	0.384	61,608.0
1,429.6	2.73	368.35	371.08	134.95	2.768	0.385	61,692.0
1,460.0	2.66	370.00	372.66	139.28	2.839	0.385	61,773.7
1,490.4	2.59	371.62	374.20	143.67	2.909	0.386	61,852.9
1,520.8	2.52	373.23	375.75	148.20	2.980	0.386	61,930.9
1,551.3	2.45	374.79	377.23	152.84	3.052	0.387	62,007.4

the "front" tilted and extended over a large portion of the model. Water and oil undergo gravity segregation with water flowing vertically toward the bottom of the model and oil flowing toward the top of the model. The displacement process is characterized by early breakthrough of water, as well as the displacement of the recoverable oil at higher WOR's. The high injection rate shown in Fig. 5.5b prevents the development of gravity segregation. A sharp front, resembling those described in Sec. 3.4, is observed.

Fig. 5.6 is a photograph showing water and oil saturations during displacement of oil from a stratified sandpack.[2] The sandpack consisted of two homogeneous layers of sand. Permeability of the top layer is seven times larger than that of the lower layer. Water injected into both layers at the left end of the sandpack travels at different rates in each layer because of the difference in permeabilities. Crossflow of water and oil from one layer to the other may occur from density differences and capillary-pressure gradients. As the displacement proceeds, water that entered the lower layer at the left end of the sandpack will flow to the upper layer because there is less resistance to flow in the higher-permeability layer.

We will examine the mechanisms that cause flow in a second direction by writing the fluid-flow equations developed in Sec. 4.2 in terms of two coordinate directions, horizontal (x) and vertical (z). The vertical axis z is positive when directed up.

Eqs. 3.14 and 3.15 describe the conservation of mass for oil and water phases at a point x,y,z. When x and z

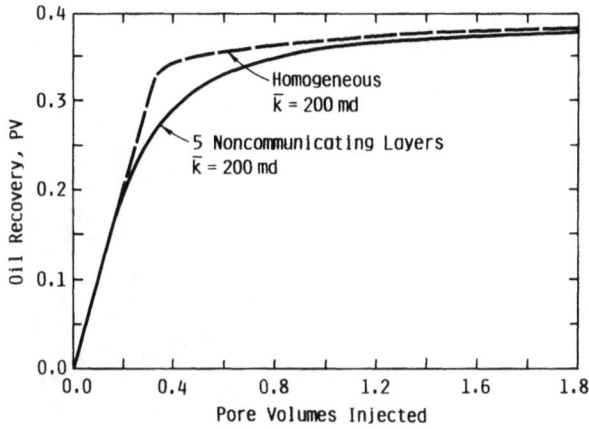

Fig. 5.2—Oil recovery for homogeneous and layered reservoirs that have equal thickness-averaged permeabilities.

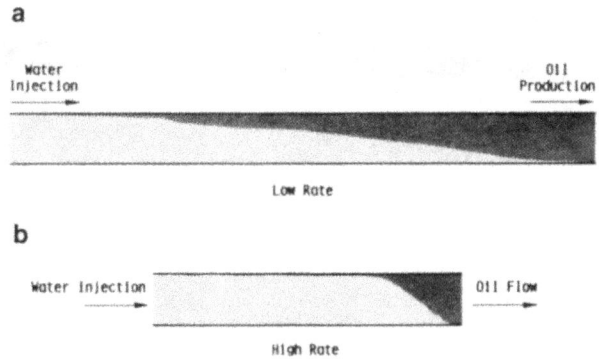

Fig. 5.5—Laboratory experiments illustrating the effect of injection rate on gravity segregation in water-floods. Part a shows the low-rate displacement and Part b shows the high-rate displacement.

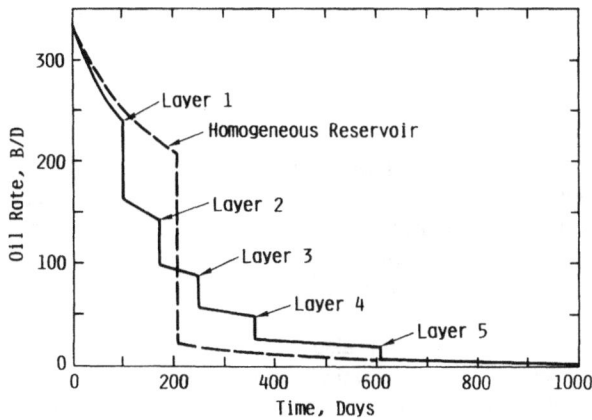

Fig. 5.3—Production rates for homogeneous and five-layer reservoirs.

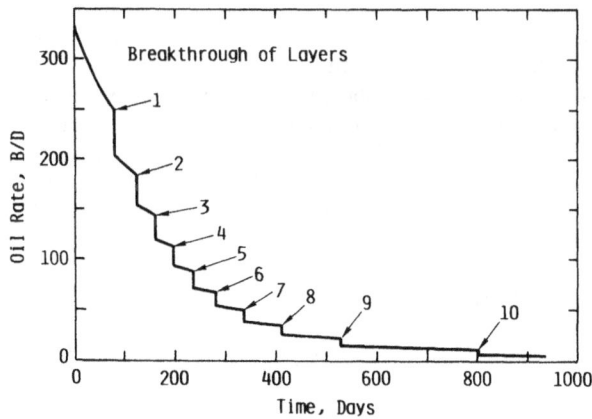

Fig. 5.4—Production rate for the reservoir when divided into 10 equal-thickness layers.

are the two dimensions, terms involving y are dropped, as in Eqs. 5.15 and 5.16.

For the oil phase,

$$\frac{\partial}{\partial x}\left(\frac{\rho_o k_{ox}}{\mu_o}\frac{\partial p_o}{\partial x}\right)+\frac{\partial}{\partial z}\left[\frac{\rho_o k_{oz}}{\mu_o}\left(\frac{\partial p_o}{\partial z}\right.\right.$$

$$\left.\left.+\rho_o g\right)\right]=\frac{\partial}{\partial t}(\rho_o S_o\phi). \quad\ldots\ldots\ldots\ldots\ldots(5.15)$$

The first quantity represents the rate that oil mass changes at x,z because of flow in the x direction. The second quantity represents the rate that oil mass changes at x,z because of flow in the z direction. The final term represents the rate that oil mass accumulates at x,z.

For the water phase,

$$\frac{\partial}{\partial x}\left(\frac{\rho_w k_{wx}}{\mu_w}\frac{\partial p_w}{\partial x}\right)+\frac{\partial}{\partial z}\left[\frac{\rho_w k_{wz}}{\mu_w}\left(\frac{\partial p_w}{\partial z}+\rho_w g\right)\right]$$

$$=\frac{\partial}{\partial t}(\rho_w S_w\phi). \quad\ldots\ldots\ldots\ldots\ldots\ldots(5.16)$$

Oil- and water-phase pressures are related through the capillary-pressure/saturation curves.

$$p_o-p_w=P_c(S_w). \quad\ldots\ldots\ldots\ldots\ldots\ldots(5.17)$$

When solved with appropriate boundary and initial conditions, Eqs. 5.15 through 5.17 yield pressure and saturation distributions that vary with time and spatial position (x,z) during a displacement process. These solutions may be obtained for some conditions by the use of scaled laboratory experiments and numerical simulation. Thus, it is important to develop some feeling for situations that produce a second velocity component as well as the effect of this component on displacement efficiency. We do this by examining approximate models of displacement processes where gravity segregation and stratified flow are present.

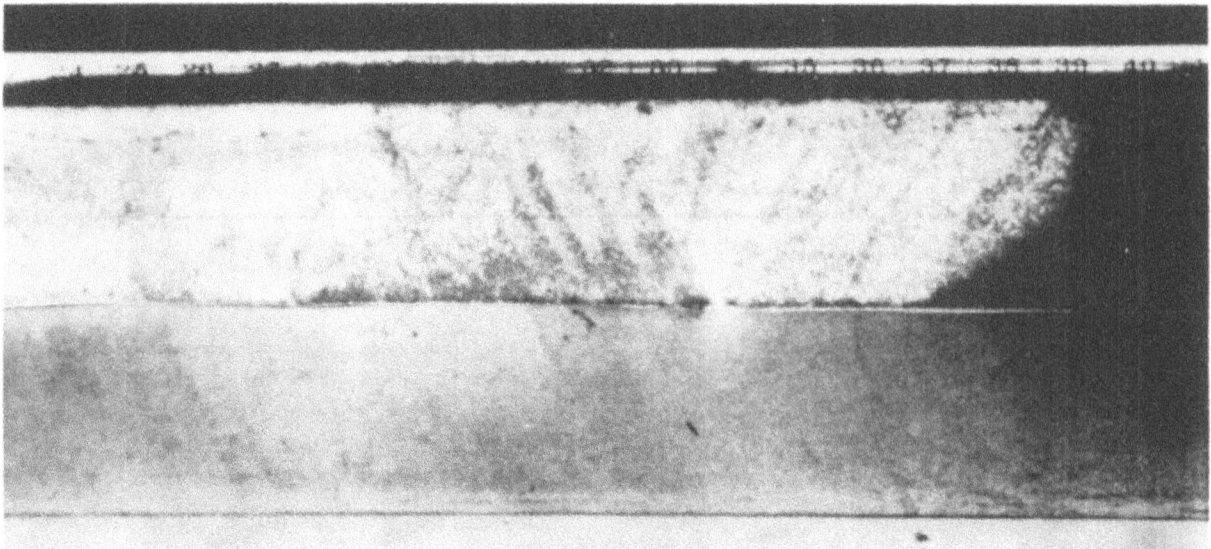

Fig. 5.6—Visualization of a displacement process where crossflow may be caused by capillary and viscous forces.[2]

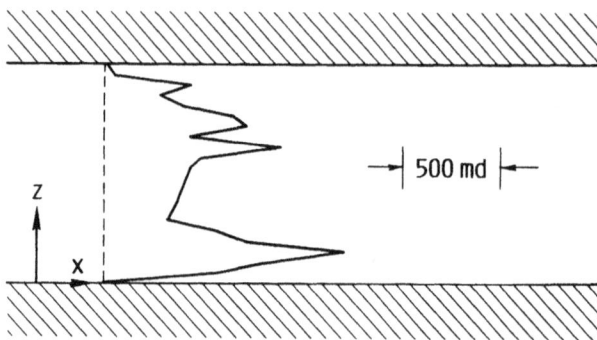

Fig. 5.7—Permeability distribution in a vertical cross section through a reservoir.

5.4 Approximation of 2D Flow With Thickness-Averaged Properties

In the linear displacement problems discussed in Sec. 5.2, vertical flows of water and oil were considered to be negligible. This is equivalent to the assumption that

$$\frac{\partial}{\partial z}\left[\frac{\rho_o k_{oz}}{\mu_o}\left(\frac{\partial p_o}{\partial z}+\rho_o g\right)\right] << \frac{\partial}{\partial x}\left(\frac{\rho_o k_{ox}}{\mu_o}\frac{\partial p_o}{\partial x}\right)$$

$$\dots\dots\dots\dots\dots\dots\dots\dots(5.18)$$

and

$$\frac{\partial}{\partial z}\left[\frac{\rho_w k_{wz}}{\mu_o}\left(\frac{\partial p_w}{\partial z}+\rho_w g\right)\right] << \frac{\partial}{\partial x}\left(\frac{\rho_w k_{wx}}{\mu_w}\frac{\partial p_w}{\partial x}\right).$$

$$\dots\dots\dots\dots\dots\dots\dots\dots(5.19)$$

Thus, the frontal-advance solution describes a displacement process where the rate of fluid movement in the x direction is much larger than in the z direction. Vertical

flows are present, but their effect on the displacement process is small.

As displacement rates decrease, we should expect to find a region where vertical flows caused by gravity forces become important, as shown in Fig. 5.5a. Gravity segregation also occurs during the displacement of a liquid by a gas. Gravity forces may be 5 to 10 times larger when gas, instead of water, is the displacing fluid. Capillary forces may be present also.

If crossflow occurs rapidly compared with fluid movement in the principal direction of flow, a modified form of the Buckley-Leverett solution can be used to estimate displacement behavior.[3] This solution leads directly to the concept of pseudo-relative-permeability curves for simulation of 2D flow in a one-dimensional (1D) model or three-dimensional (3D) flow with a 2D model.

Fig. 5.7 shows the permeability distribution in a vertical cross section through a reservoir. For the purpose of this section, consider the general case where permeability and porosity vary with Position z from the base of the reservoir. The development of the model is begun with the definition of thickness-averaged properties. The vertical dashed line on Fig. 5.7 is an arbitrary cross section located at a Distance x from the origin. A thickness-averaged permeability in the x direction is defined by Eq. 5.20.

$$\bar{k}=\frac{\int_0^h k\,dz}{\int_0^h dz}, \dots\dots\dots\dots\dots\dots(5.20)$$

where

k = the permeability of the porous rock in the x direction that varies with z,
$\bar{k}$ = thickness-averaged permeability in the x direction,
h = thickness of the formation, and
z = distance from the base of the formation.

In a similar manner, thickness-averaged porosity and water saturation are defined by Eqs. 5.21 through 5.23.

$$\bar{\phi} = \frac{\int_0^h \phi(z)dz}{h}, \quad \ldots\ldots\ldots\ldots\ldots\ldots (5.21)$$

$$S_{wz} = \frac{\int_0^h S_w \phi dz}{\int_0^h \phi dz}, \quad \ldots\ldots\ldots\ldots\ldots (5.22)$$

and

$$S_{wz} = \frac{\int_0^h S_w \phi dz}{\bar{\phi} h}, \quad \ldots\ldots\ldots\ldots\ldots (5.23)$$

where S_{wz} = thickness-averaged water saturation and $\bar{\phi}$ = thickness-averaged porosity.

Consider a material balance on the water phase for a differential element of thickness h and width dx. The fluids are incompressible. Properties (S_w, ϕ, and k) are assumed to vary with z in each cross section. Following the procedures illustrated in Sec. 3.4, we obtain Eq. 5.24.

$$-q_t \left(\frac{\partial f_{wz}}{\partial x}\right)_t = A\bar{\phi}\left(\frac{\partial S_{wz}}{\partial t}\right)_x, \quad \ldots\ldots\ldots\ldots (5.24)$$

where

f_{wz} = fractional flow of water in Plane bh,
q_t = is flow rate in the x direction passing through Plane bh, and
$A = b \times h$ = cross-sectional area perpendicular to the direction of flow.

Eq. 5.24 is simply the Buckley-Leverett equation expressed in thickness-averaged properties. If we assume $S_{wz} = S_{wz}(x,t)$ and $f_{wz} = f_{wz}(S_{wz})$, we obtain Eq. 5.25, the frontal advance equation expressed in terms of thickness-averaged parameters.

$$\left(\frac{dx}{dt}\right)_{S_{wz}} = \frac{q_t}{A\bar{\phi}}\left(\frac{\partial f_{wz}}{\partial S_{wz}}\right)_t. \quad \ldots\ldots\ldots\ldots (5.25)$$

The term on the left side of Eq. 5.25 is the velocity of the saturation S_{wz}. Thus

$$v_{S_{wz}} = \frac{q_t}{A\bar{\phi}}\left(\frac{\partial f_{wz}}{\partial S_{wz}}\right)_t. \quad \ldots\ldots\ldots\ldots (5.26)$$

When the fractional flow can be expressed as a function of S_{wz}, the velocity of each S_{wz} is constant and can be obtained from the slope of the $f_{wz} - S_{wz}$ graph. Integration of Eq. 5.25 for $0 < x < x_{S_{wz}}$ gives

$$x_{S_{wz}} = \frac{1}{A\bar{\phi}}\left(\frac{\partial f_{wz}}{\partial S_{wz}}\right)\int_0^t q_t dt. \quad \ldots\ldots\ldots (5.27)$$

Dividing both sides of Eq. 5.27 by L gives

$$\frac{x_{S_{wz}}}{L} = Q_i\left(\frac{\partial f_{wz}}{\partial S_{wz}}\right). \quad \ldots\ldots\ldots\ldots (5.28)$$

Thus, the thickness-averaged saturation S_{wz2} arrives at the end of the system ($x_{S_{wz}} = L$) when

$$Q_{i2} = \frac{1}{\left(\frac{\partial f_{wz}}{\partial S_{wz}}\right)_{S_{wz2}}}. \quad \ldots\ldots\ldots\ldots (5.29)$$

The Welge method, derived in Sec. 3.5, also applies to thickness-averaged saturations. Eq. 5.30 relates the volumetric average water saturation, $\overline{S_{wz}}$, to thickness-averaged parameters at the end of the reservoir.

$$\overline{S_{wz}} = S_{wz2} + Q_i(1 - f_{wz2}). \quad \ldots\ldots\ldots\ldots (5.30)$$

It can be seen from this development that there is an exact analogy between the Buckley-Leverett model for uniform and thickness-averaged saturations.

The introduction of thickness-averaged properties leads to a displacement model in which gross fluid movement is in one direction with the possibility of crossflow in the direction perpendicular to the principal direction of flow. Eq. 5.31 represents the fractional flow of water in the horizontal direction through the plane of thickness h located at some position x:

$$f_{wz} = \frac{\int_0^h \frac{k_w}{\mu_w}\left(\frac{dp_w}{dx}\right)dz}{\int_0^h \frac{k_w}{\mu_w}\left(\frac{dp_w}{dx}\right)dz + \int_0^h \frac{k_o}{\mu_o}\left(\frac{dp_o}{dx}\right)dz}. \quad ..(5.31)$$

To develop a 1D model that includes the effects of vertical flow through thickness-averaged properties, it is assumed that dp_w/dx and dp_o/dx do not vary with z. Thus, dP_c/dx does not vary with z. With this assumption, Eq. 5.31 becomes

$$f_{wz} = \frac{\int_0^h \frac{k_w}{\mu_w}dz}{\int_0^h \frac{k_w}{\mu_w}dz + \int_0^h \frac{k_o}{\mu_o}dz}. \quad \ldots\ldots\ldots\ldots (5.32)$$

The integrals in Eq. 5.32 may be replaced by thickness-averaged permeabilities for the water and oil phases, analogous to Eq. 5.20.

$$k_{wz} = \frac{\int_0^h k_w dz}{h} \quad \ldots\ldots\ldots\ldots\ldots (5.33)$$

and

$$k_{oz} = \frac{\int_0^h k_o dz}{h}. \quad \ldots\ldots\ldots\ldots\ldots (5.34)$$

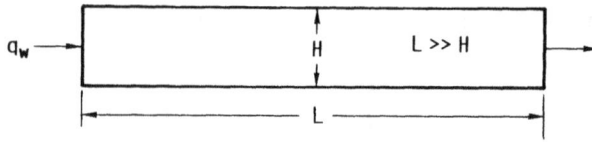

Fig. 5.8—Homogeneous linear reservoir for vertical equilibrium model.

Then Eq. 5.32 becomes

$$f_{wz} = \frac{\dfrac{k_{wz}}{\mu_w}}{\dfrac{k_{wz}}{\mu_w} + \dfrac{k_{oz}}{\mu_o}}, \quad\quad\quad\quad\quad (5.35)$$

which is identical to Eq. 3.77, which was developed for uniform properties.

Thickness-averaged relative permeabilities are defined by Eqs. 5.36 and 5.37.

$$k_{rwz} = \frac{k_{wz}}{\overline{k}} \quad\quad\quad\quad\quad\quad (5.36)$$

and

$$k_{roz} = \frac{k_{oz}}{\overline{k}}. \quad\quad\quad\quad\quad\quad (5.37)$$

Thickness-averaged relative permeabilities are also called pseudorelative permeabilities.

Application of Eq. 5.32 or 5.35 to estimate displacement performance of a waterflood requires information on the variation of k_w and k_o with z. Relative permeability curves for the reservoir rock and fluids provide a relationship between phase permeability and water saturation. Thus evaluation of k_o, k_{ro}, k_w, and k_{rw} depends on some approximation of the variation of water saturation with z during the displacement.

To approximate vertical flow in a horizontal displacement model, it must be assumed that vertical flow occurs "instantaneously," leaving a vertical saturation distribution that may be computed at various stages in the displacement. Three models to approximate vertical flow are included in this text. One method uses vertical equilibrium of capillary and gravity forces to determine the vertical saturation distribution. Vertical equilibrium dominated by gravity segregation is the basis for the second method. The third method assumes that pressures are equal in each vertical cross section.

5.4.1 Vertical Equilibrium of Capillary and Gravity Forces

Under some conditions, vertical flows of oil and water are so rapid (compared to horizontal flow) that the capillary and gravity forces in each vertical cross section are in equilibrium at every point in time during the displacement process.[4,5] Any saturation changes in a vertical cross section are assumed to occur instantaneously. A system that satisfies these conditions is said to exhibit vertical equilibrium in the displacement.

Consider the displacement of oil by water in the linear reservoir shown in Fig. 5.8. For the purposes of the development, we assume the reservoir is homogeneous with thickness h and porosity ϕ, and is thin relative to its length, L. The reservoir is at interstitial water saturation. Vertical equilibrium in Fig. 5.8 is equivalent to assuming that oil- and water-phase potentials are constant (but unequal) in a particular vertical cross section. Recalling the definition of oil- and water-phase potentials from Chap. 2,

$$\Phi_o = \int_{p_{od}}^{p_o} \frac{dp_o}{\rho_o} + g(Z - Z_d) = \text{constant}, \quad\quad (5.38)$$

or

$$\frac{\partial \Phi_o}{\partial z} = 0 = \frac{1}{\rho_o} \frac{\partial p_o}{\partial z} + g \frac{\partial Z}{\partial z} \quad\quad\quad (5.39)$$

and

$$\Phi_w = \int_{p_{wd}}^{p_w} \frac{dp_w}{\rho_w} + g(Z - Z_d) = \text{constant}, \quad\quad (5.40)$$

$$\frac{\partial \Phi_w}{\partial z} = 0 = \frac{1}{\rho_w} \frac{\partial p_w}{\partial z} + g \frac{\partial Z}{\partial z} \quad\quad\quad (5.41)$$

where z is the distance normal to the direction x, and Z is the elevation in the vertical direction above an arbitrary horizontal reference phase located at Z_d.

Eqs. 5.39 and 5.41 can also be written in terms of capillary pressure, as in Eq. 5.42.

$$\frac{dP_c}{dz} + g(\rho_o - \rho_w)\frac{\partial Z}{\partial z} = 0. \quad\quad\quad (5.42)$$

When Z and z are in the vertical plane (dip angle $=0$) and z is directed upward, $\partial Z/\partial z = +1$ and Eq. 5.42 becomes

$$\frac{dP_c}{dz} = -g(\rho_o - \rho_w). \quad\quad\quad\quad (5.43)$$

Integration of Eq. 5.42 through any vertical cross section gives Eq. 5.44.

$$P_c - P_{co} = -g(\rho_o - \rho_w)z \quad\quad\quad\quad (5.44)$$

or

$$P_c = P_{co} - g(\rho_o - \rho_w)z, \quad\quad\quad\quad (5.45)$$

where P_{co} is the value of the capillary pressure corresponding to the water saturation at the bottom of the reservoir ($z=0$). A system in vertical equilibrium has a capillary-pressure distribution that varies from P_{co} at the bottom of the cross section to $P_{ch} = P_{co} - g(\rho_o - \rho_w)h$ at the top of the reservoir. Because the capillary pressure is a unique function of water saturation along a particular saturation path (drainage or imbibition), the water-saturation distribution can be found from the capillary-pressure curve.

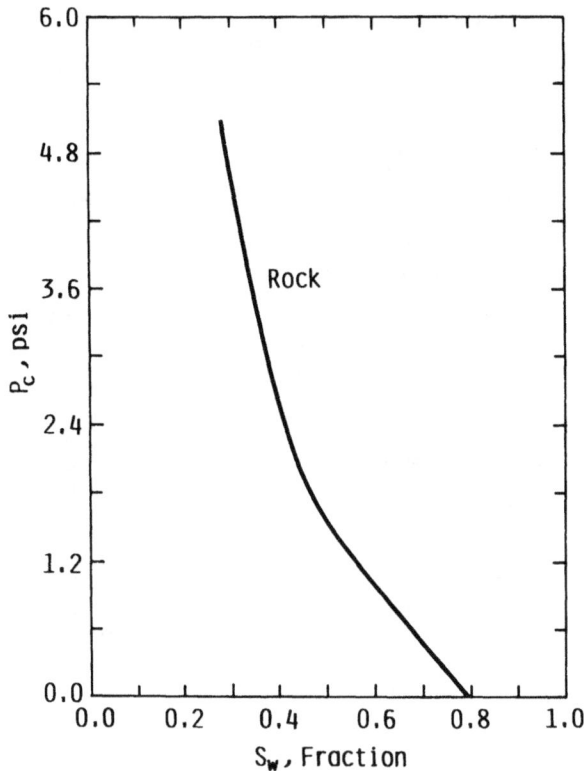

Fig. 5.9—Capillary-pressure curve for imbibition path, strongly water-wet porous media.

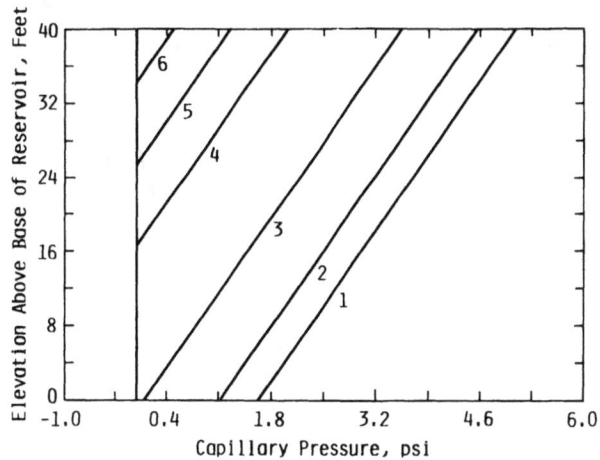

Fig. 5.10—Capillary-pressure distributions under the assumption of vertical equilibrium.

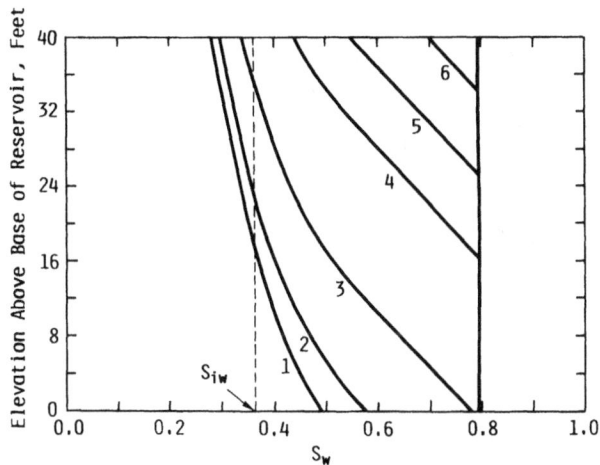

Fig. 5.11—Water-saturation distributions corresponding to capillary-pressure curves in Fig. 5.10.

The variation of capillary pressure with depth can be expressed also in terms of an arbitrary reference plane located at an elevation Z_p above the base of the reservoir. Eq. 5.46 gives the capillary pressure $P_c(Z)$ when the capillary pressure at the reference plane is P_{cp}.

$$P_c(Z) = P_{cp} - g(\rho_w - \rho_o)(Z_p - Z). \quad \ldots \ldots \ldots (5.46)$$

The capillary pressure (P_{cp}) at the reference elevation (Z_p) is called the pseudocapillary pressure.

To determine saturation distributions, one must choose a reference elevation. Although the location of the reference elevation is arbitrary, choosing a reference plane within the reservoir boundaries is desirable for computation. For the purposes of this section, the reference plane is the top of the reservoir ($Z=h$); the pseudocapillary pressure is the capillary pressure at the top of the reservoir (P_{ch}). Thus for $Z < h$,

$$P_c(Z) = P_{ch} - g(\rho_w - \rho_o)(h - Z). \quad \ldots \ldots \ldots (5.47)$$

Fig. 5.9 is a capillary-pressure curve for the imbibition path in a water-wet rock. During displacement of oil by water, the capillary pressure varies from $P_c(S_{wc})$ to $P_c(1-S_{or})$. Fig. 5.10 shows capillary-pressure distributions in a reservoir 40 ft [12 m] thick at selected values of P_{ch}. The capillary-pressure curve from Fig. 5.9 was used to prepare Fig. 5.10 with $\rho_o = 49.92$ lbm/cu ft [799.64 kg/m³] and $\rho_w = 62.4$ lbm/cu ft [999.55 kg/m³]. The capillary-pressure distributions are linear with a slope of 0.0867 psi/ft [1.9612 kPa/m].

Water-saturation distributions corresponding to the pseudocapillary pressure are presented in Fig. 5.11. For each pseudocapillary pressure, there is a unique saturation profile, and thus a unique thickness-averaged water saturation. Because relative permeabilities are unique functions of saturation on a particular path, the variation of relative permeabilities with Position Z is determined. This permits computation of thickness-averaged relative permeabilities with Eq. 5.20 and fractional flow as a function of thickness-averaged water saturations.

An important feature shown in Fig. 5.10 is the vertical displacement process resulting from the equilibrium of capillary and gravity forces. As P_{ch} decreases, an oil/water contact (OWC) eventually appears at $Z=0$ and subsequently moves vertically in the cross section. The oil phase becomes hydraulically disconnected when $P_c = P_c(1-S_{or})$, and capillary pressures can no longer be computed from Eq. 5.47. Thus the change in capillary pressure is confined to the interval $Z_{wc} < h$ where Z_{wc} is the position of the OWC. The capillary pressure for $Z < Z_{wc}$ is $P_c(1-S_{or})$. The position of the OWC is shown in Fig. 5.12.

Example 5.2 illustrates the effects of vertical equilibrium on the waterflood performance of the linear reservoir that was used in Example 3.4 with the thickness increased from 20 to 40 ft [6 to 12 m].

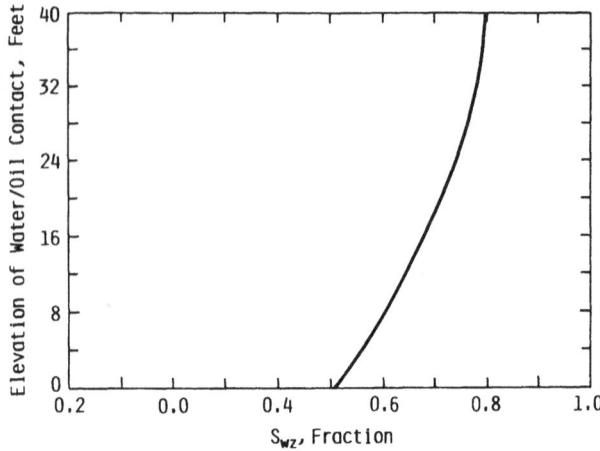

Fig. 5.12—Location of OWC as thickness-averaged water saturation varies from S_{iw} to $1-S_{or}$.

TABLE 5.8—COMPUTED DISTRIBUTIONS AT P_{ch} =3.056 psi

P_c (psi)	S_w	z (ft)	k_{rw}	k_{ro}
3.0560	0.3690	40.0000	0.0000	0.9648
2.9032	0.3775	38.2368	0.0000	0.9162
2.7504	0.3866	36.4736	0.0000	0.8661
2.5976	0.3963	34.7104	0.0001	0.8142
2.4448	0.4069	32.9472	0.0002	0.7601
2.2920	0.4185	31.1840	0.0004	0.7033
2.1392	0.4315	29.4208	0.0008	0.6429
1.9864	0.4462	27.6576	0.0017	0.5784
1.8336	0.4631	25.8944	0.0034	0.5093
1.6808	0.4826	24.1312	0.0066	0.4360
1.5280	0.5052	22.3680	0.0125	0.3600
1.3752	0.5307	20.6048	0.0231	0.2842
1.2224	0.5589	18.8416	0.0412	0.2129
1.0696	0.5889	17.0784	0.0699	0.1504
0.9168	0.6195	15.3152	0.1122	0.0996
0.7640	0.6500	13.5520	0.1704	0.0611
0.6112	0.6798	11.7888	0.2460	0.0339
0.4584	0.7088	10.0256	0.3409	0.0161
0.3056	0.7374	8.2625	0.4580	0.0058
0.1528	0.7660	6.4993	0.6021	0.0010
0.0000	0.7950	4.7361	0.7800	0.0000

p_{ch} = 3.056 psi, S_{wz} = 0.5674, k_{rwz} = 0.2016, and k_{roz} = 0.3497

Example 5.2. The reservoir used in Example 3.4 is homogeneous, with porosity of 0.15 and an absolute permeability of 0.2 darcies. Relative permeability curves for the fluid/rock combination are given as Eqs. 5.48 through 5.50.

$$k_{ro} = (1 - S_{wD})^{2.56}, \quad\quad\quad\quad\quad (5.48)$$

$$k_{rw} = 0.78 S_{wD}^{3.72}, \quad\quad\quad\quad\quad (5.49)$$

and

$$S_{wD} = \frac{S_w - S_{iw}}{1.0 - S_{or} - S_{iw}}. \quad\quad\quad\quad (5.50)$$

For the purposes of this example, the imbibition capillary-pressure curve is given by Eqs. 5.51 and 5.52.

$$P_c = \frac{2.52b}{\sqrt{k}} \quad\quad\quad\quad\quad\quad (5.51)$$

and

$$S_w = \frac{(1 - S_{or})}{1 + 1.845b}, \quad\quad\quad\quad\quad (5.52)$$

where

$\quad P_c$ = capillary pressure, psi,

$\quad k$ = absolute permeability, darcies, and base permeability for relative permeability curves, and

$\quad b$ = correlating parameter.

The densities of oil and water are 49.92 and 62.4 lbm/cu ft [799.64 and 999.55 kg/m³], respectively. Initial water saturation is 0.363 and is equal to the interstitial water saturation.

When the reservoir is in vertical equilibrium, the initial water saturation is actually the thickness-averaged saturation that corresponds to the capillary-pressure dis-

tribution at the beginning of the displacement. For this example, the reservoir thickness is 40 ft [12 m]. Thus,

$$P_c(Z) = P_{ch} - (0.0867)(40 - Z), \quad\quad\quad (5.53)$$

as long as $P_c \geq 0$. Capillary-pressure Curve 1 in Fig. 5.10 corresponds to the saturation distribution labeled 1 in Fig. 5.11. The thickness-averaged saturation of Curve 1 in Fig. 5.11 is 0.363, the capillary pressure at $Z = h$ (40 ft [12 m]) is 5.0933 psi [35.1 kPa]. The capillary-pressure distribution at initial water saturation is found by a trial-and-error process in which P_{ch} is assumed and S_{wz} is computed repeatedly until $S_{wz} = S_{wc}$. In this example, the water saturation distribution was integrated by the trapezoidal rule.

The remainder of the computations are relatively straightforward. The capillary-pressure interval from S_{wc} to $1 - S_{or}$ was subdivided into 20 increments for computation in a small program. Table 5.8 shows the distribution of capillary pressure, water saturation, and relative permeabilities to oil and water with elevation, Z, above the base of the reservoir for $P_{ch} = 3.056$ psi [21 kPa]. The reservoir is assumed to be horizontal. Thickness-averaged values of water saturation, relative permeability to oil, and relative permeability to water were computed with the trapezoidal rule. Thickness-averaged or pseudorelative permeabilities are shown at the end of Table 5.8.

When values of k_{roz} and k_{rwz} were computed, it was assumed that $k_{ro} = 1.0$ and $k_{rw} = 0$, for $S_w \leq S_{wc}$. Table 5.9 shows thickness-averaged properties corresponding to the values of P_{ch} in Col. 1.

The capillary pressure, P_{ch}, is now a pseudocapillary pressure because it refers to a particular capillary-pressure distribution. The graph of P_{ch} vs. S_{wz} in Fig. 5.13 is called a pseudo-capillary-pressure curve. The dashed line plotted in Fig. 5.13 is the actual capillary-pressure curve for the reservoir-rock/fluid interaction computed from Eqs. 5.51 and 5.52.

TABLE 5.9—THICKNESS-AVERAGED PROPERTIES AT SELECTED PSEUDOCAPILLARY PRESSURES RESERVOIR THICKNESS = 40 ft; FRACTIONAL FLOW BASED ON $\mu_o/\mu_w = 2.0$

P_{ch} (psi)	S_{wz}	k_{rwz}	k_{roz}	$f(S_{wz})$
5.0933	0.3630	0.0006	0.8797	0.0013
4.8386	0.3793	0.0017	0.8315	0.0040
4.5840	0.3982	0.0046	0.7746	0.0118
4.3293	0.4201	0.0115	0.7103	0.0313
4.0746	0.4450	0.0252	0.6411	0.0730
3.8200	0.4727	0.0495	0.5691	0.1482
3.5653	0.5032	0.0889	0.4959	0.2639
3.3106	0.5356	0.1445	0.4225	0.4061
3.0560	0.5674	0.2016	0.3497	0.5356
2.8013	0.5981	0.2588	0.2818	0.6475
2.5466	0.6278	0.3160	0.2201	0.7417
2.2920	0.6561	0.3732	0.1649	0.8190
2.0373	0.6830	0.4304	0.1169	0.8804
1.7827	0.7080	0.4874	0.0770	0.9268
1.5280	0.7306	0.5441	0.0459	0.9596
1.2733	0.7503	0.5997	0.0240	0.9804
1.0187	0.7666	0.6529	0.0106	0.9919
0.7640	0.7791	0.7012	0.0037	0.9973
0.5093	0.7879	0.7413	0.0009	0.9994
0.2547	0.7932	0.7692	0.0001	1.0000
0.0000	0.7950	0.7800	0.0000	1.0000

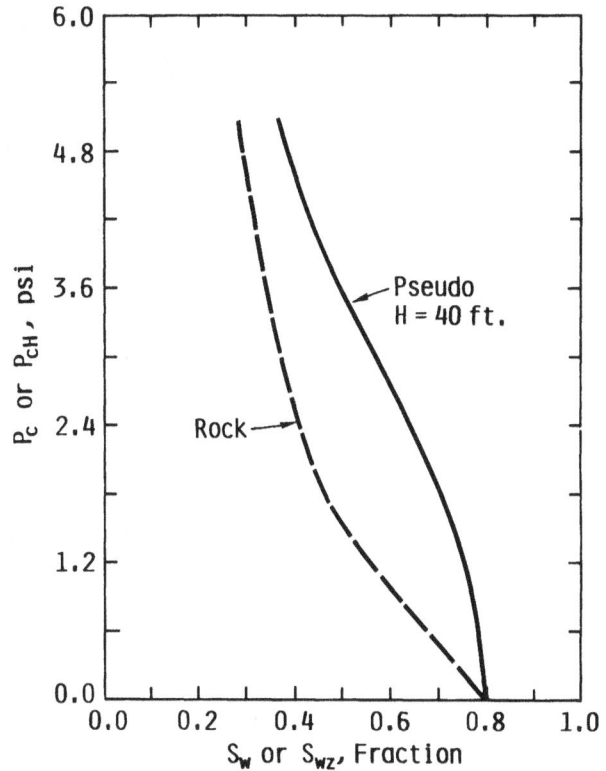

Fig. 5.13—Comparison of pseudo-capillary-pressure curve with rock-capillary-pressure curve.

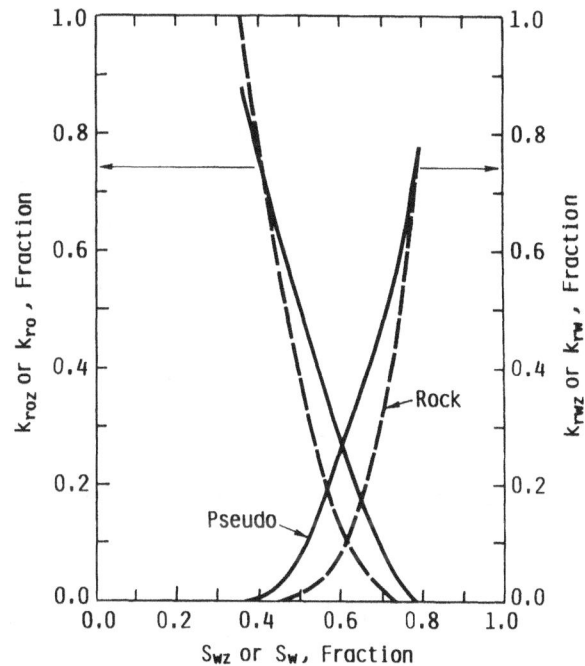

Fig. 5.14—Pseudo- and rock-relative-permeability curves.

Thickness-averaged or pseudorelative permeabilities are plotted against S_{wz} in Fig. 5.14. The dashed curves are the corresponding relative permeabilities computed from Eqs. 5.48 through 5.50.

Displacement performance is obtained by computation of fractional flow at values of S_{wz}, as in Example 3.4. Fig. 5.15 shows the fractional flow curve vs. thickness-averaged saturation for the data of Table 5.9 with $\mu_o/\mu_w = 2.0$. The dashed curve in Fig. 5.14 is the fractional flow curve from Example 3.4. The remainder of the computations follow the procedures illustrated in Chap. 3.

When vertical equilibrium is assumed in a linear model, the tangent drawn from S_{iw} to the fractional-flow curve determines the thickness-averaged breakthrough saturation. As shown in Sec. 3.8, the point of tangency is the first saturation at the outlet of the core that satisfies the material balance on the water phase.

Computed displacement performance is summarized in Table 5.10. Fig. 5.16 compares the displacement performance estimated from the frontal advance solution (Example 3.4) with the performance under the assumption of vertical equilibrium. The effect of vertical equilibrium for this set of fluid and rock properties is to increase the average permeability to oil when compared to water. Displacement efficiency increases because the fractional flow of water is reduced at the same saturation.

5.4.2 Vertical Equilibrium—Gravity Segregation

When gravity segregation occurs in a waterflood, the water runs under the oil (Fig. 5.5a). Dake[6] developed a linear displacement model based on the concept of vertical equilibrium with gravity forces dominating the displacement process. This model is equivalent to the vertical equilibrium of capillary and gravity forces when the capillary transition zone has a small thickness. Fluid flow in the vertical direction occurs instantaneously because of gravity forces. The material in this section closely follows the derivation presented by Dake.

Horizontal Reservoirs. Fig. 5.17 depicts the displacement model for gravity-segregated flow in a homogeneous, linear system. An interface forms between oil and water zones. Capillary forces are neglected, so displacement is piston-like. The change in saturation across the interface is a step function from $S_w = 1 - S_{or}$ in the water zone to S_{iw} in the oil zone. Only water flows in the water

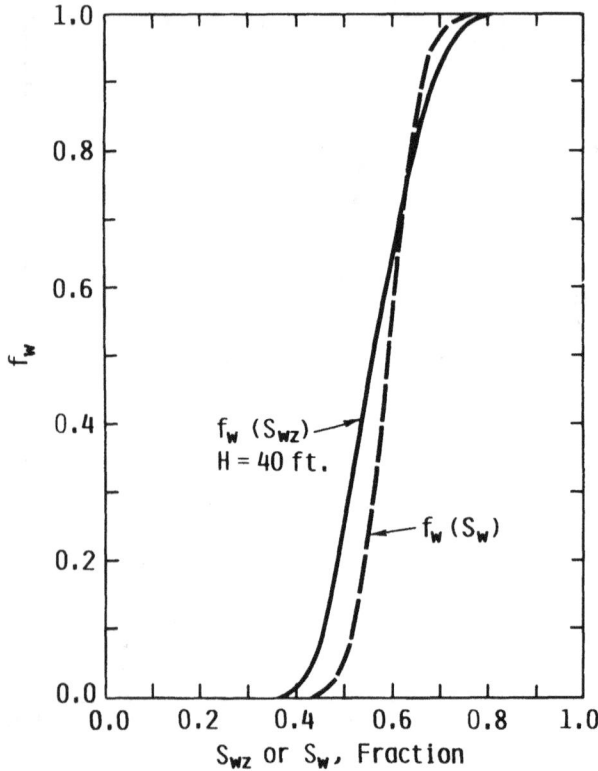

Fig. 5.15—Comparison of fractional flow curve for vertical equilibrium (pseudo) with fractional flow curve obtained from fluid/rock properties.

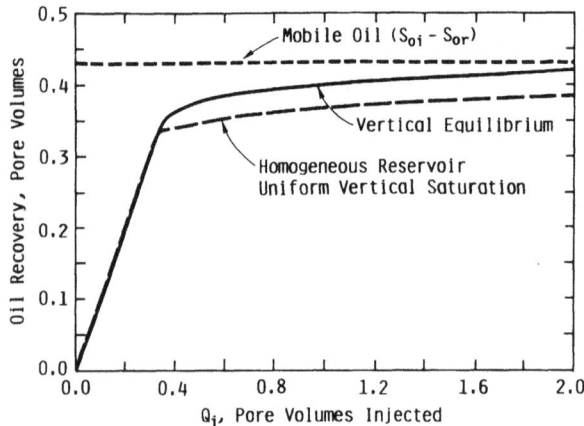

Fig. 5.16—Displacement performance of vertical equilibrium model.

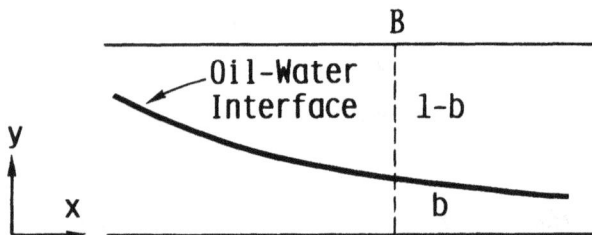

Fig. 5.17—Gravity-segregated flow in a linear homogeneous reservoir.[6]

TABLE 5.10—DISPLACEMENT PERFORMANCE FOR VERTICAL EQUILIBRIUM MODEL

S_{wz2}	f_{wz}	$\overline{S_{wz}}$	N_p (PV)	f'_{Swz} (PV)	Q_i	WOR
0.656	0.819	0.722	0.359	2.785	0.359	4.52
0.683	0.880	0.741	0.378	2.062	0.485	7.33
0.708	0.927	0.752	0.389	1.664	0.601	12.70
0.731	0.960	0.764	0.401	1.063	0.941	24.00
0.750	0.980	0.772	0.409	0.883	1.133	49.00
0.767	0.992	0.782	0.419	0.540	1.852	129.00
0.779	0.997	0.788	0.425	0.330	3.000	332.33

zone and oil in the oil zone. Vertical equilibrium is assumed so that fluids are distributed instantaneously in the vertical cross section by gravity forces.

At any vertical cross section, thickness-averaged properties are given by Eqs. 5.54 through 5.56.

$$S_{wz} = h_b(1-S_{or}) + (1-h_b)S_{iw}, \quad \ldots\ldots\ldots\ldots (5.54)$$

$$k_{roz} = (1-h_b)k_{ro}, \quad \ldots\ldots\ldots\ldots\ldots\ldots (5.55)$$

and

$$k_{rwz} = h_b k_{rw}, \quad \ldots\ldots\ldots\ldots\ldots\ldots\ldots (5.56)$$

where

k_{ro} = oil endpoint relative permeability at S_{iw},

k_{rw} = water endpoint relative permeability at S_{or}, and

h_b = fraction of the reservoir thickness flooded to residual oil saturation at a particular vertical cross section.

The parameter h_b is also the thickness-averaged dimensionless water saturation, S_{wDz}, as shown in Eq. 5.57. The value of h_b or S_{wDz} varies from 0 to 1.

$$S_{wDz} = \frac{S_w - S_{iw}}{1 - S_{or} - S_{iw}}. \quad \ldots\ldots\ldots\ldots\ldots (5.57)$$

The flow of oil and water through the vertical Cross Section B in Fig. 5.17 is given by Eqs. 5.58 and 5.59.

$$q_o = -\left(\frac{kk_{roz}A}{\mu_o}\right)\left(\frac{\partial p_o}{\partial x}\right) \quad \ldots\ldots\ldots\ldots (5.58)$$

and

$$q_w = -\left(\frac{kk_{rwz}A}{\mu_w}\right)\left(\frac{\partial p_w}{\partial x}\right). \quad \ldots\ldots\ldots\ldots (5.59)$$

Because capillary forces are neglected and the interface is almost horizontal, Eqs. 5.60 and 5.61 are the corresponding fractional flow equations.

$$f_{wz} = \frac{\dfrac{k_{rwz}}{\mu_w}}{\dfrac{k_{rwz}}{\mu_w} + \dfrac{k_{roz}}{\mu_o}} \quad \ldots\ldots\ldots\ldots\ldots\ldots (5.60)$$

or

$$f_{wz} = \frac{Mh_b}{1+(M-1)h_b}, \quad \ldots\ldots\ldots\ldots\ldots\ldots (5.61)$$

where

$$M = (k_{rw}/\mu_w)/(k_{ro}/\mu_o).$$

Expressions for displacement performance can be derived from the frontal advance theory presented in Chap. 3.

$$Q_{i2} = \frac{1}{\left(\dfrac{\partial f_{wz}}{\partial S_{wz}}\right)_{S_{wz2}}} \quad \ldots\ldots\ldots\ldots\ldots (5.62)$$

and

$$\overline{S}_{wz} = S_{wz2} + (1-f_{wz2})Q_{i2}. \quad \ldots\ldots\ldots\ldots (5.63)$$

From Eq. 5.60,

$$\left(\frac{\partial f_{wz}}{\partial S_{wz}}\right) = \left(\frac{\partial f_w}{\partial h_b}\right)\left(\frac{\partial h_b}{\partial S_{wz}}\right) \quad \ldots\ldots\ldots (5.64)$$

$$= \frac{M}{(1-S_{or}-S_{iw})[1+h_b(M-1)]^2}. \quad (5.65)$$

Thus,

$$Q_{i2} = \frac{(1-S_{or}-S_{iw})[1+h_b(M-1)]^2}{M}. \quad \ldots\ldots (5.66)$$

In this displacement model, the breakthrough saturation is S_{iw}, which is equivalent to $h_b = 0$. Therefore,

$$Q_{ibt} = \frac{1-S_{or}-S_{iw}}{M}. \quad \ldots\ldots\ldots\ldots\ldots (5.67)$$

Displacement is complete when $S_{wz2} = 1-S_{or}$ or $h_b = 1$.

Waterflood performance can be predicted by variation of h_b from 0 to 1 at the end of the system and computation of fractional flows and saturations from the appropriate equations. Note that this model is valid for $M > 1$. Problem 5.25 illustrates the computation of displacement performance with this model.

Gravity Segregation in Dipping Reservoirs. Many reservoirs have a significant dip. The gravity segregation model of Dake[6] can be adapted readily to displacement in a dipping reservoir, as shown in Fig. 5.18 by addition of the appropriate terms.

Oil and water flow rates in the direction of the dip are given by Eqs. 5.68 and 5.69.

$$q_o = -\frac{kk_{roy}A}{\mu_o}\left(\frac{\partial p_o}{\partial x} + \rho_o g \, \sin \, \alpha\right) \quad \ldots\ldots\ldots (5.68)$$

and

$$q_w = -\frac{kk_{rwy}A}{\mu_w}\left(\frac{\partial p_w}{\partial x} + \rho_w g \, \sin \, \alpha\right). \quad \ldots\ldots (5.69)$$

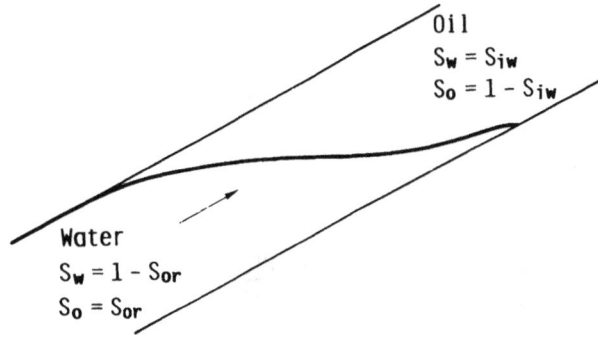

Fig. 5.18—Unstable displacement with gravity segregation in a dipping reservoir.[6]

If we assume that $\partial p_o/\partial x = \partial p_w/\partial x$, then

$$f_{wy} = \frac{\left(\dfrac{k_{rwy}}{k_{rog}}\right)\left(\dfrac{\mu_o}{\mu_w}\right) + \dfrac{kk_{rwy}A}{\mu_w q_t}(\rho_o - \rho_w)g \sin \alpha}{\left(\dfrac{k_{rwy}}{\mu_o}\right)\bigg/\left(\dfrac{k_{roy}}{\mu_w}\right) + 1},$$

$$\ldots\ldots\ldots\ldots\ldots\ldots\ldots\ldots (5.70)$$

where the subscript y refers to thickness-averaged properties in the dip-normal direction, y. Substituting for k_{rwy}, k_{roy}, we obtain

$$f_{wy} = \frac{Mh_b - h_b(h_b-1)G}{(M-1)h_b + 1}, \quad \ldots\ldots\ldots\ldots (5.71)$$

where G is the dimensionless gravity number defined by Equation 5.72 in oil-field units.

$$G = \frac{(4.9\times10^{-4})kk_{rw}A(\gamma_w - \gamma_o) \sin \, \alpha}{\mu_w q_t}. \quad \ldots (5.72)$$

Applying frontal advance concepts, we obtain

$$Q_{i2} = \frac{(1-S_{or}-S_{iw})[1+h_b(M-1)]^2}{M-G[1-2h_b+h_b^2(M-1)]}. \quad \ldots\ldots (5.73)$$

At breakthrough, $S_w = S_{wi}$ or $h_b = 0$. Thus,

$$Q_{ibt} = \frac{1-S_{or}-S_{iw}}{M-G}. \quad \ldots\ldots\ldots\ldots\ldots (5.74)$$

The remainder of the computations follow procedures described in the previous section. All oil is displaced when $h_b = 1$.

Gravity-Stabilized Displacement in Dipping Reservoirs. When displacement velocities are less than a critical value, Dietz[7] derived a relationship to predict the critical velocity required to propagate a stable interface through a linear system when gravity forces dominate. This relationship, given by Eq. 5.75, assumes piston-like displacement and negligible capillary effects, as discussed in the previous section.

$$q_c = \frac{(4.9\times10^{-1})kk_{rw}A(\gamma_w - \gamma_o) \sin \, \alpha}{\mu_w(M-1)}. \quad \ldots (5.75)$$

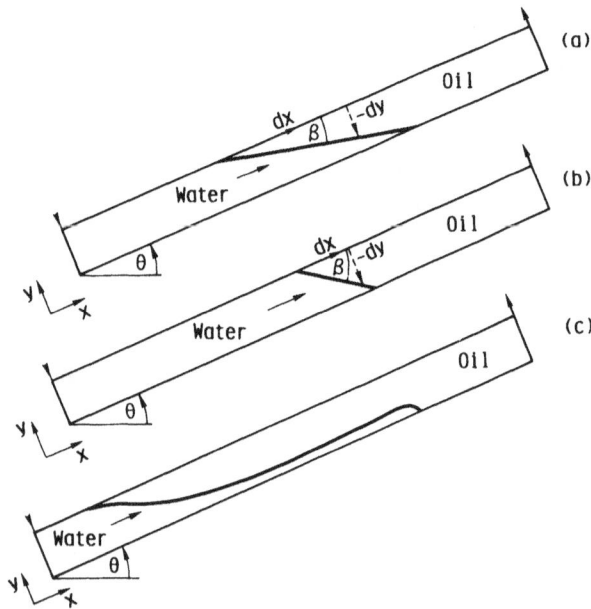

Fig. 5.19—Stable and unstable displacement in gravity-segregated displacement: (a) stable $G > M - 1$, $M > 1$, and $\beta < \theta$; (b) stable: $G > M - 1$, $M < 1$, $\beta > \theta$, and (c) unstable: $G < M - 1$.[6]

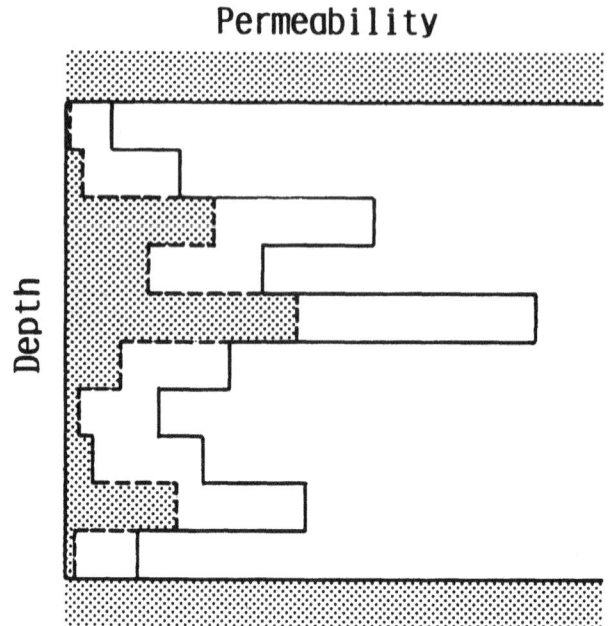

Fig. 5.20—Piston-like displacement in a linear reservoir with several communicating layers.

When the interface is stable, the velocities of oil and water are equal at every point on the interface. Thus, the interface is linear, as shown in Fig. 5.19, and is displaced at a constant velocity through the system as long as $q < q_{crit}$. The angle that represents the slope of the interface, β, can be computed from Eq. 5.76. For stable displacement, the slope must be negative.

$$\frac{dy}{dx} = -\tan \beta = \frac{(M - 1 - G)}{G} \tan \alpha. \qquad (5.76)$$

5.4.3 Displacement Performance When Pressure is Uniform in the Vertical Cross Section
A third model approximating the effects of vertical fluid flows is based on viscous forces rather than capillary and gravity forces. This model, called the uniform vertical pressure (UVP) model, has three assumptions: (1) capillary and gravity forces are negligible when compared to viscous forces (dp/dx) in the direction of flow; (2) vertical flows occur so rapidly that the pressure is uniform in any vertical cross section; and (3) the displacement is piston-like in the principal direction of flow.[3,8] The UVP model results in vertical saturation distributions that can be represented by thickness-averaged saturations and relative permeabilities. Displacement performance is computed with the same approaches as in Example 3.4.

The UVP model is applied in a linear reservoir with several communicating layers, as depicted in Fig. 5.20. Each layer may have different properties. The displacement front shown in Fig. 5.20 is continuous across the thickness h. The saturation distribution has a stair-step profile because in piston-like displacement the mobile oil saturation ($S_{oi} - S_{or}$) is displaced by the flood front. Thus, the water saturation changes abruptly from S_{iw} to $1 - S_{or}$ across the flood front in Fig. 5.20 where the shaded areas represent the portion of the reservoir that has been flooded to S_{or}.

Expressions for thickness-averaged or pseudoproperties are derived by relating fractional flow to the thickness-averaged water saturation in any vertical cross section. For example, the fractional flow of water through Cross Section x in Fig. 5.20 is given by Eq. 5.77.

$$f_{wz} = \frac{\int_A u_{wx} dz}{\int_A u_{wx} dz + \int_B u_{ox} dz}, \qquad (5.77)$$

where u_{wx} and u_{ox} are the water and oil velocities in the x direction. The integral over Region A contains all portions of the cross section that have been displaced to S_{or}, while Region B includes those parts of the cross section where only oil is flowing.

Substitution of Eqs. 2.12 and 2.13 for the water and oil velocities gives Eq. 5.78.

$$f_{wz} = \frac{\int_A \left(\frac{k_w}{\mu_w} \frac{dp_w}{dx}\right) dz}{\int_A \left(\frac{k_w}{\mu_w} \frac{dp_w}{dx}\right) dz + \int_B \left(\frac{k_o}{\mu_o} \frac{dp_o}{dx}\right) dz}. \quad (5.78)$$

When pressures are assumed to be equal at every cross section, $dp_w = dp_o = dp$. This assumption leads to Eq. 5.79, which defines the fractional flow in terms of the fraction of the cross section that has been displaced.

$$f_{wz} = \frac{\int_A \left(\frac{k_w}{\mu_w}\right)_{S_{or}} dz}{\int_A \left(\frac{k_w}{\mu_w}\right)_{S_{or}} dz + \int_B \left(\frac{k_o}{\mu_o}\right)_{S_{iw}} dz}. \qquad (5.79)$$

From Eq. 5.1,

$$S_{wz} = \frac{\int\limits_{A}\phi(S_{oi}-S_{or})dz + \int\limits_{A,B}\phi S_{iw}dz}{\bar{\phi}h} . \ \ldots\ldots (5.80)$$

Inspection of Eqs. 5.79 and 5.80 shows that f_{wz} is a function of only S_{wz}, as assumed in the development of the Buckley-Leverett solution with thickness-averaged properties. Thus, Eqs. 5.79 and 5.80 may be combined with Eq. 5.78 to compute the displacement performance of a stratified reservoir when pressure equilibrates at each cross section because of crossflow. The manner in which permeability, porosity, connate water, and residual oil saturations vary with vertical position has not been specified. No assumptions are needed if actual distributions are known. In practice, the cross section is subdivided into layers. Each layer has distinct values of porosity, permeability, thickness, and connate and residual oil saturations.

When a discrete distribution of properties is used to represent the reservoir, the flood front takes on the stair-step profile depicted in Fig. 5.20. From Eq. 5.26, we know that the flood front moves at a unique velocity in each layer. For computational purposes, it is convenient to arrange the layers in order of decreasing flood-front velocity, as shown in Fig. 5.21.

Eqs. 5.81 and 5.82 define the fractional flow and the thickness-averaged saturation when m layers have been displaced to residual oil saturation at a given position x.

$$f_{wz} = \frac{\sum\limits_{j=1}^{m}\frac{k_j k_{rwj}}{\mu_w}h_j}{\sum\limits_{j=1}^{m}\frac{k_j k_{rwj}}{\mu_w}h_j + \sum\limits_{j=m+1}^{n}\frac{k_j k_{roj}}{\mu_o}h_j} , \ \ldots (5.81)$$

$$S_{wz} = \frac{\sum\limits_{j=1}^{m}(1-S_{orj})\phi_j h_j + \sum\limits_{j=m+1}^{n}S_{wj}\phi_j h_j}{\bar{\phi}h} ,$$

$$\ldots\ldots\ldots\ldots\ldots\ldots\ldots\ldots\ldots (5.82)$$

and

$$\bar{\phi} = \frac{\sum\limits_{j=1}^{n}\phi_j h_j}{h} , \ \ldots\ldots\ldots\ldots\ldots\ldots (5.83)$$

where

k_j = base permeability for Layer j,

k_{rwj} = relative permeability to water at residual oil saturation in Layer j,

k_{roj} = relative permeability to oil at connate water saturation in Layer j,

h_j = thickness of Layer j,

h = total thickness, and

ϕ_j = porosity of Layer j.

If permeability is the only parameter that varies with thickness, the flood front will move through the layers in order of decreasing permeability. The relationship be-

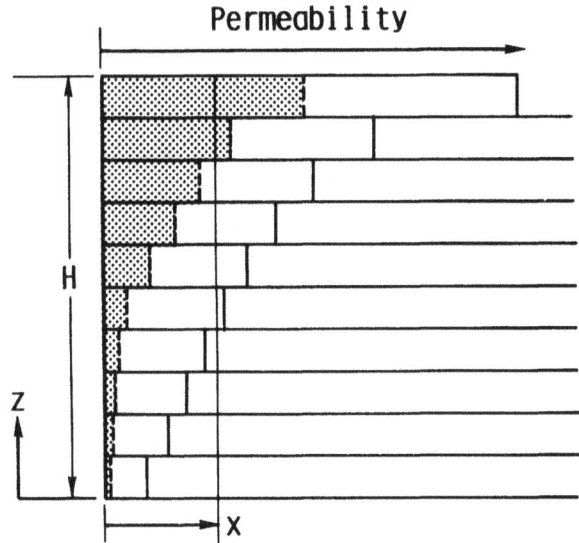

Fig. 5.21—Arrangement of layers in order of decreasing flood-front velocity.

tween fractional flow and thickness-averaged saturation can be computed from Eqs. 5.84 and 5.85.

$$f_{wz} = \frac{\sum\limits_{j=1}^{n}k_j h_j}{\sum\limits_{j=1}^{m}k_j h_j + \frac{1}{M}\sum\limits_{j=m+1}^{n}k_j h_j} \ \ldots\ldots (5.84)$$

and

$$S_{wz} = \frac{(1-S_{or})\sum\limits_{j=1}^{m}h_j}{h} + \frac{S_{iw}\sum\limits_{j=m+1}^{n}h_j}{h} . \ \ldots (5.85)$$

When k, S_{iw}, ϕ, and S_{or} vary for each layer, it is necessary to compute flood-front velocities so that layers can be ordered with $v_{Swz1} > v_{Swz2} > v_{Swzn}$ for computation of the $f_{wz} - S_{wz}$ relationship.

The flood-front velocity in Layer $m+1$ is given by Eqs. 5.86 and 5.87.

$$v_{Swzm+1} = \frac{q_t}{A\bar{\phi}}\left(\frac{\Delta f_{wz}}{\Delta S_{wz}}\right)_{m+1} \ \ldots\ldots\ldots\ldots (5.86)$$

$$= \frac{q_t}{A\bar{\phi}}\left(\frac{f_{wzm+1} - f_{wzm}}{S_{wzm+1} - S_{wzm}}\right). \ \ldots\ldots\ldots (5.87)$$

Expressions for f_{wz} and S_{wz} at m and $m+1$ are obtained from Eqs. 5.81 and 5.82. Substituting these expressions into Eq. 5.87 yields Eq. 5.88, a general expression for the velocity of the flood front in Layer $m+1$.

$$v_{Swzm+1} = \frac{q_t}{b}\left[\frac{\sum\limits_{j=1}^{m+1}\frac{k_j k_{rwj}h}{\mu_w}}{(1-S_{orm+1}+S_{iwm+1})\phi_{m+1}h_{m+1}}\right].$$

$$\ldots\ldots\ldots\ldots\ldots\ldots\ldots\ldots\ldots (5.88)$$

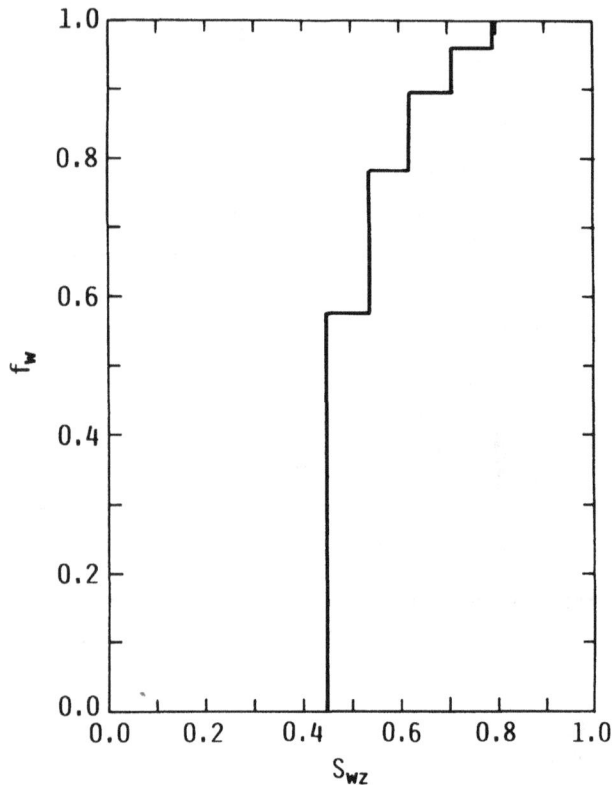

Fig. 5.22—Fractional flow curve for uniform vertical pressure model of a five-layer reservoir.

**TABLE 5.11—COMPUTATION OF f_w/S_{wz}
RELATIONSHIP FOR THE PERMEABILITY
DISTRIBUTION OF EXAMPLE 5.1 ASSUMING UNIFORM
VERTICAL PRESSURE**

Layers Flooded (n) (md)	k_j	Σk_j (flooded)	Σk_j (not flooded)	f_{wz}	S_{wz}
0		0.0	1,000.1	0	0.363
1	406.8	406.8	593.3	0.578	0.449
2	240.7	647.5	352.6	0.786	0.536
3	167.4	814.9	185.2	0.898	0.622
4	116.4	931.3	68.8	0.964	0.709
5	68.8	1,000.1	0.0	1.000	0.795

A trial-and-error procedure is used to determine the order of layers so that the flood-front velocities decrease.[8] Layers are arranged in order of decreasing values of $k_j/[\phi(1-S_{or}-S_w)]_j$. Velocities are computed from Eq. 5.88 for this arrangement of layers. Layers are reordered until an ordering is found such that

$$v_{Swz1} > v_{Swz2} \ldots v_{Swzn}.$$

Because properties (k, ϕ, S_w) change in discrete steps between layers, the fractional-flow/S_{wz} relationships also have a discrete (i.e., stair-step) change, as shown in Fig. 5.22. The WOR changed in discrete jumps as each layer experienced breakthrough. The stair steps became smaller as the number of layers increased; for some large number of layers, the f_{wz}/S_{wz} relationship can be approximated by a smooth curve.

**TABLE 5.12—COMPUTATION OF Q_i
AND OIL RECOVERY**

Layers Flooded (n) (md)	Δf_{wz}	ΔS_{wz}	$\dfrac{\Delta f_{wz}}{\Delta S_{wz}}$	Q_i	S_{wz2}	N_p (PV)
1	0.578	0.0864	6.690	0.149	0.512	0.149
2	0.208	0.0864	2.407	0.415	0.625	0.262
3	0.112	0.0864	1.296	0.771	0.701	0.338
4	0.066	0.0864	0.734	1.309	0.756	0.393
5	0.036	0.0864	0.417	2.400	0.795	0.432

Example 5.3 illustrates the prediction of waterflood performance in the linear system with five layers from Example 5.1 when crossflow occurs between layers and allows pressures to equalize at every cross section.

Example 5.3. The linear reservoir in Example 5.1 has five layers that have an equal thickness. Permeabilities are given in Table 5.1. All other parameters are constant for each layer. From Example 5.1,

$$\bar{k} = 0.200 \text{ darcies},$$
$$\phi = 0.15,$$
$$k_{rw} = 0.78 \text{ at } S_{or} = 0.205,$$
$$k_{ro} = 1.0 \text{ at } S_{iw} = 0.363,$$
$$\mu_o = 2.0 \text{ cp } [2.0 \text{ MPa·s}], \text{ and}$$
$$\mu_w = 1.0 \text{ cp } [1.0 \text{ MPa·s}].$$

Permeability is the only parameter that varies with thickness, so Eqs. 5.84 and 5.85 are used to compute the thickness-averaged water saturation (S_{wz}) and the fractional flow (f_{wz}). Table 5.11 summarizes the computed results.

The f_{wz}/S_{wz} graph is shown in Fig. 5.22. Next, the slope of the fractional-flow/S_w graph is determined to compute Q_i and the oil recovery. From Eq. 5.29,

$$Q_{i2} = \frac{1}{\left(\dfrac{\Delta f_{wz}}{\Delta S_{wz}}\right)}.$$

Oil recovery is found by calculation of S_{wz2} with Eq. 5.30 from each value of Q_{i2}. The oil recovery expressed as fraction of the PV is $S_{wz2} - S_{iw}$. Computed results are summarized in Table 5.12.

The flood-front velocity in each layer is proportional to $\Delta f_{wz}/\Delta S_{wz}$. Values of $\Delta f_{wz}/\Delta S_{wz}$ are in declining order for this example when the layers are flooded out in order of decreasing permeability, as assumed in the computation of the thickness-averaged saturation. This may not be the case when porosities and endpoint saturations (S_{iw}, S_{or}) vary, as discussed in the preceding section.

Predicted performance with the crossflow model is compared with the no-crossflow results from Example 5.1 on Fig. 5.23. Recovery from the stratified reservoir model is higher than the UVP model until approximately 1 PV. The lower recovery in the UVP model is attributed to the effects of crossflow from low-permeability layers to high-permeability layers. For $Q_i > 1$ PV, the UVP model becomes less appropriate because it is based on piston-like displacement. All the mobile oil (indicated by dashed line at 0.432 on Fig. 5.23) is displaced when the flood front breaks through the last layer.

5.5. Estimation of Vertical Displacement Efficiency With Scaled Laboratory Models

The displacement models presented in Secs. 5.2 through 5.4 were developed with certain assumptions about displacement mechanisms. These assumptions are approximated closely under some conditions. Before the development of numerical simulators, laboratory models were used to simulate displacement processes in reservoirs. When properly scaled, laboratory experiments that included displacement mechanisms not considered in other models could be conducted. For example, the Craig, Geffen, and Morse model presented in Sec. 4.4 is based on the results of scaled experiments in five-spot patterns. Analogous correlations for displacement processes involving vertical flow were developed and can be used for first estimates. In this section, an introduction to scaled-model theory and illustrations of development of scaled models to predict displacement performance when gravity segregation is an important mechanism are presented.

5.5.1 Development of Scaled Model Concepts

The idea behind scaled model theory is that displacement performance observed in a laboratory model can be scaled to simulate reservoir performance when the laboratory experiment is designed and carried out in a prescribed manner. This technique will be illustrated by the study of gravity segregation in a displacement process.

Consider the displacement of oil by water in a reservoir that has unit width, height h, and length L. The reservoir is homogeneous and isotropic. Water is injected into the reservoir at a constant rate. The reservoir is horizontal, as shown in Fig. 5.24. Boundary conditions are also indicated.

Scaling theory presented in this text is based on conversion of the partial differential equations that describe fluid flow into dimensionless form.[9,10] When this is done, dimensionless groups that determine the manner in which a laboratory experiment must be conducted to obtain results that are scalable to field dimensions are found. One requirement of scaling is geometric similarity between the model (the laboratory experiment) and the prototype (the reservoir). The geometric parameters of length (L), width (b), and thickness (h) are related as in Eqs. 5.89 and 5.90, where the subscripts m and p refer to model and prototype, respectively.

$$\frac{L_m}{h_m} = \frac{L_p}{h_p}, \quad \text{...........................} (5.89)$$

$$\frac{L_m}{b_m} = \frac{L_p}{b_p}, \quad \text{...........................} (5.90)$$

or

$$\frac{L_m}{L_p} = \frac{h_m}{h_p}. \quad \text{...........................} (5.91)$$

Next, we examine the conversion of Eqs. 5.15 through 5.17 into a dimensionless form.[9,11] Dimensionless independent and dependent variables (time, distance, and pressure) are defined by dividing each variable by an appropriate reference variable that has the same dimensions.

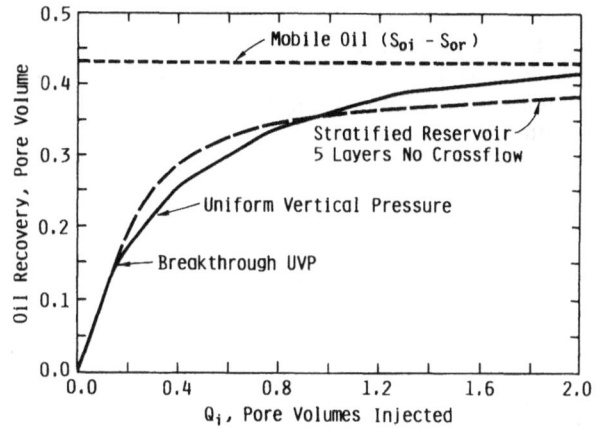

Fig. 5.23—Displacement performance for five-layer model estimated with the uniform vertical pressure model.

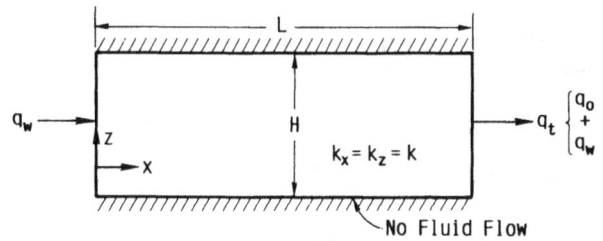

Fig. 5.24—Sketch of homogeneous reservoir.

Because reference variables may be selected arbitrarily, it is useful to choose parameters that have physical meaning. Eqs. 5.92 through 5.96 are the dimensionless spatial coordinates, time, and pressure selected for this illustration. In Eq. 5.95, the reference time is the time required to inject one PV of water at a constant volumetric injection rate of q. Thus the dimensionless time is the number of PV's of fluid injected.

$$x_D = \frac{x}{L}, \quad \text{...........................} (5.92)$$

$$z_D = \frac{z}{h}, \quad \text{...........................} (5.93)$$

$$t_D = \frac{q_t}{bhL\phi}, \quad \text{...........................} (5.94)$$

$$t_D = \frac{u_t}{L\phi}, \quad \text{...........................} (5.95)$$

and

$$p_D = \frac{p}{(q\mu_o/k_x h)}, \quad \text{...........................} (5.96)$$

where

L = length of the reservoir in Direction x,
h = thickness of the reservoir in Direction z,
b = width of the reservoir,
q = volumetric injection rate, and
u_t = darcy velocity based on water injection rate.

Saturations and relative permeabilities are dimensionless. For the purposes of this section, fluid viscosities are considered constant. Conversion of a partial derivative to dimensionless form is illustrated with the chain rule of differentiation.

$$\frac{\partial p_o}{\partial x} = \left(\frac{\partial p_o}{\partial x_D}\right)\left(\frac{\partial x_D}{\partial x}\right), \quad \ldots\ldots\ldots\ldots (5.97)$$

$$\frac{\partial p_o}{\partial x} = \frac{1}{L}\frac{\partial p_o}{\partial x_D}, \quad \ldots\ldots\ldots\ldots\ldots (5.98)$$

and

$$\frac{\partial p_o}{\partial x_D} = \left(\frac{\partial p_{oD}}{\partial x_D}\right)\left(\frac{\partial p_o}{\partial p_{oD}}\right). \quad \ldots\ldots\ldots\ldots (5.99)$$

By differentiating Eq. 5.96 expressed in terms of the oil-phase pressure and substituting for $\partial p_{oD}/\partial x_D$, we obtain

$$\frac{\partial p_o}{\partial x_D} = \left(\frac{q\mu_o}{k_x h}\right)\left(\frac{\partial p_{oD}}{\partial x_D}\right). \quad \ldots\ldots\ldots\ldots (5.100)$$

Thus when ρ_w and ρ_o are constant,

$$\frac{\partial}{\partial x}\left(\frac{\rho_o k_x k_{ro}}{\mu_o}\frac{\partial p_o}{\partial x}\right) = \frac{\rho_o q}{L^2 h}\left[\frac{\partial}{\partial x_D}\left(k_{ro}\frac{\partial p_{oD}}{\partial x_D}\right)\right].$$

$$\ldots\ldots\ldots\ldots\ldots (5.101)$$

Eqs. 5.15 and 5.16 represent the fluid flow of oil and water when gravity forces act in the z direction. When the conversions described in the previous paragraphs have been made, Eqs. 5.102 and 5.103 are the resulting dimensionless partial differential equations that describe the flow of oil and water in two dimensions with gravity forces. Verifying the derivation of Eqs. 5.102 and 5.103 is left as an exercise for the student.

Oil Phase:

$$\frac{\partial}{\partial x_D}\left(k_{ro}\frac{\partial p_{oD}}{\partial x_D}\right) + \left(\frac{L}{h}\right)^2\left(\frac{k_z}{k_x}\right)\frac{\partial}{\partial z_D}\left(k_{ro}\frac{\partial p_{oD}}{\partial z_D}\right)$$

$$+ \left(\frac{\rho_o k_z g L^2}{q\mu_o}\right)\frac{\partial k_{ro}}{\partial z_D} = \left(\frac{L}{b}\right)\frac{\partial S_o}{\partial t_D}. \quad \ldots\ldots (5.102)$$

Water Phase:

$$\frac{\partial}{\partial x_D}\left(k_{rw}\frac{\partial p_{wD}}{\partial x_D}\right) + \left(\frac{L}{h}\right)^2\left(\frac{k_z}{k_x}\right)\frac{\partial}{\partial z_D}$$

$$\times\left(k_{rw}\frac{\partial p_{wD}}{\partial z_D}\right) + \left(\frac{\rho_w k_z g L^2}{q\mu_o}\right)\frac{\partial k_w}{\partial z_D}$$

$$= \left(\frac{L}{b}\right)\left(\frac{\mu_w}{\mu_o}\right)\left(\frac{\partial S_w}{\partial t_D}\right). \quad \ldots\ldots\ldots\ldots (5.103)$$

If capillary forces are important, a relationship is needed between oil- and water-phase pressures. The function J, defined by Eq. 5.104, is frequently used for a particular saturation path. [12]

$$J(S_w) = \frac{P_c\sqrt{k/\phi}}{\sigma \cos \theta}. \quad \ldots\ldots\ldots\ldots (5.104)$$

The function J depends only on water saturation.

The dimensionless form of the capillary pressure relationship is given in Eq. 5.105:

$$P_{cD} = p_{oD} - p_{wD}$$

$$= J(S_w)\frac{\sigma \cos \theta h\sqrt{k_x \phi}}{q\mu_o}. \quad \ldots\ldots\ldots\ldots (5.105)$$

When capillary effects are considered negligible, $J(S_w)=0$ in Eq. 5.106, and $p_{oD}=p_{wD}=p_D$ in Eqs. 5.102, 5.103, 5.106, and 5.107.

The conditions required for displacement performance in the laboratory to be scalable to a reservoir can now be developed. Specifically, we desired

$$p_{oD}(x_D,z_D,t_D)\Big|_m = p_{oD}(x_D,z_D,t_D)\Big|_p, \quad \ldots (5.106)$$

$$p_{wD}(x_D,z_D,t_D)\Big|_m = p_{wD}(x_D,z_D,t_D)\Big|_p, \quad \ldots (5.107)$$

and

$$S_o(x_D,z_D,t_D)\Big|_m = S_o(x_D,z_D,t_D)\Big|_p, \quad \ldots\ldots (5.108)$$

when

$$x_{Dm} = x_{Dp}, \quad \ldots\ldots\ldots\ldots\ldots\ldots\ldots (5.109)$$

$$z_{Dm} = z_{Dp}, \quad \ldots\ldots\ldots\ldots\ldots\ldots\ldots (5.110)$$

and

$$t_{Dm} = t_{Dp}. \quad \ldots\ldots\ldots\ldots\ldots\ldots\ldots (5.111)$$

Another way to describe Eqs. 5.106 through 5.111 is that laboratory model results in dimensionless form must be equal to the field results in dimensionless form. Then, field results at x, z, and t will be scaled properly.

Laboratory data, converted to dimensionless form, represent experimental solutions to dimensionless fluid-flow equations. These equations will be identical to those obtained for a reservoir if geometric similarity is maintained.

$$\left(\frac{k_z}{k_x}\right)_m = \left(\frac{k_z}{k_x}\right)_p, \quad \ldots\ldots\ldots\ldots\ldots (5.112)$$

$$\left(\frac{\rho_o k_z g L^2}{q\mu_o}\right)_m = \left(\frac{\rho_o k_z g L^2}{q\mu_o}\right)_p, \quad \ldots\ldots\ldots (5.113)$$

$$\left(\frac{\mu_o}{\mu_w}\right)_m = \left(\frac{\mu_o}{\mu_w}\right)_p, \quad \dots \dots \dots \dots \dots (5.114)$$

$$k_{rw}(S_w)_m = k_{rw}(S_w)_p, \quad \dots \dots \dots \dots (5.115)$$

$$k_{ro}(S_w)_m = k_{ro}(S_w)_p, \quad \dots \dots \dots \dots (5.116)$$

$$\left[\frac{J(S_w)\sigma\cos\theta h\sqrt{k_x\phi}}{q\mu_o}\right]_m = \left[\frac{J(S_w)\sigma\cos\theta h\sqrt{k_x\phi}}{q\mu_o}\right]_p,$$

$$\dots \dots \dots \dots \dots \dots (5.117)$$

and

$$\left(\frac{\rho_w k_z g L^2}{q\mu_o}\right)_m = \left(\frac{\rho_w k_z g L^2}{q\mu_o}\right)_p. \quad \dots \dots \dots (5.118)$$

Eqs. 5.113 and 5.118 are usually combined as Eq. 5.119.

$$\left[\frac{k_z g L^2(\rho_o - \rho_w)}{q\mu_o}\right]_m = \left[\frac{k_z g L^2(\rho_o - \rho_w)}{q\mu_o}\right]_p.$$

$$\dots \dots \dots \dots \dots \dots (5.119)$$

There are two additional requirements for scaling. The boundary conditions in dimensionless form must be equal, and the initial conditions in dimensionless form must be equal in model and prototype.

Example 5.4 illustrates the development of scaling criteria for a reservoir where gravity segregation may be important.

Example 5.4. A homogeneous reservoir 25 ft [8 m] thick and 300 ft [91 m] long is saturated with oil and water.[13] Permeability of the reservoir is 85 md. The reservoir can be waterflooded at interstitial velocities as low as 0.047 ft/D [0.014 m/d]. It is desired to determine the effect of gravity segregation on oil displacement with a laboratory model. Other properties of the reservoir and fluids are summarized in the second column of Table 5.13.

Model properties and rates must be selected to satisfy scaling criteria in Eqs. 5.89 through 5.91 and 5.112 through 5.118. Several assumptions will be made to simplify the example. Reservoir fluids will be used in the model that will be operated at reservoir temperature. Rock type and pore structure are considered identical so that porosity, wettability, relative permeability, and capillary pressure curves coincide for model and reservoir. With these assumptions, which are substantial, the scaling requirements become

$$\frac{L}{h_m} = \frac{L}{h_p} = \frac{300}{25} = 12. \quad \dots \dots \dots \dots (5.120)$$

From Eq. 5.117,

$$\left(\frac{\sigma\cos\theta h\sqrt{k_x\phi}}{q\mu_o}\right)_m = \left(\frac{\sigma\cos\theta h\sqrt{k_x\phi}}{q\mu_o}\right)_p. \quad (5.121)$$

TABLE 5.13—PROPERTIES OF RESERVOIR AND SCALED LABORATORY MODEL

Property	Reservoir	Model
Wettability	Strongly water wet; $\cos\theta = 1$	Strongly water wet; $\cos\theta = 1$
Porosity	0.37	0.37
Interfacial tension, dynes/cm	37.0	37.0
Oil viscosity, cp	1.8	1.8
Water viscosity, cp	0.89	0.89
Density difference, g/cm^3	0.20	0.20
Permeability, darcy	0.085	212.5
Length, ft	300.0	6.00
Sand thickness, ft	25.0	0.50
Frontal velocity, ft/D	0.047	117.5
Time, t_m, days	125,000 t_m	t_m

Because

$$\left(\frac{\sigma\cos\phi}{\mu_o}\right)_m = \left(\frac{\sigma\cos\phi}{\mu_o}\right)_p, \quad \dots \dots \dots (5.122)$$

the scaling relationship is

$$\frac{h_m\sqrt{k_{xm}}}{q_m} = \frac{h_p\sqrt{k_{xp}}}{q_p}. \quad \dots \dots \dots \dots (5.123)$$

To satisfy Eq. 5.119,

$$\frac{k_{zm}L_m{}^2}{q_m} = \frac{k_{zp}L_p{}^2}{q_p}. \quad \dots \dots \dots \dots (5.124)$$

In this example, the porous medium is homogeneous and isotropic. Thus, $k_x = k_z = k$.

Eqs. 5.123 and 5.124 can be satisfied only if $q_m = q_p$. Thus

$$\frac{k_m}{k_p} = \frac{L_p}{L_m}. \quad \dots \dots \dots \dots \dots \dots (5.125)$$

If L_m or h_m is chosen, all model parameters are fixed.

Suppose the laboratory model is 6 ft [2 m] in length. From Eq. 5.120, $h_m = 0.5$ ft [0.152 m]. Also from Eq. 5.125,

$$k_m = k_p\left(\frac{L_p}{L_m}\right)^2,$$

$$= 85\left(\frac{300}{6}\right)^2$$

$$= 212,500 \text{ md}.$$

Thus the permeability of the laboratory model must be 2,500 times the reservoir permeability.

The injection rate in the model is set by the requirement that $q_m = q_p$:

$$q_p = u_{tp}\phi_p b_p h_p \quad \dots \dots \dots \dots (5.126)$$

and

$$q_m = u_{tm}\phi_m b_m h_m. \quad \dots \dots \dots (5.127)$$

Therefore,

$$u_{tm} = u_{tp}\frac{(\phi_p)(b_p)(h_p)}{(\phi_m)(b_m)(h_m)} \quad \dots \dots (5.128)$$

$$= u_{tp}(1)(50)(50)$$

$$= 2,500\, u_{pt}, \text{ and}$$

$$u_{tm} = 117.5 \text{ ft/D } [35.8 \text{ m/d}].$$

The laboratory displacement rate must be large to meet the scaling requirements. The time scale is altered also, as can be seen by a comparison of times between model and reservoir.

$$t_{Dm} = t_{Dp},$$

$$\frac{q_m t_m}{b_m h_m L_m \phi_m} = \frac{q_p t_p}{b_p h_p L_p \phi_p},$$

$$t_p = t_m\frac{(q_m)(b_p)(h_p)(L_p)(\phi_p)}{(q_p)(b_m)(h_m)(L_m)(\phi_m)} \quad \dots \dots (5.129)$$

$$= 125,000 t_m.$$

Scaling requirements between model and reservoir are summarized in Table 5.13. The parameters used in this example are identical to those used by Richardson and Perkins.[13]

Other dimensionless groups or scaling parameters can be derived by the selection of appropriate reference parameters. For example, when the reference pressure is defined by Eq. 5.130, the dimensionless scaling groups presented in Eqs. 5.131 through 5.133 can be obtained.

$$p_r = \frac{q\mu_o}{h\sqrt{k_x k_z}}, \quad \dots \dots \dots (5.130)$$

$$R_a = \frac{L}{h}\sqrt{\frac{k_z}{k_x}}, \quad \dots \dots \dots (5.131)$$

$$R_c = \frac{u_t\mu_o L}{\sqrt{k_x}\,\sigma\cos\theta}, \quad \dots \dots \dots (5.132)$$

and

$$R_d = \frac{u_t\mu_o}{\sqrt{k_x k_z}\,g\Delta\rho}. \quad \dots \dots \dots (5.133)$$

These scaling groups were developed by Goddin et al.[14] The parameter R_c represents the ratio of viscous forces to capillary forces, F_c, in the displacement. The product of R_a and R_d given as Eq. 5.134 may be viewed as the ratio of viscous forces acting in the horizontal direction to gravity forces in the vertical direction.[14]

$$R_a R_d = \frac{F_v}{F_g} = \frac{u_t\mu_o L}{k_x g\Delta\rho h}. \quad \dots \dots (5.134)$$

In strongly water-wet porous rock, capillary forces, F_c, acting in the vertical direction oppose gravity forces, F_g, that cause water to move toward the bottom of the reservoir. The term $(R_a R_d/R_c)$ given in Eq. 5.135 represents the ratio of capillary forces, F_c, to gravity forces, F_g.

$$\frac{R_a R_d}{R_c} = \frac{F_c}{F_g} = \frac{\sigma\cos\theta}{k_x g\Delta\rho h}. \quad \dots \dots (5.135)$$

The scaling groups represented by Eqs. 5.130 through 5.135 are equivalent to those presented in Eqs. 5.112 through 5.119. Selection of dimensionless scaling groups in the forms represented by Eqs. 5.134 and 5.135 permits comparison of forces causing displacement in a series of scaled-model experiments.

5.5.2 Application of Scaled Models to Predict Displacement Performance

In the previous section, the concern was with specification of laboratory parameters so that displacement performance in a laboratory model could be scaled directly to the field for a particular reservoir. Now the focus is on another aspect of the scaled-model approach. If scaling parameters in laboratory experiments are varied over the range of values anticipated in field operations, correlation of displacement performance with these parameters provides a generalized method to predict reservoir performance. This approach was used widely before the availability of numerical models.[15,16] Correlations obtained in this manner are easy to use and are often the basis for preliminary estimates of performance.

Craig et al.[16] conducted an extensive study of gravity segregation in linear drives using scaled laboratory models. Numerical values of scaling parameters were varied over a wide range by the selection of fluid properties, geometric ratios, and injection rates. Correlations were developed between volumetric sweep efficiency at breakthrough and values of the dimensionless scaling parameters. Miscible fluids were used so that capillary forces were negligible.

Fig. 5.25 illustrates the variation of volumetric sweep efficiency with the dimensionless group $(R_a R_d)$ in Eq. 5.134. This group may be interpreted in another manner. When terms are collected, it is noted that $u_t\mu_o L/k_x$ is analogous to a pressure drop in the direction of flow, while the group $g\Delta\rho h$ is a pressure change caused by gravity forces acting over the vertical distance h. Thus Craig *et al.* introduced Eq. 5.136 to lend physical meaning to the dimensionless group.

$$\frac{(\Delta p)_h}{(\Delta p)_V} = \frac{u_t \mu_o L}{k_x g \Delta \rho h}. \qquad\qquad (5.136)$$

The volumetric sweep efficiency was defined as the portion of the rock volume contacted by the injected fluid at breakthrough of the injected fluid. When the fluids are miscible, the initial saturation of the displacing fluid is zero. Therefore, the volumetric sweep efficiency is equivalent to overall recovery if it is assumed that the displacement front is sharp with no "residual oil saturation" in the swept region.

The correlations in Fig. 5.25 indicate that gravity segregation reduces volumetric sweep efficiency at breakthrough. The correlations include only volumetric sweep efficiency at breakthrough; no information is provided on the ultimate sweep efficiency. However, the remaining displaceable fluid will be recovered at higher fluid ratios (displacing/displaced) in the production after breakthrough.

Use of the mobility ratio, defined in Sec. 3.11.2 as a dimensionless scaling group, is a direct result of scaling relationships developed in the previous section for *miscible fluids*. The viscosity ratio (μ_o/μ_w or μ_d/μ_D) is specified in Eq. 5.114. When fluids are miscible, the mobility ratio is equal to the ratio of the displaced-phase viscosity to the displacing-phase viscosity.

Scaling relationships for immiscible displacement processes generally do not produce the mobility ratio as one of the parameters. In nearly all cases, the mobility ratio is an empirical correlating parameter. The one exception is the case in which displacement is complete at breakthrough of the displacing phase. The average displacing-phase saturation behind the front is $1-S_{or}$. The displacing-phase relative permeability becomes the endpoint of the relative permeability curve. Permeability curves must be scaled also to satisfy the scaling requirements of Eqs. 5.112 through 5.116. From Eqs. 5.114 through 5.116,

$$M_m = \left[\frac{k_{rw} \left|_{1-S_{or}}}{k_{ro} \left|_{S_{iw}}} \left(\frac{\mu_o}{\mu_w} \right) \right]_m$$

$$= \left[\frac{k_{rw} \left|_{1-S_{or}}}{k_{ro} \left|_{S_{iw}}} \left(\frac{\mu_o}{\mu_w} \right) \right]_p = M_p. \quad \ldots (5.137)$$

Therefore, the equivalence between M_m and M_p occurs when other scaling parameters are satisfied for this case. Perkins and Collins[10] show how relative permeability functions may be redefined to meet scaling criteria in some situations.

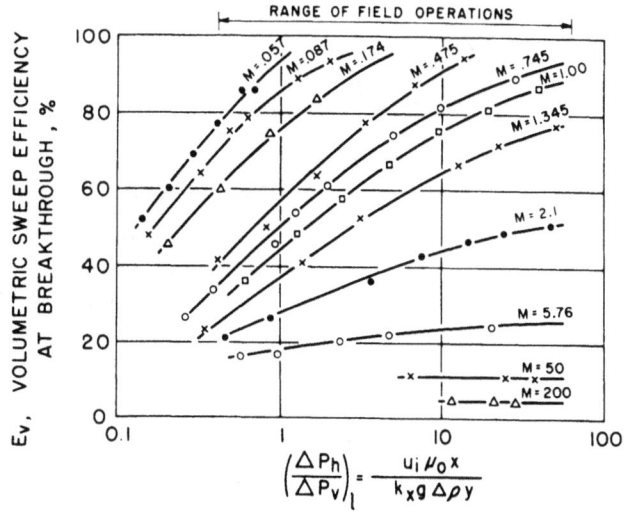

Fig. 5.25—Correlation of scaled-model experiments to determine the effect of gravity segregation on volumetric sweep efficiency in uniform linear systems.[16]

5.6 Numerical Simulation of 2D Displacement in Linear Systems Including Gravity Segregation and/or Crossflow

When fluid flow in the second direction occurs at rates that are of the same order of magnitude as flow rates in the principal direction of fluid movement, using numerical simulation to estimate displacement performance is necessary. This section illustrates the application of numerical models to predict displacement performance when vertical fluid flow caused by gravity and viscous gravity, and capillary forces is significant. It was prepared to introduce displacement mechanisms, as well as to present computed results that show the capability of numerical simulators to estimate the contribution of these mechanisms to displacement efficiency.

5.6.1 Gravity Segregation

The displacement of oil by gas in a dipping stratum illustrates gravity segregation in homogeneous reservoirs. When gas is injected at the top of a dipping reservoir, as shown in Fig. 5.26, a tilted front forms between the gas and the oil bank.[17] Displacement occurs in both x and y directions; thus the displacement process cannot be simulated by the Buckley-Leverett solution. Gas is compressible, which increases the complexity of the problem.

Numerical techniques are needed to estimate the effects of 2D flow on displacement performance. Fig. 5.27 illustrates the flow model used. In this simulation, it is convenient to use coordinate axes x and y that have been rotated the angle α with respect to the elevation direction Z. Thus $Z = -x \sin \alpha + y \cos \alpha$.

Flow equations are given by Eqs. 5.138 and 5.139.

$$\frac{\partial}{\partial x}\left[\frac{\rho_o k_o}{\mu_o} \left(\frac{\partial p_o}{\partial x} + \rho_o g \frac{\partial Z}{\partial x} \right) \right] + \frac{\partial}{\partial y}\left[\frac{\rho_o k_o}{\mu_o} \left(\frac{\partial p_o}{\partial y} \right. \right.$$

$$\left. \left. + \rho_o g \frac{\partial Z}{\partial y} \right) \right] = \phi \frac{\partial}{\partial t}(\rho_o S_o) \qquad \ldots\ldots\ldots\ldots (5.138)$$

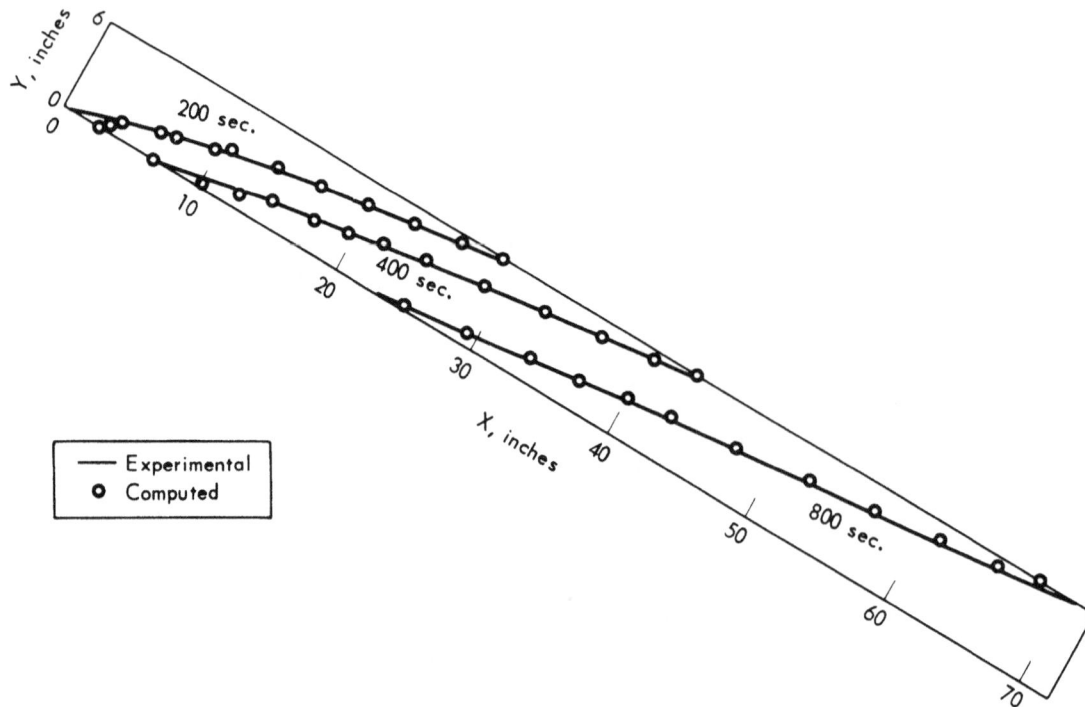

Fig. 5.26—Position of gas/oil contact during 30° displacement at 83% of the critical rate.[17]

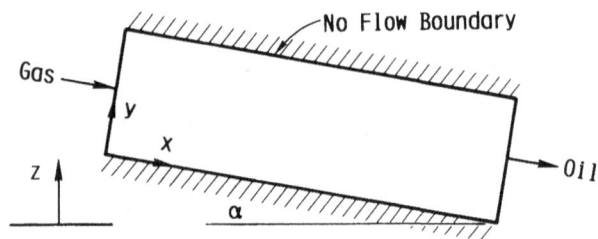

Fig. 5.27—Model of immiscible displacement of oil by gas in a dipping reservoir.

and

$$\frac{\partial}{\partial x}\left[\frac{\rho_g k_g}{\mu_g}\left(\frac{\partial p_g}{\partial x}+\rho_g g\frac{\partial Z}{\partial x}\right)\right]+\frac{\partial}{\partial y}\left[\rho_g\frac{k_g}{\mu_g}\left(\frac{\partial p_g}{\partial y}\right.\right.$$

$$\left.\left.+\rho_g g\frac{\partial Z}{\partial y}\right)\right]=\phi\frac{\partial}{\partial t}(\rho_g S_g). \quad\ldots\ldots\ldots\ldots(5.139)$$

In addition,

$$P_c=p_g-p_o, \quad\ldots\ldots\ldots\ldots\ldots\ldots\ldots(5.140)$$

$$\rho_o=\rho_o(p_o), \quad\ldots\ldots\ldots\ldots\ldots\ldots\ldots(5.141)$$

$$\rho_g=\rho_g(p_g), \quad\ldots\ldots\ldots\ldots\ldots\ldots\ldots(5.142)$$

$$k_o=k_o(S_o), \quad\ldots\ldots\ldots\ldots\ldots\ldots\ldots(5.143)$$

and

$$k_g=k_g(S_g). \quad\ldots\ldots\ldots\ldots\ldots\ldots\ldots(5.144)$$

Nitrogen was used in the model experiment, and mass transfer between phases was neglected. Oil was the wetting phase; thus the displacement process follows the drainage path. Capillary pressure and relative permeability data were obtained for this path. Fig. 5.28 shows the capillary pressure data as a function of oil saturation. There is little change in capillary pressure from $S_o=40\%$ to $S_o=100\%$. The effects of capillary forces are limited to a small interval of saturation change. Because the displacing phase is the nonwetting phase, capillary forces actually oppose displacement, in direct contrast to the water/oil displacement examples that have been considered previously in this text.

Computed and experimental locations of the gas/oil contact as a function of time are compared in Fig. 5.26 when the displacement rate was 83% of the critical rate. The critical rate was defined by Terwilliger *et al.*[18]

$$q_c=\frac{k_L A}{\mu}\Delta\rho\,\sin\,\alpha. \quad\ldots\ldots\ldots\ldots\ldots\ldots(5.145)$$

At rates above the critical rate, the displacement front is unstable, and viscous fingering or channeling is observed. At rates lower than the critical rate, the displacement front is stabilized by gravity segregation. Viscous fingering is discussed in Sec. 5.7.

5.6.2 Crossflow in Stratified Systems in the Absence of Gravity Forces

Many heterogeneous or stratified reservoirs are not adequately represented by a series of noncommunicating layers. Vertical flows or crossflows occur as fluids move from one layer to another in response to changes in fluid potentials. Consider a waterflood in the linear, stratified reservoir depicted in Fig. 5.29. The reservoir has two

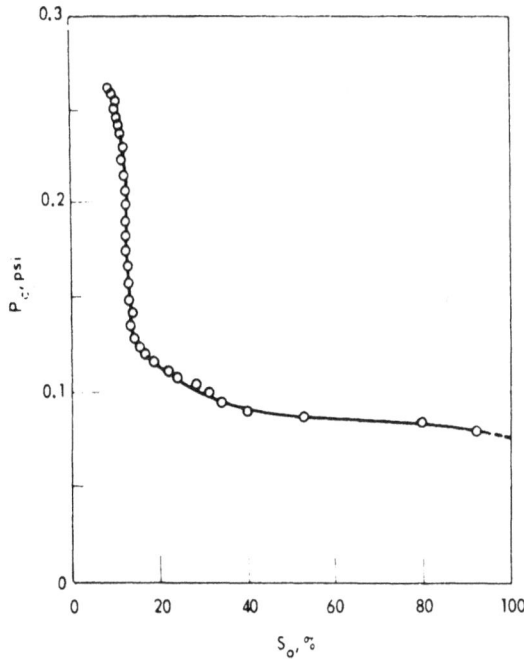

Fig. 5.28—Drainage capillary-pressure curve for 20- to 30-mesh Ottawa sand.[17]

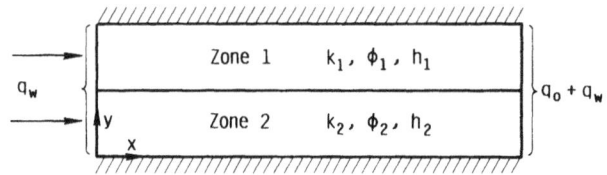

Fig. 5.29—Stratified model of a linear reservoir.

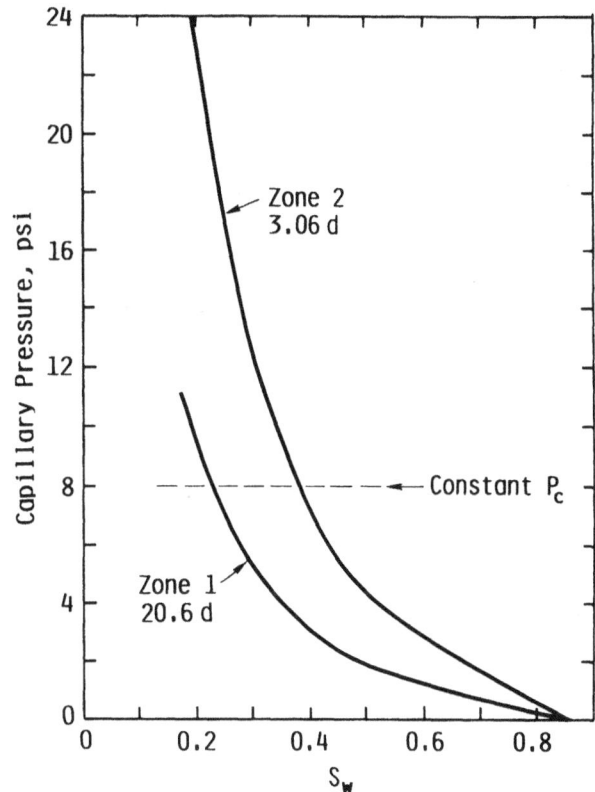

Fig. 5.30—Capillary-pressure curves for two layers that have different permeabilities.

horizontal zones or layers that are connected hydraulically so that both water and oil may move from one zone to another, depending on relative permeability relationships and potential differences. Absolute permeability of each zone is uniform and does not vary with spatial position. The permeability of Zone 1 is assumed to be larger than that of Zone 2.

When water is injected into both zones at $x=0$, several possible displacement mechanisms may be present. Because of permeability differences, the displacement front will move faster in Zone 1 than in Zone 2. Viscous crossflow may occur from Zone 2 into Zone 1 because the mobilities of oil and water in Zone 1 are greater than in Zone 2. Capillary forces may cause crossflow. If the reservoir is strongly water-wet, water from Zone 1 may imbibe into Zone 2. Some oil from Zone 2 flows countercurrent to the water entering Zone 1 where it is eventually displaced. In strongly oil-wet reservoirs, capillary forces oppose crossflow of water. Gravity segregation is superimposed on crossflows induced by viscous forces and imbibition. In the section that follows, the effect of gravity forces is neglected by consideration of displacement in stratified laboratory models that have small thicknesses and are oriented in the horizontal plane. First, the concepts of capillary equilibrium in a uniform stratified reservoir will be reviewed.

Capillary Equilibrium. Capillary forces may be important in stratified systems because changes in permeability and porosity in geologically similar rocks cause large differences in capillary pressure curves. This is seen in Eq. 5.146, the Leverett correlation,[12] which is frequently used to correlate capillary-pressure data from rocks that are from the same formation.

$$J(S_w) = \frac{P_c\sqrt{k/\phi}}{\sigma\cos\theta}. \quad \dots\dots\dots\dots\dots\dots (5.146)$$

For example, Fig. 5.30 shows the capillary-pressure curves for two unconsolidated sands that have permeabilities of 20.6 and 3.06 darcies, respectively. With the exception of $(1-S_{or})$, the capillary pressure in the lower-permeability sand is higher than that in the high-permeability sand.

Two layers with different capillary pressure characteristics are connected hydraulically at the boundary between the layers in Fig. 5.29. Pressures in the oil phases are equal, as are the pressures in the water phases. Thus, the capillary pressures in both layers are equal at this common boundary. If the system is at capillary equilibrium, the saturation distributions determined by capillary-pressure curves are not equal. A dashed line of constant capillary pressure is drawn on Fig. 5.30 to illustrate the saturation discontinuity between Layers 1 and 2 at a common boundary where the capillary pressure is P_c. When Zones 1 and 2 in Fig. 5.29 are in the horizontal plane so that gravity forces are neglected, the phase pressures are uniform within each layer. At capillary equilibrium, the two layers will have saturations S_{w1} and S_{w2} where $S_{w1} < S_{w2}$.

The average saturation of the composite system is obtained from Eq. 5.147.

$$\overline{S_w} = \frac{\phi_1 S_{w1} h_1 bL + \phi_2 S_{w2} h_2 h_2 bL}{\phi_1 h_1 bL + \phi_2 h_2 bL}. \quad \ldots \ldots (5.147)$$

When Zones 1 and 2 are oriented in the vertical plane, the pressure distribution is affected by gravity. At capillary equilibrium, $\Phi_{o1} = \Phi_{o2}$ and $\Phi_{w1} = \Phi_{w2}$. Saturations will vary somewhat throughout the cross section.

Effects of crossflow in stratified systems (as in Fig. 5.28) will be illustrated for the case where both layers are in the horizontal plane so that gravity forces are negligible. Coordinate axes are now x and y.

The displacement process in two dimensions is represented by the equivalent forms of Eqs. 5.15 and 5.16 for each zone.

Zone 1: $h_2 \leq y \leq h_2 + h_1$.

$$\frac{\partial}{\partial x}\left(\frac{\rho_o k_{o1}}{\mu_o}\frac{\partial p_{o1}}{\partial x}\right) + \frac{\partial}{\partial y}\left(\frac{\rho_o k_{o1}}{\mu_o}\frac{\partial p_{o1}}{\partial y}\right)$$

$$= \frac{\partial}{\partial t}(\rho_o S_{o1}\phi_1), \quad \ldots \ldots (5.148)$$

$$\frac{\partial}{\partial x}\left(\frac{\rho_w k_{w1}}{\mu_w}\frac{\partial p_{w1}}{\partial x}\right) + \frac{\partial}{\partial y}\left(\frac{\rho_w k_{w1}}{\mu_w}\frac{\partial p_{w1}}{\partial y}\right)$$

$$= \frac{\partial}{\partial t}(\rho_w S_{w1}\phi_1), \quad \ldots \ldots (5.149)$$

$$k_{w1} = k_{w1}(S_{w1}), \quad \ldots \ldots (5.150)$$

$$k_{o1} = k_{o1}(S_{w1}), \quad \ldots \ldots (5.151)$$

and

$$P_{c1} = P_{c1}(S_{w1}). \quad \ldots \ldots (5.152)$$

Zone 2: $0 \leq y \leq h_2$.

$$\frac{\partial}{\partial x}\left(\frac{\rho_o k_{w2}}{\mu_o}\frac{\partial p_{o2}}{\partial x}\right) + \frac{\partial}{\partial y}\left(\frac{\rho_o k_{o2}}{\mu_o}\frac{\partial p_{o2}}{\partial y}\right)$$

$$= \frac{\partial}{\partial t}(\rho_o S_{o2}\phi_2), \quad \ldots \ldots (5.153)$$

$$\frac{\partial}{\partial x}\left(\frac{\rho_w k_{w2}}{\mu_w}\frac{\partial p_{w2}}{\partial x}\right) + \frac{\partial}{\partial y}\left(\frac{\rho_w k_{w2}}{\mu_w}\frac{\partial p_{w2}}{\partial y}\right)$$

$$= \frac{\partial}{\partial t}(\rho_w S_{w2}\phi_2), \quad \ldots \ldots (5.154)$$

$$k_{w2} = k_{w2}(S_{w2}), \quad \ldots \ldots (5.155)$$

$$k_{o2} = k_{o2}(S_{w2}), \quad \ldots \ldots (5.156)$$

and

$$P_{c2} = P_{c2}(S_{w2}). \quad \ldots \ldots (5.157)$$

Zones 1 and 2 meet at the boundary where $y = h_2$. Because these layers are connected hydraulically at the boundary $y = h_2$,

$$p_{w1} = p_{w2},$$

$$p_{o1} = p_{o2}, \text{ and}$$

$$P_{c1} = P_{c2}.$$

Continuity of oil and water flows across the boundary is required also. Thus at $y = h_2$,

$$k_{o1}\frac{\partial p_{o1}}{\partial y} = k_{o2}\frac{\partial p_{o2}}{\partial y} \quad \ldots \ldots (5.158)$$

and

$$k_{w1}\frac{\partial p_{w1}}{\partial y} = k_{w2}\frac{\partial p_{w2}}{\partial y}. \quad \ldots \ldots (5.159)$$

The mathematical description of the problem is completed by the specification of boundary and initial conditions. No-flow boundaries are assumed at $y = 0$ and $y = h$. Thus at $y = 0$, $0 \leq x \leq L$,

$$\frac{\partial \Phi_{o2}}{\partial y} = \frac{\partial p_{o2}}{\partial y} = 0 \quad \ldots \ldots (5.160)$$

and

$$\frac{\partial \Phi_{w2}}{\partial y} = \frac{\partial p_{w2}}{\partial y} = 0. \quad \ldots \ldots (5.161)$$

At $y = h$, $0 \leq x \leq L$,

$$\frac{\partial \Phi_{o1}}{\partial y} = \frac{\partial p_{o1}}{\partial y} = 0 \quad \ldots \ldots (5.162)$$

and

$$\frac{\partial \Phi_{w1}}{\partial y} = \frac{\partial p_{w1}}{\partial y} = 0. \quad \ldots \ldots (5.163)$$

Boundary conditions at $x=0$ and $x=L$ are more difficult to formulate. If the injection rate per unit width is constant with respect to time, then at $x=0$,

$$\frac{q}{b} = \int_0^h u_{w2}(y) \Big|_{x=0} dy + \int_{h_2}^h u_{w1}(y) \Big|_{x=0} dy \quad \dots (5.164)$$

or

$$\frac{q}{b} = \int_0^{h_2} \left[-\left(\frac{k_{w2}}{\mu_2}\right)\left(\frac{\partial p_{w2}}{\partial x}\right) \Big|_{x=0} \right] dy$$

$$+ \int_{h_2}^h \left[-\left(\frac{k_{w1}}{\mu_w}\right)\left(\frac{\partial p_{w1}}{\partial x}\right) \Big|_{x=0} \right] dy. \quad \dots\dots(5.165)$$

At $x=L$, the flow rates are constant for an incompressible system. Thus

$$\frac{q_p}{b} = \frac{(q_o + q_w)}{b} \Big|_{x=L},$$

$$\frac{q_p}{b} = \int_0^{h_2} (u_{o2} + u_{w2}) \Big|_{x=L} dy + \int_{h_2}^h (u_{o1} + u_{w1}) \Big|_{x=L} dy,$$

$$\dots\dots\dots\dots\dots\dots\dots\dots\dots(5.166)$$

and

$$\frac{q_p}{b} = \int_0^{h_2} -\left(\frac{k_{w2}}{\mu_w}\frac{\partial p_{w2}}{\partial x} + \frac{k_{o2}}{\mu_o}\frac{\partial p_{o2}}{\partial x}\right) \Big|_{x=L} dy$$

$$+ \int_{h_2}^h -\left(\frac{k_{w1}}{\mu_w}\frac{\partial p_{w1}}{\partial x} + \frac{k_{o1}}{\mu_o}\frac{\partial p_{o1}}{\partial x}\right) \Big|_{x=L} dy.$$

$$\dots\dots\dots\dots\dots\dots\dots\dots\dots(5.167)$$

Other boundary conditions used at $x=0$ and $x=L$ include specification of potentials or pressures in oil and water phases as constants or a function of time. When the potential difference is constant, the flow rates vary with time. Initial saturations are assumed to be at capillary equilibrium at the beginning of the displacement.

The mathematical problem defined by Eqs. 5.148 through 5.167 is formidable. Complete solutions were not possible before development of numerical techniques. Numerical solutions are discussed in the next section. Ref. 13 contains a limited amount of experimental data that were obtained from scaled-model experiments. Those results are useful in providing some feel for the displacement mechanisms.

TABLE 5.14—DATA FOR COMPUTING STRATIFIED MODEL

Constants	
Length, in.	72
Width, in.	6
Thickness, in.	3/8
Flow rate, cu in./min	0.61
Viscosity of oil, poise	0.017
Viscosity of water, poise	0.009
Porosity	0.357
Permeability fine sand, darcies	3.06
Permeability coarse sand, darcies	20.6
Initial water saturation	0.198

5.6.3 Immiscible Displacement in a Uniform Stratified Reservoir with Crossflow Caused by Capillary and Viscous Forces

Simulation of oil displacement from a uniform stratified reservoir with crossflow was first reported in a classic paper by Douglas et al.[2] They were able to solve the problem described by Eqs. 5.148 through 5.167 in the previous section. Properties of their stratified system are shown in Table 5.14. Gravity forces were not important because the laboratory results were obtained in a horizontal model that was 3/8 in. [0.953 cm] thick.

Fig. 5.31 shows computed saturation and velocity profiles at one instant of time during the simulation. Line segments in Fig. 5.31 indicate the direction and magnitude of oil- and water-velocity vectors. Water is represented by the broken line. Curves of constant water saturation, shown in Fig. 5.31, indicate the relative displacement efficiency from each zone. The large difference in water saturations between the two zones at the right side of the model reflects different initial water saturations corresponding to capillary equilibrium at the beginning of the displacement. The initial water saturation in Zone 1 was 0.13, while the initial water saturation in Zone 2 was 0.27.

Fig. 5.32 shows a comparison of calculated vs. experimental oil recovery for the stratified laboratory model. The agreement is excellent to the point of water breakthrough but poor thereafter. This example, however, clearly demonstrated the potential of numerical techniques in solving displacement problems.

Numerical solutions permit investigation of the effects of different operating parameters. Figs. 5.33 and 5.34 show saturation contours and crossflow rates for two mobility ratios ($M_{\bar{S}}=0.6$ and $M_{\bar{S}}=0.21$). The mobility ratio is defined in terms of the saturations in the loose layer. Crossflows of water and oil were larger at the lower mobility ratio. Other saturation profiles are presented in Refs. 14 and 19. Crossflow rates were highest near the saturation front in the loose layer where the largest capillary-pressure gradients were found.

Recovery efficiencies for the base case with $M_{\bar{S}}=0.6$, along with two limiting cases in which crossflow was neglected, are shown in Fig. 5.35. The uniform system is a reservoir that has an average permeability based on permeability thickness as defined by Eq. 5.168. Displacement in this system was predicted with $\bar{k}$ in the frontal advance solution.

$$\bar{k} = \Sigma \frac{k_j h_j}{h}. \quad \dots\dots\dots\dots\dots\dots(5.168)$$

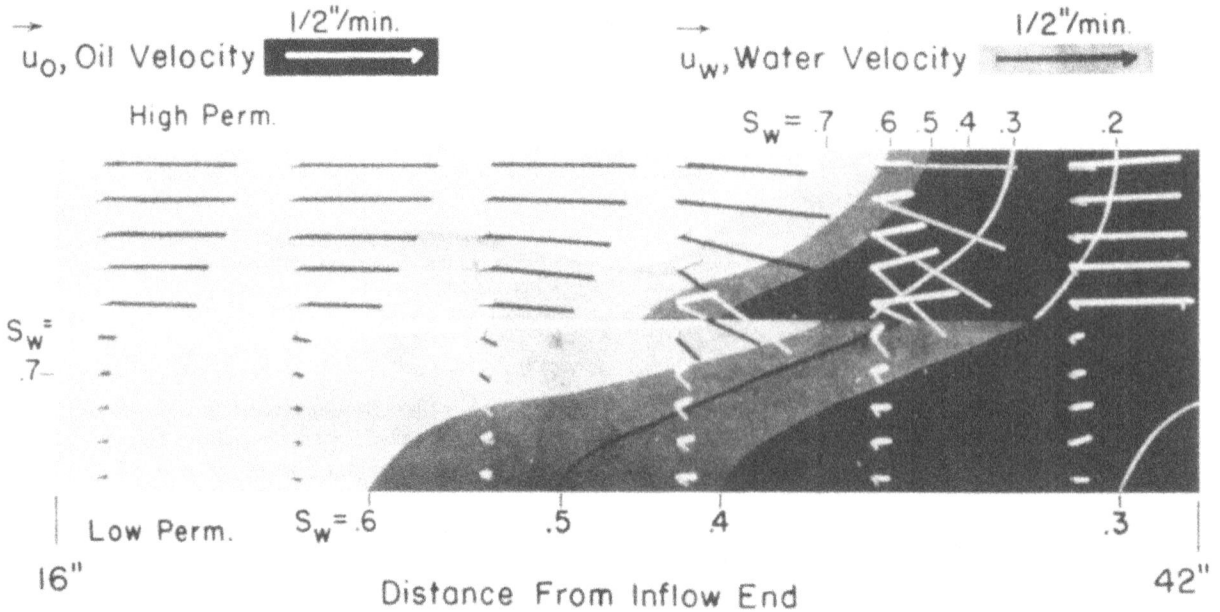

Fig. 5.31—Computed saturation and velocity distributions in the stratified model.[2]

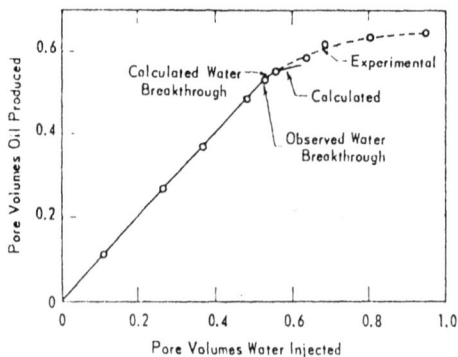

Fig. 5.32—Oil recovery from stratified model.[2]

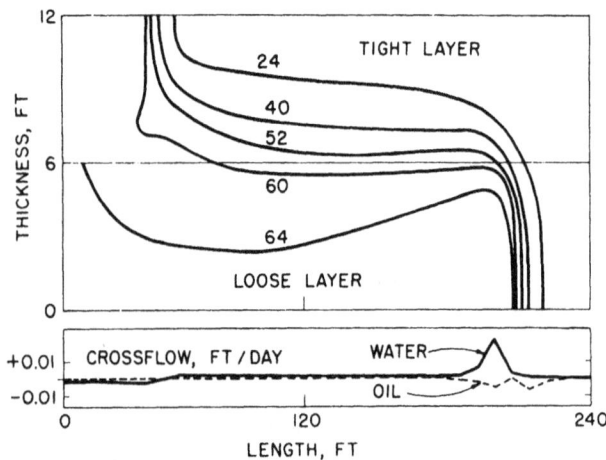

Fig. 5.33—Saturation contours and crossflow rates before breakthrough for $M_s = 0.6$.[14]

In the no-crossflow case, the two layers were not connected hydraulically. Displacement performance of each layer was calculated from the frontal advance solution. Recoveries from each layer were combined at equal points in time, as discussed earlier, to obtain the total recovery for the reservoir. Oil recovery and water injection in Fig. 5.35 are reported in terms of recoverable hydrocarbon PV $(S_{oi} - S_{or})$.

Performance of the base case is between the computed performance for no crossflow and the uniform system. Thus, for this system, the predicted performance is bounded by two cases that can be computed from the frontal advance solution. Eq. 5.169 defines a crossflow index that quantifies the effect of crossflow on oil recovery.

$$\text{Crossflow index} = \frac{N_{pcf} - N_{pncf}}{N_{pu} - N_{pncf}}, \quad \ldots\ldots\ldots (5.169)$$

where

N_{pu} = oil recovery from uniform system with average permeability,

N_{pcf} = oil recovery from layered system with crossflow, and

N_{pncf} = oil recovery from stratified system with no crossflow.

A crossflow index of 1.0 means that the performance of the layered system with crossflow is identical to that of the uniform system. Fig. 5.36 illustrates the effect of the layer permeability ratio (R_k) on the crossflow index with other parameters fixed at base-case values. The layer ratio for the base case was 4.0.

Of several dimensionless groups investigated in numerical simulations, the mobility ratio shown in Fig. 5.37 appeared to have the largest effect on crossflow, particularly at low values of the ratio.[14]

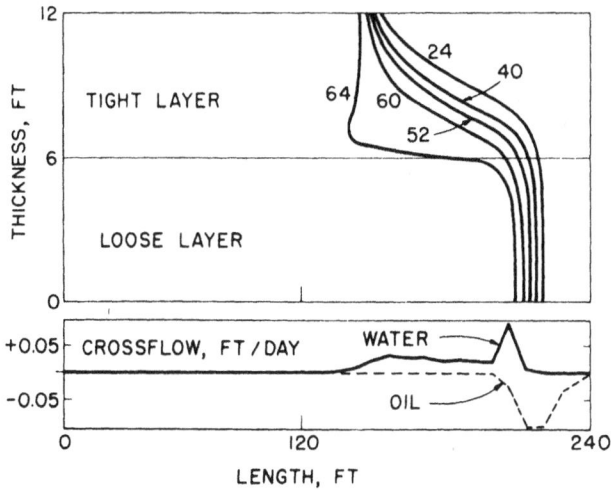

Fig. 5.34—Saturation contours and crossflow rates for $M_s = 0.2$.[14]

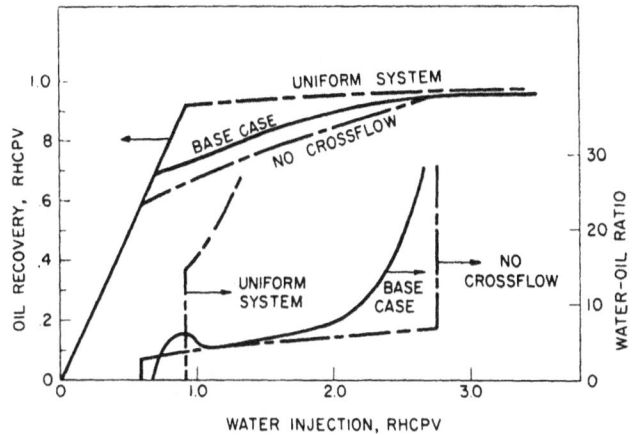

Fig. 5.35—Oil recovery and WOR for $M_s = 0.6$.[14]

5.6.4 Limitations of Numerical Simulation

Although numerical simulators are capable of solving complex two-phase flow problems, such as those described in this section, they have limitations. One computational problem involves the evaluation of the mobilities of oil and water phases as saturations change during the numerical simulation. Fig. 5.38 compares computed saturation profiles during a linear waterflood for two mobility weighting schemes with the Buckley-Leverett solutions for three weighting schemes when the mobility ratio is favorable ($M_{\bar{s}} < 1$) and unfavorable ($M_{\bar{s}} > 1$).[20] Fig. 5.39 compares computed water saturation profiles for Case 5, Ref. 19, using two different mobility computation schemes.[20] The secondary saturation hump reported by Goddin[19] at 50/50 mobility weighting is absent when a different mobility weighting scheme is used. These examples show that numerical simulators do not necessarily produce unique solutions.

5.7 Viscous Fingering

The displacement theory developed in Chap. 3 does not represent a phenomenon that has been observed when the viscosity ratio is much larger than 1.0. Fig. 5.40 shows cross sections through a sandpack (1.9 in. [4.8 cm] in diameter) that was waterflooded at an oil/water viscosity ratio of 102.5.[21] White spots show the paths taken by the water.

This phenomenon, called viscous fingering, was first reported by Engleberts and Klinkenberg[22] for immiscible displacement processes. The same phenomenon, called channeling, was observed by Hill,[23] who studied miscible displacement in packed beds when the viscosity of the displacing fluid was much less than that of the displaced fluid. In gravity drainage experiments similar to those depicted in Fig. 4.7, a stabilized zone formed only when the displacement rate was less than the gravity drainage reference rate defined by Eq. 5.170.

$$u_c = \frac{q_c}{A} = \frac{k_L}{\mu} \Delta\rho \sin \alpha. \qquad (5.170)$$

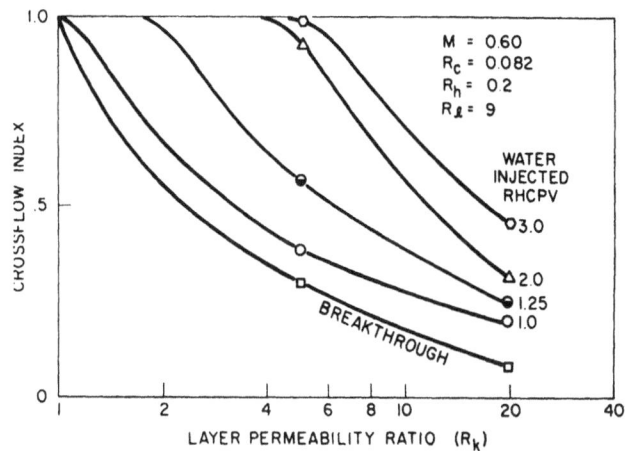

Fig. 5.36—Effect of layer permeability ratio on crossflow index.[14]

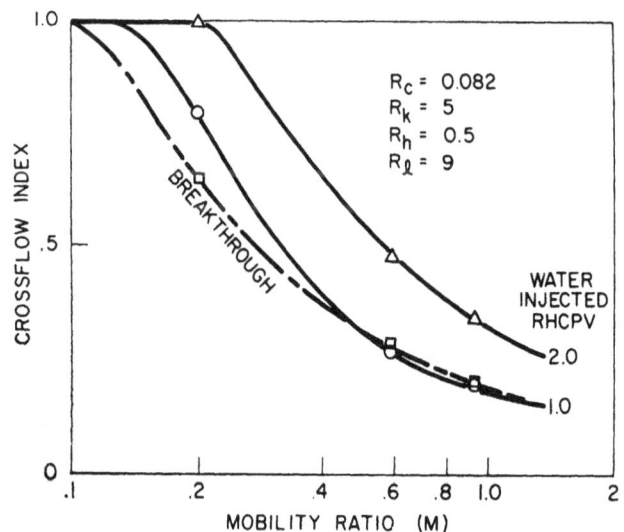

Fig. 5.37—Effect of mobility ratio (M) on crossflow index.[14]

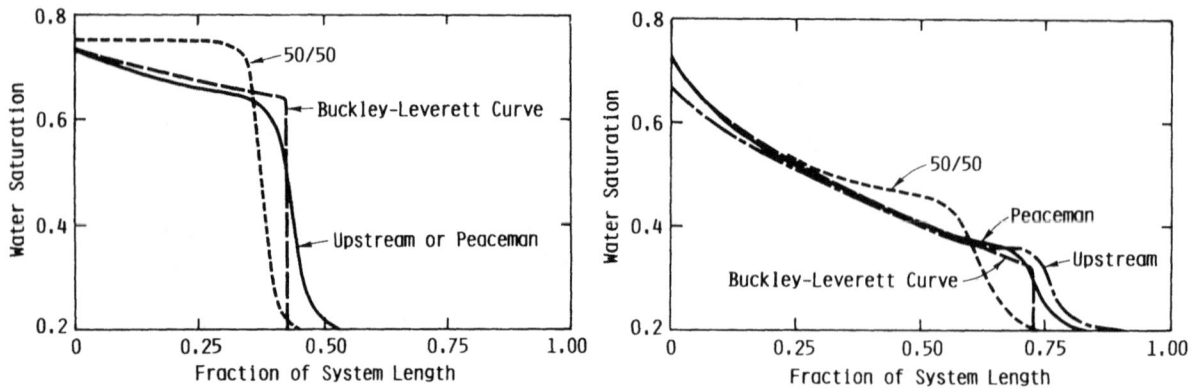

Fig. 5.38—Computed saturation profiles with different mobility weighting schemes for a linear waterflood at favorable and unfavorable mobility ratios after 0.2 PV injection.[20]

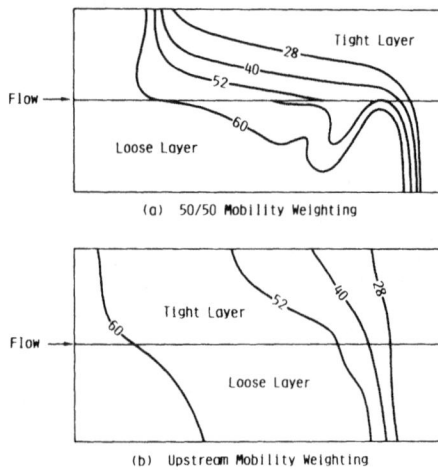

Fig. 5.39—Effects of mobility weighting schemes on saturation contours for Case 5B of Ref. 14 after 0.275 PV injection.[20]

The displacement front was not stabilized at displacement rates larger than the gravity drainage rate. The lack of stabilization was caused by viscous fingering of the displacing gas through the oil. Fingering is characterized by 2D (or possibly 3D) flow in random, unpredictable paths. Velocities vary with position as well as time. The following section provides an overview of viscous fingering.

The initiation and growth of viscous fingers is believed to be caused by instabilities at the interface between the displacing fluid and the displaced fluid whenever the viscosity of the displacing fluid is much less than that of the displaced fluid.[23,24] The development of a method to predict the onset of viscous fingering in a gravity-stabilized displacement is attributed to Hill. Although the development was based on miscible displacement in a vertical packed column, extension to immiscible displacement in a porous column follows from the development. Hill postulated the initiation of a small finger, at a flat interface separating a fluid with Viscosity 2 and Density 2, which displaced a fluid with Viscosity 1 and Density 1. The displacement rate u was constant. Fig. 5.41 illustrates a finger of length z extending from the interface AB.[23] The finger cannot propagate or grow if $p_1 > p_2$. Thus, the criterion for stable flow is $p_1 - p_2 > 0$.

Expressions for p_2 and p_1, obtained by assuming that Darcy's law describes flow of liquid and gas between Planes AB and CD, are given by Eqs. 5.171 and 5.172.

$$p_1 = p_o + g\rho_1 \Delta z - \frac{\mu_1 u \Delta z}{k_1} \qquad \ldots\ldots\ldots\ldots (5.171)$$

and

$$p_2 = p_o + g\rho_2 \Delta z - \frac{\mu_2 u \Delta z}{k_2}. \qquad \ldots\ldots\ldots\ldots (5.172)$$

Because $p_1 - p_2 > 0$ for stable flow,

$$g(\rho_1 - \rho_2)\Delta z - \left(\frac{\mu_1}{k_1} - \frac{\mu_2}{k_2}\right) u \Delta z > 0. \qquad \ldots\ldots (5.173)$$

Therefore, stable flow will occur when

$$u < \frac{g(\rho_1 - \rho_2)}{\left(\dfrac{\mu_1}{k_1} - \dfrac{\mu_2}{k_2}\right)}. \qquad \ldots\ldots\ldots\ldots\ldots (5.174)$$

If $k_1 = k_2 = k$,

$$u < \frac{kg(\rho_1 - \rho_2)}{(\mu_1 - \mu_2)}. \qquad \ldots\ldots\ldots\ldots\ldots (5.175)$$

Eq. 5.175 is identical to Eq. 5.170 when the viscosity of the displacing phase is omitted, which is reasonable when the displacing phase is a gas. Eq. 5.176 defines the critical velocity for onset of viscous fingering for a porous reservoir.[7] The angle ZZ' is between the vertical axis and the normal to the interface in the direction of flow.

$$u_c \leq \frac{g(\rho_1 - \rho_2) \cos (ZZ')}{\left(\dfrac{\mu_1}{k_1} - \dfrac{\mu_2}{k_2}\right)}. \qquad \ldots\ldots\ldots\ldots (5.176)$$

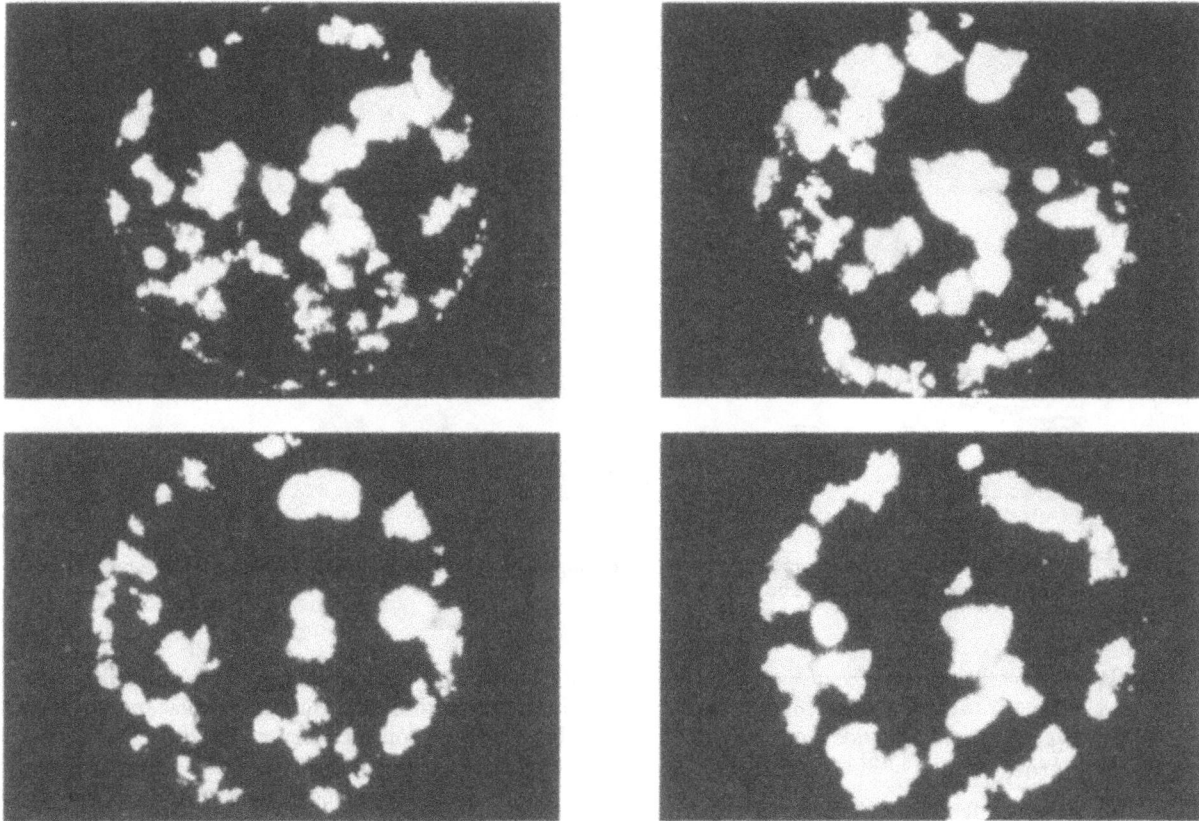

Fig. 5.40—Viscous fingers in an oil-wet system. [21]

Although criteria for initiation of fingers can be defined through the approach outlined in the previous paragraphs, there are differing viewpoints on what happens after fingers develop.

Some data show that viscous fingers tend to disperse as they move through a porous medium. Fig. 5.42 illustrates the fingering pattern at two stages of displacement of oil by water in a glass-bead pack. [25] Initial water saturation was nearly immobile, and the viscosity ratio (μ_o/μ_w) for the displacement was 10.0. Fingers that formed at the entrance of the glass-bead pack deteriorated into a zone of graded saturation. Similar results have been reported for Berea sandstone from the interpretation of X-ray shadowgraphs. At high viscosity ratios (>50), viscous fingering occurred rapidly during the initial part of the displacement, and stabilized as water imbibed into bypassed regions.

The dispersion of fingers into graded saturation zones suggests that capillary forces may oppose the development of fingers and may dampen their propagation in strongly water-wet reservoir rock. However, the opposite situation exists during a gas-displacement process or when a strongly oil-wet reservoir is waterflooded. Capillary-pressure gradients oppose the development of fingers when the nonwetting fluid displaces a wetting fluid. Thus, once a finger develops, there is no tendency for the dampening of the finger, as might occur when a wetting fluid displaces the nonwetting fluid.

It is useful to identify some situations in which viscous fingering has not been observed even though viscosity ratios were much larger than 1.0. In Chap. 2, the effect

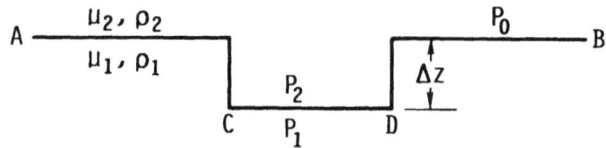

Fig. 5.41—Model of interface for stability analysis. [23]

of viscosity ratio on relative permeability curves was examined. All deviations from uniform flow in such methods as the Pennsylvania State U. (PSU) or unsteady-state methods are incorporated into relative permeability curves that are derived from the experimental data. Viscous fingering in permeability measurements would appear as a variation of relative permeability with viscosity.

Relative permeability curves for displacement of nonwetting fluid by the wetting fluid are usually independent of viscosity ratio. The data presented in Fig. 2.42 were obtained at viscosity ratios of 1.8 and 151. [26] Data obtained on Lloydminster crude are shown in Fig. 5.43. Viscosity ratios investigated include 1.0, 292, 646, and 1,430. [27,28] In contrast, changes have been reported in relative permeability curves with viscosity ratio when a nonwetting fluid displaced a wetting fluid. These differences in displacement performance under similar conditions have been attributed to viscous fingering. [29]

Although several models have been proposed to describe viscous fingering during immiscible displacement in porous media, [21,24,30,31] no model has gained acceptance as a tool for displacement performance when vis-

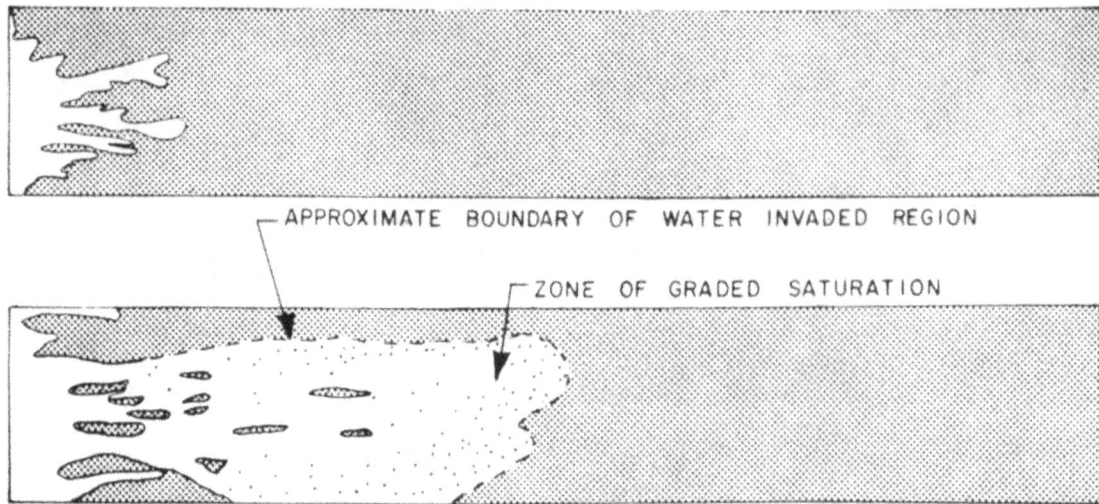

Fig. 5.42—Fingering pattern and zone of graded saturation developed in a bead-pack model. Initial water saturation near irreducible minimum. Viscosity ratio = 10.0. [25]

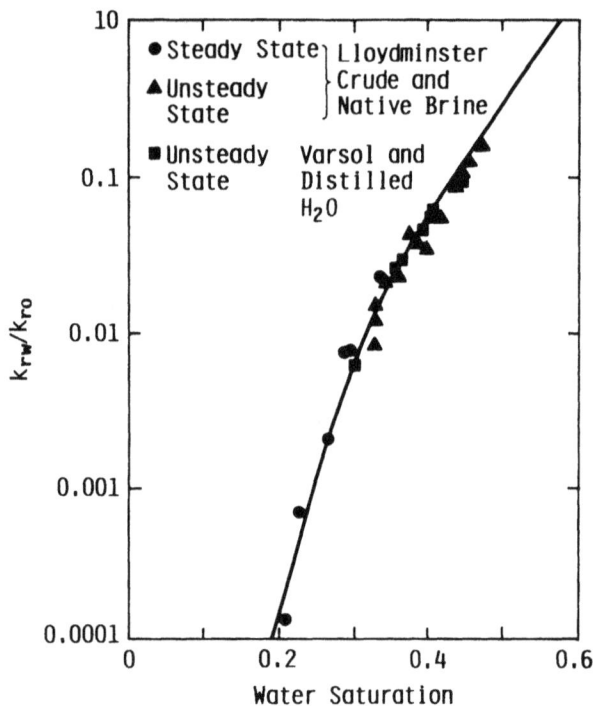

Fig. 5.43—Relative permeability ratio for tank Sandpacks I and II. [27]

of technology is generally referred to as mobility control. Mobility control is important in micellar, polymer, miscible, and CO_2 oil recovery processes.

Viscous fingering may be desirable and perhaps even necessary for the application of thermal oil recovery processes in heavy oil reservoirs. Viscosities of some heavy oils are so large that injection of steam or air would not be possible at the rates required to sustain displacement if the displacement rates were limited to those considered attainable without viscous fingering.

5.8 Estimating Waterflood Performance With 3D Models and Reservoir Simulators

In previous sections of this chapter, we examined displacement mechanisms and methods of estimating the macroscopic displacement efficiency for 2D immiscible displacement processes in vertical and horizontal planes. Reservoirs are 3D geologic deposits that have properties that may vary throughout the deposit. Frequently, the amount of variation is not known until revealed by analysis of displacement performance. Thus, simulation of displacement performance may be limited by knowledge of reservoir properties as well as available reservoir models.

Methods of estimating displacement efficiency in three dimensions may be subdivided into two classifications. One classification, called layered models, considers the heterogeneous (vertical) reservoir to be composed of noncommunicating layers. Displacement performance in each layer is computed from a 2D model. Results from each layer are combined to simulate the total reservoir. The second classification contains numerical models or reservoir simulators. Included are pseudo-3D simulators and models based on complete solution of the fluid-flow equations in three spatial coordinates.

5.8.1 Layered Models

When a reservoir is approximated by a series of noncommunicating layers, the displacement performance of *any* pattern flood may be estimated by summing the contributions of each layer at equal points in time as illustrated in Sec. 5.2 for a linear system. It makes no difference

cous fingering occurs. Craig[32] expanded the dilemma by concluding that nonuniform displacement fronts similar to those attributed to viscous fingering could form in a reservoir for no other reason than nonuniformities such as microscopic variations in porosity and permeability.

In most EOR processes, viscous fingering decreases the macroscopic displacement efficiency because high oil saturations remain in regions bypassed by the displacing fluid. The displacing fluid is expensive and in some cases cannot be reinjected. For these reasons, many EOR processes incorporate schemes that attempt to eliminate or to minimize viscous fingering at certain interfaces. This area

**TABLE 5.15—PERMEABILITY DATA—
WELL MP 110[38]**

Sample Number	Depth	Permeability to Air (md)
1	608	450
2	609	283
3	610	407
4	611	650
5	612	730
6	613	430
7	614	900
8	615	500
9	616	440
10	617	420
11	618	400*
12	619	381
13	620	358*
14	621	324
15	622	565
16	623	714
17	624	1,212
18	625	315
19	626	591

*Missing cores, arithmetic average of adjacent samples.

**TABLE 5.16—SUBDIVISION OF
RESERVOIR INTO LAYERS BASED
ON PERMEABILITY THICKNESS**

Layer	Core Samples	k_j (md)	h_j (ft)
1	1 to 3	380	3
2	4 to 6	603	3
3	7 to 9	613	3
4	10 to 12	401	3
5	13 to 15	414	3
6	16 to 19	944	4

what type of displacement model is used. Furthermore, it is not necessary to use the same displacement model, rock properties, fluid properties, or rock/fluid relationships for each layer.

If permeability is the only parameter that varies between layers, the method described in Sec. 5.2 and illustrated in Example 5.1 may be used to extend displacement predictions for one layer to estimation of the performance of a multilayer reservoir. This method is valid whether the flooding pattern is five-spot, linear, staggered line drive, or another pattern based on an arbitrary well arrangement.

Application of Layered Models to Simulate Displacement Performance. Once the areal simulator is chosen for a pattern, several decisions must be made. Availability of high-speed digital computers and programs has made simulation of multiple layers possible both in terms of simulation cost and time required to produce meaningful results. The engineer must decide whether there is enough reservoir information to justify multilayer simulation. Then it is necessary to determine the number of layers required to produce an acceptable simulation of the displacement process.

Selection of layers is controlled by the engineer's perception of how permeability varies with depth and areal location within the reservoir. The first step in developing this perception is to identify distinct geological zones that are mappable between wells. These zones should be treated as layers when they are present. In many cases a detailed geological and engineering analysis is necessary to identify these zones. This subject is covered in more detail in Sec. 7.6.

Other methods of selecting layers are used when geological subdivision is not possible. Four ways of subdividing reservoirs into layers are presented in this section. Other methods have been proposed.[33-37]

Table 5.15 contains permeability data from core plugs taken at 1-ft [0.305-m] intervals.[38] Permeabilities are missing from two depths because of broken cores, leaving 17 permeability samples. All other properties will be considered uniform for the purposes of this section. The

core analysis is assumed to be representative of the entire reservoir.

One method is based on subdividing the reservoir into nineteen 1-ft [0.305-m] layers that have permeabilities corresponding to the core samples. Estimates of the permeabilities for the two missing intervals are required. Permeabilities within parentheses in Table 5.15 are the arithmetic averages of adjacent samples. The thickness-averaged permeability for the data in Table 5.15 is 530 md.

In the second method, an arbitrary number of layers that have permeabilities averaged on permeability thickness (kh) is selected. Table 5.16 shows the permeability distribution for 6 layers that were obtained by finding the average permeability of the samples included in each layer with Eq. 5.177.

Because core samples were available (or a value was assumed) at 1-ft [0.305-m] increments, the kh product is identical to the product of the arithmetic average k and the net thickness.

$$\bar{k}_j = \frac{\sum_{i=1}^{n} k_i h_i}{\sum_{i=1}^{n} h_i}, \quad \ldots\ldots\ldots\ldots\ldots\ldots\ldots (5.177)$$

where

$\bar{k}_j$ = average permeability based on thickness,
k_i = permeability of Core Sample i,
h_i = thickness represented by Core Sample i,
        and
n = number of core samples in Layer j.

In Eq. 5.177, h_i is the thickness of Layer j.

The two methods presented in the preceding paragraphs or some variation of them have been used widely to subdivide reservoirs into layers because of their simplicity. However, these methods have a major deficiency that is related to the interpretation of core analysis data. Each method assumes that the core sample is representative of the interval above and below the sampling depth. The actual variation of permeability with vertical position between sampling points is never known. Core samples are seldom taken at the intervals required to describe these variations. Large changes in permeability occur over short intervals of distance. All these factors introduce uncertainty into the interpretation of core data.

Although arithmetic average permeabilities were computed for Core Samples 11 and 13, the values may be incorrect. It is not known from inspection of the core

**TABLE 5.17—FREQUENCY DISTRIBUTION OF
PERMEABILITY DATA**

Permeability to Air (md)	Number of Samples With Larger Permeability	Cumulative Frequency Distribution ($\% > k_j$)
1,212	0	0
900	1	5.8
730	2	11.8
714	3	17.7
650	4	23.5
591	5	29.4
565	6	35.3
500	7	41.2
450	8	47.1
440	9	52.9
430	10	58.8
420	11	64.7
407	12	70.6
381	13	76.5
324	14	82.4
315	15	88.2
283	16	94.1

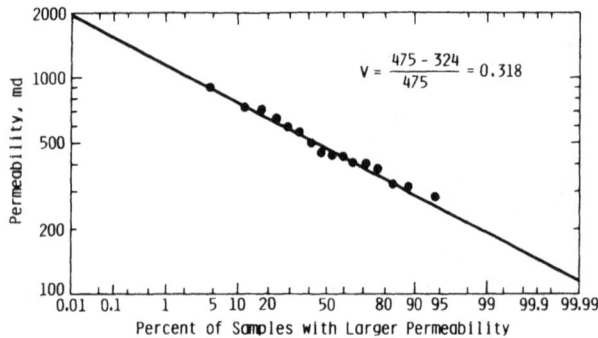

Fig. 5.45—Characterization of reservoir heterogeneity by permeability variation.

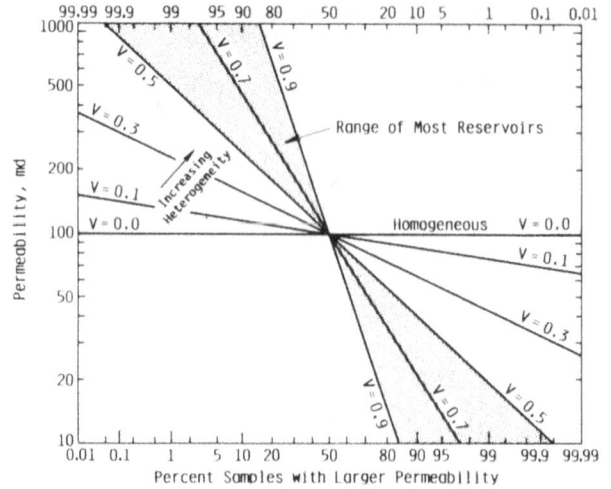

$$V = \frac{475 - 324}{475} = 0.318$$

Fig. 5.44—Plot of permeability data on log-normal paper.

analysis whether the permeability between Core Samples 17 and 18 decreases linearly with distance as assumed in the computation of the arithmetic average for Samples 11 and 13 in some other manner. Considering the changes in permeability of other samples over short distance intervals, one could argue that the permeabilities of Samples 11 and 13 could be much higher or lower than computed from a simple arithmetic average. Core samples represent a small portion of the interval in a particular well and an even smaller portion of a reservoir. Subdivision methods that do not compensate for sample variation may not be an adequate representation of the reservoir.

The third method of subdividing a reservoir into layers uses an empirical correlation of permeability data. Intuitively, this method seems to provide the closest representation of permeability changes caused by geological processes. Law[39] found that permeabilities in most reservoirs have a log-normal distribution. That is, the geologic processes that create permeability in reservoir rocks appear to leave permeabilities distributed around the geometric mean. If it is assumed that there are enough core samples to establish the true shape of the distribution curve, subdivision of the reservoir into layers with the distribution curve should give a more accurate representation of permeability distribution within the depth interval.

Example 5.5 shows how to plot permeability data to determine whether the data are log-normally distributed.

Example 5.5. The fit of a set of permeability data to a log-normal distribution can be done numerically or graphically. The graphical approach will be used in the example for the data of Table 5.15.

To evaluate the permeability distribution it is necessary to determine the cumulative frequency distribution of the raw data. Permeability data are log-normally distributed if the graph of permeability vs. cumulative frequency distribution on log-normal probability paper can be approximated by a straight line. Permeability data from Table 5.15 are arranged in descending order in the first column of Table 5.17. Cumulative frequency distribution, the fraction of the samples with permeabilities greater than the particular sample, is in Col. 3.

Fig. 5.44 is a plot of the permeability frequency distribution on log-normal probability paper. A straight line was drawn through the data with more weight placed on points in the central portion of the plot. The midpoint of the permeability distribution (50th percentile) is the log mean permeability. The log mean permeability for the data in Table 5.17 is 475 md.

The straight line drawn through the data is also a measure of the dispersion or the heterogeneity of the reservoir rock. Dykstra and Parsons[34] recognized this important feature and introduced the permeability variation, which characterizes a particular distribution. Eq. 5.178 defines the permeability variation, V, used in petroleum engineering.

$$V = \frac{k_{50} - k_{84.1}}{k_{50}}. \qquad \ldots \ldots \ldots \ldots \ldots \ldots (5.178)$$

The term V is also called the Dykstra-Parsons coefficient of permeability variation. For the data in Fig. 5.44 the permeability variation is

$$V = \frac{475 - 324}{475} = 0.32.$$

Permeability variation is an excellent descriptor of reservoir heterogeneity. A homogeneous reservoir has a permeability variation that approaches zero while an extremely heterogeneous reservoir would have a permeability variation approaching 1.0. Fig. 5.45 illustrates how reservoir heterogeneity changes with permeability variation for a sample that has a log mean permeability of 100 md. The reservoir described by the data in Table 5.17 and in Fig. 5.44 is considered to have low heterogeneity.

Subdivision of a reservoir with a log-normal permeability distribution was first introduced indirectly by Dykstra and Parsons,[34] who developed a method to compute the displacement efficiency of a linear reservoir subdivided into n layers. The displacement model assumed piston-like displacement (i.e., no oil displacement in the region behind the front). Displacement performance was computed for a range of mobility ratios for a reservoir that consisted of 50 layers of equal thickness. Correlations of displacement performance in core plugs with permeability variation (Eq. 5.155) at different WOR's were obtained to compensate for the assumption of piston-like displacement. The Dykstra-Parsons method is illustrated in Problem 5.28.

Another variation of the Dykstra-Parsons approach is more useful for computations.[40] The reservoir is subdivided into some number of layers that have equal thickness. Permeability of each layer is assigned from the log-normal permeability distribution, which is presumed to be representative of the reservoir. Thus, a reservoir is approximated by n hypothetical layers that have the same permeability variation and geometric mean permeability as the available core data. The number of layers is determined at the time of simulation for a particular problem.

Example 5.6 illustrates the procedure used to subdivide a reservoir into n equal-thickness layers using the log-normal permeability distribution.

Example 5.6. The permeability distribution shown in Fig. 5.44 will be assumed to represent the cross section. Subdividing the reservoir into 10 layers that have the same permeability variation is desired. The permeability variation of 10 layers will be the same as the original layers if the permeabilities are chosen from the line drawn through the data in Fig. 5.44 at the proper cumulative frequency distribution. This is done by choosing the permeability of the layer at the midpoint of the interval.

The frequency distribution is divided into 10 increments, as shown in Table 5.18. Permeability for each layer is selected at the midpoint of the interval. The permeabilities selected from Fig. 5.44 are presented in Col. 3 and can be checked for internal consistency by noting that the geometric mean permeability defined by Eq. 5.179 is equal to the log mean permeability.

$$k = \sqrt[n]{k_1 k_2 k_3 \ldots k_n}. \quad \ldots\ldots\ldots\ldots\ldots\ldots (5.179)$$

For the permeabilities in Table 5.18, $k = 473.3$ md, which is in good agreement with the value obtained from the graph.

Estimates of kh are sometimes available from pressure buildup or falloff tests. If kh is known, it is possible to subdivide a reservoir into layers that have (1) a specified

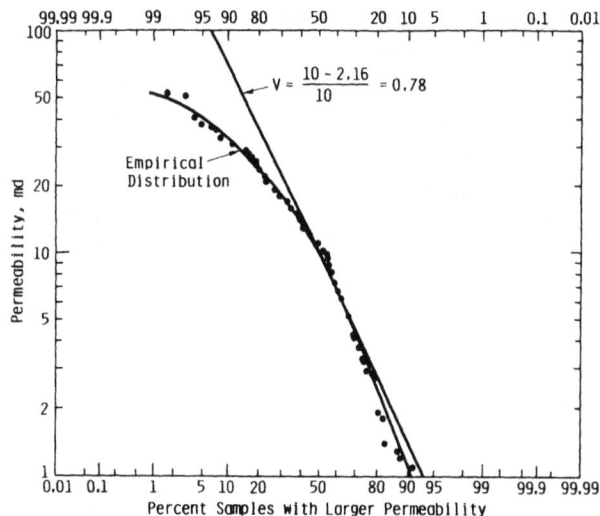

Fig. 5.46—Correlation of permeability data on log-normal probability paper.

TABLE 5.18—PERMEABILITY DISTRIBUTION OF 10 LAYERS THAT HAVE THE SAME PERMEABILITY VARIATION AS THE ORIGINAL DATA

Layer	Midpoint of Frequency Distribution	Permeability (md)	Permeability (md) $\bar{k} = 530$ md
1	5	880	936.7
2	15	700	740.8
3	25	610	644.3
4	35	550	576.3
5	45	495	521.4
6	55	452	473.2
7	65	410	428.1
8	75	366	382.9
9	85	320	333.0
10	95	254	263.4

permeability variation and (2) a specified kh. When a reservoir is simulated as a series of equal-thickness layers, kh is approximated by Eq. 5.180.

$$\bar{k}h = \sum_{i=1}^{n} k_i h_i, \quad \ldots\ldots\ldots\ldots\ldots\ldots (5.180)$$

where $\bar{k}$ is thickness-averaged permeability. Subdivision of the reservoir into 10 equal-thickness layers that have a kh product of 10,070 md-ft [3069 md-m], or a thickness-averaged permeability of 530 md, can be done by iteration (or trial and error). The permeability distribution in Col. 4 of Table 5.18 was obtained in this manner. In general, the thickness-averaged permeability, $\bar{k}$, is not equal to k, the geometric mean permeability. The availability of kh data provides another constraint on the subdivision of a reservoir into layers.

Subdivision of a reservoir into layers based on the log-normal permeability distribution should be done with care. When the permeability variation is high, one layer can have an unreasonably large permeability. The layer can dominate the performance of a simulated waterflood because it provides a path for water to flow without displac-

TABLE 5.19—COMPARISON OF PERMEABILITY DISTRIBUTION FOR 10 EQUAL-THICKNESS LAYERS

Layer	Interval	Midpoint of Interval	Log Normal $v=0.78$	Empirical Correlation
1	0 to 10	5	116.0	40.6
2	10 to 20	15	47.5	27.8
3	20 to 30	25	27.4	20.8
4	30 to 40	35	17.6	15.8
5	40 to 50	45	12.0	11.8
6	50 to 60	55	8.4	8.5
7	60 to 70	65	5.5	5.5
8	70 to 80	75	3.5	3.3
9	80 to 90	85	2.0	1.7
10	90 to 100	95	0.9	0.5
$\bar{k}$			24.1	13.6

TABLE 5.20—MINIMUM NUMBER OF EQUAL-THICKNESS LAYERS REQUIRED TO OBTAIN PERFORMANCE OF A 100-LAYER FIVE-SPOT WATERFLOOD AT PRODUCING WOR'S ABOVE 2.5[40]

(Confidence Level: Mean Square Difference ≤1% Sweep)

Mobility Ratio	Permeability Variation							
	0.1	0.2	0.3	0.4	0.5	0.6	0.7	0.8
0.05	1	1	2	4	10	20	20	20
0.1	1	1	2	4	10	20	100	100
0.2	1	1	2	4	10	20	100	100
0.5	1	2	2	4	10	20	100	100
1.0	1	3	3	4	10	20	100	100
2.0	2	4	4	10	20	50	100	100
5.0	2	5	10	20	50	100	100	100

ing much oil. An example of this situation is shown in Fig. 5.46, where permeability data from a well are plotted on log-normal probability paper. The upper line represents the log-normal distribution obtained by fitting the data with a straight line weighing the central points most heavily. The permeability variation is 0.78, indicating a very heterogeneous reservoir. The solid line in Fig. 5.46 is an empirical correlation of the data on log-normal probability paper.

If the log-normal permeability distribution is subdivided into equal-thickness layers, as in Example 5.6, there will be at least one layer that has a permeability significantly higher than the permeability on the empirical distribution curve. For example, if the reservoir is subdivided into 10 equal-thickness layers, the permeability of Layer 1 (at the fifth percentile) is 116.0 md, the log-normal distribution, compared to 40.6 md for the empirical distribution. The high-permeability layer obtained from the assumed log-normal distribution can dominate the performance of a simulated waterflood.

The preferred method for subdividing the reservoir into layers is to use the empirical correlation as the basis for selecting permeabilities of layers. This can be done when there are sufficient data points to define the empirical distribution.

Table 5.19 contains permeability distributions obtained from Fig. 5.46 with the log-normal and the empirical correlation. Included in Table 5.19 are the thickness-averaged permeabilities.

Because it is possible to subdivide a vertical cross section into an arbitrary number of layers, criteria for deter-

mining when the number of subdivisions is adequate need to be developed. Time and simulation cost, important factors that depend on specific situations, are recognized but neglected in the following discussion. When the number of layers is variable, a simulation is judged to be acceptable if further subdivision of the reservoir into layers produces a change in displacement performance that is within some predetermined tolerance. The tolerance may be set on any one of several key variables. For example, agreement of cumulative-production/PV-injected curves to within 1 to 2% could be one criterion. Another measure of agreement is the mean-squared difference defined by Eq. 5.181. The number of subdivisions would be determined when the mean-squared difference between cumulative oil production from two different subdivision schemes was within some tolerance (e.g., 1%).

$$(\Delta N_p)^2 = \frac{\sum_{i=1}^{n} (N_{pk} - N_{pk+1})^2_{Q_{iD}}}{\sum_{i=1}^{n} i}, \quad \ldots \ldots (5.181)$$

where the subscript k is an index on the number of layers.

In Eq. 5.181, the differences in N_p are computed at the same values of Q_i. A guide for determining the number of layers for a five-spot model is presented in Ref. 42. In this case, the Craig *et al.* model was used (see Sec. 4.4) with relative-permeability curves selected so that displacement was piston-like at all mobility ratios. The performance of a 100-layer model was taken as the standard for comparison.

Craig compared WOR and volumetric sweep efficiency at breakthrough for equal-thickness layers of 1, 2, 3, 4, 5, 10, 20, and 50 layers; results were obtained with 100 layers. The minimum number of layers required to approximate the performance of the 100-layer model was determined as a function of mobility ratio and permeability variation. The criterion for agreement was that the mean-squared error in sweep efficiency at equal WOR was less than 1%. Tables 5.20 through 5.22 summarize results of these simulations and provide a guide to selection of number of layers for five-spot patterns.

Similar correlations could be developed for other displacement models (Higgins and Leighton), specific relative permeability curves, and other patterns. The important feature of the approach used by Dykstra and Parsons[34] and Craig[40] is that it uses information on the probable variation of permeability over the entire vertical cross section that is not accounted for in other models.

5.8.2 Pseudomodels

The concept of pseudo-relative-permeability curves was introduced in Sec. 5.4 to account for vertical fluid flows when the assumptions of vertical equilibrium or uniform vertical pressure were applicable. By the use of pseudo-relative-permeability curves in place of "rock" curves in five-spot, streamtube, or numerical simulation, it is possible to approximate the effect of vertical flow caused by capillary, gravity, and viscous forces on waterflood performance with relatively simple models. Problems 5.16 and 5.23 illustrate simulation of waterflood performance in a five-spot pattern using pseudo-relative-permeability concepts.

TABLE 5.21—MINIMUM NUMBER OF EQUAL-THICKNESS LAYERS REQUIRED TO OBTAIN PERFORMANCE OF A 100-LAYER FIVE-SPOT WATERFLOOD AT PRODUCING WOR'S ABOVE 5[40]

(Confidence Level: Mean Square Difference ≤1% Sweep)

Mobility Ratio	Permeability Variation							
	0.1	0.2	0.3	0.4	0.5	0.6	0.7	0.8
0.05	1	1	2	4	5	10	10	20
0.1	1	1	2	4	10	10	10	100
0.2	1	1	2	4	10	10	20	100
0.5	1	2	2	4	10	10	20	100
1.0	1	2	3	4	10	10	20	100
2.0	2	3	4	5	10	10	50	100
5.0	2	3	5	10	20	100	100	100

TABLE 5.22—MINIMUM NUMBER OF EQUAL-THICKNESS LAYERS REQUIRED TO OBTAIN PERFORMANCE OF A 100-LAYER FIVE-SPOT WATERFLOOD AT PRODUCING WOR'S ABOVE 10[40]

(Confidence Level: Mean Square Difference ≤1% Sweep)

Mobility Ratio	Permeability Variation							
	0.1	0.2	0.3	0.4	0.5	0.6	0.7	0.8
0.05	1	1	1	2	4	5	10	20
0.1	1	1	1	2	5	5	10	20
0.2	1	1	2	3	5	5	10	20
0.5	1	1	2	3	5	5	10	20
1.0	1	1	2	3	5	10	10	50
2.0	1	2	3	4	10	10	20	100
5.0	1	3	4	5	10	100	100	100

5.8.3 Numerical Solutions

Three-dimensional simulation of displacement processes may be required when there are significant changes in reservoir properties (permeability, porosity, and thickness) or saturations across the areal or vertical extent of a reservoir or when crossflow or gravity segregation are potentially important because of the structural position of the reservoir and/or wells. Solution of fluid-flow equations in three dimensions, including the effects of rock and fluid compressibility, is possible with numerical techniques for almost any reservoir geometry. Examples of numerical simulations are presented in Refs. 41 through 45. Solutions of displacement problems in three dimensions have been developed with two techniques. These techniques will be illustrated briefly in this section with an example from the literature.

Fig. 5.47 shows a 3D model of half of a typical well arrangement on the flank of a large reservoir.[46] The model has been subdivided into gridblocks in x, y, and z directions for numerical solution. Production wells were drilled at the crest of the dome. The original OWC is indicated by the dashed line. A large aquifer underlies the oil zone, but the water drive from the aquifer is limited because of a "tar barrier" that effectively reduces permeability in the shaded region in Fig. 5.47. The tar barrier is 100 ft [30.48 m] thick. Table 5.23 shows the vertical zonation of the reservoir as well as fluid properties at two pressures. Formation thickness and vertical stratification were considered uniform throughout the reservoir. Water Injection Wells 1 and 2 were used to supplement the natural bottomwater drive.

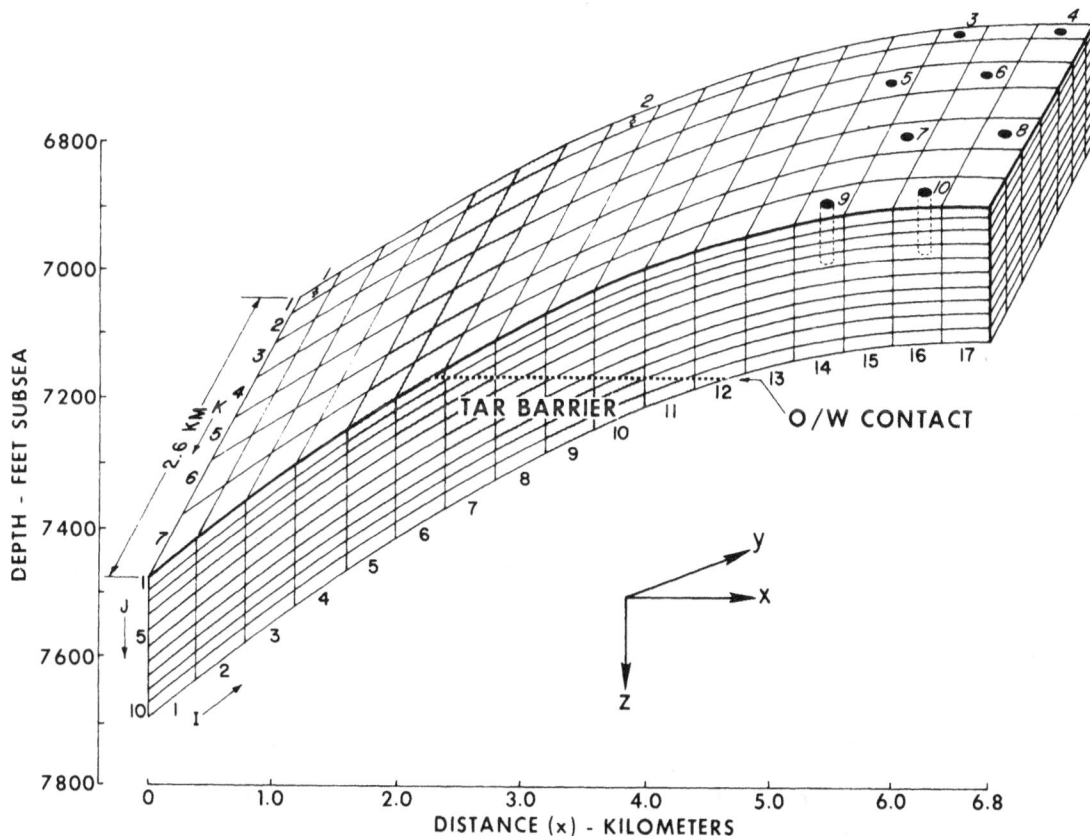

Fig. 5.47—Three-dimensional model of typical pattern on flanks of reservoir.[46]

TABLE 5.23—RESERVOIR ZONATION AND FLUID PROPERTIES[46]

Cycle	Thickness ft (Δz)	Thickness Percent of Total	Porosity-Thickness % of Total	Horizontal Permeability-Thickness % of Total
1	16.6	7.8	8.60	3.9
2	20.0	9.4	12.75	17.1
3	20.0	9.4	12.75	17.1
4	24.4	11.6	15.50	20.9
5	23.3	10.9	13.08	14.5
6	23.3	10.9	13.08	14.5
7	21.3	10.0	6.06	3.0
8	21.3	10.0	6.06	3.0
9	21.3	10.0	6.06	3.0
10	21.3	10.0	6.06	3.0
Totals	212.8	100.0	100.00	100.0

Rock compressibility, 3.0×10^{-6}/psi

	Fluid Properties			
	Oil		Water	
	2,000 psi	4,000 psi	2,000 psi	4,000 psi
FVF, RB/STB	1.5730	1.5231	1.0397	1.0397
Viscosity, cp	0.312	0.412	0.434	0.434
g/cm^3	0.6456	0.6682	1.1047	1.1123

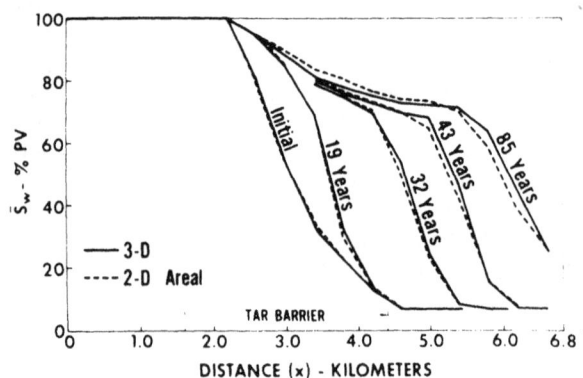

Fig. 5.48—Average water saturation distributions calculated using 3D and 2D areal models.[46]

The portion of the reservoir represented by the model in Fig. 5.47 was simulated with two numerical techniques. One technique solved the two-phase fluid-flow equations in three dimensions. The second technique is actually a 2D areal (x-y) model that approximates the 3D solution by use of dynamic pseudopermeability functions. These functions are a family of pseudo-relative-permeability curves that vary with saturation and velocity. They are derived from other numerical solutions and account for the vertical flow component in the 2D areal model. The 2D areal model had the same 17×7 areal grid and well locations as shown in Layer 1 ($J=1$) of Fig. 5.46. Background literature on pseudofunctions can be found in Refs. 5, 8, and 46.

Results of several simulations are shown in Figs. 5.48 and 5.49. Fig. 5.48 shows the average water saturation at different distances along the x axis. Water saturation contours (averaged over thickness in the 3D model) are presented in Fig. 5.49 for several times. These results show the capabilities of numerical models to simulate displacement processes. In this example, the 2D areal model, with dynamic pseudopermeability correlations, gave results comparable to those obtained with the 3D simulator

at a savings of computer time by a factor of 40. Both models, however, simulate a displacement process that is 3D because of variations in reservoir properties, initial fluid saturations, structural position of the reservoir, and location of injection and production wells.

5.9 Summary

The models presented in this chapter represent solutions of the fluid-flow equations for well-defined physical models of reservoirs. In every case, assumptions are made concerning the variation of reservoir properties and fluid saturations with spatial position. Selection of the model to predict displacement performance for a particular reservoir assumes that these variations can be approximated by the assumptions incorporated into the model. There are few guidelines other than experience in determining how closely the actual reservoir must conform to the mathematical model to yield acceptable results. In some cases, 2D and 3D numerical simulators show that what appears to be a complicated problem can be simulated with a 1D model. In other cases, the use of simple models, even if the only tools available, can produce results that are worthless. Coats[47] offers some suggestions on model selection.

Problems

5.1. The linear reservoir studied in Problem 3.10 was homogeneous with an absolute permeability of 23 md and a porosity of 0.206. In Problem 5.1, you will consider the reservoir to have two noncommunicating layers with properties given in Table 5.24.

The average porosity and the average absolute permeability based on thickness are identical to the values used for the uniform reservoir in Problem 3.10. For the purposes of this problem, assume that the relative permeability curves and initial and residual oil saturations are the same for each layer. When these assumptions are approximated closely, the procedure presented in Sec. 5.2 and illustrated in Example 5.1 can be used to estimate the production performance of the reservoir.

Using the computed results in Table 5.25, estimate the performance of the two-layer system when a constant pressure drop of 2,000 psia [14 MPa] is maintained across the system. The results should be presented in the same column headings as Part d of Problem 3.10 up to a WOR of 50. The time increment for the combined results should be 365 days.

When performance is desired for a few times or layers, hand computations are reasonable. In most cases, multilayer calculations should be done with a short computer program that does the necessary scaling and linear interpolation. The values in Table 5.25 were generated by the computer program that was written to solve Problem 3.10. These data are saved easily in a file that can be read at a later time for use in multilayer calculations.

Plot oil production rate vs. time. Compare the results of Problem 3.10 with Problem 5.1 by plotting the oil rate for the homogeneous reservoir (Problem 3.10) on the same graph.

5.2. The reservoir depicted in Fig. 5.50 consists of two noncommunicating zones. These zones have quite different properties, as indicated in Table 5.26. The plan is to waterflood the reservoir by the injection of water in wells located along the lease boundary as depicted in Fig. 5.50. Predictions of waterflood performance for each zone are given in Tables 5.27 and 5.28 when the pressure drop between injection and production wells is 600 psi [4 MPa]. Demonstrate your understanding of waterflood concepts by estimating the waterflood performance of the composite system after 30, 60, 90, and 180 days of injection at a constant pressure drop of 600 psi [4 MPa]. You are requested to determine the following: (1) oil production rate, B/D; (2) volume of water injected, barrels; (3) cumulative oil recovery, barrels; (4) water injection rate, B/D; and (5) WOR.

5.3. The predicted performance of a waterflood in a 12.66-acre [51 233-in.^{2}] five-spot pattern is given in Table 5.29. The reservoir is 20 ft [6.1 m] thick, and has a porosity of 0.10 and a mean permeability of 150 md. There was no initial gas saturation. The performance in Table 5.29 was obtained with the Higgins-Leighton streamtube model assuming the reservoir was homogeneous and was operated under a constant pressure drop.

Reassessment of the reservoir identified two noncommunicating layers with properties shown in Table 5.30.

Using the predicted performance in Table 5.29, determine the following: (1) at what time will the water production begin in the two-layered reservoir? (2) what is the cumulative oil recovery when water production begins? (3) estimate the injection rate 3 years after the beginning of the flood, (4) estimate the cumulative oil recovery from the pattern after 3 years of water injection, and (5) when will a WOR of 10 occur in the flood?

5.4. A narrow reservoir (2,000 ft [610 m] wide by 3 miles [5 km] long) will be waterflooded with existing wells that are spaced 660 ft [201 m] apart. There are three wells per row across the width of the reservoir. The reservoir is fractured with the direction of fractures approximately perpendicular to the axis of the reservoir. The reservoir will be flooded with a linedrive pattern by aligning rows of injection and production wells parallel to the fracture

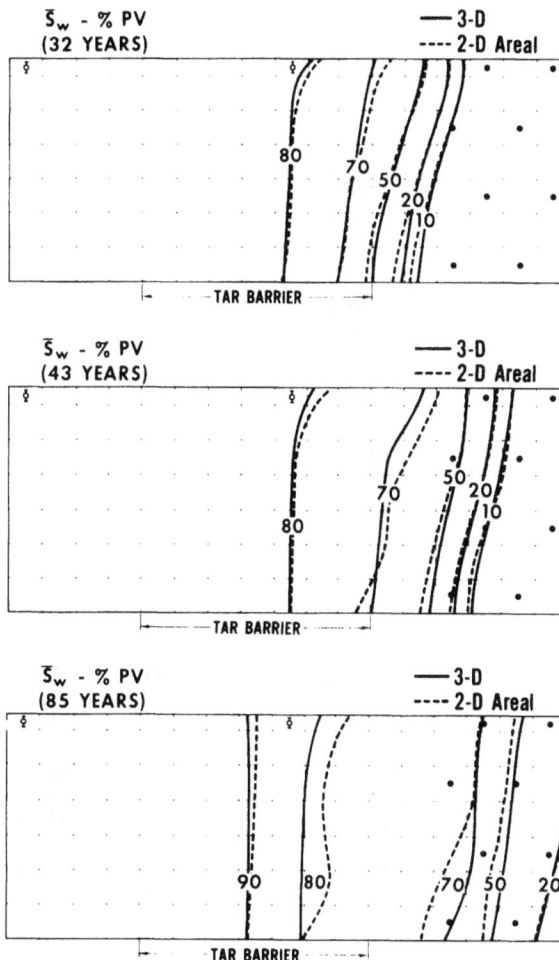

Fig. 5.49—Water saturation contours at 32, 43, and 85 years calculated from 3D and 2D areal models.[46]

TABLE 5.24—RESERVOIR PROPERTIES FOR PROBLEM 5.1

Layer	Thickness (ft)	Absolute Permeability (md)	Porosity
1	15	10	0.17
2	5	62	0.314

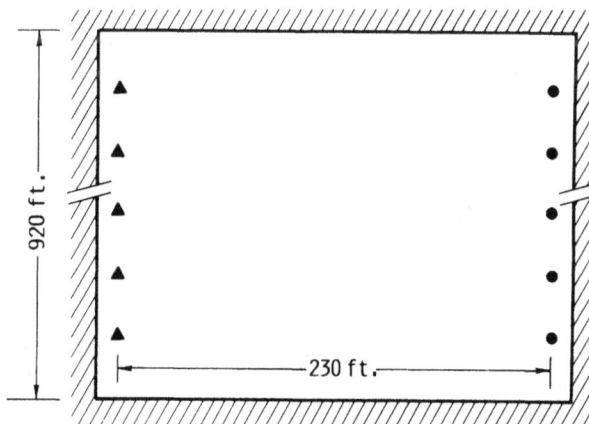

Fig. 5.50—Arrangement of injection and production wells in an edgewater flood.

TABLE 5.25—LINEAR WATERFLOOD
WITH CONSTANT PRESSURE DROP
$\Delta p = 2,000$ psia

Time (days)	q_o	q_w (B/D)	q_t (B/D)	WOR	Q_i (PV)	N_p/N (OOIP)	N_p (bbl)
0.0	51.85	0.0	51.85	0.0	0.0	0.0	0.0
96.2	52.59	0.0	52.59	0.0	0.014	0.021	5,023.0
191.0	53.35	0.0	53.35	0.0	0.027	0.043	10,045.9
284.5	54.13	0.0	54.13	0.0	0.041	0.064	15,068.9
376.6	54.94	0.0	54.94	0.0	0.055	0.086	20,091.8
467.3	55.77	0.0	55.77	0.0	0.068	0.107	25,114.8
556.7	56.63	0.0	56.63	0.0	0.082	0.129	30,137.8
644.7	57.51	0.0	57.51	0.0	0.096	0.150	35,160.7
731.4	58.42	0.0	58.42	0.0	0.110	0.172	40,183.7
816.7	59.36	0.0	59.36	0.0	0.123	0.193	45,206.7
900.6	60.33	0.0	60.33	0.0	0.137	0.215	50,229.6
983.2	61.33	0.0	61.33	0.0	0.151	0.236	55,252.6
1,064.4	62.37	0.0	62.37	0.0	0.164	0.258	60,275.5
1,144.2	63.44	0.0	63.44	0.0	0.178	0.279	65,298.5
1,222.7	64.55	0.0	64.55	0.0	0.192	0.301	70,321.5
1,299.9	65.70	0.0	65.70	0.0	0.205	0.322	75,344.4
1,375.6	66.89	0.0	66.89	0.0	0.219	0.344	80,367.4
1,450.0	68.13	0.0	68.13	0.0	0.233	0.365	85,390.4
1,523.1	69.41	0.0	69.41	0.0	0.246	0.387	90,413.3
1,594.7	70.74	0.0	70.74	0.0	0.260	0.408	95,436.3
1,665.1	10.73	61.39	72.12	5.72	0.274	0.430	100,459.2
1,782.2	10.14	65.03	75.17	6.41	0.297	0.435	101,680.5
1,907.8	9.57	68.83	78.39	7.19	0.324	0.440	102,917.7
2,042.5	9.02	72.77	81.79	8.07	0.353	0.446	104,168.4
2,186.7	8.49	76.86	85.35	9.05	0.386	0.451	105,430.2
2,341.0	7.98	81.11	89.09	10.16	0.423	0.457	106,700.9
2,506.2	7.50	85.51	93.01	11.40	0.464	0.462	107,978.6
2,682.9	7.03	90.06	97.09	12.81	0.509	0.468	109,261.6
2,872.0	6.59	94.77	101.36	14.39	0.560	0.473	110,548.4
3,074.4	6.16	99.63	105.79	16.18	0.618	0.479	111,837.5
3,291.2	5.75	104.65	110.40	18.20	0.682	0.484	113,127.6
3,523.7	5.36	109.82	115.18	20.49	0.753	0.490	114,417.7
3,773.0	4.99	115.15	120.14	23.10	0.833	0.495	115,706.7
4,040.9	4.63	120.64	125.27	26.06	0.923	0.501	116,993.5
4,329.0	4.29	126.28	130.57	29.43	1.023	0.506	118,277.3
4,639.3	3.97	132.07	136.04	33.28	1.136	0.512	119,557.4
4,974.0	3.66	138.02	141.68	37.69	1.262	0.517	120,833.0
5,335.6	3.37	144.12	147.49	42.75	1.405	0.522	122,103.3
5,727.2	3.10	150.36	153.46	48.57	1.566	0.528	123,367.9
6,152.0	2.84	156.76	159.59	55.29	1.747	0.533	124,626.0
6,613.9	2.59	163.30	165.89	63.07	1.952	0.539	125,877.3
7,117.4	2.36	169.98	172.34	72.10	2.184	0.544	127,121.1
7,667.8	2.14	176.81	178.95	82.62	2.447	0.549	128,357.0
8,271.2	1.94	183.77	185.71	94.94	2.747	0.554	129,584.7
8,934.8	1.74	190.87	192.61	109.42	3.089	0.560	130,803.6

TABLE 5.26—PROPERTIES OF ZONES

	Zone 1	Zone 2
Permeability to oil at S_{iw}, md	116.6	32.8
Permeability to water at S_{or}, md	28.2	4.07
S_{iw}	0.45	0.356
S_{or}	0.37	0.426
Thickness, ft	7	8
Porosity, fraction	0.229	0.174
Oil viscosity, cp	22.7	22.7
Water viscosity, cp	1.0	1.0

trend. With this well arrangement, the system can be approximated by linear displacement between rows of injection and production wells.

There are three distinct noncommunicating zones in the reservoir. Properties of the zones are given in Table 5.31. Although the fluids in each zone are identical ($\mu_o = 5$ cp

[5 mPa·s], $\mu_w = 0.5$ cp [0.5 mPa·s]), the relative permeability curves are different primarily because the initial water saturations vary with rock permeability because of pore size distribution. There is no initial gas saturation.

The waterflood will be operated at constant pressure drop by controlling the injection rate. Bottomhole pressure in injection wells is limited to 1,700 psig [12 MPa] to avoid further fracturing. Production wells will be operated with 200 ft [61 m] of fluid over the pump. Pumps are set about 50 ft [15 m] above the productive interval. Assume that the relative permeability curves are given by Eqs. 5.182 and 5.183. Other data are presented in Table 5.31.

$$k_{ro} = (1 - S_{wD})^{1.5} \quad \dots \dots \dots \dots \dots (5.182)$$

and

$$k_{rw} = 0.3 S_{wD}^3, \quad \dots \dots \dots \dots \dots (5.183)$$

where S_{wD} is given by Eq. 3.111.

**TABLE 5.27—WATERFLOOD PERFORMANCE OF ZONE 1,
CONSTANT PRESSURE DROP = 600 psi**

Time (days)	q_o	q_w (B/D)	q_t (B/D)	WOR	Q_i (PV)	N_p (PV)	N_p (bbl)
0.0	97.27	0.0	97.27	0.0	0.0	0.0	0.0
5.3	102.49	0.0	102.49	0.0	0.009	0.009	528.7
10.3	108.30	0.0	108.30	0.0	0.018	0.018	1,057.3
15.0	114.81	0.0	114.81	0.0	0.026	0.026	1,586.0
19.5	122.16	0.0	122.16	0.0	0.035	0.035	2,114.7
23.7	130.51	0.0	130.51	0.0	0.044	0.044	2,643.3
27.6	140.08	0.0	140.08	0.0	0.053	0.053	3,172.0
31.2	151.16	0.0	151.16	0.0	0.061	0.061	3,700.6
34.6	164.16	0.0	164.16	0.0	0.070	0.070	4,229.3
37.7	179.59	0.0	179.59	0.0	0.079	0.079	4,758.0
40.5	100.47	97.77	198.24	0.97	0.088	0.088	5,286.6
41.6	96.69	106.35	203.04	1.10	0.091	0.089	5,401.7
42.9	93.04	115.14	208.18	1.24	0.096	0.091	5,522.5
44.3	89.52	124.12	213.64	1.39	0.101	0.093	5,648.0
45.7	86.10	133.28	219.39	1.55	0.106	0.096	5,777.2
47.3	82.78	142.61	225.39	1.72	0.112	0.098	5,909.5
49.0	79.54	152.08	231.63	1.91	0.118	0.100	6,044.1
50.7	76.39	161.70	238.09	2.12	0.125	0.102	6,180.5
52.6	73.30	171.45	244.75	2.34	0.132	0.105	6,318.2
54.5	70.28	181.31	251.59	2.58	0.140	0.107	6,456.9
56.5	67.33	191.27	258.60	2.84	0.148	0.109	6,596.1
58.6	64.44	201.32	265.76	3.12	0.158	0.111	6,735.5
60.9	61.61	211.45	273.06	3.43	0.168	0.114	6,874.9
63.2	58.84	221.64	280.49	3.77	0.178	0.116	7,014.0
65.6	56.13	231.89	288.02	4.13	0.189	0.118	7,152.6
68.1	53.48	242.18	295.66	4.53	0.202	0.121	7,290.5
70.7	50.88	252.49	303.37	4.96	0.215	0.123	7,427.5
73.5	48.35	262.82	311.16	5.44	0.229	0.125	7,563.4
76.3	45.86	273.15	319.01	5.96	0.244	0.127	7,698.2
79.3	43.44	283.47	326.91	6.53	0.260	0.130	7,831.6
82.4	41.07	293.77	334.84	7.15	0.277	0.132	7,963.6
85.7	38.76	304.03	342.79	7.84	0.295	0.134	8,093.9
89.1	36.51	314.24	350.75	8.61	0.315	0.136	8,222.6
92.7	34.32	324.39	358.71	9.45	0.336	0.138	8,349.6
96.5	32.18	334.47	366.65	10.39	0.358	0.140	8,474.7
100.4	30.11	344.46	374.57	11.44	0.383	0.142	8,597.8
104.6	28.09	354.36	382.45	12.61	0.409	0.144	8,718.9
109.0	26.14	364.14	390.28	13.93	0.437	0.146	8,837.9
113.6	24.25	373.80	398.05	15.42	0.467	0.148	8,954.8
118.5	22.42	383.33	405.74	17.10	0.500	0.150	9,069.4
123.8	20.65	392.71	413.36	19.02	0.535	0.152	9,181.8
129.3	18.94	401.93	420.87	21.22	0.573	0.154	9,291.7
135.3	17.30	410.98	428.29	23.75	0.615	0.156	9,399.3
141.6	15.73	419.86	435.59	26.69	0.661	0.157	9,504.5
148.5	14.22	428.54	442.76	30.14	0.711	0.159	9,607.1
155.9	12.78	437.02	449.79	34.21	0.766	0.161	9,707.2
164.0	11.40	445.28	456.68	39.06	0.826	0.162	9,804.7
172.8	10.09	453.32	463.41	44.93	0.893	0.164	9,899.5
182.6	8.85	461.13	469.98	52.11	0.969	0.165	9,991.7
193.4	7.68	468.70	476.38	61.03	1.054	0.167	10,081.1
205.6	6.58	476.01	482.59	72.35	1.151	0.168	10,167.8
219.5	5.55	483.06	488.61	87.03	1.262	0.170	10,251.7
235.5	4.59	489.84	494.43	106.60	1.392	0.171	10,332.7

Estimate the waterflood performance of this reservoir that to a WOR of 100. Assume that the waterflood begins at the same time in each pattern. The computed performance should be presented in the format of Table 5.7.

5.5. A reservoir that consists of two noncommunicating zones is to be waterflooded by the injection of water at a constant rate. Relative-permeability curves and initial and residual oil saturations are the same in each zone and are represented by values presented in Example 3.5. Permeability, porosity, and thickness are different for each zone. Thus, the relationship between Q_i given in Fig. 3.16 and $\overline{\lambda}^{-1}$ represents the linear waterflood in each zone. Derive a method for computing the displacement performance of a linear waterflood in this reservoir at constant injection rate. In your derivation, let k_1 =permeability to oil at connate water saturation in the high-permeability zone, ϕ_1 =porosity of the high-permeability zone, and h_1 =thickness of the high-permeability zone.

Hint. The displacement performance of each zones where there is no crossflow between zones is determined only by the number of PV's of fluid that have been injected into the zone. The amount of fluid injected in a given time increment is proportional to the average flow resistance of that zone during a time increment.

5.6. Dykstra and Parsons[34] presented a model for estimation of the displacement performance of a layered

**TABLE 5.28—WATERFLOOD PERFORMANCE OF ZONE 2,
CONSTANT PRESSURE DROP = 600 psi**

Time (days)	q_o	q_w (B/D)	q_t (B/D)	WOR	Q_i (PV)	N_p (PV)	N_p (bbl)
0.0	31.27	0.0	31.27	0.0	0.0	0.0	0.0
20.4	32.47	0.0	32.47	0.0	0.012	0.012	650.1
40.0	33.76	0.0	33.76	0.0	0.025	0.025	1,300.1
58.9	35.16	0.0	35.16	0.0	0.037	0.037	1,950.2
77.0	36.68	0.0	36.68	0.0	0.050	0.050	2,600.2
94.3	38.33	0.0	38.33	0.0	0.062	0.062	3,250.3
110.9	40.15	0.0	40.15	0.0	0.074	0.074	3,900.3
126.7	42.14	0.0	42.14	0.0	0.087	0.087	4,550.4
141.7	44.34	0.0	44.34	0.0	0.099	0.099	5,200.4
156.0	46.79	0.0	46.79	0.0	0.112	0.112	5,850.5
169.5	25.36	24.15	49.51	0.95	0.124	0.124	6,500.5
173.0	24.38	25.67	50.05	1.05	0.127	0.126	6,588.0
176.9	23.43	27.20	50.62	1.16	0.131	0.127	6,679.9
181.0	22.49	28.74	51.23	1.28	0.135	0.129	6,775.5
185.5	21.58	30.29	51.87	1.40	0.139	0.131	6,874.4
190.3	20.69	31.84	52.53	1.54	0.144	0.133	6,976.1
195.5	19.82	33.41	53.22	1.69	0.149	0.135	7,080.2
200.9	18.97	34.97	53.94	1.84	0.155	0.137	7,186.3
206.7	18.14	36.54	54.68	2.01	0.161	0.139	7,294.0
212.9	17.33	38.12	55.44	2.20	0.167	0.141	7,403.2
219.4	16.53	39.69	56.23	2.40	0.174	0.143	7,513.5
226.3	15.76	41.27	57.03	2.62	0.182	0.145	7,624.7
233.6	15.00	42.84	57.84	2.86	0.190	0.147	7,736.6
241.3	14.27	44.41	58.68	3.11	0.198	0.150	7,849.0
249.4	13.55	45.98	59.53	3.39	0.207	0.152	7,961.7
257.9	12.85	47.54	60.39	3.70	0.217	0.154	8,074.6
267.0	12.17	49.10	61.26	4.04	0.228	0.156	8,187.5
276.5	11.50	50.64	62.14	4.40	0.239	0.158	8,300.3
286.6	10.86	52.18	63.04	4.81	0.251	0.160	8,412.9
297.2	10.23	53.71	63.94	5.25	0.264	0.163	8,525.1
308.5	9.62	55.23	64.85	5.74	0.278	0.165	8,636.8
320.4	9.03	56.74	65.77	6.29	0.293	0.167	8,748.0
333.1	8.45	58.23	66.69	6.89	0.308	0.169	8,858.5
346.5	7.90	59.72	67.61	7.56	0.326	0.171	8,968.3
360.8	7.36	61.18	68.54	8.31	0.344	0.173	9,077.2
376.0	6.84	62.63	69.47	9.15	0.364	0.175	9,185.2
392.3	6.34	64.06	70.41	10.10	0.386	0.177	9,292.3
409.7	5.86	65.48	71.34	11.17	0.409	0.179	9,398.3
428.3	5.40	66.87	72.27	12.39	0.435	0.181	9,503.2
448.4	4.95	68.25	73.20	13.78	0.463	0.183	9,607.0
470.0	4.53	69.60	74.13	15.37	0.493	0.185	9,709.5
493.5	4.12	70.93	75.05	17.22	0.527	0.187	9,810.8
519.0	3.73	72.24	75.97	19.37	0.563	0.189	9,910.8
546.8	3.36	73.53	76.89	21.89	0.604	0.191	10,009.3
577.4	3.01	74.79	77.79	24.87	0.649	0.193	10,106.5
611.1	2.67	76.02	78.69	28.44	0.699	0.194	10,202.2
648.7	2.36	77.23	79.59	32.74	0.756	0.196	10,296.4
690.7	2.06	78.41	80.47	38.02	0.820	0.198	10,389.0
738.1	1.79	79.56	81.34	44.57	0.893	0.200	10,480.1
792.3	1.53	80.68	82.21	52.85	0.978	0.201	10,569.5
854.9	1.29	81.77	83.06	63.54	1.076	0.203	10,657.3
928.4	1.07	82.83	83.90	77.68	1.193	0.205	10,743.4
1,016.2	0.86	83.86	84.72	96.96	1.334	0.206	10,827.7
1,123.6	0.68	84.85	85.54	124.25	1.508	0.208	10,910.3

reservoir by subdivision of the reservoir into n noncommunicating layers with no crossflow between layers. A constant pressure drop is maintained across the layers. Piston-like displacement was assumed for the waterflood. Thus, as each layer is flooded out, production goes from 100% oil to 100% water. The WOR is a direct measure of the fraction of the vertical cross section that is flooded. The term coverage, denoted by C, was defined as the fraction of the reservoir depleted by water. Eq. 5.184 is an expression for the coverage when the thickness of each layer is h_i.

$$C = \sum_{i=1}^{j} \frac{h_i}{h} + \sum_{i=j+1}^{n} \alpha_i \frac{h_i}{h}. \quad \ldots\ldots\ldots\ldots (5.184)$$

In Eq. 5.184, the n layers are arranged in order of descending permeability with Layer j representing the most recent layer flooded out. The term α_i is given by Eq. 5.185.

$$\alpha_i = \frac{M - \sqrt{M^2 + \frac{k_i}{k_j}(1 - M^2)}}{M - 1}, \quad \ldots\ldots\ldots\ldots (5.185)$$

where M is the mobility ratio based on endpoints of the relative permeability curves.

Compute the displacement performance of the five-layer reservoir presented in Example 5.1 using Eqs. 5.184 and 5.185. Compare the computed performance with the results of Example 5.1 where the displacement is not piston-like. Is it possible to define coverage for a linear waterflood when piston-like displacement does not occur?

5.7. In 1949, Stiles[33] proposed one of the first methods of predicting the vertical sweep efficiency of a linear waterflood operating at constant pressure drop in a layered reservoir when there was no crossflow between layers. The Stiles model assumes piston-like displacement with unit mobility ratio. With these assumptions, the coverage (fraction of the vertical cross section swept to S_{or} by water) is given by Eq. 5.186.

$$C = \sum_{i=1}^{j} \frac{h_i}{h} + \sum_{i=j+1}^{n} \frac{\frac{k_i}{k_j}h_i}{h}, \quad \ldots\ldots\ldots(5.186)$$

where the subscript j represents the last layer flooded out and the index, n, is the number of layers. The WOR corresponding to the floodout of Layer j is given by Eq. 5.187.

$$F_{wo} = \frac{\sum_{i=1}^{j} k_i h_i}{\sum_{i=j+1}^{n} k_i h_i}. \quad \ldots\ldots\ldots\ldots(5.187)$$

Compute the waterflood performance of Example 5.1 using the Stiles method, and compare with the results of Example 5.1 using frontal advance theory and the Dykstra-Parsons model given by Eqs. 5.184 and 5.185.

5.8. Estimate the waterflood performance of a 10-acre [40 469-m^2] five-spot pattern in a reservoir that has five noncommunicating zones. The waterflood will be operated at a constant pressure drop of 1,000 psi [7 MPa] until the WOR reaches 50. Relative permeability data are represented by Eqs. 5.188 and 5.189. Reservoir rock and fluid properties are given in Tables 5.32 and 5.33.

$$k_{ro} = (1 - S_{wD})^{2.5} \quad \ldots\ldots\ldots\ldots(5.188)$$

and

$$k_{rw} = 0.1 S_{wD}^{1.5}, \quad \ldots\ldots\ldots\ldots(5.189)$$

where S_{wD} is given by Eq. 3.111.

5.9. Rework Problem 5.4 using the Higgins-Leighton model[48] with direct line-drive geometry. The reservoir will not be considered to be fractured. Effective wellbore radius is 1.0 ft [0.305 m]. How would you expect the waterflood performance under direct line-drive pattern to differ from that computed using the linear displacement model in Problem 5.4?

5.10. Core analyses from four wells located on the corners of an area about 1,500 ft [457 m] by 1,500 ft [457 m] are given in Table 5.34. Determine the Dykstra-Parsons coefficient of permeability variation (V) for this set of data. Subdivide the reservoir into 10 equal-thickness

TABLE 5.29—PREDICTED PERFORMANCE—WATERFLOOD OF HOMOGENEOUS RESERVOIR FIVE-SPOT PATTERN, PATTERN AREA 12.66 ACRES

W_i (bbl)	Time (days)	Injection Rate (B/D)	WOR	N_p (STB)
4,466	141.77	38.7	0.0	2,524
9,425	283.6	39.4	0.0	7,461
14,563	425.3	40.7	0.0	12,600
19,843	567.1	41.7	0.0	17,879
25,221	708.9	42.3	0.0	23,258
30,680	850.6	42.9	0.0	28,716
36,203	992.4	43.3	0.0	34,239
41,785	1,134.2	44.0	0.0	39,821
47,426	1,276.0	44.7	0.0	45,426
53,250	1,417.7	48.4	0.46	51,006
59,535	1,559.5	51.3	0.63	55,291
66,297	1,701.3	53.3	0.453	58,219
77,177	1,843.1	54.1	1.49	61,002
80,246	1,984.8	58.1	3.89	63,757
95,328	2,268.4	59.6	4.63	66,521
110,787	2,551.9	61.0	4.94	69,190
126,638	2,835.5	63.5	5.82	71,829
143,247	3,119.0	65.5	16.03	72,991
160,192	3,402.6	66.7	17.77	73,938
177,966	3,686.1	67.84	19.53	74,813

TABLE 5.30—RESERVOIR PARAMETERS—HETEROGENEOUS CASE

Layer	Thickness (ft)	Porosity	Permeability (md)
1	10	0.08	50
2	10	0.12	250

TABLE 5.31—PROPERTIES OF RESERVOIR ROCK FOR ZONES 1 THROUGH 3, PROBLEM 5.4

Zone	Thickness (ft)	k_o/S_{iw} (md)	S_{iw} (fraction)	S_{or} (fraction)	ϕ (fraction)
1	5	500	0.20	0.30	0.25
2	10	50	0.40	0.30	0.20
3	5	100	0.35	0.30	0.23

TABLE 5.32—RESERVOIR AND FLUID PROPERTIES

S_{oi}	0.76
S_{or}	0.30
S_g	0.16
B_o, bbl/STB	1.30
B_w, bbl/STB	1.05
r_w, ft	0.75
μ_o, cp	3.5
μ_w, cp	0.8

TABLE 5.33—PROPERTIES OF LAYERS

Layer	Thickness (ft)	Permeability (md)	Porosity (fraction)
1	5.00	406.8	0.245
2	5.00	240.7	0.222
3	5.00	167.4	0.215
4	5.00	116.4	0.205
5	5.00	68.8	0.185

TABLE 5.34—PERMEABILITY DISTRIBUTION FOR FOUR WELLS, EL DORADO FIELD, PROBLEM 5.10[38]

Well MP-101		Well MP-102		Well MP-104		Well MP-105	
Depth (ft)	Permeability to Air (md)	Depth (ft)	Permeability to Air (md)	Depth (ft)	Permeability to Air (md)	Depth (ft)	Permeability to Air (md)
617.0	0.480	605.5	455.000	654.0	70.000	614.0	170.000
617.5	294.000	606.0	525.000	655.0	302.000	615.0	69.000
618.0	137.000	606.5	195.000	656.0	96.000	616.0	0.200
618.5	231.000	607.0	101.000	657.0	500.000	617.0	73.000
619.0	172.000	607.5	387.000	658.0	84.000	618.0	380.000
619.5	480.000	608.0	95.000	659.0	700.000	619.0	315.000
620.0	560.000	608.5	180.000	660.0	760.000	620.0	48.000
620.5	445.000	609.0	37.000	661.0	392.000	621.0	0.080
621.0	131.000	609.5	51.000	662.0	414.000	622.0	61.000
621.5	362.000	610.0	540.000	663.0	420.000	623.0	270.000
622.0	435.000	610.5	5.000	664.0	600.000	624.0	250.000
622.5	260.000	611.0	185.000	665.0	309.000	625.0	415.000
623.0	14.000	611.5	660.000	666.0	740.000	626.0	42.000
623.5	366.000	612.0	655.000	667.0	18.000	627.0	Broken
624.0	560.000	612.5	380.000	668.0	486.000	628.0	215.000
624.5	487.000	613.0	10.000	669.0	127.000	629.0	285.000
625.0	75.000	613.5	265.000	670.0	Broken	630.0	Broken
625.5	Broken	614.0	26.000	671.0	670.000	631.0	Broken
626.0	Broken	614.5	1,620.000	672.0	5.000	632.0	455.000
626.5	600.000	615.0	185.000	673.0	11.000		
627.0	460.000	615.5	775.000				
627.5	Broken	616.0	121.000				
628.0	780.000	616.5	345.000				
628.5	560.000	617.0	2.000				
629.0	580.000	617.5	Broken				
629.5	1,045.000	618.0	850.000				
630.0	225.000	618.5	Broken				
630.5	600.000	619.0	145.000				
631.0	20.100	619.5	800.000				
631.5	465.000	620.0	980.000				
632.0	2.880	620.5	850.000				
632.5	Broken	621.0	885.000				
633.0	920.000	621.5	810.000				
633.5	1,570.000	622.0	540.000				
634.0	380.000	622.5	805.000				
634.5	630.000	623.0	380.000				
		623.5	69.000				
		624.0	Broken				
		624.5	530.000				
		625.0	1,120.000				

layers assuming that (1) the permeability distribution follows the log-normal distribution, and (2) the permeability distribution is empirical, as discussed in Sec. 5.8.1.

5.11. Permeability and porosity data from Well Hiram 17WI are presented in Table 5.35.[49] Determine whether the permeabilities are distributed log-normally. Select a method to subdivide the reservoir into 10 equal-thickness layers, and determine the permeability and porosity of each layer. To determine porosity, develop a correlation between permeability and porosity by plotting permeability vs. porosity on semilog paper. Permeability should be plotted on the log scale.

5.12. A reservoir is to be waterflooded on 10-acre [40 468-m^2] spacing with five-spot patterns. The reservoir is 30 ft [9 m] thick and has a porosity of 0.25. Relative permeability data are given by Eqs. 5.190 and 5.191. Other parameters are given in Table 5.36.

$$k_{ro} = 0.70(1 - S_{wD})^{1.5} \quad \ldots\ldots\ldots\ldots\ldots (5.190)$$

and

$$k_{rw} = 0.068(S_{wD})^{1.2}, \quad \ldots\ldots\ldots\ldots\ldots (5.191)$$

where S_{wD} is given by Eq. 3.111. The reservoir is heterogeneous with a geometric mean permeability of 275 md and a permeability variation of 0.7.

Estimate the waterflood performance of this pattern using the Higgins-Leighton program.[48] The flood will be operated at a constant pressure drop of 500 psi [3.4 MPa]. The radii of production and injection wells are 0.5 ft [0.15 m]. For this calculation, assume that there is no crossflow between layers. Thus, heterogeneity can be approximated by subdividing the reservoir thickness into equal-thickness layers. Simulate waterflood performance for 5, 10, and 20 layers using the log-normal permeability distribution to assign permeability to each layer. Compare the computed results.

5.13. In Problem 5.12, the oil-viscosity and relative permeability data lead to a favorable mobility ratio. Waterflood performance is almost piston-like with little oil flow

TABLE 5.35—PERMEABILITY AND POROSITY DATA FROM CORE ANALYSIS, HIRAM NO. 17 WI WELL,[49] PROBLEM 5.11

Sample Number	Depth (ft)	Permeability to Air (md)	Porosity
1	2,880.0 to 2,881.0	1,271.0	28.9
2	2,881.0 to 2,882.0	1,239.0	28.5
3	2,882.0 to 2,883.0	1,184.0	28.1
4	2,883.0 to 2,884.0	1,891.0	28.8
5	2,884.0 to 2,885.0	1,500.0	27.9
6	2,885.0 to 2,886.0	1,271.0	29.2
7	2,886.0 to 2,887.0	1,565.0	29.0
8	2,887.0 to 2,888.0	1,325.0	29.7
9	2,888.0 to 2,889.0	967.0	27.4
10	2,889.0 to 2,890.0	717.0	27.8
11	2,890.0 to 2,891.0	728.0	28.0
12	2,891.0 to 2,892.0	554.0	22.2
13	2,892.0 to 2,893.0	130.0	20.3
14	2,893.0 to 2,894.0	218.0	21.5
15	2,894.0 to 2,895.0	466.0	25.5
16	2,895.0 to 2,896.0	684.0	24.9
17	2,896.0 to 2,897.0	600.0	27.2
18	2,897.0 to 2,898.0	336.0	23.7
19	2,898.0 to 2,899.0	150.0	21.9
20	2,899.0 to 2,900.0	277.0	22.0
21	2,900.0 to 2,901.0	78.0	19.4
22	2,901.0 to 2,902.0	101.0	17.4
23	2,902.0 to 2,903.0	82.0	18.4
24	2,903.0 to 2,904.0	82.0	16.7
25	2,904.0 to 2,905.0	49.0	16.9
26	2,905.0 to 2,906.0	36.0	17.1
27	2,906.0 to 2,907.0	23.0	15.9
28	2,907.0 to 2,908.0	20.0	16.5
29	2,908.0 to 2,909.0	<0.1	13.0
30	2,909.0 to 2,910.0	56.0	16.8
31	2,910.0 to 2,911.0	49.0	17.3
32	2,911.0 to 2,912.0	26.0	17.8
33	2,912.0 to 2,913.0	33.0	17.8
34	2,913.0 to 2,914.0	26.0	15.6
35	2,914.0 to 2,915.0	36.0	17.4
36	2,915.0 to 2,916.0	42.0	17.3
37	2,916.0 to 2,917.0	33.0	16.8
38	2,917.0 to 2,918.0	39.0	16.6
39	2,918.0 to 2,919.0	52.0	17.2
40	2,919.0 to 2,920.0	56.0	16.9
41	2,920.0 to 2,921.0	33.0	15.2
42	2,921.0 to 2,922.0	46.0	16.1
43	2,922.0 to 2,923.0	36.0	17.4
44	2,923.0 to 2,924.0	29.0	14.8
45	2,924.0 to 2,925.0	33.0	15.7
46	2,925.0 to 2,926.0	23.0	15.7
47	2,926.0 to 2,927.0	33.0	15.6
48	2,927.0 to 2,927.5	17.0	14.4
	2,927.5 to 2,930.0	Lost Core	

TABLE 5.36—RESIDUAL SATURATIONS AND FLUID VISCOSITIES, PROBLEM 5.12

S_{or}	0.36
S_{iw}	0.41
μ_o, cp	0.88
μ_w, cp	0.83

TABLE 5.37—RESERVOIR AND PROPERTIES DATA FOR SLAUGHTER ESTATE UNIT[50]

Reservoir Data

Layer	Thickness (ft)	Permeability (md)	Porosity (fraction)
1 (Top layer)	18.1	0.75	0.087
2	21.9	2.1	0.114
Impermeable layer			
3	12.5	9.6	0.134
4	13.4	11.6	0.169
5	9.3	16.0	0.109

Fluid Properties and Initial Saturations

Oil viscosity, cp	2.28
Water viscosity, cp	0.7
Oil density, lbm/cu ft	51.4
Water density, lbm/cu ft	62.4
Oil FVF, RB/STB	1.108
Water FVF, RB/STB	1.0
Dissolved GOR, scf/STB	250
Initial water saturation	0.087
Initial gas saturation	0.188
Residual oil saturation	0.311

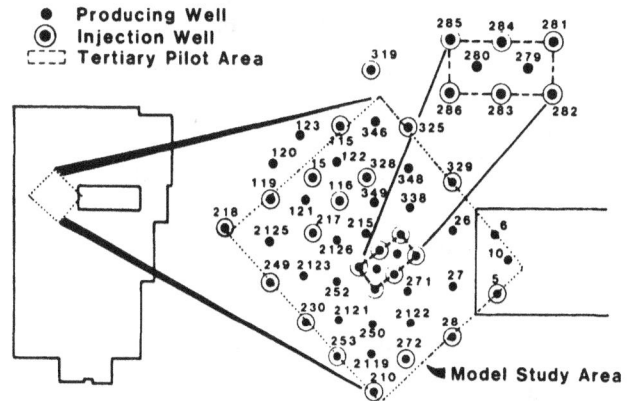

Fig. 5.51—Slaughter Estate unit model study and pilot area.[50]

after breakthrough in regions swept by water. Investigate the effect of mobility ratio on waterflood performance by simulating waterfloods for oil viscosities of 2, 5, 10, and 20 cp [2, 5, 10, and 20 mPa·s]. For this case, subdivide the reservoir into five equal-thickness layers, as in Problem 5.12, using the log-normal distribution with a geometric mean permeability of 275 md and a permeability variation of 0.7.

5.14. A pilot waterflood was conducted in the Slaughter Estate unit[50] in two adjacent five-spot patterns that had a total area of 12.3 acres [49 777 m²]. The pilot pattern shown in Fig. 5.51 consists of six injection wells and two production wells and will be considered to be con-

fined. The productive interval is the San Andres dolomite, which was subdivided into five discrete layers on the basis of a study of core and log data. Properties of the layers are given in Table 5.37.

The reservoir was produced by solution gas drive before water injection in the pilot area. A gas saturation of 0.188 was estimated to exist in the pilot area when the waterflood began. Estimate the waterflood performance of the pilot pattern using the Higgins-Leighton model. Compare your results to actual performance of the pilot waterflood in Ref. 50. Identify differences between predicted and observed performance, and discuss possible reasons for the differences.

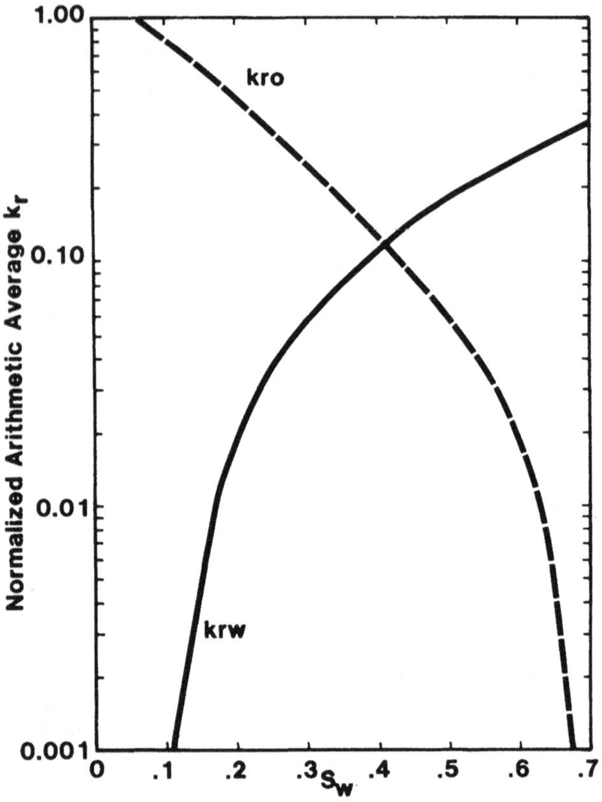

Fig. 5.52—San Andres Zone V water/oil relative permeability curves.[50]

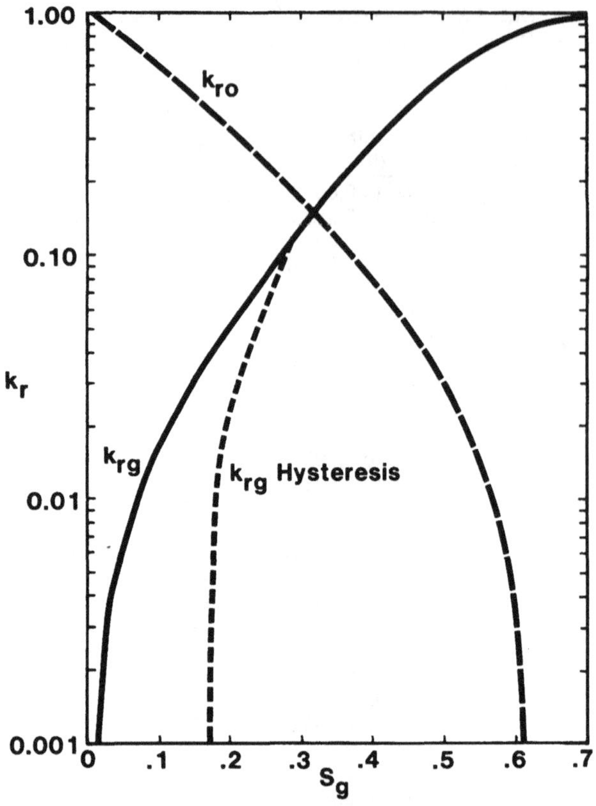

Fig. 5.53—San Andres Zone V oil/gas permeability and gas relative permeability hysteresis curves.[50]

TABLE 5.38—FLUID AND ROCK PROPERTIES FOR PROBLEM 5.17

S_{or}	0.25
S_{iw}	0.30
μ_o, cp	2.5
μ_w, cp	1.0
ρ_o, g/cm^3	0.85
ρ_w, g/cm^3	1.01
k, md	10
σ, dynes/cm	30
ϕ	0.20

Assume that the pressure drop between injection and production wells is constant at 2,750 psi [19 MPa]. Rock and fluid properties are given also in Table 5.37. Relative-permeability data are presented in Figs. 5.52 and 5.53.

5.15. Rework Problem 5.14 assuming that the injection rate was held constant at 425 B/D [68 m^3/d].

5.16. A set of pseudo- or thickness-averaged relative permeability data were developed in Sec. 5.4.1 for a waterflood in a reservoir under the assumption of vertical equilibrium. Use the data presented in Table 5.9 to compute the waterflood performance of the five-spot pattern in Example 4.4. Change the reservoir thickness to 40 ft [12 m] to conform to the assumptions used to generate thickness-averaged relative permeability curves. Compare predicted performance to that obtained by assuming uniform fluid saturations in the vertical cross section.

5.17. The effect of vertical equilibrium on displacement performance is related directly to the thickness of the reservoir. You are requested to develop thickness-averaged relative permeability curves for a homogeneous reservoir under the assumption of vertical equilibrium. Curves for reservoir thicknesses of 5, 10, 20, and 50 ft [1.5, 3.05, 6.1, and 15.2 m] will be developed. Capillary pressure and relative permeabilities for the reservoir rock are given by Eqs. 5.192 through 5.194.

$$k_{ro} = 0.404(1-S_{wD})^{1.77}, \quad\ldots\ldots\ldots\ldots (5.192)$$

$$k_{rw} = 0.189 S_{wD}^{3.70}, \quad\ldots\ldots\ldots\ldots\ldots (5.193)$$

and

$$P_c = 0.12\sigma\left(\frac{\phi}{k}\right)^{0.5}(1-S_{wD})^2, \quad\ldots\ldots\ldots\ldots (5.194)$$

where S_{wD} is given by Eq. 3.111. Properties of oil and water are found in Table 5.38.

Compare thickness-averaged relative permeability curves by plotting relative permeability vs. saturation. What is the impact of vertical equilibrium on displacement efficiency of a waterflood?

In developing thickness-averaged relative permeability curves, it is necessary to define a datum for capillary-pressure distributions. The location of the datum is arbitrary. For this problem, the datum plane will be the top of the reservoir. When capillary equilibrium is assumed, the initial water saturation is not uniform. The interstitial water saturation in Table 5.38 is a thickness-averaged saturation corresponding to the initial capillary-pressure distribution, found by trial and error, as illustrated in Example 5.2. This is easily done with a short computer program. In computing thickness-averaged relative permeability, remember that $k_{rw} = 0$ and $k_{ro} = 0.404$ when $S_w < S_{iwz}$.

TABLE 5.39—BASIC DATA
FOR A TWO-LAYER MODEL[52]

Length (L), ft	625
Thickness	
h_1, ft	10
h_2, ft	10
Width, ft	1,000
Horizontal permeability to oil	
at interstitial water saturation	
k_{x1}, md	150
k_{x2}, md	50
Porosity	0.2
Oil viscosity, cp	0.9
Water viscosity, cp	0.65
Interstitial water saturation	0.16
Residual oil saturation	0.2
Interfacial tension, dynes/cm	24.5
Oil density, g/cm^3	0.8
Water density, g/cm^3	1.0

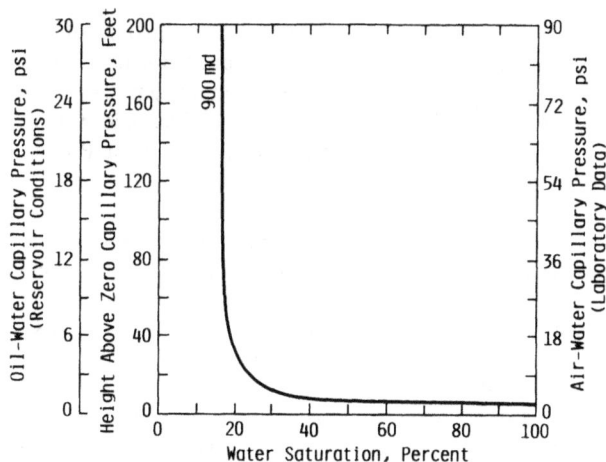

Fig. 5.54—Capillary-pressure distributions when transition zone is narrow.[51]

5.18. The transition zone under capillary equilibrium is defined as the width of the region where the water saturation changes from $S_w = 100$ to $S_w = S_{iw}$. When the capillary pressure curve is flat (as depicted in Fig. 5.54), the transition zone is narrow relative to the thickness of the reservoir. Under these conditions, the thickness-averaged relative permeability curves become linear functions of thickness-averaged saturation.

What type of displacement process is implied by the assumption of negligible width of the transition zone?

Hint. It may be useful to compute thickness-averaged relative permeability curves as in Problem 5.17 to determine how parameters vary with reservoir thickness. A steep capillary-pressure curve can be obtained by selecting a large value of the permeability in Eq. 5.194.

5.19. Table 5.39[52] contains basic data for the two-layer model shown in Fig. 5.55. Relative permeability and capillary-pressure relationships are given by Eq. 5.195 through 5.198.

$$k_{ro} = 0.9(1 - S_{wD})^{2.2}, \quad \dots \dots \dots \dots (5.195)$$

$$k_{rw} = 0.4 S_{wD}^{1.5}, \quad \dots \dots \dots \dots \dots (5.196)$$

and

$$P_c = 0.7\sigma \left(\frac{\phi}{k}\right)^{0.5} (1 - S_{wD})^4, \quad \dots \dots \dots (5.197)$$

where

$$S_{wD} = \frac{S_{iw} - S_{or}}{1 - S_{or} - S_{iw}},$$

σ = interfacial tension, dynes/cm,
ϕ = porosity, fraction,
k = permeability to oil at connate water saturation, darcies, and
P_c = capillary pressure, psi [kPa].

Using the center of the reservoir as the datum, construct pseudo-capillary-pressure and pseudo-relative-permeability curves for this two-layer system assuming vertical equilibrium. Neglect crossflow caused by viscous forces.

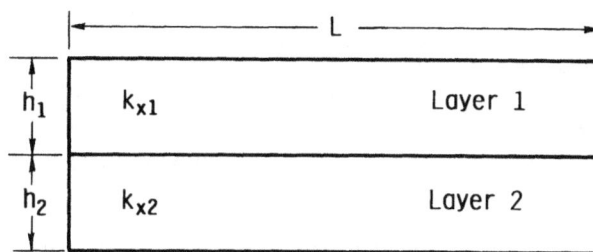

Fig. 5.55—Two-layer model for Problem 5.19.

5.20. Interchange Layers 1 and 2 in Problem 5.19. What effect does the order of layers have on the pseudo-relative-permeability and pseudo-capillary-pressure curves?

5.21. Suppose the vertical permeability between layers in the reservoir described in Problem 5.1 is so high that crossflow occurs quickly and the assumptions of uniform vertical pressure are valid. Estimate the waterflood performance using the fractional flow curve derived for uniform vertical pressure. What are the effects of crossflow on waterflood performance when compared to the no-crossflow calculations in Example 5.1?

5.22. Derive pseudo-relative-permeability curves for the waterflood of the reservoir described in Example 5.5 assuming complete crossflow between each layer and neglecting capillary and gravity forces.

5.23. Predict the waterflood performance of the 10-acre [40 468-m^2] five-spot pattern in Example 4.1 when the reservoir is subdivided into five equal-thickness layers, as in Example 5.3. Crossflow is instantaneous so that the assumption of uniform vertical pressure is valid.

5.24. Compute the pseudorelative permeabilities for oil and water using the data given in Example 5.5 when gravity segregation dominates vertical fluid flow. Compare pseudo-relative-permeability curves with rock curves.

5.25. Determine the waterflood performance of the reservoir described in Example 3.4 under the conditions of gravity segregation with vertical equilibrium. Compare your results with those obtained from Example 3.4 by plotting oil recovery expressed as a fraction of PV vs. PV's injected.

5.26. A linear reservoir that has a dip of 30° will be waterflooded by injection into downdip wells. The reservoir has the same properties and dimensions as used in Example 3.4. Assume gravity segregation occurs. Specific gravities of the oil and water are 0.8 and 1.05, respectively. (1) Determine the maximum injection rate if the displacement front is stabilized by gravity. (2) Predict waterflood performance to a WOR of 100 when the injection rate is 1.5 times the critical rate determined in (1). (3) Compare displacement performance under gravity segregation with performance predicted by the frontal advance theory by plotting oil recovery and WOR vs. PV's injected.

5.27. The gravity segregation model presented in Sec. 5.4.2 does not contain a method of predicting pressure drop when the flow rate is constant or flow rate when the pressure drop is fixed. Using frontal advance theory, develop a method of predicting pressure drop and/or flow rate in a horizontal reservoir when gravity segregation controls the displacement process.

5.28. The Dykstra-Parsons method of estimating waterflood performance of a stratified reservoir is based on a series of correlations between permeability variation, endpoint mobility ratios, and recovery performance for corefloods on California reservoir rock. These correlations are presented as Figs. 8.1 through 8.4 in Ref. 32. Determine the recovery (fraction of oil displaced) using the Dykstra-Parsons correlation for WOR = 1, 5, 25, and 100. Compare the results with those computed in Example 5.1. Why would you expect the results to be different? Would you expect the Dykstra-Parsons correlation to be valid for pattern waterfloods?

Nomenclature

A = cross-sectional area perpendicular to direction of flow, bh, sq ft [m²]

b = width of reservoir, ft [m]

f_w = fractional flow of water

Δf_w = incremental change in fractional flow of water

f_{wj} = fractional flow of water in Layer j

f_{wz} = thickness-averaged fractional flow of water

F_c = $\alpha \cos\theta$ in Eq. 5.135, a measure of capillary forces

F_g = $k_x g \Delta\rho h$ in Eq. 5.134, a measure of gravity forces

F_v = $u_t \mu_o L$ in Eq. 5.134, a measure of viscous forces

g = gravitational constant, ft/sec² [m/s²]

G = gravity number defined by Eq. 5.72, dimensionless

h = thickness of reservoir, ft [m]

h_j = thickness of Layer j, ft [m]

h_m = thickness of model, ft [m]

h_p = thickness of prototype, ft [m]

h_1 = thickness of Layer 1, ft [m]

h_2 = thickness of Layer 2, ft [m]

$J(S_w)$ = Leverett function, dimensionless

k = absolute permeability, md

$\bar{k}$ = thickness-averaged permeability, md

k_b = base permeability, md

k_j = permeability of Layer j, md

k_{ox} = permeability to oil in x direction, md

k_{oz} = thickness-averaged oil permeability, md

k_{ro} = relative permeability to oil

k_{roz} = pseudo- or thickness-averaged relative permeability to oil

k_{rw} = relative permeability to water

k_{rwz} = pseudo- or thickness-averaged relative permeability to water

k_{wx} = permeability to water in the x direction, md

k_{wz} = thickness-averaged permeability to water, md

k_x = permeability in x direction, md

k_z = permeability in z direction, md

L = length of reservoir, ft [m]

L_m = length of model, ft [m]

L_p = length of prototype, ft [m]

M_m = mobility ratio in model, dimensionless

M_p = mobility ratio in prototype, dimensionless

N_p = cumulative oil produced, bbl [m³]

N_{pcf} = oil recovery from layered system with crossflow, bbl [m³]

N_{pj} = cumulative oil produced from Layer j, bbl [m³]

N_{pncf} = oil recovery from layered system with crossflow, bbl [m³]

N_{pu} = oil recovery from uniform system with average permeability, bbl [m³]

p_D = dimensionless pressure

Δp_h = pressure change in horizontal direction, psi [kPa]

p_I = pressure at injection well, psi [kPa]

p_o = oil-phase pressure, psi [kPa]

p_{od} = oil-phase pressure at horizontal datum located at Elevation Z, psi [kPa]

p_{oD} = dimensionless oil-phase pressure

p_p = pressure in model, psi [kPa]

p_r = dimensionless reference pressure defined by Eq. 5.130

Δp_V = pressure change in vertical direction

p_w = water-phase pressure, psi [kPa]

p_{wd} = water-phase pressure at horizontal datum located at Elevation Z_d, psi [kPa]

p_{wD} = dimensionless water-phase pressure

P_c = capillary pressure, psi [kPa]

P_{ch} = capillary pressure Elevation h, psi [kPa]

P_{co} = capillary pressure at bottom of reservoir ($Z=0$), psi [kPa]

P_{cp} = pseudocapillary pressure at reference elevation, Z_p, psi [kPa]

q = flow rate, B/D [m^3/d]

q_c = critical rate for gravity-stabilized displacement, B/D [m^3/d]

q_{ob} = oil production rate in base problem, B/D [m^3/d]

q_{oj} = oil production from Layer j, B/D [m^3/d]

q_t = total flow rate, B/D [m^3/d]

q_{tj} = total flow rate in Layer j, B/D [m^3/d]

q_{wb} = water production rate in base problem, B/D [m^3/d]

q_{wj} = water production rate from Layer j, B/D [m^3/d]

Q_i = PV's injected, fraction

Q_{i2} = PV's injected corresponding to saturation S_{w2}, fraction

Q_{ib} = PV's injected in base model, fraction

Q_{ij} = PV's injected in Layer j, fraction

R_a = dimensionless group, defined by Eq. 5.131

R_c = dimensionless group, defined by Eq. 5.132

R_d = dimensionless group, defined by Eq. 5.133

R_k = layer permeability ratio

R_l = reservoir effective length/thickness ratio

S_{iw} = interstitial water saturation where the water phase is immobile under the applied pressure gradient, fraction of PV

S_o = oil saturation, fraction of PV

S_{or} = residual oil saturation, fraction of PV

S_w = water saturation, fraction of PV

$\overline{S_w}$ = volumetric-averaged water saturation

S_{wD} = dimensionless water saturation $= (S_w - S_{iw})/(1 - S_{or} - S_{iw})$

S_{wf} = frontal advance water saturation at breakthrough, fraction of PV

S_{wj} = water saturation of Layer j, fraction of PV

S_{wz} = thickness-averaged water saturation

$\overline{S_{wz}}$ = volumetric-averaged thickness-averaged water saturation

ΔS_{wz} = incremental change in thickness-averaged saturation

S_{wz2} = thickness-averaged water saturation at end of system, fraction of PV

t = time, days

t_b = time for base calculation, days

t_D = dimensionless time

t_j = time in Layer j, days

t_m = time in model, days

t_p = time in prototype, days

u = darcy velocity, ft/D [m/d]

u_{ox} = darcy velocity of oil phase in x direction, ft/D [m/d]

u_{tm} = darcy velocity in model, ft/D [m/d]

u_{tp} = darcy velocity in prototype, ft/D [m/d]

u_{wx} = darcy velocity of water phase in x direction, ft/D [m/d]

v_{Swz} = frontal velocity of thickness-averaged saturation S_{wz}, ft/D [m/d]

v_t = total interstitial velocity, ft/D [m/d]

V = Dykstra-Parsons coefficient of permeability variation, dimensionless

V_{pj} = pore volume of Layer j, bbl [m^3]

x_{Swz} = x location of thickness-averaged saturation, ft [m]

x_D = dimensionless distance in x direction

z_D = dimensionless distance in z direction

z_p = z coordinate in prototype, ft [m]

Z = elevation relative to a horizontal datum plane, ft [m]

Z_d = elevation of horizontal datum plane, ft [m]

Z_p = reference elevation where pseudocapillary pressure is defined, ft [m]

λ_r = relative total mobility, cp^{-1} [Pa·s^{-1}]

μ_o = viscosity of oil, cp [Pa·s]

μ_D = viscosity of displacing phase, cp [Pa·s]

μ_w = viscosity of water, cp [Pa·s]

ρ_o = density of oil, lbm/cu ft [kg/m^3]

ρ_w = density of water, lbm/cu ft [kg/m^3]

ϕ = porosity, fraction

ϕ_b = porosity base computation, fraction

ϕ_j = porosity of Layer j, fraction

ϕ_m = porosity of model, fraction

ϕ_p = porosity of prototype, fraction

Φ_o = oil-phase potential

Φ_w = water-phase potential

References

1. Jordan, J.K., McCardell, W.M., and Hocott, C.R.: "Effect of Rate on Oil Recovery by Waterflooding," Humble Oil and Refining Co., Houston (Aug. 1956).
2. Douglas, J. Jr., Peaceman, D.W., and Rachford, H.H.: "A Method for Calculating Multi-Dimensional Immiscible Displacement," Trans., AIME (1959) 216, 297-308.
3. Hiatt, W.N.: "Injected-Fluid Coverage of Multi-Well Reservoirs with Permeability Stratification," Drill. and Prod. Prac., API (1958) 165, 165-94.
4. Coats, K.H., Dempsey, J.R., and Henderson, J.H.: "The Use of Vertical Equilibrium in Two-Dimensional Simulation of Three-Dimensional Reservoir Performance," Soc. Pet. Eng. J. (March 1971) 63-71; Trans., AIME, 251.
5. Coats, K.H. et al.: "Simulation of Three-Dimensional, Two-Phase Flow in Oil and Gas Reservoirs," Soc. Pet. Eng. J. (Dec. 1967) 377-88; Trans., AIME, 240.
6. Dake, L.P.: Fundamentals of Reservoir Engineering, Elsevier Scientific Publishing Co., New York City (1978) 372-90.
7. Dietz, D.N.: "A Theoretical Approach to the Problem of Encroaching and By-Passing Edge Water," Proc., Akad. Wetenschappen, Amsterdam, V56-B (1953) 83.
8. Hearn, C.L.: "Simulation of Stratified Waterflooding by Pseudo Relative Permeability Curves," J. Pet. Tech. (July 1971) 805-13.
9. Rapoport, L.A.: "Scaling Laws for Use in Design and Operation of Water-Oil Flow Models," J. Pet. Tech. (Sept. 1955) 143-50; Trans., AIME, 204.
10. Perkins, F.M. Jr. and Collins, R.E.: "Scaling Laws for Laboratory Flow Models of Oil Reservoirs," J. Pet. Tech. (Aug. 1960) 69-71; Trans., AIME, 219.
11. Geertsma, J., Croes, G.A. and Schwarz, N.: "Theory of Dimensionally Scaled Models of Petroleum Reservoirs," J. Pet. Tech. (June 1956) 118-27; Trans., AIME, 207.
12. Leverett, M.C.: "Capillary Behavior in Porous Solids," Trans., AIME (1941) 142, 152-69.
13. Richardson, J.G. and Perkins, F.M. Jr.: "A Laboratory Investigation of the Effect of Rate on Recovery of Oil by Waterflooding," J. Pet. Tech. (April 1957) 114-21; Trans., AIME, 210.
14. Goddin, C.S. Jr. et al.: "A Numerical Study of Waterflood Performance in a Stratified System With Crossflow," J. Pet. Tech. (June 1966) 765-71; Trans., AIME, 237.
15. Henley, D.H., Owens, W.W., and Craig, F.F. Jr.: "A Scale-Model Study of Bottom-Water Drives," J. Pet. Tech. (Jan. 1961) 90-98; Trans., AIME, 222.

16. Craig, F.F. Jr. *et al.*: "A Laboratory Study of Gravity Segregation in Frontal Drives," *J. Pet. Tech.* (Oct. 1957) 275-81; *Trans.*, AIME, **210**.

17. Blair, P.M. and Peaceman, D.W.: "An Experimental Verification of a Two-Dimensional Technique for Computing Performance of Gas-Drive Reservoirs," *Soc. Pet. Eng. J.* (March 1963) 19-27; *Trans.*, AIME, **228**.

18. Terwilliger, P.L. *et al.*: "An Experimental and Theoretical Investigation of Gravity Drainage Performance," *Trans.*, AIME (1951) **192**, 285-96.

19. Goddin, C.S. Jr.: "Two-Dimensional Flow of Two Immiscible Incompressible Fluids in a Stratified Porous Medium," PhD dissertation, U. of Michigan, Ann Arbor, MI (1965).

20. Huppler, J.D.: "Numerical Investigation of the Effects of Core Heterogeneities on Waterflood Relative Permeabilities," *Soc. Pet. Eng. J.* (Dec. 1970) 381-91; *Trans.*, AIME, **249**.

21. Peters, E.J. and Flock, D.L.: "The Onset of Instability During Two-Phase Immiscible Displacement in Porous Media," *Soc. Pet. Eng. J.* (April 1981) 249-58.

22. Engelberts, W.F. and Klinkenberg, L.J.: "Laboratory Experiments on the Displacement of Oil by Water from Packs of Granular Materials," *Proc.*, Third World Pet. Cong., The Hague (1951) Part II, 544.

23. Hill, S.: "Channeling in Packed Columns," *Chem. Eng. Sci. I* (1952) No. 6, 247-53.

24. Chuoke, R.L., van Meurs, P., and van der Poel, C.: "The Instability of Slow, Immiscible, Viscous Liquid-Liquid Displacements in Permeable Media," *Trans.*, AIME (1959) **216**, 188-94.

25. Perkins, T.K. and Johnson, O.C.: "A Study of Immiscible Fingering in Linear Models," *Soc. Pet. Eng. J.* (March 1969) 39-45.

26. Richardson, J.G.: "The Calculation of Waterflood Recovery From Steady-State Relative Permeability Data," *J. Pet. Tech.* (May 1957) 64-66; *Trans.*, AIME, **210**.

27. Scott, G.R., Collins, H.N., and Flock, D.L.: "Improving Waterflood Recovery of Viscous Crude Oils by Chemical Control," *J. Cdn. Pet. Tech.* (Oct.-Dec. 1965) 243-51.

28. Collins, H.N.: "High Viscosity Crude Oil Displacement in a Long Unconsolidated Sand Pack Using a Native Brine," MS thesis, U. of Alberta, Edmonton, Alta., Canada.

29. Mungan, N.: "Interfacial Effects in Immiscible Liquid-Liquid Displacement in Porous Media," *Soc. Pet. Eng. J.* (Sept. 1966) 247-53; *Trans.*, AIME, **237**.

30. Smith, C.R.: *Mechanics of Secondary Oil Recovery*, Krieger Publishing Corp., New York City (1975) 210.

31. Vossoughi, S.: "Viscous Fingering in Immiscible Displacement," PhD dissertation, U. of Alberta, Edmonton, Alta., Canada (1976).

32. Craig, F.F. Jr.: *The Reservoir Engineering Aspects of Waterflooding*, Monograph Series, SPE, Richardson, TX (1971) **5**.

33. Stiles, W.E.: "Use of Permeability Distribution in Waterflood Calculations," *Trans.*, AIME (1949) **186**, 9-13.

34. Dykstra, H. and Parsons, R.L.: "The Prediction of Oil Recovery by Waterflood," *Secondary Recovery of Oil in the United States*, API (1950) 160-74.

35. Warren, J.E. and Price, H.S.: "Flow in Heterogeneous Porous Media," *Soc. Pet. Eng. J.* (Sept. 1961) 153-69; *Trans.*, AIME, **222**.

36. Testerman, J.D.: "A Statistical Reservoir Zonation Technique," *J. Pet. Tech.* (Aug. 1962) 889-93; *Trans.*, AIME, **225**.

37. Warren, J.E. and Cosgrove, J.J.: "Prediction of Waterflood Behavior in a Stratified System," *Soc. Pet. Eng. J.* (June 1964) 149-57; *Trans.*, AIME, **231**.

38. Coffman, C.L. and Howell, W.D.: "El Dorado Micellar-Polymer Demonstration Project," First Annual Report, BERC/TPR-75/1, Bartlesville Energy Center, Bartlesville, OK (Oct. 1975) II-38.

39. Law, J.: "A Statistical Approach to the Interstitial Characteristics of Sand Reservoirs," *Trans.*, AIME (1944) **155**, 202-22.

40. Craig, F.F. Jr.: "Effect of Reservoir Description on Performance Predictions," *J. Pet. Tech.* (Oct. 1970) 1239-45.

41. Peaceman, D.W.: *Fundamentals of Numerical Reservoir Simulation*, Elsevier Scientific Publishing Co., New York City (1977).

42. Aziz, K. and Settari, A.: *Petroleum Reservoir Simulation*, Applied Science Publishers Ltd., London (1979).

43. Thomas, G.W.: *Principles of Hydrocarbon Reservoir Simulation*, second edition, IHRDC, Boston, MA (1981).

44. Critchlow, H.B.: *Modern Reservoir Engineering—A Simulation Approach*, Prentice Hall, Englewood Cliffs, NJ (1977).

45. *Numerical Simulation*, Reprint Series, SPE, Richardson, TX (1973) **11**.

46. Jacks, H.H., Smith, O.W.J.E., and Mattax, C.C.: "The Modeling of a Three-Dimensional Reservoir With a Two-Dimensional Reservoir Simulation—The Use of Dynamic Pseudo Functions," *Soc. Pet. Eng. J.* (June 1973) 175-85.

47. Coats, K.H.: "Use and Misuse of Reservoir Simulation Models," *J. Pet. Tech.* (Nov. 1969) 1391-98.

48. Leighton, A.J. and Higgins, R.V.: "Improved Method To Predict Multiphase Waterflood Performance for Constant Rates or Pressures," RI8055, USBM (1975).

49. Johnson, J.P. and Burtch, F.W.: "North Stanley Polymer Demonstration Project," First Annual Report, BERC/TPR-78/19, Natl. Technical Information Service, U.S. DOE (Oct. 1976).

50. Ader, J.C. and Stein, M.H.: "Slaughter Estate Unit Tertiary Miscible Gas Pilot Reservoir Description," *J. Pet. Tech.* (May 1984) 837-45.

51. Amyx, J.W., Bass, D.M., and Whiting, R.L.: *Petroleum Reservoir Engineering*, McGraw-Hill Book Co. Inc., New York City (1960).

52. Lake, L.W. and Yokoyama, Y.: "The Effects of Capillary Pressure on Immiscible Displacements in Stratified Porous Media," paper SPE 10109 presented at the 1981 SPE Annual Technical Conference and Exhibition, San Antonio, Oct. 5-7.

General Reference

Bentsen, R.G.: "Scaled Fluid-Flow Models with Permeabilities Differing from That of the Prototype," *J. Cdn. Pet. Tech.* (July-Sept. 1976) 46-52.

SI Metric Conversion Factors

acre $\times$ 4.046 873	E+03	= m^2
bbl $\times$ 1.589 873	E−01	= m^3
cp $\times$ 1.0*	E−03	= Pa·s
cu ft $\times$ 2.831 685	E−02	= m^3
cu in. $\times$ 1.638 706	E+01	= cm^3
dyne/cm $\times$ 1.0*	E+00	= mN/m
ft $\times$ 3.048*	E−01	= m
in. $\times$ 2.54*	E+00	= cm
lbm $\times$ 4.535 924	E−01	= kg

*Conversion factor is exact.

Chapter 6
Waterflood Design

6.1 Introduction

The design of a waterflood involves both technical and economic considerations. Economic analyses are based on estimates of waterflood performance. These estimates may be rough or sophisticated depending on the requirements of a particular project and the philosophy of the operator. This chapter presents methods of estimating waterflood performance for economic analyses. It is organized in order of increasing complexity beginning with first-pass estimates with simple methods and ending with an introduction to the capability of reservoir simulators to evaluate waterflood designs.

6.2 Factors Constituting a Design

The five steps in the design of a waterflood are as follows.
  1. Evaluation of the reservoir, including primary production performance.
  2. Selection of potential flooding plans.
  3. Estimation of injection and production rates.
  4. Projection of oil recovery over the anticipated life of the project for each flooding plan.
  5. Identification of variables that may cause uncertainty in the technical analysis.

Other components of a waterflood design are beyond the scope of this text. Technical analysis of a waterflood produces estimates of the volumes of fluids and rates. Those estimates are used also for sizing equipment and fluid-handling systems. It is necessary to identify a source of water for injection that is compatible with connate fluids as well as with reservoir rock. Design includes arrangements for proper disposal of produced water.

6.3 Reservoir Description

The purposes of a reservoir description in waterflood design are (1) to define the areal and vertical extent of the reservoir, (2) to describe quantitatively the variation in rock properties—such as permeability and porosity within the reservoir, (3) to determine the primary production mechanism, including estimates of the oil remaining to be produced under primary operation, (4) to estimate the distribution of the oil resource in the reservoir, and (5) to

evaluate fluid properties required for predicting waterflood performance. The data and interpretations that are obtained in developing a reservoir description make up many of the input data for the waterflood design.

Methods of developing a reservoir description are presented in Chap. 7. The concern in Chap. 6 is the use of a reservoir description in waterflood design. The following information is usually available from a reservoir description.

Reservoir Characteristics
  1. Areal and vertical extent of producing formation.
  2. Isopach maps of gross and net sand.
  3. Correlation of layers and other zones.
Reservoir Rock Properties
  1. Areal variation of average permeability, including directional trends derived from geological interpretations.
  2. Areal variation of porosity.
  3. Reservoir heterogeneity—particularly the variation of permeability with thickness and zone.
Reservoir Fluid Properties
  1. Gravity, FVF, and viscosity as a function of reservoir pressure.
Primary Producing Mechanism
  1. Identification of producing mechanisms—such as fluid expansion, solution-gas drive, or water drive.
  2. Existence of gas caps or aquifers.
  3. Estimation of oil remaining to be produced under primary operations.
  4. Pressure distribution in the reservoir.
Distribution of Oil Resources in Reservoir at Beginning of Waterflood
  1. Trapped-gas saturation from solution-gas drive.
  2. Vertical variation of saturation as a result of gravity segregation.
  3. Presence of mobile connate water.
  4. Areas already waterflooded by natural water drive.
Rock/Fluid Properties
  1. Relative permeability data for the reservoir rock. Minimum data are endpoints of relative permeability curves.

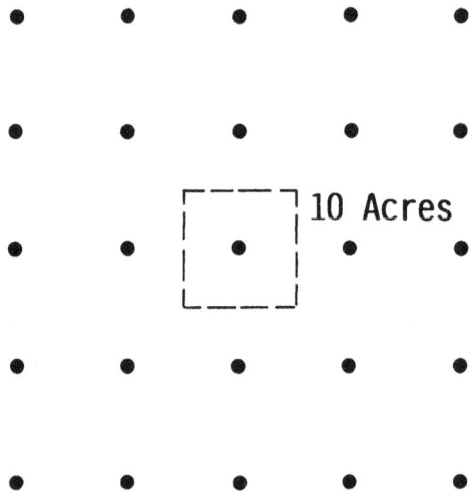

Fig. 6.1a—Regular well spacing during primary production period—10-acre spacing.

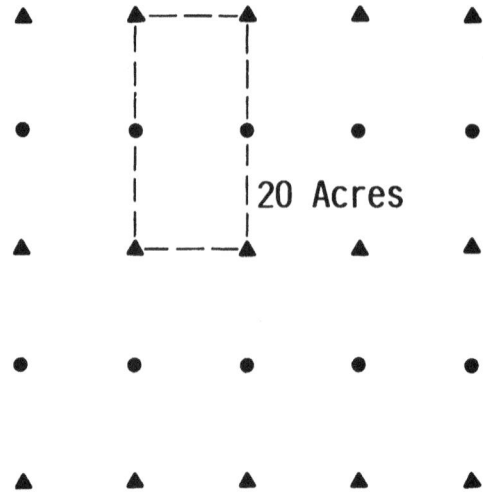

Fig. 6.1b—Linedrive spacing on 20-acre spacing.

6.4 Selection of Potential Flooding Plans

Selection of the waterflooding plan is determined by factors that are often unique to each reservoir. In some reservoirs, the waterflood may be done with edge wells to form a peripheral flood. This is called pressure maintenance when water injection supplements declining reservoir energy from solution-gas drive or an aquifer of limited extent. Pressure maintenance often begins while the reservoir is still under primary operation to maintain maximum production rates. Pattern flooding, an alternative to pressure maintenance, may be selected because reservoir properties will not permit waterflooding through edge wells at desired injection rates. In pattern flooding, injection and withdrawal rates are determined by well spacing as well as reservoir properties. Pattern size becomes a variable that is considered in economic analyses.

The selection of possible waterflooding patterns depends on existing wells that generally must be used because of economics. Pattern selection is constrained by the loca-

tions of production wells. Many fields are developed for primary production on a uniform well spacing as in Fig. 6.1A. If only existing wells are used, the options are limited to a linedrive, five-spot, or nine-spot pattern as shown in Figs. 6.1B through 6.1D. Infill drilling of injection wells can be used to reduce the spacing, as in Fig. 6.2. Any of these patterns can be used to waterflood a reservoir, but final selection of spacing and pattern type, when there are several possibilities, is determined by comparison of the economics of alternative flooding schemes.

Surface or subsurface topology and/or the use of slant-hole drilling techniques may result in production or injection wells that are nonuniformly located, as shown in Fig. 6.3.[1] In these situations, the flooding pattern—i.e., the region affected by the injection well—could be different for every injection well. Some reservoirs are small and are developed for primary production with a limited number of wells. When economics are marginal, one producer may be converted to an injection well, as in Fig. 6.4. The flooding pattern would not be uniform.

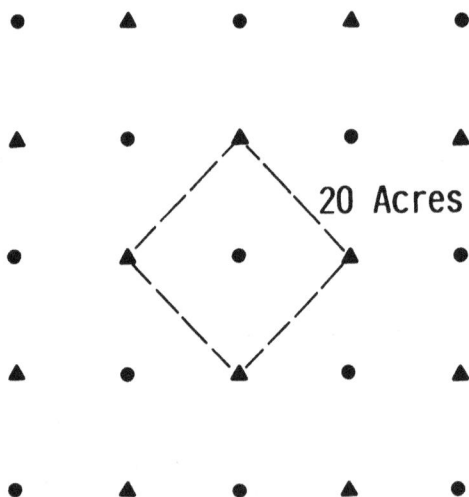

Fig. 6.1c—Five-spot pattern on 20-acre spacing.

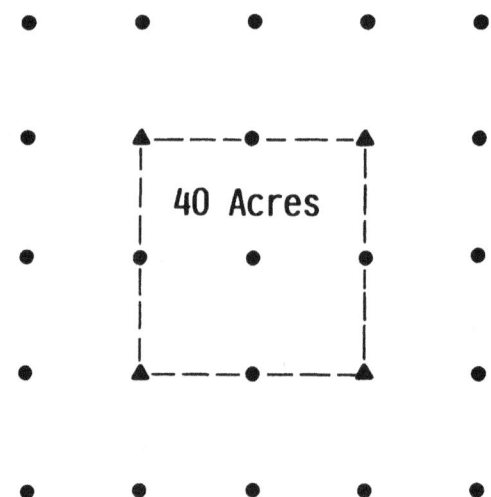

Fig. 6.1d—Nine-spot pattern on 40-acre spacing.

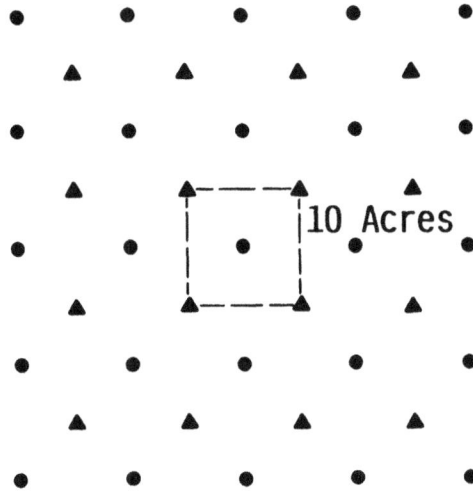

Fig. 6.2—Infill drilling to create 10-acre spacing.

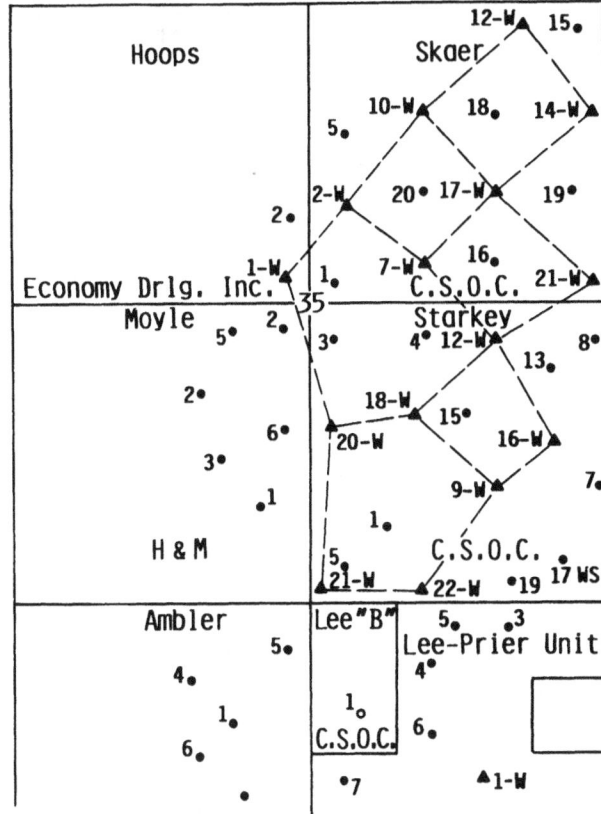

Fig. 6.3—Waterflood pattern developed with irregular spacing.[1]

Reservoir characteristics—such as gas cap, water drive, or a fault—may limit the options that can be considered. Edge or peripheral drives may be more appropriate than the pattern flood, particularly when there is structural dip as illustrated in Fig. 6.5.[2] A pronounced directional permeability trend can control the arrangement of flooding patterns. The original waterflood pattern for the North Burbank field was a five-spot with lines of injectors and producers oriented along the southwest/northeast direction.[3] It has east-west jointing or fracturing that leads to an effective permeability in the east-west direction that is five times the effective permeability in the north-south direction.[3] When this geological feature was recognized, the waterflood was developed as a line drive by drilling infill wells for injection, injecting water in east-west rows of wells, and producing alternate rows of production wells as shown in Fig. 6.6. Although the surface arrangement of wells in Fig. 6.6 is a five-spot, the subsurface fluid movement approximates a line drive because of directional permeability. A sound program that couples geological evaluation with engineering analysis often leads to better selection of flooding patterns and improved waterflooding performance.

6.5 Injection Rates

In Chaps. 4 and 5, we found that oil recovery correlates with the cumulative volume of water injected. Injection rate is a key economic variable in the evaluation of a waterflood. When a waterflood is conducted in an established area, there may be data or correlations based on operating experience. Typically, injection rates are correlated in terms of injectivities as barrels per day per acre foot, barrels per day per net foot of sand, or barrels per day per net foot per pounds per square inch. Specific values are dependent on reservoir rock properties, fluid/rock interactions, spacing, and available pressure drop. Comparable values would be expected under similar reservoir and operating conditions.

It is often possible to estimate injection rates from relatively simple equations when rates are not known. Two situations are of interest in waterflooding operations. If water injection is initiated before a mobile-gas saturation develops, the system may be treated as if it were liquid-filled. Another case is the depleted reservoir where a

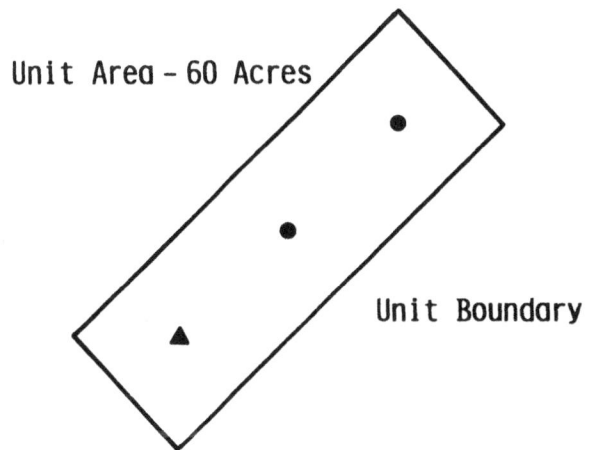

Fig. 6.4—Three-well flooding arrangement.

**TABLE 6.1—EXACT EXPRESSIONS FOR INJECTION RATES IN FULLY DEVELOPED PATTERNS
AT UNIT MOBILITY RATIO (Ref. 4 after Ref. 39)**

Direct Line Drive

$$i = \frac{3.541 \, kh(\Delta p)}{\mu\left(\ln \dfrac{a}{r_w} + 1.571\dfrac{d}{a} - 1.838\right)}$$

$$\frac{d}{a} \geq 1 \quad \dotfill \quad 6.1$$

Staggered Line Drive

$$i = \frac{3.541 \, kh(\Delta p)}{\mu\left(\ln \dfrac{a}{r_w} + 1.571\dfrac{d}{a} - 1.838\right)} \quad \dotfill \quad 6.2$$

Five-Spot

$$i = \frac{3.541 \, kh(\Delta p)}{\mu\left(\ln \dfrac{d}{r_w} - 0.619\right)} \quad \dotfill \quad 6.3$$

Seven-Spot

$$i = \frac{4.72 \, kh(\Delta p)}{\mu\left(\ln \dfrac{d}{r_w} - 0.569\right)} \quad \dotfill \quad 6.4$$

Nine-Spot

$$i = \frac{3.541 \, kh(\Delta p)_{i,c}}{\dfrac{1+R}{2+R}\left(\ln \dfrac{d}{r_w} - 0.272\right)\mu} \quad \dotfill \quad 6.5$$

$$i = \frac{7.082 \, kh(\Delta p)_{i,s}}{\left[\dfrac{3+R}{2+R}\left(\ln \dfrac{d}{r_w} - 0.272\right) - \dfrac{0.693}{2+R}\right]\mu} \quad \dotfill \quad 6.6$$

R = ratio of producing rate of corner well to side well,

$(\Delta p)_{i,c}$ = pressure difference between injection well and corner well, and

$(\Delta p)_{i,s}$ = pressure difference between injection well and side well.

*Units in these equations are barrels per day, darcies, feet, pounds per square inch, and centipoise.

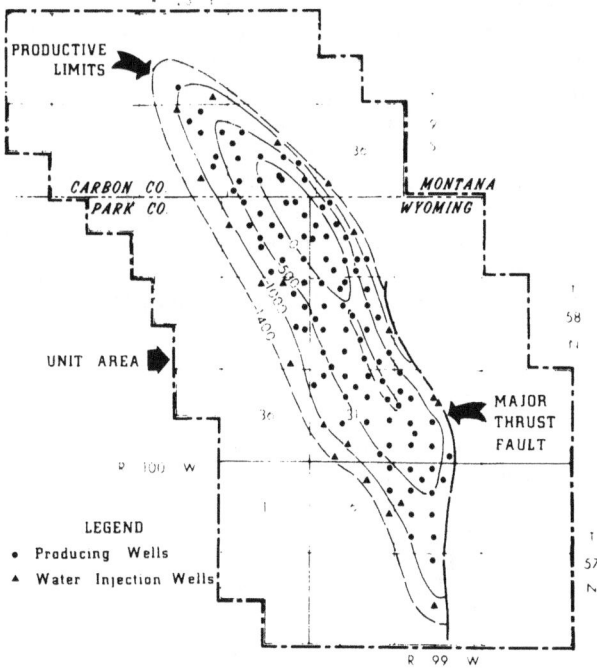

Fig. 6.5—Edge waterflood or peripheral flood.[2]

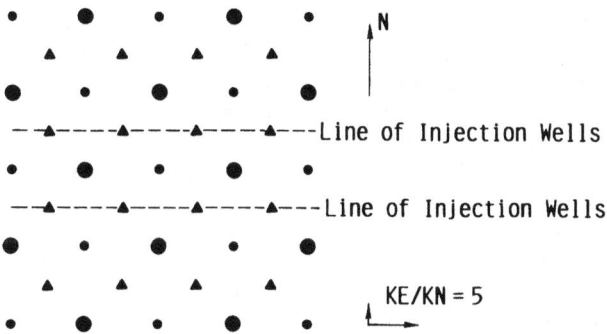

Fig. 6.6—Waterflood where subsurface flooding pattern is controlled by orientation of fractures that have an east-west trend. Original waterflood injection wells are large solid dots. Infill wells in the final flooding plan are solid triangles.[6]

mobile-gas saturation develops during primary production by solution-gas drive. In these reservoirs, initial injection rates decline rapidly as the mobile gas is displaced. Because controlling rates are those for liquid-filled systems, we begin our discussion with these systems.

6.5.1. Exact Expressions for Injection Rates in Liquid-Filled Patterns—$M=1$

Injection rates (and thus production rates) may be computed from analytical expressions for pattern floods when the mobility ratio is 1.0 and the fluids are incompressible. Table 6.1 summarizes the flooding patterns and corresponding equations.[4] The expressions for injection rate serve several purposes even if the mobility ratio is not 1.0. When gas saturation is immobile (or zero), the initial injection rate can be calculated with oil considered to be the only fluid flowing and with the oil mobility

$(k_o/\mu_o)S_{iw}$ in the injection-rate equation. Another limiting condition is the injection rate when areal sweep efficiency is 100%. At this point, water is the only fluid flowing, and the injection rate may be computed with the water mobility at residual oil $(k_w/\mu_w)S_{or}$ substituted into the injection-rate equation. Thus it is possible to estimate injection rates at 0 and 100% sweep for pattern floods. These are limiting rates for the displacement process irrespective of the mobility ratio of the actual flood.

If the mobility ratio is 1.0, the injection rate remains constant throughout the displacement. A mobility ratio less than 1.0 may be viewed as a displacement process where the injected fluid is less mobile than the displaced fluid. The injection rate decreases as the volume of the reservoir swept by the injected fluid increases. The lowest injection rate corresponds to the injection rate of water at residual oil saturation. An analogous situation exists when the mobility ratio is greater than 1.0. As the flood progresses, injection rates increase when the pressure drop is constant because the injected fluid has less resistance to flow (i.e., is more mobile) than the displaced fluid. Maximum injection rates occur when the entire pattern has been displaced to residual oil saturation.

Injection rates computed from Eqs. 6.1 through 6.6 assume that the reservoir has uniform properties. This is not usually true in the region around the wellbore. Permeability is often reduced from the migration of fines, precipitation of constituents in the oil (such as paraffins and asphaltenes), deposition of scale, the filtration of particulate matter from poor-quality inspection water, and swelling of clays caused by water sensitivity. The permeability of the wellbore may be increased by acidization or small fracture treatments.

Injection rates can be estimated when there is reduction or increase in permeability in the immediate area of the wellbore by introducing the skin factor into Eqs. 6.1 through 6.6. The skin model represents all permeability changes near the wellbore by a hypothetical zone of radius r_s that has a permeability of k_s where $r_s > r_w$. Both r_s and k_s are included in the skin factor, s, defined by Eq. 6.7:

$$s = \ln\left(\frac{r_s}{r_w}\right)\left(\frac{k}{k_s} - 1\right). \quad\ldots\ldots\ldots\ldots\ldots\ldots (6.7)$$

When the permeability of the wellbore changes, the skin factor is introduced into Eqs. 6.1 through 6.6 as illustrated in Eq. 6.8 for a five-spot pattern:

$$i = \frac{3.541kh\Delta p}{\mu\left(\ln\dfrac{d}{r_w} + \dfrac{s_i + s_p}{2} - 0.619\right)}, \quad\ldots\ldots\ldots (6.8)$$

where s_i is the skin factor for the injection well and s_p is the skin factor of the production well. The skin factor is positive when there is permeability reduction or wellbore damage and negative when the permeability of the region around the wellbore is larger than the average reservoir permeability. Methods of determining the skin factor from pressure buildup and falloff analyses are given in Refs. 5 and 6.

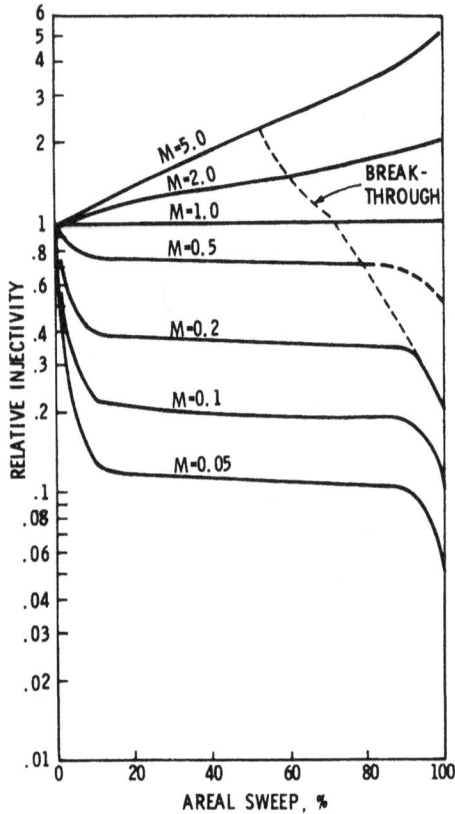

Fig. 6.7—Correlation of conductance ratio (relative injectivity) with areal sweep efficiencies at selected mobility ratios for miscible displacement in a five-spot pattern. [6]

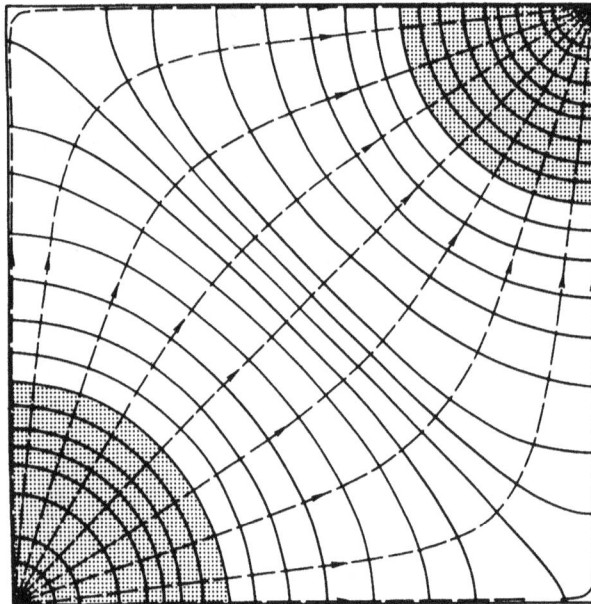

Fig. 6.8—Isopotentials and streamlines in a quadrant of homogeneous five-spot pattern.

Most floods do not have unit mobility ratios, and thus injection rates change with the area of the pattern that is swept. Fig. 6.7 shows a correlation[7] for the change in the conductance with area swept (E_A) and mobility ratio, M, for miscible displacement in a five-spot pattern from the data of Caudle and Witte[8] presented in Fig. 4.10. The conductance or relative injectivity defined by Eq. 6.9 is the ratio of the injection rate during miscible displacement to the injection rate computed from Eq. 6.3 (Table 6.1), assuming the displaced phase is flowing in the pattern as a single phase under the same pressure drop.

$$\gamma = \frac{i}{i_b}. \quad\dots\dots\dots\dots\dots\dots\dots\dots\dots\dots\dots (6.9)$$

Recall from Chap. 3 that the mobility ratio for a miscible displacement, M, process is as follows:

$$M = \frac{\left(\dfrac{k}{\mu}\right)_d}{\left(\dfrac{k}{\mu}\right)_D}$$

$$M = \frac{\mu_d}{\mu_D}.$$

Injection rates in a five-spot pattern are fairly constant after the displacing front moves away from the injection wellbore. The large change in conductance at small values of E_A is because much of the flow resistance is found in the region immediately around the injection well. A similar rapid change in injection rate is observed as E_A approaches 100% as a result of changes in the flow resistance at the producing well.

Immiscible displacement processes involve the flow of two phases as well as the presence of residual saturations of one or more phases in swept and unswept portions of the reservoir. In the next section, a general approach to estimate injection rates in an immiscible displacement process for uniform and nonuniform patterns is developed.

6.5.2 Injection Rates for Pattern Floods in Liquid-Filled Systems—$M \neq 1$

Injection rates can be estimated from models that approximate fluid flow in pattern floods. These models rely on characteristics of fluid movement near injection and production wells for homogeneous systems that can be seen in Fig. 6.8.[9] Fig. 6.8 shows isopotentials for a quadrant of a five-spot when the mobility ratio is 1.0—i.e., single-phase flow. Flow is radial in about 23% of the pattern area around the production and injection wells. About 90% of the potential drop (or pressure drop) occurs in this region. A similar situation exists for other flooding patterns—such as the direct line drive and nine-spot.

In a waterflood, a radial flood front forms when water is injected into a well, even when the mobility ratio is not unity. Injection rates change because the flow resistance behind the flood front either increases or decreases depending on the mobility of the fluids in the displaced region. Although the fraction of the pattern area where the fluid flow is radial varies with mobility ratio, the flow resistance for a significant portion of the flood occurs in the radial flow region.

When relative permeability curves and fluid properties fall within certain ranges, the displacement process is piston-like—that is, the breakthrough or flood-front saturation is essentially $1-S_{or}$, and there is little oil displacement in the swept area after the flood front arrives. Piston-like displacement occurs whenever a straight line can be drawn on the fractional flow curve from the point representing initial water saturation to $1-S_{or}$ without intersecting the curve. In piston displacement, the mobility ratio is given by Eq. 3.207.

$$M = \frac{\left(\dfrac{k_{rw}}{\mu_w}\right)_{S_{or}}}{\left(\dfrac{k_{ro}}{\mu_o}\right)_{S_{iw}}}$$

Data to compute this mobility ratio are usually obtained from flood pot tests because $(k_{rw})S_{or}$ and $(k_{ro})S_{iw}$ are the endpoints of the relative permeability curves.

The method for calculating injection rates approximates the pattern area with radial sections or a combination of radial and linear sections whose surface area is equal to the pattern area.[4] Injection rates in the pattern are estimated by combining radial and linear flow equations. This method is illustrated with the development of an approximate equation for the injection rate in a five-spot pattern for $M \neq 1$. Then, the method will be extended for use with other patterns—such as a line drive.

6.5.2.1. Approximate Model for Five-Spot.
Fig. 6.9 shows one-fourth of a five-spot approximated with two segments of a circle that have radii r_{ep} and r_{ei}, respectively. According to our model assumptions, the pattern area is equal to the area of the approximating flow segments. Thus, when $r_{ei} = r_{ep} = r_e$, then $r_e = d/\sqrt{\pi}$.

The fluid-flow model assumes that flow is steady, incompressible, and radial from the injection well to r_{ei}, the outer radius of the injection segment; then fluid flows radially from the outer radius of the production segment, r_{ep}, to the production well. The equivalent flow system is depicted in Fig. 6.10.* In the approximate system, the dotted areas are included in both injection and production segments, while the cross-hatched areas are excluded.

The location of the flood-front saturation, r_f, can be computed with material balance. Neglecting the radius of the wellbore,

$$W_i = (\pi r_f^2 \phi h)(\overline{S}_{wf} - S_{iw}). \quad \ldots \ldots \ldots \ldots \ldots (6.10)$$

Thus

$$r_f = \sqrt{\frac{W_i}{\pi \phi h(\overline{S}_{wf} - S_{iw})}}, \quad \ldots \ldots \ldots \ldots \ldots (6.11)$$

for $r_f < r_e$, when the displacement is piston-like, $\overline{S}_{wf} = 1 - S_{or}$. Then,

$$r_f = \sqrt{\frac{W_i}{\pi \phi h(S_{oi} - S_{or})}}, \quad \ldots \ldots \ldots \ldots \ldots (6.12)$$

when $S_{iw} = 1 - S_{oi}$.

*Personal communication with W.E. Brigham, Stanford U. (1980).

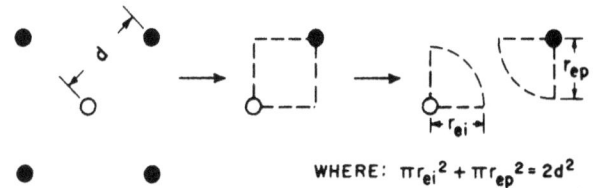

Fig. 6.9—Approximate model for fluid flow in a five-spot pattern.[4]

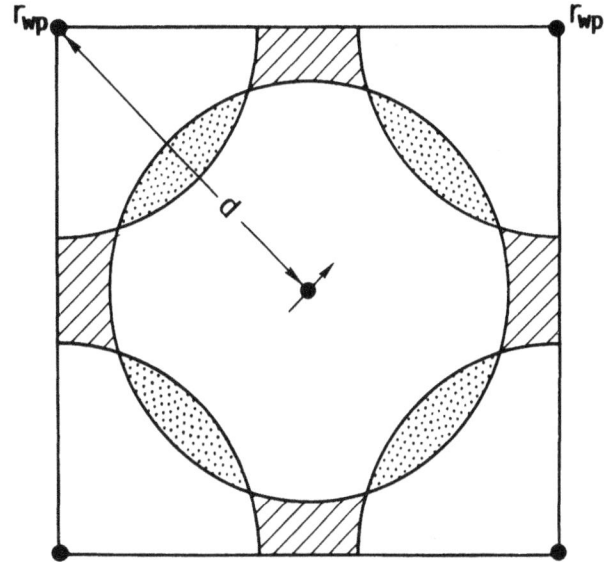

Fig. 6.10—Radial flow segments in approximating fluid flow in a five-spot pattern (after Brigham).

The expression for the flow rate is obtained by considering the system as resistances in series. For steady-state radial flow in a porous rock with radii r_1 and r_2,

$$p_1 - p_2 = \frac{i \ln \dfrac{r_2}{r_1}}{\left(\dfrac{k}{\mu}\right) 2\pi h}. \quad \ldots \ldots \ldots \ldots \ldots (6.13)$$

Thus

$$p_w - p_p = (p_w - p_f) + (p_f - p_e) + (p_e - p_p) \quad \ldots (6.14)$$

or

$$p_w - p_p = \frac{i \ln \dfrac{r_f}{r_w}}{\left(\dfrac{k_w}{\mu_w}\right)_{S_{or}} 2\pi h} + \frac{i \ln \dfrac{r_e}{r_f}}{\left(\dfrac{k_o}{\mu_o}\right)_{S_{iw}} 2\pi h}$$

$$+ \frac{i \ln \dfrac{r_e}{r_{wp}}}{\left(\dfrac{k_o}{\mu_o}\right)_{S_{iw}} 2\pi h}. \quad \ldots \ldots \ldots \ldots \ldots (6.15)$$

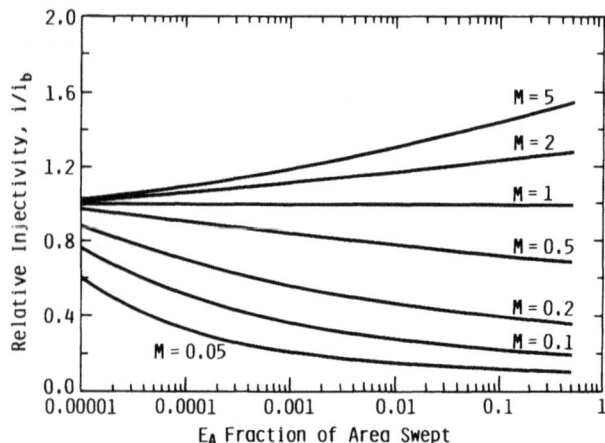

Fig. 6.11—Variation of conductance (i/i_b) with mobility ratio.

Solving for the injection rate,

$$i = \frac{2\pi h(p_w - p_p)}{\dfrac{\ln\dfrac{r_f}{r_w}}{\lambda_w} + \dfrac{\ln\dfrac{r_e}{r_f}}{\lambda_o} + \dfrac{\ln\dfrac{r_e}{r_{wp}}}{\lambda_o}}, \quad \dots \dots (6.16)$$

and

$$i = \frac{2\pi\lambda_o h(p_w - p_p)}{\dfrac{1}{M}\ln\dfrac{r_f}{r_w} + \ln\dfrac{r_e}{r_f} + \ln\dfrac{r_e}{r_{wp}}}. \quad \dots \dots \dots (6.17)$$

Eq. 6.17 may be compared with the exact solution (Eq. 6.3) at $M=1$ to check the effects of the assumptions used in developing the model. When $M=1$ and $r_w = r_p$, Eq. 6.17 becomes

$$i = \frac{2\pi\lambda_o(p_w - p_p)}{\ln\left(\dfrac{r_e^2}{r_w^2}\right)}. \quad \dots \dots \dots (6.18)$$

Because

$$r_e = \frac{d}{\sqrt{\pi}},$$

Eq. 6.18 may be written in terms of d as in Eq. 6.19:

$$i = \frac{\pi\lambda_o h(p_w - p_p)}{\ln\left(\dfrac{d}{r_w}\right) - 0.572}. \quad \dots \dots \dots (6.19)$$

If variables are expressed in oilfield units (pounds per square inch, centipoise, days, feet, barrels per day, darcies), the corresponding relationship is Eq. 6.20.

$$i = \frac{3.541\lambda_o h\Delta p}{\ln\left(\dfrac{d}{r_w}\right) - 0.572}. \quad \dots \dots \dots (6.20)$$

Eq. 6.20 differs from the exact solution by the presence of 0.572 in the denominator rather than 0.619. The log term dominates the denominator in Eq. 6.20 for all problems of practical interest. Thus the model is a close approximation to the exact solution.

It is useful to express the injection rate given by Eq. 6.20 in terms of a conductance to compare the effects of the mobility ratio on injection rates.[8] The conductance, γ, defined by Eq. 6.9 is the ratio of the injection rate during the flood to the injection rate if oil were injected into the formation at the same pressure drop.

$$\gamma = \frac{i}{i_b} = \frac{2\left[\ln\left(\dfrac{d}{r_w}\right) - 0.572\right]}{\dfrac{1}{M}\ln\dfrac{r_f}{r_w} + \ln\dfrac{r_e}{r_f} + \ln\dfrac{r_e}{r_w}}. \quad \dots \dots (6.21)$$

Because

$$E_A = \frac{\pi(r_f^2 - r_w^2)}{2d^2} \quad \dots \dots \dots (6.22)$$

and

$$r_f = d\sqrt{\frac{2E_A}{\pi}} \quad \dots \dots \dots \dots (6.23)$$

when $r_f \gg r_w$, and after rearranging Eq. 6.21,

$$\gamma = \frac{2\left[\ln\left(\dfrac{d}{r_w}\right) - 0.572\right]}{\dfrac{1}{M}\ln\left(\dfrac{d\sqrt{2E_A/\pi}}{r_w}\right) + \ln\left(\dfrac{d}{r_w\sqrt{2E_A/\pi}}\right)}$$

$$\dots \dots \dots \dots (6.24)$$

for $E_A \leq 0.5$. The conductance can be expressed also in terms of W_i with Eq. 6.10.

Fig. 6.11 illustrates the change in conductance with mobility ratio (M) and E_A for a waterflood in a 10-acre [40 469-m^2] five-spot, where $d=467$ ft [142 m] and $r_w = 0.5$ ft [0.15 m]. The computations show that most of the rate changes occur when the areal sweep is less than 0.1. Thereafter, the conductances remain relatively constant to $E_A = 0.5$—the limit of the model. Effects of mobility ratio on injection rates can be observed early in the waterflood. Thus injection rates based on short pilot tests or early waterflood performance can be misleading when projected to the entire flood.

6.5.2.2 Injection Rate Model for Staggered Line Drive in Liquid-Filled Systems—$M=1$.
Injection rates for other pattern floods may be estimated with the combination of radial and linear segments to approximate the pattern area. The derivation presented in this section is from Brigham.*

Fig. 6.12 shows a portion of a field developed with a staggered linedrive flooding pattern. The pattern area is outlined by the dashed line. Fluid flow is approximated

---

*Personal communication with W.E. Brigham, Stanford U. (1980).

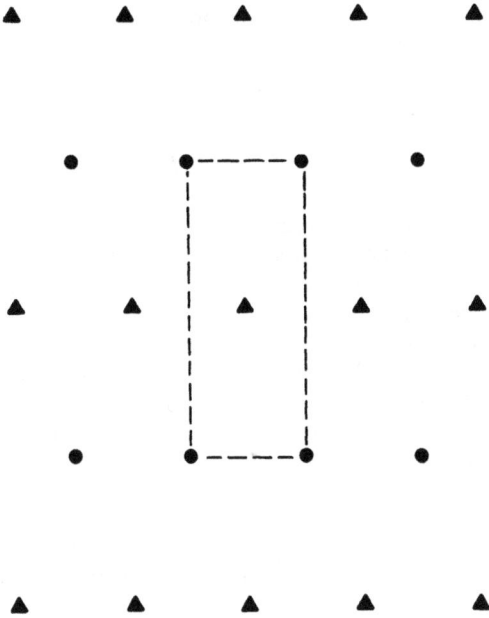

Fig. 6.12—Staggered linedrive model.

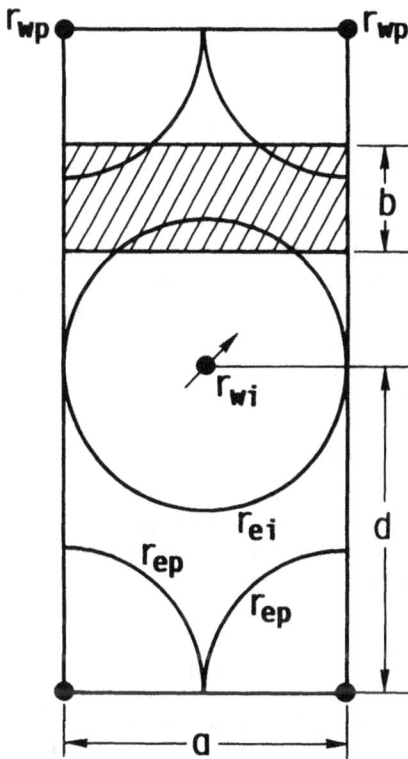

Fig. 6.14—Radial/linear flow segments in an approximate model of staggered linedrive pattern.

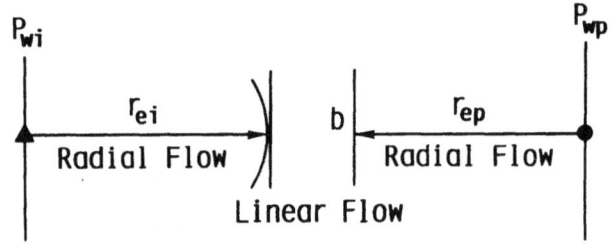

Fig. 6.13—Approximate model of a staggered linedrive pattern (after Brigham).

with the assumption that radial flow exists around injection and production wells in as much of the area as possible without overlapping, as shown in Fig. 6.13. The remainder of the pattern area (cross-hatched) is divided into two linear segments that have Width a and Length b. Length b is determined by equating the area of the model to the pattern area that is $2da$. The pattern area equals radial flow area around injection well plus linear flow area between injection and production wells plus radial flow area around production wells.

From Fig. 6.13,

$$2da = \pi r_{ei}^2 + 2(ab) + \pi r_{ep}^2. \qquad (6.25)$$

When rates are equal in each well, $r_{ei} = r_{ep} = a/2$. Substitution into Eq. 6.25 yields, after some manipulation,

$$b = d - \frac{\pi a}{4}. \qquad (6.26)$$

The equivalent one-dimensional (1D) fluid-flow model for one-half of the pattern is depicted in Fig. 6.14.

The total pressure drop $(p_{wi} - p_{wp})$ takes place over three separate flow segments. Thus

$$p_{wi} - p_{wp} = (p_{wi} - p_{ei}) + (p_{ei} - p_{ep}) + (p_{ep} - p_{wp}).$$

$$\qquad (6.27)$$

Substituting appropriate expressions for Δp into Eq. 6.27, we obtain

$$\frac{i^*\mu \ln\left(\frac{r_{ei}}{r_{wp}}\right)}{\pi k h} + \frac{i^*\mu b}{kha} + \frac{i^*\mu \ln\left(\frac{r_{ep}}{r_{wp}}\right)}{\pi k h}$$

$$= p_{wi} - p_{wp}. \qquad (6.28)$$

Solving for i^*, the injection rate for one-half the pattern, gives Eq. 6.29.

$$i^* = \frac{\pi k h (p_{wi} - p_{wp})}{\mu\left[\ln\left(\frac{r_{ei}}{r_{wi}}\right) + \frac{\pi b}{a} + \ln\left(\frac{r_{ep}}{r_{wp}}\right)\right]}. \qquad (6.29)$$

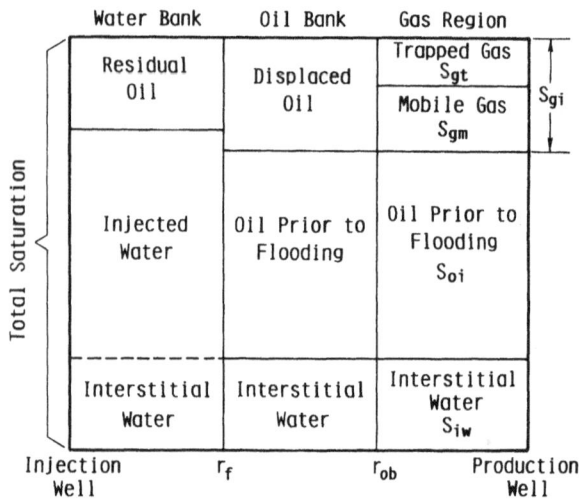

Fig. 6.15—Saturation distribution during the waterflood of a depleted reservoir when trapped gas redissolves (after Prats et al. [10]).

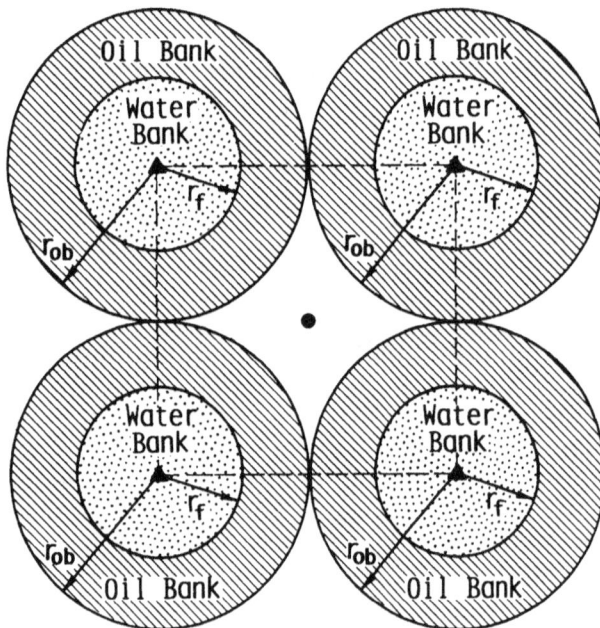

Fig. 6.16—Interference of oil banks in the waterflood of a five-spot pattern that has a uniform initial gas saturation (after Prats et al. [10]).

With the substitutions $i = 2i^*$, $r_w = r_{wi} = r_{wp}$, and $r_{ei} = r_{ep} = a/2$, and with some rearrangement, the expression for the injection rate for the full pattern is Eq. 6.30.

$$i = \frac{\pi k h(p_{wi} - p_{wp})}{\mu \left(\ln \dfrac{a}{r_w} + 1.571 \dfrac{d}{a} - 1.927 \right)} . \qquad \ldots \ldots \ldots (6.30)$$

In oilfield units, Eq. 6.30 is multiplied by 3.535. The exact solution for the injection rate [4] is

$$i = \frac{3.541 k h(p_{wi} - p_{wp})}{\mu \left(\ln \dfrac{a}{r_w} + 1.571 \dfrac{d}{a} - 1.838 \right)} . \qquad \ldots \ldots (6.31)$$

The difference between the approximate and exact expressions for the injection rate is 0.089 in the denominator. This amount is small and has little effect on computed injection rates.

As in the case of the five-spot pattern, an approximate fluid-flow model properly accounts for flow geometry when $M = 1$. Now we extend the model to include mobility ratios other than 1. For piston-like displacement, the injection rate is given by Eq. 6.32:

$$i = \frac{3.541 \lambda_o h(p_{wi} - p_{wp})}{\left(\dfrac{1}{M} - 1 \right) \ln \left(\sqrt{\dfrac{r_f}{r_w}} \right) + \ln \dfrac{a}{r_w} + 1.571 \dfrac{d}{a} - 1.927} ,$$

$$\ldots \ldots \ldots \ldots \ldots \ldots \ldots \ldots \ldots \ldots \ldots \ldots \ldots (6.32)$$

where $r_f \leq a/2$, and

$$r_f = \sqrt{\frac{2 d a E_A}{\pi}} .$$

In this section it has been shown how relatively simple fluid-flow models can be developed to estimate injection rates in pattern floods. The agreement between approximate models and the exact equations for unit mobility ratio is excellent for five-spot and staggered linedrive patterns. Problem 6.7 illustrates the development of similar equations for the nine-spot pattern. The agreement between approximate and exact equations suggests that approximate models for many types of patterns can be constructed with a combination of radial and linear flow segments.

6.5.3 Depleted Reservoirs

Waterfloods are frequently initiated after some of the oil has been produced by solution-gas drive. A mobile-gas saturation is present when water injection begins. In Sec. 3.9 a displacement model for waterflooding an oil reservoir that has an initial gas saturation was introduced. Those concepts can be incorporated into the approximate fluid-flow models to estimate injection rates.

When there is an initial gas saturation and sufficient oil saturation for an oil bank to form (Sec. 3.9), the displacement process is represented by Fig. 6.15. [10] The oil bank displaces the mobile gas, leaving a trapped-gas saturation, S_{gt}. At usual waterflood pressures, the trapped-gas saturation redissolves in the oil. During the fill-up period, mobile gas is displaced by the oil bank. Gas mobility is large in the region ahead of the oil bank. There is little pressure drop between the oil bank and the producing well; consequently, oil production is negligible.

Fluid movement in the oil and water banks during the fill-up period is radial until the oil banks from adjacent patterns meet or "interfere" with each other. Fig. 6.16 illustrates the position of the oil bank in a five-spot pattern at interference. [10]

Eq. 6.33 describes the injection rate during fill-up to the interference point for a five-spot when the flow resistance is neglected ahead of the oil bank.

$$i = \frac{2\pi\lambda_o(p_{wi} - p_{wp})}{\frac{1}{M}\ln\left(\frac{r_f}{r_w}\right) + \ln\left(\frac{r_{ob}}{r_f}\right)} , \quad \ldots\ldots\ldots (6.33)$$

where

r_{ob} = radius of the oil bank, $r_w \le r_{ob} \le d/\sqrt{2}$, and

r_f = radius of the flood-front saturation.

Both r_{ob} and r_f may be defined by a material balance on the injected water.

$$W_i = \pi(r_f^2 - r_w^2)(\bar{S}_w - S_{iw})h\phi. \quad \ldots\ldots\ldots (6.34)$$

Solving for r_f gives

$$r_f = \sqrt{\frac{W_i}{\pi\phi h(\bar{S}_w - S_{iw})} + r_w^2} . \quad \ldots\ldots\ldots (6.35)$$

The volume of water injected to fill-up is equal to the volume of gas displaced by the oil bank as the initial gas saturation, S_{gi}, is reduced to the trapped-gas saturation, S_{gt}. A material balance yields Eq. 6.36 for r_{ob}:

$$r_{ob} = \sqrt{\frac{W_i}{\pi\phi h(S_{gi} - S_{gt})} + r_w^2} . \quad \ldots\ldots\ldots (6.36)$$

At interference, $r_{ob} = d/\sqrt{2}$. Eq. 6.36 can be used to compute the volume of water required to reach interference. If the injection rate is constant, the interference time can be computed from W_i. Neglecting the wellbore radius,

$$W_{ii} = \frac{\pi d^2}{2}\phi h(S_{gi} - S_{gt}). \quad \ldots\ldots\ldots (6.37)$$

Usually, fill-up occurs in a relatively short time after interference. The volume of water injected at fill-up is given by Eq. 6.38.

$$W_{if} = 2d^2\phi h S_{gi}, \quad \ldots\ldots\ldots (6.38)$$

when the trapped gas is redissolved in the oil with negligible change in volume. At fill-up, r_f is obtained by substituting Eq. 6.38 into Eq. 6.35 to obtain

$$r_f = \sqrt{\frac{2d^2 S_{gi}}{\pi(\bar{S}_w - S_{iw})} + r_w^2} . \quad \ldots\ldots\ldots (6.39)$$

If $r_f < d$ for a five-spot, the injection rate after fill-up is given by Eqs. 6.19 and 6.20. Example 6.1 illustrates the estimation of injection rate in a 5-acre [20 235-m²] five-spot where the initial gas saturation is 0.10 and the trapped-gas saturation is zero.

Example 6.1

An estimate of the injection rate is needed for a waterflood in a 5-acre [20 235-m²] five-spot pattern. The reservoir has been depleted by solution-gas drive and has an initial gas saturation of 0.10. From laboratory correlations (Figs. 2.56 and 2.57), the gas saturation trapped by the oil bank is estimated to be 0.04. Flood pressure is expected to cause the trapped gas to redissolve in the oil, leaving the residual gas saturation equal to zero. Reservoir and fluid/rock properties are given in Table 6.2. A pressure difference of 330 psi [2.3 MPa] is maintained between the injection well and the production well.

Solution. Injection rates can be estimated with the approximate model until interference, $(r_{ob} \le d/\sqrt{2})$, where d is the distance between the injection well and the production well. When $r_{ob} \ge d/\sqrt{2}$, flow ceases to be radial and the oil-bank flood front moves at different velocities toward the production well.

Because the point of interference determines the last time when the approximate model is valid, computations begin there. The distance between the injection well and the production well is computed as follows.

$$d = \sqrt{\frac{(5 \text{ acres})(43,560 \text{ sq ft/acre})}{2}}$$

$$= 330 \text{ ft.}$$

The interference occurs when $r_{ob} = d/\sqrt{2}$ or 233.33 ft [71.12 m]. From Eq. 6.37, the volume of water injected at interference is

$$W_{ii} = \frac{\pi}{2}(330 \text{ ft})^2(0.15)(15 \text{ ft})(0.10)$$

$$= 38,488 \text{ cu ft}$$

$$= 6,855 \text{ bbl.}$$

Eq. 6.33 is used to compute the injection rate for $0 \le W_i \le 6,855$ bbl [1090 m³]. We assume $S_w = 1.0 - S_{or} = 0.7$. From Eq. 6.35,

$$r_f = \sqrt{\frac{(6,855)(5.615)}{\pi(0.15)(15.0)(0.70 - 0.27)} + (0.5)^2}$$

$$= 112.5 \text{ ft.}$$

TABLE 6.2—RESERVOIR AND FLUID/ROCK PROPERTIES

Pattern area, acres	5
Thickness, ft	15
Radius of injection and production wells, ft	0.5
S_{iw}	0.27
S_{or}	0.30
Porosity	0.15
Permeability to liquid (base permeability), darcies	0.203
k_{ro} at S_{iw}	0.7
k_{rw} (at S_{or})	0.15
Viscosity of oil, cp	2.0
Viscosity of water, cp	1.0

In oilfield units (darcies, centipoise, feet, pounds per square inch, and barrels per day), Eq. 6.33 is

$$i = \frac{7.082k_b \left(\frac{k_{ro}}{\mu_o}\right) h(p_{wi} - p_{wp})}{\frac{1}{M} \ln \frac{r_f}{r_w} + \ln \frac{r_{ob}}{r_f}} \quad \ldots \ldots (6.40)$$

The mobility ratio, M, is computed from the endpoints of the relative permeability curves.

$$M = \frac{\left(\frac{k_{rw}}{\mu_w}\right)_{S_{or}}}{\left(\frac{k_{ro}}{\mu_o}\right)_{S_{iw}}}$$

$$= \frac{\left(\frac{0.15}{1.0}\right)}{\left(\frac{0.7}{2.0}\right)}$$

$$= 0.43.$$

Substituting into Eq. 6.40 gives i at interference:

$$i = \frac{(7.082)(0.203)\left(\frac{0.7}{2.0}\right)(15)(330)}{\frac{1}{0.43}\ln\frac{112.5}{0.5} + \ln\frac{233.3}{112.5}}$$

$$= \frac{2,491}{13.32}$$

$$= 186.9 \text{ B/D.}$$

Injection rates from the beginning of injection to interference can be computed easily. Table 6.3 contains injection rates for 10 equal increments of W_i, $0 \le W_i \le W_{ii}$.

The initial injection rate is quite high (2,300 B/D [366 m³/d]) but declines rapidly with W_i as the space occupied by gas is filled with water and oil banks. High initial injection rates that decline rapidly (or lower rates at small pressure drops) should be expected in a waterflood of a depleted reservoir.

TABLE 6.3—INJECTION RATES TO INTERFERENCE

W_i (bbl)	r_f (ft)	r_{ob} (ft)	i (B/D)	Time (days)
0.1	0.66	1.02	2,298.4	0.00
686.5	36.59	73.79	233.2	0.54
1,370.9	50.33	104.36	216.8	3.59
2,056.4	61.64	127.81	208.2	5.81
2,741.8	71.17	147.58	202.5	10.15
3,427.3	79.57	166.00	198.3	13.57
4,112.7	87.17	180.75	196.0	17.06
4,798.2	94.15	196.23	192.3	20.60
5,483.7	100.65	208.71	190.0	24.18
6,169.1	106.76	221.37	188.1	27.81
6,854.6	112.53	233.35	186.3	31.47

Continued injection of water leads to fill-up of the gas space. The volume of water injected to fill-up is obtained from Eq. 6.38.

$$W_{if} = \frac{(2)(330)^2(0.15)(15)(0.10)}{5.615}$$

$$= 8,728 \text{ bbl.}$$

At fill-up, the radius of the water bank is given by Eq. 6.35.

$$r_f = \sqrt{\frac{(8,728)(5.615)}{\pi(0.15)(15)(0.70 - 0.27)} + (0.5)^2}$$

$$= 127 \text{ ft.}$$

The entire pattern is 100% liquid-saturated so that Eq. 6.17 now describes the injection rate from fill-up to the radius where the assumptions made in developing Eq. 6.17 are not valid—that is, $r_f = d/\sqrt{2}$ is the maximum value of r_f where Eq. 6.17 is valid. Thus

$$r_f = \frac{330}{\sqrt{\pi}}$$

$$= 186.2 \text{ ft.}$$

Solving for W_i with Eq. 6.34 gives

$$W_i = \frac{\pi(186.2)^2(0.70 - 0.27)(15)(0.15)}{5.615}$$

$$= 18,768 \text{ bbl.}$$

Therefore, the injection rate for $8,728 \le W_i \le 18,768$ may be estimated from Eq. 6.17.

Applying Eq. 6.17 with $r_e = 186.2$ ft [56.8 m] and $r_f = 127.0$ ft [39 m] gives the injection rate at the instant fill-up occurs. Thus

$$i_f = \frac{(7.082)(0.203)(0.7/2.0)(15)(330)}{\frac{1}{0.43}\ln\left(\frac{127.0}{0.5}\right) + \ln\left(\frac{186.2}{127.0}\right) + \ln\left(\frac{186.2}{0.5}\right)}$$

$$= \frac{2,491}{19.18}$$

$$= 129.9 \text{ B/D.}$$

Between interference and fill-up, the injection rate decreased from 186.9 to 129.9 B/D [29.7 to 20.6 m³/d] as W_i changed from 6,855 to 8,728 bbl [1090 to 1388 m³]. The time between interference and fill-up corresponds to the injection of 1,873 bbl [298 m³] of water at a declining rate. Little error would accrue if an average rate for this interval equal to (186.9+129.9)/2 or 158.4 B/D [25.2 m³/d] were assumed.

As noted previously, the approximate model for the liquid-filled system applies for $r_f \leq d/\sqrt{\pi}$. At $r_f = d/\sqrt{\pi}$,

$$i = \cfrac{2{,}491}{\cfrac{1}{0.43} \ln \cfrac{186.2}{0.5} + \ln \cfrac{186.2}{0.5}}$$

$$= 126.5 \text{ B/D.}$$

The injection rate remains essentially constant between fill-up and $W_i = 18{,}768$ B/D [2984 m^3/d]. Fig. 6.17 illustrates the variation of injection rate with cumulative water injected for $0 \leq W_i \leq 18{,}768$. The three intervals (interference, fill-up, and model limits) are indicated on the figure.

The injection-rate decline in the early stages of injection is caused by two mechanisms. The high mobility of the gas phase has been discussed. The injected water has a low mobility compared with oil ($M=0.43$). In this example, the injection rate would decline rapidly from the beginning of injection because the oil bank is displaced by water, which has a lower mobility. Because mobility in the immediate vicinity of the wellbore dominates fluid flow, the change in injection rate would be significant. The dashed curve on Fig. 6.17 shows the injection rate if the initial gas saturation were zero.

6.5.4 Incorporation of Two-Phase Flow Behind the Flood Front in Injection-Rate Models

Most waterfloods do not behave as the piston-like displacements assumed in deriving the injection-rate models in Secs. 6.5.2 and 6.5.3. The water saturation behind the flood front is less than $1-S_{or}$. Both water and oil flow in the swept area and the resistance to flow should include the contributions of each phase.

In this section, we show how two-phase flow in the swept region can be incorporated into the injection-rate models developed in Secs. 6.5.2 and 6.5.3. Then, a range of variables to establish when the assumption of piston-like displacement is an acceptable approximation is examined. The frontal advance equation can be used to determine saturation distributions for the portion of the waterflood that is radial.

6.5.4.1 Injection Rate for Radial Waterflood With Two-Phase Flow Behind the Flood Front. The total flow rate (per unit thickness) in the waterflooded region at any radius r where $r_w < r < r_f$ is given by Eqs. 6.41 and 6.42.

$$q_t = -2\pi r k(\lambda_{ro} + \lambda_{rw})\frac{dp}{dr} \quad \ldots\ldots\ldots\ldots (6.41)$$

and

$$q_t = -2\pi r k \lambda_r \frac{dp}{dr}. \quad \ldots\ldots\ldots\ldots\ldots (6.42)$$

Because fluids are incompressible, q_t is constant at every r. Eq. 6.42 can be integrated from r_w to r_f to obtain an

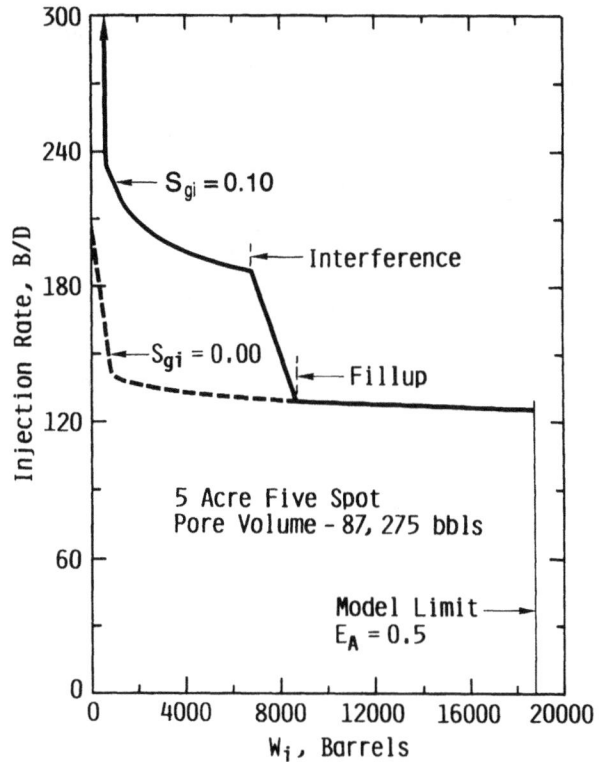

Fig. 6.17—Injection rate during waterflood of a five-spot pattern that has a mobile gas saturation.

expression for the pressure drop across the waterflooded region.

$$p_w - p_f = \frac{q_t}{2\pi k} \int_{r_w}^{r_f} \lambda_r^{-1} dr, \quad \ldots\ldots\ldots\ldots (6.43)$$

where r_f is the radius of the flood-front saturation and p_f is the pressure at r_f.

To evaluate Eq. 6.43, it is necessary to know the relationship between λ_r^{-1} and r.

Expressions necessary for these computations will be developed from the frontal-advance solution in Sec. 6.5.4.2.

An average apparent viscosity for a radial system is defined in Eq. 6.44.

$$\overline{\lambda_{rf}^{-1}} = \cfrac{\displaystyle\int_{r_w}^{r_f} \lambda_r^{-1} dr}{\displaystyle\int_{r_w}^{r_f} \cfrac{dr}{r}}. \quad \ldots\ldots\ldots\ldots (6.44)$$

Substituting Eq. 6.44 into Eq. 6.43, we obtain

$$p_w - p_f = \frac{q_t}{2\pi k} \overline{\lambda_{rf}^{-1}} \ln \frac{r_f}{r_w}. \quad \ldots\ldots\ldots (6.45)$$

When Eq. 6.45 is substituted into Eq. 6.14, the following expression is obtained for the approximate injection rate in a five-spot.

$$i = \frac{2\pi kh(p_w - p_p)}{\overline{\lambda_{rf}^{-1}} \ln \dfrac{r_f}{r_w} + \dfrac{\ln \dfrac{r_e}{r_f}}{\lambda_{ro}} + \dfrac{\ln \dfrac{r_e}{r_w}}{\lambda_{ro}}} . \quad \dots (6.46)$$

The total mobility ratio, M_t, is defined in Eq. 6.47.

$$M_t = \frac{\overline{(\lambda_{ro} + \lambda_{rw})}}{(\lambda_{ro} + \lambda_{rw})_{S_{iw}}}, \quad \dots\dots\dots\dots (6.47)$$

where $\overline{\lambda_{ro} + \lambda_{rw}}$ is behind the flood front.

$$M_t = \frac{1}{\overline{\lambda_{rf}^{-1}} \lambda_{ro}}. \quad \dots\dots\dots\dots\dots (6.48)$$

Then Eq. 6.46 becomes

$$i = \frac{2\pi h \lambda_{ro} k(p_w - p_p)}{\dfrac{1}{M_t} \ln \dfrac{r_t}{r_w} + \ln \dfrac{r_e}{r_f} + \ln \dfrac{r_e}{r_w}}, \quad \dots\dots\dots (6.49)$$

which is identical to Eq. 6.17 when M is replaced by M_t. Eq. 6.49 includes the correct flow resistance for two-phase flow in the waterflooded region.

6.5.4.2 Evaluation of Average Mobility for a Radial System.
To compute the average mobility from Eqs. 6.44 and 6.48, we use the frontal-advance solution in radial coordinates to find the saturation at each r for $r_w < r < r_f$. When the saturation distribution is known, λ_{ro} and λ_{rw} are determined and the integral in Eq. 6.43 or 6.44 can be computed easily with numerical or graphical methods. This procedure is illustrated in the following section.

The frontal-advance solution was developed in Chap. 3 for a linear waterflood. An analogous solution in radial coordinates can be derived following the same approach. The derivation of the radial form of the frontal-advance solution is given as Problem 3.18. Equations derived from that problem are used here.

The radial form of the frontal-advance equation is Eq. 6.50.

$$\left[\frac{d(r^2)}{dt}\right]_{S_w} = \frac{q_t}{\pi \phi h} \left(\frac{\partial f_w}{\partial S_w}\right)_t . \quad \dots\dots\dots (6.50)$$

The location of a saturation r_{Sw} is found by integrating Eq. 6.52 between r_w and r_{Sw} and 0 to t to obtain Eq. 6.51.

$$r_{Sw}^2 - r_w^2 = \frac{f'_{Sw}}{\pi \phi h} \int_0^t q_t dt \quad \dots\dots\dots\dots (6.51)$$

or

$$r_{Sw}^2 = r_w^2 + \frac{W_i f'_{Sw}}{\pi \phi h}. \quad \dots\dots\dots\dots (6.52)$$

When W_i is constant, Eqs. 6.51 and 6.52 give the location of each saturation. As in a linear waterflood, saturations range from the flood-front saturation, S_{wf}, to $S_w = 1 - S_{or}$. The flood-front saturation, S_{wf}, is identical for linear and radial systems and is obtained as described in Chap. 3 for the linear case. It can also be shown that the average saturation behind the flood front, $\overline{S_{wf}}$, is constant and can be computed from Eq. 6.53.

$$\overline{S_{wf}} = S_{wf} + \frac{1 - f_{Swf}}{f'_{Swf}}. \quad \dots\dots\dots\dots (6.53)$$

$\overline{S_{wf}}$ for a radial waterflood is the same as $\overline{S_{wf}}$ for a linear waterflood for the frontal-advance solution.

Example 6.2 illustrates the use of the frontal-advance equation to compute injection rates for a five-spot pattern.

Example 6.2
This example illustrates the computation of injection rates in a five-spot pattern with the frontal-advance solution to determine the average mobility in the flooded region. For this example, the parameters are given in Table 6.4.

Relative permeability relationships are given by Eqs. 6.54 and 6.55.

$$k_{ro} = (1 - S_{wD})^2 \quad \dots\dots\dots\dots (6.54)$$

and

$$k_{rw} = 0.1 S_{wD}^2, \quad \dots\dots\dots\dots (6.55)$$

where

$$S_{wD} = \frac{S_w - S_{iw}}{1 - S_{or} - S_{iw}}.$$

Note that S_w is a dimensionless water saturation that varies from 0 at S_{iw} to 1.0 at $S_w = 1.0 - S_{or}$. At S_{iw}, $k_{ro} = 1.0$, while at $S_w = 1 - S_{or}$, $k_{rw} = 0.1$. The base

TABLE 6.4—RESERVOIR ROCK AND FLUID PROPERTIES

Pattern area, acres	10
Wellbore radius, ft	0.5
Thickness, ft	1.0
Water viscosity, cp	1.0
Oil viscosity, cp	30.0
S_{iw}	0.3
S_{or}	0.25

permeability for the relative permeability relationships is the permeability to oil at interstitial water saturation. The mobility ratio based on the endpoints of the relative permeability curves is 3.0 for Example 6.2.

The approximate model is valid for $E_A < 0.5$, which is equivalent to $r_f < r_e$. For the 10-acre [40 469-m^2] five-spot pattern,

$$d = \sqrt{\frac{10 \text{ acres}(43,560 \text{ sq ft/acre})}{2}} = 466.7 \text{ ft.}$$

From the definition of r_e for a five-spot,

$$r_e = \frac{d}{\sqrt{\pi}}$$

$$= 263.3 \text{ ft.}$$

Injection rates were computed at 20 different flood-front positions by subdividing the interval between r_w and r_e into equal increments of ln r. For each r_f (i.e., r_{Swf}), the value of W_i is computed from Eq. 6.56.

$$W_i = \frac{(r_f^2 - r_w^2)\phi\pi h}{f'_{Swf}}. \qquad (6.56)$$

The integral that must be evaluated to compute the average apparent viscosity is given in Eq. 6.57.

$$\overline{\lambda_{rf}^{-1}} = \frac{\displaystyle\int_{r_w}^{r_f} \lambda_r^{-1} \frac{dr}{r}}{\ln\left(\dfrac{r_f}{r_w}\right)}. \qquad (6.57)$$

At any W_i, Eq. 6.56 relates r to S_w through f'_{Sw} by use of Eq. 6.52. Thus λ_r^{-1} can be evaluated for each value of r, $r_w \le r \le r_f$. For computations, it is convenient to subdivide the interval from $S_w = 1 - S_{or}$ to S_{wf} into 50 increments. At each r_f, there are 50 values of r between r_w and r_f where λ_r^{-1} is known. Numerical integration of Eq. 6.57 can be done by any suitable technique. The trapezoidal rule was used to obtain the results in this example.

Table 6.5 contains values of E_A, $\overline{\lambda_{rf}^{-1}}$, M_t, and i/i_b computed for 20 values of r_f between r_w and r_e.

The computed results reveal two important features of a waterflood in any pattern. First, the total mobility ratio, M_t, varies with areal sweep. Second, the largest changes occur in a relatively small area around the wellbore.

6.5.4.3 Evaluating the Assumption of Piston-Like Displacement.
In this section an investigation is made of when the total mobility ratio, M_t, should be used to estimate injection rates. Relative permeability data are required for the computation of M_t. For the purposes of this discussion, assume that Eqs. 3.143 and 3.144 describe the relative permeability relationship in terms of the dimensionless water saturation, S_{wD}.

$$k_{ro} = \alpha_1(1 - S_{wD})^m$$

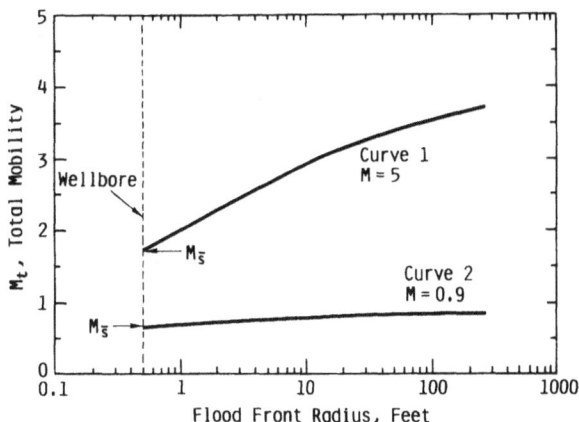

Fig. 6.18—Variation of the total mobility (M_t) with flood-front radius.

and

$$k_{rw} = \alpha_2 S_{wD}{}^n,$$

where

$$S_{wD} = \frac{S_w - S_{iw}}{1 - S_{or} - S_{iw}}.$$

The mobility ratio based on the endpoints of the relative permeability curves is given by Eq. 6.58:

$$M = \left(\frac{\alpha_2}{\alpha_1}\right)\left(\frac{\mu_o}{\mu_w}\right). \qquad (6.58)$$

Fig. 6.18 illustrates the change of M_t with r_f for $M = 0.9$ and $M = 5$. These curves were computed with values of $\alpha_1 = 1.0$, $\alpha_2 = 0.1$, $n = 2$, $m = 2$, and $r_w = 0.5$

TABLE 6.5—INJECTION RATES BASED ON AVERAGE APPARENT VISCOSITY FLOOD FRONT POSITION COMPUTED FROM RADIAL FRONTAL ADVANCE SOLUTION

r_f (ft)	E_A	$\overline{\lambda_{rf}^{-1}}$ (cp)	M	i/i_b
0.68398	0.000	20.536	1.461	1.008
0.93566	0.000	19.549	1.535	1.018
1.27995	0.000	18.595	1.613	1.029
1.75092	0.000	17.703	1.695	1.043
2.39520	0.000	16.889	1.776	1.058
3.27654	0.000	16.165	1.856	1.074
4.48218	0.000	15.531	1.932	1.092
5.13145	0.000	14.982	2.002	1.111
8.38760	0.001	14.510	2.067	1.132
11.47392	0.001	14.108	2.127	1.153
15.69589	0.002	13.763	2.180	1.175
21.47140	0.003	13.468	2.227	1.198
29.37206	0.006	13.214	2.270	1.222
40.17988	0.012	12.995	2.309	1.248
54.96457	0.022	12.803	2.343	1.274
75.18947	0.041	12.635	2.374	1.301
102.85638	0.076	12.486	2.403	1.330
140.70366	0.143	12.354	2.428	1.360
192.47734	0.267	12.235	2.452	1.391
263.30177	0.500	12.129	2.473	1.424

TABLE 6.6—TOTAL MOBILITY RATIO (M_t) AND M_S COMPUTED AT ASSUMED VALUES OF M WITH $m=2$ AND $n=2$ IN RELATIVE-PERMEABILITY CORRELATIONS

M	M_t	M_S
0.1	0.1	0.095
0.3	0.293	0.26
0.5	0.482	0.40
0.7	0.666	0.53
0.9	0.844	0.64
1.0	0.932	0.69
3.0	2.470	1.3
5.0	3.71	1.7
10.0	5.96	2.2

TABLE 6.7—VARIATION OF CONDUCTANCE (i/i_b) WITH OIL RELATIVE-PERMEABILITY CURVES (m) AT $M=1.0$

	M_t	M_S	Difference (%)
$m=1$	1.000	1.000	0.0
$m=2$	0.965	0.932	3.5
$m=3$	0.900	0.818	10.0
$m=4$	0.824	0.701	17.5
$m=5$	0.750	0.600	26.0

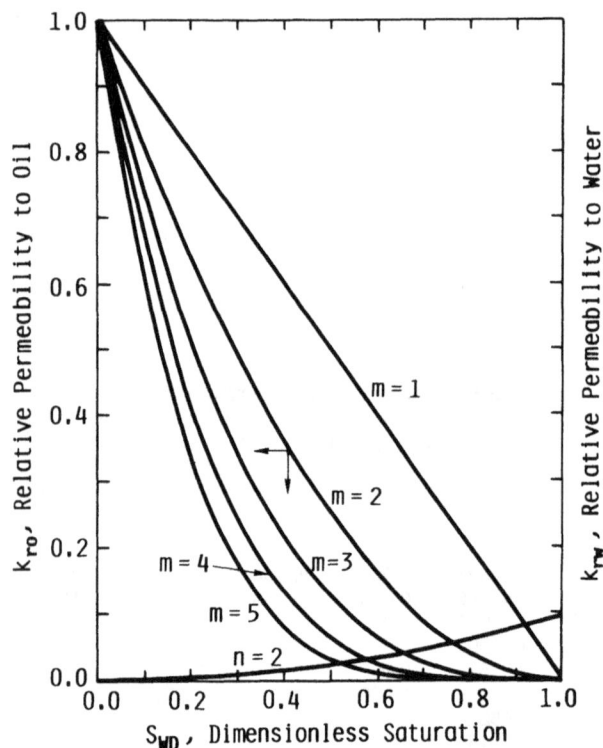

Fig. 6.19—Relative permeability curves for different values of the exponent (m) of the oil permeability correlation.

ft [0.15 m]. Curve 1 in Fig. 6.18 shows the variation of M_t with r_f for $M=5.0$, which corresponds to an oil/water viscosity ratio of 50 in Eq. 6.58. Total mobility increases by about 70% for $r_f < 20$ ft [6 m]. The change in M_t, however, is about 23% for $20 < r_f < 250$ ft [76 m]. Curve 2 shows the variation of M_t with r_f for $M=0.9$ ($\mu_o/\mu_w = 9.0$). The total mobility is about 0.68 at $r_f = 0.7$ ft [0.2 m] and increases to 0.8 when $r_f = 11.5$ ft [3.5 m]. There is little change in M_t for $r_f > 11.5$ ft [3.5 m]. Also indicated on Fig. 6.18 are values of $M_{\bar{S}}$, the mobility ratio based on the average water saturation behind the flood front computed from Eq. 3.208.

$$M_{\bar{S}} = \frac{\left(\dfrac{k_{rw}}{\mu_w}\right)_{\bar{S}_{wf}}}{\left(\dfrac{k_{ro}}{\mu_o}\right)_{S_{iw}}}.$$

This mobility ratio is a good approximation for M_t at the beginning of injection.

The examples in Fig. 6.18 lead to two observations. First, the radii where the total mobility varies appreciably are small for all mobility ratios. Thus the total mobility ratio can be considered constant in the radial flow segment of approximate models. Second, the value of M_t may be quite different from M or $M_{\bar{S}}$. In general, $M_t < M$, as seen in Table 6.6 where values of M_t computed at $r_f = 263.3$ ft [80.3 m] are compared.

For this set of relative permeability curves, there is good agreement between M_t and M for $M < 1.0$. M could be used to compute injection rates with little difference between those obtained with M_t. Large differences are found at $M = 10.0$. Included in Table 6.6 are values of $M_{\bar{S}}$. This mobility ratio accounts for the radial flow geometry only when $M < 0.5$.

The agreement of M_t and M for $M < 1.0$ reflects the particular set of parameters used in the relative permeability correlations. Table 6.7 shows M_t when $M = 1.0$ for five different oil relative permeability curves obtained by varying the exponent m in Eq. 3.143 from $m = 1$ to $m = 6$.

Relative permeability curves used to obtain these results are shown in Fig. 6.19. The water relative permeability curve ($n = 2$) was the same for each case. These computations illustrate the variation of m with permeability curve to evaluate M. Injectivities shown in Table 6.7 correspond to $E_A = 0.5$ in a 10-acre [40 469-m^2] pattern. The differences between $M_{\bar{S}}$ and M_t as well as M and M_t are small for oil relative permeability curves characterized by $m \leq 3$. For oil relative permeability curves with $m > 3$, M becomes a progressively poorer approximation. The value of M_t, however, is bounded by M and $M_{\bar{S}}$ for all cases.

6.6 Estimation of Waterflood Performance

Waterflood design involves both technical and economic considerations. To make an economic evaluation, it is necessary to estimate fluid injection and production rates and to make a projection of oil production (or recovery) for the anticipated life of the project for each flooding plan. These estimates, along with the well layout for the waterflood, provide sufficient technical data to estimate investment requirements, operating costs, and income for a proposed waterflood. These are the data required for economic analyses with discounted cash flow or other approaches.

6.6.1 Approximate or First-Pass Estimates of Waterflood Performance

Estimates of waterflood recovery, production rates, and production time curves can be made in a relatively short time with simple models of the displacement process. One approach to illustrate the procedure is presented here. Craig[11] provides an extensive survey of approximate and empirical methods.

Estimation of Waterflood Recovery by Material Balance. Waterflood recovery may be estimated from a material balance with core data and estimates of the sweep efficiency. The reservoir is considered to be subdivided into two parts—the volume swept by the waterflood and the unswept volume. With this representation, the volume of oil displaced from the reservoir is given by Eq. 6.59.

$$N_d = \frac{E_V}{B_o}(\Delta S_o)V_p, \dots\dots\dots\dots (6.59)$$

where

E_V = fraction of the reservoir volume that is swept by the injected water when the economic limit is reached,

ΔS_o = change in average oil saturation within the swept volume, and

N_d = oil displaced from the volume swept by the waterflood.

The displacement process is assumed to be piston-like so that the oil saturation in the swept region is the average residual saturation determined from flood pot tests. Thus $\Delta S = S_{o1} - S_{or}$, where S_{o1} is the volumetric average oil saturation in the reservoir at the beginning of the waterflood.

The volume of oil displaced usually is not equal to the volume of oil produced. In depletion-drive reservoirs, a gas saturation exists when waterflooding begins. Flooding pressures are high enough in most waterfloods to force the gas back into solution. If all portions of the reservoir are connected hydraulically, some of the oil displaced by the waterflood will fill the pore space occupied previously by the gas saturation in the unswept portion of the reservoir. This part of the displaced oil is not recovered. Eq. 6.60 gives the oil remaining in the reservoir when the unswept pore space is resaturated to the initial oil saturation.

$$N_r = \frac{E_V S_{or} V_p}{B_o} + \frac{(1-E_V)S_{oi}V_p}{B_o}, \dots\dots (6.60)$$

where the first term represents the oil remaining in the swept volume and the second term is the oil remaining in the unswept zone.

Oil recovered by waterflooding (N_{pw}) is computed with Eq. 6.61.

$$N_{pw} = [S_{o1} - E_V S_{or} - (1-E_V)S_{oi}]\frac{V_p}{B_o}, \dots (6.61)$$

where S_{o1} is the oil saturation at the beginning of the waterflood.

Eq. 6.62 expresses the waterflood recovery in terms of the original oil in place (OOIP) and the oil produced during primary operations.

$$N_{pw} = (N-N_p) - N\frac{B_{oi}}{B_o}\left[1 + E_V\left(\frac{S_{or}}{S_{oi}}-1\right)\right], \dots\dots\dots\dots (6.62)$$

where

N_{pw} = oil potentially recoverable by waterflooding, STB,

N = initial oil in place, STB,

N_p = oil produced during primary operations, STB, and

B_{oi} = initial FVF.

Eqs. 6.60 through 6.62 may be altered to include a residual gas saturation if sufficient information is available.

Prediction of waterflood recovery requires estimation of the residual oil saturation and volumetric sweep efficiency at the point the flood is terminated. Residual oil saturation is determined from flood pot tests on the reservoir rock or inferred from data available from similar reservoirs. Volumetric sweep efficiency may be estimated by three methods. Two methods are empirical and rely on the availability of waterflood performance or experience in similar reservoirs. In the first approach, E_V is computed from the performance of completed waterfloods with Eq. 6.61 and then assumed to apply to the new waterflood. This approach is widely used to obtain preliminary estimates of waterflood recovery.

The second method assumes that estimates of vertical and areal sweep efficiencies can be made from interpretation of core analyses, well logs, fluid and rock properties, and simplified displacement models. Volumetric sweep efficiency is assumed to be the product of the areal sweep efficiency (E_A) and the vertical sweep efficiency (E_I) as in Eq. 6.63.

$$E_V = E_A E_I, \dots\dots\dots\dots (6.63)$$

where

E_I = fraction of the reservoir cross section that has been displaced by the injected water and

E_A = fraction of the reservoir area within the vertical portion of the reservoir that has been swept to residual oil saturation.

Areal sweep efficiencies are estimated for the particular flooding pattern with scaled models—such as those described in Sec. 6.6.2. Values of E_I may be estimated from core analyses and fluid/rock properties with a correlation like the Dykstra-Parsons method,[12] which correlates E_I with permeability variation and mobility ratio (M) for several WOR's. In many cases, E_I is estimated from interpretation of injectivity profiles in water-injection wells in the reservoir to be flooded or a similar reservoir.

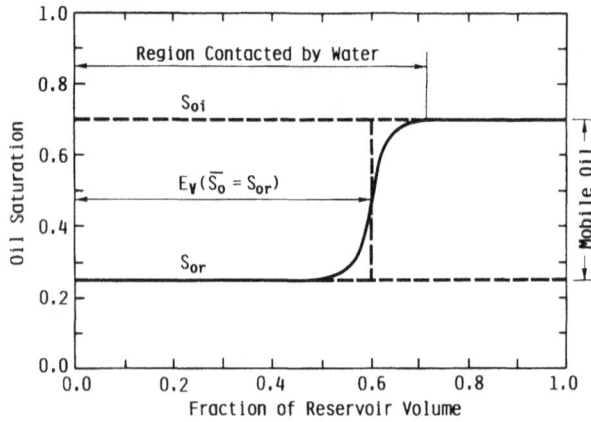

Fig. 6.20—Oil saturation distribution at $E_V = 0.6$ when the displacement process is piston-like.

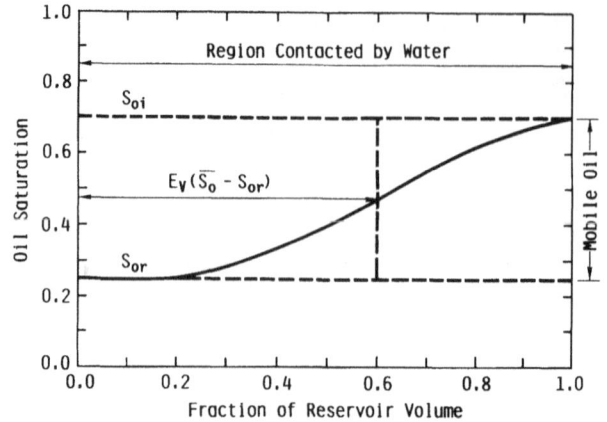

Fig. 6.21—Oil saturation distribution at $E_V = 0.6$ when the displacement process is not piston-like.

Volumetric sweep efficiency ranges from 0.1 for heterogeneous reservoirs to 0.7+ for homogeneous reservoirs with good flooding characteristics. There is uncertainty in E_V and judgments must be made. Some precautions should be exercised when using values of volumetric sweep efficiency to estimate oil recovery. In Eqs. 6.59 through 6.62, the volumetric sweep efficiency is defined by the way that the oil saturation in the swept region is specified. In these equations, the oil saturation in the swept region is assumed to be S_{or}. The validity of this approximation depends on the properties of the fluids and rock. If the displacement front is steep, as depicted in Fig. 6.20, the average oil saturation in the swept region is approximately S_{or}. If the saturation profile is gradual, as shown in Fig. 6.21, however, the average saturation in the swept region is larger than S_{or} and E_V increases. Figs. 6.20 and 6.21 have the same volume of oil displaced from the swept region. These examples are a reminder that every volumetric sweep efficiency has a corresponding definition of oil saturation in the swept zone. Volumetric sweep efficiencies obtained from other reservoirs will be similar when displacement processes

produce similar saturation profiles and have the same definition of oil saturation in the swept zone. Otherwise, large differences may occur.

The third method of estimating volumetric sweep efficiency uses correlations based on simulation of waterflood performance in pattern floods.[39] Figs. 6.22 and 6.23 show correlation of volumetric sweep efficiency for five-spot waterfloods at WOR's of 25 and 50. These figures were prepared by computing waterflood performance for a 100-layer five-spot model with the Higgins-Leighton streamtube model. Permeability of each layer was determined from the log-normal distribution following the procedure presented in Chap. 5. There was no crossflow between layers. For the purpose of these calculations, the relative permeability correlations described in Chap. 3 were used with $\alpha_1 = 0.402$, $\alpha_2 = 0.248$, $m = 2.06$, and $n = 2.33$. Values of S_{oi} and S_{or} were 0.7 and 0.25, respectively. FVF's were assumed to be 1.0. Computations were made at endpoint mobility ratios (M) of 0.1, 1.0, 10.0, 20.0, and 100.0 by selection of the appropriate oil/water viscosity ratio. Although Figs. 6.22 and 6.23 were developed for a specific set of relative permeability

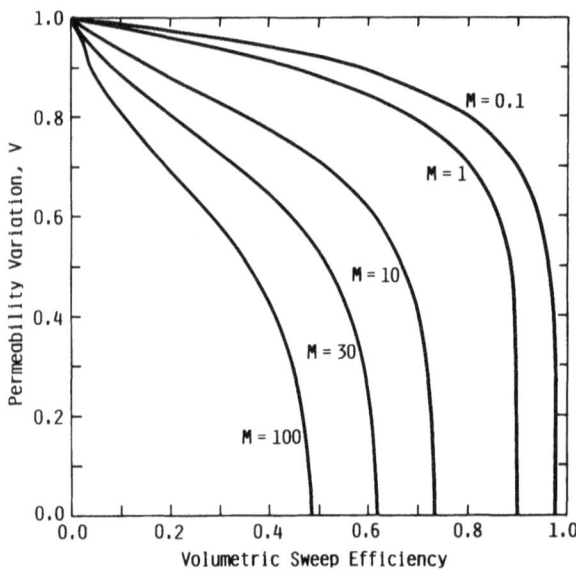

Fig. 6.22—Correlation of volumetric sweep and permeability variation for WOR = 25.[39]

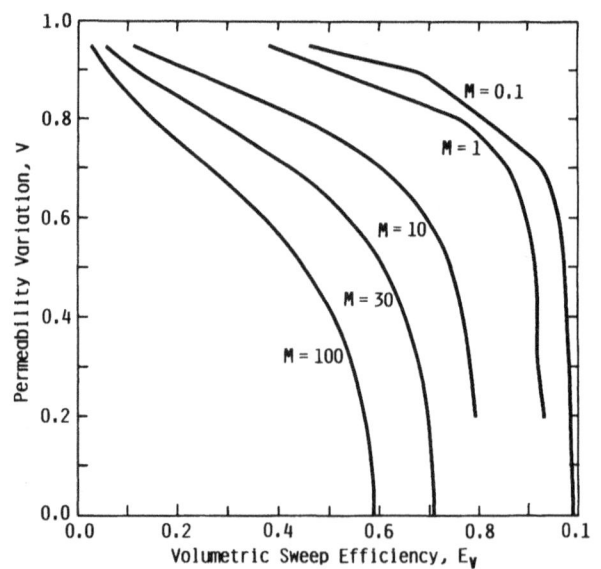

Fig. 6.23—Correlation of volumetric sweep and permeability variation for WOR = 50.[39]

data, they can be used to obtain approximate values of E_V for five-spot patterns when data are limited. It would be possible to prepare a set of type curves correlating E_V with permeability variation (V) for other patterns as well as floods where pseudorelative permeability curves were used to approximate effects of crossflow on waterflood performance.

In many areas, there are empirical or rule-of-thumb estimates of waterflood recovery developed from years of operating experience. One rule of thumb that has been widely used (and abused) assumes that a good waterflood will produce the same recovery in barrels per acre-foot as in primary. Another is an average recovery (e.g., 150 bbl/acre-ft [0.2 m³/m³]) for waterfloods in a particular formation. While these rules are sometimes used for estimation of waterflood recovery, the engineer should remember that *all* estimates must satisfy the material-balance equation for reasonable ranges of residual saturations and sweep efficiencies. Those estimates that do not meet the material-balance requirement should be viewed with skepticism.

Production Rates. To complete the approximate model, it is necessary to estimate production rates as a function of time. Rough bounds can be placed on the length of a flood. Field experience provides information on the capacity of production wells under primary conditions. Information may be available on injection rates in comparable reservoirs. If reservoir rock and fluid data are available, the methods introduced in Sec. 6.4 can be used to compute injection rates. Review of waterflood predictions in Chaps. 3 and 4 and in other references show that waterfloods require 1 to 2 PV's of water injection for recovery of the mobile oil. If the PV of the reservoir is known and all injected water is assumed to stay in the reservoir (frequently a poor assumption), it is possible to estimate the length of the flood by dividing the total volume of water injected (e.g., 1.5 PV) by the average injection rate.

The production rate/time relationship is found empirically. The oil rate, $q_o(t)$, is required to produce the estimated waterflood reserves (N_{pw}) in the estimated flood life (t_l). Mathematically,

$$N_{pw} = \int_0^{t_l} \frac{q_o(t)dt}{B_o}. \qquad (6.64)$$

Expressions for $q_o(t)$ are obtained from operating experience of similar floods. An example of the construction of an empirical model is presented in the following section.

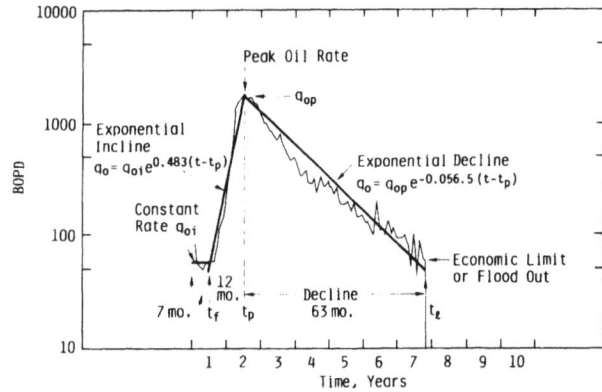

Fig. 6.24—Production response of fully developed waterflood with water input into all injection wells at the beginning of the flood (after Ref. 13).

Fig. 6.24 is the production response of a fully developed waterflood in which the water-injection rate was constant at 3,900 B/D [620 m³/d] water.[13] Water injection began in Year 1. The first production response or "oil kick" occurred 7 months after water injection started. The oil rate averaged 60 B/D [10 m³/d] oil during this period. Production peaked at 1,760 B/D [280 m³/d] oil 19 months into the flood and declined to 50 B/D [8 m³/d] oil when the flood ended after 82 months of injection.

The production/time curve shown in Fig. 6.22 can be represented empirically by dividing the flood life into three intervals:

t_f = time for the initial production response,
t_p = time at which the oil rate reaches a peak, and
t_l = flood life (time at which the production rate is at the economic limit).

For purposes of description, the time interval from 0 to t_f is termed the fill-up time, the time interval from t_f to t_p is the inclining rate period, and the time interval from t_p to t_l is the declining rate period.

An empirical model of the production response is developed by finding the mathematical relationships that best approximate the rate/time curve in the time interval. One model of the production curve in Fig. 6.24 that was constructed by assuming a constant rate, q_{oi}, from 0 to t_f, an exponential increase in oil rate to q_{op} for $t_f<t<t_p$, and an exponential decline from q_{op} to q_{ol} for $t_p<t<t_l$. With these assumptions, the mathematical expressions in Table 6.8 represent the rate/time relationship. The factor 30.4 is the average number of days per month while

TABLE 6.8—RATE EXPRESSIONS FOR OIL PRODUCTION IN APPROXIMATE WATERFLOOD MODEL

Period	Length (months)	Rate (BOPD)	Cumulative Production for Time Interval
Fill-up	t_f	q_{oi}	$q_{oi}t_f\,(30.4)$
Inclining rate	t_p-t_f	$q_{oi}e^{I(t-t_f)}$	$\dfrac{(q_{op}-q_{oi})\,30.4}{I}$
Declining rate	t_l-t_p	$q_{op}e^{-D(t-t_p)}$	$\dfrac{(q_{op}-q_{oi})\,30.4}{D}$

TABLE 6.9—RATE EXPRESSIONS FOR DECLINE CURVE ANALYSIS (after Ref. 16)

Type	Rate Where $q = q_{oi}$ at $t = 0$	Cumulative Production From $t = 0$
Coordinate	$q = -mt + q_{oi}$	$q_{oi}t - \dfrac{mt^2}{2}$
Exponential	$q = q_{oi}e^{-Dt}$	$\dfrac{q_{oi} - q_o}{D}$
Hyperbolic	$q = q_{oi}(1 + ntD)^{-1/n}$	$\dfrac{q_{oi}(q_{oi}^{1-n} - q^{1-n})}{D(1-n)}$
Harmonic	$q = \dfrac{q_{oi}}{(1 + Dt)}$	$\dfrac{q_i}{D} \ln\left(\dfrac{q_i}{q}\right)$

coefficients I and D are expressed in units of t^{-1}. Equations for the cumulative oil production during a particular time interval were obtained by integrating the corresponding rate equation as in Eqs. 6.65 through 6.68.

$$N_{pw}(t_2) - N_{pw}(t_1) = \int_{t_1}^{t_2} \frac{q_o(t)dt}{B_o}. \qquad \ldots\ldots\ldots (6.65)$$

For the declining rate period with $B_o = 1.0$,

$$\Delta N_{pw} = \frac{q_{op}}{B_o} \int_{t_p}^{t_1} e^{-D(t-t_p)} dt, \qquad \ldots\ldots\ldots\ldots (6.66)$$

$$\Delta N_{pw} = \frac{q_{op}}{DB_o} [1 - e^{-D(t-t_p)}], \qquad \ldots\ldots\ldots\ldots (6.67)$$

and

$$\Delta N_{pw} = \frac{1}{DB_o}(q_{op} - q_{ol}). \qquad \ldots\ldots\ldots\ldots\ldots (6.68)$$

The "eyeball" fit shown in Fig. 6.24 yielded values of 0.483 months^{-1} and 0.0565 months^{-1}, respectively, for I and D. Cumulative production based on the correlation is 1,039,838 bbl [165 321 m^3] oil compared with 1,029,186 bbl [163 628 m^3] oil actually produced. Although the agreement is good, there are compensating differences between the fitted and actual curves in the inclining and declining rate periods. Because of the logarithmic rate scale, the eye fit leads to more oil production than observed in the declining rate period and less oil in the inclining period. This problem could be eliminated with computer-based computation schemes or with the selection of other rate expressions to fit the data.

TABLE 6.10—RULE-OF-THUMB ESTIMATES FOR OKLAHOMA WATERFLOODS

Time	PV
t_f	time required to inject 0.104 PV of water
t_p	time required to inject 0.23 PV of water
t_l	time required to inject 1.25 PV of water

Decline curves are widely used to estimate oil production. Four commonly used decline-curve equations are given in Table 6.9. These equations lend themselves to graphical analysis. Various curve-fitting programs are available for programmable calculators and digital computers. An engineer may try many different relationships before arriving at a satisfactory choice. Refs. 13 through 21 cover decline-curve analysis.

Prediction of waterflood performance with an empirical model like the one described in the previous section requires some method of determining that rate expressions are representative of a particular reservoir and finding a method of determining the constants in the equations.

All empirical models must satisfy the material balance—that is, if the waterflood reserves (including primary) are estimated to be N_{pw}, then

$$N_{pw} = \int_o^{t_f} \frac{q_o dt}{B_o} + \int_{t_l}^{t_p} \frac{q_o dt}{B_o} + \int_{t_p}^{t_l} \frac{q_o dt}{B_o}, \qquad \ldots\ldots\ldots (6.69)$$

where the first integral represents fill-up, the second integral incline, and the third integral decline. We assume that t_f, t_p, and t_l can be estimated from operating experience for a specific type of reservoir. The correlations of Bush and Helander[13] developed for average conditions in sandstone reservoirs in Oklahoma that were depleted by primary production are summarized in Table 6.10. All times are based on the average water-injection rate for the flood. Injection rates can be estimated with the methods presented in Sec. 6.5.2.

For the purpose of this section, injection rates will be estimated by assuming an injectivity of I_{AF} (barrels per day per acre-foot) sand. Thus

$$t_l = I_{AF}Ah, \qquad \ldots\ldots\ldots\ldots\ldots\ldots\ldots\ldots (6.70)$$

$$t_l = (7,758A\phi h)(1.25 \text{ B/D/acre-ft}),$$

$$= \frac{9,698\phi}{Ah} \text{days}.$$

Flood life computed from Eq. 6.70 assumes that the average injection rate is q_w throughout the project. If the number of wells varies over the flood, as is frequently the case, it is necessary to estimate the injection rate/time relationship.

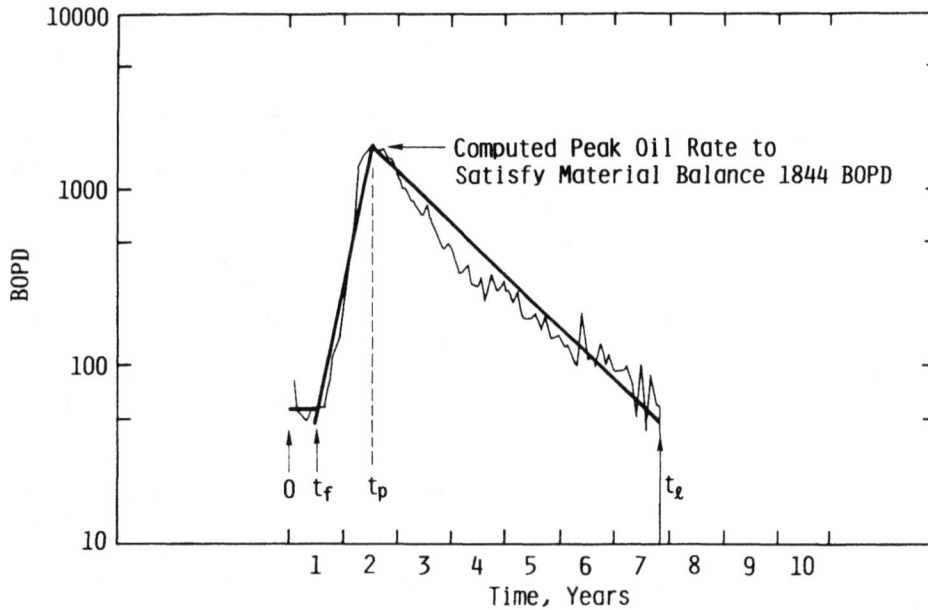

Fig. 6.25—Match of empirical model of waterflood response with actual performance of a waterflood. [13]

Production after fill-up is estimated by extrapolating the decline curve from the beginning of water injection to t_f. If exponential incline and declines are assumed, the rate/time curve after fill-up is found by solving Eq. 6.71 for q_{op}.

$$\Delta N_{pw} = \frac{(q_{op}-q_{oi})(t_p-t_f)}{B_o \ln\left(\dfrac{q_{op}}{q_{oi}}\right)} + \frac{(q_{op}-q_{ol})(t_l-t_p)}{B_o \ln\left(\dfrac{q_{op}}{q_{ol}}\right)},$$

$$\dotfill (6.71)$$

where

N_{pw} = estimated oil production for $t_f \rightarrow t_l$, and
q_{oi} = oil rate at beginning of the incline, B/D.

The oil rate at flood-out is determined by the economic limit. Eq. 6.71 can be solved for q_{op} by trial and error or by Newton's method.

The waterflood in Fig. 6.24 was fitted to exponential incline and decline by solving for q_{op} with the parameters listed in Table 6.11.

The value of q_{op} was 1,844 B/D [293 m³/d] oil compared with 1,760 B/D [280 m³/d] oil observed in the flood. Fig. 6.25 shows the match between actual and empirical production rates.

Estimation of waterflood performance with an empirical model depends on initiation of the waterflood in the entire reservoir within a short period of time. If injection wells are put on line at staggered intervals throughout the project, the field decline will be concealed by the response to new wells. There would be no basis for the uniform decline assumed in the development of this model. It may be possible to estimate project life from an average oil production rate and the oil recovery from material-balance calculations. Other methods of estimating ultimate recovery include "cut-cum" plots [22] and extrapolation at oil/water ratio (or WOR) vs. cumulative oil production. [23]

The methods described in this section have been used for design purposes in reservoirs where the operating costs are low, flood life is relatively short, and/or economics are clearly favorable. Hundreds of waterfloods have been installed with little more engineering than described in the previous paragraphs. The majority of these floods were initiated and completed before the development of reservoir models presented in Chap. 5. Thus it is not known whether other flooding plans could have been selected or oil recovery could have been improved if tools available now were used to plan and to conduct the waterfloods.

First-pass or approximate methods serve as an order-of-magnitude check on oil recovery, injection, and production rates. Small operators without engineering staffs or with limited staffs still use the method to plan waterfloods. The method is inadequate, however, for reservoirs where reservoir heterogeneity, aquifers, and gas caps introduce large uncertainties in waterflooding performance and, consequently, the economic evaluation.

6.6.2 Engineering Approaches—Scaled-Model Correlations

Waterflood design with an engineering approach is based on the simulation of waterflood performance with some type of displacement model. The model chosen should be capable of simulating the principal features of the reservoir and fluids.

TABLE 6.11—PARAMETERS USED IN MATCH OF FIG. 6.25

t_f, months (days)	7 (213)
t_p, months (days)	19 (578)
t_l, months (days)	82 (2,493)
q_{oi}, B/D	60
q_{ol}, B/D	50
N_{pw}, STB	1,142,232

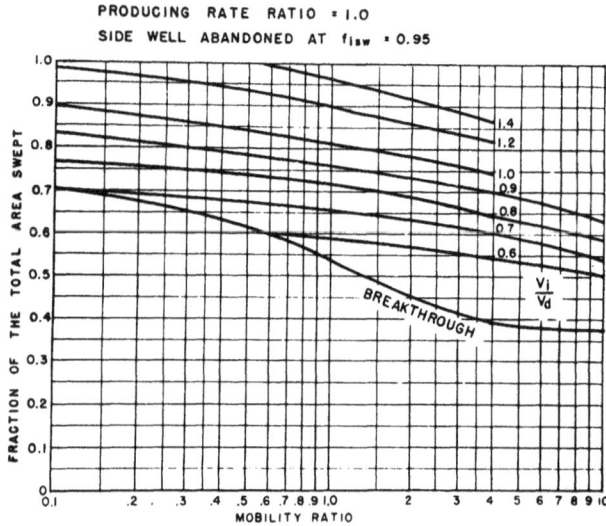

Fig. 6.26—Sweepout pattern efficiency as a function of mobility ratio for the nine-spot pattern at various displaceable volumes injected.[25]

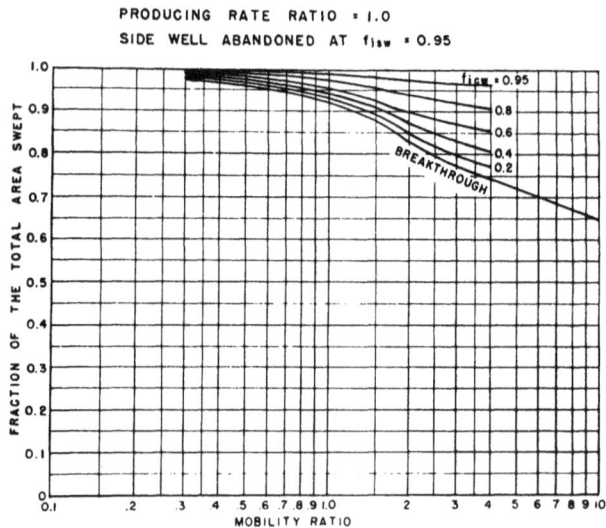

Fig. 6.27—Sweepout pattern efficiency as a function of mobility ratio for the nine-spot pattern at various corner-well producing cuts (f_{cw}).[25]

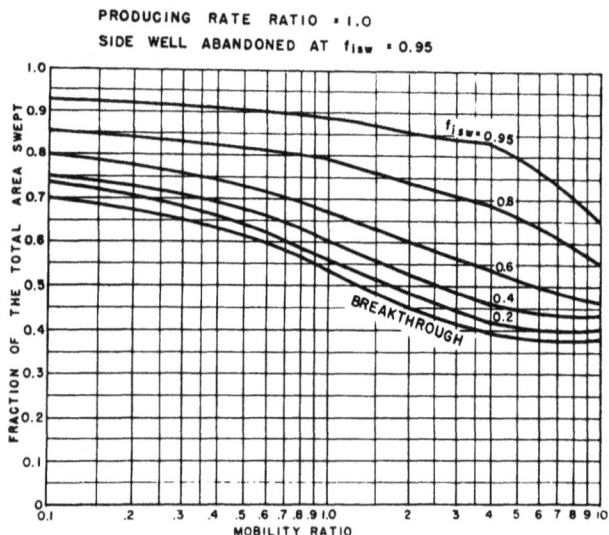

Fig. 6.28—Sweepout pattern efficiency as a function of mobility ratio for the nine-spot pattern at various side-well producing cuts (f_{Sw}).[25]

The reservoir engineer is sometimes requested to prepare estimates of waterflood recovery on relatively short notice. A decision must be made about the amount of effort required to produce the estimate. The selection of a displacement model can become a complex process involving time constraints, capabilities of the particular engineer and/or the technical staff, accuracy needed for the decision process, and, finally, the risk the producer is willing to take considering the uncertainties that are present when a displacement model is selected.

The only scaled-model correlation that accounts for immiscible displacement is the Craig, Geffen, Morse (CGM) five-spot model.[24] There are no comparable correlations for linedrive, staggered linedrive, or nine-spot patterns. As mentioned in Chap. 4, correlations from scaled-model experiments with miscible fluids have been developed. These correlations are comparable to waterfloods that behave as piston-like displacements—that is, the average water saturation in the displaced area, $\bar{S}_w$, is equal to $1-S_{or}$. There is no displacement of oil from the swept area after the flood front arrives.

Figs. 6.26 through 6.28 are correlations of displacement performance for a nine-spot flood from scaled-model displacement experiments with miscible fluids.[25] Correlations for a range of mobility ratios (M) are presented in terms of areal sweep efficiency, E_A, displaceable hydrocarbon PV's (V_d), and the fraction of the displacing fluid in the produced fluid (f_D). Eqs. 3.207, 6.72, and 6.73 define these parameters for immiscible displacement.

$$M = \frac{\left(\dfrac{k_{rw}}{\mu_w}\right)_{S_{or}}}{\left(\dfrac{k_{ro}}{\mu_o}\right)_{S_{oi}}},$$

where

E_A = area of pattern swept to a residual oil saturation of S_{or},

$$V_d = \frac{W_i}{V_p(1-S_{iw}-S_{or})}, \quad \ldots\ldots\ldots\ldots\ldots\ldots (6.72)$$

and

$$f_D = f_w. \quad \ldots\ldots\ldots\ldots\ldots\ldots\ldots\ldots\ldots\ldots\ldots (6.73)$$

Example 6.3 illustrates the use of these correlations to estimate waterflood performance for a 10-acre [40 469-m^2] nine-spot pattern.

Example 6.3

A waterflood is to be conducted in a nine-spot pattern on 10-acre [40 469-m^2] spacing. Side and corner wells will be operated so that rates are the same. Porosity is 0.2 and the mobility ratio, M, based on the endpoints of the relative permeability curves is 1.0. The reservoir is liquid-saturated with an initial oil saturation of 0.7. Residual saturation is 0.26. Prepare the recovery Q_i and WOR/recovery curves with the miscible correlations of Figs. 6.26 through 6.28.

Solution. Assume unit thickness so that V_p is the PV per unit thickness. Then

$$V_p = (10 \text{ acres})(7,758 \text{ bbl/acre-ft})(1.0 \text{ ft})(0.2)$$

$$= 38,790 \text{ bbl}.$$

The correlation in Fig. 6.26 relates E_A to V_d, the number of displaceable hydrocarbon PV's injected. From Eq. 6.72,

$$V_d = \frac{W_i}{V_p(1.0 - S_{iw} - S_{or})}$$

$$= \frac{W_i}{17,456}$$

Because $Q_i = W_i/V_p$,

$$Q_i = V_d(1 - S_{iw} - S_{or}).$$

Oil recovery at each V_d is computed by assuming piston-like displacement. Thus

$$N_p = E_A(S_{oi} - S_{or})V_p.$$

Table 6.12 summarizes the computations for N_p assuming that B_o is 1.0.

The WOR during the displacement is found by combining the water production from side and corner wells. In a fully developed nine-spot pattern, there are three net production wells for each injection well. Because there are two side wells per corner well, the water cut from the pattern is

$$f_p = \frac{q_c f_{cw} + 2q_s f_{sw}}{q_c + 2q_s}, \quad \dots \dots \dots \dots \dots (6.74)$$

where

q_c = production rate of corner well, B/D,
q_s = production rate of side well, B/D,
f_{cw} = fractional flow of water in corner well, dimensionless, and
f_{sw} = fractional flow of water in side well, dimensionless.

From the problem statement, $q_c = q_s$ so that $q_t = 3q_s$. Thus

$$f_p = \frac{(f_{cw} + 2f_{sw})}{3}. \quad \dots \dots \dots \dots \dots (6.75)$$

Figs. 6.27 and 6.28 give f_{cw} and f_{sw} for a nine-spot pattern when $q_s = q_c$ and the side well is shut in at $f_{sw} = 0.9$. Fractional flows from the side and corner wells are summarized in Table 6.13.

Rate of production is determined by the water injection rate and the water cut (q_w/q_t) at both wells. In this example, the side well is shut in at 95% water cut when the areal sweep efficiency is about 89%. Breakthrough into the corner well occurs at $E_A = 92\%$. The injection rate until the corner well is shut in can be computed from

TABLE 6.12—ESTIMATION OF PRODUCTION PERFORMANCE, 10-ACRE NINE-SPOT PATTERN

V_d	E_A	$N_p (B_o = 1)$ (bbl)	N_p (PV)	Q_i (PV)
0.53	0.53	9,251	0.239	0.239
0.60	0.59	10,211	0.263	0.270
0.70	0.66	11,521	0.297	0.315
0.80	0.72	12,568	0.324	0.360
0.90	0.76	13,266	0.342	0.405
1.00	0.82	14,226	0.367	0.450
1.20	0.90	15,710	0.405	0.540
1.40	0.97	16,845	0.434	0.630

TABLE 6.13—COMPUTATION OF WOR

E_A Pattern	f_{sw}	f_{cw}	f_p Pattern	WOR Pattern
0.53	BT*	0.0	BT	0.00
0.56	0.20	0.0	0.13	0.15
0.60	0.40	0.0	0.23	0.36
0.67	0.60	0.0	0.40	0.67
0.79	0.80	0.0	0.53	1.14
0.89	0.95	0.0	0.63	1.73
0.92	SI	BT	BT	0.00
0.93	SI	0.20	0.20	0.25
0.94	SI	0.40	0.40	0.66
0.96	SI	0.60	0.60	1.50
0.98	SI	0.80	0.80	4.00
0.99	SI	0.95	0.95	19.00

*BT = breakthrough; SI = shut in.

Eq. 6.5 in Table 6.1. Problem 6.7 illustrates the development of an approximate model to estimate injection rates when $M \neq 1$. After the corner well is shut in, the injection rate at $M = 1$ is computed from Eq. 6.3 for the five-spot flood.

Correlations based on miscible experiments are easy to use. A waterflood prediction can be done in 1 or 2 hours by an engineer who understands the method. There are differences, however, between the displacement performance predicted by a miscible correlation and those for immiscible displacement. Fig. 6.29 shows the oil-recovery curve as a function of PV's of water injected for the Caudle and Witte correlation[5] and the CGM correlation for the problem in Example 4.1. Cumulative oil recovery for both models will ultimately be identical at large Q_i. The Caudle and Witte correlation predicts earlier oil production and later arrival of the displacement front than actually observed. Thus, when it is necessary to use miscible displacement models for predicting waterflood performance, the prediction will be optimistic.

In some situations, an engineer must decide whether to use the scaled-model correlation from immiscible experiments (CGM), to use a scaled-model correlation obtained from displacement experiments with miscible fluids, or to use a more sophisticated model. One factor of the decision process is an evaluation of the assumptions made in each model against known variations in reservoir and fluid properties.

The displacement models described in this section assume uniform spacing of wells in a reservoir that is areally homogeneous. The effects of reservoir heterogeneity can be evaluated only when the reservoir is visualized as a series of noncommunicating layers with the methods discussed in Sec. 5.8.1. Fluids are incompressible and fluid

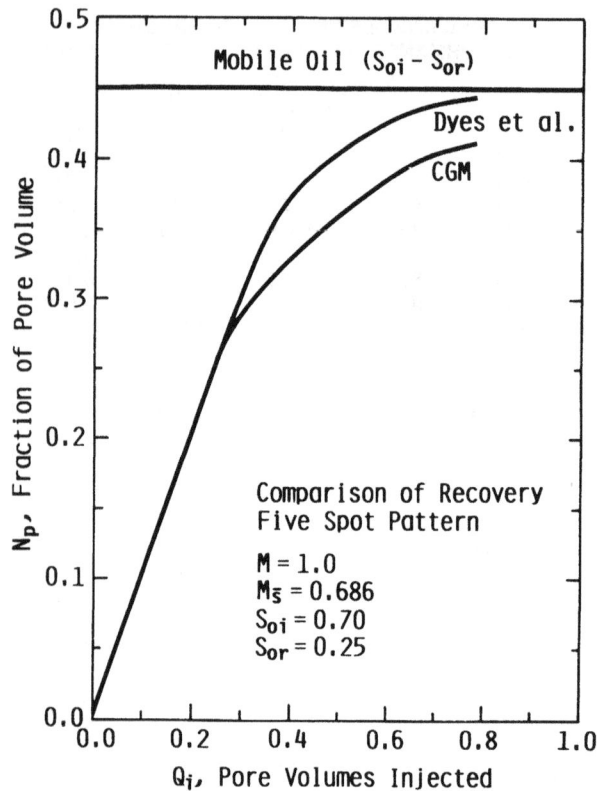

Fig. 6.29—Comparison of waterflood performance in a five-spot pattern from miscible models with predictions from the CGM correlation.

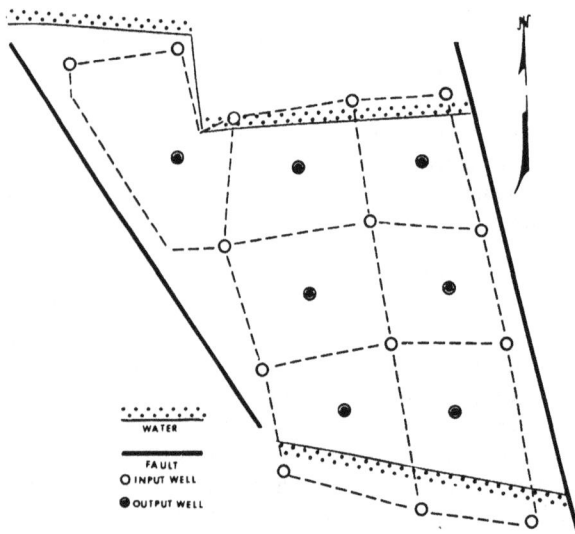

Fig. 6.30—Diagram of reservoir showing waterflood pattern.[27]

TABLE 6.14—RESERVOIR DATA

Oil gravity, °API	30
Oil viscosity, cp	1.7
Water viscosity, cp	0.42
FVF	1.05
Bubblepoint pressure, psi	1,800
Original solution GOR, cu ft/bbl	400
Reservoir temperature, °F	160
Reservoir pressure at 4,250-ft datum, psi	70
Original reservoir pressure, psi	1,900
Air permeability	
Average, md	265
Range, md	25 to 600
Coefficient of variation (V)	0.628
Porosity	
Average, %	29
Range, %	20 to 32
Sand thickness, ft	155
Reservoir volume, acre-ft	15,200
Interstitial water saturation, %	30
Residual oil saturation, %	20

6.6.3 Computer Models for Design of Waterfloods

A computer model is necessary for waterflood design when the assumptions required to use approximate models (Sec. 6.6.1) or correlations based on scaled-model results (Sec. 6.6.2) cause large uncertainties in the estimated flood performance. The level of uncertainty that is acceptable is a matter of engineering and management judgment and will not be addressed here.

This section introduces the capabilities of computer models to assist in the design of a waterflood. Models will be described in terms of the types of problems for which they are best suited. Our emphasis is on identifying key assumptions that separate one type of model from another.

Computer models are separated into two categories in this text. The streamtube model presented in Sec. 4.5 is a computer model in the sense that access to a computer is necessary to use the model. All other computer models are referred to as numerical simulators because they are based on the numerical solution of the partial differential equations describing two- and three-phase flow in a porous rock in one, two, or three spatial coordinates. In this section, some of the capabilities of computer models to assist in the design of a waterflood are presented.

6.6.3.1 Streamtube Models. The theory of streamtube models was introduced in Sec. 4.5 and the five-spot pattern was used to illustrate concepts. A computer program is available in FORTRAN language that permits estimation of waterflood performance in five-spot, linedrive, staggered linedrive, and seven-spot patterns.[26] These are the patterns for which shape factors (G_j) have been published. Heterogeneity is included by subdivision of the pattern into a specified number of noncommunicating layers. The program simulates a waterflood in a depleted reservoir with the displacement model shown in Sec. 3.9.

The capability of a streamtube model to match waterflood performance is shown in Example 6.4.[27]

Example 6.4

Fig. 6.30 shows a reservoir in the Dominguez field in the Los Angeles basin that was waterflooded with a nominal five-spot flooding pattern. The reservoir is bounded by faults on the east and west sides and edge water on the north and south sides. Reservoir data are summarized in Table 6.14.

properties vary with saturation but not with pressure or composition. These models do not have injection-rate correlations that are applicable for immiscible displacement. Consequently, either a constant injection rate is assumed or the injection rate is estimated with the procedures in Sec. 6.5.

Models that relax some or all of the assumptions described in the previous paragraph have been developed for use in waterflood design and analysis. The capabilities of these models are examined in Sec. 6.6.3.

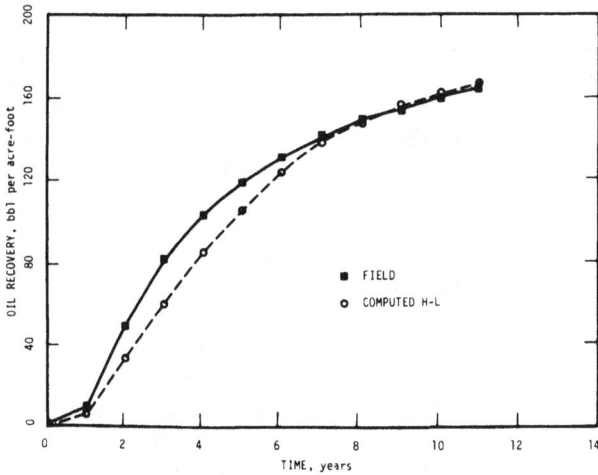

Fig. 6.31—Performance of waterflood compared with match obtained with the Higgins-Leighton streamtube model. [27]

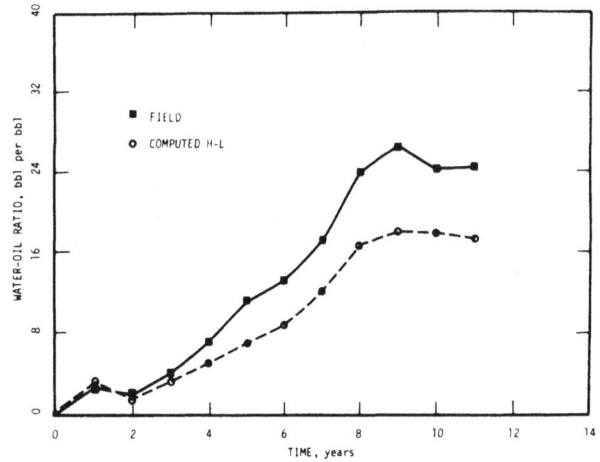

Fig. 6.33—WOR vs. time calculated with Higgins-Leighton streamtube model with the first set of parameters. [27]

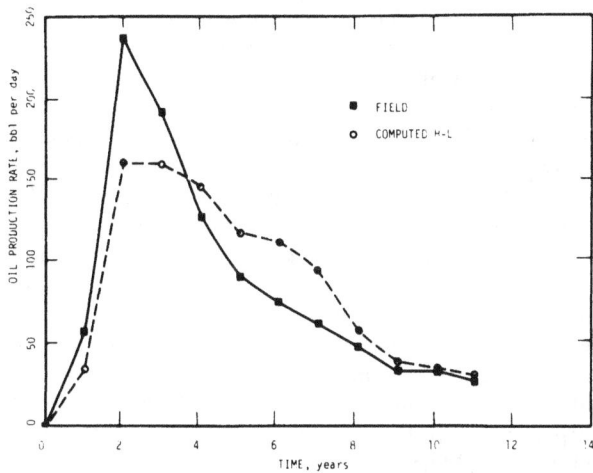

Fig. 6.32—Oil production rates for the Dominquez waterflood—first match. [27]

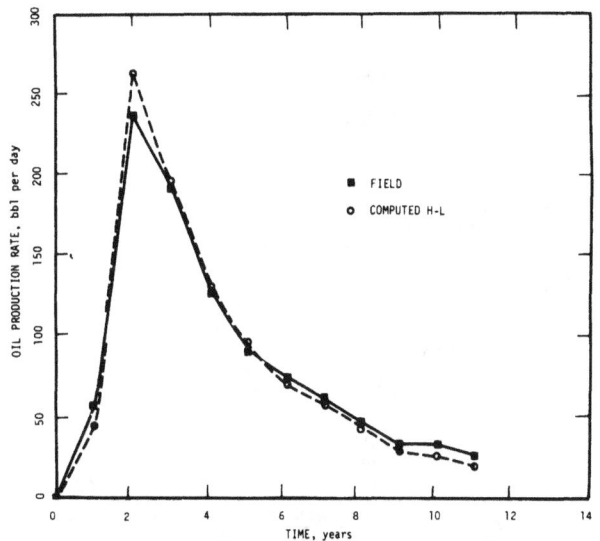

Fig. 6.34—Comparison of oil production rates computed with revised parameters. [27]

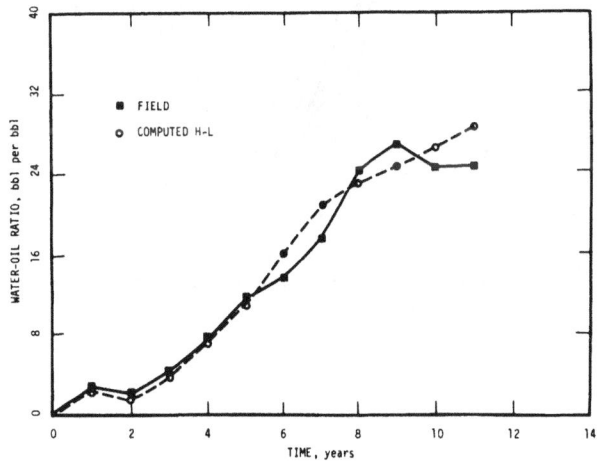

Fig. 6.35—Comparison of WOR at different times with field performance with revised parameters. [27]

The production responses shown in Figs. 6.31 through 6.33 were estimated with the streamtube model and the computer program in Ref. 21. Although the five-spot patterns were irregular, they were simulated as seven regular five-spot patterns with equal reservoir volume. The net thickness of 155 ft [47 m] was subdivided into five layers of equal thickness with permeabilities of 663, 312, 187, 110, and 52 md. The comparisons shown in Figs. 6.34 and 6.35 were obtained by selecting the set of relative permeability curves that gave the best match and assuming that the initial water saturation was 0.39.

The agreement shown in Figs. 6.31 and 6.32 is adequate for design purposes and verifies the capability of the model to simulate performance. Proof of prediction requires *a priori* selection of relative permeability curves, which was not done in this case. Because of the ease and low cost of simulation, however, it is possible to investigate a number of different relative permeability curves to determine the sensitivity of waterflood performance to uncertainty in this variable. In a design application, large uncertainty might justify a data acquisition program to obtain suitable relative permeability curves.

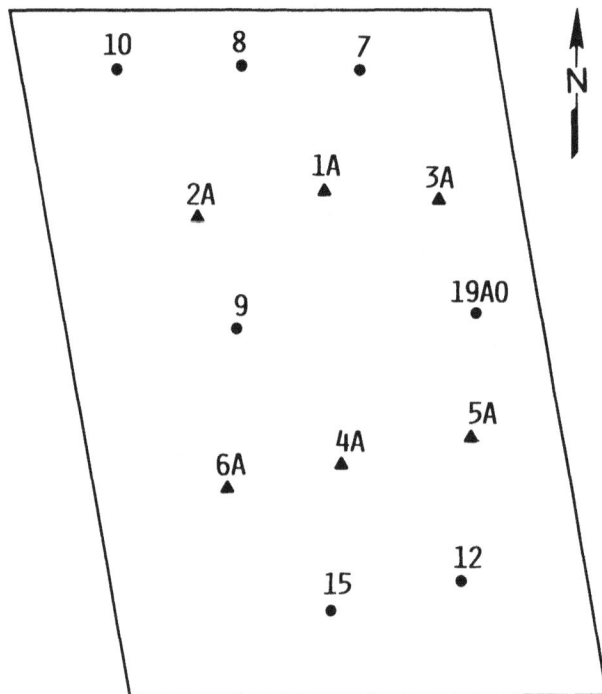

Fig. 6.36—Well arrangement in the Keown unit.[28]

Fig. 6.37—Streamtubes for the Keown unit assuming the reservoir is homogeneous and isotropic (after Refs. 28 and 29).

Streamtube models are easy to use when shape factors are available, as is the case for five-spot, linedrive, staggered linedrive, and seven-spot patterns. It is necessary to generate streamtubes and shape factors for all other well configurations. Fig. 6.36 illustrates a linedrive displacement pattern in a reservoir where injection and production rates vary considerably. Fig. 6.37 is one set of streamtubes developed for this displacement pattern with the techniques and computer programs in Refs. 28 and 29. Refs. 29 through 31 present techniques for determining streamlines, streamtubes, and shape factors for arbitrary well arrangements.

Finally, the limitations of streamtube models are summarized. In streamtube models, fluids are incompressible and there are no changes of fluid properties with pressure or composition. Gravity and capillary forces are neglected. Reservoir heterogeneity (vertical) is considered by subdividing a reservoir into noncommunicating layers. Each layer is considered to be homogeneous. Streamtubes are based on the flow of a single phase in homogeneous media with injection and production wells operating at fixed rates. In waterfloods, production wells are shut in when WOR's become excessive. Shutting in a production well forces fluids flowing through streamtubes collected by that well to flow to other wells. New streamlines form; consequently, the original streamtubes no longer exist. Although new streamtubes can be located with the methods described in this section and in Refs. 29 through 32, the saturation distribution and location of the flood front would not be known. Streamtube models would probably not be effective in this situation. Streamtubes usually are generated from displacement patterns or models that have unit mobility ratio. This would appear to be a serious limitation. Martin et al. [33] illustrate difficulties of streamtube models in predicting waterflood response of an isolated five-spot pattern at favorable mobility ratios. Martin and Wegner,[34] however, show that streamlines generated at unit mobility ratios are approximated closely in confined patterns, except for mobility ratios near 0.1.

6.6.3.2 Reservoir Simulators. Reservoir simulators are complex computer programs that simulate multiphase displacement processes in two or three dimensions. Simulators solve the appropriate fluid-flow equations (described in Chaps. 3 and 4) by numerical techniques to produce estimates of saturation distributions, pressure, and flows of each phase at discrete points in a reservoir. Fluid and rock properties may vary with position. Thus it is possible to simulate waterflood performance in nearly any reservoir if there is sufficient information to establish parameters required by the simulator.

Selection of reservoir simulation to design and/or manage waterfloods is governed by several considerations. Foremost is the assessment of whether the simulation has the potential of improving recovery from the flood when compared with simpler methods of estimating performance described earlier in this chapter. Large reservoirs are prime candidates for reservoir simulation because small changes in field operation can have sizable effects on oil recovery. Other reservoirs are candidates because of peculiar combinations of fluid and rock properties. Another factor influencing the possibility of reservoir simulation is the availability of sufficient data to use the simulator.

Technical considerations control the type of simulation that is required. Many reservoirs have complex, and sometimes unknown, distributions of rock properties (such as porosity and permeability). Parameters in reservoir simulators can be adjusted to account for known or hypothesized variations. Previously described models have limited capabilities to account for differences other than reservoir heterogeneity through layered models. Fluid saturations in reservoirs change with position because of variations in fluid/rock properties associated with geology as well as equilibrium of gravity and capillary forces. Only limiting cases (i.e., vertical equilibrium discussed in Chap. 5) can be examined with these models. Most consider the variation of fluid properties with pressure and composition. Cases between these extremes require numerical simulation. Reservoirs that have gas/oil contacts and/or water-oil contacts (WOC's) associated with an aquifer also require numerical simulation. Thus the engineer's assessment of the reservoir coupled with an estimate of potential benefits of simulation are used to select a reservoir simulator to predict waterflood performance.

The topic of reservoir simulators is beyond the scope of this text. Reservoir simulation is an important tool, and a case history of a reservoir simulation is included in the next section to illustrate some of the possibilities of the approach. Other case histories and information on reservoir simulation are found in Refs. 35 through 37.

Example 6.5

The West Seminole field shown in Figs. 6.38 and 6.39a produces from the San Andres formation in west Texas at an average depth of 5,100 ft [1555 m].[37,38] Fig. 6.39 is a cross section of the reservoir revealing a large dome on the west end and a smaller dome on the east end of the structure. A large gas cap covers most of the field, and there is a WOC near the bottom of the reservoir. The reservoir is geologically complex and heterogeneous, as illustrated by the cross section in Fig. 6.40 showing the permeability distribution.

The reservoir was discovered in 1948 and was developed on approximately 40-acre [161 880-m^2] well spacing. Reservoir pressure declined rapidly during primary production leading to fluid injection projects in an attempt to arrest pressure decline. Reinjection of produced gas began in 1964 with limited results. A peripheral waterflood was initiated in 1969–71 in the wells shown in Fig. 6.41. Peripheral wells were completed at or below the WOC to avoid possible loss of oil into the gas cap, as depicted in Fig. 6.42. The peripheral water injection program, however, did not produce a significant change in reservoir pressure.

During 1973–75, fifteen 40-acre [161 880-m^2] five-spot patterns were developed in the main dome as shown in Fig. 6.41. Data from new wells provided more insight into the geology of the reservoir.[38] Several geological units that correlated across the entire field, suggesting that vertical barriers to flow might exist, were identified. These interpretations, combined with other data, led to a decision to use reservoir simulation to investigate the possibility of vertical movement of oil into the gas cap under the existing waterflooding program.

Fig. 6.43 shows the subdivision of the study region into gridblocks used in a three-dimensional (3D), three-phase

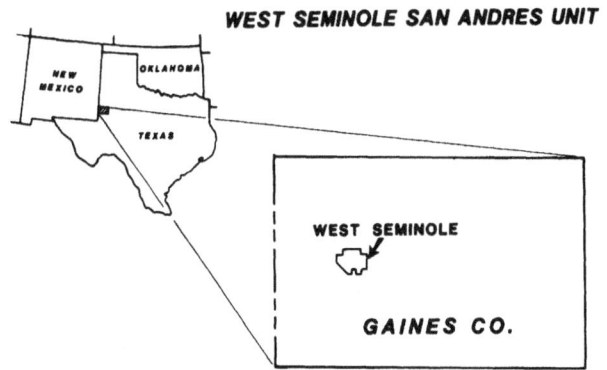

Fig. 6.38—Field location map.[37]

WEST SEMINOLE FIELD

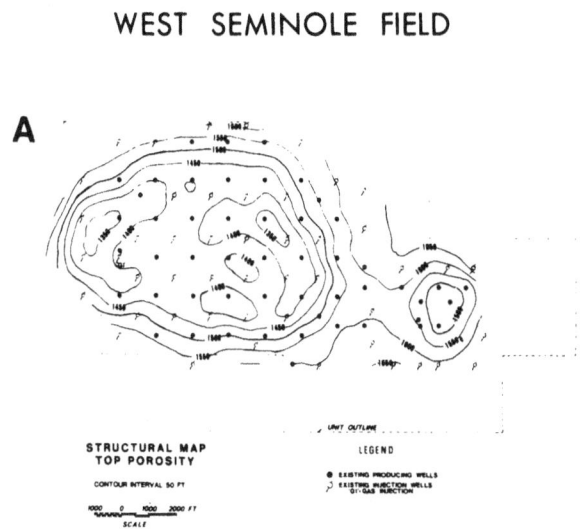

Fig. 6.39a—Structure map on top of porosity.

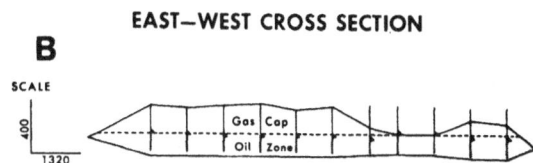

Fig. 6.39b—Schematic cross section through the field.

black-oil reservoir simulator. The reservoir was subdivided vertically into six layers that approximately matched the geological units.

Simulation of waterflood performance was done by adjusting the parameters of the reservoir simulator with the available reservoir data for the period 1948–78. History matches for the simulations are shown in Fig. 6.44. The predicted performance of the 40-acre [161 880-m^2] five-spot patterns is shown in Fig. 6.45.

The reservoir simulator provided information that would otherwise not be available with less sophisticated models. For example, it was not possible to match field performance unless vertical permeabilities were on the order of 10^{-3} to 10^{-4} md. Thus the simulation indicated strong barriers to vertical flow of fluids between layers. Simulation of peripheral water injection indicated that

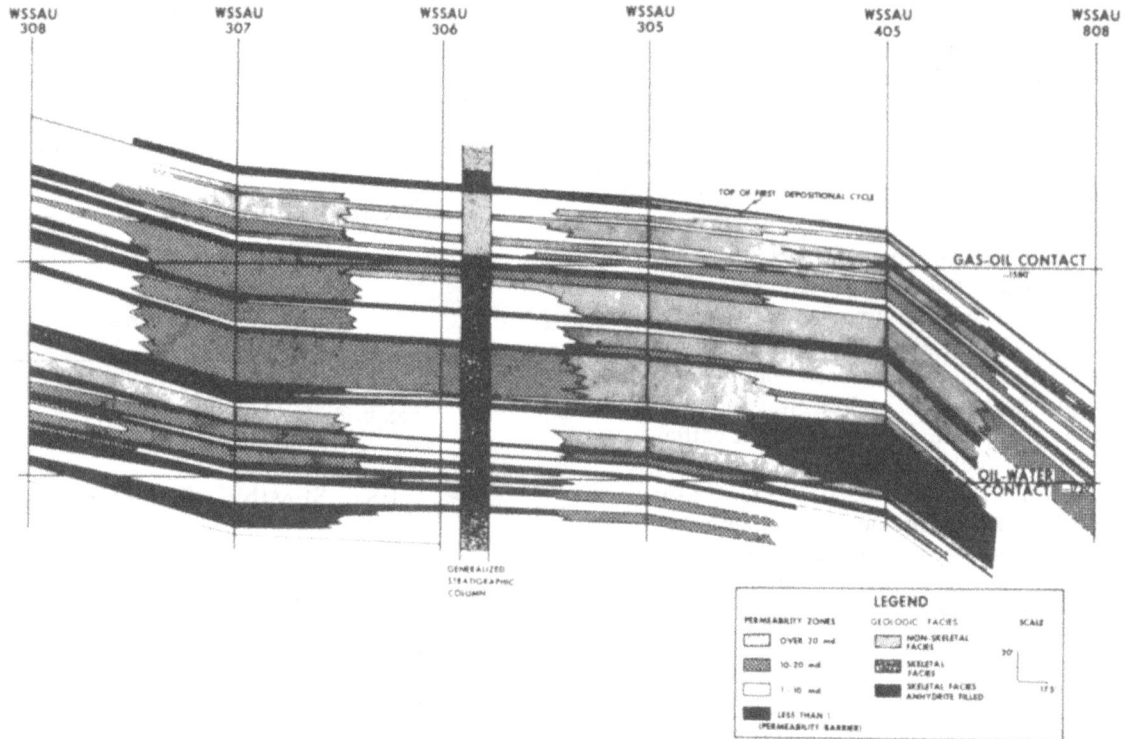

Fig. 6.40—Cross section showing permeability distribution.[37]

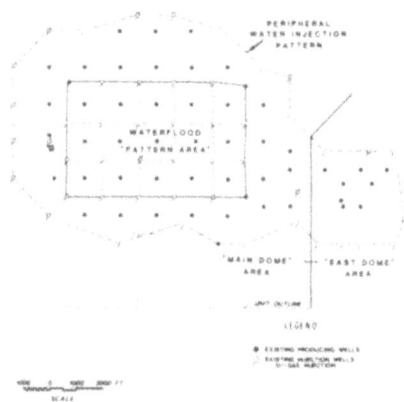

Fig. 6.41—Unit outline, well plan, and study area designations.[37]

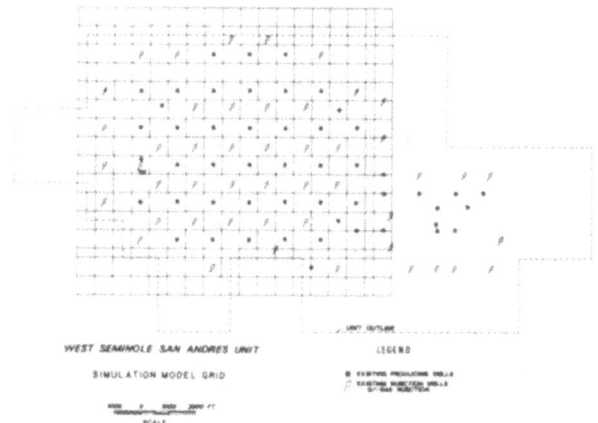

Fig. 6.43—Areal grid used in the simulation model.[37]

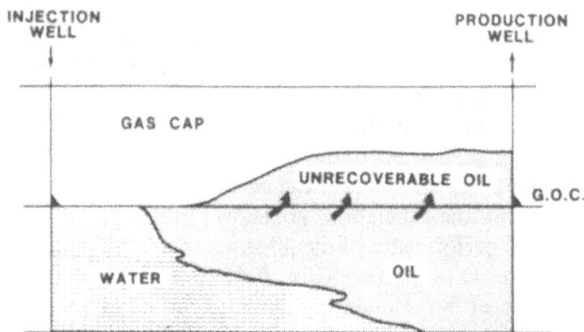

Fig. 6.42—Possible loss of oil to the gas cap.[37]

(1) the injection interval was separated from the reservoir by several "tight" barrier zones, and (2) there was not sufficient continuity in the injection interval to permit water to move over the distances required to affect the main productive zones.

Further simulation studies showed that waterflood recovery could be increased and oil movement into the gas zone could be decreased by a reservoir management plan. The plan involved drilling enough production wells to convert the 40-acre [161 880-m^2] five-spots to 40-acre [161 880-m^2] inverted nine-spot patterns. Infill drilling, combined with recompletion of some peripheral injection wells and balancing of water-injection/reservoir-voidage rates, was the most desirable operating scheme. The development of this reservoir management scheme could not be done without reservoir simulators. Complex geology, coupled with the nature of the reservoir, precludes effective waterflood design and analysis by other techniques. It is likely that there will be some deviation between the prediction and the actual performance. If the differences are small, minor adjustments in reservoir parameters by history match can be done to produce a revised projection. In some cases, the differences between predicted and actual performance may be significant. Restudy and simulation may produce new factors controlling the waterflood in the reservoir. Even the sophisticated reservoir simulator has some measure of uncertainty associated with the predicted results. Reservoir simulation, however, may be the only tool that can account for the major variables affecting waterflood performance.

6.6.4 Uncertainty in Technical Analyses

Prediction of waterflood performance inevitably involves making estimates of parameters that are not well known or perhaps not known at all. Rather than accept a single answer as the design choice, it is always a good procedure to identify key parameters that control the waterflood performance and to determine the sensitivity of the proposed design to reasonable uncertainties in these parameters. Sensitivity analyses are common in economic studies and should be used in a similar manner in technical analyses. A simple approach is to hold all variables constant and determine the effect of a single variable on performance. Sophisticated methods involving Monte Carlo simulation methods are used also. Again, the choice of the method to study sensitivity is measured against the degree of uncertainty in the data and the value of the expected outcome.

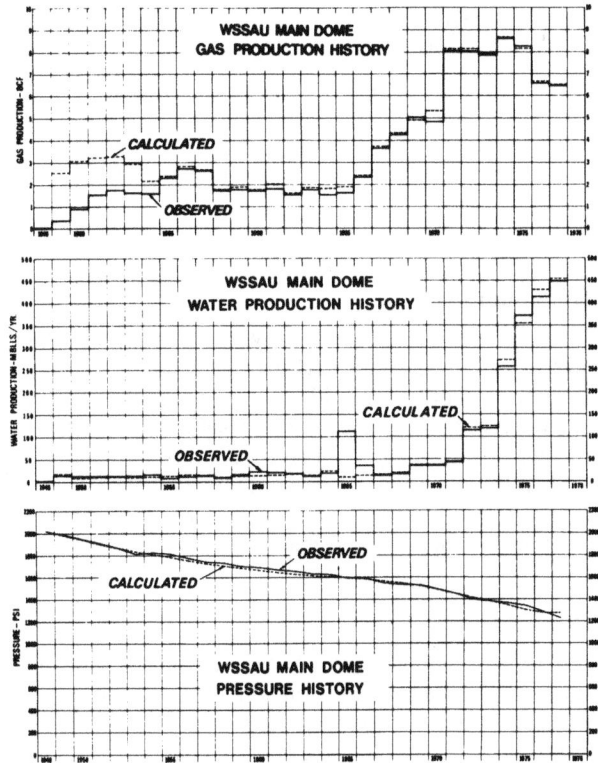

Fig. 6.44—Pressure and production history match on main dome.[37]

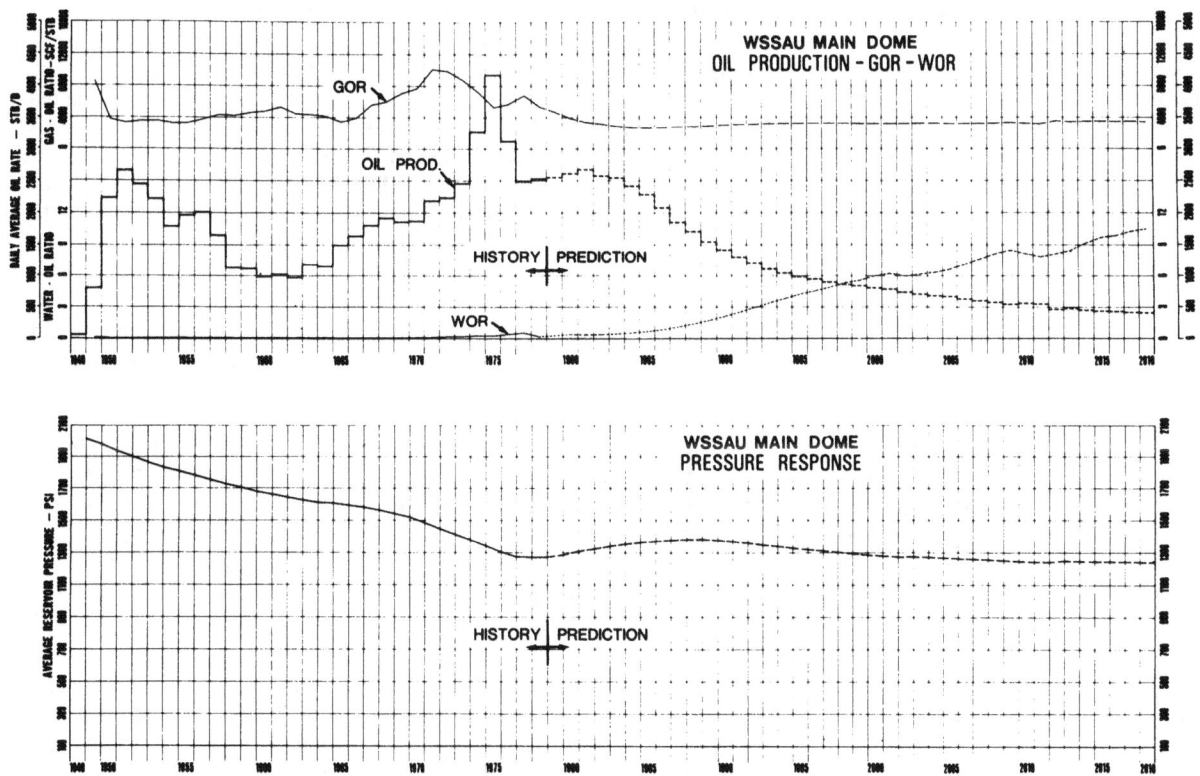

DETAILS OF PERFORMANCE UNDER CURRENT OPERATIONS

Fig. 6.45—Predicted field performance with continued operation of 40-acre five-spot pattern.[37]

TABLE 6.15—RESERVOIR ROCK AND FLUID PROPERTIES

k, md	10
h, ft	50
ϕ, fraction	0.15
Δp, psi	2,000
r_w, ft	0.75
μ_o, cp	2.75
μ_w, cp	1.0
S_{iw}	0.25
S_{or}	0.30

6.6.5 Summary

The reservoir engineer has many models available to design and/or to study a waterflood. Selection of the specified model requires matching the model to the reservoir and the economic potential. This chapter is developed so that waterflood design can be approached at the level of sophistication commensurate with the complexity of the reservoir and the decisions that must be made concerning the development and management of a waterflood.

Problems

6.1. In a confined five-spot pattern, there is one injection well for each production well. What is the ratio of injection to production wells if reservoirs of 160, 640, and 2,560 acres [64.8, 259, and 1036 ha] are developed on 20-acre [8.1-ha] five-spot spacing? Discuss the impact of using a representative five-spot pattern to predict waterflood performance.

6.2. One consideration in the selection of a waterflooding pattern is the rate at which water can be injected or oil produced. Consider a reservoir that was drilled with 40-acre [161 880-m^2] spacing for primary depletion. For the purpose of this problem, it is assumed that the waterflood will be conducted at unit mobility ratio. Determine injection rates for development of the field with direct linedrive, staggered linedrive, five-spot, and nine-spot patterns. Properties of the reservoir are given in Table 6.15. Relative permeability data are represented by Eqs. 6.76 and 6.77.

$$k_{ro} = 0.404(1 - S_{wD})^{1.77} \quad \dots \dots \dots \dots \dots \dots (6.76)$$

and

$$k_{rw} = 0.189 S_{wD}^{3.70}, \quad \dots \dots \dots \dots \dots \dots \dots \dots (6.77)$$

where S_{wD} is given by Eq. 3.111.

6.3. In Sec. 6.5.2, an approximate method for prediction of injection rate in a five-spot pattern when the mobility ratio is not unity was developed. This method is valid for areal sweep efficiency to about 50%. Compute the injection rates for $d/r_w = 50$ and 1,000 for a mobility ratio of 0.1. Compare your results with the relative injectivity determined from the Caudle and Witte correlation presented in Fig. 4.10. Why are the results different?

6.4. Rework Problem 6.3 for a mobility ratio of 2.0.

6.5. The waterflood performance of a five-spot pattern was simulated with the Higgins-Leighton model in Example 4.4. Compare injection rates predicted by the approximate method in Sec. 6.5.2.1 with the values in Table 7.

6.6. Estimate the injection rate as a function of PV's injected into each zone for the direct linedrive waterflood in Problem 5.4 with the assumption that the reservoir is homogeneous. The wellbore radius is 0.5 ft [0.15 m].

6.7. Derive an approximate formula for the injection rate in an inverted nine-spot following the procedures described in Sec. 6.5.1 for a staggered linedrive pattern. The ratio of corner-well production to side-well production rate is 1.5. In developing this model, assume that the dashed dividing line shown in Fig. 6.46 separates the region drained by the side well from the region drained by the corner well. The ratio of angle θ_1 to $\theta_1 + \theta_2$ is the fraction of the injection rate that is produced by the side well in the symmetric unit. Compare your equation with the exact expression in Table 6.1.

6.8. Following the derivation of Problem 6.7 and discussion in Sec. 6.5.2, derive an expression for the injection rate when the mobility ratio is not unity.

6.9. A reservoir has been produced by depletion gas drive to a pressure of about 5 psig [35 kPa]. The reservoir covers about 270 acres [1 092 656 m^2] and is completed on 10-acre [40 469-m^2] spacing. Reservoir data are summarized in Tables 6.16 and 6.17. It is planned to waterflood the reservoir with five-spot patterns by infill drilling.

The reservoir has three hydrocarbon intervals: a gas sand (5.8 ft [1.8 m]), a depleted oil sand (3.7 ft [1.1 m]), and an oil sand (20.5 ft [6.2 m]). All intervals are open in both injection and production wells. Assume that there is no vertical communication between the zones. Endpoint relative permeabilities are available but relative permeability relationships are not known. The waterflood will be operated at a constant pressure drop of 900 psig [6

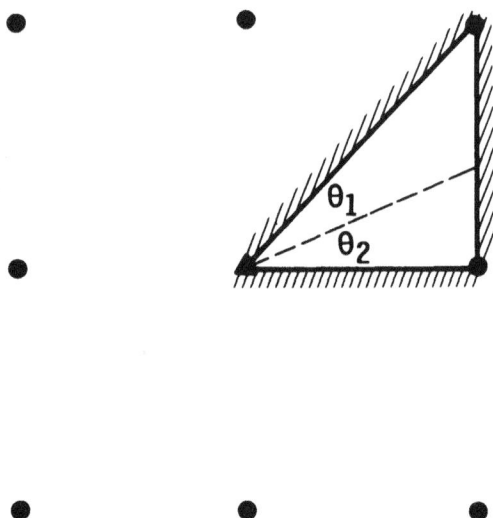

Fig. 6.46—Inverted nine-spot pattern with element of symmetry.

TABLE 6.16—PERMEABILITY AND THICKNESS DATA

Zone	Sand	Average Permeability (md)	Thickness (ft)
1	Gas	117.7	0.94
2	Gas	52.8	2.50
3	Gas	15.0	2.36
4	Depleted oil	86.2	1.40
5	Depleted oil	23.7	2.30
6	Oil	157.5	1.14
7	Oil	99.1	2.19
8	Oil	64.5	4.42
9	Oil	41.0	4.41
10	Oil	21.2	5.36
11	Oil	4.3	2.98

TABLE 6.17—PROPERTIES OF RESERVOIR AND FLUIDS

Zone	Gas Sand	Depleted Oil Sand	Oil Sand
Net sand, ft	5.80	3.70	20.50
Porosity	0.20	0.20	0.20
S_{iw}	0.35	0.35	0.35
S_o (beginning of flood)	0.11	0.21	0.51
S_g (beginning of flood)	0.54	0.44	0.14
S_{or}	0.11	0.21	0.17
k_{ro} at S_{iw}	0.00	0.00	0.30
k_{rw} at S_{or}	0.50	0.10	0.10
k_{rg}	0.50	0.30	0.10
Oil gravity, °API	36	—	—
Viscosity, cp			
Water	1		
Oil	8		
Gas	0.015		
Oil FVF	1.00		
Radius of wells, ft	0.25		

TABLE 6.18—RESERVOIR AND FLUID PROPERTIES

Area, acres	206
Net oil sand, ft	25.5
Porosity	0.21
S_{iw}	0.32
S_o (beginning of flood)	0.47
S_g (beginning of flood)	0.21
S_{or}	0.10
k_{ro} at S_{iw}	0.30
k_{rw} at S_{or}	0.10
k_{rg}	0.05
Oil gravity, °API	32
Viscosity, cp	
Oil	4.0
Water	0.7
Gas	0.011
Oil FVF	1.05
Radius of wells, ft	0.33

TABLE 6.19—PERMEABILITY AND THICKNESS DATA

Sample	Average Permeability (md)	Thickness (ft)
1	305.0	1.47
2	142.0	5.46
3	73.0	4.80
4	37.0	4.79
5	19.0	3.23
6	9.3	1.67
7	3.5	2.64
8	0.8	1.45

TABLE 6.20—SUMMARY OF AVERAGE MONTHLY PRODUCTION OIL RATE (B/D)

Month	Year 1	Year 2	Year 3	Year 4
January	1,480	1,230	890	820
February	1,380	1,300	930	865
March	1,350	1,120	920	800
April	1,220	1,200	850	820
May	1,356	1,160	890	805
June	1,420	1,120	920	750
July	1,290	1,040	855	720
August	1,235	1,060	900	780
September	1,140	1,030	830	810
October	1,145	960	850	770
November	1,165	1,050	900	780
December	1,120	960	880	735

MPa]. Estimate the oil production rate as a function of time until the water cut is 98%. The flood will be operated at sufficient pressure that trapped gas goes into solution.

6.10. Properties of a depleted reservoir are summarized in Table 6.18. The reservoir located at a depth of 2,500 ft [762 m] consists of a single oil sand with a net thickness of 25.5 ft [7.8 m]. The producing area is 206 acres [833 656 m²]. The field was developed on 20-acre [80 938-m²] spacing. It is planned to waterflood the field on 20-acre [80 938-m²] spacing by drilling infill wells to create five-spot patterns. Assume injection pressure is limited by a fracture gradient of 1.0 psi/ft [22.6 kPa/m] of depth. Production wells are operated with a backpressure of about 50 psig [345 kPa] on the formation. Predict the waterflood performance to a WOR of 50. Permeability distribution averaged from core data is given in Table 6.19.

6.11. Use the empirical method discussed in Sec. 6.6.1 to estimate waterflood performance for the reservoir described in Problem 6.10, assuming that the average water injection rate for the field is 3,500 B/D [556 m³/d]. The economic limit is 60 B/D [10 m³/d]. Field production at the beginning of the waterflood was 70 B/D [11 m³/d].

6.12. A reservoir has been under waterflood for 10 years. The water injection rate has been constant at about 8,800 B/D [1399 m³/d] for the majority of the flood. Average monthly oil production rates are given in Table 6.20 for a 3-year period. Estimate the volume of oil that is likely to be recovered if the waterflood continues to an economic limit of 100 B/D [16 m³/d].

Cumulative Oil Production (bbl)	Cumulative Water Production (bbl)	WOR	Water Cut (%)
578,721	989,884	2.86	74.12
625,969	1,125,844	2.66	72.66
668,434	1,228,335	2.74	73.25
708,392	1,351,496	2.81	73.77
747,166	1,449,794	2.62	72.40
783,119	1,547,547	2.90	74.33
817,313	1,652,934	3.49	77.73
851,608	1,786,533	3.92	79.66
883,612	1,912,601	4.16	80.61
913,345	2,043,149	4.29	81.09
942,872	2,166,711	4.51	81.85
970,186	2,299,502	4.58	82.07
995,613	2,408,179	3.80	79.17
1,019,513	2,486,967	3.36	77.09
1,042,851	2,567,133	3.58	78.18
1,064,115	2,646,741	3.75	78.96
1,085,278	2,726,349	3.75	78.96
1,105,484	2,802,008	3,83	79.32
1,125,173	2,879,338	3.97	79.90
1,143,977	2,954,966	4.10	80.38
1,162,268	3,031,288	4.30	81.12
1,179,333	3,106,946	4.42	81.56
1,196,770	3,183,858	4.45	81.67
1,213,680	3,259,934	4.59	82.11
1,230,110	3,336,846	4.78	82.71
1,245,573	3,412,504	5.05	83.46
1,260,378	3,489,567	5.24	83.97
1,274,708	3,565,135	5.37	84.30
1,288,681	3,641,495	5.48	84.56
1,302,373	3,716,610	5.63	84.92
1,315,576	3,792,970	5.81	85.31
1,328,531	3,868,500	6.01	85.74
1,338,694	3,931,995	7.36	88.03
1,353,054	4,048,927	10.26	91.12
1,365,762	4,209,753	12.86	92.79
1,378,049	4,370,400	11.41	91.94
1,391,308	4,501,216	9.76	90.70
1,404,097	4,624,502	10.04	90.94
1,417,196	4,761,104	12.26	92.46
1,429,750	4,938,977	14.48	93.54
1,442,352	5,125,427	15.64	93.99
1,454,413	5,324,717	16.29	94.22

TABLE 6.21—PRODUCED OIL AND WATER DATA—PROBLEM 6.13

Cumulative Oil Production (bbl)	Cumulative Water Production (bbl)	WOR	Water Cut (%)
191,530	1,928	0.01	1.00
225,831	2,210	0.01	1.00
258,476	2,681	0.01	1.00
351,898	3,605	0.01	1.40
625,352	6,186	0.01	0.80
1,223,175	15,655	0.02	2.20
1,790,610	35,690	0.05	4.50
2,312,541	71,650	0.08	7.30
2,789,068	138,519	0.19	16.30
3,161,755	254,315	0.36	26.20
3,528,379	404,335	0.43	30.20
3,874,513	595,084	0.63	38.50
4,223,013	833,440	0.73	42.10
4,585,502	1,150,953	0.97	49.20
4,951,254	1,543,150	1.08	51.90
5,312,933	1,976,370	1.27	56.00
5,643,467	2,439,496	1.54	60.60
5,964,412	2,957,634	1.65	62.30
6,256,792	3,510,597	1.99	66.60
6,558,588	4,132,590	2.16	68.40
6,871,200	4,833,724	2.29	69.60
7,163,971	5,513,439	2.42	70.80
7,420,640	6,262,036	3.10	75.60
7,686,932	7,064,046	2.92	74.50
7,971,668	7,905,123	3.12	75.70
8,258,932	8,824,849	3.44	77.50
8,530,956	9,816,896	3.76	79.00
8,821,112	10,868,298	3.81	79.20
9,113,009	12,034,986	3.67	78.60
9,398,438	13,066,711	3.88	79.50
9,684,807	14,127,484	3.63	78.40
9,992,441	15,314,703	4.13	80.50
10,282,572	16,594,479	4.56	82.00
10,541,951	17,809,107	4.68	82.40
10,760,920	18,881,768	5.17	83.80
10,969,148	20,163,482	5.94	85.60
11,197,873	21,618,316	6.81	87.20
11,423,393	23,188,377	7.47	88.20
11,624,745	24,695,808	8.09	89.00
11,803,789	26,181,679	7.20	87.80
11,951,950	27,490,350	9.42	90.40

TABLE 6.22—OIL AND WATER PRODUCTION DATA—PROBLEM 6.14*

*Data from I. Ershaghi, USC, 1985.
Initial oil in place = 54,696,000 bbl.

6.13. Estimation of waterflood performance in a fully developed waterflood with no major changes in operation may be done by plotting WOR vs. recovery on semilog paper and extrapolating a curve through the data to a WOR corresponding to the economic limit. A set of data from an ongoing waterflood is presented in Table 6.21. Plot these data with WOR on the log scale and estimate the total recovery when the WOR = 25.

6.14. A method to estimate waterflood recovery from water-cut/cumulative-production data was proposed by Ershaghi and Omoregie.[23] The method, developed with the Buckley-Leverett theory, is based on representation of relative permeability relationships by Eq. 6.78 in the region where data are available.

$$\frac{k_o}{k_w} = ae^{bS_w}. \quad \quad \quad \quad \quad (6.78)$$

When this correlation is valid, the oil recovery expressed as a fraction of the OOIP is given by Eq. 6.79.

$$E_R = \frac{1}{b(1-S_{wi})}\left[\ln\left(\frac{1}{f_w}-1\right)-\frac{1}{f_w}\right]$$
$$-\frac{1}{(1-S_{wi})}\left(S_{wi}+\frac{1}{b}\ln A\right), \quad \quad (6.79)$$

where $A = a\mu_w/\mu_o$.

If E_R is plotted vs. $-\{\ln[(1/f_w)-1]-(1/f_w)\}$, the groups $1/[b(1-S_{wD})]$ and $1/(1-S_{wi})[S_{wi}+(1/b)\ln A]$ can be found graphically or with a least-squares program. Recovery at higher water cuts is determined by extrapolation.

Table 6.22 contains production and recovery data for a fully developed waterflood that is expected to operate with no major changes. With Ershaghi and Omoregie's method, estimate the oil recovery to a WOR of 20. Compare your estimate to one you obtain by plotting WOR vs. recovery as described in Problem 6.13.

TABLE 6.23—RESERVOIR AND FLUID PROPERTIES

S_o	0.558
S_{or}	0.144
S_g	0.242
S_{iw}	0.200
k_{rw} at S_{or}	0.10
k_{ro} at S_{iw}	0.133
k_{rg}	0.148
μ_o, cp	3.5
μ_w, cp	0.87
μ_g, cp	0.013
ϕ	0.176

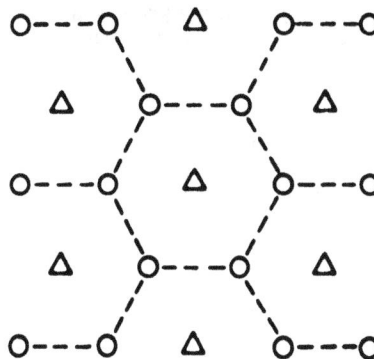

Fig. 6.47—Inverted seven-spot pattern.

6.15. Reservoir and fluid properties are given in Tables 6.23 and 6.24 for a representative 20-acre [80 938-m^2] five-spot pattern that is to be waterflooded. Assume that the waterflood will be conducted at a constant pressure drop of 1,000 psi [6.9 MPa]. You are to predict the waterflood performance to a WOR of 100. Only relative permeability endpoints are available and thus the shape of the relative permeability curves is not known. You can determine the effects of this uncertainty on your predicted results, however, by making several runs in which the relative permeabilities are assumed to fit the dimensionless relationships in Eqs. 3.143 and 3.144.

$$k_{ro} = \alpha_1 (1 - S_{wD})^m$$

and

$$k_{rw} = \alpha_2 S_{wD}{}^n,$$

where S_{wD} is given by Eq. 3.111.

Select appropriate values of m and n and evaluate the effects of the uncertainty in these parameters on the estimated waterflood performance.

6.16. An oil field will be waterflooded in an inverted seven-spot pattern as shown in Fig. 6.47.*

A fully developed seven-spot pattern has two producers for every injector.

1. With the concepts developed in Sec. 6.5, develop an approximate equation for injectivity in this system for the case where the injection well radius is different from the producing well radii and where the mobility ratio is 1.0.

2. Based on Eq. 6.4 (Table 6.1), what is the *exact* equation for Part 1?

3. If the mobility ratio is not 1.0, how should the equation in Part 2 be changed to be correct? You do not have to derive it in detail; merely indicate how and why it must be changed, and write it down.

*Personal communication with W.E. Brigham, Stanford U. (1980).

TABLE 6.24—PERMEABILITY AND THICKNESS DISTRIBUTION

Layer	k_w (md)	Thickness (ft)
1	1,636	0.88
2	524	4.88
3	188	6.10
4	60	6.05
	Total	17.91

4. In the system sketched in Fig. 6.47, the distance between wells is 1,000 ft [305 m] and the injection well radii are 10 ft [3 m].* (They have been fractured.) The producing well radii are 0.25 ft [0.08 m]. The effective permeability to oil at irreducible water saturation is 100 md, and the initial (irreducible) water saturation is 10%.

The average water saturation behind the front is 56.3%. The formation thickness is 10 ft [3 m], the porosity 20%, the oil viscosity 1.0 cp [1.0 mPa·s], the water viscosity 0.5 cp [0.5 mPa·s], and the water relative permeability 0.25. The system is liquid-filled at the time injection is started.

If the pressure drop from injectors to producers is 3,000 psi [21 MPa], what will be the injection rate into a well at the time 50,000 bbl [7949 m^3] of water have been injected?

6.17. An undersaturated reservoir is considered to have a shape as shown in Fig. 6.48.* Reservoir and aquifer properties are given in Table 6.25. PVT data for the reservoir are presented in Fig. 6.49.

The reservoir has a connecting aquifer that is suspected of extending 60,000 ft [18 288 m] beyond the WOC. To maintain the production and pressure profile as presented in Table 6.26, it might be considered necessary to inject water at the WOC. On the basis of the data given, determine whether such a water-injection program is required, and if it is, the water-injection profile for the first 2 years.

*Personal communication with W.E. Brigham, Stanford U. (1980).

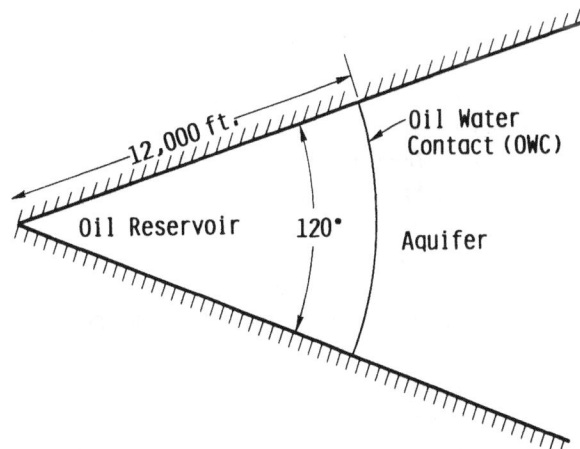

Fig. 6.48—Reservoir with edgewater drive in Problem 6.17.

**TABLE 6.25—RESERVOIR AND
AQUIFER PROPERTIES FOR 6.17**

Reservoir Properties*	
Stock-tank oil in place, STB	530×10^6
Interstitial water saturation, S_{iw}	0.05
Reservoir and Aquifer Properties	
Thickness of reservoir, ft	100
Permeability, md	300
Viscosity of water, cp	0.55
Porosity	0.20
Water compressibility, psi^{-1}	3.0×10^{-6}
Effective rock compressibility, psi^{-1}	4.0×10^{-6}
Water FVF	1.0

*The PVT data are presented in Fig. 6.47.

**TABLE 6.26—RESERVOIR PRESSURE HISTORY DURING
PRIMARY PRODUCTION UNDER AN ACTIVE
WATER DRIVE—PROBLEM 6.17**

Time (years)	Average Reservoir Pressure* (psia)	Cumulative Oil Production** (10^6 STB)
0	4,200	0
0.5	3,900	6
1	3,700	20
1.5	3,540	35
2	3,400	50
2.5	3,300	65
3	3,200	80

*It can be assumed that the pressure decline at the OWC is
the same as that in the oil reservoir.
**Although there might be some water production, the quantity
will probably be negligible for the first 5 years.

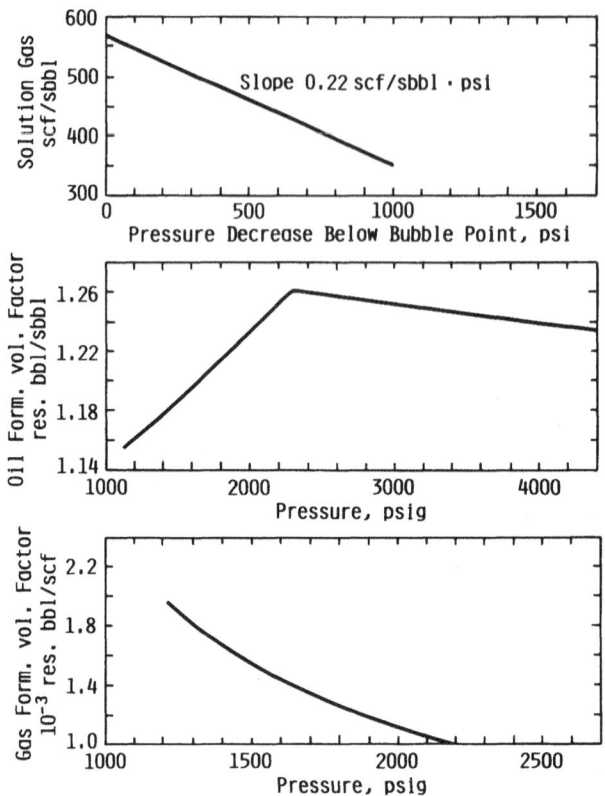

Fig. 6.49—PVT properties for reservoir fluids.

Nomenclature

a = distance between producing wells, or constant in Eq. 6.78, ft [m]

A = constant in Eq. 6.79

b = length of segment—staggered linedrive model, or constant in Eq. 6.78, ft [m]

B_o = oil FVF, STB/bbl [stock-tank m^3/m^3]

B_{oi} = initial oil FVF, STB/bbl [stock-tank m^3/m^3]

d = distance from injection well to producing well, ft [m]

D = decline rate in exponential rate equation, month^{-1}

E_A = areal sweep efficiency, fraction

E_I = vertical sweep efficiency, fraction

E_R = waterflood recovery, fraction OOIP

E_V = volumetric sweep efficiency, fraction

f_{cw} = fractional flow of corner well—nine-spot pattern, fraction

f_p = fractional flow of producing well, fraction

f_{sw} = fractional flow of side well—nine-spot pattern, fraction

f_w = fractional flow of water, fraction

f'_{Sw} = derivative of f_w with respect to S_w

G_j = shape factor in Higgins-Leighton model, ft^{-1} [m^{-1}]

h = thickness of formation, ft [m]

i = injection rate, B/D [m^3/d]

i^* = injection rate for half of a staggered linedrive pattern

i_b = injection rate at unit mobility when oil is flowing under the applied pressure drop at S_{iw}, B/D [m^3/d]

I = incline rate in exponential rate equation, month^{-1}

I_{AF} = injectivity, B/D-acre-ft [m^3/d·m^3]

k = permeability, darcies

k_o = permeability to oil, darcies

k_w = permeability to water, darcies

M = endpoint mobility ratio based on viscosity ratio in miscible flooding

$M_{\bar{S}}$ = mobility ratio based on average water saturation behind the flood front

M_t = total mobility ratio—mobility of fluids behind the front to mobility of fluids ahead of front

N = initial oil in place, STB [stock-tank m^3]

N_p = oil produced by primary, STB [stock-tank m^3]

N_{pw} = oil produced by waterflood, STB [stock-tank m^3]

N_r = oil remaining after the reservoir has been waterflooded, STB

p_e = pressure in reservoir at effective radius, r_e, psi [kPa]

p_{ei} = pressure in reservoir at effective radius, r_{ei}, psi [kPa]

p_f = pressure in reservoir at radius r_f, psi [kPa]

p_p = pressure in producing well, psi [kPa]

p_w = pressure in well, psi [kPa]

p_{wi} = pressure in injection well, psi [kPa]

p_{wp} = pressure in producing well, psi [kPa]

p_1 = pressure at radius r_1, psi [kPa]

p_2 = pressure at radius r_2, psi [kPa]

q_c = flow rate from corner well, B/D [m^3/d]

q_o = oil production rate, B/D [m^3/d]

q_{oi} = initial oil production rate or oil production rate at beginning of waterflood, B/D [m^3/d]

q_{ol} = oil production rate at economic limit, B/D [m^3/d]

q_{op} = peak oil production rate, B/D [m^3/d]

q_s = flow rate in side well, B/D [m^3/d]

q_t = total production/injection rate, B/D [m^3/d]

q_w = water production rate, B/D [m^3/d]

Q_i = PV's of water injected, dimensionless

r_e = effective radius, ft [m]

r_{ei} = effective radius of injection well, ft [m]

r_{ep} = effective radius of production well, ft [m]

r_f = radius of flood-front saturation, ft [m]

r_{ob} = radius of oil bank, ft [m]

r_w = wellbore radius, ft [m]

r_{wi} = radius of injection well, ft [m]

r_{wp} = radius of production well, ft [m]

r_{Swf} = radius of flood-front saturation, ft [m]

r_1 = radius of Position 1, ft [m]

r_2 = radius of Position 2, ft [m]

s_i = skin factor for injection well

s_p = skin factor for production well

S_{gi} = initial gas saturation, fraction

S_{gm} = mobile gas saturation, fraction

S_{gt} = trapped gas saturation, fraction

S_{iw} = interstitial water saturation where water phase is immobile under the applied pressure gradient, fraction

S_o = oil saturation, fraction

S_{oi} = initial oil saturation, fraction

S_{or} = residual oil saturation, fraction

S_w = water saturation, fraction

$\overline{S}_w$ = average water saturation

$\overline{S}_{wD}$ = dimensionless water saturation, fraction

$\overline{S}_{wf}$ = average water saturation behind the flood front, fraction

t_f = fill-up time, days

t_1 = time to reach economic limit, days

t_p = time to peak oil production rate, days

V_d = number of displaceable hydrocarbon PV's = $Q_i/(1 - S_{or} - S_{iw})$

V_p = pore volume, bbl [m^3]

W_i = cumulative volume of water injected, bbl [m^3]

W_{if} = cumulative volume of water injected to fill-up, bbl [m^3]

W_{ii} = cumulative volume of water injected to interference, bbl [m^3]

α_1 = relative permeability to oil at interstitial water saturation, dimensionless

α_2 = relative permeability to water at residual oil saturation, dimensionless

γ = relative injectivity or conductance

λ_d = mobility of displaced fluid, darcy/cp [darcy/Pa·s]

λ_D = mobility of displacing fluid, darcy/cp [darcy/Pa·s]

λ_o = mobility of oil phase, darcy/cp [darcy/Pa·s]

λ_r = relative mobility, cp^{-1} [$Pa·s^{-1}$]

λ_r^{-1} = apparent viscosity, cp [Pa·s]

$\overline{\lambda}_{rf}^{-1}$ = average apparent viscosity for radial system, cp [Pa·s]

λ_{ro} = relative mobility of oil phase, cp^{-1} [$Pa·s^{-1}$]

λ_{rw} = relative mobility of water phase, cp^{-1} [$Pa·s^{-1}$]

λ_w = mobility of water phase, darcy/cp [darcy/Pa·s]

μ_d = viscosity of displaced phase, cp [Pa·s]

μ_D = viscosity of displacing phase, cp [Pa·s]

μ_o = oil viscosity, cp [Pa·s]

μ_w = water viscosity, cp [Pa·s]

Superscripts

m = oil exponent relative permeability correlation

n = water exponent relative permeability correlation

References

1. Multizone Waterflooding Operations, Lansing-Kansas City Limestones, Kansas, *Proc.*, SPE/AIME, Great Bend and Wichita Sections Meeting (1970) 72.
2. Wayhan, D.A. and McCaleb, J.A.: "Elk Basin Madison Heterogeneity—Its Influence on Performance," *J. Pet. Tech.* (Feb. 1969) 153-59.
3. Hunter, Z.A.: "Progress Report, North Burbank Unit Waterflood—January 1, 1956," *Drill. and Prod. Prac.*, API (1956) 262-73.
4. Deppe, J.C.: "Injection Rates— The Effect of Mobility Ratio, Area Swept and Pattern," *Soc. Pet. Eng. J.* (June 1961) 81-91; *Trans.*, AIME, **222**.
5. Earlougher, R.C. Jr.: *Advances in Well Test Analysis*, Monograph Series, SPE, Richardson, TX (1977) **5**.
6. Lee, W.J.: *Well Testing*, Textbook Series, SPE, Richardson, TX (1982) **1**.
7. Craig, F.F. Jr.: "Effect of Reservoir Description on Performance Predictions," *J. Pet. Tech.* (Oct. 1970) 1239-45.
8. Caudle, B.H. and Witte, M.D.: "Production Potential Changes During Sweepout in a Five-Spot System," *J. Pet. Tech* (Dec. 1959) 63-65; *Trans.*, AIME, **216**.
9. Hurst, W.: "Determination of Performance Curves in Five-Spot Waterflood," *Pet. Eng.* (1953) **25**, B40-46.
10. Prats, M. *et al.*: "Prediction of Injection Rate and Production History for Multifluid Five-Spot Floods," *J. Pet. Tech.* (May 1959) 98-105; *Trans.*, AIME, **216**.
11. Craig, F.F. Jr.: *The Reservoir Engineering Aspects of Waterflooding*, Monograph Series, SPE, Richardson, TX (1971) **3**, 78-95.
12. Dykstra, H. and Parsons, R.L.: "The Prediction of Oil Recovery by Waterflood," *Secondary Recovery of Oil in the United States*, API, Dallas (1950) 160-74.

13. Bush, J.L. and Helander, D.P.: "Empirical Prediction of Recovery Rate in Waterflooding Depleted Sands," *J. Pet. Tech.* (Sept. 1968) 933–43.

14. Arps, J.J.: "Analysis of Decline Curves," *Trans.*, AIME (1945) **160**, 228–47.

15. Shea, G.B., Higgins, R.V., and Lechtenberg, H.J.: "Decline and Forecast Studies Based on Performances of Selected California Oil Fields," *J. Pet. Tech.* (Sept. 1964) 959–65.

16. Ramsay, H.J. Jr. and Guerrero, E.T.: "The Ability of Rate-Time Decline Curves to Predict Production Rates," *J. Pet. Tech.* (Feb. 1969) 139–41.

17. Slider, H.C.: "A Simplified Method of Hyperbolic Decline Curve Analysis," *J. Pet. Tech.* (March 1968) 235–36.

18. Meehan, D.N.: "Exponential, Harmonic Declines Programmed for Calculators," *Oil and Gas J.* (May 28, 1980) 95–100.

19. Meehan, D.N.: "Hyperbolic Oil Production Decline Analysis Programmed," *Oil and Gas J.* (June 9, 1980) 52–56.

20. Hom, D.H.: "Calculator Program Determines Decline Curves," *Oil and Gas J.* (Aug. 3, 1981) 102–04.

21. Meehan, D.N.: "Three New Programs Aid Decline Curve Analysis By Calculator," *Oil and Gas J.* (April 20, 1981) 90–96.

22. Hollo, R.: "Program Handles Two Decline-Curve Types," *Oil and Gas J.* (Sept. 7, 1981) 154–56.

23. Ershaghi, I. and Omoregie, O.: "A Method for Extrapolation of Cut vs. Recovery Curves," *J. Pet. Tech.* (Feb. 1978) 203–04.

24. Craig, F.F. Jr., Geffen, T.M., and Morse, R.A.: "Oil Recovery Performance of Pattern Gas or Water Injection Operations from Model Tests," *J. Pet. Tech.* (Jan. 1955) 7–15; *Trans.*, AIME, **204**.

25. Kimbler, O.K., Caudle, B.H., and Cooper, H.E. Jr.: "Areal Sweepout Behavior in a Nine-Spot Injection Pattern," *J. Pet. Tech.* (Feb. 1964) 199–202; *Trans.*, AIME, **231**.

26. Leighton, A.J. and Higgins, R.V.: "Improved Method to Predict Multiphase Waterflood Performance for Constant Rates or Pressures," RI 8055, USBM (1975).

27. Higgins, R.V. and Leighton, A.J.: "Matching Calculated With Actual Waterflood Performance by Estimating Some Reservoir Properties," *J. Pet. Tech.* (May 1974) 501–06.

28. Diaz Correa, P.A.: "A Stream-Tube Model for Simulation of Fluid Flow in Porous Media," MS thesis, U. of Kansas, Lawrence (1980).

29. Vossoughi, S.: "TORP Stream Tube Model," Tertiary Oil Recovery Project, U. of Kansas, Lawrence (July 1980).

30. Wang, G.C.: "Streamline Model Studies of Viscous Waterfloods," PhD dissertation, U. of Texas, Austin (1970).

31. James, L.J.K.: "An Image-Well Method for Bounding Arbitrary Reservoir Shapes in the Streamline Model," PhD dissertation, U. of Texas, Austin (1972).

32. Parsons, R.W.: "Directional Permeability Effects in Developed and Unconfined Five-Spots," *J. Pet. Tech.* (Apr. 1972) 487–94; *Trans.*, AIME, **253**.

33. Martin, J.C., Woo, P.T., and Wegner, R.E.: "Failure of Stream Tube Methods To Predict Waterflood Performance of an Isolated Inverted Five-Spot at Favorable Mobility Ratios," *J. Pet. Tech.* (Feb. 1973) 151–53.

34. Martin, J.C. and Wegner, R.E.: "Numerical Solution of Multiphase, Two-Dimensional Incompressible Flow Using Stream-Tube Relationships," *Soc. Pet. Eng. J.* (Oct. 1979) 313–23; *Trans.*, AIME, **267**.

35. *Numerical Simulation*, Reprint Series, SPE, Richardson, TX (1973) **11**.

36. Fanchi, J.R. and Harpole, K.J.: "Boast: A Three-Dimensional, Three-Phase Black Oil Applied Simulation Tool (Version 1.1), Volume I: Technical Description and Fortran Code," DOE/BC/10033-3, U.S. DOE (Sept. 1982).

37. Harpole, K.J.: "Improved Reservoir Characterization—A Key to Future Reservoir Management for the West Seminole San Andres Unit," *J. Pet. Tech.* (Nov. 1980) 2009–19.

38. Barrett, D.D., Harpole, K.J., and Zaaza, M.W.: "Reservoir Data Pays Off: West Seminole San Andres Unit, Gaines County, Texas" paper SPE 6738 presented at the 1977 SPE Annual Technical Conference and Exhibition, Denver, Oct. 9–12.

39. Hirasaki, G.J., Morra, F. Jr., and Willhite, G.P.: "Estimation of Reservoir Heterogeneity from Waterflood Performance," paper SPE 13415 available from SPE, Richardson, TX.

SI Metric Conversion Factors

$$
\begin{array}{rlll}
\text{acre} & \times\ 4.046\ 873 & \text{E}+03 & = \text{m}^2 \\
\text{acre-ft} & \times\ 1.233\ 489 & \text{E}+03 & = \text{m}^3 \\
{}^{\circ}\text{API} & 141.5/(131.5+{}^{\circ}\text{API}) & & = \text{g/cm}^3 \\
\text{bbl} & \times\ 1.589\ 873 & \text{E}-01 & = \text{m}^3 \\
\text{cp} & \times\ 1.0* & \text{E}-03 & = \text{Pa}\cdot\text{s} \\
\text{cu ft} & \times\ 2.831\ 685 & \text{E}-02 & = \text{m}^3 \\
\text{ft} & \times\ 3.048* & \text{E}-01 & = \text{m} \\
{}^{\circ}\text{F} & ({}^{\circ}\text{F}-32)/1.8 & & = {}^{\circ}\text{C} \\
\text{psi} & \times\ 6.894\ 757 & \text{E}+00 & = \text{kPa}
\end{array}
$$

*Conversion factor is exact.

Chapter 7
The Role of Reservoir Geology in the Design and Operation of Waterfloods

7.1 Introduction

Chaps. 3 through 6 focused on microscopic and macroscopic displacement efficiency of waterflooding in porous rocks and the design of waterfloods. The effects of reservoir heterogeneity on waterflood performance were illustrated by the development of displacement models for common geological variations, such as a layered reservoir with and without crossflow. Throughout this book the reader has been reminded that reservoir rocks were created by geological processes. It should not be surprising, then, that knowledge of these processes could help determine how important engineering properties, such as permeability and porosity, vary at points throughout the reservoir.

This chapter will emphasize the importance of geology in the design of fluid displacement processes and will present a methodology for developing an adequate reservoir model. Much of the methodology relies on geological concepts and principles that may not be familiar to engineers. It is necessary for engineers to understand the concepts used to develop a geologic model and the limitations that are inherent in the process.

7.2 Field Cases Illustrating the Importance of Reservoir Description

Waterflooding has been practiced widely throughout the world. There are many examples of waterflood experience that illustrate the importance of developing a sound geological model (i.e., a reservoir description) as part of the waterflood design. The approach taken in this chapter builds on this experience by examining waterfloods that are in progress or have been completed.

East Texas Field. Petroleum engineers often are introduced to waterflooding when material balance concepts are covered in reservoir engineering. Water-drive reservoirs are those associated with an aquifer. Part or all of the reservoir energy may be supplied by the aquifer. The East Texas field shown in Fig. 7.1 is a classic example of a strong, natural-water-drive reservoir underlain by a huge aquifer. A map showing the pressure decline in the aquifer as a result of oil production from the East Texas and Mexia pools is shown in Fig. 7.2. Fig. 7.3 is a west/east cross section showing the movement of the oil/water contact after 25 years of production.[4]

The natural water drive is effective in this East Texas field for several reasons: (1) the amount of energy available in the aquifer was large and was supplemented with saltwater injection into watered-out wells; (2) the aquifer has good hydraulic contact with the reservoir; (3) the East Woodbine sand is quite uniform with good continuity; and (4) the sand exhibits intermediate wettability with residual oil saturation of about 0.15.[5] The uniformity of the Woodbine sand throughout the reservoir means volumetric sweep will be on the order of 80 to 90% when the reservoir is abandoned.

Intisar "A" Field. The selection of a waterflooding plan often is limited by the geology of the reservoir. In this example, unique geologic properties permitted the use of a bottomwater drive to deplete the reservoir.

The Intisar reservoir complex in Fig. 7.4 was discovered in 1967.[6] Oil was found in carbonate reefs at depths of 8,900 to 10,000 ft [2713 to 3048 m]. Fig. 7.5 shows a cross section through the Intisar "A" Reef reservoir. This reservoir had a gross oil column of 1,002 ft [305 m] at the thickest point. Log analyses indicated that the oil column was essentially continuous from the oil/water contact (OWC) to the top of the reef. A field structure map for Intisar "A" (Fig. 7.6) shows the shape of the reservoir. Reservoir properties are summarized in Table 7.1.

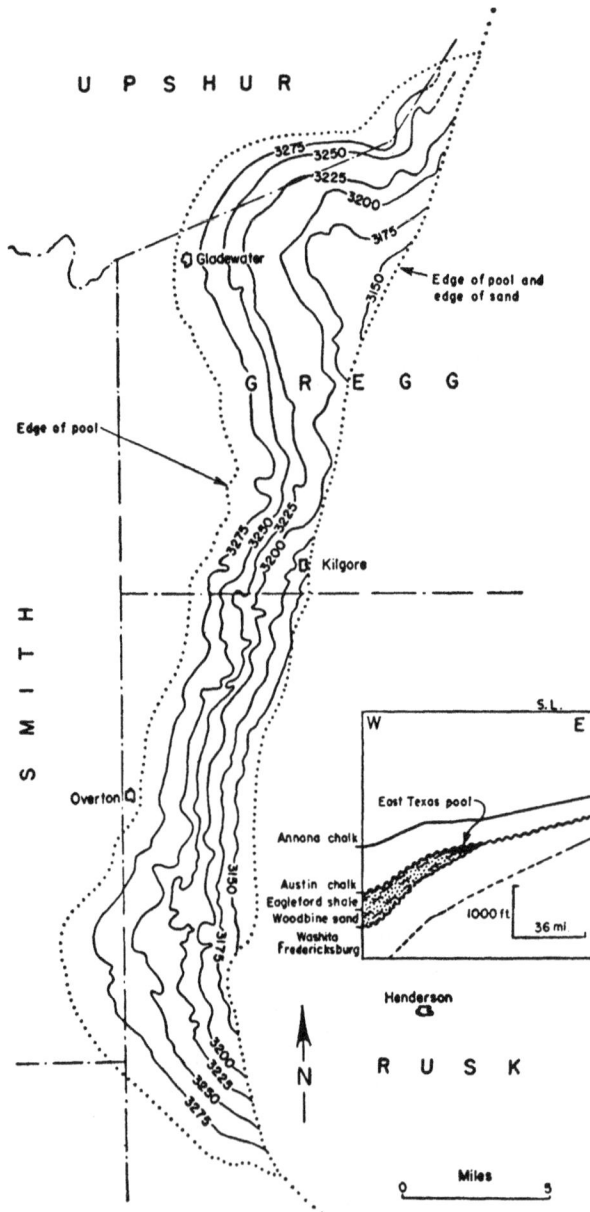

Fig. 7.1—Structure on top of the Woodbine (Cretaceous producing sand in the East Texas pool). The intersection of two unconformity planes along the east boundary marks the edge of the trap (Ref. 1, Fig. 8-3, redrawn from Ref. 2).

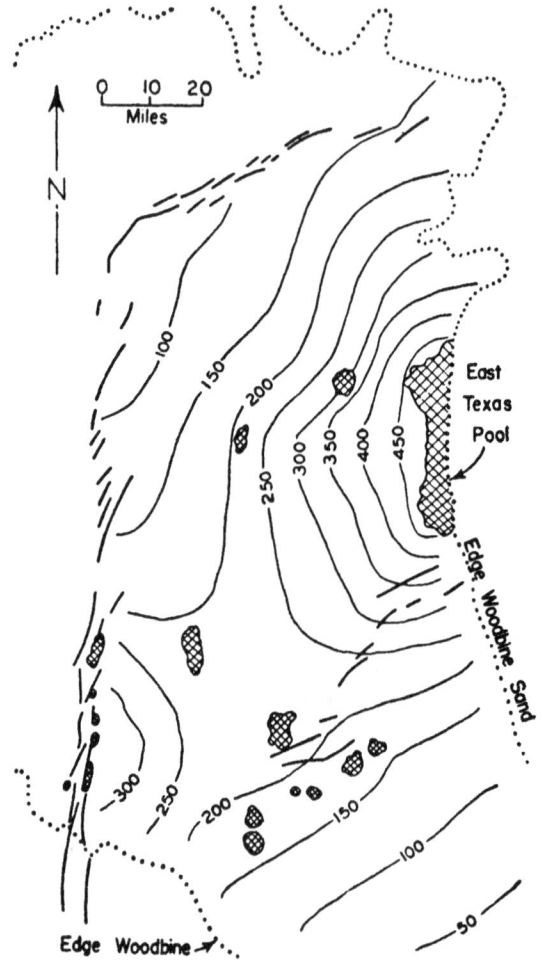

Fig. 7.2—Map showing the decline in pressure by contours (isobars) in East Texas–Tyler basin as a result of the production of oil in the East Texas and Mexia pools. Contour interval is 50 psi (Ref. 1, Fig. 10-17, redrawn from Ref. 3).

Primary recovery from the reservoir was estimated to be low because the oil was highly undersaturated. Although an OWC was present at the base of the reef, there was no evidence of a natural water drive. Reservoir energy was thought to be limited to fluid and rock expansion and to solution gas drive.

A bottomwater injection program was selected for pressure maintenance in this field. Water was injected below the OWC in the 29 wells shown in Fig. 7.6. The applicability of a bottomwater flood depends on the changes in rock properties within the reservoir that are associated with geology. For a bottomwater flood to be effective, the reservoir must have good communication in the horizontal and vertical directions with no barriers to vertical flow.

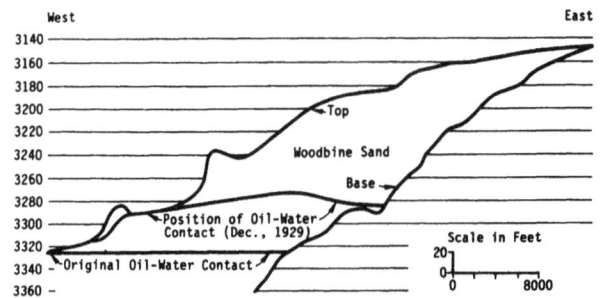

Fig. 7.3—West/east cross section, East Texas field. [4]

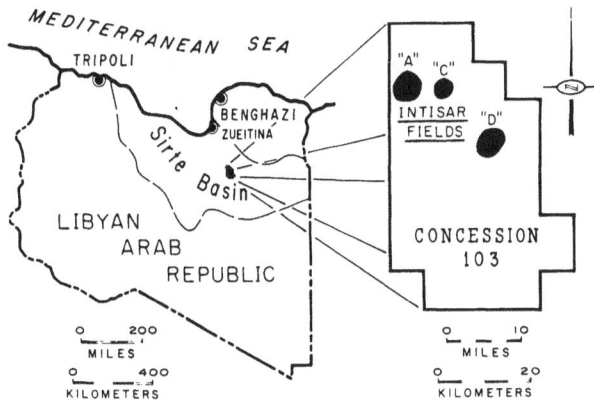

Fig. 7.4—Location map.[6]

The reservoir consists of three major carbonate units (Fig.7.5). Although porosity and permeability vary between the units, permeability barriers to vertical flow were not detected from core or log analysis. The ratio of vertical to horizontal permeability was about 0.75, indicating good communication in both horizontal and vertical directions.

The reservoir pressure history for the three intervals in the Intisar "A" field is shown in Fig. 7.7. Reservoir pressure declined rapidly with fluid withdrawal before water injection. By July 1968, the reservoir pressure had decreased from 4,532 to about 3,700 psig [31 to about 26 MPa] when the cumulative production was 40×10^6 bbl [6×10^6 m^3]. Water injection began in July 1968 and pressure decline was arrested in about 2 years. Fig. 7.7 also includes reservoir pressures computed in Dec. 1982 using a reservoir simulator.

Field performance through the end of 1983 is shown in Fig. 7.8. At that time, the field had produced

683×10^6 bbl [109×10^6 m^3] oil, and 1.17×10^9 bbl [2×10^8 m^3] water had been injected. Ultimate recovery is estimated to be 750×10^6 bbl [119×10^6 m^3], which is almost 50% of the stock-tank oil originally in place.

Elk Basin Madison Waterflood. Production performance can indicate that the geological model of the reservoir is not adequate. The Madison reservoir in the Elk basin anticline is shown in Figs. 7.9 and 7.10.[7] Initial wells were completed open hole through the entire section (Fig. 7.9). Performance during the first 10 years was characteristic of a strong water drive in that a small pressure decline was observed. For this reason, the entire vertical section was assumed to be under active water drive. The reservoir was considered homogeneous with tight streaks that were believed not to correlate.

The operating plan for the reservoir was based on this reservoir description. Water influx rates were estimated to be sufficient to maintain a fluid withdrawal rate of 25,000 B/D [3795 m^3/d] while maintaining reservoir pressure above the bubblepoint (1,140 psi [7860 kPa] at −700 ft [−213 m] datum). Production would be taken from the wells located near the top of the structure as the water moved updip, displacing the oil. Edge wells were planned as needed to ensure good areal-sweep efficiency and to accelerate production.

As development proceeded, interpretation of well logs led to the identification of four distinct rock types in the cross section, which were designated Zones A, B, C, and D. These zones were believed to be still under active water drive but perhaps not in communication. The first indication of an inadequate reservoir description came during 1958–61, when several wells were drilled and completed in single zones. Production rates from wells completed in Zone A, as well as bottomhole pressure (BHP), declined with time. In contrast, wells completed in Zones B, C, or D responded as expected under a water

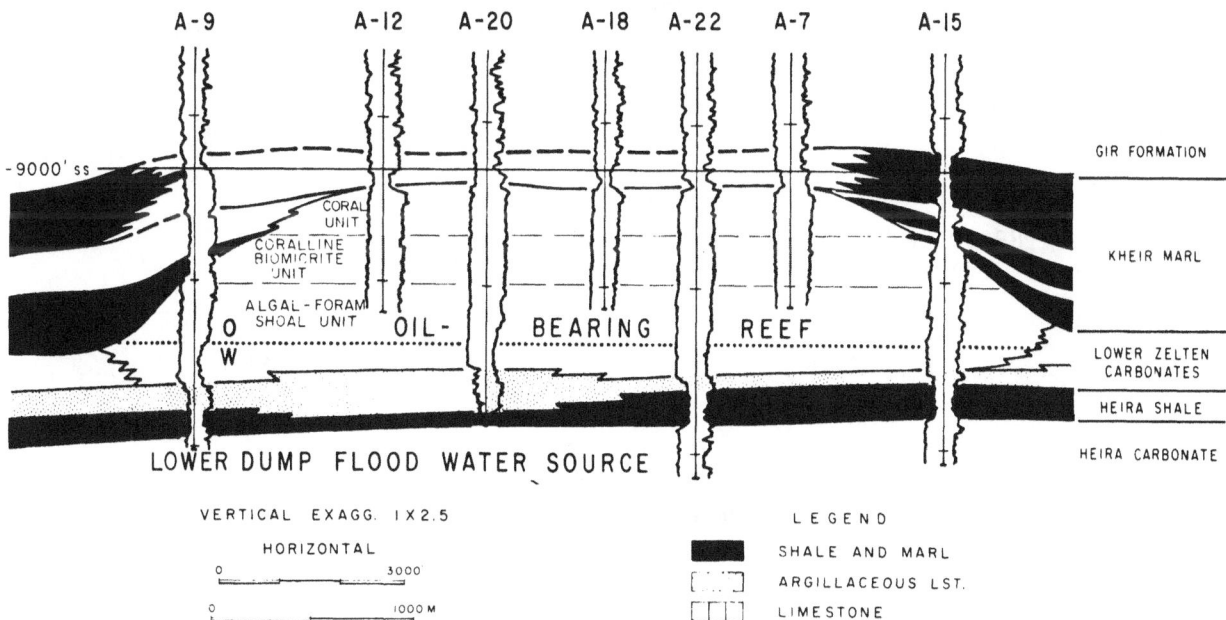

Fig. 7.5—Intisar "A" field cross section.[6]

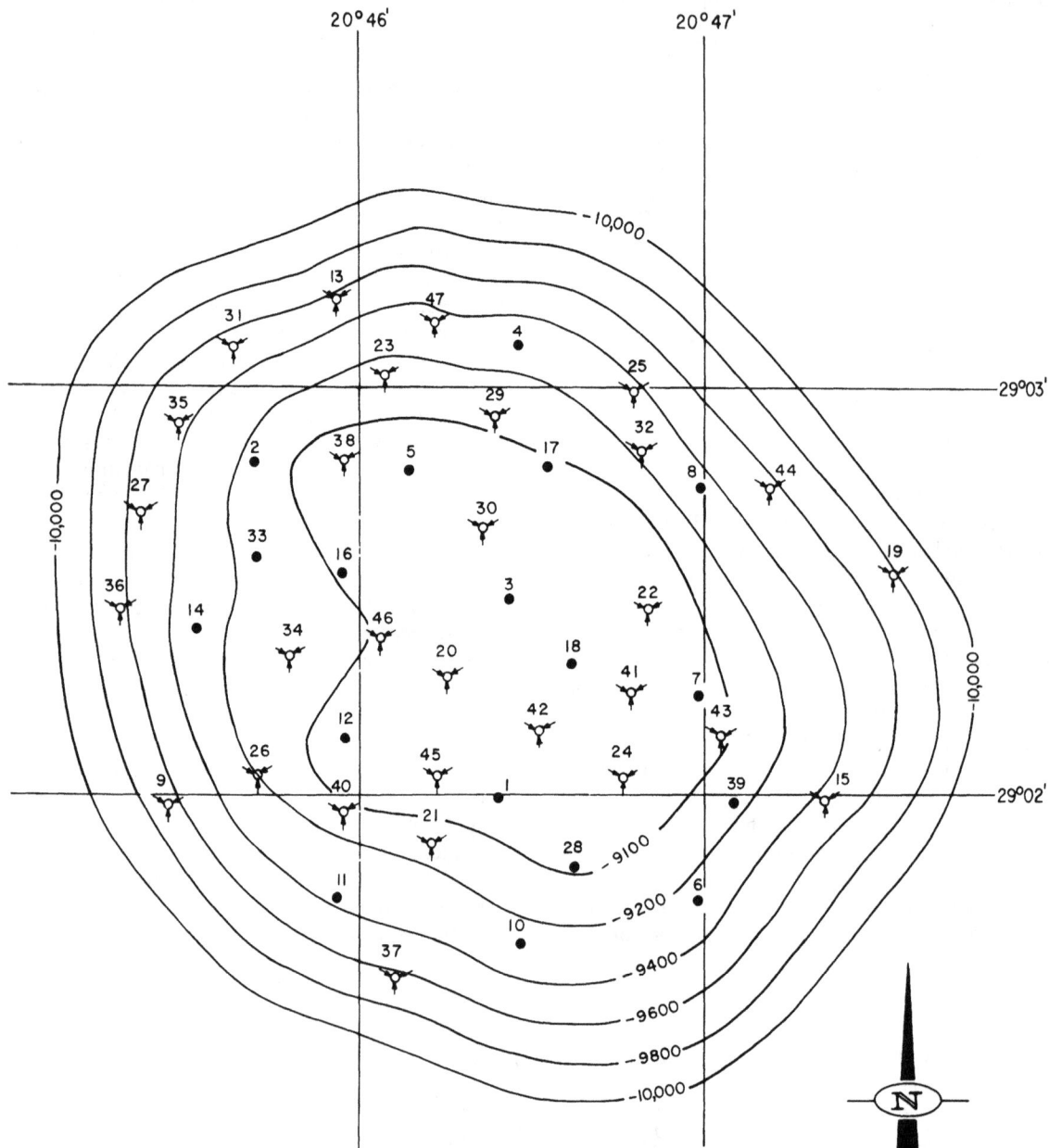

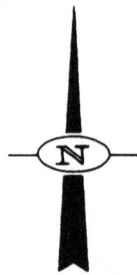

Fig. 7.6—Intisar "A" field structure map.[6]

TABLE 7.1—RESERVOIR PROPERTIES, INTISAR "A" FIELD[6]

Average horizontal permeability to air, md	26
Average net porosity	20
Connate water, %	12
Oil gravity, °API	45
Initial reservoir pressure at datum, psig	4,532
Saturation pressure, psig	3,240
Initial solution GOR, scf/STB	1,336
Initial oil FVF, RB/STB	1.81
Reservoir temperature, °F	237
Oil viscosity at initial pressure, cp	0.15

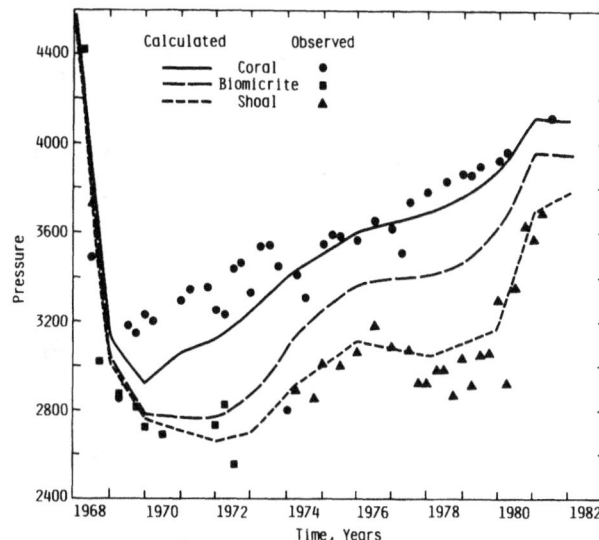

Fig. 7.7—Pressure history during depletion of Intisar "A" (Personal communication with C.L. DesBrisay, Occidental Petroleum Co., 1984).

drive. Increased fluid withdrawals accelerated the decline of production rates with accompanying increases in GOR. Although part of the produced water was reinjected into interior Zone A injection wells, response to water injection was temporary.

The geology of this reservoir was revealed from an intensive data gathering program in which 3,000 ft [914 m] of core was cut in 10 wells. Analysis of the core, log, and production data led to a revised interpretation of the reservoir. Figs. 7.11 and 7.12 show reservoir zonation concepts interpreted from core and log data. The study shows a strongly developed Karst topography surface with associated carbonate dissolution that causes a high degree of variability in permeability and vertical/lateral continuity. Zone A is characterized by high permeability (caused by the downward-percolating groundwater in Karst development), low lateral continuity of the pay zones, and a lack of a natural water drive. Zones B, C, and D are characterized by lower permeability (because of the lack of dissolution below the paleo water table), a higher degree of lateral continuity, and a strong natural water drive.

Zone A is separated from the other zones by an areally continuous solution breccia, which represents the approximate location of the paleo water table. This solution breccia zone, having good lateral continuity and being present over the entire area, effectively separates Zone A from the lower three zones, thereby giving two distinctly different reservoirs.

Major subzones were found within Zones A and B but were somewhat less continuous, were separated by local

permeability barriers, and still had to be mapped separately.

The revised reservoir description was used to help interpret the observed production data. Individual well analysis for the period 1958–61 revealed that the reservoir pressure in Zone A declined from 2,264 to 700 psi [16 to 5 MPa] by 1963. Pressure in the other zones remained high. It was concluded that the production performance of Zone A was by solution gas drive, while Zones B, C, and D were under a natural water drive. Table 7.2 summarizes the characteristics of the zones and subzones of the Elk Basin Madison Reservoir.

The revised reservoir description, combined with results of the initial water injection program, was used to alter the water-injection program and to drill new producing wells in underdeveloped areas. In the initial water-injection program, water breakthrough was rapid in interior wells and caused scaling problems, which resulted in production rate declines in production wells. Water injection into Zone A was shifted to peripheral wells. Injection into Zone B was discontinued. Fig. 7.13 shows

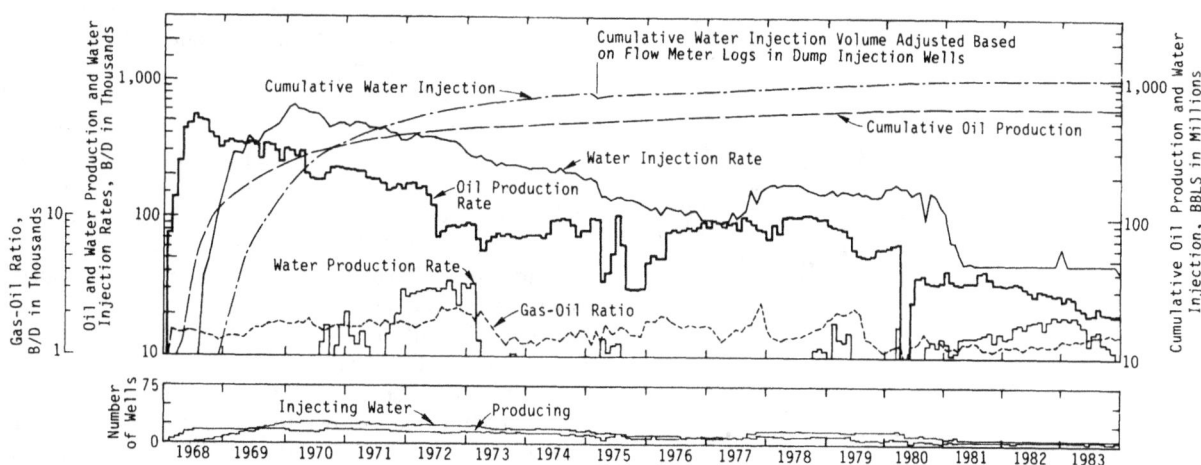

Fig. 7.8—Field performance—Intisar "A" (Personal communication with C.L. DesBrisay, Occidental Petroleum Co., 1984).

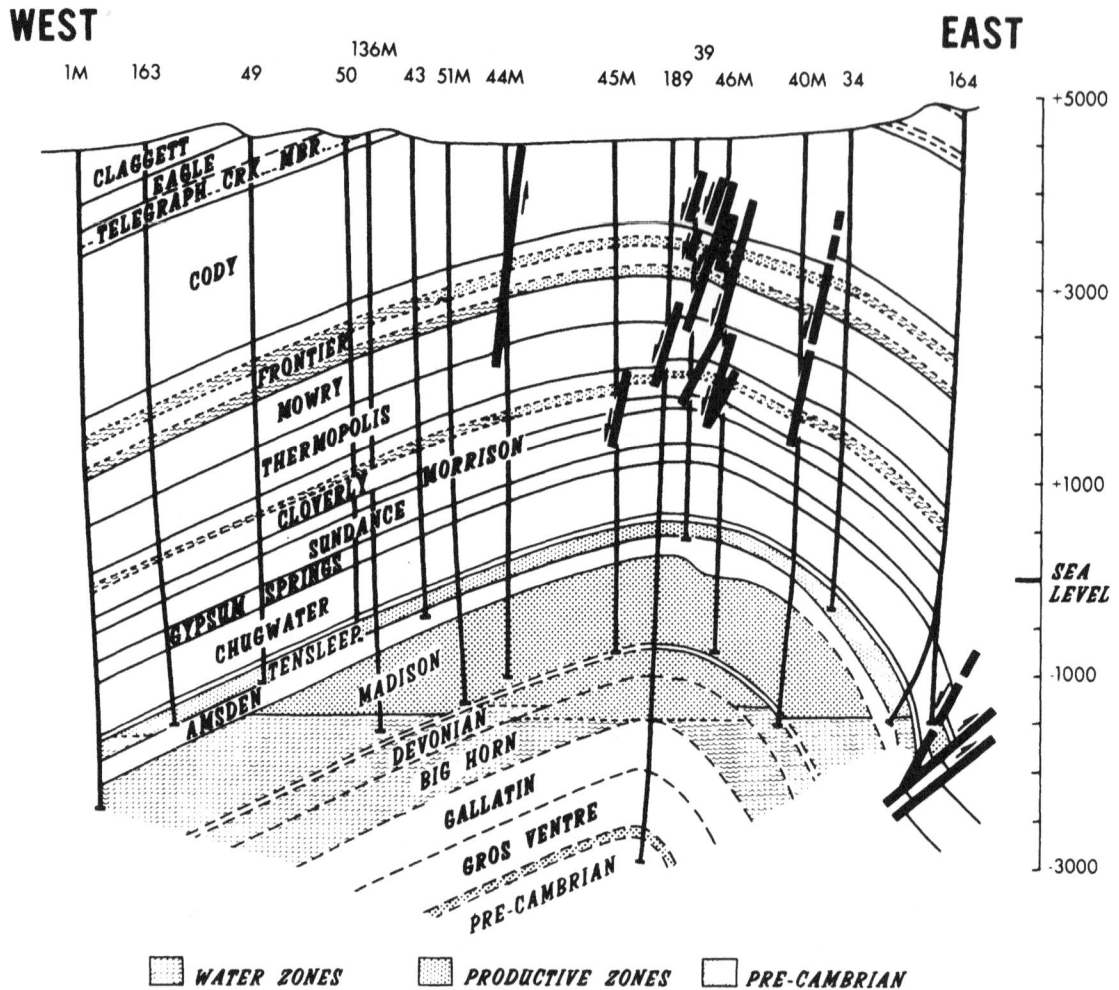

Fig. 7.9—West/east cross section of Elk Basin anticline showing pay horizons.[7]

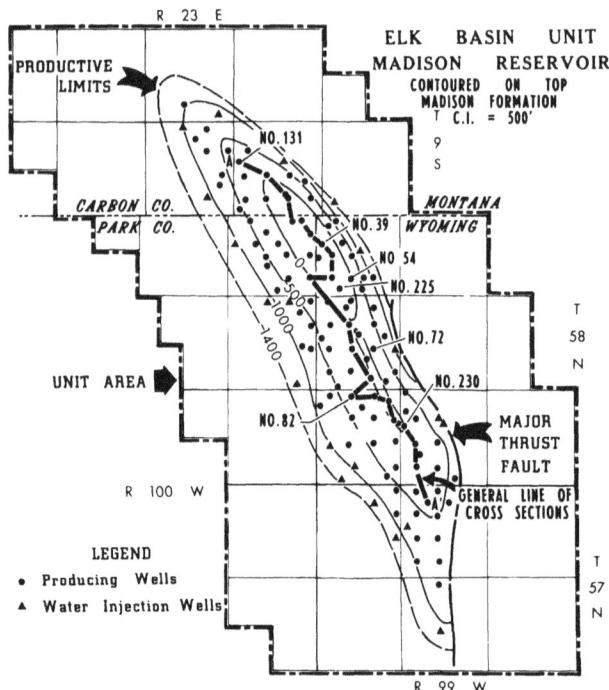

Fig. 7.10—Plat showing Elk Basin Madison structure and location of wells.[7]

the performance history of the Elk Basin Madison that resulted from this analysis. This resulted in an increase in production of 10,000 B/D [1590 m^3/d] and an increase in ultimate recoverable reserves of 62×10^6 bbl [10 m^3] or 8%.[8]

The Elk Basin Madison reservoir illustrates the importance of obtaining extensive reservoir data during field development so that reservoir geology can be defined as soon as possible and incorporated into waterflooding plans. There has been lateral application of these findings to other Madison producing fields in Wyoming.

Denver Unit Waterflood. The Denver Unit waterflood in the Wasson San Andres field of west Texas, shown in Fig. 7.14, illustrates how geological concepts were used to redesign a waterflood.[9,10] Discovered in 1936, this field produces from the San Andres carbonate interval at a depth of about 5,000 ft [1524 m]. The productive interval varies from 300 to 500 ft [91 to 152 m] thick. Primary development was completed on 40-acre [161 875-m^2] spacing in the early 1940's. The primary producing mechanism was solution gas drive.

Waterflooding began in 1964. The initial design was the peripheral flood shown in Fig. 7.15. Water was injected below the OWC, as depicted in Fig. 7.16. Because the reservoir was considered continuous, it was assumed that injection of water at or below the OWC would create an edgewater and/or bottomwater drive. Under this

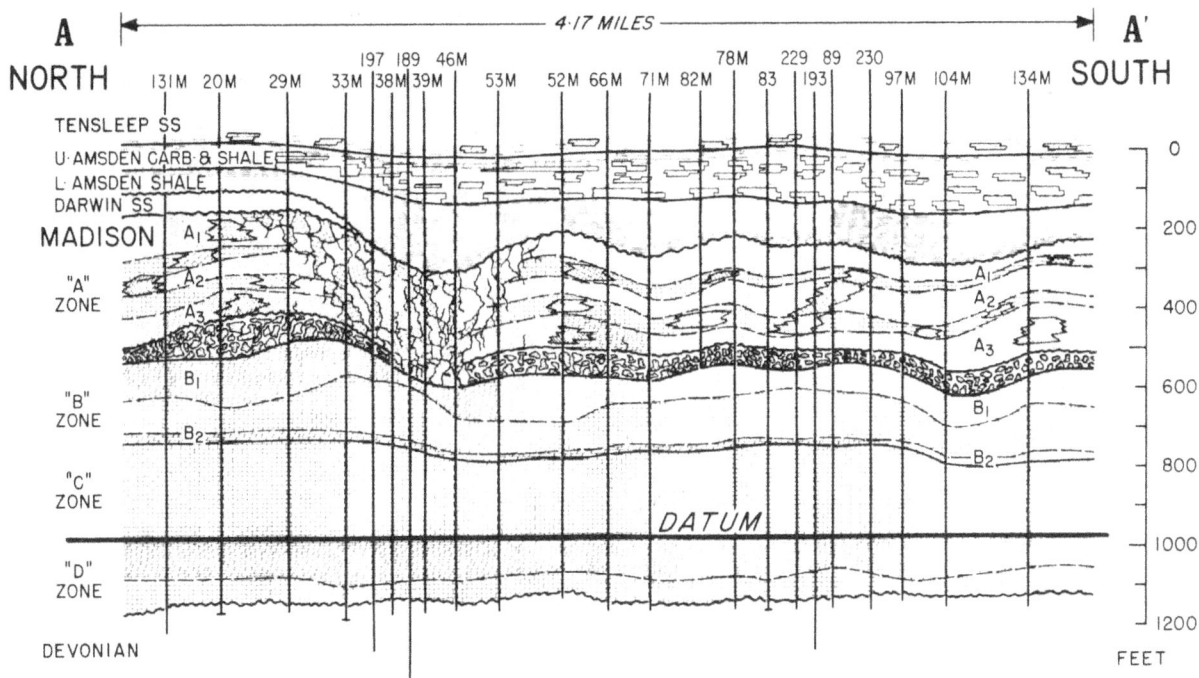

Fig. 7.11—North/south stratigraphic cross section of Elk Basin Madison reservoir showing Karst development, a collapsed sinkhole on the north end, and lateral continuity (Personal communication with J.A. McCaleb, Amoco Production Co., 1983). [7]

geological model of the reservoir, injected water would move laterally and vertically throughout the productive formation.

Performance of the peripheral waterflood was less than expected. Injectivity was low because edge wells selected for water injection often had the poorest-quality reservoir rock. Production wells located 3 to 4 miles [5 to 6 km] from the injection wells did not respond to water injection. Water input volumes were so low that response was slow and limited to the first row of production wells. Inadequate injectivity and poor reservoir transmissibility over long distances caused the peripheral flood to fail. These results indicated that pattern flooding would be required.

Geological concepts of the reservoir also changed during—and perhaps as a result of—the initial waterflood response. Detailed geological studies indicated that the pay interval could be divided into 10 discrete zones (Fig. 7.17). Zones were mapped vertically and laterally over distances of several well locations. Permeability barriers were identified between zones that would restrict the amount of crossflow. The most important discovery from these studies was that some pay members were not continuous over large distances and would not be flooded on the 40-acre [161 875-m^2] spacing selected for the waterflood.

A new geologic model evolved in which the reservoir was represented as a series of continuous and discontinuous pay zones (Fig. 7.18). Continuous pay was defined as the portion of the total net pay that is connected hydraulically between two wells as at the well spacing existing in the field. Discontinuous pay is the fraction of the total net pay considered to be connected to a single well. Further study led to infill drilling on 20-acre [80 938-m^2] spacing to increase the fraction of continuous pay under waterflood. The field subsequently was developed on

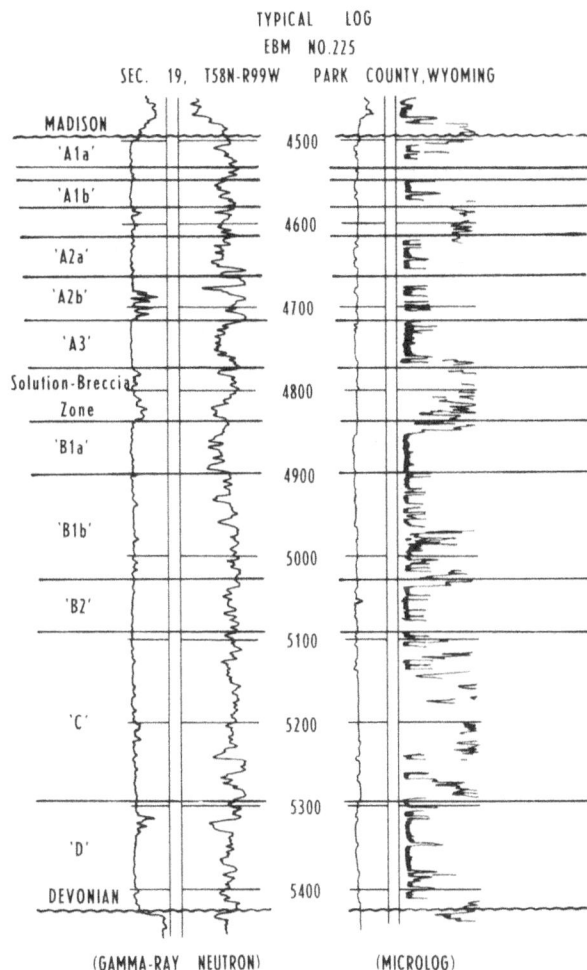

Fig. 7.12—Typical Elk Basin Madison well log showing subzones. [7]

TABLE 7.2—SUMMARY OF ELK BASIN MADISON ZONE CHARACTERISTICS[7]

Zone	Average Net Pay (ft)	Gross Pay Sections (ft)	Porosity (%)	Permeability (md)	Lithology of Pay	Recovery Mechanism
A_{1a}	8	25	13.6	48	Fine-to-medium grained dolomite	
A_{1b}	10	20	12.1	71		
A_{2a}	19	38	12.4	47	Medium-grained dolomite	Solution gas drive artificial waterflood
A_{2b}	9	45	11.3	73		
A_3	22	62	12.1	368	Medium-grained dolomite with vugs	
				Solution Breccia Zone		
B_{1a}	37	65	11.5	5		
B_{1b}	22	100	10.0	3	Fine crystalline dolomite	
B_2	6	70	11.5	6		Natural water drive
C	7	215	10.3	3	Microcrystalline dolomite	
D	48	140	11.2	11	Fine crystalline dolomite	

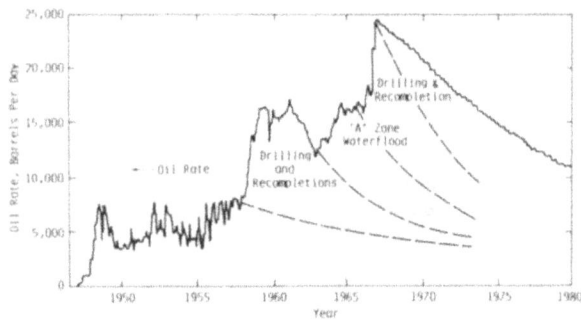

Fig. 7.13—Performance of Elk Basin Madison field after waterflooding plans were revised. [8]

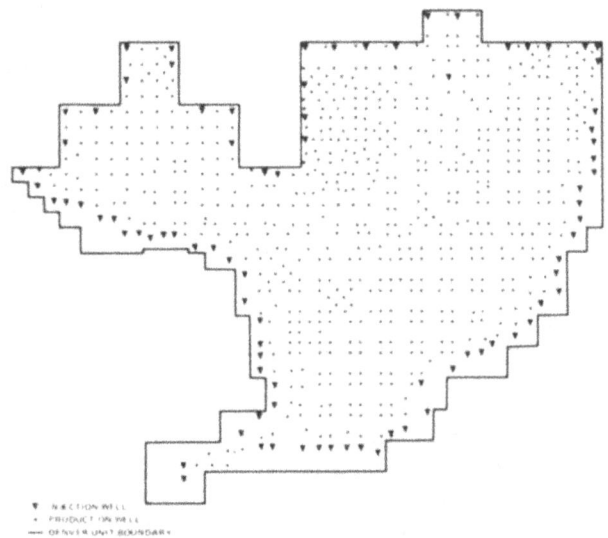

Fig. 7.15—Original peripheral waterflood pattern. [10]

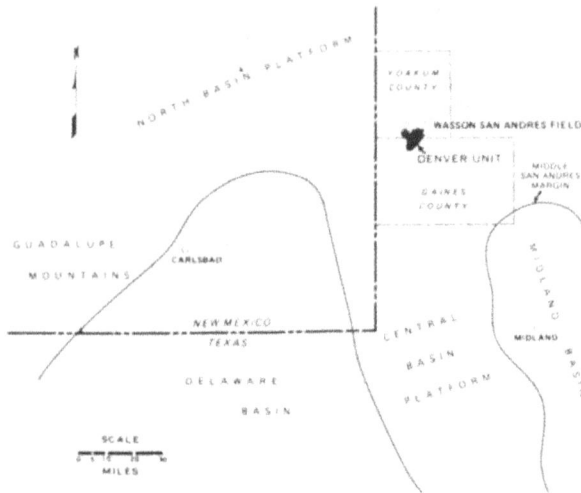

Fig. 7.14—Location map—Denver Unit, Wasson San Andres field. [9]

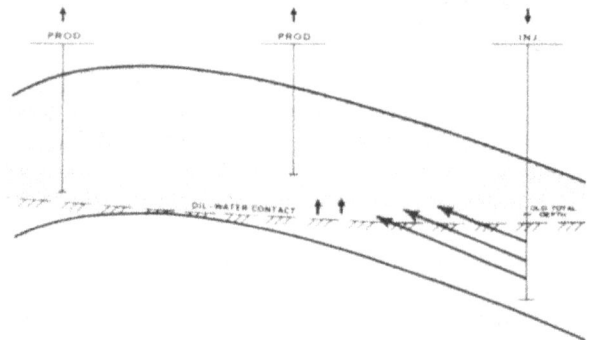

Fig. 7.16—Old geologic concept of continuous pay. [9]

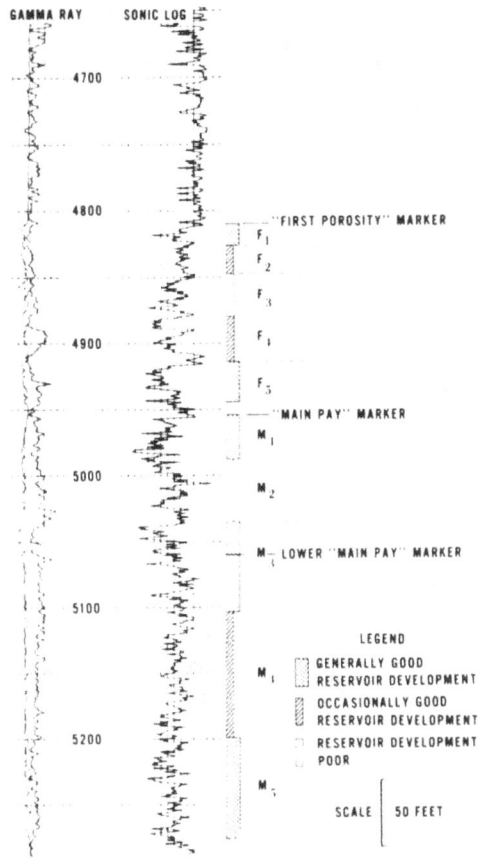

Fig. 7.17—Zonal subdivisions of San Andres reservoir.[9]

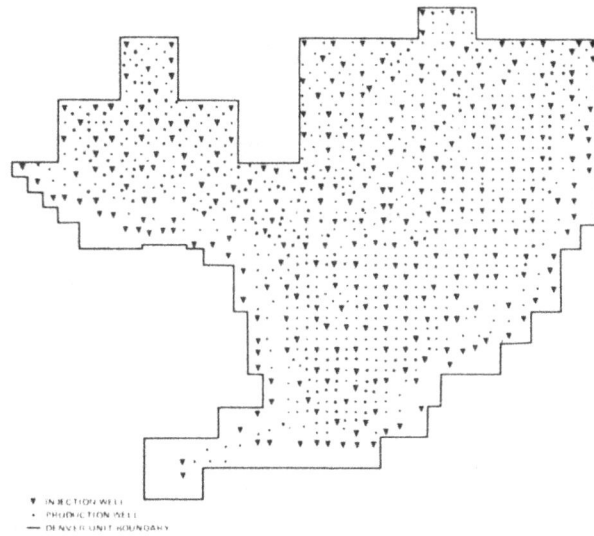

Fig. 7.19—1979 project status.[10]

about 20-acre [80 938-m^2] spacing with an inverted nine-spot pattern. Fig. 7.19 shows the waterflood pattern as of 1979. Production performance from the waterflood is presented in Fig. 7.20. Further background on the Wasson San Andres unit is found in Refs. 5 and 6.

Jay-Little Escambia Creek Waterflood. The Jay-Little Escambia Creek (Jay-LEC) waterflood illustrates how an extensive geological study performed during development of the field was integrated into the waterflood design for a large field.[11-14] The Jay-LEC field was discovered in 1970 and is shown in Fig. 7.21. Production is from the Smackover carbonate and Norphlet sand formations at a

depth of about 15,400 ft [4694 m]. The crest of the reservoir is at a subsea depth of 15,100 ft [4603 m]. An OWC is located at a subsea depth of 15,480 ft [4718 m]. Fig. 7.22 is a typical well log from the field. More than 90% of the oil in place is in the first Smackover formation. Table 7.3 summarizes rock, fluid, and reservoir properties.

The reservoir contains a volatile crude oil that is sour. Because the oil is highly undersaturated ($p_i - p_{BP} = 5,020$ psi [35 MPa]), a rapid decline in reservoir pressure would be expected unless a natural water drive were present. Although an OWC exists, an early reservoir study indicated that there was little evidence that a natural water drive would be an effective source of reservoir energy. Thus, a pressure maintenance program was required to increase oil recovery. Waterflooding was selected from among several possible processes.

An extensive coring program provided the basis for unitization of the field as well as development of a reservoir description for waterflood design.[14] The Smackover formation in the Jay-LEC field is layered uniformly.

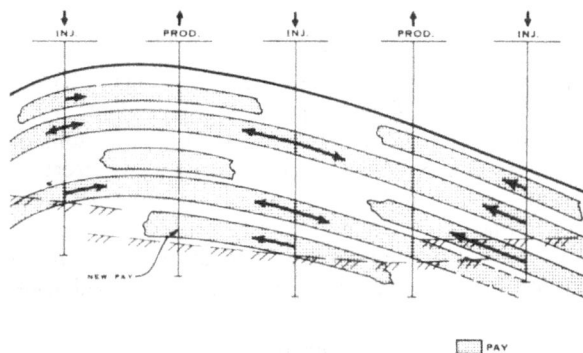

Fig. 7.18—Current geologic concept—noncontinuous pay.[9]

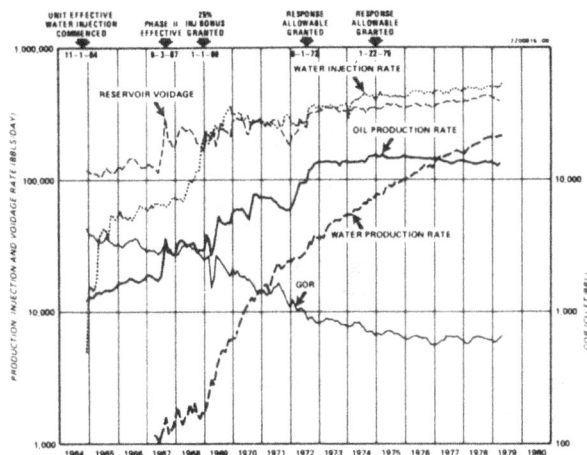

Fig. 7.20—Project performance curves—Denver unit.[10]

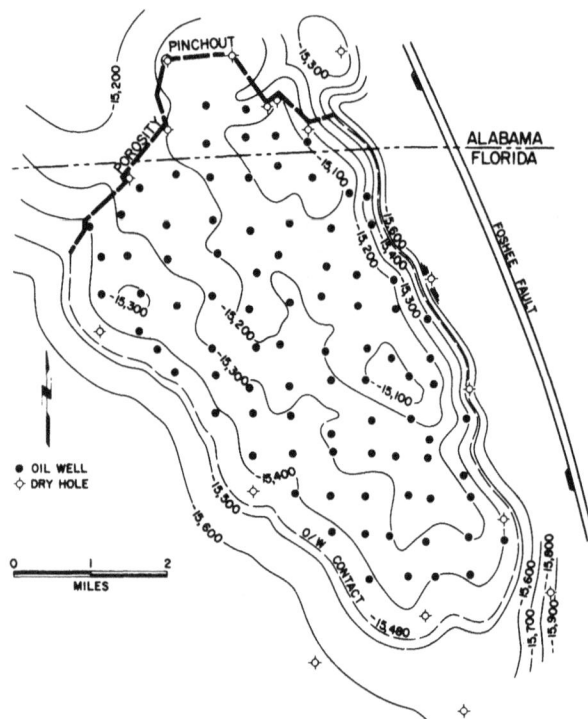

Fig. 7.21—Structure map, top of Smackover Jay-LEC fields.[11]

TABLE 7.3—SUMMARY OF ROCK AND FLUID PROPERTIES, RESERVOIR PROPERTIES, AND PRODUCTION/INJECTION DATA[15]

Rock and Fluid Properties	
Porosity, %	14.0
Permeability, md	35.4
Water saturation, %	12.7
Oil FVF, RB/STB	1.76
Oil viscosity, cp	0.18
Oil gravity, °API	51
Solution GOR, scf/STB	1,806
Hydrogen sulfide content, mol%	8.8
Mobility ratio (water/oil)	0.3

Reservoir Properties	
Datum, ft subsea	15,400
Original pressure, psia	7,850
Current pressure, psia	5,750
Saturation pressure, psia	2,830
Temperature, °F	285
Productive area, acres	14,415
Net thickness, ft	95
OOIP, million STB	728

Production/Injection (Jan. 1, 1981)	
Oil production rate, 1,000 STB/D	90
Cumulative oil production, million STB	296
Water injection, thousand B/D	250
Cumulative water injection, million bbl	524

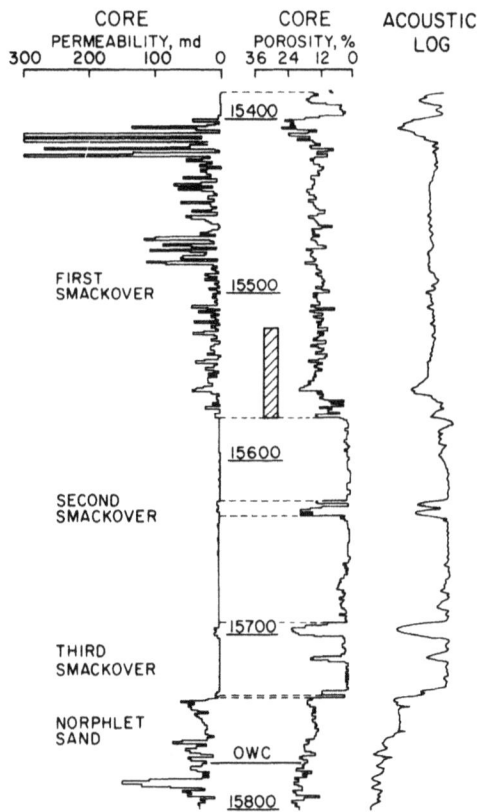

Fig. 7.22—Typical well log, Exxon-Bray 10-4, Jay-LEC fields.[11]

Correlation of high- and low-permeability zones was possible over large areas of the field. This meant that thin, tight streaks, which had limited or no vertical flow capability, could inhibit crossflow between layers. Properties within individual layers were considered homogeneous. Best continuity of the reservoir occurs in the thick central area that parallels the long axis of the reservoir where net pay was greater than 200 ft [61 m]. Porosity pinchouts exist across the northern part of the structure where the porous dolomite changes to dense limestone. The northern area of the reservoir is highly stratified. The net pay thins into low-permeability zones separated by dense limestone along the flanks.

The First Smackover formation was selected to be flooded first because geological correlations indicated that the formation was massive and interconnected, and because the mobility ratio was quite favorable (0.3). The waterflooding plan was developed by use of a two-dimensional (2-D) waterflooding simulator to compare alternative flooding schemes. Four waterflood plans were evaluated: (1) peripheral flood, (2) five-spot pattern, (3) a 3:1 staggered linedrive pattern, and (4) a combination of peripheral wells and five-spot patterns. Figs. 7.23 and 7.24 show three of the proposed flooding plans.

The peripheral flood was rejected from the results of the 2D simulator. Water injection capacity and reservoir transmissibility on the north, west, and east sides were low. It was not possible to maintain reservoir pressure in the central portion of the field even though high-capacity injection wells could be used along the south flank of the reservoir.

Pattern arrangement for the remaining three waterflooding plans was selected to take advantage of reservoir geology. Injection wells were located on the outside in the five-spot plan, assuming that water influx and coning

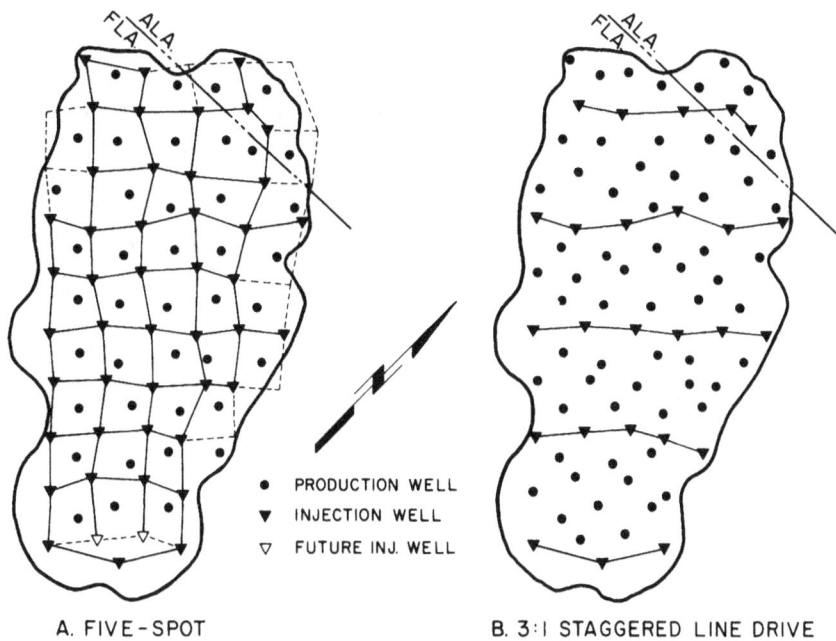

Fig. 7.23—Repeating patterns 2D areal reservoir model, Jay-LEC fields.[11]

would limit the effectiveness of these wells as producers. The 3:1 staggered linedrive pattern was oriented on the short axis of the reservoir because the best reservoir continuity was believed to exist along the long axis of the reservoir. The combination plan in Fig. 7.24 was conceived to match the flood plan to reservoir geology. Stratified sections along the north, west, and east flanks of the reservoir were developed on five-spot patterns, while a peripheral flood was placed along the southern flank. The thick, central portion of the reservoir was flooded as a 3:1 staggered linedrive.

Projected waterflood recovery from the 3:1 staggered linedrive plan was more than 200×10^6 bbl [32×10^6 m^3]. The 3:1 plan yielded 9.8×10^6 bbl [1.6×10^6 m^3] incremental oil recovery over the five-spot plan and 14.4×10^6 bbl [2.3×10^6 m^3] over the combination pattern. Selection of the waterflooding plan was based on technical and economic considerations with some allowance for geologic uncertainty.[11] Economic potential of the 3:1 staggered linedrive pattern was also larger because fewer wells were required and less water was produced.

The 3:1 staggered linedrive provided allowance for poor flood performance from unanticipated variations in reservoir properties. In the 3:1 staggered linedrive plan, flood fronts may be adjusted on the basis of the response of the first row of producing wells to obtain closure on the center row of producers. The 3:1 pattern had other advantages to respond to uncertainties in flood performance, including the ability for additional injection wells to be drilled if needed and the conversion to a five-spot flood by connecting center-row producers to injection wells if reservoir transmissibility was low or if vertical sweep efficiency was poor because of reservoir stratification.

Details of the Jay-LEC waterflood are presented in several papers.[11-15] Fig. 7.25 shows the waterflood response from the Jay-LEC unit.

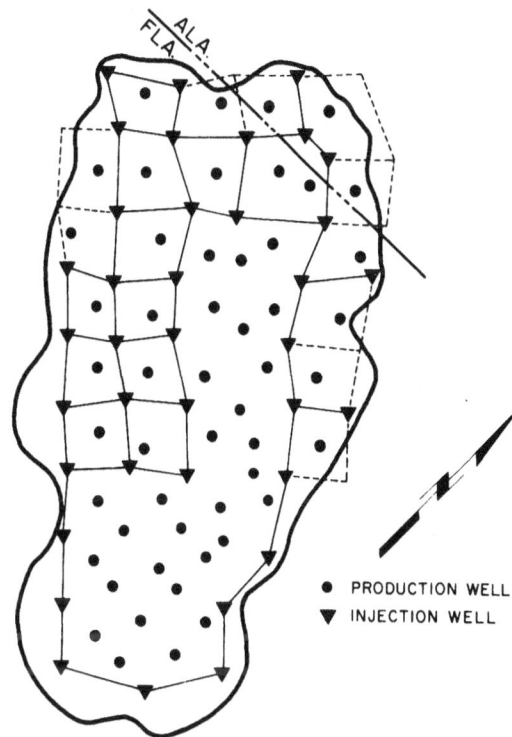

Fig. 7.24—Combination pattern, 2D areal reservoir model, Jay-LEC fields.[11]

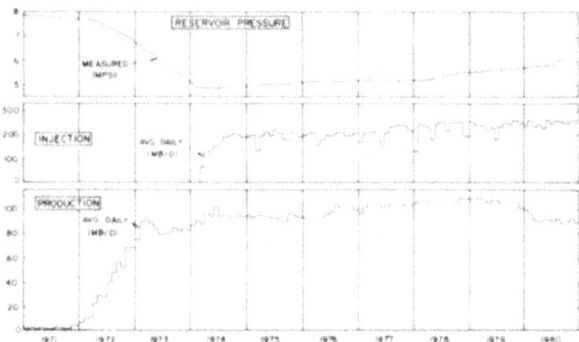

Fig. 7.25—Jay-LEC unit performance. [15]

Judy Creek Waterflood. The Judy Creek field in central Alberta produces from a Devonian reef. [16,17] The field was discovered in 1959 and covers a surface area of 47 sq miles [122 km^2]. Maps and cross sections illustrating facies, rock properties, and fluid distributions are shown in Fig. 7.26. Original oil in place (OOIP) was estimated at 830×10^6 bbl [132×10^6 m^3]. Because the field is not connected to a large aquifer, a peripheral waterflood was initiated in 1962 to maintain the reservoir pressure. By 1973, pressure distributions in the field indicated that the waterflood was ineffective. It was concluded that permeability barriers existed within the reservoir that prevented communication between the peripheral wells and parts of the reservoir.

A combined engineering and geological study, which revealed the complex structure of the reservoir, was used to prepare a revised waterflooding plan. Fig. 7.27 is a cross section showing the complex lithofacies in the Judy Creek field. The term "facies" is used to describe any areally restricted point of a designated stratigraphic unit that exhibits characteristics significantly different from other parts of the unit. [18] The reservoir was subdivided into three stratigraphic units—S3, S4, and S5—that corresponded to three periods of reef growth. Correlation cross sections similar to Fig. 7.28 were prepared to correlate porous units throughout the field. Porous intervals in Fig. 7.28 were judged to be continuous (fluid connected) vertically or horizontally. Discontinuous beds in Units S3 and S4 are not shaded. Isopachs of net porosity and continuous porosity for Units S3, S4, and S5 are shown in Fig. 7.29. The original distribution of oil and water is shown in Fig. 7.30. A small water zone was present in Unit S4 while Unit S3 was about half filled with water.

Water was injected into the reef through bottom and peripheral wells in Units S3 and S4. The rate of water advance is mapped in Fig. 7.31. Injected water flowed through permeable rocks on the outer reef rim, bypassing the interior regions of Units S4 and S5. Bypassing or lack of communication was also evident from pressure surveys. In early 1973, a pressure gradient of 2,000 psi [14 MPa] existed across the pool. Pressures in injection wells were 1,000 psi [7 MPa] above the original reservoir pressure while interior well pressures were below the bubblepoint pressure. Fig. 7.32 shows the isobaric map in March 1974.

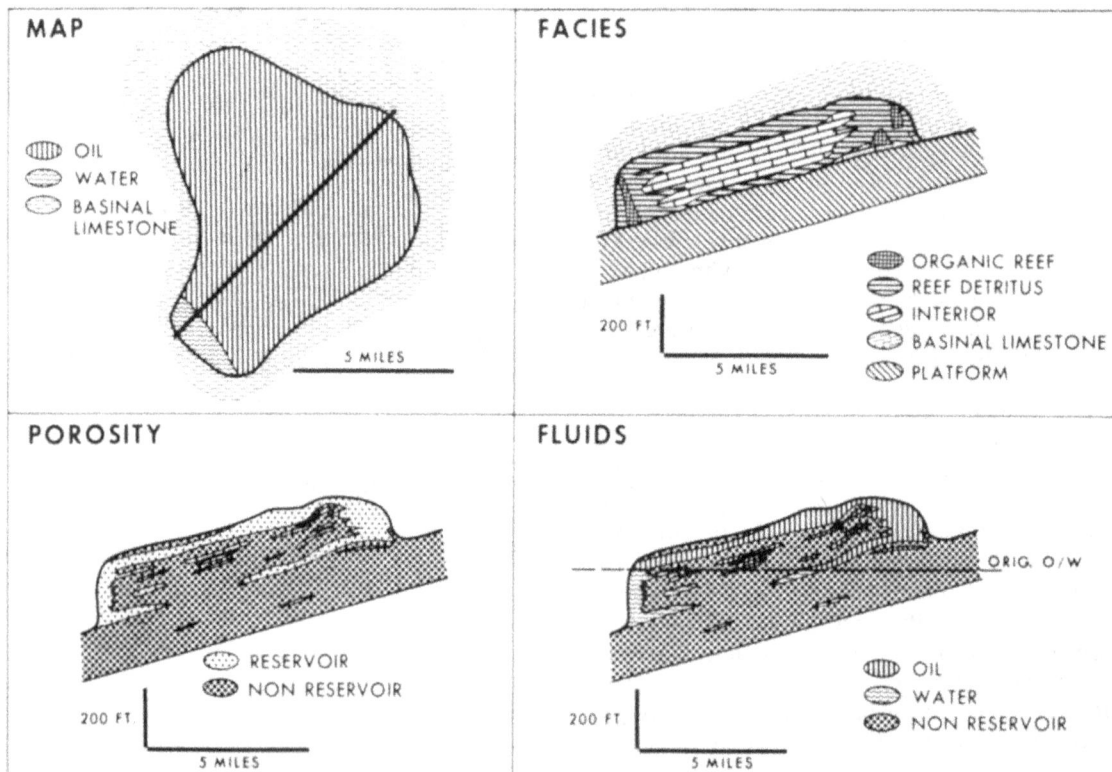

Fig. 7.26—Judy Creek oilfield map and cross sections illustrating lithofacies, porosity, and fluid distributions along Line A-A1. [16]

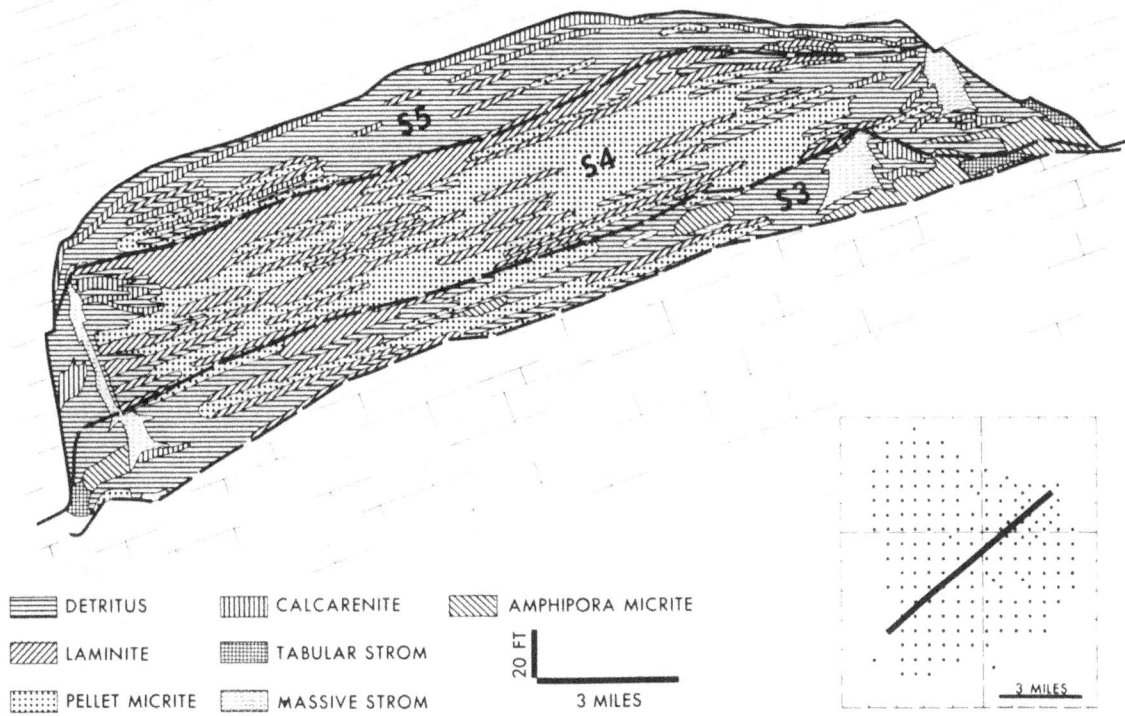

DETRITUS   CALCARENITE   AMPHIPORA MICRITE

LAMINITE   TABULAR STROM

PELLET MICRITE   MASSIVE STROM   20 FT   3 MILES

Fig. 7.27—Cross section showing complexity of lithofacies, Judy Creek field.[16]

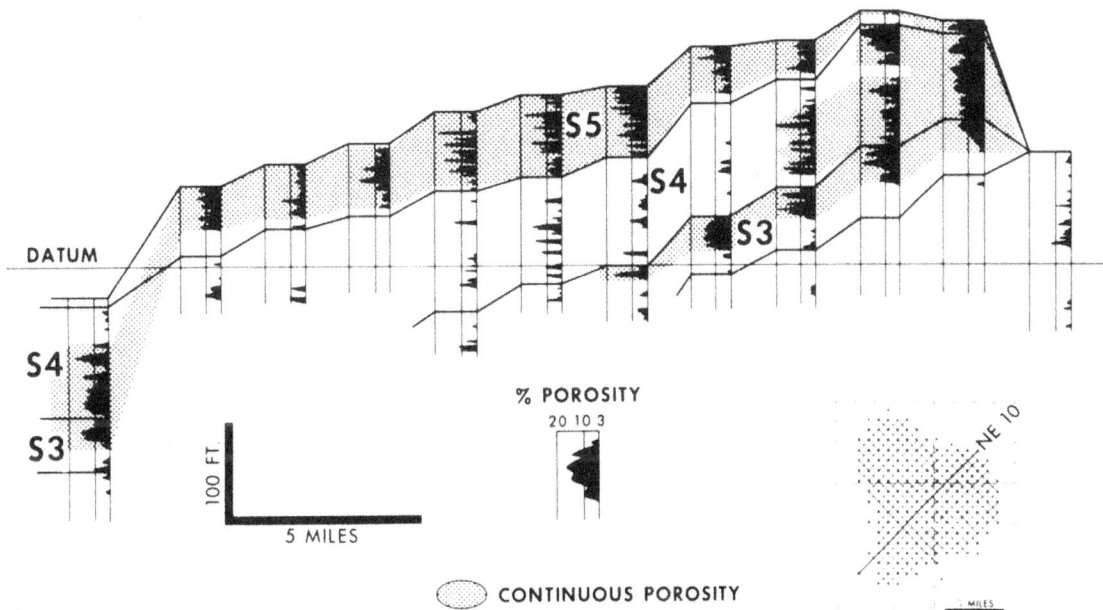

DATUM

S4

S3

% POROSITY
20 10 3

100 FT

5 MILES

NE 10

CONTINUOUS POROSITY

MILES

Fig. 7.28—Example of a correlation cross section, Judy Creek field (Ref. 16, after Ref. 17).

238

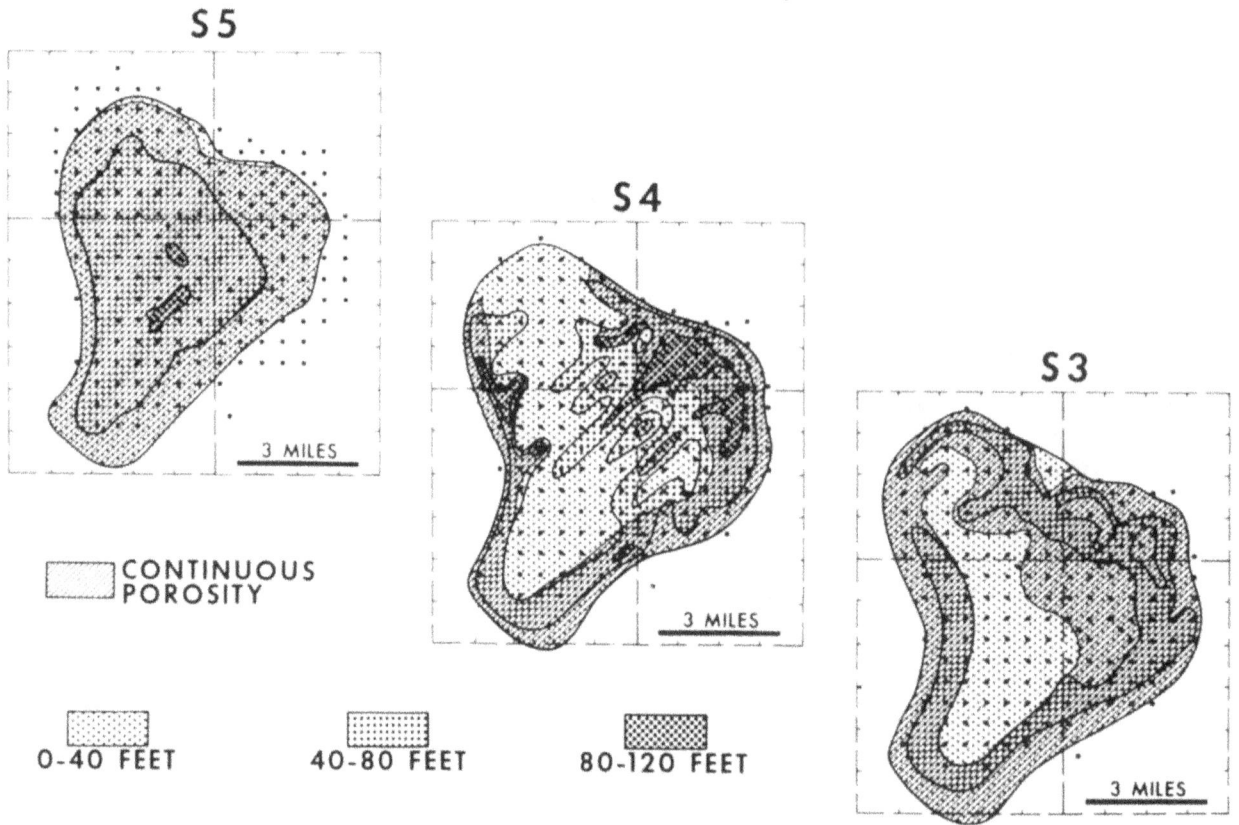

Fig. 7.29—Isopach of net porosity and zones of continuous porosity in Units S3, S4, and S5, Judy Creek field (Ref. 16, after Ref. 17).

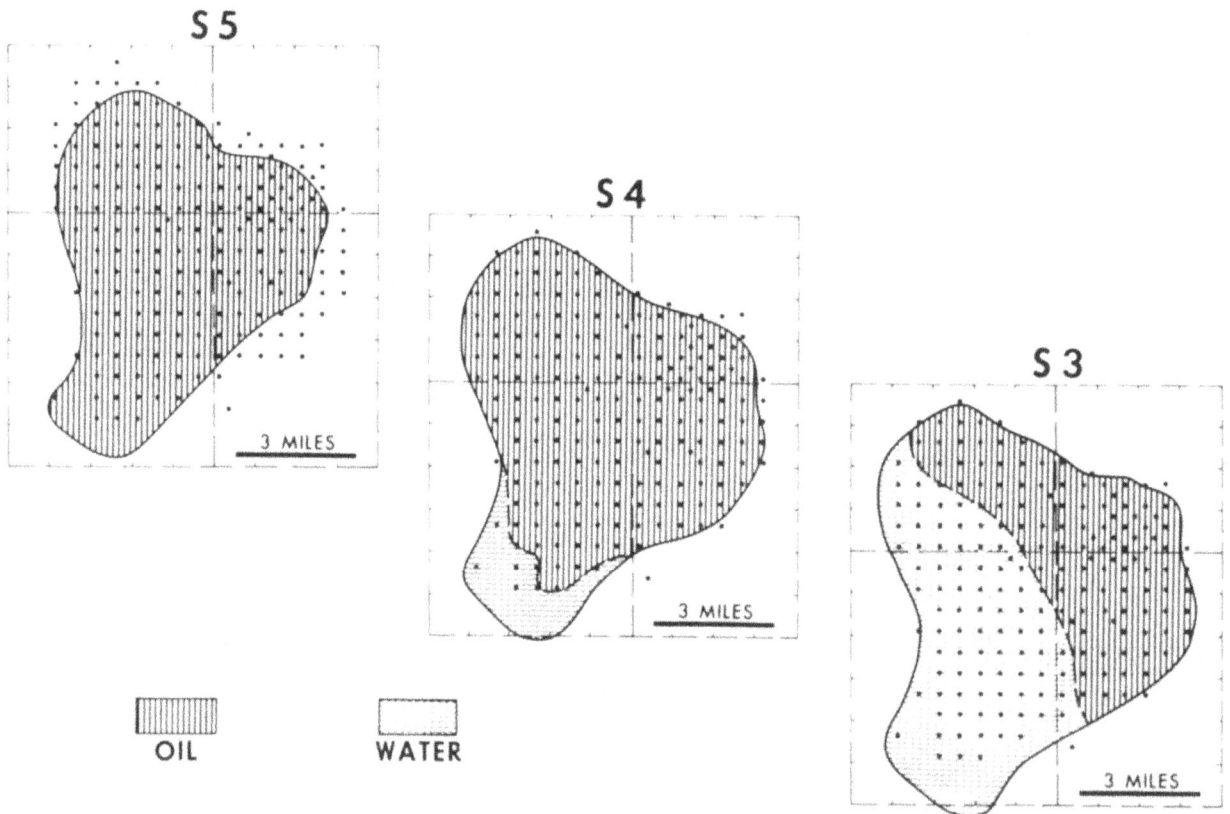

Fig. 7.30—Maps showing original distribution of oil and water in Units S3, S4, and S5, Judy creek field (Ref. 16, after Ref. 17).

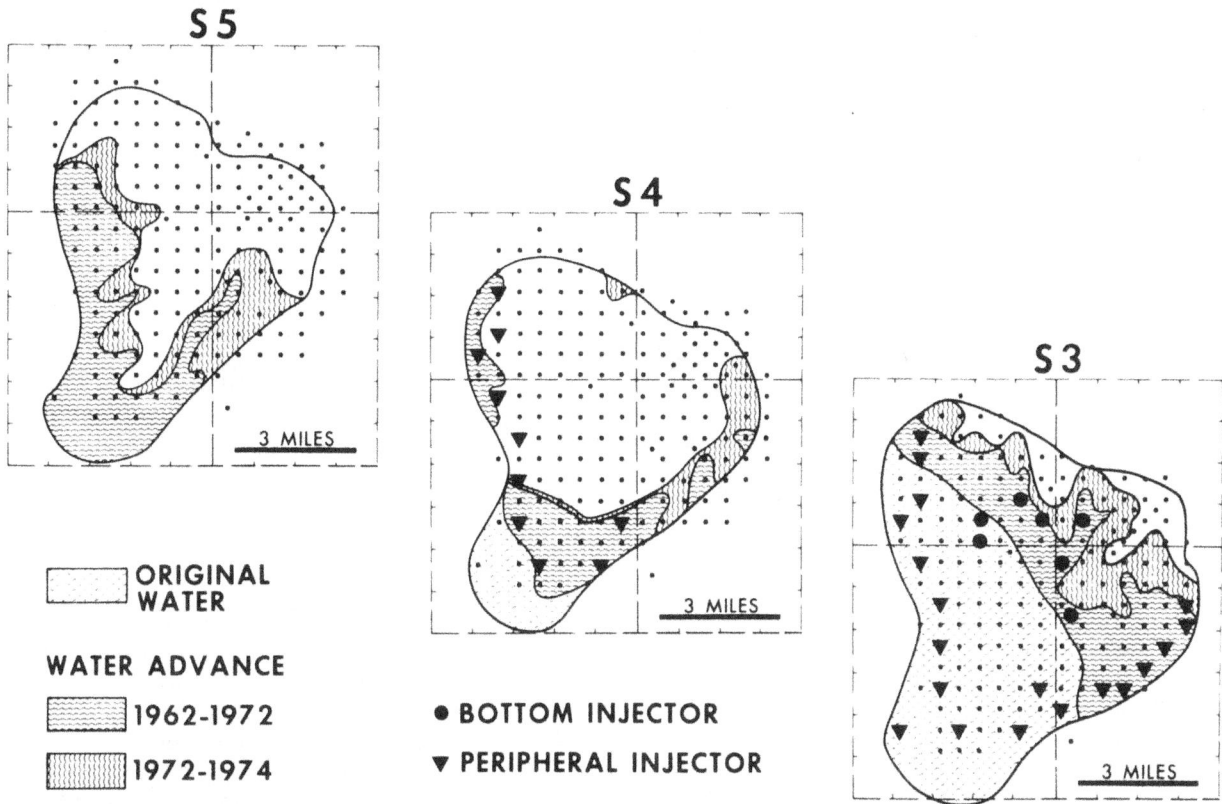

S5

S4

S3

ORIGINAL WATER

WATER ADVANCE

1962-1972

1972-1974

● BOTTOM INJECTOR

▼ PERIPHERAL INJECTOR

3 MILES

Fig. 7.31—Maps showing rate of water advance peripheral and bottomwater flood in Units S3, S4, and S5, Judy Creek waterflood (Ref. 16, after Ref. 17).

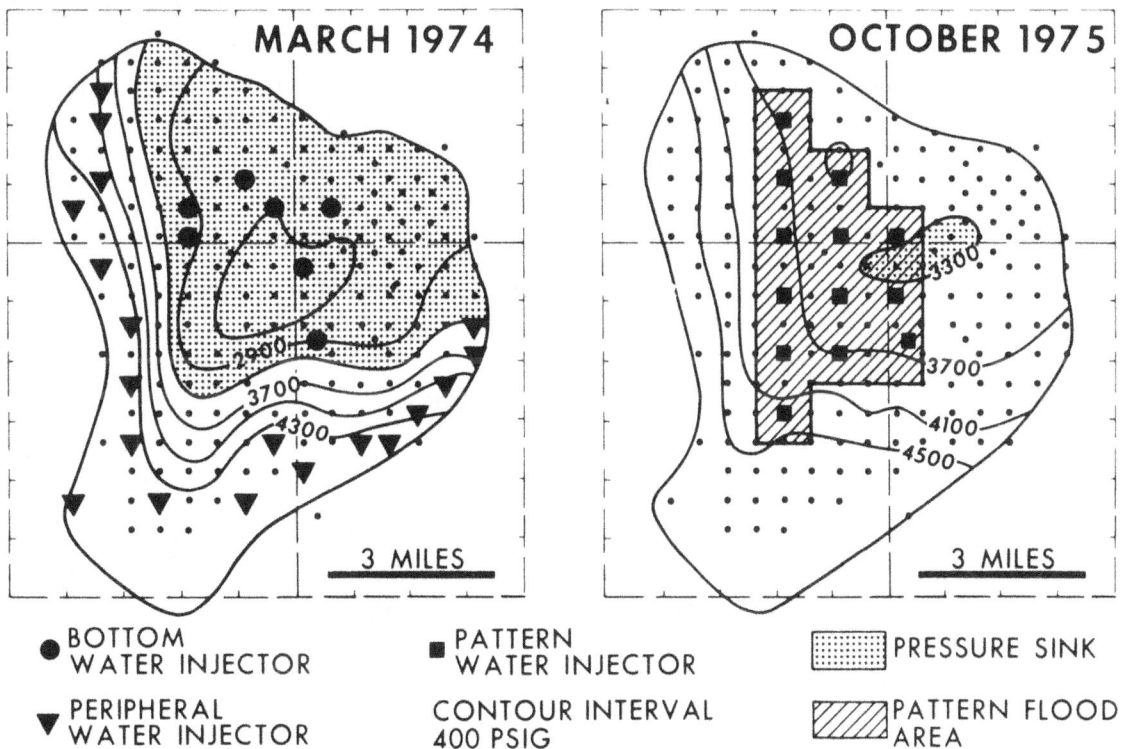

MARCH 1974

OCTOBER 1975

3300

3700

4100

4500

2900

3700

4300

3 MILES

● BOTTOM WATER INJECTOR

▼ PERIPHERAL WATER INJECTOR

■ PATTERN WATER INJECTOR

CONTOUR INTERVAL 400 PSIG

PRESSURE SINK

PATTERN FLOOD AREA

Fig. 7.32—Isobaric maps showing the result of conversion from peripheral to pattern flood, Judy Creek field (Ref. 16 after Ref. 17).

TABLE 7.4—SAMPLING OF RESERVOIR VOLUME FROM CORE AND LOG ANALYSIS

Well Spacing (acres/well)	Fraction of Reservoir Volume Sampled by by 6-in. Core ($\times 10^6$)	Fraction of Reservoir Volume Sampled Log Penetration Radius 5 ft ($\times 10^4$)
2.5	1.80	7.21
5.0	0.90	3.61
10.0	0.45	1.80
20.0	0.225	0.90
40.0	0.112	0.451
80.0	0.056	0.225
160.0	0.028	0.113

The detailed engineering and geological evaluation led to two major conclusions: (1) a pattern waterflood should be installed to flood Unit S5; and (2) discontinuous beds in Unit S4 were not waterflooded effectively by peripheral and bottomwater injection. A pattern waterflood was installed in 1974 with wells completed in all porous intervals in Units S3, S4, and S5. By Oct. 1975, the pressure distribution improved considerably, as can be seen in Fig. 7.31. The combined geological/engineering study also resulted in the opening of several zones in Unit S4 in wells behind the flood front that were thought to be flooded out. Many wells were reactivated by the opening up of porous regions defined from detailed geologic analysis.

7.3 Reservoir Description

Design of a waterflood or any other process in which oil is displaced from porous rock is based on a specified geologic model of the reservoir. For example, it is essential to determine whether the reservoir or part of the reservoir is connected hydraulically to an aquifer that might supply all or part of the reservoir energy for a natural waterflood. Volumetric sweep efficiency depends on contacting as much reservoir volume as is possible. Flooding plans depend on *knowledge* and/or *hypothesis* of how pore space is connected between wells as well as within the reservoir.

A *reservoir description* is a conceptual model that describes the spatial distribution of fluid and rock properties and saturations for reservoir and nonreservoir rock within the gross thickness and areal extent of the porous rock that constitutes the reservoir.

A reservoir description is always constructed from limited information. Important data are often missing. Much of the available information is obtained during drilling and completion of wells. Even here the volume of the reservoir sampled by core and log analyses is an infinitesimal fraction of the total reservoir volume, as illustrated in Table 7.4.

A second source of information is data collected during primary depletion. This information represents the response of the reservoir to production. We may also have pressure buildup, falloff, or interference tests that provide data on the average flow characteristics in the vicinity of a production well.

The task of developing a reservoir description is depicted in Table 7.5. The goal of this task is to describe the reservoir and the variation of properties (such as permeability, porosity, and fluid saturation) with spatial position throughout the reservoir from data collected from a few sample points, primarily the wells. Averaged values are acceptable as long as the distributions are consistent with lateral and vertical continuity in the reservoir. We are interested also in the properties of nonreservoir rock because these rocks act as barriers for vertical flow. The reservoir model is obtained by *extrapolation* of these data to the interwell area with geologic and engineering analysis, experience, and assumptions. Until recently, however, much of the model development was performed by geologists and/or engineers working independently.

Historically, the description of a reservoir was left to the geologist, who usually developed maps defining reservoir boundaries and gross and net sand. Cross sections showing principal structural markers were prepared. In some cases, stratigraphic and structural cross sections were made of identifiable rock zones within the reservoirs. Detailed descriptions of core materials and cuttings often accompanied the geological report. Geologists were interested primarily in rock units and viewed their responsibilities to be completed when rock units were described. There was little interest or incentive in relating geological descriptions to such engineering properties as permeability and porosity.

The reservoir engineer had the responsibility of developing possible waterflood plans and predicting the performance of these plans. Maps produced by the geological section were combined with core data to estimate permeability and porosity distributions throughout the reservoir

TABLE 7.5—FLOW PATH DEPICTING DEVELOPMENT OF A RESERVOIR DESCRIPTION

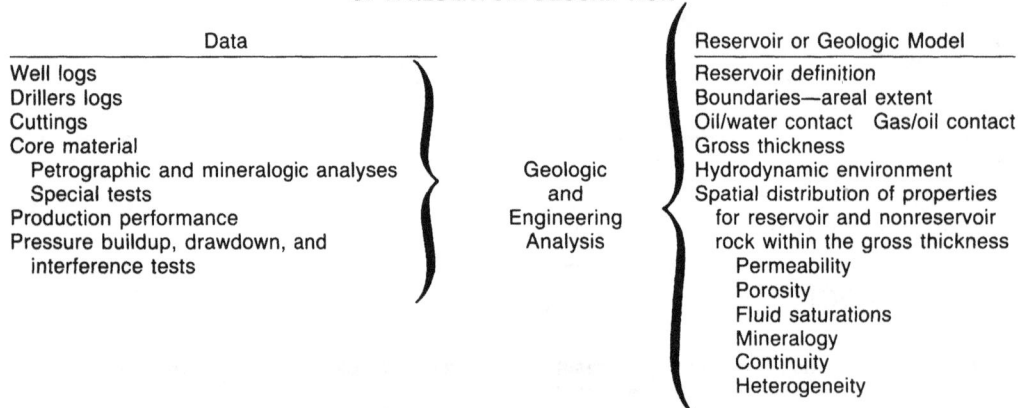

Data		Reservoir or Geologic Model
Well logs		Reservoir definition
Drillers logs		Boundaries—areal extent
Cuttings		Oil/water contact Gas/oil contact
Core material		Gross thickness
Petrographic and mineralogic analyses	Geologic	Hydrodynamic environment
Special tests	and	Spatial distribution of properties
Production performance	Engineering	for reservoir and nonreservoir
Pressure buildup, drawdown, and	Analysis	rock within the gross thickness
interference tests		Permeability
		Porosity
		Fluid saturations
		Mineralogy
		Continuity
		Heterogeneity

for simulation of the waterflood. There was little communication between geologists and engineers once the geological report was prepared and transmitted to the appropriate office. It was not unusual for a geologic report to contain more information than the reservoir engineer could use because of limited geologic background or the capability of reservoir simulators. The consistency or adequacy of engineering and geological interpretations was seldom questioned until a reservoir did not respond in the manner anticipated; then both engineers and geologists became involved in the search for explanations of performance. More often than not, reservoir performance was linked to geologic assumptions made in constructing the reservoir description.

The development of reservoir descriptions evolved from the need to improve predictions of reservoir simulation models, management of existing waterfloods, and design of other EOR processes. [19,20] Development of a reservoir description requires close interaction between geologists and engineers. Geological assumptions and interpretations must be measured against the reservoir performance reflected by production history and pressure tests. Physical properties used in reservoir simulators should be consistent with geological interpretations. [21] This process is more effective when geologists and engineers define the scope of the investigation at the beginning, including points where interaction will occur. [20]

The process of developing a reservoir description from geologic studies involves two major tasks. [22] First, the reservoir rock and nonreservoir rocks must be identified and mapped. Then, properties (i.e., permeability and porosity) of the rocks must be determined (or estimated) and mapped in three dimensions.

A geologic study involves determining the depositional environment from analysis of cores, logs, and cuttings as well as reconstructing the history of diagenesis that caused the distribution of porosity and permeability in the reservoir and nonreservoir rocks. Information derived from control points (the wells) is used to select models of deposition and diagenesis for the entire reservoir area, thereby permitting prediction of properties in the interwell area where little information is available. Construction of these models is an iterative process in which conceptual models of deposition and diagenesis are proposed and tested against models that have been developed from studies of recent sedimentary bodies and ancient deposits. [23,24] Secs. 7.4 and 7.5 summarize depositional models for sandstones and carbonate reservoirs. Sec. 7.6 presents a procedure for developing a reservoir description.

7.4 Depositional Environment for Sandstone Reservoirs

Recognition of the environment of deposition is a principal step in a geologic study because the distribution and continuity of the reservoir and nonreservoir rock usually are related to depositional conditions. [21,23] Most is known about sandstone reservoirs that have been studied extensively. Figs. 7.33 and 7.34 illustrate continuity and thickness patterns for sandstones deposited under two different depositional environments. Fig. 7.33 is a correlation section in the Bradford sandstone in western Pennsylvania. [20] Sandstone and shale units in this deposit correlated over

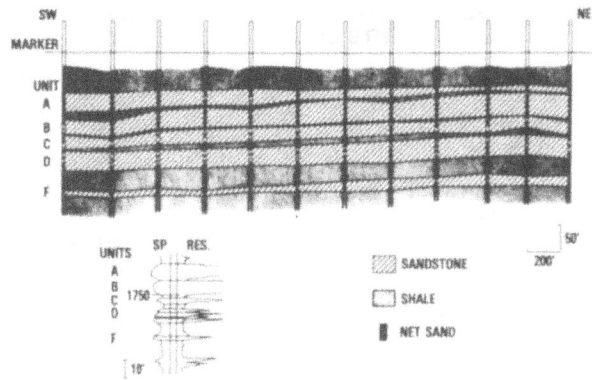

Fig. 7.33—Correlation section showing continuity pattern of the Bradford sandstone in the Sage Lease. [20]

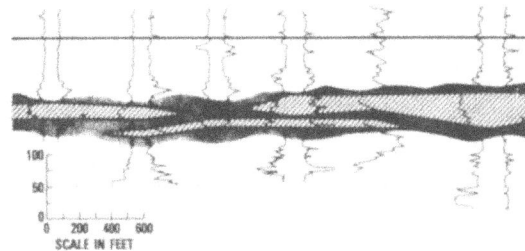

Fig. 7.34—Correlation section of the Robinson sandstone showing continuity pattern in the Robinson main field. [20]

large distances. This sandstone is typical of a marine deposit in which the grain material was deposited in seawater. In marine deposits, sandstone and shale rock units correlate over distances of thousands of feet to a few miles.

The Robinson sandstone in southeastern Illinois in Fig. 7.34 originated from a river that deposited sediments in an extensive alluvial valley or plain. [25] Continuity of individual sand units shown in the correlation section in Fig. 7.34 is variable and depends on location. Thin sand units (10 to 20 ft [3 to 6 m]) usually extend over distances of about 1,000 ft [305 m], while thick sands (50 ft [15 m]) are correlatable for distances of a few thousand feet in some parts of the Robinson main field. [25]

Sandstone reservoirs are most likely to be created under three depositional environments: continental, transitional, and marine. [23] A simplified classification of sandstone sedimentary models corresponding to depositional environment is shown in Table 7.6. [23] Refs. 20 and 26 provide extensive background material.

7.4.1 Rock Properties

Clastic sediments, which comprise sandstone and shale deposits, were deposited in aqueous or subaerial windblown environments. Reservoir-rock quality and pore space are determined by the characteristics of the granular material that was deposited and alteration (diagenesis) of the material after deposition. Table 7.7 summarizes these factors. [26]

TABLE 7.6—CLASSIFICATION OF MODERN ENVIRONMENTS OF SAND DEPOSITION[23]

Continental

Alluvial (fluvial)
Alluvial fan
Braided stream
Meandering stream (includes flood basins between meander belts)
Aeolian (can occur at various positions within continental and transitional environments)

Transitional (Shallow Marine)

Deltaic
Birdfoot-lobate (fluvial dominated)
Cuspate-arcuate (wave and current dominated)
Estuarine (with strong tidal influence)
Coastal Interdeltaic
Barrier island (includes barrier islands, lagoons behind barriers, tidal channels, and tidal deltas)
Chenier plain (includes mud flats and cheniers)
Transgressive marine

Marine*
Deep marine
Outer shelf
Inner shelf

*Note: Sediments deposited in shallow marine environments, such as deltas and barrier islands, are included under the transitional group of environments above.

TABLE 7.7—GEOLOGIC PROCESSES AFFECTING RESERVOIR ROCK QUALITY[26]

	Depositional	Diagenetic Processes
Texture {	Grain size	Compaction
	Sorting	
	Grain shape	Cementation
	Grain packing	
	Grain roundness	Solution
Mineral composition		Fracturing

The sorting coefficient, S_o, is a measure of the dispersion of grain size around the mean value and is calculated from Eq. 7.2.

$$S_o = \sqrt{\frac{d_{p25}}{d_{p75}}}, \quad \dots\dots\dots\dots\dots\dots (7.2)$$

where d_{p25} is the particle size (millimeters) at the 25th percentile and d_{p75} is the particle size (millimeters) at the 75th percentile. A sorting coefficient of 1 indicates homogeneous deposit with uniform sand-grain size, while a sorting coefficient >3 has an extremely wide distribution of grain sizes.

Diagenetic processes (compaction, cementation, solution, and fracturing) alter the pore space and, consequently, the porosity and permeability. Changes in porosity resulting from these processes are large and are usually in the direction of lower porosities and permeabilities. However, Sneider et al.[27] observed from their experience that grain size and sorting are principal variables in correlating porosity, permeability, and pore structure of reservoir rocks even though the initial sedimentary deposit was subjected to severe diagenesis. Trends expected between texture and pore space are summarized in Table 7.8.

In the same reservoir environment, both permeability and porosity of the reservoir rock would be expected to follow the trends predicted from the size and distribution of the original sediments.

The initial porosity of unconsolidated sands is controlled primarily by grain-size distribution (sorting coefficient, S_o) as shown in Fig. 7.35[27,28] Permeability varies with both sorting and grain size. Sorting can be determined by sieving unconsolidated sediments through a series of standard sieves or by comparing the sample with comparators of known size and sorting.[27] Fig. 7.36 shows the cumulative size distribution of a sample in terms of particle diameter. Also plotted on Fig. 7.36 is particle diameter expressed in "phi units," where

$$\phi = -\log_2 \text{ (particle diameter in inches [millimeters])}.$$

$$\dots\dots\dots\dots\dots\dots\dots\dots (7.1)$$

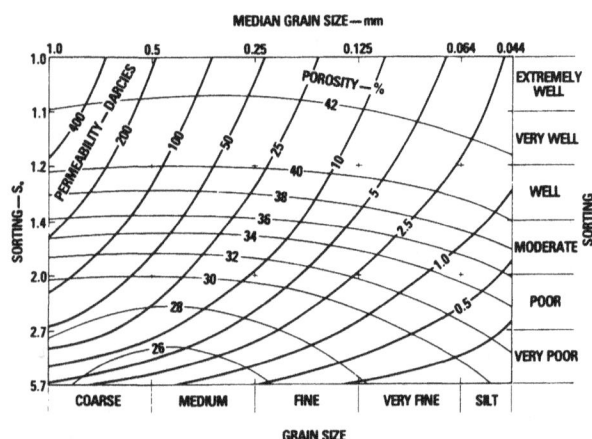

Fig. 7.35—Porosity and permeability of artificially mixed and wet-packed sand of different sizes and sortings (Ref. 27 after Ref. 28).

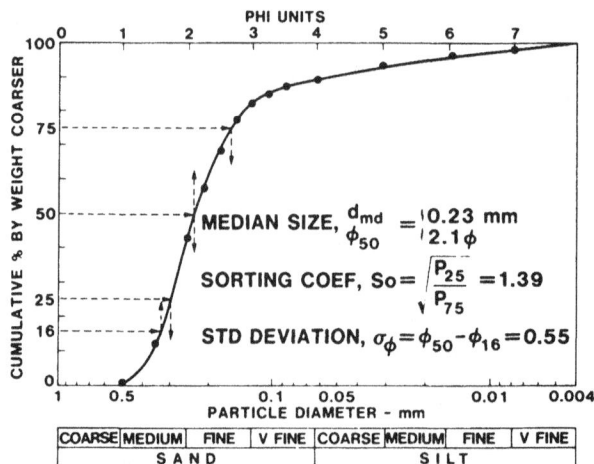

Fig. 7.36—Cumulative size curve showing textural parameters.[27]

7.4.2 Continental Environment

Sands of continental (nonmarine) origin are alluvial (derived from streams or rivers) or aeolian (windblown). Petrographic analysis is used to determine whether a particular sediment is marine or nonmarine but may not be adequate.[1] Alluvial deposits are most abundant and will be the only deposit considered here. LeBlanc[23] describes characteristics of aeolian deposits and reservoirs. Fig. 7.37 illustrates idealized alluvial sand patterns deposited as alluvial fans, meandering streams, and braided streams.[29,30] Alluvial fans are formed by rivers or streams that deposit their sediment load downslope of a confined valley or canyon.[31] They are deposited typically in front of mountains of high relief as shown in Parts a and d of Fig. 7.37, which describes the formation in terms of two models.[23] Some alluvial fans are built by braided streams as they flow into a canyon or valley. Braided stream deposits are often coarse sediments. In the lower fan, some braided stream deposits have good lateral and vertical continuity. When the braided streams contain substantial amounts of fine-grained sediments, however, the channels may have poor lateral and vertical continuity.[23]

Other alluvial fans form when floods depositing a sheet of sand over the entire fan surface occur. Sands deposited under these conditions often have good lateral continuity. Because of the high energy content of the flowing water, fine sediments are winnowed from the deposit, leaving coarser grains with little clay.

Sands deposited from fluvial (river) systems are the most common type of continental sandstone.[23] Parts b, c, f, and g of Fig. 7.37 show possible sand deposits created by a meandering stream that flows through a flood plain. Part e of Fig. 7.37 depicts a deposit formed by a braided stream. Meandering stream deposits have been studied extensively and are well-known reservoirs for hydrocarbons.

Fig. 7.38 shows the main features of a meandering river and its flood plain.[31] In Part a of Fig. 7.38, two meandering streams have deposited sand bodies in the flood plain, which are stacked and discontinuous in the vertical section. Details of a particular depositional cycle are presented in Part b of Fig. 7.38. Three types of deposits in Part b of Fig. 7.38 are important in the analysis of petroleum reservoirs. Two (vertical accretion and channel fill) form reservoir boundaries or barriers to flow. Vertical accretion deposits consist of fine sand and mud that spread over the flood plain when flooding occurs. The stream meanders across the flood plain, leaving abandoned channels. Channel abandonments occur in two ways.[23] A river may cut off a single loop (formation of an oxbow) in the stream, as in Part f of Fig. 7.37, or a significant part of the meander belt, as in Part g of Fig. 7.37.[23] Abandonment is caused by gradual clogging of the channel with sediment that reduces stream flow or by a sudden change in the flow path following a flood. When abandonment is sudden, the abandoned channel eventually is filled with fine overbank muds and other organic material forming the channel deposit (clay plug) depicted in Part b of Fig. 7.38.[31] Abandonment removes a channel from active sedimentation so that it eventually becomes covered with clay and silt when flooding occurs. These deposits are barriers to fluid flow.

Oil has been found in a number of channel deposits.[32-34] One of the earliest mapped channel deposits

TABLE 7.8—RELATIONSHIPS BETWEEN TEXTURE AND PORE SPACE[27]

Texture		Pore Space	
Grain Size	Sorting	Porosity	Pore Size
Very fine	Well to very well	Highest	Predominantly fine and very fine pores
Grain size increase ↓	Sorting poorer ↓	Porosity decrease ↓	Numbers of medium and large pores and average pore size increase ↓
Gravel	Poor	Lowest	Large, medium, and fine pores

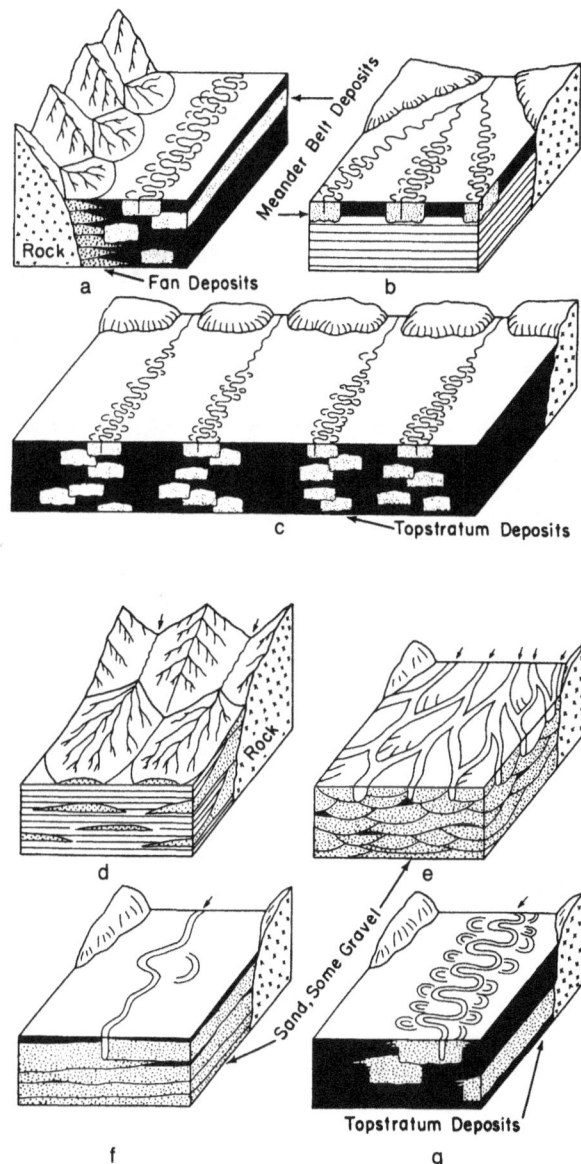

Fig. 7.37—Contrasting idealized alluvial sand patterns (Ref. 29, Fig. 11-13, after Ref. 30).

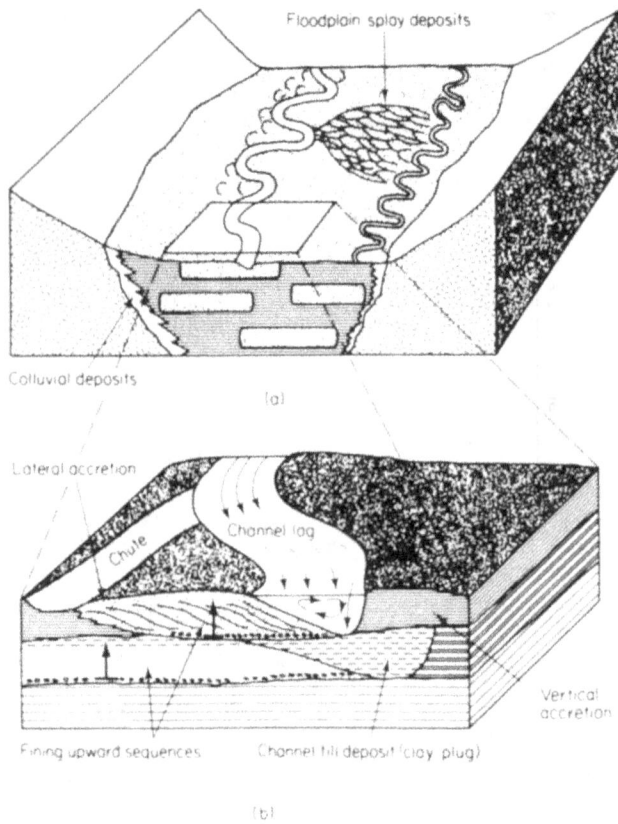

Fig. 7.38—Block diagrams showing main features of a meandering river and its floodplain (Ref. 31, Fig. 19-4).

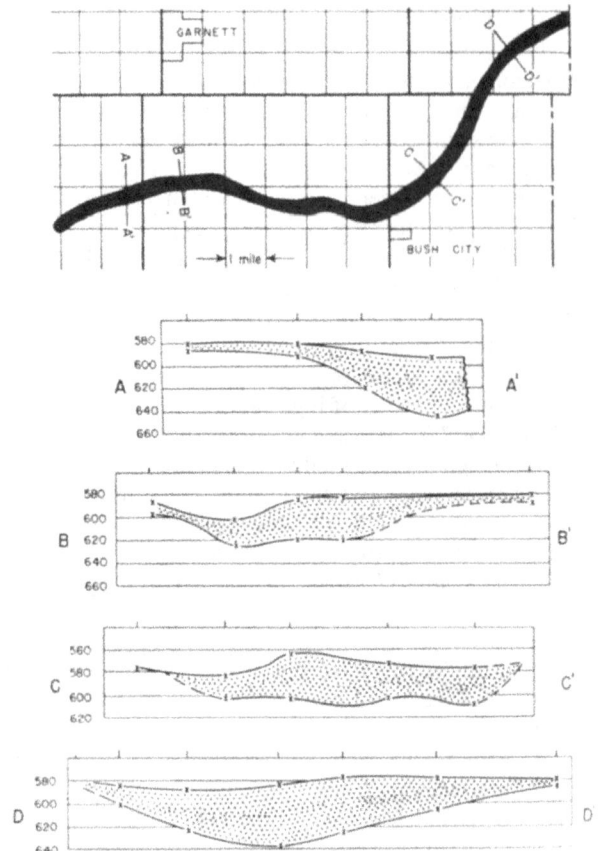

Fig. 7.39—Map and sections of the Bush City pool, Anderson County, eastern Kansas, characteristic of a
number of pools in this region. The sinuous pattern and the central thickening of the sand sug-
gest that this is a sand-filled channel deposit. Oil
fills the channel (Ref. 1, Fig. 7-20, redrawn from
Ref. 34).

is the Bush City shoestring in southeastern Kansas, which
was discovered in 1923. A map showing the location and
cross sections through this deposit is given in Fig. 7.39.
The Bush City shoestring consists of a sand body that is
about 13 miles [21 km] long and ¼ mile [0.4 km] wide,
which was deposited in a channel cut into the Cherokee
shale.

Meandering streams also deposit sediment on the inner bank of a channel bend to form a point bar. This
deposit, built by lateral accretion (Part b of Fig. 7.38),
has a distinct internal structure. Fig. 7.40 shows the secondary flow pattern in a meandering stream and the internal structure created by sediments deposited on the
bank.[31] Coarser sediment is deposited in deeper water,
while fine sediments are deposited on the edge. This
produces a characteristic fining-upward sequence consist-
ing of (1) a poorly sorted basal section of gravel, sand,
and salt; (2) a zone of trough crossbedded sand; (3) a zone
of horizontal and small-scale crossbedded and silty sands;
and (4) a zone of fine-grained, crossbedded silts and sands
with small ripples.[23,31] The channel migrates by erod-
ing the outer band, and the point bar continues to grow.
Meandering streams are subject to flood cycles, which
deposit clays and silts on the top of the point bar and create a discontinuity that is an upper seal. However, discontinuities may also occur within the point bar. At the
end of a flood cycle, the velocity of the floodwaters approaches zero, permitting fine silts and clays to settle on
the sloping surface of the point bar.[23] This layer, termed
a clay drape, is an effective permeability barrier. The
sloped lines within the point bar in Part b of Fig. 7.38

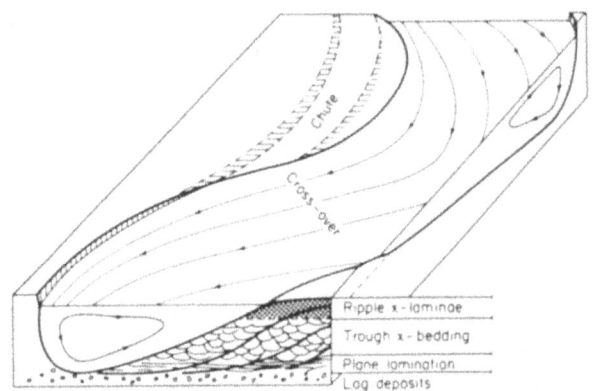

Fig. 7.40—Model of a meandering river, showing secondary
flow pattern and internal structure of point-bar
deposits (Ref. 31, Fig. 19-5).

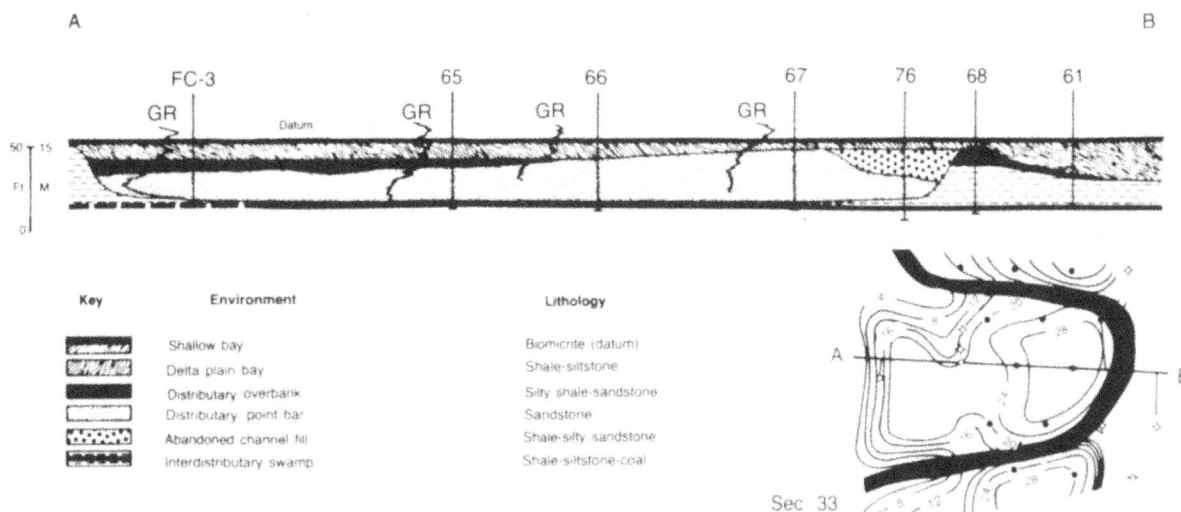

Fig. 7.41—Stratigraphic cross section of a sandstone reservoir of point-bar origin, Eastburn field, MO.[35]

represent episodes of clay-drape formation. According to LeBlanc,[23] clay drapes ranging from 0.125 in. to 2 ft [0.32 cm to 0.61 m] thick are observed primarily in the upper half of the point-bar sequence. Clay drapes may be detected from self-potential (SP) or gamma ray logs by the appearance of serrations in the log.

Point-bar deposits also form in deltaic deposits when the distributary channel meanders. Fig. 7.41 is a cross section through one meander loop of the Eastburn field, Vernon County, MO.[35] Fig. 7.42 shows the sequence of meander loops that span the field.

Sand bodies deposited from river systems appear to have general characteristics. For example, width of the deposit is confined to the meander belt, which may be 15 to 20 times the width of the channel.[23] Also, the thickness of a point bar or channel from a meandering stream is limited to the depth of the channel in which it was formed.[23,31] However, these characteristics are not universally applicable. Parts c and g of Fig. 7.37 depict alluvial deposits that are stacked and thus are thicker than the meandering stream that created them. Fig. 7.43 represents three different sandstone geometry models of sandstones deposited in a river channel. Because of stacking, vertical continuity depends on intersection of different sand bodies as well as the type of deposit at the intersections. Unique patterns

of lateral continuity are associated with the stacking model. Harris and Hewitt[20] propose that these sandstone models should be used to interpret and to predict continuity and thickness patterns in sandstone reservoirs. The thickness and continuity pattern of the Cypress sandstone in the Loudon field, which is shown in Fig. 7.44, contains belt, continuous sheet, and discontinuous sheet geometries. This sandstone was formed in a shallow-water deltaic environment.[21] Recovery efficiencies from belt and continuous sheet regions in the southernmost part of the field were better than from the discontinuous sheet geometry representing the northern part of the reservoir. Thus, recovery efficiency agreed qualitatively with performance that could be predicted from the study of cross sections and maps of the field.

7.4.3 Transitional

7.4.3.1 Deltaic Environment. Sandstones of deltaic origin are the most common reservoirs for hydrocarbons.[23,26] A delta is a deposit built when a river discharges its sediment load into a large body of water, usually the sea. Fig. 7.45 depicts depositional environments typical of modern deltas.[23] Deltaic sands are deposited within the river, which becomes a distributary channel, and along the front of the delta.

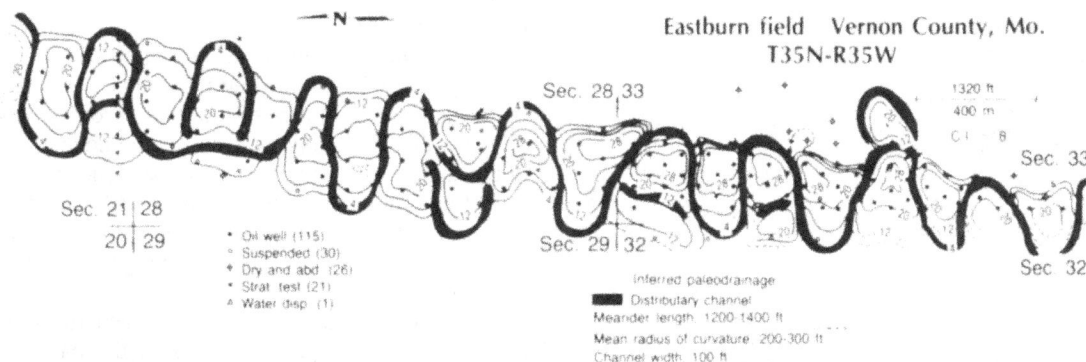

Fig. 7.42—Cross-sand thickness of point-bar reservoir, Eastburn field, MO.[35]

MAPS

BELT

CONTINUOUS SHEET

DISCONTINUOUS SHEET

CROSS SECTIONS

VERTICAL STACKING

LATERAL STACKING

ISOLATED STACKING

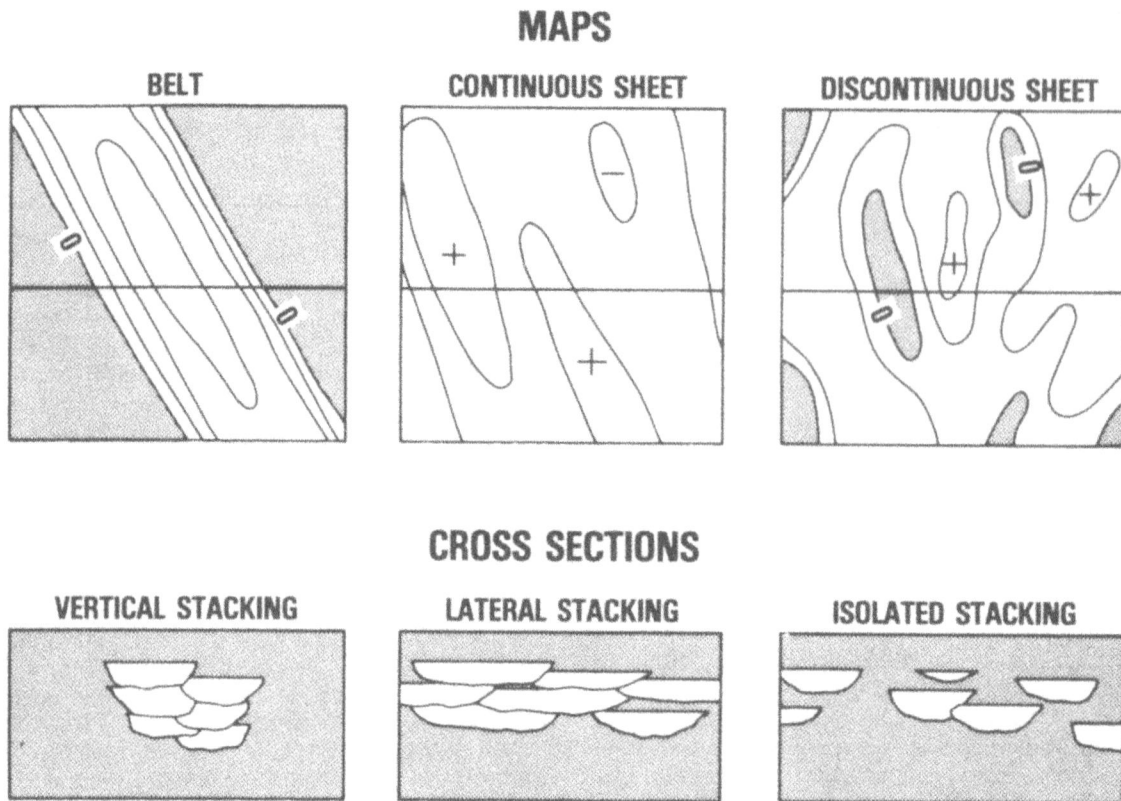

Fig. 7.43—Principal types of sandstone reservoir geometries.[20]

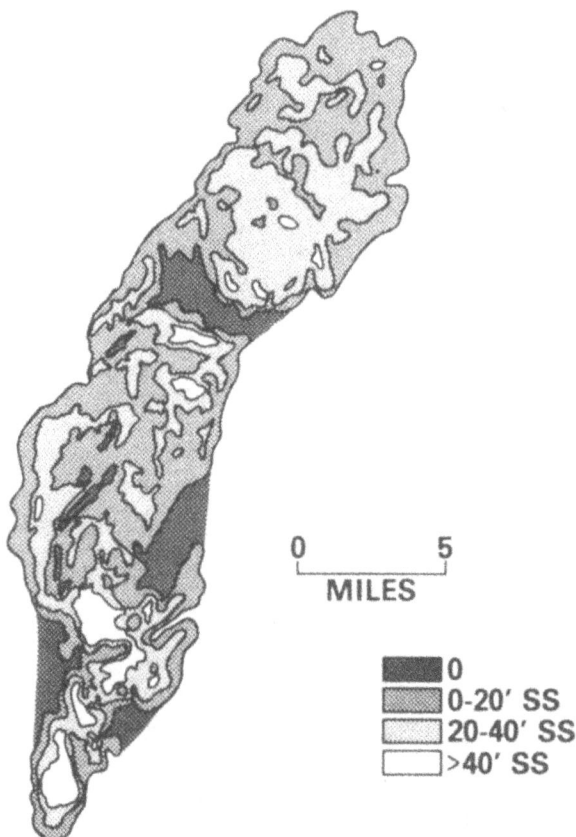

0   5

MILES

0
0-20' SS
20-40' SS
>40' SS

Fig. 7.44—Thickness pattern of the Cypress (Weiler) sandstone in the Loudon field.[20]

Cross sections through a typical delta building into a marine basin are shown in Fig. 7.46.[26] The delta progrades (builds seaward) as sand is deposited at the mouth of the distributary channel. Coarse sand deposited at the mouth of a distributary channel forms river mouth bars that may be dispersed by sea currents to form the sheet-like fringe sands depicted in Fig. 7.45. Finer sediment is transported farther out into the sea where it is dispersed by shore currents and settles slowly to form the prodelta muds and clay shown in Figs. 7.45 and 7.46. The delta sequence builds outward on top of marine sediments.

The distributary channel cuts down through the fringe sand and silts as the delta progrades seaward. In long deltas, the distributary channel may cut into the marine shale. Sand deposited in the distributary channel is highly crossbedded. Multiple distributary channels are common (as shown in Fig. 7.45), with channel initiation and abandonment occurring in a cyclic pattern (Fig. 7.47). Most modern deltas have many abandoned distributary channels.[23]

As a channel is abandoned, it is filled with sediments. The type of sediments filling the channel is determined by the abruptness of the abandonment process. A gradually abandoned channel will have sand added to the top of the channel fill, while a rapidly abandoned channel will have fine sediments on the top of the fill. Channels can be filled up to 90% with sand or clay/silt.[26] A clay plug covers the channel sand deposit and provides an effective seal.

Modern deltas may be placed in two general classes that characterize the dispersal energy of the basin and the amount of sediment.[26] Table 7.9 presents this classification as proposed by Sneider et al.[26] When the dispersal energy of the basin is high, fines (silt and clay) are

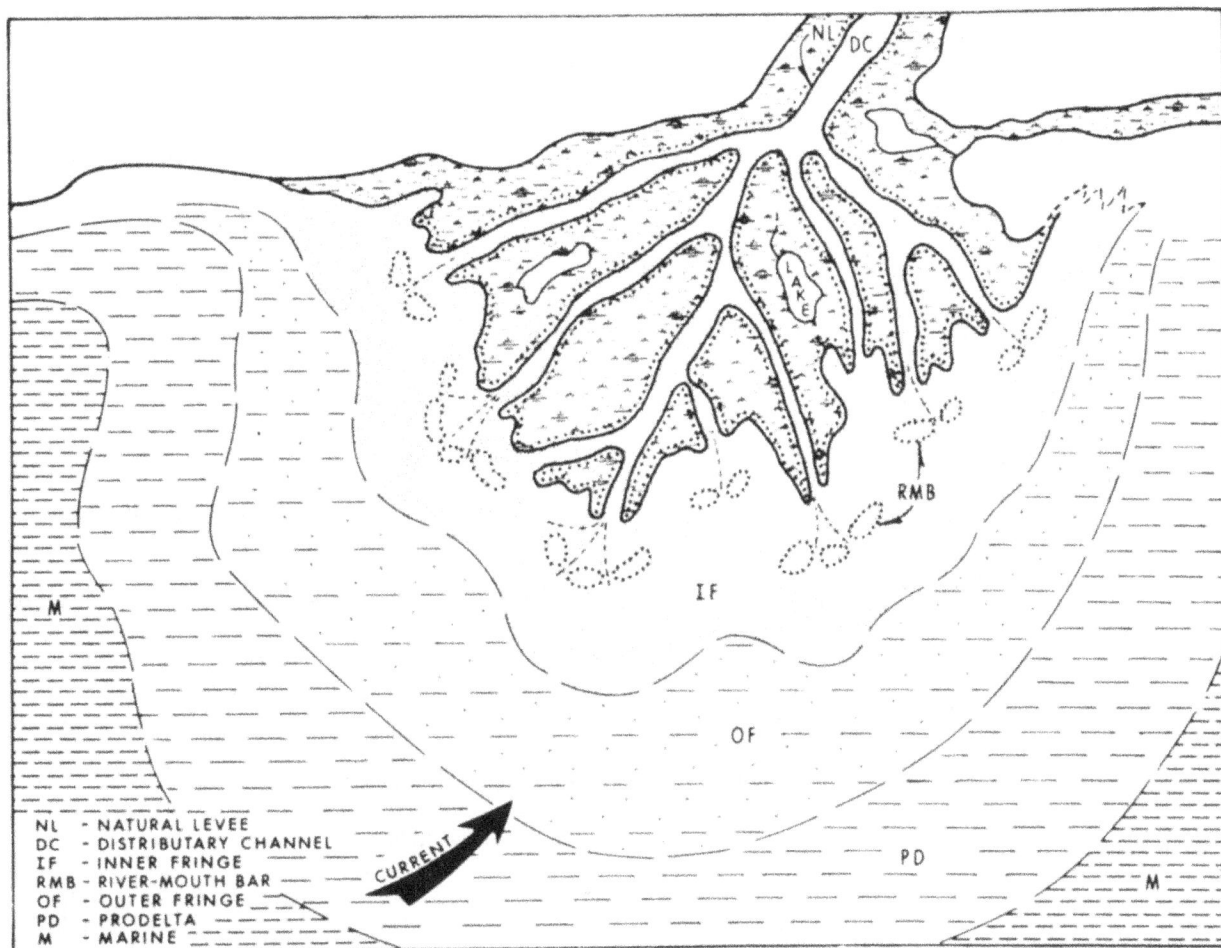

NL - NATURAL LEVEE
DC - DISTRIBUTARY CHANNEL
IF - INNER FRINGE
RMB - RIVER-MOUTH BAR
OF - OUTER FRINGE
PD - PRODELTA
M   - MARINE

Fig. 7.45—Depositional environments typical of modern deltas.[23]

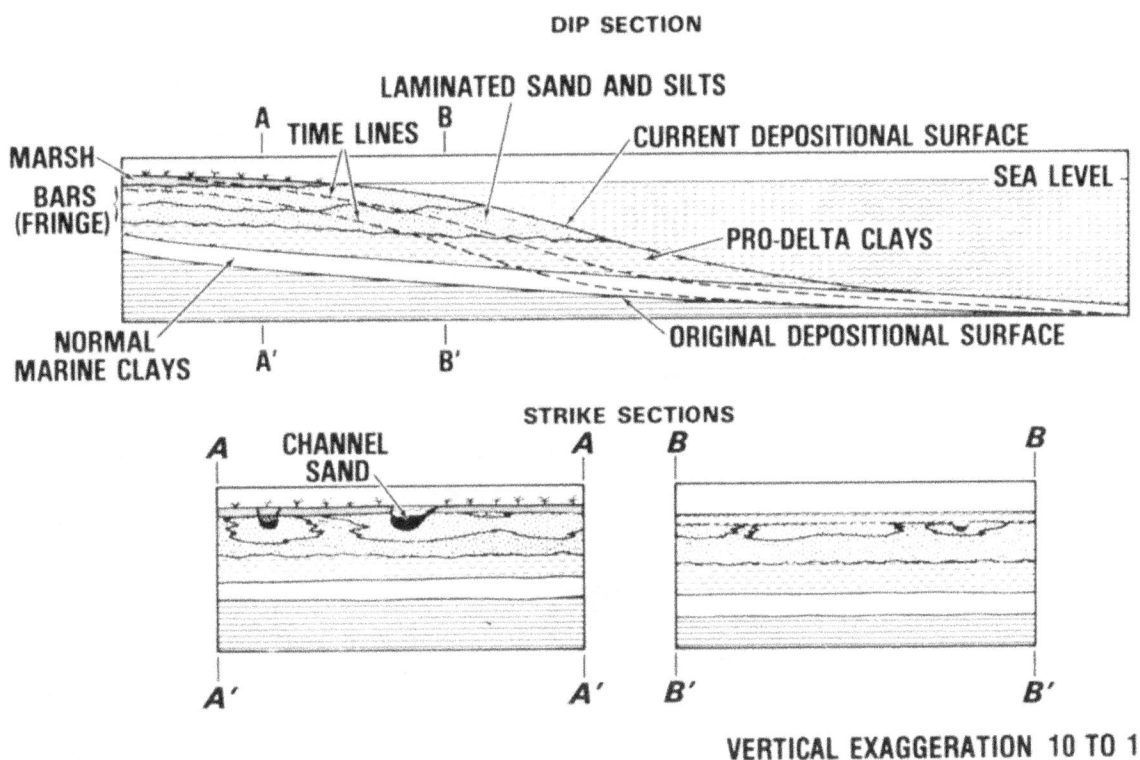

DIP SECTION

LAMINATED SAND AND SILTS

A TIME LINES B

CURRENT DEPOSITIONAL SURFACE

SEA LEVEL

MARSH

BARS
(FRINGE)

PRO-DELTA CLAYS

NORMAL
MARINE CLAYS

A'   B'

ORIGINAL DEPOSITIONAL SURFACE

STRIKE SECTIONS

A CHANNEL   A   B   B
   SAND

A'   A'   B'   B'

VERTICAL EXAGGERATION 10 TO 1

Fig. 7.46—Delta growth for dip and strike sections.[26]

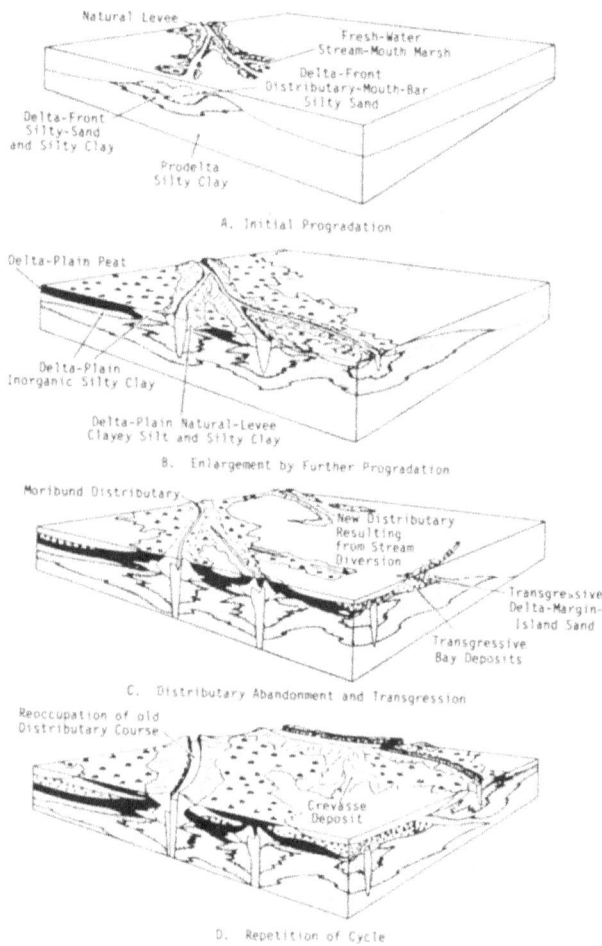

Fig. 7.47—Development of delta sequence in the Mississippi delta.[31]

separated (winnowed) from coarser sediments and transported away from the delta by the current. Coarse sediments that settle on the delta surface are clean and well sorted and have excellent reservoir properties. In low-energy basins, both sand and fines settle together so that the delta builds on a muddy or clay environment.

The deltaic sequence may be recognized from detailed petrographic analysis of cores. Fig. 7.48 depicts several types of modern deltaic deposits.[26] Regions not containing a distributary channel have a characteristic sequence: (1) a top, marsh-derived layer of the deltaic plain that contains marine clay, silt, or coal with scattered reservoir sands, (2) fringe sands, (3) basal prodelta muds, and (4) marine clays. Fringe sands may be subdivided in the vertical sequence into an inner bar (reservoir sand with some intervening clay/silt layers) and the outer bar (reservoir sand with many clay/silt layers).

A vertical section through a distributary channel will show the channel fill cutting through the deltaic sequence. The vertical sequence encountered in a cross section through a distributary channel depends on the depth of the channel at the position where the cross section is constructed. Fig. 7.48 shows distributary channels that rest on inner and outer bars and cut into the marine clay. Thus, a vertical section through the distributary channel could show all or part of the deltaic sequence.

Fig. 7.49 is a core description graph for an interval from the Loudon reservoir.[21] The interval is part of the Cypress formation that contains the Weiler sandstone. Examination of cores and fossils led to the identification of the vertical sequence of the lithologic units as having deltaic origin.[21] Six lithologic units are shown in order of deposition. These are prodelta, shale (Lithologic Unit 1),

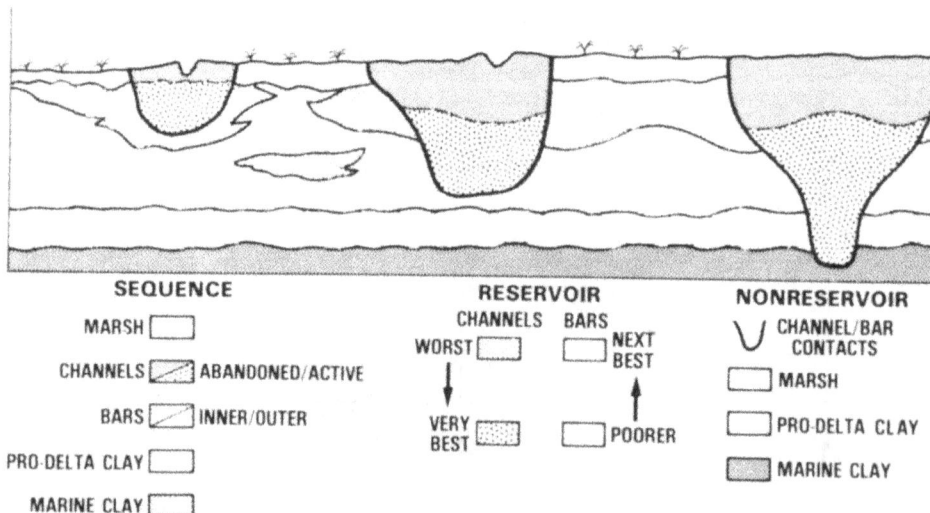

Fig. 7.48—Deltaic environment reservoir character and continuity.[26]

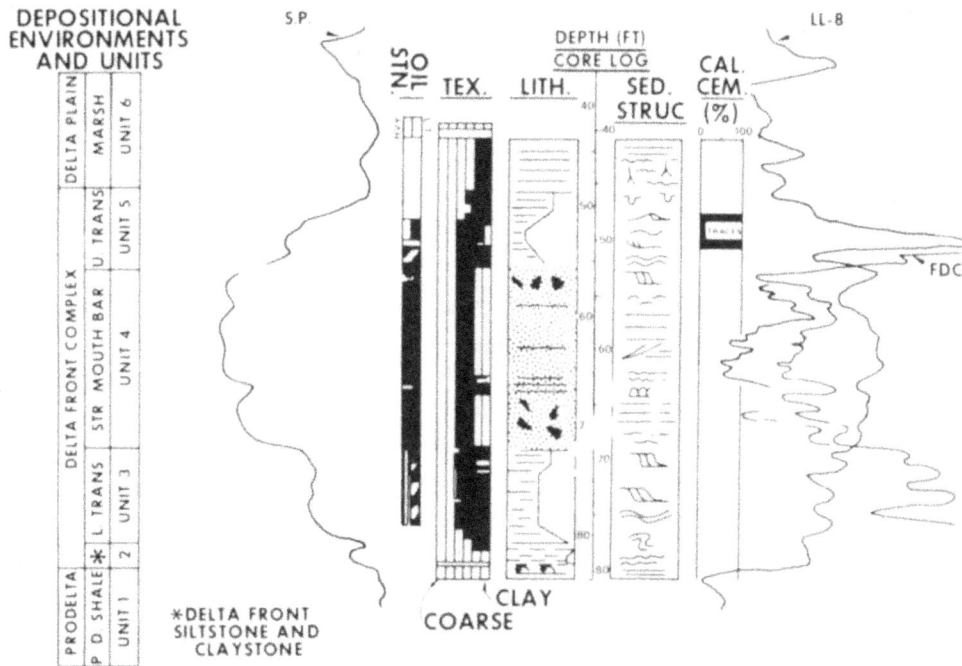

Fig. 7.49—Example core—description graph showing results of rock studies.[21]

delta front shale siltstone and sandstone (Lithologic Unit 2 through 5), and delta/plain marsh (Lithologic Unit 6). Fig. 7.50 shows the location of this vertical section relative to the units in a modern delta.[20] The identification of this deposit as having deltaic origin is based on studies of the sequence recent and ancient deltas.

Studies of deltaic environments also led to generalizations concerning continuity and properties of reservoir sands.

Reservoir Quality. Permeability and porosity of deltaic deposits follow the trends discussed in Sec. 7.4.1. Grain size and sorting correlate with pore size, porosity, and permeability in unconsolidated to slightly consolidated

sands. Similar correlations are observed in reservoir sands that have moderate cementation and compaction. Sneider *et al.*[26] conclude that the most important parameters controlling reservoir rock in most unfractured reservoirs are the size and sorting of the grains. Applying these correlations to channel and bar deposits produces the conceptual relationships between position in the vertical sequence and reservoir quality presented in Fig. 7.51, which shows that the best reservoir-quality sand in channel deposits is near the bottom of the deposit. In bar deposits, the best reservoir-quality sand is at the top of the deposit. The conceptual relationships in Fig. 7.51 are reversed in some cases by diagenesis.

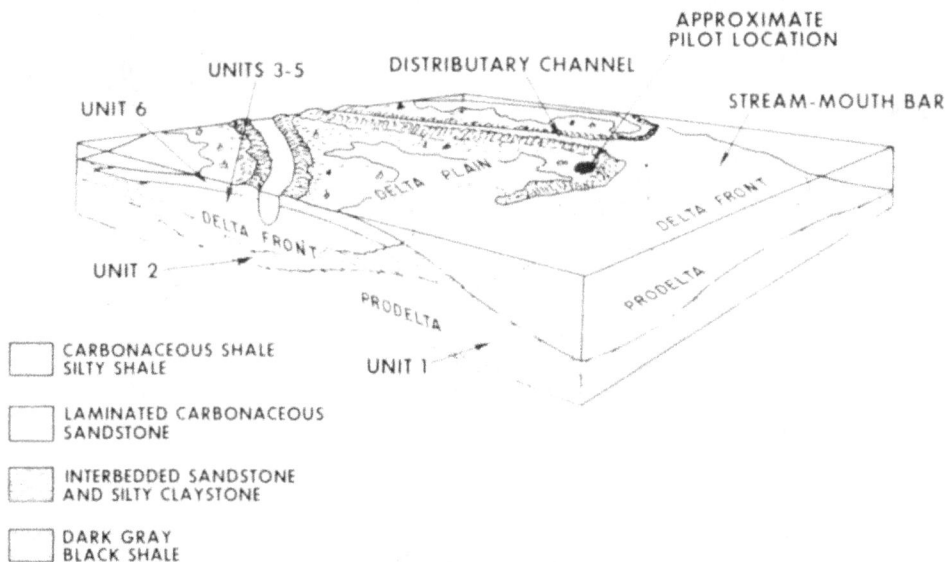

Fig. 7.50—Block diagram showing the vertical and areal distribution of units in a typical modern delta.[21]

Type	Texture		Pore Space		"Capillarity"		Continuity
	Grain Size	Sorting	Porosity	Pore Size	Permeability	Sw	
Top	Finest	Best	Highest	Very Fine	Lowest	Highest	Deteriorates Upward
CHANNELS	↑	↑	↑	↑	↑	↑	
Bottom	Coarsest	Poorest	Lowest	Large	Highest	Lowest	Best
Top	Coarsest	Best	Highest	Large	Highest	Lowest	Best
BARS	↓	↓	↓	↓	↓	↓	
Bottom	Finest	Poorest	Lowest	Very Fine	Lowest	Highest	Deteriorates Downward

Fig. 7.51—Deltaic reservoir conceptual relationships.[26]

Fig. 7.52 presents idealized permeability profiles and log responses for deltaic bar and channel deposits. Correlation of permeability with grain size is apparent. The raggedness of the profiles reflect interbedding of silt and clay in the vertical sequence. Parts A, B, C, and D of Fig. 7.52 are typical of high-energy sand delta deposits. Parts E and F of Fig. 7.52 show numerous permeability breaks found in deltaic deposits in low-energy mud deltas.

Reservoir Continuity. The type of delta affects the continuity of the deposited bar (fringe), channels, and units. High-energy deltas have sand bars with good lateral continuity except in the region cut by a distributary channel. Vertical continuity is good. The few channels are meandering, creating point-bar deposits within the meander

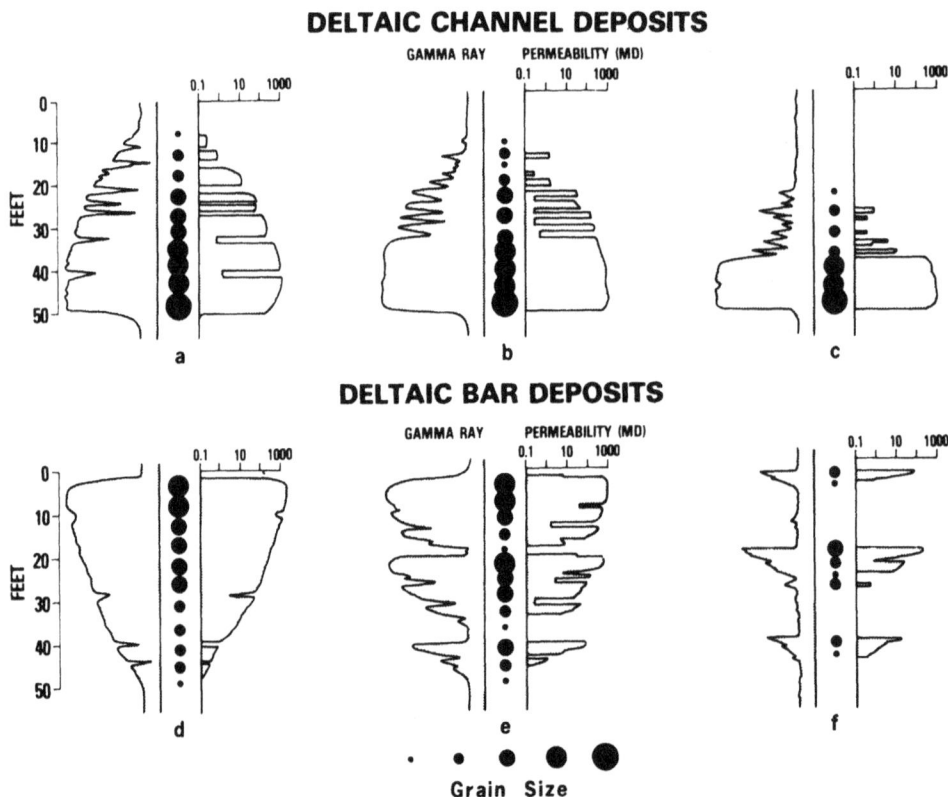

Fig. 7.52—Idealized permeability profiles and log responses.[26]

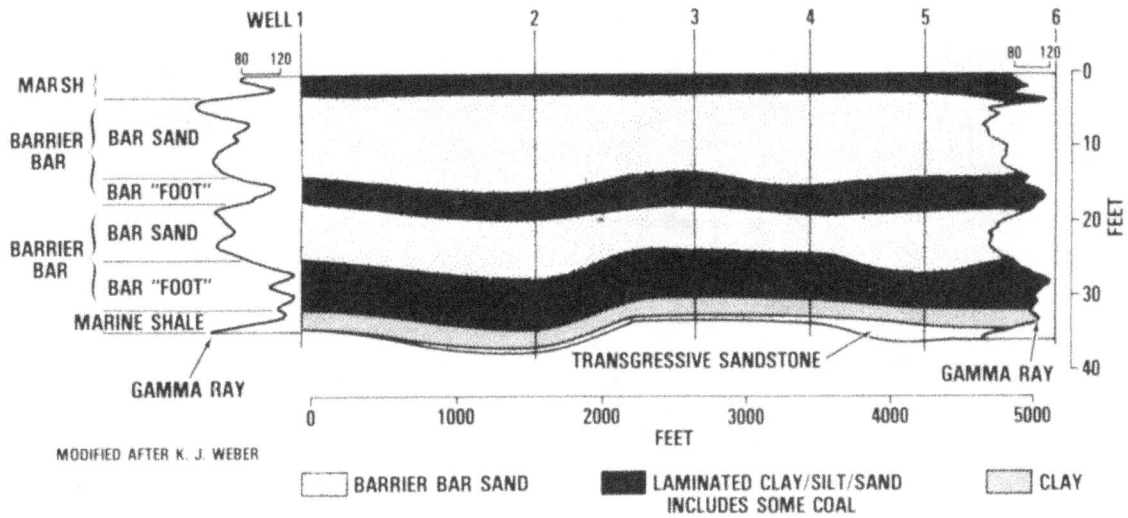

Fig. 7.53—Continuity in sand-delta system, Afiesere field, Niger delta (Ref. 26, after Ref. 36).

belt. As shown in Fig. 7.51, the sand deposit has the characteristic fining-upward sequence. Best continuity is found at the bottom of the channel where coarsest sediments are deposited. Silt/clay interbeds are deposited in increasing amounts and thicknesses as the top of the channel is approached.

The intersection of the channel and bar/fringe sands is a potential region of reservoir discontinuity. Some channel deposits are in direct contact with bar/fringe sands with good hydraulic connections. Others are separated from fringe sands by a layer of fine-grained sediments that may be effective barriers to flow.[23,26] Figs. 7.53 and 7.54 illustrate continuity in the sand delta system found in the Afiesere field of the Niger delta.[26]

Reservoir continuity in bar and channel sands deposited in a low-energy (mud) environment is controlled by

the numerous floods that carried sediments into the delta and by the absence of silt/clay removal (winnowing) between floods. Silt/clay interbeds or layers of large areal extent are formed. Mud deltas have many distributary channels that cut through the delta sequence. Thus, reservoir sands are less continuous.

Figs. 7.55 and 7.56 illustrate sand continuity in two mud deltas in south Louisiana.[26]

7.4.3.2 Barrier Island Sands. Barrier island deposits are long, narrow sand bodies that formed parallel to ancient coastlines. A model of a recent barrier island complex is shown in Fig. 7.57.[31] A lagoon separates the barrier island from the shore or coastal plain. The origin of the barrier island still is being debated, but the island must have emerged from the sea. Longshore currents transport

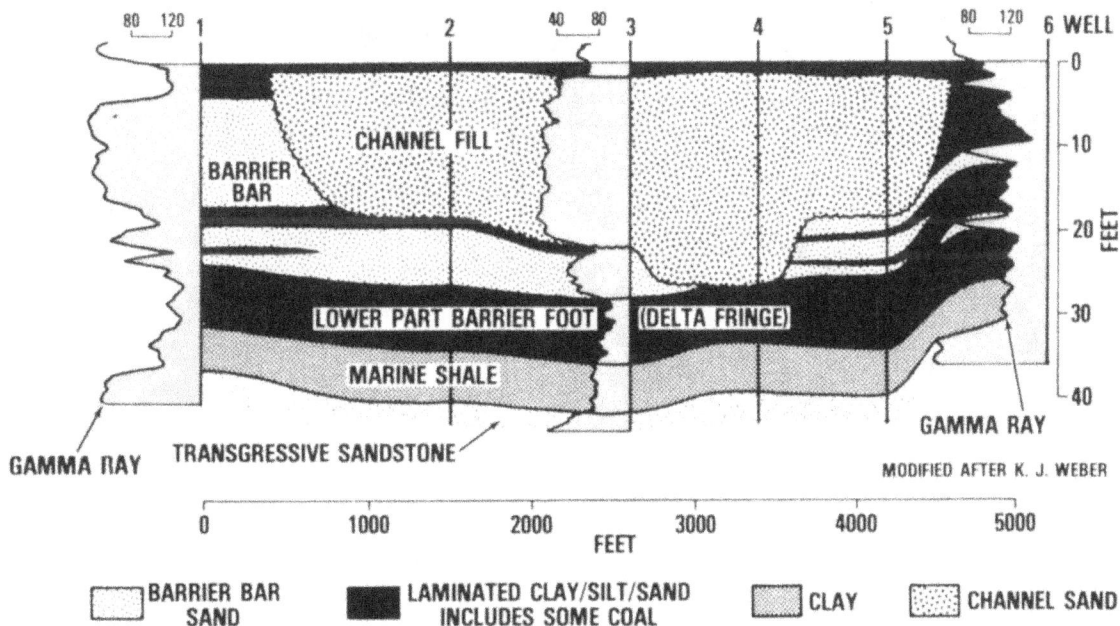

Fig. 7.54—Continuity in sand-delta system, Afiesere field, Niger delta (Ref. 26, after Ref. 36).

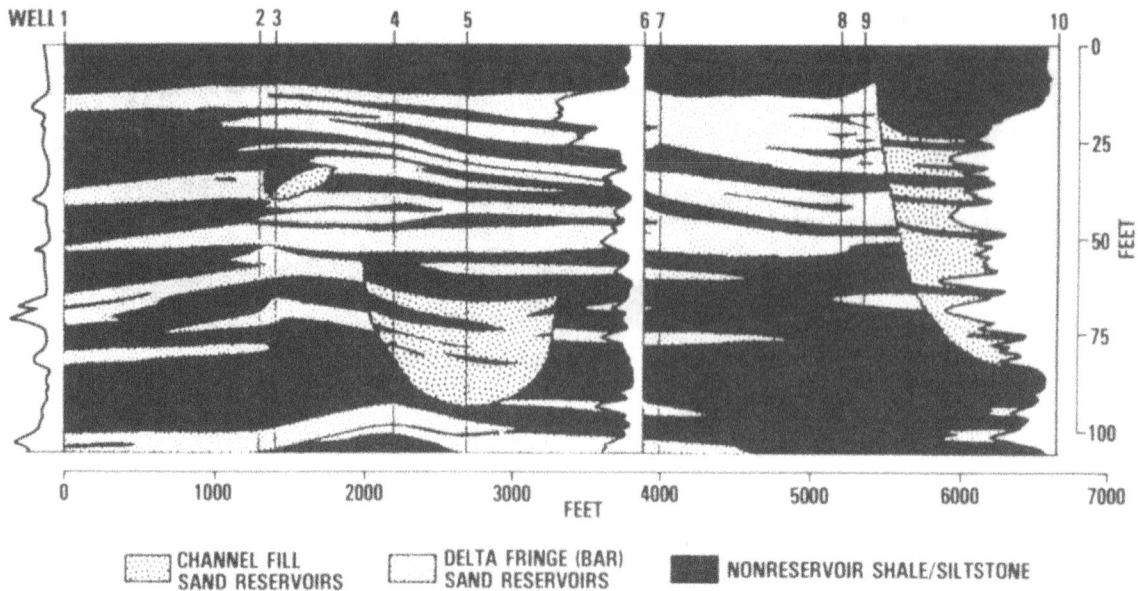

Fig. 7.55—Continuity in mud-delta system in south Louisiana Miocene.[26]

sediments parallel to the shore. The barrier island may be cut by a tidal channel that creates a depositional environment of its own along the axis of the channel.

Barrier island deposits are recognized by a characteristic vertical sequence of sedimentary structures, the presence of a lagoonal facies in adjacent areas, and some evidence that the barrier island emerged from the sea.[31] Because

these deposits are a small percentage of sands in a sedimentary basin, they are hard to find and to identify.[23] They make excellent petroleum reservoirs, however, because they are bounded by impermeable silts and clays, thereby providing an effective stratigraphic trap, and because there is little clay in the deposit and excellent reservoir continuity.

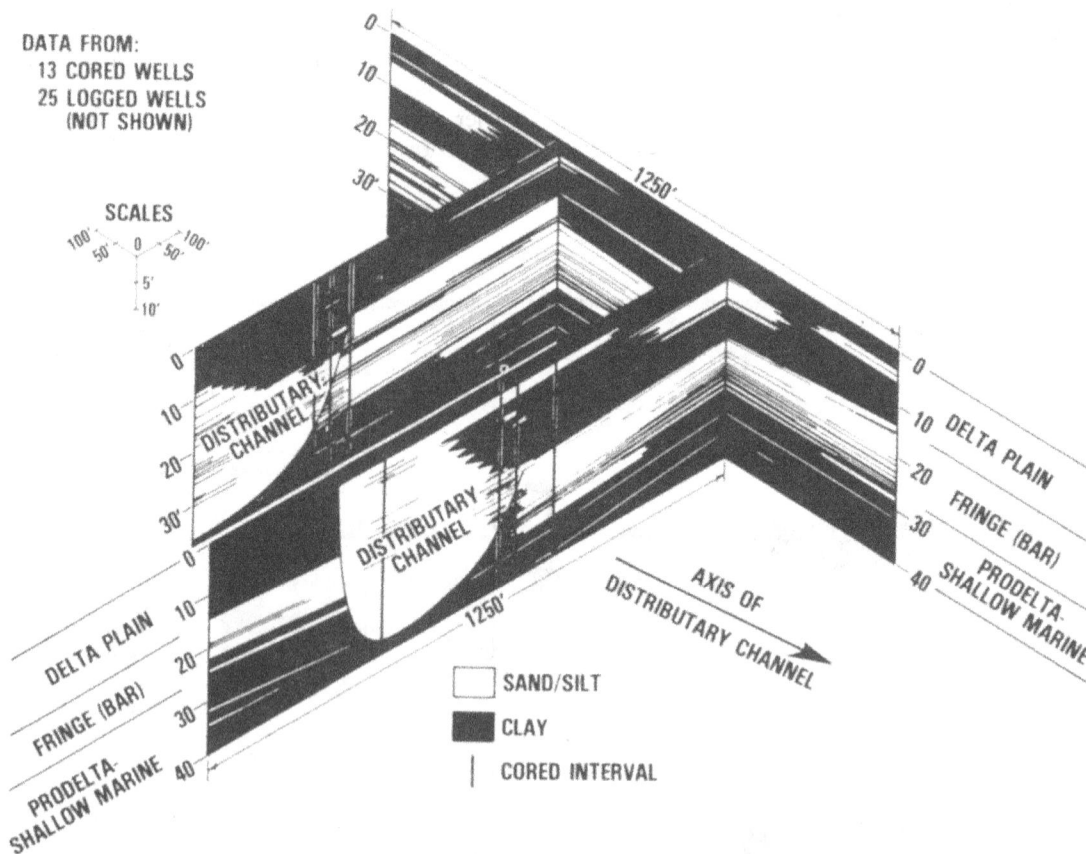

Fig. 7.56—Geometry and continuity of reservoir and nonreservoir unit for recent Lafourche Mississippi delta.[26]

The vertical sedimentary sequence consists of four layers. The lowest layer comprises interbedded sands, silts, and clays and is overlain by a thick zone of highly burrowed sand. The third layer contains horizontal, laminated sands characteristic of the beach and upper-shore facies. The top layer has sediments that were above sea level and may be crossbedded. Sedimentation on a recent barrier island is depicted in the cross section in Fig. 7.58.[23] The coarsest material is found at the top of the sequence and the finest materials at the bottom. This trend is followed in Fig. 7.59,[26] a vertical sequence in a barrier-bar deposit found in the Elk City field, and in Fig. 7.60, which depicts barrier island sandstone and lagoonal facies in the Bell Creek field.[39] The distinct shale/sand boundaries of the barrier island facies show up as pronounced breaks on the SP log. Thus, the shape of the SP log often is used qualitatively to identify possible barrier deposits.

7.4.3.3 Transgressive Marine Sandstones

Sands deposited as the sea advances or transgresses on the land are referred to as transgressive marine sands. Transgressions occur because of rising seas, subsidence of the coastal plain, abandonment of a delta, or a combination of these effects during a period of constant sea level.[23] Transgressive sandstones deposited during delta abandonment are particularly important because there are recognizable geologic markers associated with deltaic deposits. As the sea moves inland, a small portion of the upper deltaic sequence is removed by wave erosion and deposited as a thin blanket of sand (<10 ft [3 m]) over the deltaic sequence. Cementation of these sands occurs rapidly after burial because of the large amount of shell material in the sand. These sands have wide area distribution and good continuity but are not important as reservoirs. They usually give large spikes on resistivity logs. LeBlanc[23] notes that deltaic facies could not be correlated if transgressive beds did not occur in the vertical section.

7.4.4 Deep Marine Sandstones

Turbidite sandstones are the principal reservoir rocks formed in a deep marine environment. A turbidite sequence consists of many layers of sandstones separated by interbedded shales. Turbidites are sediments deposited by turbidity currents, which are periodic, pulse-like flows of sediment. Although causes of turbidity currents in the deep marine environment are not fully understood, the geologic record indicates that each flow deposited an individual sand bed as the flow moved out of its submarine canyon or valley onto a fan where the current dissipated as it spread across the basin floor (Fig. 7.61).[31] According to Pettijohn et al.,[29] turbidites are perhaps the most continuous individual beds known because they are deposited during short-lived pulses and are not reworked by later currents.

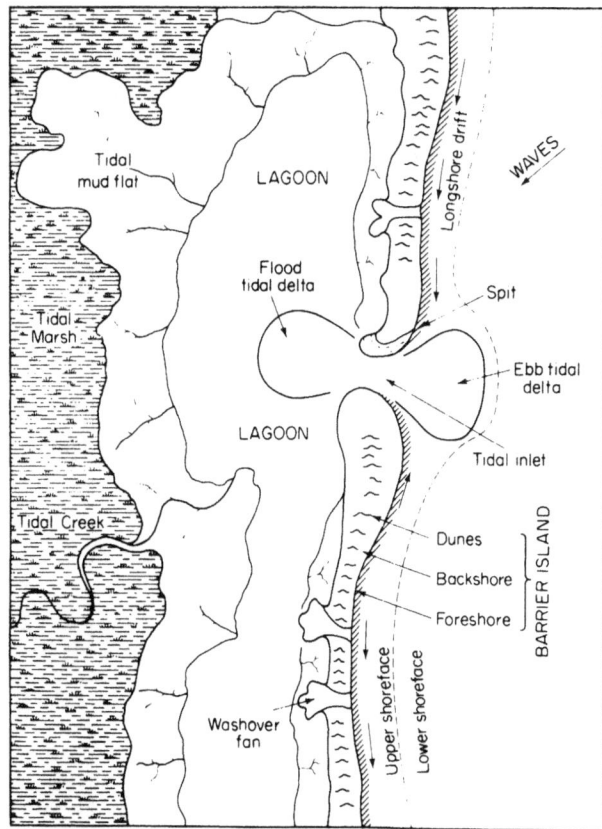

Fig. 7.57—Main components of Barrier Island model (Ref. 31, Fig. 19-14).

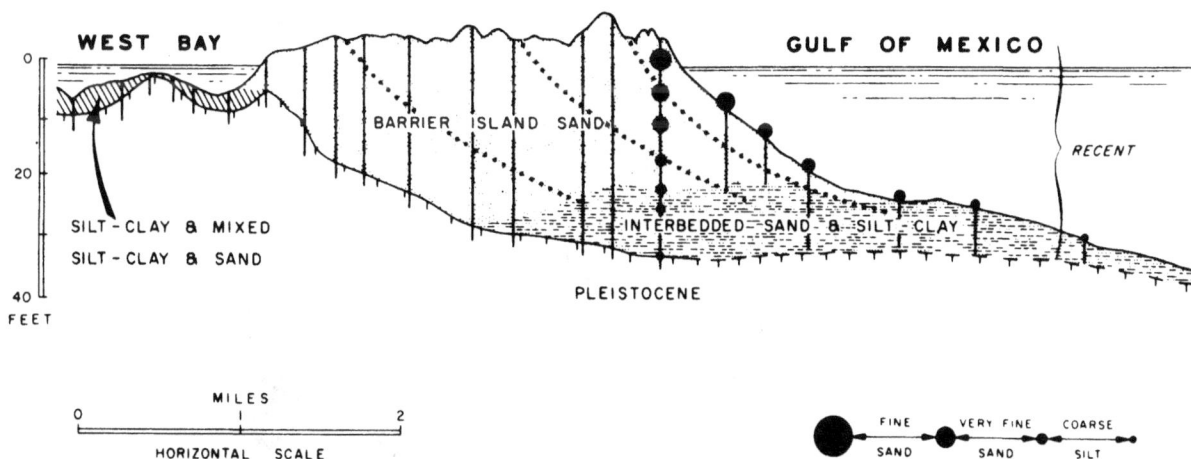

Fig. 7.58—Section across the Galveston barrier island showing relationship of barrier sands to lagoonal and marine fine-grained sediments (Ref. 23, after Ref. 37).

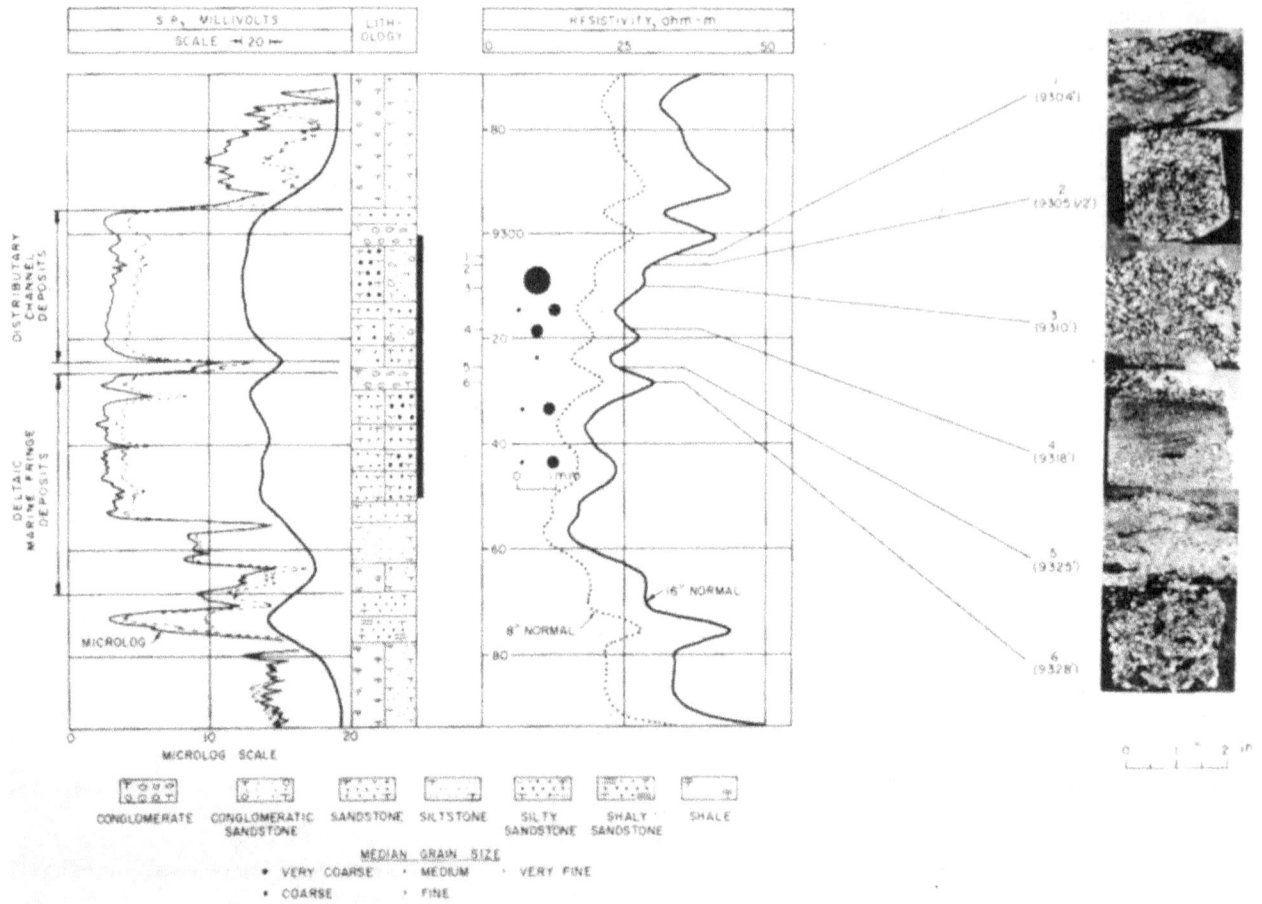

Fig. 7.59—Vertical sequence in barrier-bar complex from Elk City field.[38]

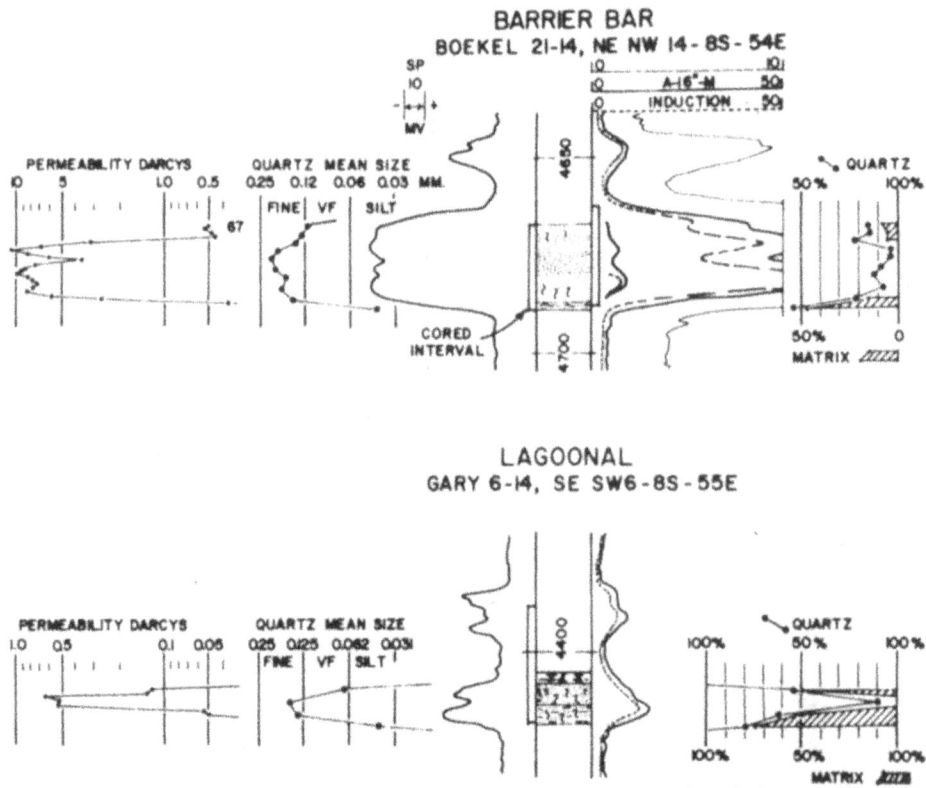

Fig. 7.60—Barrier island and lagoonal facies in Bell Creek Field (Ref. 30, Fig. 19-42).

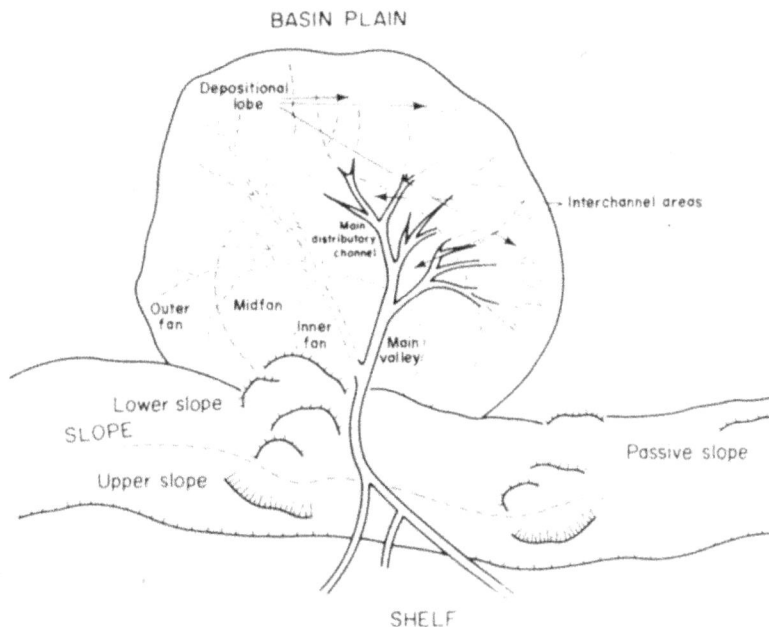

Fig. 7.61—Submarine fan model of Mutti and Ricci Lucchi (Ref. 31, Fig. 19.42, from Ref. 41).

The facies model for submarine fans based on modern and ancient fans is described in detail in several references.[40-42] Submarine fans in Fig. 7.61 are formed as turbidity currents dissipate. Each turbidity current episode disperses a large volume of sediment through active fans. When the turbidity current begins, the coarser material is concentrated at the head and the bottom of the current.[43] Thus, coarser material moves faster than finer sediments that remain in the rear and upper parts of the flow. As turbidity currents emerge from canyons or valleys and spread out over the fans, velocities decrease,

leading to the formation of thick, coarse beds near the canyon or valley outlet with thin fine-grain beds deposited at the extremities of the fan. An ideal vertical sequence generated by a turbidity current episode, termed the Bouma sequence, is shown in Fig. 7.62.[31] This sequence has a characteristic decrease of grain size from bottom to top, referred to as fining upward, and is found primarily on the outer fan.

Submarine fans develop from many episodes of turbidity currents. The fan is subdivided into three regions: the inner (upper) fan, the middle fan, and the outer fan. As

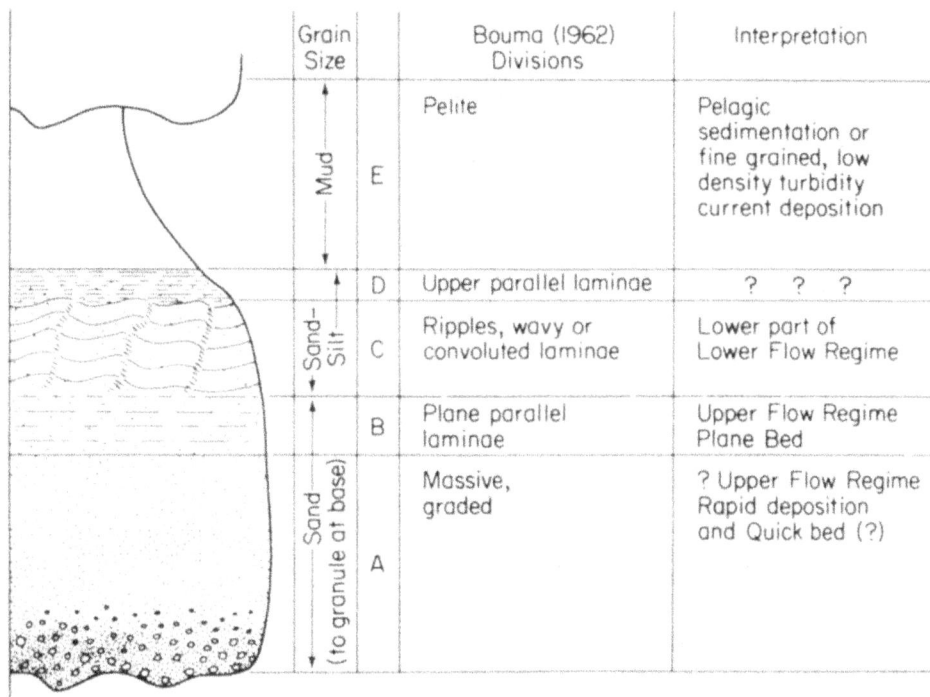

Fig. 7.62—Ideal sequence of structures in a turbidite bed with interpretation (Ref. 31, Fig. 5.7, after Ref. 44 with interpretation, after Refs. 45–47).

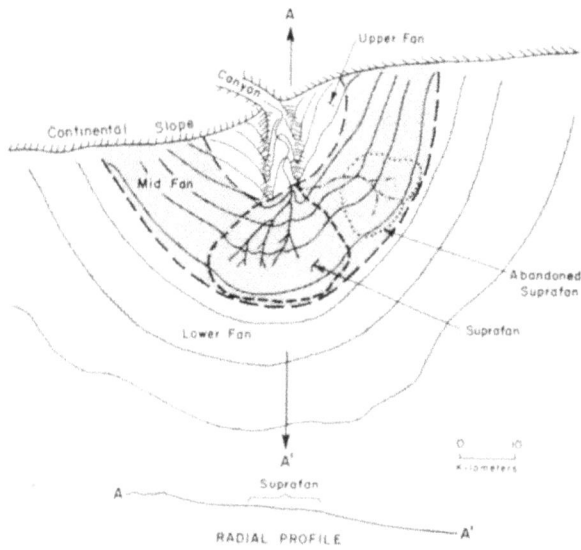

Fig. 7.63—Model for submarine fan growth showing active and abandoned lobes (suprafans). Courtesy International Human Resources Development Corp.

progradation occurs, the middle fan builds over the outer fan, and the inner fan is on top of the middle fan. Generally, the coarsest sediments are deposited in the distributary channels in the inner fan. Deposition of finer sand and silt is rapid on the middle fan near the end of the main distributary channel,[31] forming a lobe of sediment or a suprafan that builds until the distributary channel shifts and begins to form a new suprafan. The distributary channel outlined with dashed lines in Fig. 7.61 is an abandoned channel. Fig. 7.63, showing active and abandoned lobes, is a model for submarine fan growth. When the distributary channel shifts, the old suprafan receives limited sediments, mainly clays and similar fines transported by overflows from the new distributary channel.

The vertical sequence through a submarine fan was interpreted in a classic paper by Mutti and Ricci Lucchi.[41] Two main fan-channel sequences were identified. They were (1) a thickening and coarsening-upward sequence of facies that were associated with the outer fan and progradation of suprafan lobes and (2) thinning and fining-upward sequences typical of channel fills in the lower slope and inner or middle fan.[31] Fig. 7.64 shows this vertical sequence using five of the seven facies proposed by Mutti and Ricci Lucchi.[31]

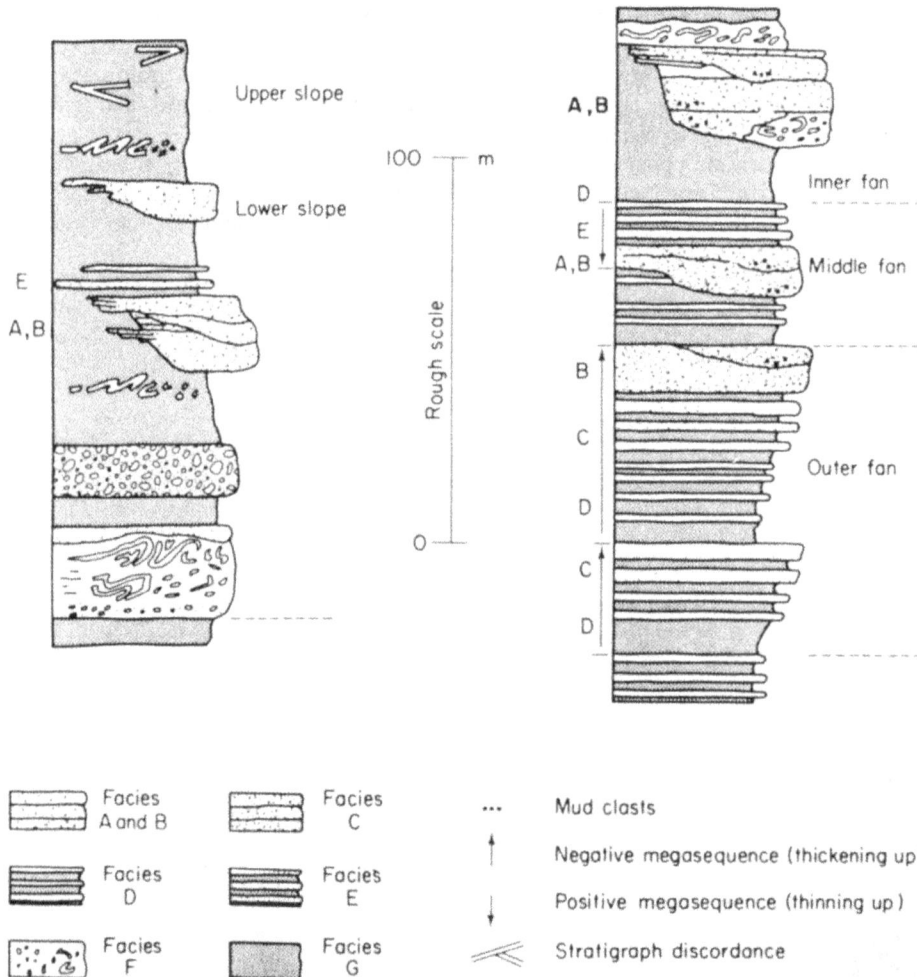

Fig. 7.64—Typical vertical sequences of facies developed in slope and submarine fan environments (Ref. 31, Fig. 19–43).

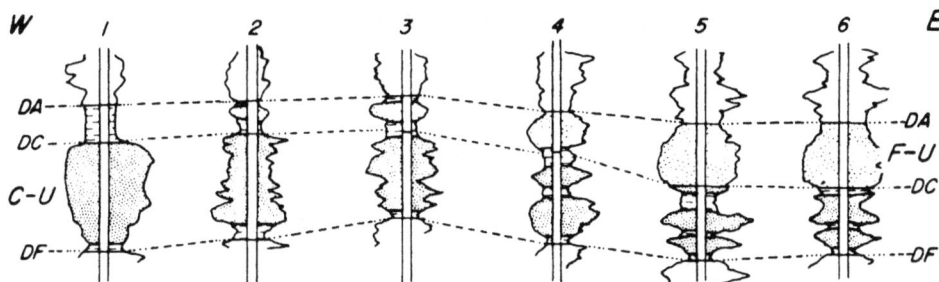

Fig. 7.65—Revision of electric logs of lower Pliocene Repetto formation, Ventura Basin, CA. Courtesy International Human Resources Development Corp.

Several oilfields produce from turbidite deposits.[48] The lower Pliocene Repetto formation of the Ventura field in California was identified as a turbidite deposit.[42,49] Fig. 7.65 is a correlation of electric logs showing coarsening-upward deposits characteristic of a suprafan lobe and fining-upward trend of individual turbidite beds.

7.5 Depositional Environment of Carbonate Reservoirs

The depositional environment of carbonate reservoirs differs from sandstone clastic sedimentation in that all sediments were produced by chemical and biochemical processes within a sedimentary basin of an ancient sea. Most carbonate sediments are formed in an environment of warm, shallow, clear, marine water.[50] The basic processes of carbonate sedimentation produce facies patterns that are recognizable in the geological record throughout the world. Fig. 7.66 illustrates three major depositional environments for carbonate sediments.[51] They are basin

(a deep low-energy zone), slope or shelf margin (a shallow high-energy zone), and shelf or backreef (a shallow low-energy zone). Typical carbonate bodies found in these environments are also indicated in Fig. 7.66.

The three depositional environments may be divided into the nine subenvironments or belts shown in Table 7.10.[50] Because of environmental differences (temperature, slope, geologic age, and water energy), not all facies are found in a particular deposit. The general pattern, however, is a model for prediction of geographic distribution of rock types.[50]

Most carbonate sediments are deposited on the shelf and shelf margins as carbonate ramps, platforms, and banks (Figs. 7.67 and 7.68). Table 7.11 contains definitions of terms that describe the shape of these large carbonate bodies.[50]

Characteristic shapes of five types of carbonate deposits are shown in Fig. 7.69. The small biohermal reefs have high relief above the seafloor. Biostromal reefs are sheet-like deposits with abrupt edges that can extend over

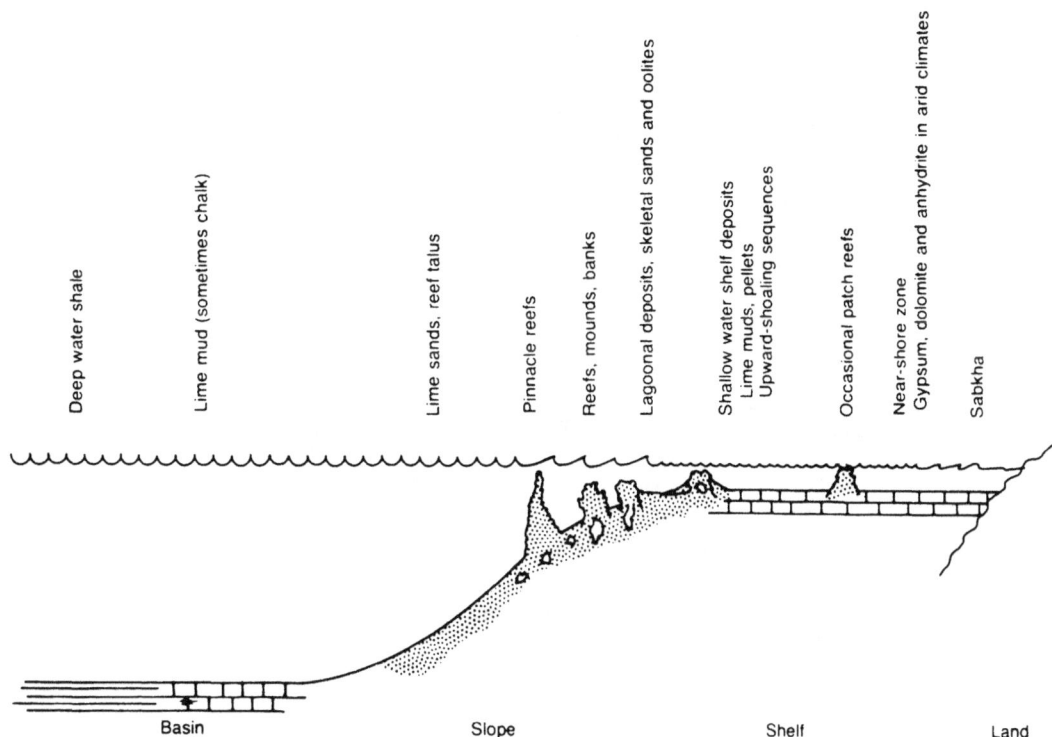

Fig. 7.66—Carbonate depositional environments and characteristic rock types (Ref. 51, Fig. 9-8).

TABLE 7.10—IDEALIZED SEQUENCE OF STANDARD FACIES BELTS (Ref. 50 after Refs. 52 and 53)

Sealed Cross Section

Diagrammatic cross section

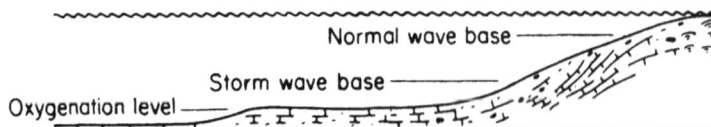

Facies Number	1	2	3	4
Facies	*Basin* (euxinic or evaporite) a) Fine clastics b) Carbonates c) Evaporites	*Open shelf* (undaform) Open marine neritic a) Carbonates b) Shale	*Toe of slope carbonates*	*Foreslope* a) Bedded fine-grain sediments with slumps b) Foreset debris and lime sands c) Lime mud masses
Lithology	Dark shale or silt, thin limestones (starved basin); evaporite fill with salt	Very fossiliferous limestone interbedded with marls; well-segregated beds	Fine-grain limestone; cherty in some cases	Variable, depending on water energy upslope; sedimentary breccias and lime sands
Color	Dark brown, black, red	Gray, green, red, brown	Dark to light	Dark to light
Grain type and depositional texture	Lime mudstones; fine calcisiltites	Bioclastic and whole fossil wackestones; some calcisiltites	Mostly lime mudstone with some calcisiltites	Lime silt and bioclastic wackestone-packstone; lithoclasts of various sizes
Bedding and sedimentary structures	Very even mm lamination; rhythmic bedding; ripple cross lamination	Thoroughly burrowed; thin to medium; wavy to nodular beds; bedding surfaces show diastems	Lamination may be minor; often massive beds; lenses of graded sediment; lithoclasts and exotic blocks. Rhythmic beds	Slump in soft sediments; foreset bedding; slope bioherms; exotic blocks
Terrigenous clastics admixed or interbedded	Quartz silt and shale; fine-grain siltstone; cherty	Quartz silt, siltstone, and shale; well segregated beds	Some shales, silt, and fine-grain siltstone	Some shales, silt, and fine-grain siltstone
Biota	Exclusively nektonic-pelagic fauna preserved in local abundance on bedding planes	Very diverse shelly fauna preserving both infauna and epifauna	Bioclastic detritus derived principally from upslope	Colonies of whole fossil organisms and bioclastic debris

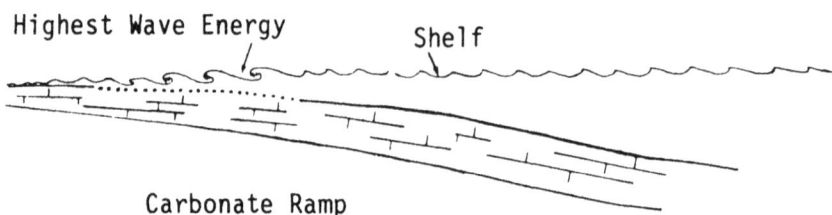

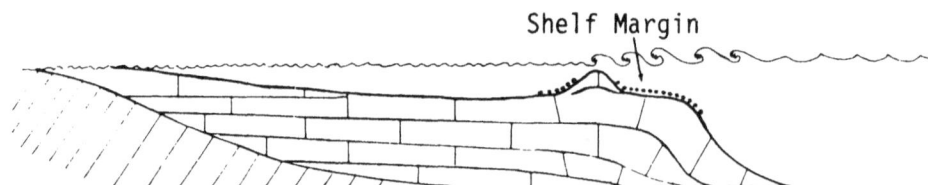

Fig. 7.67—Definition of carbonate ramps and platforms (Ref. 50, Fig. II-2).

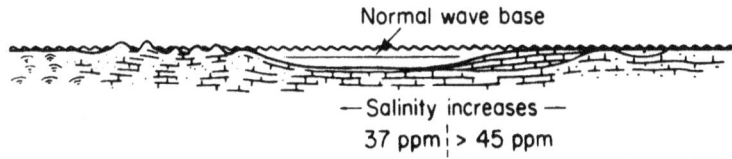

Normal wave base

— Salinity increases —
37 ppm ⋮ > 45 ppm

Facies Number	5	6	7	8	9
Facies	*Organic (ecologic) reef* a) Boundstone mass b) Crust on accumulation of organic debris and lime mud; bindstone c) Bafflestone	*Sands on edge of platform* a) Shoal lime sands b) Islands with dune sands	*Open platform* (normal marine, limited fauna) a) Lime sand bodies b) Wackestone mudstone areas, bioherms c) Areas of clastics	*Restricted platforms* a) Bioclastic wackestone, lagoons and bays b) Litho-bioclastic sands in tidal channels c) Lime mud-tide flats d) Fine clastic units	*Platform evaporites* a) Nodular anhydrite and dolomite on salt flats b) Laminated evaporite in ponds
Lithology	Massive limestone-dolomite	Calcarenitic-oolitic lime sand or dolomite	Variable carbonates and clastics	Generally dolomite and dolomitic limestone	Irregularly laminated dolomite and anhydrite, may grade to red beds
Color	Light	Light	Dark to light	Light	Red, yellow, brown
Grain type and depositional texture	Boundstones and pockets of grainstone; packstone	Grainstones well sorted; rounded	Great variety of textures; grainstone to mudstone	Clotted, pelleted mudstone and grainstone; laminated mudstone; coarse lithoclastic wackestone in channels	
Bedding and sedimentary structures	Massive organic structure or open framework with roofed cavities; lamination contrary to gravity	Medium to large scale crossbedding; festoons common	Burrowing traces very prominent	Birdseye, stromatolites, mm lamination, graded bedding, dolomite crusts on flats. Cross-bedded sand in channels	Anhydrite after gypsum; nodular, rosettes, chickenwire, and blades, irregular lamination; carbonate caliche
Terrigenous clastics admixed or interbedded	None	Only some quartz sand admixed	Clastics and carbonates in well-segregated beds	Clastics and carbonates in well-segregated beds	Windblown, land-derived admixtures; clastics may be important units
Biota	Major frame-building colonies with ramose forms in pockets; in-situ communities dwelling in certain niches	Worn and abraded coquinas of forms living at or on slope; few indigenous organisms	Open marina fauna lacking (e.g., echinoderms, cephalopods, brachiopods); mollusca, sponges, forams, algae abundant; patch reefs present	Very limited fauna, mainly gastropods, algae, certain foraminifera (e.g., miliolids) and ostracods	Almost no indigenous fauna, except for stromatolitic algae

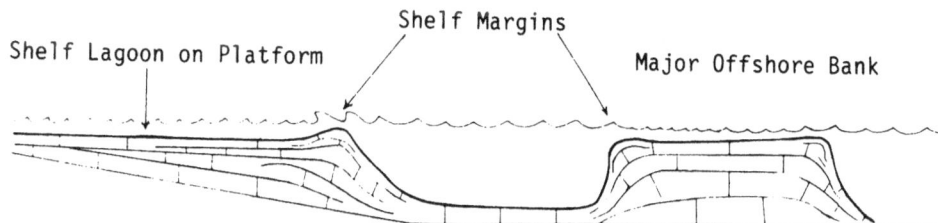

Shelf Margins

Shelf Lagoon on Platform

Major Offshore Bank

Fig. 7.68—Definition of carbonate platforms, shelf margins, and offshore banks (Ref. 50, Fig. II-3).

TABLE 7.11—MAJOR CARBONATE DEPOSITS[50]

Carbonate ramps:	Huge carbonate bodies built away from positive areas with gentle slope.
Carbonate platform:	Huge carbonate bodies built up with a more or less horizontal top and abrupt shelf margins.
Major offshore banks:	Complex carbonate buildups of great size and thickness well offshore from coastal ramps or platforms.

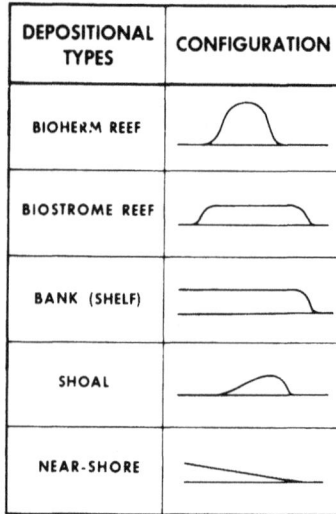

Fig. 7.69—Shapes of carbonate deposits.[16]

hundreds of square miles. Carbonate banks are thick, tabular deposits formed by the buildup of carbonate sediments across a basin during extended periods of constant sea level. Depositing is usually cyclic, and mapping depositional units over areas as large as thousands of square miles is possible. Shoal carbonates are mound-like deposits. The near-shore deposits include beach, tidal flat, and supratidal deposits, which are deposited cyclically as the sea level rises and falls.

Carbonate Sediments. Primary porosity of carbonates is related to depositional environment through grain-size distribution. Sedimentary materials that make up a carbonate reservoir were derived primarily from the basin where deposited. Little silicate detritus is present. Carbonate sediments range from microscopic particles (lime muds) to sand-size particles of calcium carbonate. Folk[54,55] developed a classification scheme in which carbonate rocks are considered as a mixture of sand-size particles of calcium carbonate and carbonate mud. In this classification scheme, grain materials (intraclasts, skeletal grains, pelloids, and oolites) termed allochems form the framework, while lime muds (sparry calcite or micrite) make up the remainder of the rock. Fig. 7.70 illustrates Folk's classification of limestones.[43] Micrite is the term used to describe microcrystalline lime mud. Particle size is less than 5 μm. Sparry calcite is coarse calcium-carbonate cement. Table 7.12 contains descriptions of carbonate allochems.[56]

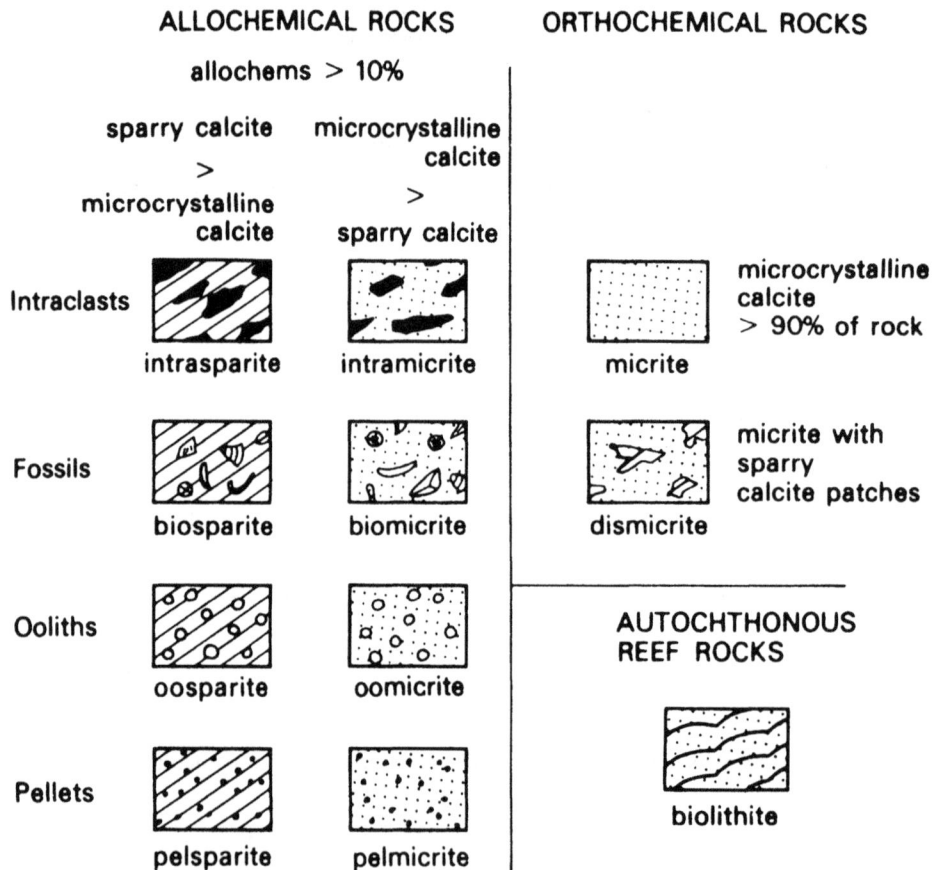

Fig. 7.70—Folk's classification of limestones. Application of terminology to the main varieties of limestone (Ref. 43, Fig. 8.2).

GRADE SIZE SCALES

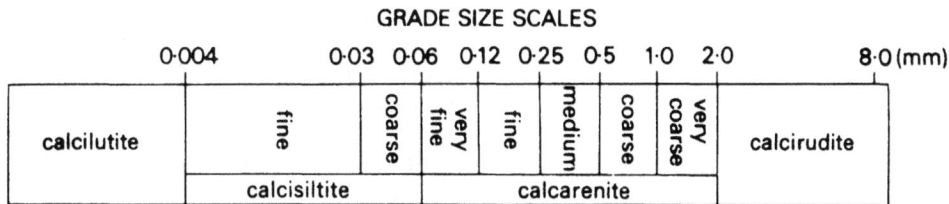

Fig. 7.71—Classic limestone classification by grain size (Ref. 43, Fig. 8.1).

With this classification scheme, the depositional process may be visualized as a competition between the rate of production of grain materials and carbonate muds within the sedimentary basin.[31] Thus, carbonate sediments may be found with any ratio of grain material, depending on the environment at the time of deposition. Some geologists use a classification scheme based on grain size. The term calcarenite refers to a carbonate deposit with grain size between 0.0024 and 0.0787 in. [0.06 and 2.0 mm] (Fig. 7.71).[43] When the deposit contains fossils, complex names are created to reflect fossil content. For example, foraminiferal calcarenite and argillaceous molluscan micrite are names given to two fundamental rock types in the Zelten field, Libya.[20]

Dunham[57] developed a widely used classification scheme based on depositional texture. Dunham's scheme, shown in Table 7.13, describes carbonate rock in terms of the grain material that was able to remain at the site of deposition. The presence or absence of lime mud (micrite) is a measure of the wave energy during deposition. For example, mudstone is primarily micrite and was deposited in a calm sea, while grainstone is mud-free with muds removed by a high-energy environment. Many carbonate rocks lack sedimentary structures (such as cross-bedding or preferential grain orientation), indicating little influence of hydrodynamics on deposition. Therefore, the size and shape of the carbonate deposits depend on the area of deposition, the length of time the depositional process continued, and the wave energy during deposition.[31]

Depositional Models. Depositional models of carbonate rocks are inferred from studies of ancient and modern rocks with the facies models discussed earlier. Two examples are included to illustrate application of these concepts, a regressive shoreline model and an organic reef. Other models are found in Refs. 31, 50, and 56.

Regressive Shoreline. Fig. 7.72 shows the geological provinces in the Permian basin of west Texas and New

TABLE 7.12—VARIETIES OF CARBONATE ALLOCHEMS[56]

Intraclasts

fragments of penecontemporaneous, generally weakly consolidated carbonate sediment that have been eroded from the sea bottom and redeposited.[56]

Detrital Skeletal Grains

crinoid, brachiopods, foraminifera, ostracods, etc., may be whole or fragmented, commonly transported after death.

Pellets (Peloids)

a rounded or oval mass of structureless micrite (0.001 to 0.006 in. in diameter) that may be fecal in origin.[55]

Oolites

a spherical to ellipsoidal body 0.01 to 0.079 in. in diameter, which may or may not have a nucleus and has concentric or radial structure or both.

Pisolite

a spherical or subspherical, accretionary body more than 0.079 in. in diameter.

Mexico, a major oil-producing region. The basin contains two major carbonate platforms, as well as extensive carbonate shelves. Locations of three of the many reservoirs in the Permian basin are indicated on Fig. 7.72. The Means field (San Andres) and the Robertson field (Clearfork) are on the shelf margin, while the Fullerton field (Clearfork) formed on the shelf away from the margin.[58] Sediments were deposited as limestone in the shallow shelf environment with vertical sections several hundred feet thick. Most of the limestone was converted subsequently to dolomite. The lithological changes observed in the vertical sections were interpreted as a series of transgressions and regressions of the sea during deposition. This depositional environment resulted in highly stratified reservoirs.

TABLE 7.13—CLASSIFICATION OF CARBONATE ROCKS ACCORDING TO DEPOSITIONAL TEXTURE FROM DUNHAM. (Ref. 50, Fig. I-5)

Depositional Texture Recognizable				Depositional Texture Not Recognizable
Original Components Not Bound Together During Depositions			Original components were bound together during deposition...as shown by intergrown skeletal matter lamination contrary to gravity, or sediment-floored cavities that are roofed over by organic or questionably organic matter and are too large to be interstices.	Crystalline carbonate
Contains Mud (particles of clay and fine silt size)		Lacks mud and is grain-supported		
Mud-Supported		Grain-supported		(Subdivide according to classifications designed to bear on physical texture or diagenesis.)
Less than 10% grains	More than 10% grains			
Mudstone	Wackestone	Packstone	Grainstone	Boundstone

Deposition in a regressive environment is depicted in the models of Fig. 7.73. In this sequence, the sea is regressing with the barrier reef/oolite bank overlain progressively by lagoonal (subtidal), algal flat (intertidal), and sabkha (supratidal) regions. Land finally covers the entire sequence. The deposition regions may be classified as follows.[40]

Supratidal   Above-normal high tide, but within the range of spring and storm tides.

Intertidal   Between normal high and low tides.

Subtidal   Open marine and lagoons permanently below low tide.

Two characteristics of carbonate deposition are indicated on Fig. 7.73. Subsidence must occur to build thick carbonate sections. In addition, the adjacent land mass must have been arid during the time of carbonate deposition or must be some distance away from the shelf margin because there is little silicate detritus within the deposit.

If the sea level remains relatively high, the zone of clastic influx is limited (as indicated in Part C of Fig. 7.73) and the carbonate sequence continues to build toward the basin. Transgression of the sea causes relocation of major

Fig. 7.72—Permian basin geological provinces.[58]

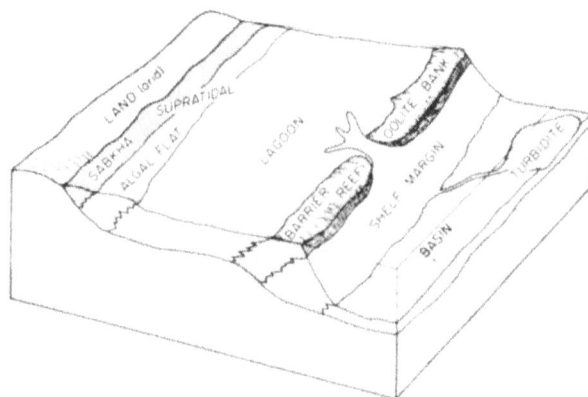

(a)

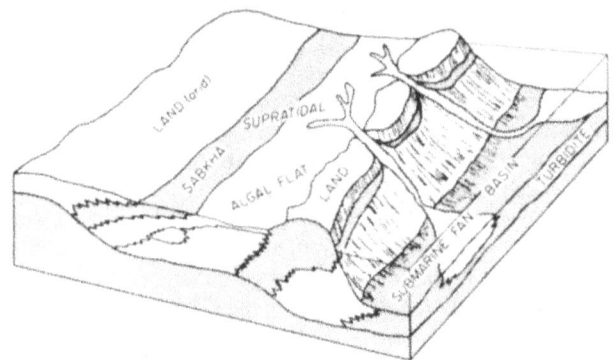

(c)

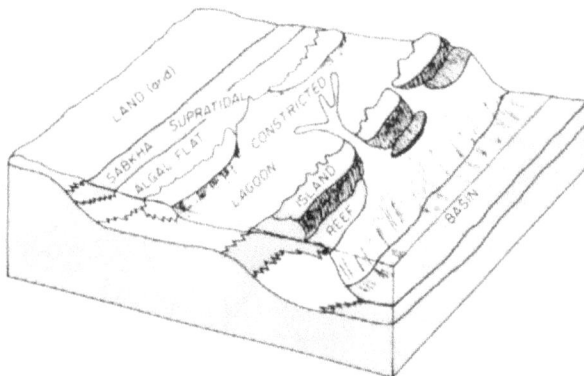

(b)

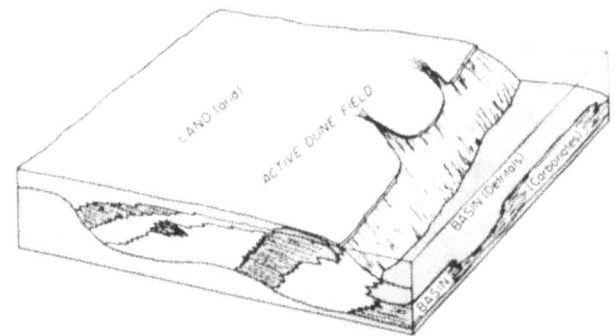

(d)

Fig. 7.73—Depositional environments related to sea-level change or sedimentation, northern Permian basin, southeastern New Mexico, and west Texas (Ref. 30, Fig. 19-40, after Ref. 59).

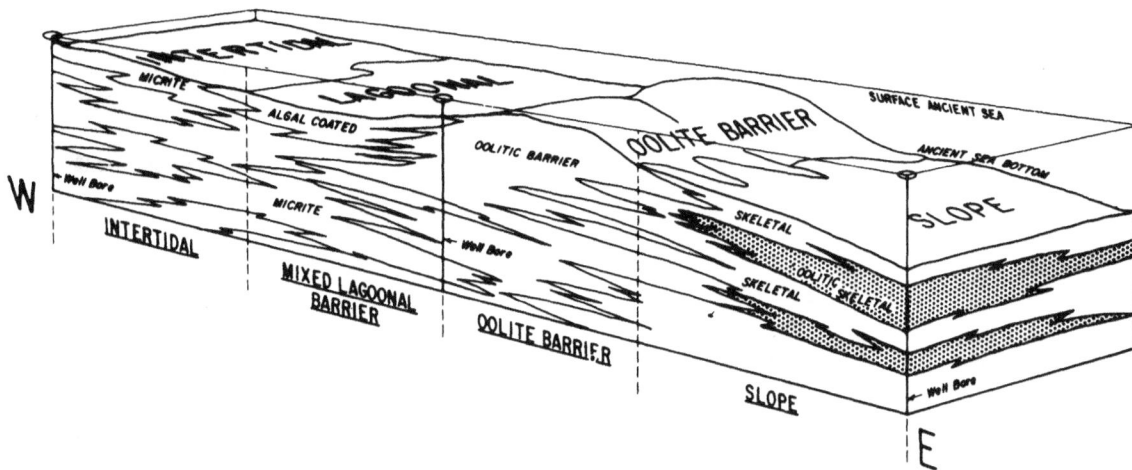

Fig. 7.74—Means field (San Andres) schematic. [58]

facies belts toward the shore and thus interfingering of carbonate facies. A schematic block diagram through the Means field in Fig. 7.74 illustrates the distribution of carbonate facies. Deposition in the Means field is typical of an intertidal bank sequence. [58] Best porosity is generally in the oolitic facies. A skeletal facies having good porosity (secondary) was deposited on the front of the oolite bank. The lagoonal facies has many thin porous zones interbedded with shales and carbonate muds. The intertidal region on the shelf side of the lagoon is composed of anhydrite and micritic dolomite. Porosity in this region is poor.

Fig. 7.74 also illustrates an important factor in the analysis of carbonate reservoirs. Identification and positioning of the carbonate facies establishes the deposition cycle for a particular basin. Thus, identification of the facies in a particular well determines the location of the well in a basin once the depositional model is established. Interpretation of facies in a vertical cross section often requires construction of complex conceptual models. Fig. 7.75 is a stratigraphic model of cyclic and reciprocal sedimentation in the Permian reef complex of New Mexico, showing several clastic/carbonate depositional cycles. [50,60] Facies lines are tilted seaward because of regression of the sea and progradation of clastic sediments. A well drilled through this vertical sequence would pass through a complex sequence of time lines.

Organic Reefs. Organic reef deposits are important reservoir rocks throughout the world. A reef is a ridge or mound-like structure that contains a high percentage of skeletal material. [16] Organic reefs form on the shelf margin or on platforms as depicted in Fig. 7.76 and may grow to heights ranging from a few hundred to a thousand feet above the platform. Reefs vary in areal extent ranging from a surface area of less than 1 sq mile [3 km^2] to tens of sq miles. Fig. 7.76 is a cross section illustrating the different facies across a typical reef complex. [56] The reef contains a core, termed the reef flat, which acts as a wave-resistant framework around which the reef develops. [56] The reef flat is an area of carbonate sand deposition just leeward of the actively growing reef.*

Carbonate reefs have many different shapes. These are given in Table 7.14. [50,56]

Fig. 7.77, a generalized cross section across a Canadian Devonian reef, indicates the organisms found in each depositional environment. [56] Identification of the skeletal remains is a principal factor in determining depositional environment.

The Swan Hills field is one of several reef reservoirs in Alta., Canada. This and several other reservoirs formed on the Beaver Hill Lake platform (see Fig. 7.78). The reservoir rock consists of a biohermal reef that formed

*Personal communication with W.J. Ebanks Jr., 1983.

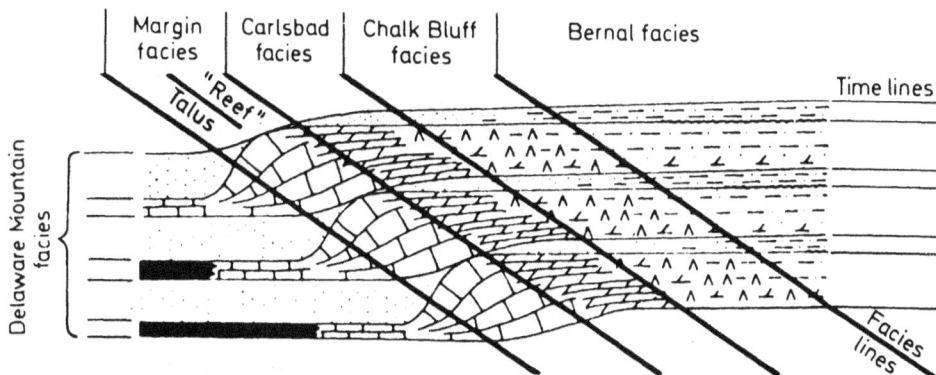

Fig. 7.75—Stratigraphic model of cyclic and reciprocal sedimentation in Permian reef complex, New Mexico (Ref. 50, Fig. II-21, from Ref. 60).

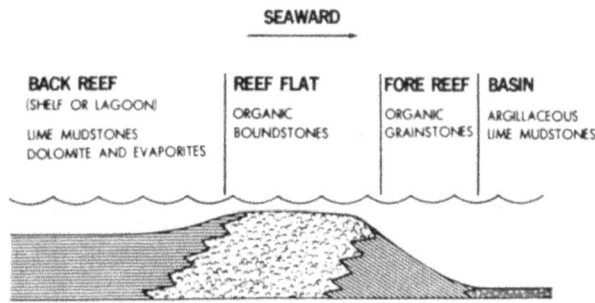

Fig. 7.76—Generalized cross section illustrating the different facies across a reef complex (Ref. 56, Fig. 6-1).

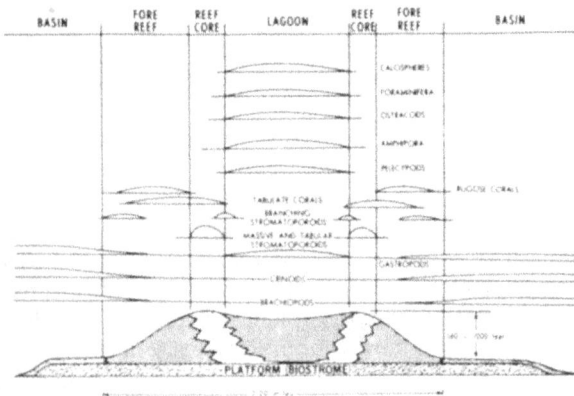

Fig. 7.77—Generalized cross section across a Canadian Devonian reef with paleontology (Ref. 56, Fig. 6-12).

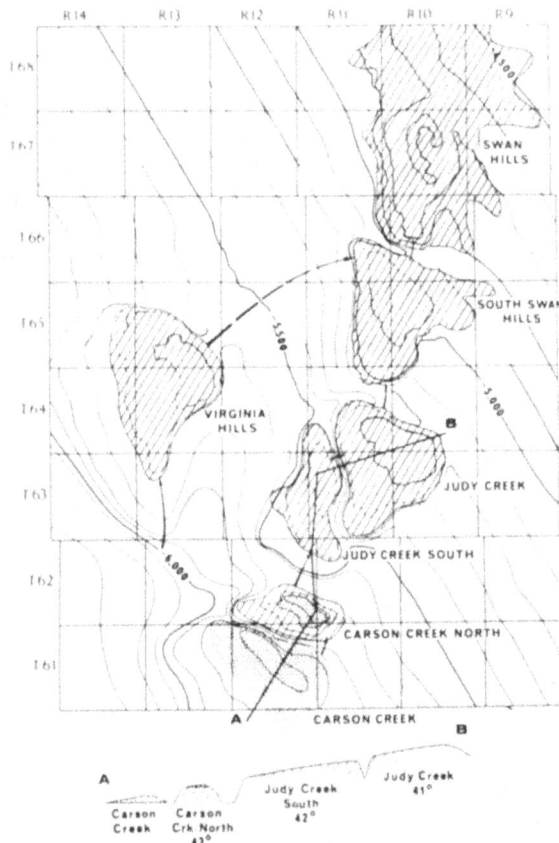

Fig. 7.78—Differential entrapment along the Swan Hills reef trend, northern Alta., Canada.[61]

TABLE 7.14—CARBONATE REEFS[50,56]

Atoll reef:	Ring-like organic accumulation in offshore or oceanic position surrounding a lagoon of variable depth
Pinnacle reef:	A conical or steep-sided upward-tapering reef several hundred feet high and usually less than 4 miles in diameter
Patch reef:	Isolated, organic frame constructed buildup
Fringe reef:	Curvilinear reef built directly out from the coast
Barrier reef:	Curvilinear reef separated from the coastline by a carbonate shelf or wide lagoon

on top of a biostrome. Fig. 7.79 is an isopach showing biohermal and biostromal facies. Cross Section A-A[1] in Fig. 7.80 illustrates the position of the different carbonate facies in the reef complex.

The depositional history of the Swan Hills formation can be described in terms of six major changes of sedimentation and environment shown in Fig. 7.81.[62] The formation is divided into two members on the basis of color and morphologic characterization—the lower or dark brown lime member and the upper, light brown lime member. The dark brown lime member was deposited during Stages I to III while Stages IV to VI cover the period when the deposition of the light brown lime occurred.

In Stage I, the Beaver Hill Lake sea transgressed westward from the basin, creating an environment for organic growth. A coral zone developed in the upper part of the basal beds, creating a rigid base for further development. A stromatoporoidal zone developed along the northeast rim in areas of high wave energy. Stromatoporoids are the organisms that were principal contributors of skeletal particles to the sediment and builders of the reef. Continued growth of these regions occurred in Stage II under a small but persistent rise in sea level. A broad shelf lagoon formed behind the stromatoporoid reef front where a brecciated zone developed in quieter waters. On the west side, amphipora beds were deposited in quiet semistagnant waters of the lagoon.

Development of the dark brown lime terminated at the end of Stage III as the carbonate bank on the west emerged, coupled with the reworking of the upper part of the bank that created a thin, fragmented zone on this part of the platform. Deposition of the light brown lime began in Stage IV with the first unit covered on the flanks and near the top with porous calcarenite.

In Stage V, a "table reef" developed on the beds deposited in Stage IV. Slow subsidence led to the formation of a circular stromatoporoidal reef atoll that enclosed a central lagoon. Growth during Stage V was probably terminated by emergence. The last stage of development (Stage VI) occurred after the deposit became submerged. Renewed growth of amphipora and deposition of pelletoid and unfossiliferous carbonate mud beds occurred. Strong erosion and emergence prevented upward growth of the organic reef, causing calcarenite beds and coarse reef rubble to form on the top and upper slopes of the buildup. A sudden increase in water depth drowned the reef complex at the end of Stage VI.

The Swan Hills formation is a complex carbonate deposit that contains at least 10 definable stratigraphic units.[63] There are seven major porous intervals and numerous other minor porous intervals. The rock is predominantly limestone, so it was possible to recognize many of the original depositional environments.

Fig. 7.82 is a reservoir-quality profile showing the sub-division of the reservoir into six zones (A, B, C, C', D, and dark brown). This well encountered all major rock facies except the foreslope and open marine. The deposi-tional environment of each zone inferred from core descriptions is described in more detail in Ref. 63.

Reservoir quality varies throughout the deposit as can be seen in Table 7.15. Highest permeability and porosity usually are found in the reef rim deposits. Reservoir qual-ity is controlled by the water energy at the time of depo-sition and by the position within the biohermal reef.

Rock Properties. The initial pore structure of carbonate sediments is similar to clastic sediments but is more var-ied because of the diversity of sediments. Because car-bonate sediments are soluble and brittle, diagenetic processes often alter the properties of the original deposit and leave complex pore systems that may bear little resem-blance to the original pore space.[53]

Porosity in carbonate systems is categorized as primary or secondary by origin. Primary (or matrix) porosity refers to intergranular pore space between grain material and/or skeletal materials that often have internal (cellu-lar) porosity. Primary porosity is controlled by grain size and shape and by the nature of skeletal organisms. Secon-dary porosity is created by diagenetic changes. Table 7.16 summarizes some of the important types of porosity.[56] An extensive classification scheme for carbonate systems is found in Ref. 64. Jodry[65] also presents a classifica-tion scheme.

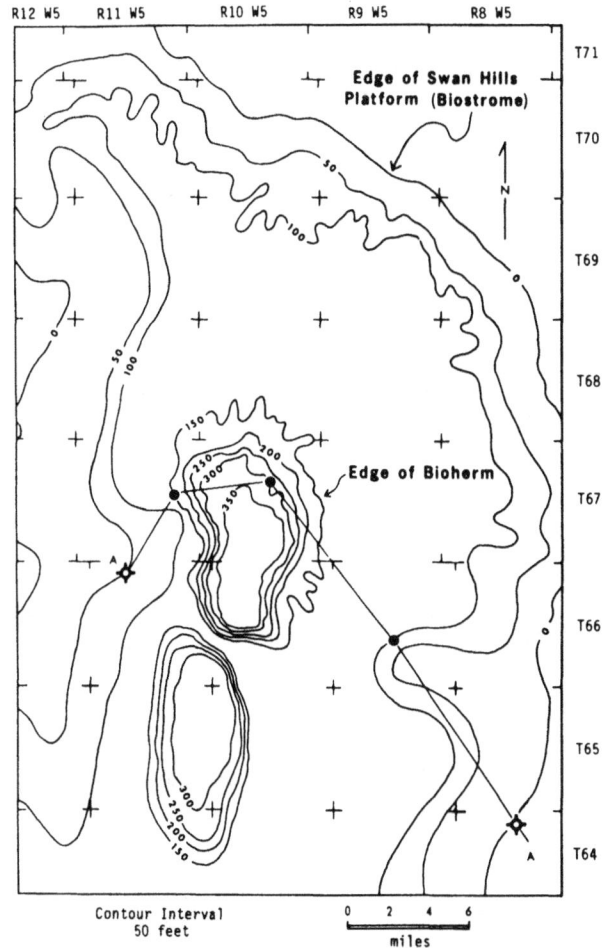

Fig. 7.79—Isopach map of the Swan Hills member (biohermal and biostromal facies) of the Beaver Hill Lake formation (Ref. 56, Fig. 6-13).

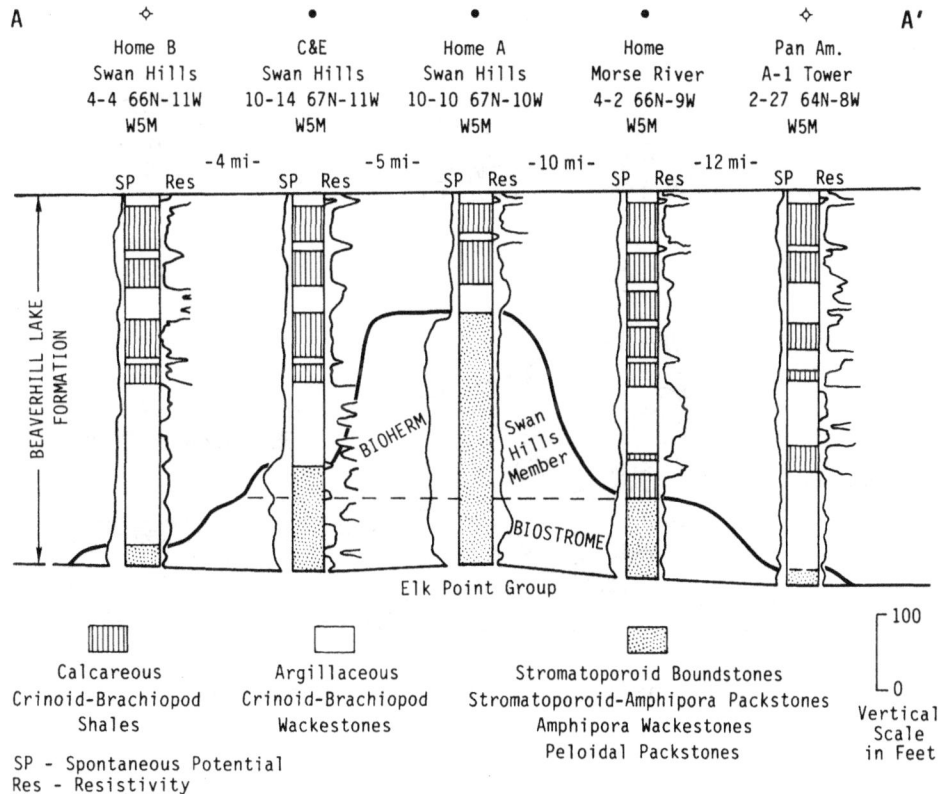

Fig. 7.80—Electric log cross section across Swan Hill reef, Alta., Canada (Ref. 56, Fig. 6-14).

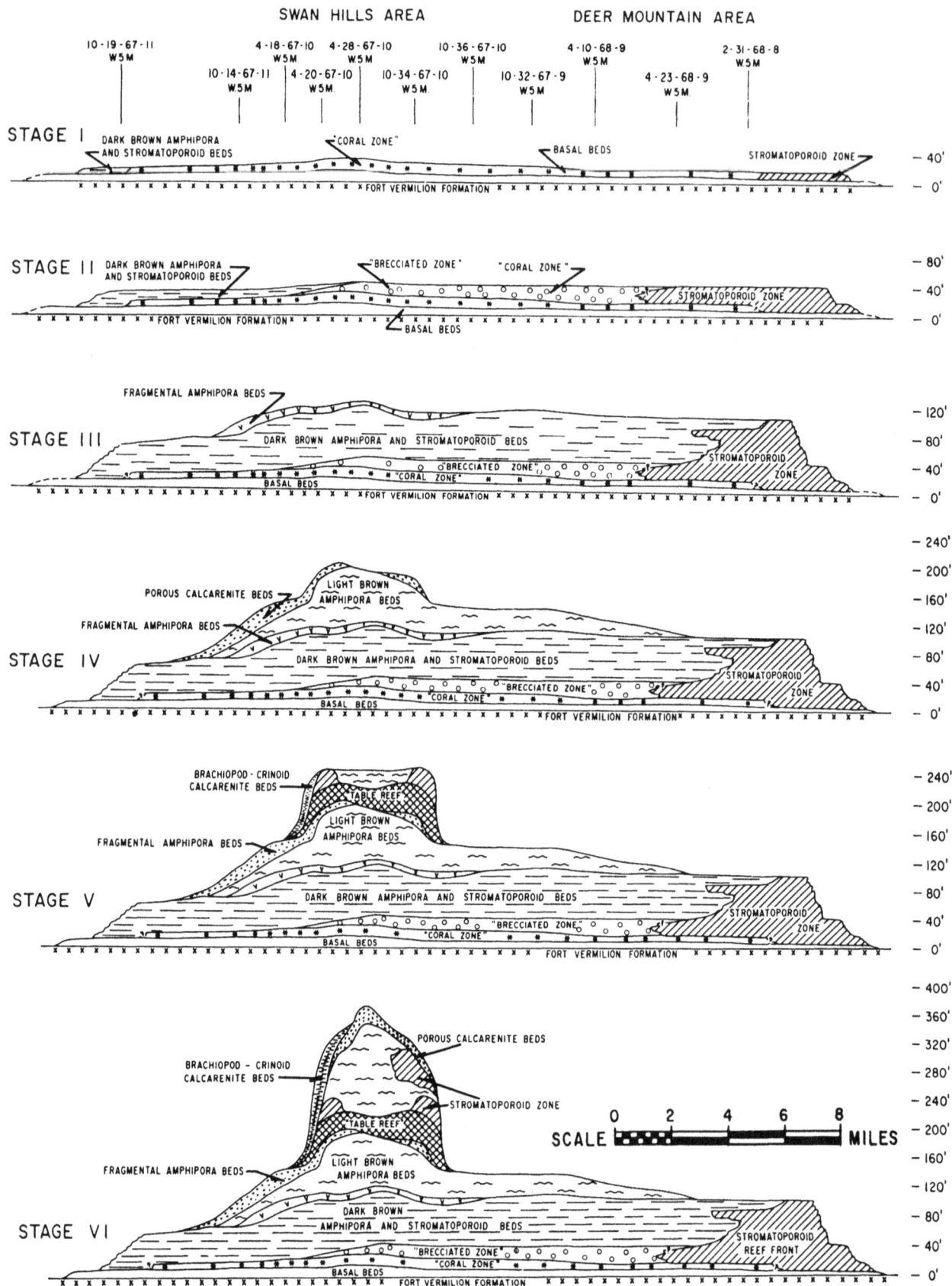

Fig. 7.81—Schematic diagram of the Swan Hills reef growth.[62]

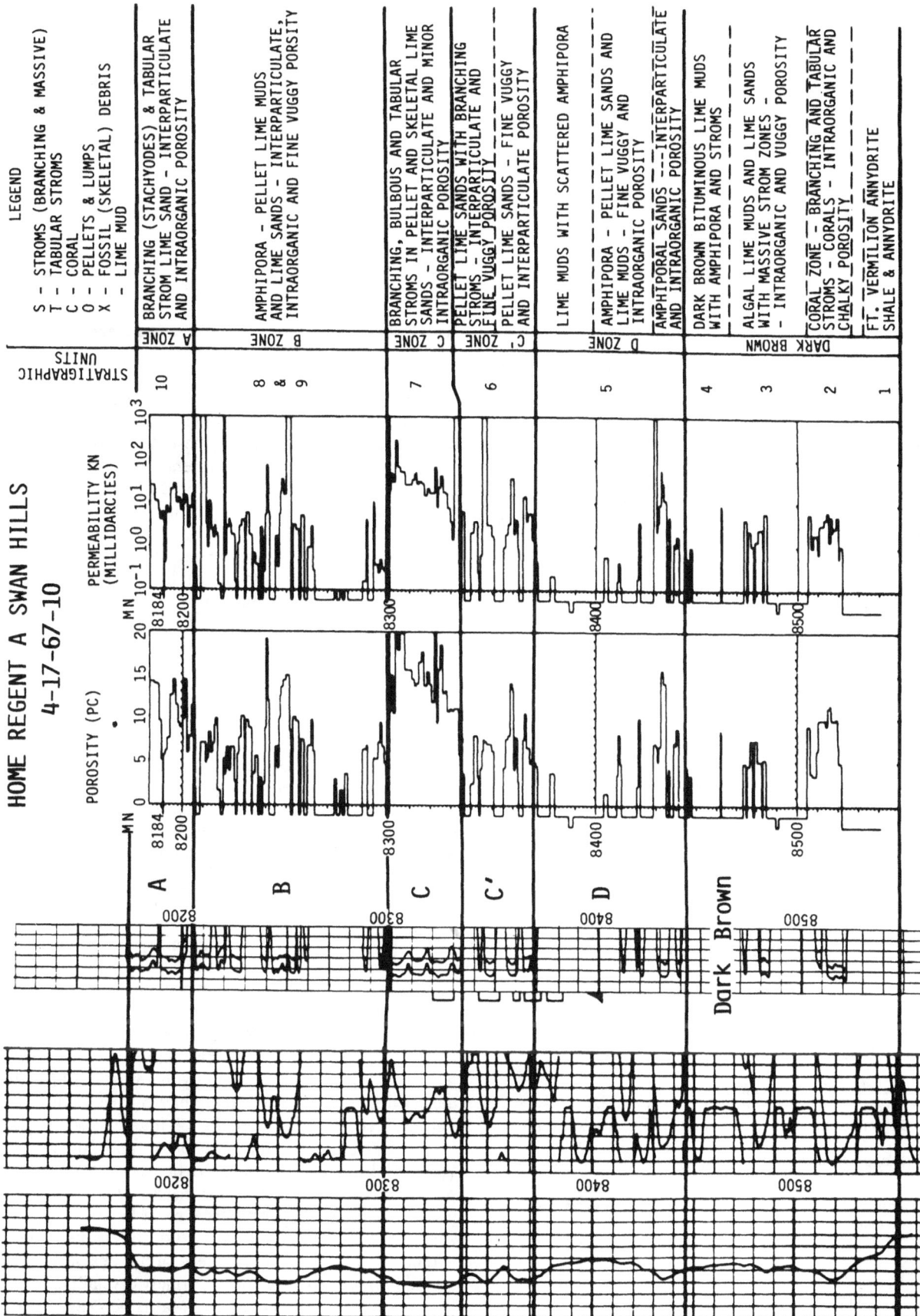

Fig. 7.82—Reservoir quality profile of Home Regent A Swan Hills.[63]

TABLE 7.15—PARAMETERS OF THE POROUS INTERVALS [63]

	Light Brown Lime					Dark Brown Lime*
	A	B	C	C'	D	Coral
A, acres × 1,000	11	22	25	23	43	39
h net, ft	23	46	27	12	15	7
h gross, ft	30	100	30	20	60	100**
k—reef rim, md	33	44	23	19	26	5
k—reef core, md	33	16	18	6	6	5
ϕ—reef rim, %	10.0	10.3	9.5	6.7	7.0	7.0
ϕ—reef core, %	10.0	8.0	9.0	6.2	6.3	7.0

*Porous intervals in the dark brown lime member other than the coral zone have been excluded because the areal extent of these zones appears to be small and much of the pay seems to be ineffective because of low permeability.
**Total dark brown lime thickness, coral zone averages 12 ft gross thickness.

TABLE 7.16—TYPES OF POROSITY— CARBONATE SEDIMENTS [56]

Primary Porosity

Intergranular—pores between carbonate allochems in a noncemented or partially cemented grainstone.

Organic—pores developed within the skeleton of an organism (for example, the uncemented cells in a stromatoporoid) or holes between organisms.

Secondary Porosity

Intercrystalline—pores between carbonate crystals (for example, sucrosic dolomite and chalky limestone).

Vuggy—pores (vugs) created by leaching of the carbonate rock, may cut across several original grains. If the vug is small, it is sometimes referred to as pin-point porosity.

Moldic porosity—implies what was occupying the vug (i.e., fossilmoldic and oomoldic).

Fractures—fissures or cracks in the carbonate rock. Fractures are favorable because they increase both porosity and permeability; however, they can be unfavorable because they can cause channeling.

TABLE 7.17—SECONDARY POROSITY OF CARBONATE RESERVOIRS [16]

Process	Favorable Effects	Unfavorable Effects
Fracturing joints breccia	Increase k	Increase channeling
Leaching	Increase k and ϕ	
Dolomitization	Increase k	Can also decrease ϕ and k
Recrystallization	May increase pore size and k	Decrease ϕ and k
Cementation by Calcite Dolomite Anhydrite Pyrobitumen Quartz	—	Decrease ϕ and k

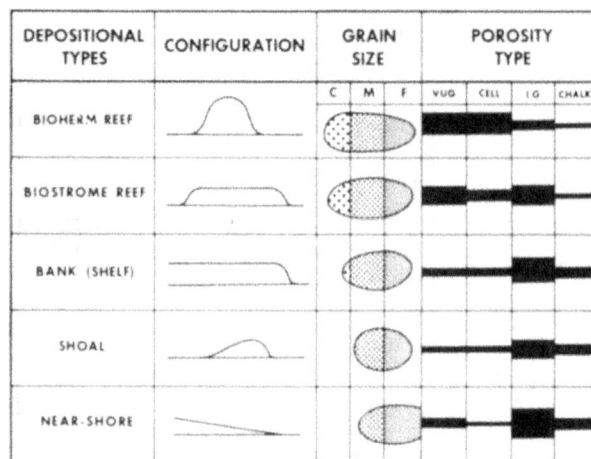

Fig. 7.83—Primary porosity of carbonate reservoirs. [16]

Several processes are known to alter the pore structure of carbonate sediments after deposition. Ref. 66 includes an extensive discussion of diagenesis of carbonate rocks. Five common processes are presented in Table 7.17. [16] Included in Table 7.17 are effects of these processes on the porosity and permeability of the original carbonate sediment.

Carbonate sediments are very susceptible to leaching because calcium carbonate is soluble in water. Leaching increases both porosity and permeability. Dolomitization occurs when calcium is replaced by magnesium.

$$2\ CaCO_3 + Mg^{++} \rightarrow CaMg(CO_3)_2 + Ca^{++} \quad \ldots (7.3)$$

Calcite   Dolomite

Because dolomite is denser than calcite, dolomitization usually causes an increase in porosity and permeability. Jodry[65] observes that dolomitization enlarges pore throats and improves both capillary properties and permeability of the rock. Jardine et al.[16] compared undolomitized reefs and dolomitized reefs in western Canada. Porosity was slightly higher in dolomitized reefs. Average permeability in undolomitized reefs was 68 md, while dolomitized reefs had an average permeability of 800 md. The effects of dolomitization were attributed to development of solution vugs and to extensive fracturing.

Fracturing is common in carbonate rocks because the rocks are brittle. Although fracturing slightly increases porosity, it heavily impacts permeability. Some reservoirs have such low matrix permeability that fractures provide the permeability necessary to produce oil. Felsenthal and Ferrell[67] cite several examples of fractured reservoirs and provide an excellent discussion of fluid flow in carbonate rocks. Recrystallization is a process that commonly results in coarser crystals in the matrix than the original particles of micrite. Recrystallized rocks have a "chalky" appearance and a fine pore structure that is not visible without the aid of a microscope. Careful analysis is required because severely leached, finely crystalline rocks also have a "chalky" appearance. Matrix porosity is high, but permeabilities are quite low.

Cementation occurs by several processes and reduces both permeability and porosity.

Reservoir Continuity. Carbonate rocks are extremely heterogeneous, and several types of porosity and corresponding ranges of permeability are found commonly within a particular depositional unit. Fig. 7.83 illustrates the distribution of porosity types in five types of carbonate deposits.[16] Rock types are arranged in order of decreasing energy in the depositional environment. Also shown in Fig. 7.83 is the grain-size distribution. There is a decrease in average grain size, porosity, and permeability with decreasing energy of deposition. Table 7.18 summarizes characteristics of rock properties within each deposit.

The dominant feature of carbonate reservoirs is heterogeneity. Reservoir geology is usually far more complex than initially assumed. Continuity of carbonate reservoirs is variable and best determined by correlating permeability and porosity with facies and mapping the facies across the deposit. The complexity of carbonate facies is shown in Fig. 7.84, a map of 13 facies identified in a pinnacle reef.[68] Units grouped according to similarity of flow properties are shown in Fig. 7.85. Fig. 7.86 is a correlation section through the Zelten field, Libya.[20] Heterogeneity in the two reservoir facies—coralgal micrite and discocyclina-foraminiferal calcarenite—as well as seven other facies is evident. Five types of porosity were identified from the study of the cores: intergranular, intragranular, leached skeletal, leached micrite, and vuggy. Rock properties for each lithologic class were determined by correlating core analysis, log analysis, and well test data. Porosities and permeabilities determined from data analysis are included in Fig. 7.86. Core analyses are represented as small dots. Flowmeters were used to obtain average permeability over a particular interval. Permeabilities derived from flowmeter tests were consistently higher than permeabilities based on core analysis. This is a common observation in carbonate deposits that have fractures or solution voids and is often used as a diagnostic tool to detect presence of fractures.[67]

TABLE 7.18—CHARACTERISTICS OF ROCKS IN CARBONATE DEPOSITS[16]

Depositional Type	Porosity and Permeability
Biohermal reef:	Porosity and permeability highest in skeletal rim. Interior has lower permeability and interbedding of porous/nonporous sediments. Poor vertical permeability.
Biostromal reef:	Highest porosity toward seaward edge but porosity throughout reef. Highly stratified with $k_h > k_v$ by an order of magnitude.
Carbonate bank:	Porosity confined to narrow rim near seaward edge or in local patches. Permeability and porosity low unless altered by diagenesis.
Shoal carbonates:	Intergranular.
Near shore:	Thin sequences, strongly stratified with limited areal extent. High porosity but permeability less than 1 md.

7.6 Development of a Reservoir Description

The reservoir description is the foundation for designing, operating, and evaluating the performance of a waterflood as well as any other EOR project. Selection of the flooding plan and the model used to estimate flooding performance is determined in large measure by the reservoir description. The development of a reservoir description should begin early in the life of a reservoir. Because developing a reservoir description requires investment of the time of many people and expenses to obtain and process data, the scope of a reservoir description is weighted against anticipated benefits. This is often a subjective judgment.

Reservoir descriptions may be developed in a number of ways. A systematic method formalized by Harris[21] is described here. A key feature in the method is the integra-

GENETIC UNITS (FACIES)

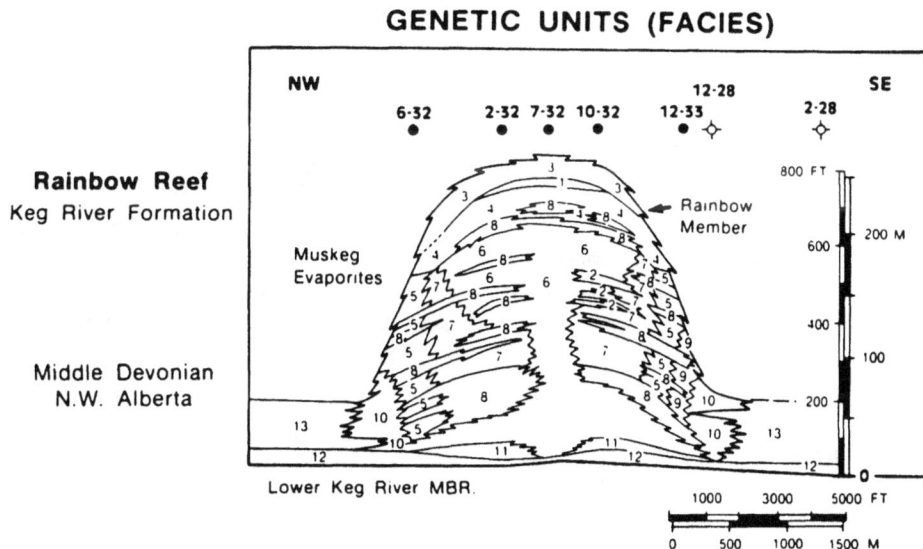

Fig. 7.84—Sedimentary facies of a pinnacle reef in the Rainbow Member of the Keg River formation, northern Alberta (Ref. 68, after Ref. 69).

FLOW UNITS

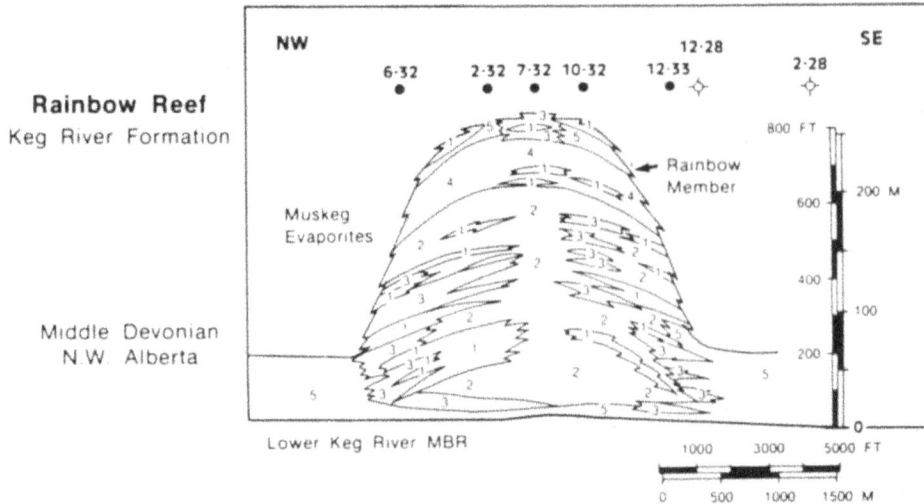

Fig. 7.85—Facies of the same Rainbow reef as in Fig. 7.84, showing units in which the flow of fluids would be expected to be similar. Facies have been grouped to reduce complexity for numerical simulation.[68]

tion of geology and engineering in the process of forming the reservoir description. In this approach, a geologist/engineer team is formed and areas of responsibility are defined before the study is begun. Integration and iteration continue throughout the study until a reservoir description that is consistent with project objective, geological interpretation, and engineering performance is obtained.

Table 7.19 outlines this method in terms of four major

studies. These are (1) rock studies to define lithology, depositional environment of the reservoir, and correlations of rock properties; (2) framework studies that establish the structure, continuity of reservoir and nonreservoir rock, and gross thickness; (3) reservoir quality studies to determine the variation of rock properties (permeability, porosity) and the areal extent of each reservoir zone; and (4) integration studies that yield maps of PV and transmissibility across the reservoir. Table 7.19

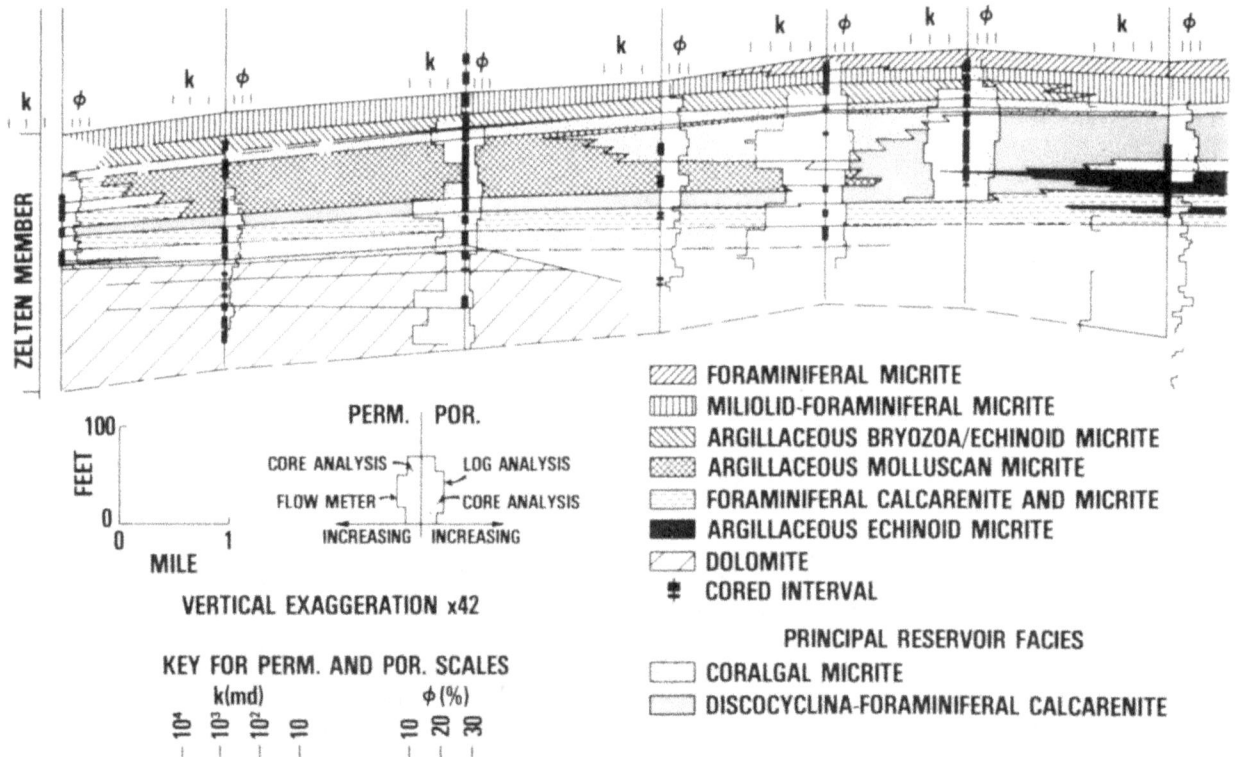

Fig. 7.86—Correlation section showing distribution of fundamental rock types and porosity/permeability profiles in the Zelten Field, Libya.[20]

TABLE 7.19—DEVELOPMENT OF A RESERVOIR DESCRIPTION[20,21]

Activity	Input Parameters	Objectives	Responsibility
Rock studies	Cores	Core descriptions—including mineralogy	Geologist
	Cuttings	Lithology (facies)	Geologist
	Core analysis	Depositional environment	Geologist
	Drillers logs	Correlations of core data for each	
	Well logs	productive facies	Engineer/Geologist
		Porosity $- k_h$	
		$k_h - k_v$	
		$S_w - k_h$	
		Log porosity—core porosity	
Framework studies	Core analyses	Structure map	Geologist/Geophysicist
	Core description graphs	Delineation of aquifer/gas cap	Geologist
	Well logs	Principal reservoir zones	
	Depositional model	Areal and vertical extent of principal	
		reservoir/nonreservoir zones	Geologist/Engineer
	Correlations of rock properties	Continuity of reservoir and	
		nonreservoir zones	Geologist/Engineer
	Pulse tests	Cross sections	
	Seismic data	Fence or panel diagrams	
Reservoir quality studies	Framework maps	Area distribution of permeability, porosity,	
	Core analyses	net sand and water saturation for	
	Well logs	each reservoir zone	Engineer/Geologist
	Correlations of rock properties	Contour maps of permeability, porosity,	
	Well tests—pulse, buildup,	water saturation, and net sand	Geologist/Engineer
	and falloff		
Integration studies	Reservoir description from	Reservoir maps	Engineer/Geologist
	rock and reservoir quality	PV	
	studies	Hydrocarbon PV	
		Transmissibility (k_h)	
	Mathematical model		
	simulating production/	History match of pressure/production	Engineer
	pressure primary production	Response by adjusting rock parameters	
	waterflood	Consistent with geological model	Geologist

depicts input parameters, objectives, and principal responsibilities for the four studies. Harris[21] and Harris and Hewitt[20] give examples illustrating the method. A number of other studies follow a similar approach although it is not formally presented in the same way.

Rock Studies. The rationale for estimating properties in the interwell area is based on identification of the depositional environment. Typically, this is done by geologists who study cores, cuttings, thin sections, and core analyses to prepare core description graphs like those shown in Figs. 7.49 and 7.87. The vertical sequence of sedimentation is studied to establish the depositional ori-

gin based on the conceptual models discussed in Secs. 7.4 and 7.5. Rock studies include correlation of rock properties (permeability, porosity, and capillary pressure) with rock type. Horizontal permeability often is correlated with rock type. Other correlations that may be developed include horizontal permeability vs. vertical permeability and horizontal permeability vs. connate water saturation.

Fig. 7.88 is a porosity/permeability crossplot that shows the properties of different rock types in the Loudon field.[21] Permeability/porosity correlations must be prepared with particular attention given to mineralogy. Correlation of permeability of the Rotliegendes sandstone from

Eastburn sandstone, composite log, IF 117 - IF 102A

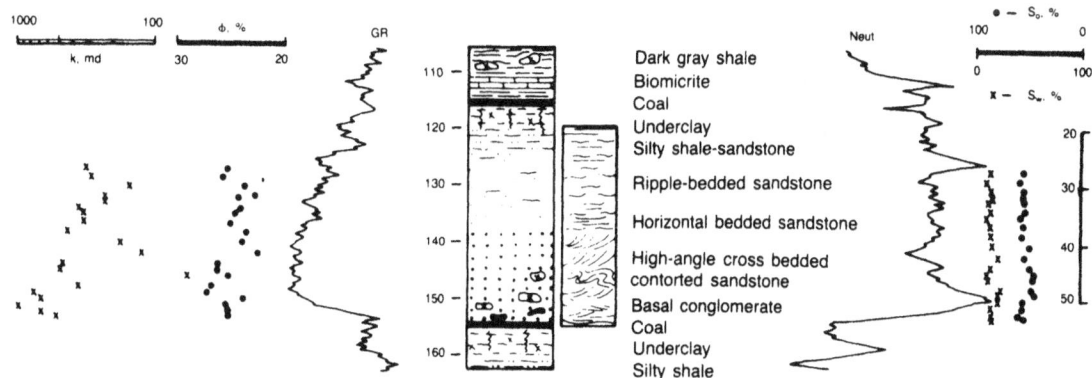

Fig. 7.87—Reservoir quality profile, Eastburn sandstone.[35]

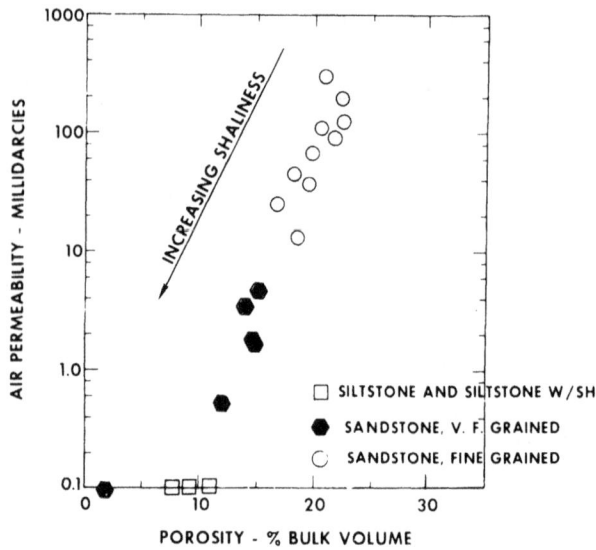

Fig. 7.88—Porosity vs. permeability crossplot for various rocks aids identification of reservoir rocks. [21]

the North Sea province with porosity is shown in Fig. 7.89. [68] The kaolinite-cemented sandstone is more permeable than illite-cemented sandstone. Correlation was obtained when clay mineralogy was identified and data were grouped accordingly.

Permeability/porosity crossplots may produce distinct separation of facies. Fig. 7.90 is a cross section through the Judy Creek field showing the distribution of three facies-controlled porosity groups. [16]

Group I rocks consist of reef framework and reef detritus; they occur in organic-reef and shallow-water facies. Group II rocks consist of reef detritus and algae laminates; they are found in shoals and back reefs. Group III rocks are composed of organic debris and pellets in a lime/mud matrix.

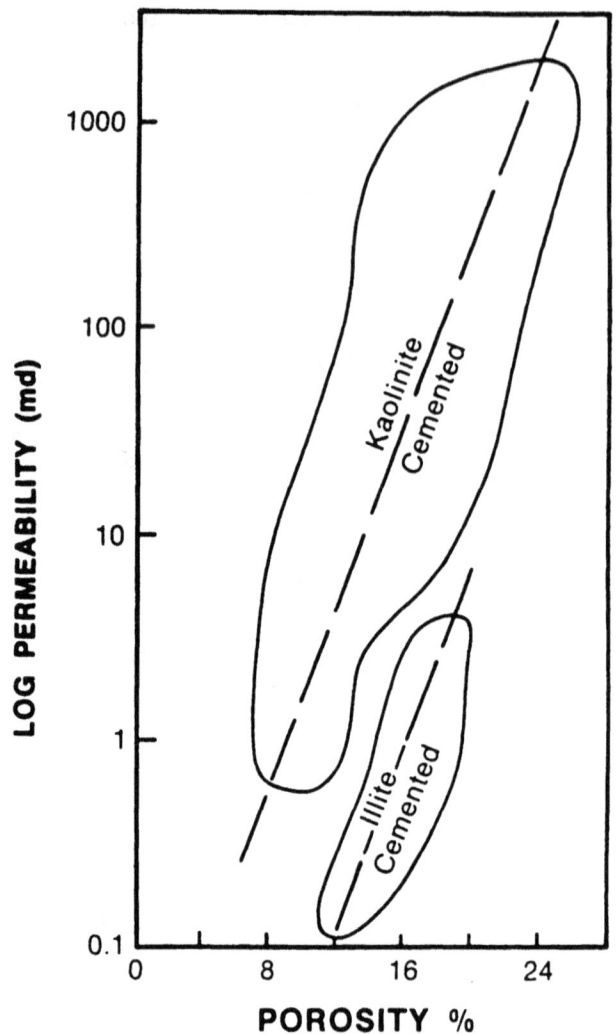

Fig. 7.89—Sandstones of the Rotliegendes formation in the North Sea in which discrete families of properties relating to the types of authigenic clay cement are evident (Ref. 68, after Ref. 70).

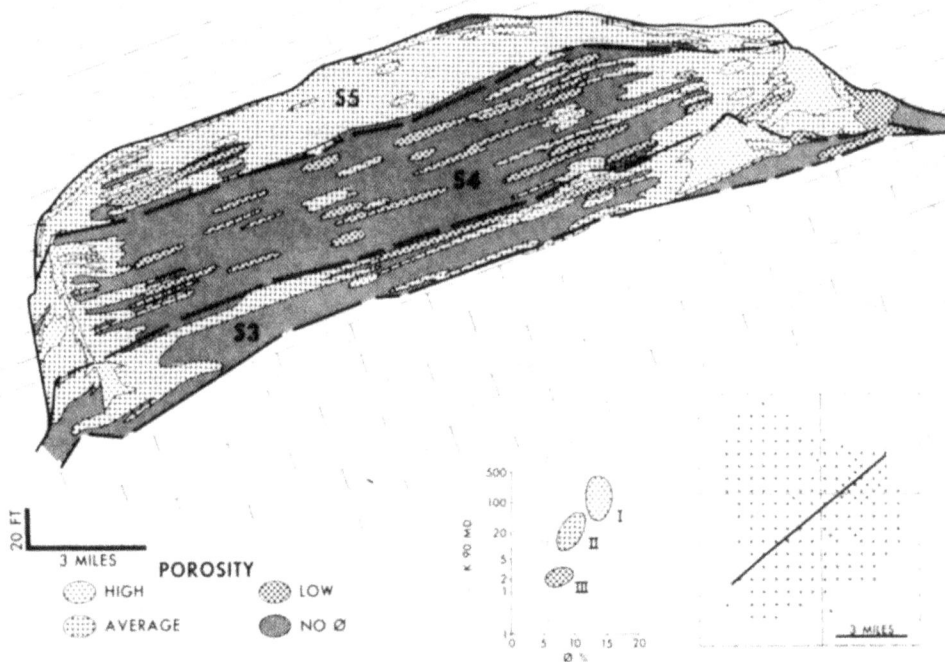

Fig. 7.90—Cross section showing the distribution of three facies-controlled porosity groups. [16]

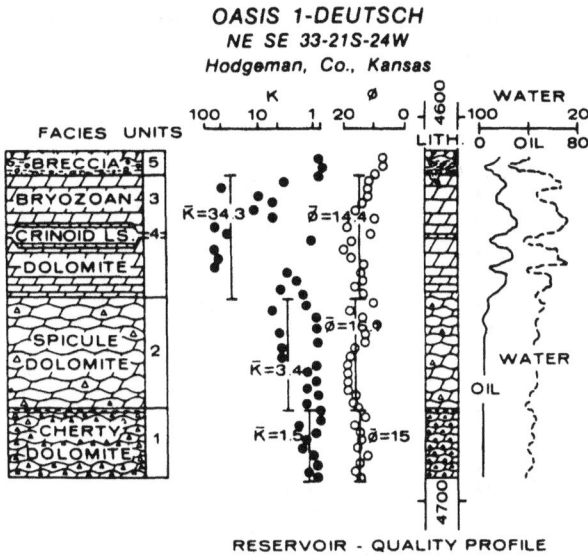

Fig. 7.91—Vertical sequence of carbonate facies in Bindley field, KS, plotted to show their different reservoir rock properties and their relation to accumulation of oil.[71]

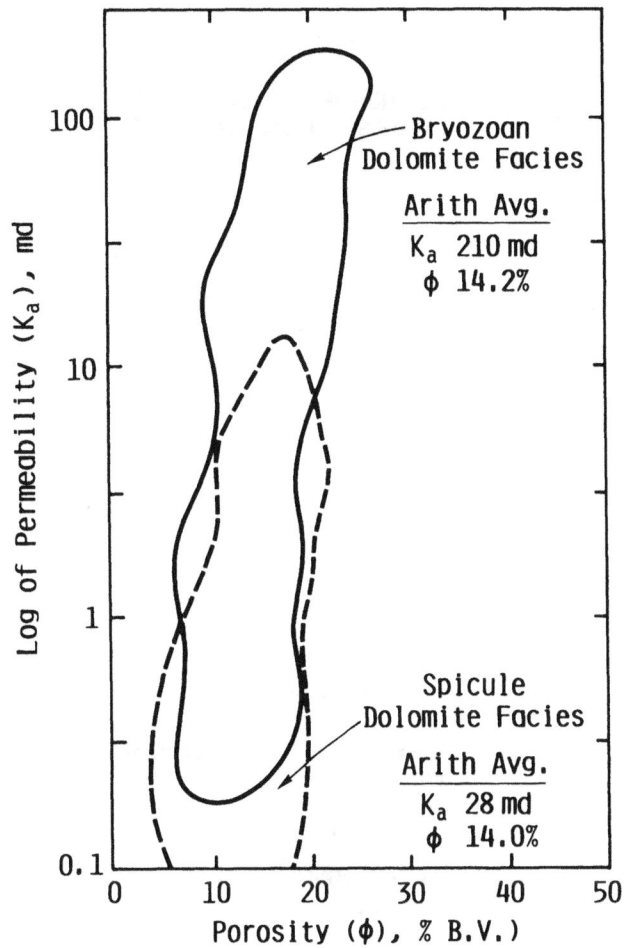

Fig. 7.92—Plot of log permeability to air vs. porosity in Mississippian dolomites from Bindley field, KS. Courtesy W.J. Ebanks.

Caution is required when a correlation is developed from core data. Different facies may have different types of pore systems. For example, Ebanks[71] identified two distinct carbonate facies—bryozoan dolomite and spicule dolomite—in Mississippian dolomites from the Bindley field, KS. A core description graph is shown in Fig. 7.91.

The permeability/porosity correlation (see Fig. 7.92) is weak for two reasons. First, small changes in porosity are associated with large changes in permeability. Second, the correlations for each facies overlap throughout much of the porosity interval. As indicated in fig. 7.92, large differences in permeability occur because facies change when porosities are essentially equal. In this case, detailed studies of the pore structure revealed that bryozoan dolomite facies had a well-connected network of vuggy pores. Most were connected effectively through intercrystalline pores, while the spicule dolomite had isolated pores in a matrix of fine crystalline dolomite.

Engineers and geologists frequently find that core data are limited and that well logs must be used to obtain estimates of rock properties. A common approach is to correlate permeability with porosity from the available core data, as illustrated in the previous section. Then, permeabilities are determined for wells lacking core analyses by the development of correlations between a porosity log and porosities from core data.

Three types of porosity logs are used in reservoir analysis: acoustic, neutron, and density logs. Although each of these logs responds primarily to porosity, other formation characteristics (such as lithology, type, and amount of fluids) influence the interpretation of log response to obtain porosity.[72] Table 7.20 summarizes factors affecting the response of the three logs in clean carbonate rocks and sand/shale sequences.

Because other factors affect the porosities indicated by logs, correlations between log and core properties are developed by comparing log and core data on a foot-by-foot basis. It may be necessary to shift core data a few feet to obtain the best correlation with log data.[27] Core data

TABLE 7.20—FACTORS AFFECTING LOG RESPONSE IN CLEAN CARBONATE ROCKS AND SAND/SHALE SEQUENCES[72]

Rock Type	Log Type	Factors Affecting Log Response
Carbonate	Density Neutron	Matrix, total porosity, amount, and type of fluid in pore space
	Acoustic	Matrix, primary porosity, amount, and type of fluid in pore space and compaction
Clastic sediments (sand/shale)	Density	Effective porosity + clay effect − depending on shale density, + light hydrocarbon, gas effects
	Acoustic	Effective porosity, clay effect + compaction, + light hydrocarbon, gas effect
	Neutron	Effective porosity + clay effect, + light hydrocarbon, gas effect, or + heavy oil effect

should be corrected for the effects of overburden pressure where appropriate.[73]

Two examples showing the development of correlations from log data are presented to illustrate the technique. Fig. 7.93 shows the correlation of sonic travel time with

CORRELATION OF SONIC TRAVEL TIMES FROM LOGS WITH CORE POROSITIES

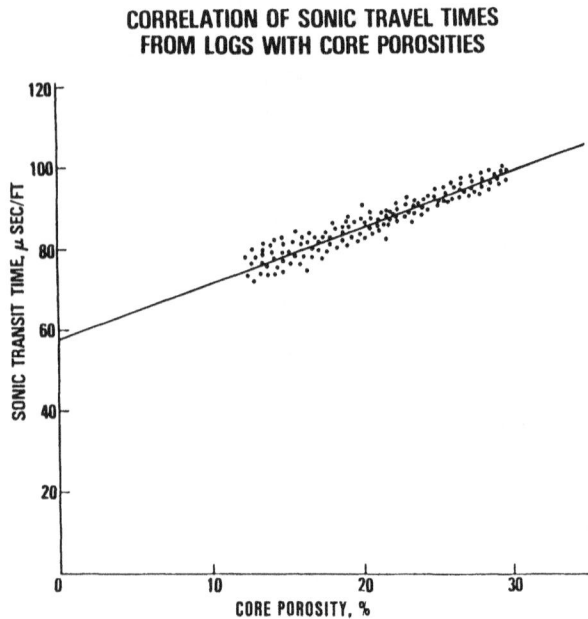

Fig. 7.93—Correlation of sonic travel times from logs with core porosities.[27]

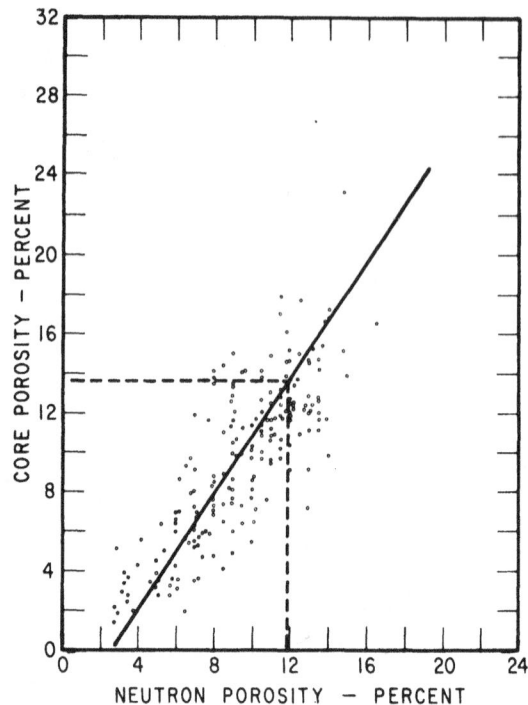

Fig. 7.95—Core porosity vs. neutron porosity for Means field.[58]

POROSITY-PERMEABILITY RELATION

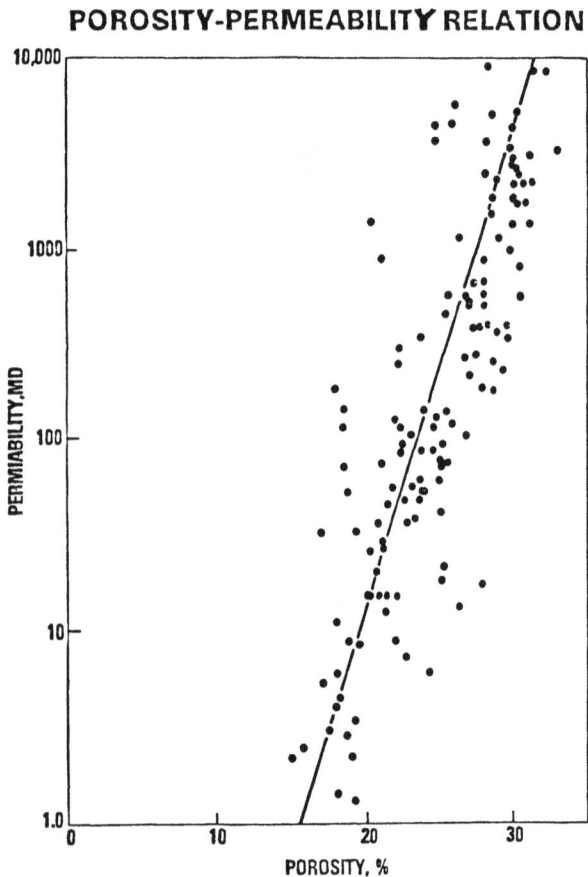

Fig. 7.94—Correlation of permeability with porosity for Jay field.[27]

core porosities from the Jay field in Florida.[27] Correlation of permeability and porosity from core data is shown in Fig. 7.94.

In the Means field, a small percentage of wells were cored, so developing correlations from the available logs was necessary.[58] Many of the logs were 15 to 25 years old, offering a formidable challenge. Gamma ray neutron logs constituted the majority of the logs. Neutron log response is sensitive to hole size and condition, logging tool, and logging company.[58] Thus a separate correlation was developed for each well. The most common procedure used to develop the correlation was to identify two endpoints on the gamma ray response and to fit the logarithmic scale expected for the neutron log response to these values. This is a widely used method of interpreting neutron log data when calibration or scale headings are missing or uncertain. For the Means field, the neutron zero was assumed to be 100% porosity, whereas a dense streak was assigned a porosity of 1 to 1½. Lithology was incorporated into the correlation when the shale cutoff was considered to be 25% of the total gamma ray deflection from the baseline. Fig. 7.95 is a typical correlation of core porosity with log porosity from a well in the Means field.

The composite correlation for all cored wells is shown in Fig. 7.96. When data were reduced to one correlation, the neutron porosity was slightly less than the core porosity. The correlation of core permeability with core porosity is shown in Fig. 7.97.

Well log analyses are crucial in developing a reservoir description in most fields. In addition to porosity logs discussed in the previous section, gamma ray and SP logs are used widely for the correlation of lithology. Determination of lithology and mineralogy may be possible by cross correlation of any two or three porosity logs.[74] In

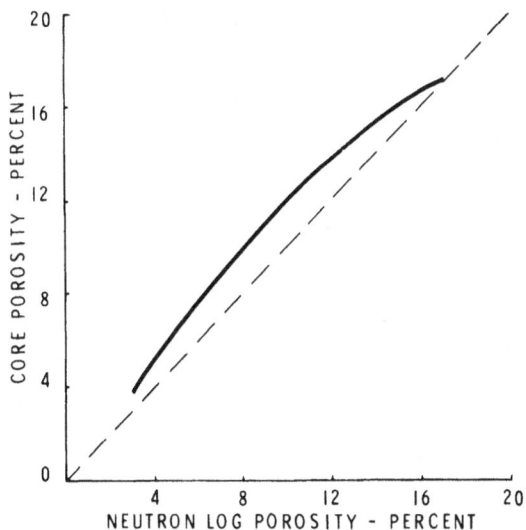

Fig. 7.96—Composite correlation of core porosity with neutron porosity for Means field.[58]

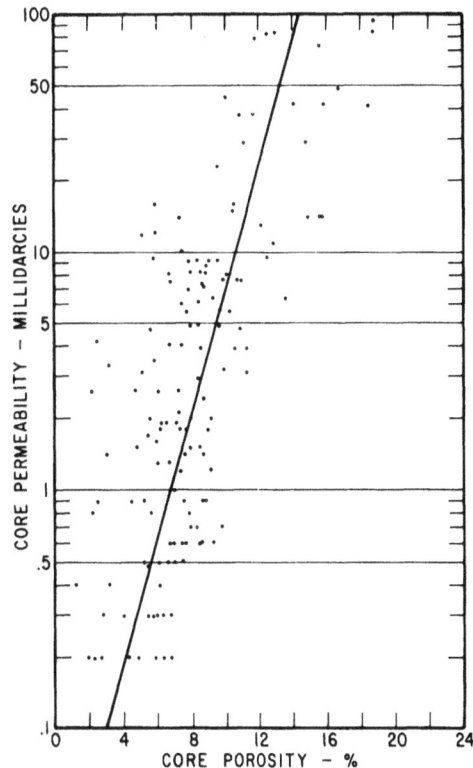

Fig. 7.97—Core permeability vs. core porosity at Means field.[58]

some situations, micrologs give good resolution between permeable and impermeable strata. Resistivity logs are usually interpreted to obtain estimates of water and oil saturations but may provide other information.

When core analyses are limited, as is often the case, calibrating logs against core data to extrapolate limited core data across the field is essential.

Sneider *et al.*[38] illustrate how rock types, sequences of rock types, and porosity were determined by interpretation of SP, 8-in. [20-cm] normal resistivity and micrologs for the Elk City field. Fig. 7.98 shows typical log-response/rock-type calibrations. The SP deflection from the shale baseline distinguishes the sandstone and conglomerates from the impermeable siltstones, calcareous shales, and limestones. Microlog response showed positive separation opposite all permeable formations. Microlog resolution, when free from artifacts, is typically on the order of 1 ft [0.3 m].* The "hashy" microlog response was characteristic of very silty, shaly, or cemented rocks; thus there was no separation. Subsequent correlation of micrologs with core analyses revealed that microlog separation occurred opposite rocks that had permeabilities of 0.5 md or greater.

The 8-in. [20-cm] normal measures resistivity in the region around the wellbore that is flushed with filtrate. Resistivity is primarily a function of rock pore structure and mud filtrate. Studies of the 8-in. [20-cm] normal resistivity curves in the sandstone and conglomerates led to correlation of grain size with resistivity. That is, resistivity increased with increasing grain size. Correlation of grain size with resistivity was attributed to sorting. Fig. 7.99 shows the relationships between sorting, grain size, and porosity in some Elk City sandstones and conglomerates. Fine-grained rocks have the highest porosity and the best sorting. Thus, low formation factors and resistivities were observed. Porosity decreases as grain size increases and sorting becomes poorer. These factors cause the formation factor and resistivity to increase. Correlation of grain size and porosity also enabled correlation of total porosity with permeability for distinct genetic sand units.

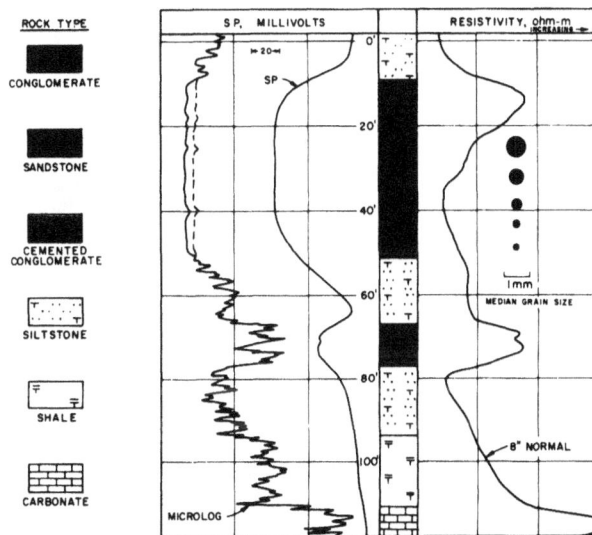

Fig. 7.98—General relations between rock type and log response in the Elk City field.[38]

*Personal communication with J.H. Doveton, Kansas Geological Survey, 1983.

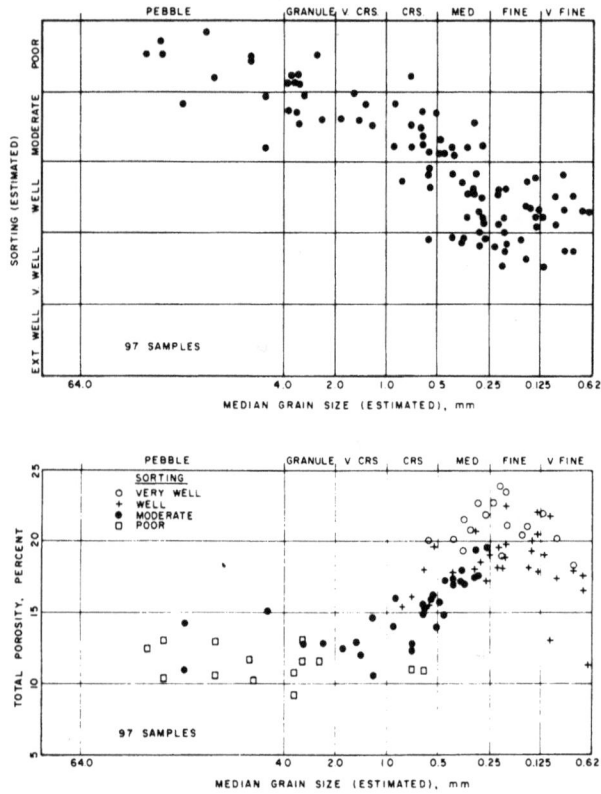

Fig. 7.99—Relationship among sorting, grain size, and porosity in some Elk City sandstones and conglomerates.[38]

Permeability varies with position and direction. Values of permeability reported in most core analyses are measured on core plugs cut horizontally. The orientation of the core and, thus, the core plug relative to the coordinate axis is not usually known. Because crossflow must be considered in reservoir simulators, determining vertical permeabilities on a few cores to obtain the ratio of horizontal to vertical permeability is necesary. The ratio is considered constant within the same facies because of the limited amount of data usually available. Ratios may vary from 1 to 2 for unconsolidated sands to 100 or more for consolidated rocks. Vertical permeability may be determined also from well tests, as discussed later.

Measurement of permeabilities with core plugs should be done with specific attention given to sedimentary structures (such as crossbedding and other geologic features in the core). Fig. 7.100 illustrates two examples of these effects.[8] Cores A and B in Fig. 7.100 have festoon crossbedding, a common sedimentary feature in sandstone reservoirs that have a fluvial origin. The visible bedding planes intersect the horizontal axis at an angle. Permeability was measured parallel and perpendicular to the bedding planes as well as in the usual horizontal direction. Directional permeability on the scale of the core sample is indicated because the permeability parallel to the crossbedding is more than four times greater than the permeability perpendicular to the crossbeds. Horizontal permeability was lower because of the composite effects of these directional trends. Vertical permeability determined from core plugs would not indicate correctly the

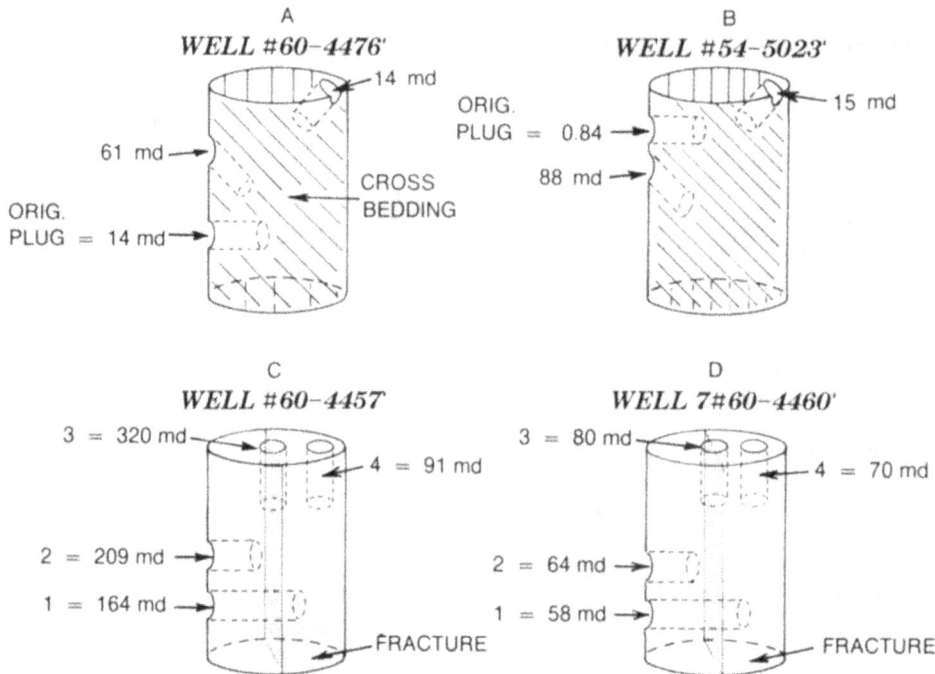

Fig. 7.100—Core permeability studies.[8]

directional permeability present in the crossbeds. Permeabilities measured on core plugs may not be useful if the size of the core plug is small relative to the geologic structure.

Core Samples C and D in Fig. 7.100 are from a homogeneous sandstone with high permeability and porosity. The samples contain small, vertical features that probably contributed to the high permeability attributed to the interval from the interpretation of waterflood performance. Close inspection of the fractures revealed the presence of very fine sand to silt-size quartz and an asphaltic residue along the fractures. Permeabilities measured in the horizontal direction decreased in Plugs C1 and D1, which cut across the fracture. Locations of vertical core plugs are indicated also. Vertical permeabilities were higher than horizontal permeabilities in this rock, as might be expected from the fracture orientation, but the differences were not large. The lower permeability in Plug D4 was attributed to smaller grain size and to the effects of localized shear.

When data are limited, model calculations have been used to determine the k_h/k_v ratio.[75] Weber[76] presents a method to model permeability in a crossbedded sand with discontinuous clay and silt intercalations. Halderson and Lake[77] proposed a statistical method of determining the horizontal and vertical permeabilities in reservoirs with random shales that cannot be correlated between wells.

Correlations of other properties may also be developed by the calibration of resistivity logs against water saturations from oil-based cores and water saturations from capillary pressure measurements on cores.[27,78]

Framework Studies. Framework studies determine the number and distribution of reservoir zones. The continuity of reservoir and nonreservoir rocks is an important objective. Framework studies begin by mapping the gross structure from well and seismic data to define the areal and vertical extent of the deposit. Structure maps like the map shown in Fig. 7.101 developed for the Elk City field usually are generated early in the framework studies.[38]

Identification of an aquifer and estimation of aquifer size may be done at this point if adequate data are available. This is important because the size of the aquifer (if any) relative to the reservoir is a measure of the capacity of the aquifer to maintain reservoir pressure under primary production. Table 7.21 contains estimates of aquifer capacity when the effective compressibility of the aquifer is 6×10^{-6} psi^{-1} [41×10^{-6} kPa^{-1}].

The principal activity in framework studies is the determination of area and vertical limits and continuity of reservoir and nonreservoir zones. Core description graphs, if available, are used to identify principal reservoir and nonreservoir zones. Correlation between wells is performed by the integration of the depositional model with core description graphs and well logs. Reservoir zonation is portrayed on cross sections that are prepared for principal regions of the field. Examples of framework studies in three fields are presented in this section. Other examples are found in Refs. 79 and 80.

In sandstone reservoirs, the principal nonreservoir rock is usually shale. Study of shale units provides guidelines for use in determining continuity. Richardson *et al.*[81] summarized some aspects of shale continuity in terms of distributions and dimensions of major and minor shale barriers. Major shale barriers are those that separate distinct

sand units in a field, while minor shale barriers are interbedded within a sand unit.

Fig. 7.102 shows the continuity of major shale (silt) intercalations in sandstone deposits from studies of outcrop samples having known depositional environments.[76]

Major shale barriers (see Fig. 7.103) may be continuous or discontinuous. Continuous shales extend over large distances (0.19 mile [0.3 km]) and have such low vertical permeability that the adjacent sands behave as separate reservoirs. Discontinuous shales are barriers to fluid flow in local areas. The continuity of major shale barriers may be determined qualitatively by the determination of the depositional environment (see Table 7.22).[81]

Minor shale barriers (intrasand shales) are depicted in Fig. 7.104.[82] These shales are thin (0.98 to 1.97 ft [0.3 to 0.6 m]) and are identified through core analysis because they are too thin to resolve with well logs. Intrasand shales usually do not correlate between wells and appear to be scattered randomly within the sand matrix.[77]

Continuity of minor shales can be estimated from a reference model that resembles the depositional area. Wadman *et al.*[84] estimated the areal extent of minor shales deposited in a braided stream environment in the Sadlerochit sand, Prudhoe Bay unit, using a study of 17 modern braided and straight stream deposits. Results of this study showed that stream channel width was about 40 times the channel depth and that the deposited shales had an average length-to-width ratio of 3. By assuming that the average minor shale thickness at 1.8 ft [0.55 m] in the Sadlerochit was equal to the channel depth at the time of deposition, Wadman *et al.*[84] estimated the shale width at 75 ft [23 m] and shale length about 200 ft [61 m]. A statistical method of representing lateral continuity and spatial distribution of intrasand shales is described by Halderson and Lake.[77]

Loudon Field. Fig. 7.105 shows the location of a pilot site (0.625 acres [2529.296 m^2]) in the Loudon field in central Illinois.[21] The field is on an anticline that has a northeast/southwest axial trend with 1 to 2° flank dips and a 165-ft [50-m] closure relative to the Cypress datum. The Weiler sandstone reservoir interval is in the Cypress formation of the upper Mississippian (Chester Age). A core description graph from this field is shown in Fig. 7.49. The deposition model was determined to be deltaic and the location of the pilot area is on the deltaic fringe as depicted in Fig. 7.50.

Correlation of reservoir and nonreservoir zones between wells was done for the pilot area and the 160 acres [647 500 m^2] (including the pilot) by systematic analysis of core descriptions and well logs with the depositional model as a guide. Several deltaic cycles occur in the upper Mississippi strata in Illinois separated by shallow-water carbonates that have large areal extent. These carbonates were excellent correlation markers regionally and in the pilot area.

Fig. 7.106 illustrates the method used to correlate intervals between Wells 1 and 2. Depositional environments and units analogous to those presented in Fig. 7.49 are not shown but were incorporated in the analysis. Wells were correlated within the interval defined by the limestone datum at the top of the Cyprus formation and a limestone marker below the formation. Correlation numbers indicated on Fig. 7.106 identify intervals in each well that were judged to have the same depositional unit based on

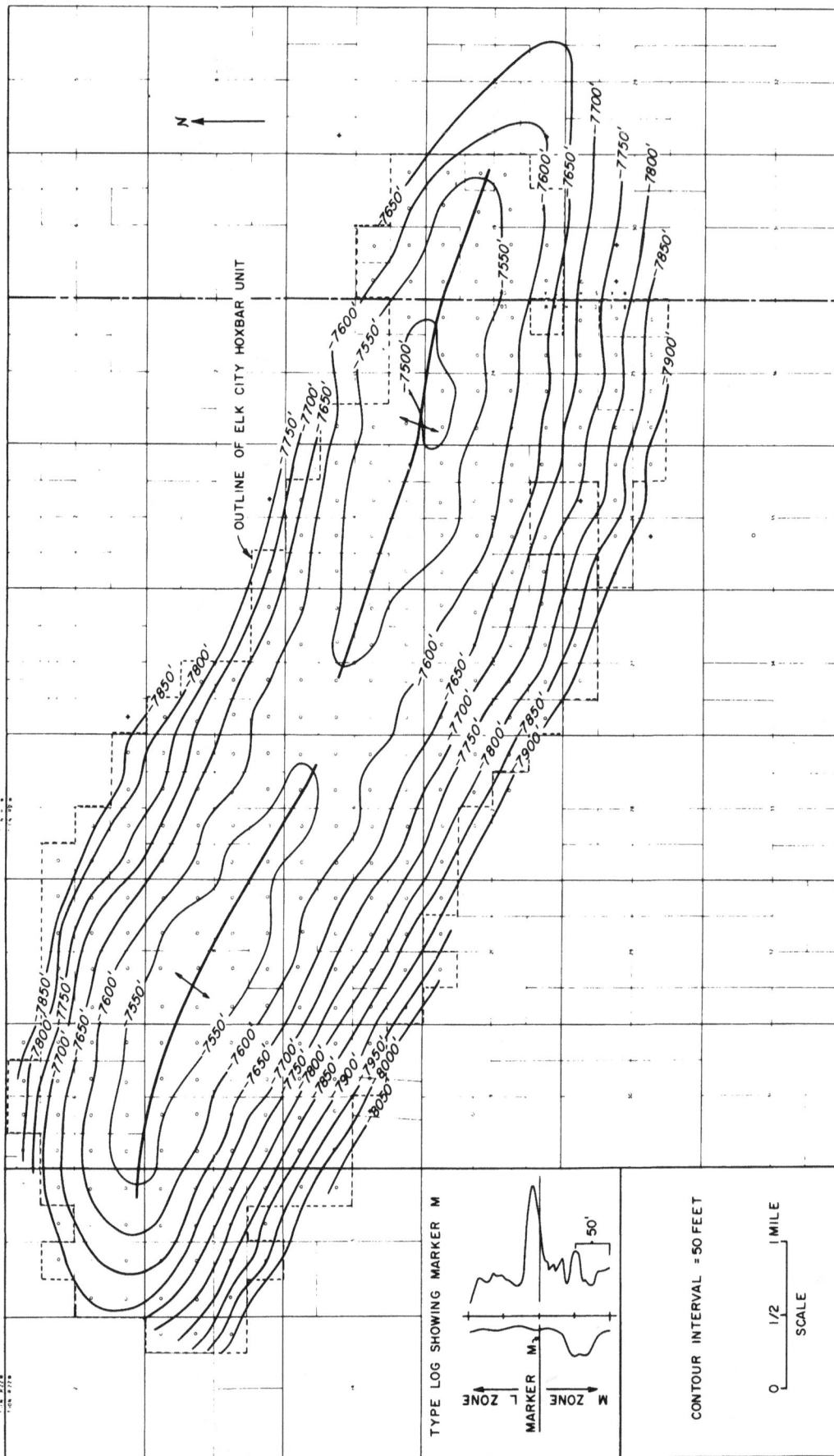

Fig. 7.101—Structure contour map on Marker M, Elk City field, OK.[38]

**TABLE 7.21—CAPACITY OF AQUIFERS
TO MAINTAIN PRESSURE[27]**

Aquifer Size/Reservoir Size (PV/PV)	Water Efflux* (Reservoir PV/1,000 psi pressure drop in aquifer)
5	0.03
10	0.06
50	0.3
100	0.6

*$C_{f+w} = 6 \times 10^{-6}$ psi^{-1}.

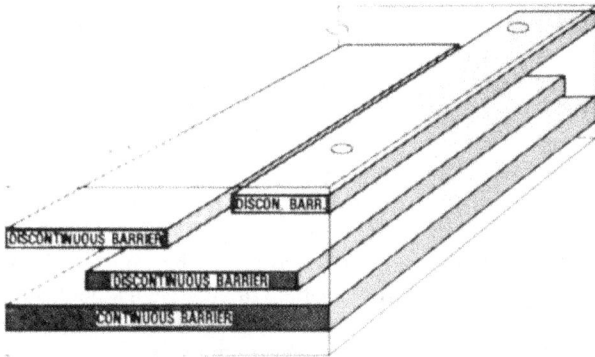

Fig. 7.102—Continuity of shale (silt) intercalations as a function of depositional environment (Ref. 76, after Refs. 82, 83).

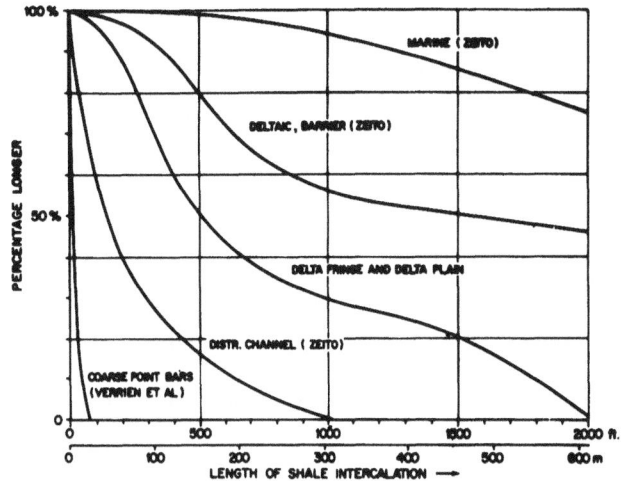

Fig. 7.103—Major shale barriers (intersand) defined by correlation.[81]

TABLE 7.22—ASSOCIATION OF SHALE CONTINUITY AND ENVIRONMENT OF DEPOSITION[81]

Principal Depositional Environment*	General Shale Continuity in Subenvironment		Some Considerations in Shale Barrier Description
	Continuous	Discontinuous	
Dunes (Aeolian)		Interdune playas and wadi (stream) deposits	Shales rare
Alluvial fan	Lower fan "sheets"	Upper fan channels	Channel spacing and gradients, debris flow deposits
Braided stream		Channel braids and lateral secondary channels	Channel depth (based on fundamental bed thicknesses), valley width
Meandering stream	Flood basin (remnant in meandering belt complexes)	Top part of point bars and abandoned channels	Channel depth (based on fundamental bed thicknesses), valley width
Upper deltaic plain	Flood basin (remnants; as above)	Upper point bar, abandoned channels, channel braids	Channel depth, coastal plain vs. valley confinement
Lower deltaic plain	Interdistributary areas	Distributary channels	Channel depth, spacing of distributaries
Delta fringe	Areas marginal to river-mouth bars	River-mouth bars	Size of feeder river, wave, tidal, and strength of currents along the shore
Beaches—barrier island	Lagoons, shoreface	Tidal inlet channel, shoreface	Shales uncommon in lower, rare in upper parts
Tidal flat	High flat	Low flat and channels	Tide range
Submarine fan	Lower fan "sheet"	Upper fan	Feeder canyon size, particle-size range

*Based on LeBlanc.[23]

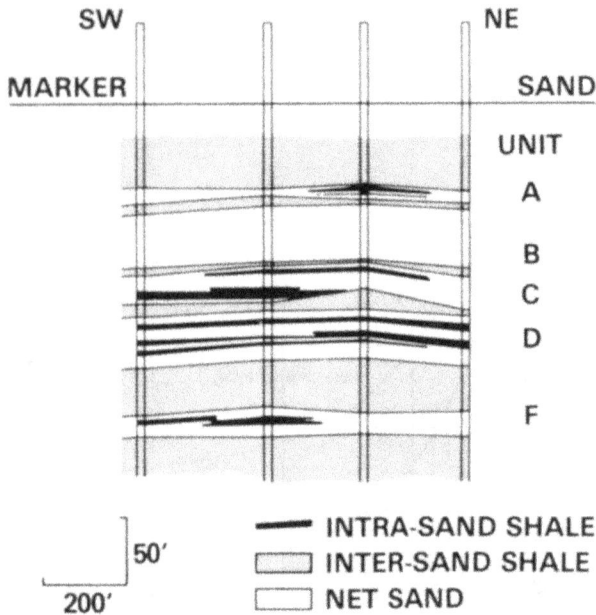

Fig. 7.104—Example of intrasand and intersand shale barriers. [81]

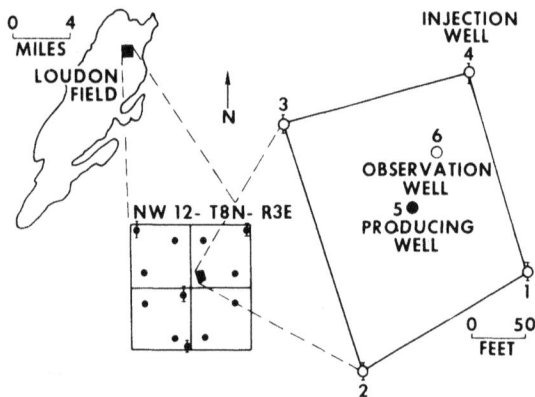

Fig. 7.105—Location and plan of the pilot-test site in the Loudon field, IL. [21]

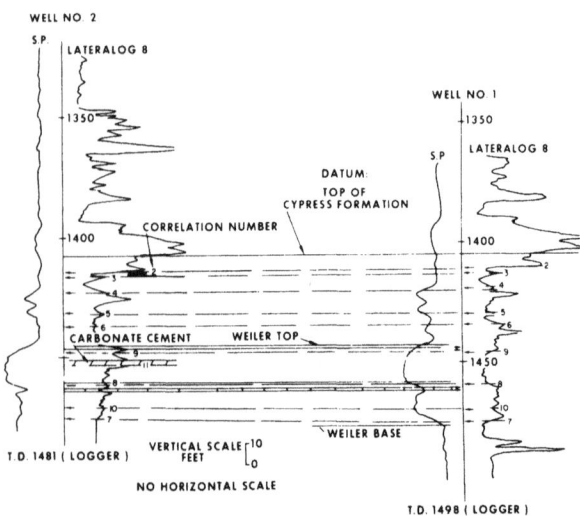

Fig. 7.106—Example of well-log correlation in the pilot indicating detail required for sand continuity studies. [21]

core description or similarity of log response to a well where the core description was known.

Parallel correlation lines shown in Fig. 7.106 were characteristic of the pilot area and the 160-acre [647 500-m^2] area surrounding the pilot. Thus, reservoir beds were inferred to have a nearly horizontal, tabular geometry. Correlation of particular units were mapped to show areal continuity. Fig. 7.107 is the gross thickness map of Unit 4, which is the principal reservoir unit in the pilot area. Note that this map includes both reservoir and nonreservoir zones and must be adjusted to reflect net pay.

Interpretation of the correlations between wells revealed that continuity of reservoir and nonreservoir zones varied with well spacing. Two scales of continuity were identified for shale breaks in the pilot area: (1) those with radii between 200 and 1,000 ft [61 and 305 m] (Correlation No. 8 in Fig. 7.106) and those with radii less than 50 ft [15 m] (determined from correlation between Wells 5 and 6 that were about 50 ft [15 m] apart). The carbonate cement noted in Well 2 was found in five other wells but did not correlate between wells.

Correlations developed with this method assume there are no discontinuities in the correlatable units between wells. Other data are needed to assess the validity of this assumption for a particular field or an area in a field. For example, pressure pulse tests can often be used to evaluate gross continuity between two wells. [85-88] In the Loudon field, it was not possible to obtain a measurable response between one of two well pairs tested at a well spacing of 10 acres [40 469 m^2]. Thus, extrapolation of well log/core description correlations on this spacing is questionable. Well testing is an important tool that can yield quantitative data on reservoir continuity in the interwell area and should be included as part of the framework activities.

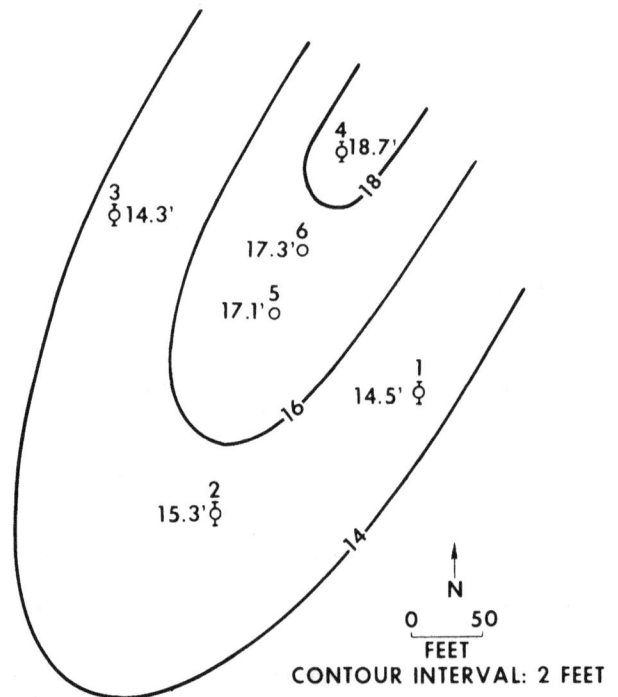

Fig. 7.107—Thickness map of the reservoir and nonreservoir rocks in Unit 4. [21]

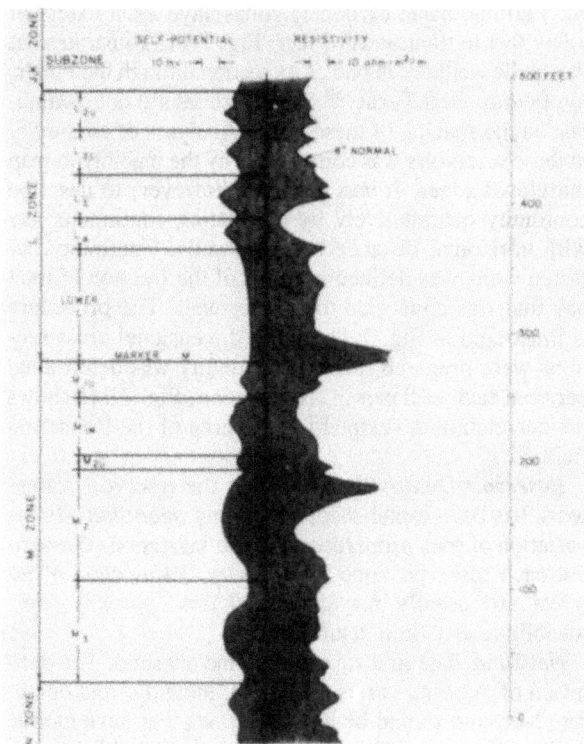

Fig. 7.108—Idealized log of Zones L and M, Elk City field.[38]

Elk City Field. The Elk City field is a 100×10^6 bbl $[16 \times 10^6 \ m^3]$ field located in the Anadarko basin in southwestern Oklahoma. The field produces from the Pennsylvanian sandstone and conglomerates. Productive intervals are interbedded with nonproductive siltstones, shales, and carbonates. Sneider *et al.*[38] conducted an extensive geological and petrophysical study of this field to develop a reservoir description during the planning stage of a waterflood.

The field was developed with 310 wells on 40-acre $[161 \ 875\text{-}m^2]$ spacing over an area of about 25 sq miles $[65 \ km^2]$. The interval studied included Zones L and M, which were the two major reservoirs. Core analyses (1,700 ft [518 m]) were available from 26 wells. Most wells were logged with a combination of SP log, 8- and 16-in. [20- and 41-cm] normal resistivity logs, 24-ft [7-m] lateral log and microlog. A gamma ray neutron log was run instead of the microlog in 16 wells. The relatively small number of cored wells placed a premium on calibration of well logs with core data.

Fig. 7.108 is a type log of Zones L and M. A dense marine limestone marker (Marker M) separates Zones L and M and was used as a datum to hang cross sections. Core studies of Zones L and M were used to determine depositional environment following procedures described in previous sections of this text. Four distinct sand bodies were identified: (1) barrier bar, (2) alluvial channel, (3) deltaic distributary channel, and (4) deltaic marine fringe. Mapping of these bodies in Zones L and M was a key to developing the reservoir description.

Correlation was done by the preparation of correlation sections similar to Fig. 7.109. Markers L, LI, LII, LIII, MI, MII, and N were identified as marine shales and limestones from core analyses and could be recognized on logs. Because they had marine origins, these markers were expected to correlate over large areas of the field. This was observed, as is shown on Fig. 7.109. In some areas, however, markers were absent. Markers LII and LIII are missing from Wells 1-15-8 and 1-14-5 in Fig. 7.109. The discontinuity in these markers usually was identified as a channel sand deposit. The width of the deposit was estimated by arbitrarily locating the discontinuities midway between Wells 1-15-8/1-15-7 and 1-14-5/1-14-6.

Correlation of the marine markers also provided the basis for subdivision of the L and M producing zones. Subzones shown in Fig. 7.109 and the type log (Fig. 7.108) were separated by impermeable strata over most of the

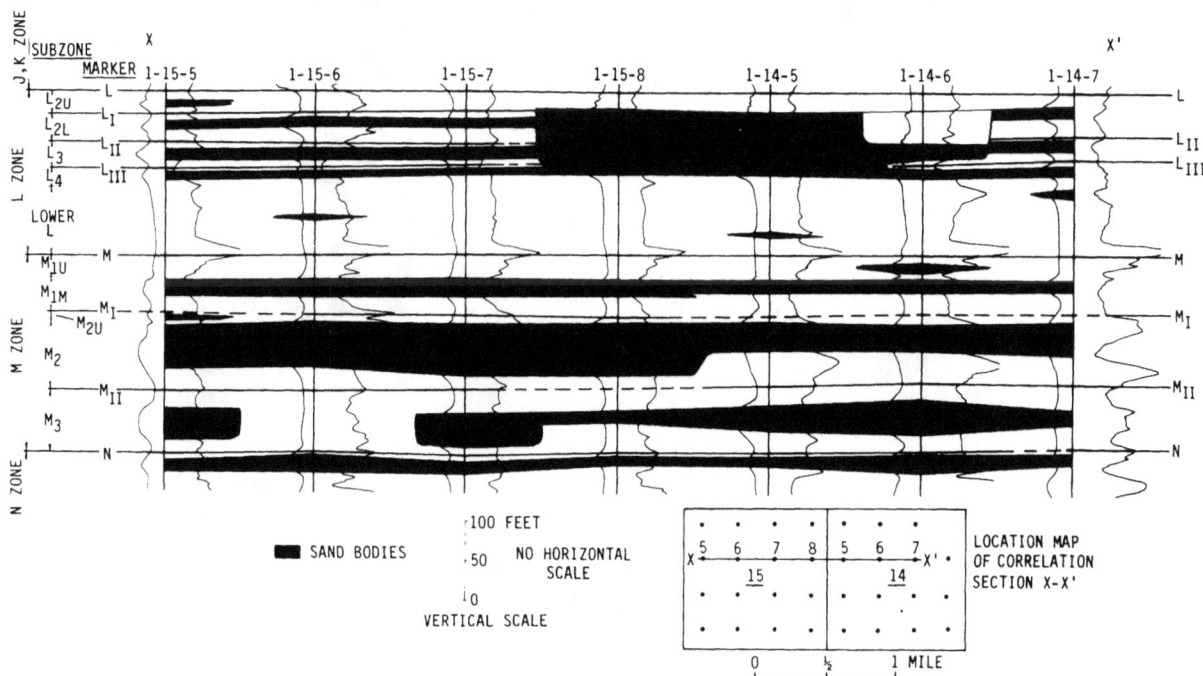

Fig. 7.109—Correlation cross section, Elk City field.[38]

field. The discontinuities in sand deposits in Zones L2L, L3, and L4 identified in Wells 1-15-8 and 1-14-5 were interpreted as a channel cutting through the zones that was filled with gravel. These subzones were believed to be hydraulically connected as a result of the channel deposit. Also shown on Fig. 7.109 are sand bodies in Subzones L2U, L, M1U, and M3 that were believed not to be continuous throughout the cross section shown. These correlations led to the identification of six separate, major reservoir zones and three minor zones that would act as separate reservoirs during waterflooding.

The Elk City study is an excellent example of a complex depositional environment. Fig. 7.110, which is a map showing the distribution of genetic types of sand bodies in the L3 subzone, also shows the complexity of the environment. The areal distribution of the genetic sandbodies can be visualized with panel diagrams. Fig. 7.111 depicts the distribution of reservoir and nonreservoir rock in the region cut by the alluvial channel. Fig. 7.112 is a sequence of cross sections through the barrier bar deposit.

The study of the Elk City field illustrates how sand body genesis can be combined with core and log analyses to estimate the distribution of reservoir and nonreservoir rock. The depositional model depicted in Fig. 7.110 will be used also in reservoir quality studies to estimate net sand.

Permian Basin Carbonates. Continuity of reservoir zones can be described quantitatively when data are limit-

ed. Permian basin carbonate zones have areal extent of a few feet to thousands of feet. Fig. 7.113 illustrates net pay at the wellbore in one cross section through the Fullerton field in west Texas.[58] Several zones are not continuous on the spacing of these wells. Analyses of continuity in these reservoirs was complicated by the inability to map individual zones. It was possible, however, to describe continuity quantitatively by correlating continuous pay with horizontal distance between wells. Continuity between wells was defined in terms of the fraction of total pay that was connected to another well. The procedure is illustrated in Fig. 7.114. Two-dimensional cross sections were prepared, and the continuity was determined between each well pair in cross section. Fig. 7.115 shows the correlation developed for one area of the Robertson field.[89]

Reservoir Quality Studies. Once the reservoir framework has been established, describing *quantitatively* the variation of rock properties and fluid saturations throughout each reservoir zone is necessary. Parameters of interest are usually net sand thickness, porosity, permeability, and fluid saturations.

Net Sand. The first step is to define net sand. The definition of net sand varies with field, location, and operator. Net sand should be those intervals that have mobile hydrocarbon saturation and sufficient permeability to allow flow of fluids under the expected pressure or potential gradients. Often, a lower limit of absolute permeability

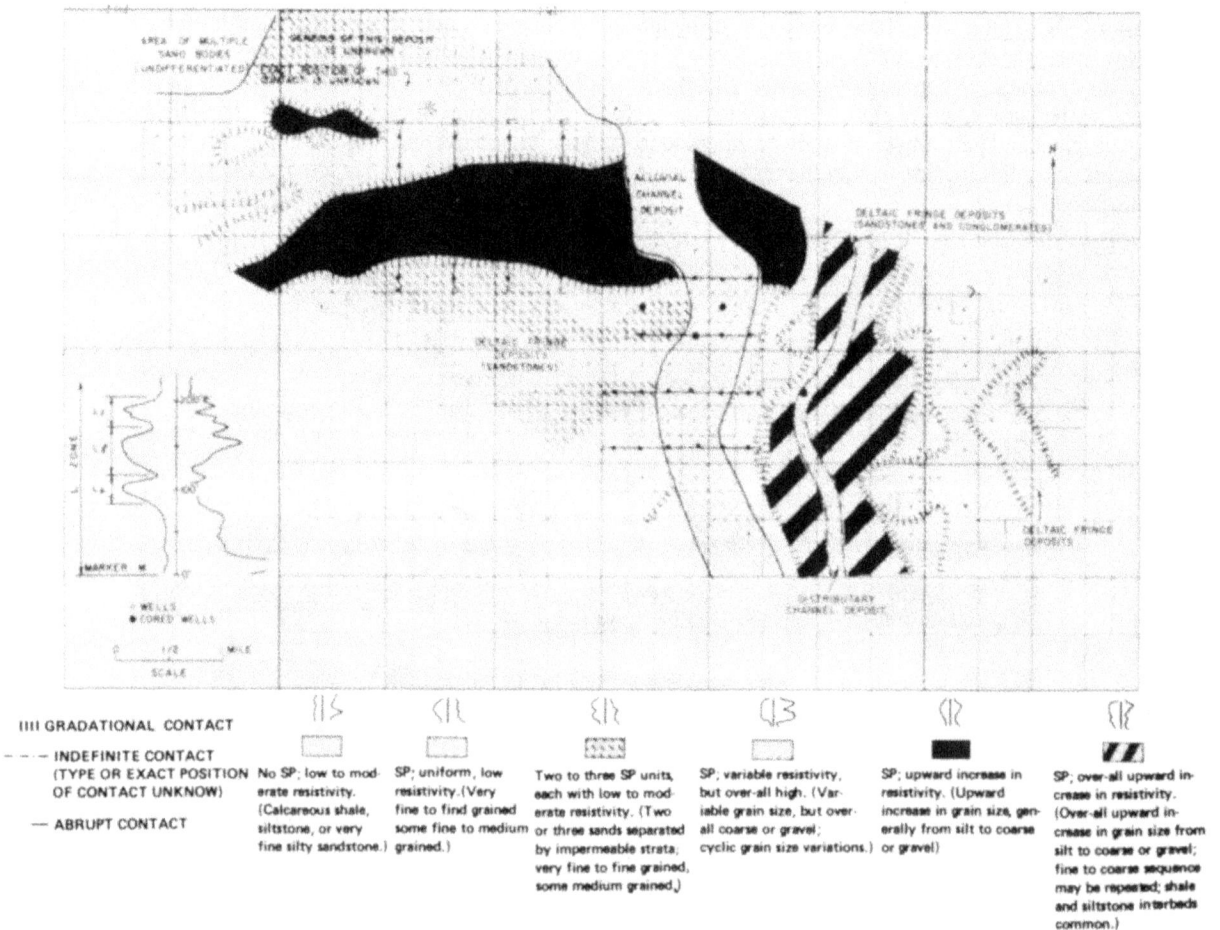

Fig. 7.110—Map showing distribution of sequence of rock types, Subzone L3 with distribution of genetic types of sand bodies.[38]

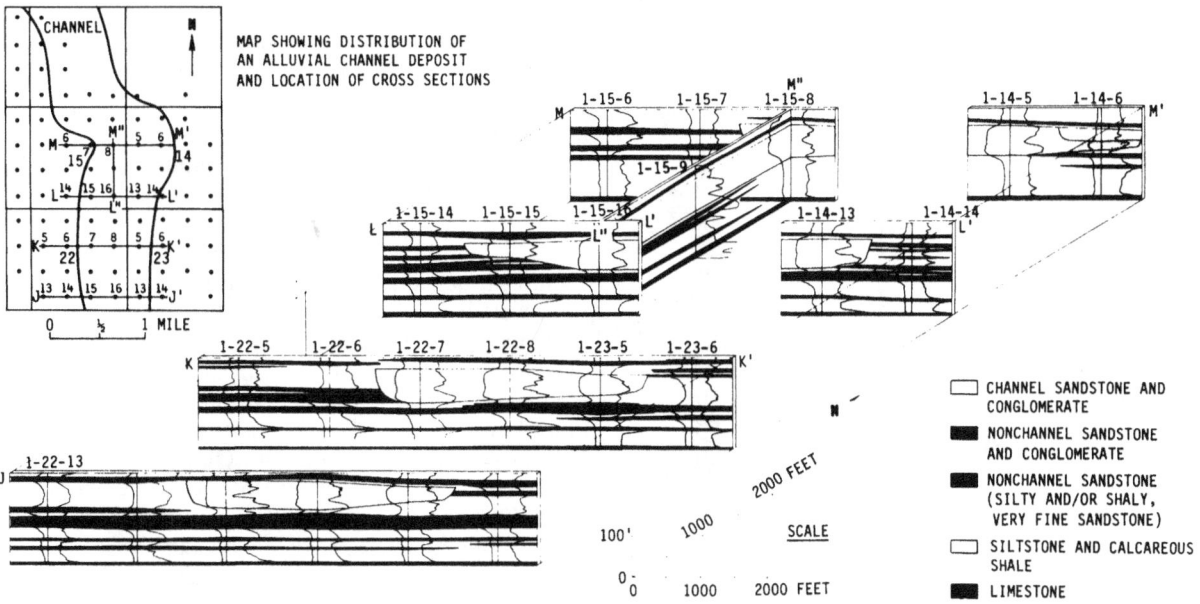

Fig. 7.111—Areal distribution of a portion of an alluvial channel deposit and cross sections through the deposit.[38]

is chosen, implying that fluids cannot be produced from sections that have lower permeability. In the pilot area of the Loudon field, a permeability of 10 md was selected as the cutoff for defining net sand because this value was considered a practical limit for a surfactant flood that was to be tested there.[21] Sneider *et al.*[38] selected a permeability cutoff of 0.5 md to define reservoir rock based on positive microlog separation. A permeability cutoff of 0.1 md is commonly used in the carbonate formations of the Permian basin in west Texas. Some engineers use a cutoff porosity. This is equivalent to a cutoff permeability when a permeability/porosity correlation is available. Net sand is sometimes determined from SP and gamma ray logs as the interval above the shale base line, corrected for adjacent bed effects. The method used is arbitrary but should be checked against volumetric consistency.

Some fields are not suited to setting a cutoff permeability for determining net pay. In the Robertson field, a Permian carbonate reservoir, the correlation between permeability and porosity was poor.[58] Core samples with porosities less than 2% had permeabilities greater than 0.1 md, the conventional permeability cutoff, while other cores had permeabilities less than 0.1 md when porosities were as high as 8%.

George and Stiles[58] developed an empirical technique to determine net pay to improve oil-in-place calculations and to obtain a reasonable distribution of net pay in the field. Two net pay cutoffs were defined. Actual pay was defined as the net thickness of core samples with permeability greater than 0.1 md, the permeability cutoff. Apparent pay was defined as the net thickness of core samples with porosity greater than a specific cutoff. A plot of ap-

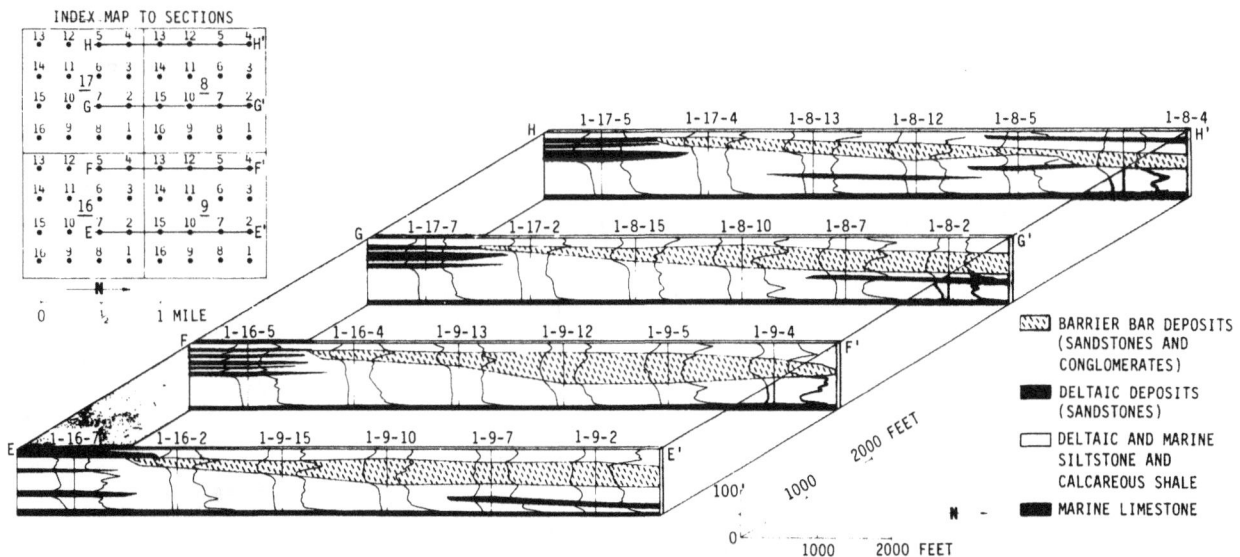

Fig. 7.112—Cross sections through a barrier bar deposit.[38]

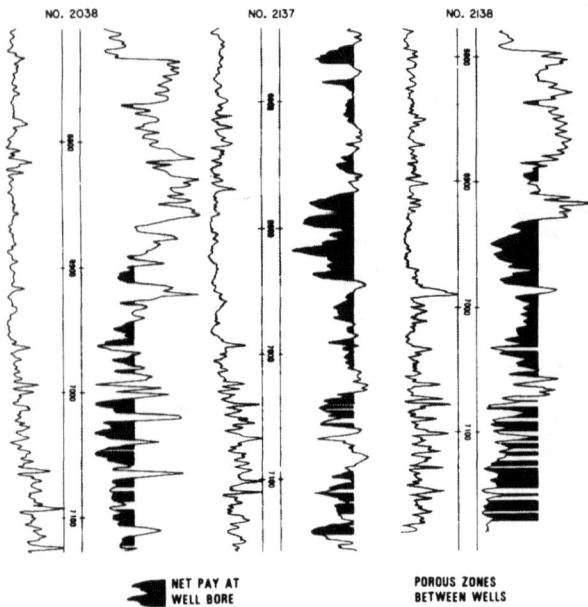

Fig. 7.113—North/south cross section at Fullerton field.[58]

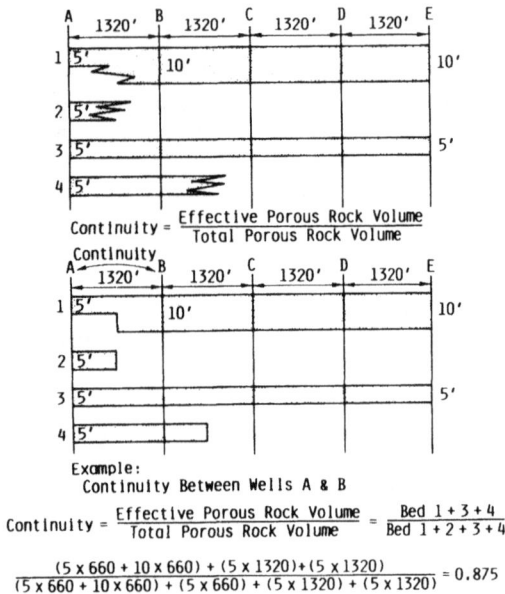

$$\text{Continuity} = \frac{\text{Effective Porous Rock Volume}}{\text{Total Porous Rock Volume}}$$

Example:
Continuity Between Wells A & B

$$\text{Continuity} = \frac{\text{Effective Porous Rock Volume}}{\text{Total Porous Rock Volume}} = \frac{\text{Bed } 1+3+4}{\text{Bed } 1+2+3+4}$$

$$\frac{(5 \times 660 + 10 \times 660) + (5 \times 1320) + (5 \times 1320)}{(5 \times 660 + 10 \times 660) + (5 \times 660) + (5 \times 1320) + (5 \times 1320)} = 0.875$$

Fig. 7.114—Computation of reservoir continuity from connected sand.[89]

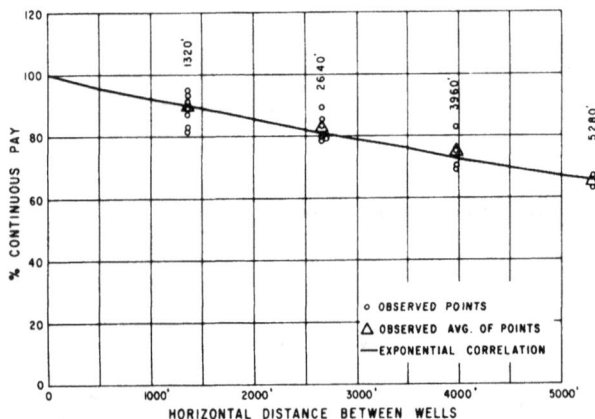

Fig. 7.115—Correlation of reservoir continuity with distance between wells, Robertson Upper Clearfork field.[89]

parent pay expressed in porosity feet vs. porosity cutoff for the Robertson Upper Clearfork unit is shown in Fig. 7.116. The porosity cutoff for the OOIP calculations of 4.2% is the porosity where the apparent pay and actual pay were equal.

The approach outlined in the previous paragraph did not allocate net pay properly across the field. The porosity cutoff of 4.2%, which satisfied volumetric balances, led to the assignment of no net pay to wells that had produced significant quantities of oil because porosities in those wells were below the cutoff. Thus it was necessary to develop an empirical method to allocate net pay. The ratio of actual pay (0.1 md cutoff) to apparent pay (at a specific porosity cutoff) is plotted vs. porosity. The correlation shown in Fig. 7.117 represents data from the Robertson Upper Clearfork unit. Of those core samples that had porosity of 8%, about 75% would be pay, while about 50% of the samples with a porosity of 1% would be pay. Allocation of net pay to individual wells was performed by the selection of a low porosity cutoff, the subdivision of the porosity distribution in a well into discrete intervals, and the assignment of the appropriate ratio of apparent pay to actual pay from Fig. 7.117 to each interval. This approach enables a more realistic evaluation of net pay than would have been obtained with the conventional permeability cutoff and illustrates the level of effort that may be needed to develop a consistent definition of net pay when data are poor, are missing, or do not correlate well.

Permeability and/or porosity cutoffs are used widely because the required data are available or can be estimated as described in the previous paragraphs. These cutoffs are approximate because they do not always determine whether an interval is productive or not. Another approach to defining net sand is possible when sufficient core material is available throughout the reservoir to obtain capillary pressure curves and thin sections characteristic of rock types.

It is well established that capillary pressure curves vary with rock type, as illustrated in Fig. 7.118.[38] Use of this information in defining productive intervals is illustrated by Jodry[65] in a study of Montana/North-Dakota carbonate limestones. In the Montana/North-Dakota area,

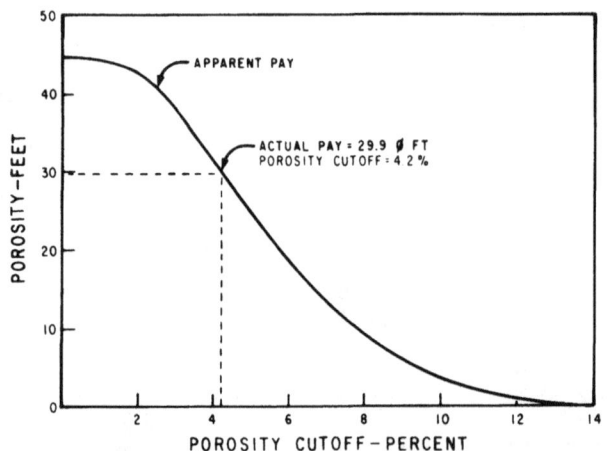

Fig. 7.116—Correlation of actual pay with apparent pay, Robertson Upper Clearfork unit.[58]

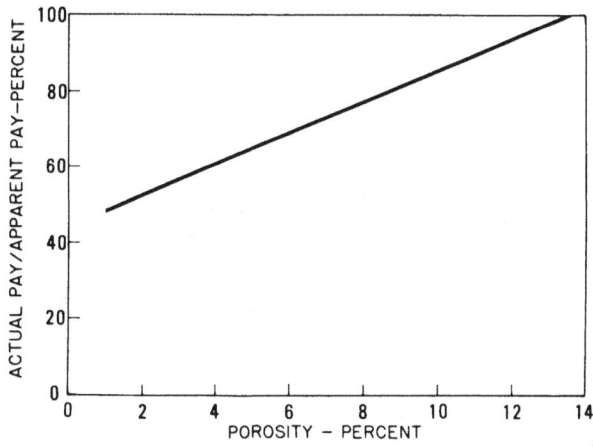

Fig. 7.117—Correlation of net pay with actual pay.[58]

rocks with good porosity and permeability frequently were found to produce only water. There was little correlation between oil-productive intervals and porosity or permeability cutoffs. Study of thin sections and mercury capillary pressure curves from hundreds of wells led to correlation of oil-productive regions with characteristic families of capillary pressure curves and mean pore radii. This approach indirectly incorporates pore geometry and size as a parameter that defines productive intervals. A productive interval must have sufficient hydrocarbon saturation to be mobile. Thus, fluid saturations are involved also in selecting net pay cutoffs, as discussed later in this section.

Definition of net sand allows determination of net sand in a well for each discrete zone. Net thickness isopachs may be prepared, as in Fig. 7.119 for individual zones. A map of the depositional model often is used as a

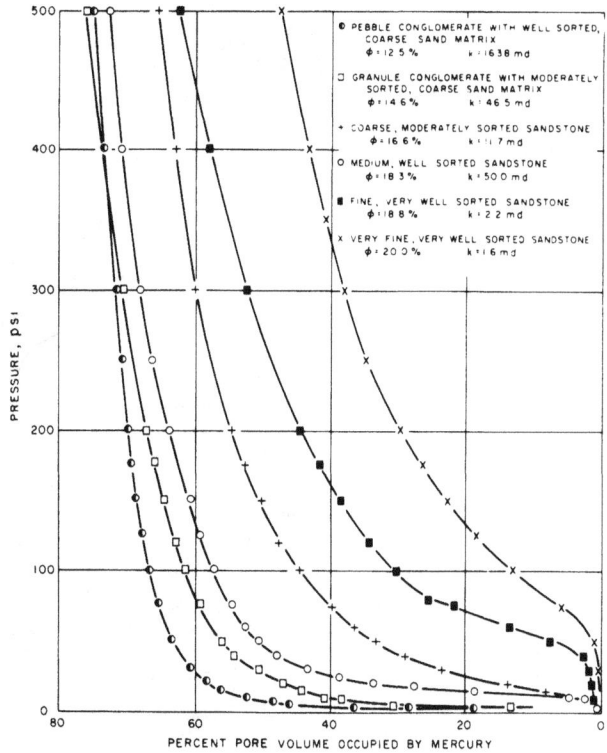

Fig. 7.118—Capillary-pressure curves of typical reservoir rock types.[38]

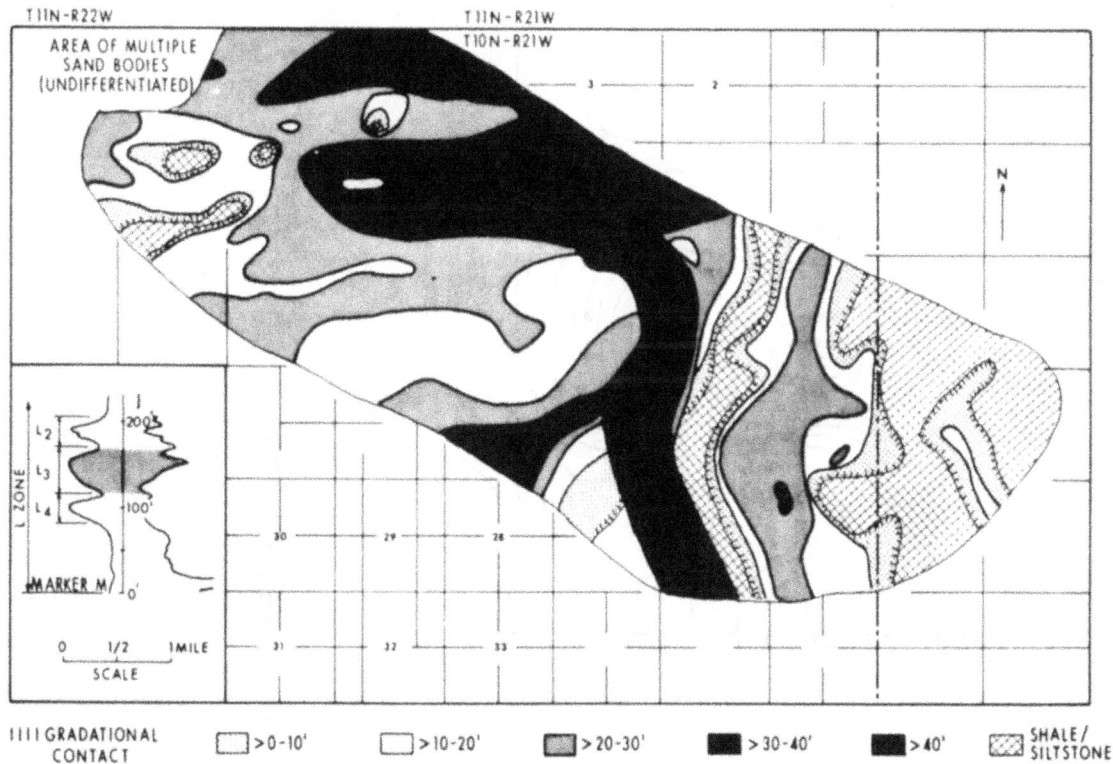

Fig. 7.119—Thickness map of net sand, Subzone L3, Elk City field.[38]

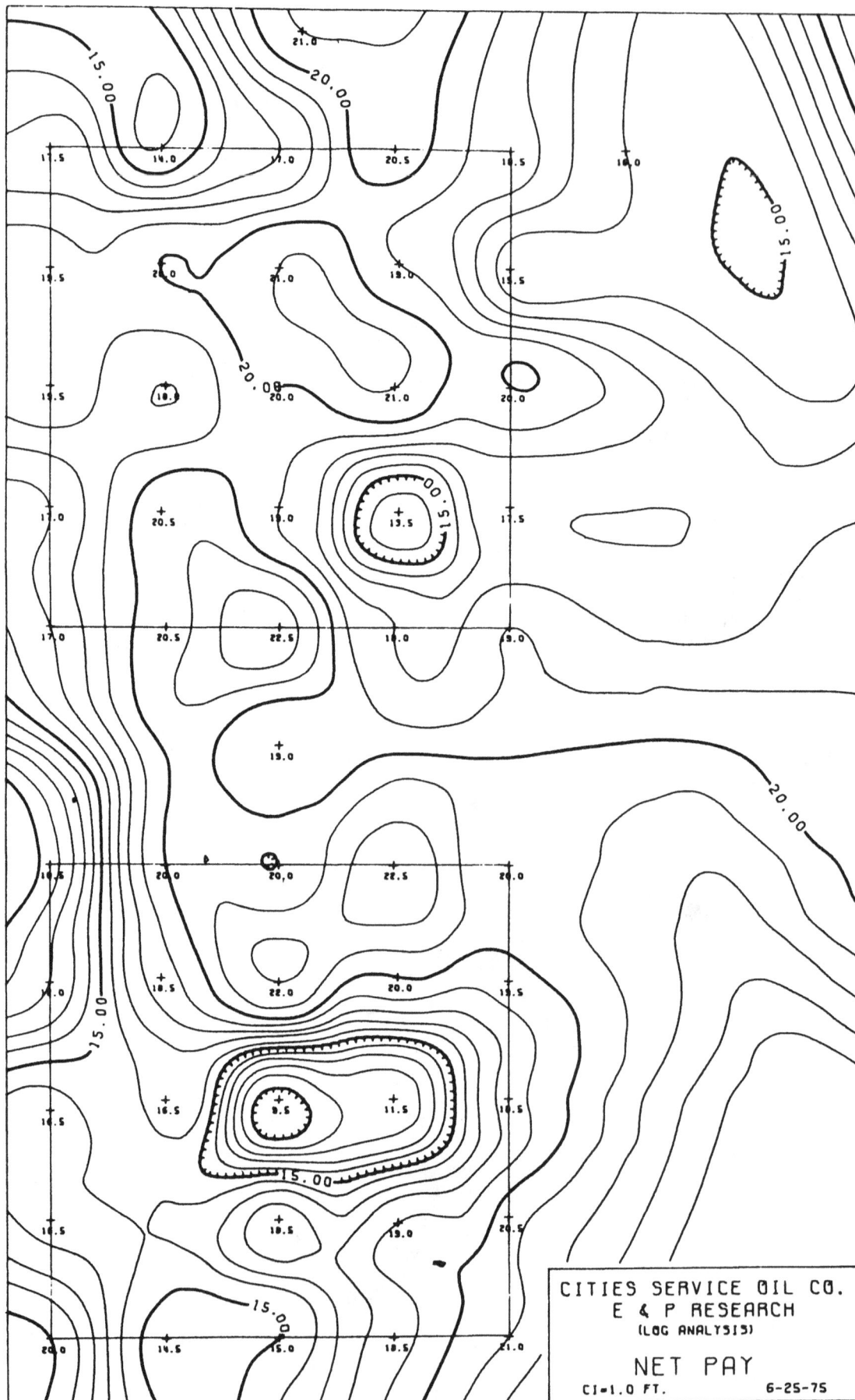

Fig. 7.120—Computer-generated contour map of net pay, El Dorado pilot area, KS.[90]

"guide" in drawing contour lines, particularly in geologically complex regions.[21] A certain amount of judgment, experience, and intuition is inherent in this process when contour lines are drawn manually. Computer programs that prepare isopachs according to fixed algorithms are available. Fig. 7.120 is a typical contour map prepared by computer analyses.[90] These programs seldom incorporate depositional considerations and should be used only when such additional information on geology is not available.

Fluid Saturations. Fluid saturations usually are estimated for each productive interval from interpretation of well logs and cores. While it is common practice to run one or more logs in wells when drilled, the expense of coring and analysis often results in a limited coring program. There may be no cores, so estimates of saturation must be based on log analysis.

When core analyses are available, fluid saturations rarely add up to one because of loss of fluids by expansion or flushing during coring operations. Other considerations are thermal contraction as the core is brought from the reservoir to the surface and changes in porosity when the core is removed from the net confining pressure found in the reservoir. The engineer is always faced with the problem of what to do with the vacant space. Is vacant space oil, gas, or water? Unfortunately, there are no easy answers to this question. Multiple approaches must be used to develop a consistent method of determining fluid saturations.

TABLE 7.23—TYPICAL CORE ANALYSIS RESULTS: IMPERIAL SIMMONS 67[*][93]

Sample No Core 2	Depth (ft)	Brine	Oil	Coring Fluid	Porosity (%)	Rock Type[**]
9	3,169.50 to 69.75	60.0	39.4	0.6	1.8	G
10	70.75 to 71.25	48.5	43.2	8.3	5.7	G
11	71.25 to 71.65	61.4	37.5	1.1	3.9	G
12	71.65 to 72.10	32.6	66.3	1.1	6.4	G
13	72.25 to 72.60	32.4	66.5	1.1	7.5	G
14	72.60 to 72.90	43.5	52.5	4.0	7.2	G
15	73.05 to 73.50	36.8	50.2	13.0	5.3	G
16	73.50 to 73.95	33.8	51.3	14.9	4.4	G
17	73.95 to 74.40	27.8	54.2	18.0	5.9	V
18	74.40 to 75.15	24.2	54.6	21.2	5.4	V
19	75.15 to 75.60	12.4	70.2	17.4	7.4	V
20	75.60 to 76.00	13.6	64.5	21.9	5.6	V
21	76.00 to 76.45	43.6	30.0	26.4	2.7	V
23	77.10 to 77.35	20.9	45.2	34.9	6.2	V

*Saturations, % PV
**G = granular and V = vuggy

The preferred approach is to calibrate well logs with core data. Calibration is done by the determination of a cored interval where the water saturation and porosity are known. Generally, this interval is nearly 100% saturated with formation water, so the true formation resistivity can be obtained directly. Porosity logs are calibrated against core porosities that have been adjusted for the effects of net confining pressure. Water saturations are computed from resistivity logs with Archie's equation or other appropriate equations. Fig. 7.121 is an example of a correlation between resistivity and hydrocarbon-filled PV, which was derived from log calculations.[75] This correlation was used to assign initial hydrocarbon saturations to the productive intervals representative of the geologic interval used to develop the correlation.

Coring techniques[91] (such as pressure core barrel or sponge coring[92]) have been developed to minimize or to allow determination of the effects of fluid expansion on saturations obtained from core analyses. Special precautions may be necessary to obtain cores with minimum contamination. Willmon[93] describes such a coring program to evaluate the displacement efficiency in the Redwater field, Alta., Canada, which produced under a strong bottomwater drive. The productive interval is the Leduc limestone formation, an Upper Devonian reef. Cores were taken with a pressure core barrel and a low-invasion oil-based drilling fluid tagged with tritiated hexadecane to evaluate mud filtrate invasion. The pressure core barrel was frozen before removal from the core barrel to prevent fluid expansion loss. Table 7.23 shows core analyses from the oil zone unadjusted for displacement caused by the drilling fluid. Because the mud filtrate was thought to displace a comparable volume of mobile oil, the water saturations in Table 7.23 were considered to be connate water. Note that connate saturation varies widely with rock type (granular or vuggy) and porosity.

Further core analyses were conducted on cores taken from the Redwater reservoir to determine capillary pressure curves for both imbibition and drainage cycles with the restored state technique on preserved and weathered cores. Results of these tests are summarized in Table 7.24. When all data were grouped according to rock type, corre-

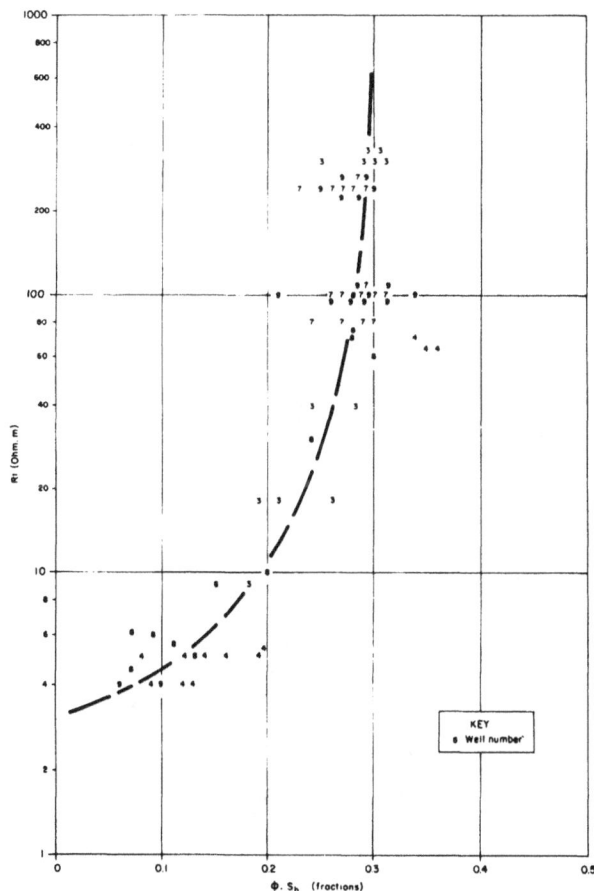

Fig. 7.121—Trend of resistivity vs. hydrocarbon-filled PV.[75]

TABLE 7.24—CAPILLARY PRESSURE TEST RESULTS[93]

Plug	Type*		Porosity (%)	k_{N2} (md—H or V)	S_{cw} at P_c (% PV)	(psi)	S_{wr} at P_r (% PV)	(psi)	Recovery (% HCPV)	Spontaneous Imbibition of Brine (% PV)
NP	H	Granular	17.8	61.6	22.0	39	49.1	56	37.0	3.74
P	H	Vuggy	17.5	49.7	16.0	56	30.4	56	63.8	4.56
P	H	Vuggy	18.2	118.6	13.9	39	35.9	56	58.3	9.70
P	H	Vuggy	13.9	917.0	19.1	39	35.3	56	56.3	12.82
P	H	Granular	12.6**	5.0**	38.1	39	34.5	56	44.2	2.72
P	V	Granular	22.4	114.1	4.7	†	37.8	56	60.3	7.05
P	V	Granular	18.7	44.0	25.6	†	41.1	56	44.7	2.94
P	V	Granular	19.0	67.4	21.5	†	22.2	56	71.7	2.28
P	V	Granular	18.4	39.2	26.7	†	29.2	56	60.1	3.60
NP	V	Granular	11.3	32.4	16.3	56	29.7	56	64.5	0.00
NP	V	Granular	18.7	159.4	11.9	39	22.9	56	74.0	0.97
NP	V	Granular	10.2	57.2	29.1	56	23.3	56	67.2	1.05
NP	H	Granular	12.0	315.1	13.9	39	16.1	56	81.4	1.50
NP	H	Granular	11.6	43.2	33.8	56	25.7	56	61.1	1.92
NP	H	Granular	22.0	527.0	16.8	39	20.0	56	76.0	1.14
NP	H	Granular	11.1	5.6	22.0	39	49.1	56	37.0	3.74
NP	H	Granular	17.8	61.6	26.4	39	(20.2)	‡	(72.5)	0.93

*P = preserved, NP = nonpreserved, V = vertical, and H = horizontal.
**Estimated.
†S_{cw} established by flushing.
‡Extrapolated.

lations of interstitial water saturation with porosity were developed for the two principal rock types in this reservoir. These correlations, shown in Figs. 7.122 and 7.123, permit assignment of average interstitial water saturation to individual positions in the reservoir.

When core data are available but interpretation of logs is difficult because of thin bed or other effects, it may be possible to estimate initial saturations by reconstructing the history that the core went through during the coring process and by making assumptions about saturation changes that are consistent with the properties of the drilling fluid, FVF's of oil and water, and the temperature and pressure of the reservoir. This is always a difficult method. For example, if the core was taken with an oil-based drilling fluid, the vacant space is not likely to be water saturation unless water is mobile and is displaced by filtrate loss. Thus, the water saturation is considered to be interstitial water. This can be confirmed by independent measurement of interstitial water saturation from capillary pressure tests. Coring is frequently performed with water-based coring fluids. When no precautions have been taken to control fluid loss, the water saturation usually will be higher than the true saturation. Fluid saturations can be estimated from core analyses in this case only, assuming the reservoir was at interstitial saturation and using capillary pressure, S_w, or porosity, S_{iw}, correlations. There may be significant uncertainty in the values of S_w. Sometimes average interstitial water saturation can be estimated by material balance if water production during a depletion drive is negligible.

Also, water saturation distributions have been estimated by drainage capillary pressure data in reservoirs that have oil/water or gas/oil contacts by assuming the reservoir is in capillary equilibrium at the time of discovery. This approach assumes that the reservoir was originally water wet when the oil first saturated the rock and that changes in wettability that occurred after original migration did not alter the saturation distribution or the capillary pressure curve. Drainage capillary pressure curves

are used to assign saturations as illustrated in Problem 2.11 for a two-layered reservoir. As the number of discrete layers with unique capillary curves increases, applying this method without deriving an averaging technique to represent the data becomes increasingly difficult. As a result, assignment of initial water saturation distributions based solely on capillary equilibrium and capillary pressure curves is not universally accepted.

In depletion drive reservoirs, a gas saturation is present at the beginning of a waterflood. The average saturation in the vicinity of each well can be estimated from gas/oil relative permeability curves using the GOR at the end of primary production. Gas saturations may also be estimated by material balance if interstitial water saturation is known.

It should be apparent that porosity, S_w, correlations for distinct rock types provide a porosity cutoff that corresponds to a water saturation where the oil is immobile. Determination of this saturation requires relative permeability data or generalized correlations. Fortunately, in most core analyses, endpoint saturations and water permeability can be determined quite accurately from flood pot tests.

Porosity and Permeability. Porosities for each zone are determined from core analysis or porosity logs, as discussed earlier. Contour maps of porosity can be generated in the same manner as described for net sand. Fig. 7.124 is a porosity map for the Chesney-Hegberg pilot area of the El Dorado field in Kansas.

Permeabilities are assigned to each zone in a well from core analyses, if available, or from permeability/porosity correlations developed in rock studies. Permeability may also be assigned by the development of correlations between permeability derived from core analyses and log-derived water saturations, as in Fig. 7.125. Permeability maps are often constructed for individual zones (Fig. 7.126). Assignment of permeabilities for a particular well can be checked against values of kh obtained from pressure buildup/falloff analyses. Fig. 7.127 compares the re-

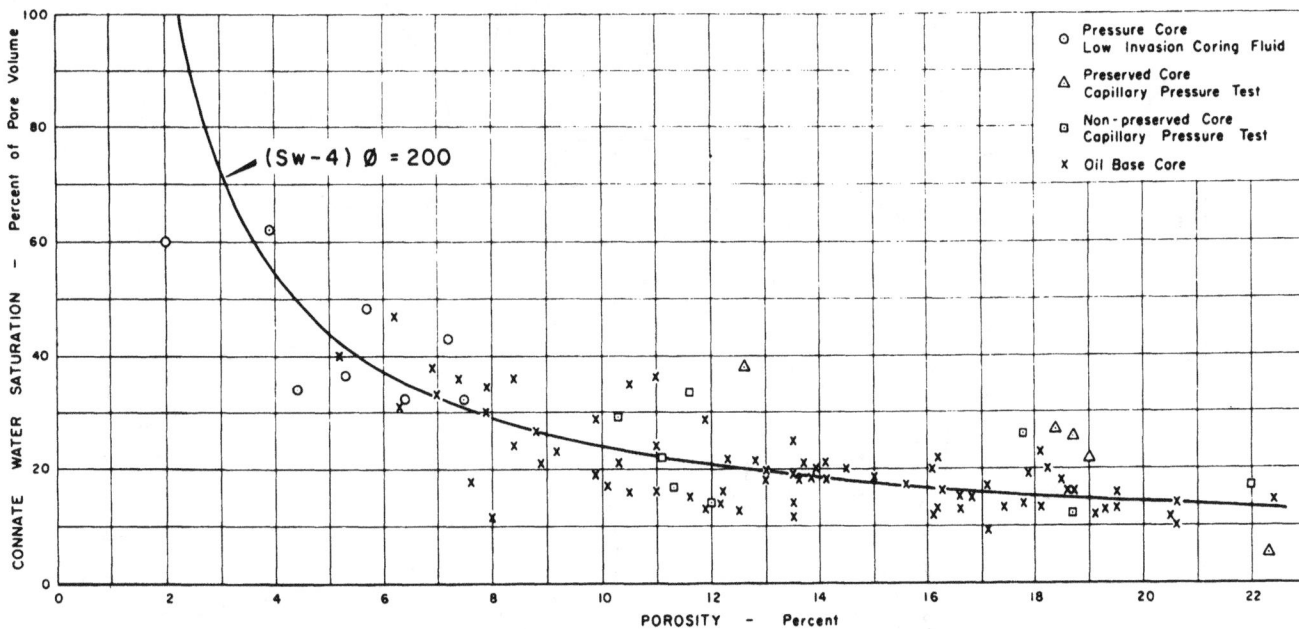

Fig. 7.122—Correlation of interstitial water saturation with porosity, granular-type rock. [93]

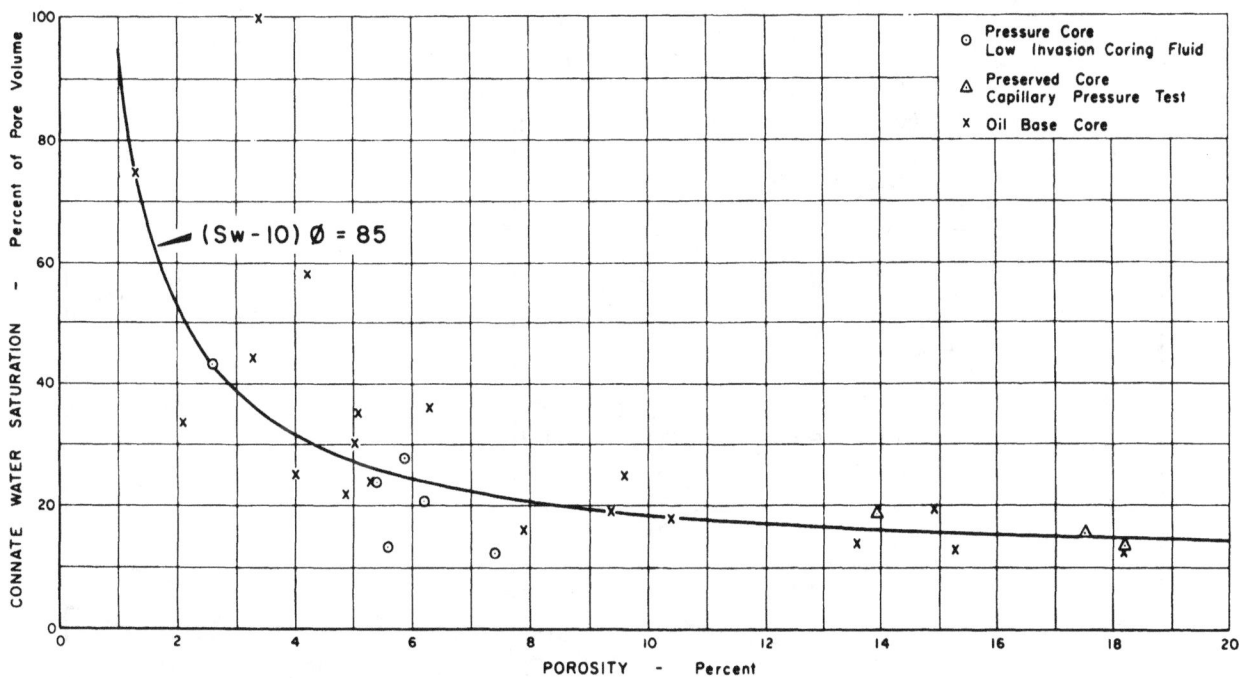

Fig. 7.123—Correlation of interstitial water saturation with porosity, vuggy-type rock. [93]

Fig. 7.124—Porosity map for pilot area, El Dorado field, KS.[90]

sults of tests from the Jay-LEC field.[27] Good correlations between pressure buildup results and core analyses were obtained when kh was computed from core analyses through arithmetic averaging. When values of kh do not agree, it is necessary to adjust k or h, or both.

Once the permeability distribution is determined within a specific geologic zone, subdividing the zone further for simulation purposes may be necessary. Subdivision based on the Dykstra-Parsons coefficient of permeability variation was presented in Sec. 5.8.1. This method does not retain information on the orientation of one layer to another. When crossflow can occur between naturally occurring layers, it is necessary to subdivide the zone into intervals that are physically adjacent. While these zones may sometimes be selected by eye from permeability breaks, in many cases there are no obvious separations between layers.

When there is considerable variability in reservoir properties within facies, a different subdivision technique is proposed by Hearn et al.[80] The reservoir is subdivided into "flow units." Flow units are regions in the sedimentary sequence that are judged to control the movement of injected and produced fluids within the reservoir. They are selected with the choice of certain ranges of porosity and permeability. Selection is subjective but is used to segregate contiguous regions of the reservoir by their capability to transmit fluids laterally and vertically. Application of the flow unit concept is illustrated through a detailed study of the Hartzog Draw field.[80]

Testerman[94] describes a statistical reservoir zonation technique that retains the actual location of the strata within the reservoir. The procedure involves zonation of permeability data from the top to the bottom of the interval. Statistical tests are used to minimize the variation of permeability within zones and to maximize the variation in permeability between zones. After all the wells have been zoned, a second comparison is made to correlate zones from well to well throughout the reservoir. Zones in adjacent wells are considered to be continuous (i.e., connected) if the difference in mean permeability of two zones in adjoining wells is less than the difference expected from variations of measurements within zones. Although Testerman used the arithmetic mean permeability to compare zones, analogous comparisons can be made with geometric mean permeabilities that are more consistent with the distribution found in many porous rocks. Because there are no geological parameters in the statistical evaluation, judgment is necessary to determine whether the derived zonation is consistent with the geological model.

As noted in Rock Studies, permeability varies with position and direction. In that section, directional permeability referred to variations in permeability that were observed on the scale of a core sample. Directional permeability usually refers to a pronounced permeability trend within the reservoir or a relatively large portion of a reservoir. At one time, reservoirs were thought to be fairly homogeneous aerially, and directional trends were considered unique in certain reservoirs. As detailed knowledge of reservoirs and depositional environments increases, it is perhaps more appropriate to consider all reservoirs as heterogeneous with some directional permeability.

Directional permeability is an important factor in the development of a reservoir description for a waterflood or other fluid displacement processes. Identification of

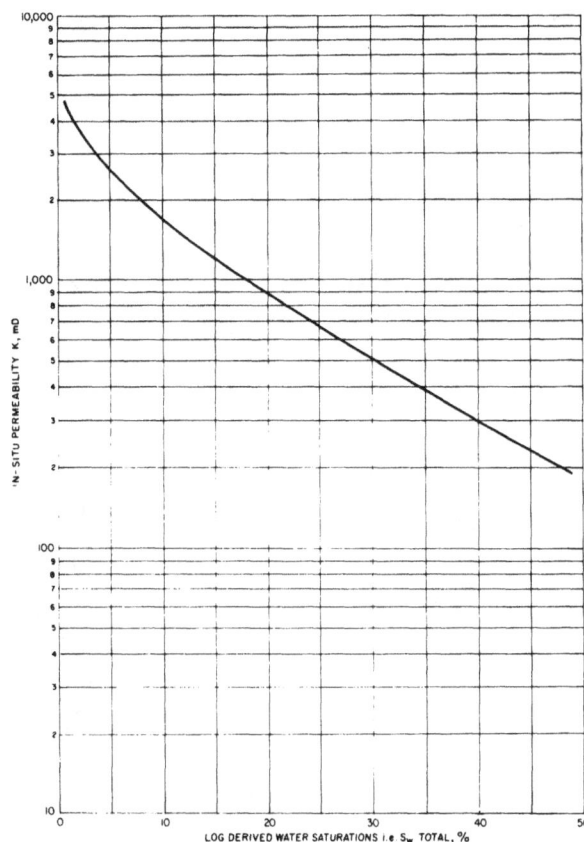

Fig. 7.125—Trend of permeability vs. water saturation for Well 8.[75]

directional permeability and mapping of trends often use data from several sources. Historically, directional permeability often was discovered in the early stages of a pilot flood when water unexpectedly broke through into some producing wells and never seemed to reach others. One of the earliest examples of a directional permeability trend caused by natural fractures in the North Burbank field was discussed in Chap. 6. The trend was discovered early in the waterflood development (1951) and was confirmed by use of water-soluble dyes as a tracer.[51] A subsequent study by Hagen[95] suggested that the subsurface fractures were part of an extensive joint system that was visible on the ground and from the air. The average direction of the joints as interpreted from aerial photographs was N70°E, which agrees closely with the observed subsurface directional permeability trend.

The Booch sand unit in southern Oklahoma produces from a channel deposit. Fig. 7.128 shows the net sand isopach of the Booch sand unit.[96] Wells 12, 17, and 21 were selected for injection in the pilot flood because these wells were located along the axis of the sand body, and their directional permeability, if any, was assumed to be parallel to the axis of the deposit. Injection water from Well 12 broke into Well 11 within 30 days after the beginning of injection. Within 9 months of injection, production from Well 15 went from 1 to 2 BOPD [0.2 to 0.3 m³ d oil] and 0 to 76 BWPD [0 to 12 m³/d water]. Oil production increased to 4 B/D [0.6 m³/d], and water production decreased to 8 B/D [1.3 m³/d] immediately after Well 17 was shut in. Preferential movement of water in the east/west direction in the vicinity of Wells 12 and 17 indicates directional permeability in those areas but not its cause.

Fig. 7.126—Permeability distribution, El Dorado pilot area, KS. [90]

A different situation occurred in the Spraberry trend area in Texas where reservoirs produced from a tight, fractured sand. Oriented cores in the Spraberry trend area indicated that directional permeability should exist. Results of a pilot waterflood showed preferential movement of water through a major fracture system that was oriented in a northeast/southwest direction.[97] Fig. 7.129 illustrates streamlines traced by water moving from injection Wells B-2, B-4, B-6, and B-10 in the pilot area after 2 years of injection. From this match of computed and observed water breakthrough, the ratio of permeability along the major fracture axis to that perpendicular to the major axis was found to be about 144:1.

Directional permeability may be determined in three ways: (1) interpretation of deposition and sedimentary structures, (2) well testing, and (3) analysis of oriented cores. Experience with certain types of geologic formations indicates when to anticipate directional permeability. Pressure pulse testing is a powerful tool in evaluating directional permeability and kh between well pairs because existing wells are used with minimal disruption of normal operations.[85-88] In reservoirs where the method can be used, pulse testing[98,99] can be performed at any stage of primary or secondary operation. Testing of a well pair may be done in a few hours to a few days. It is also possible to determine vertical permeability in a well through vertical pulse testing. This technique should be considered whenever the determination of directional permea-

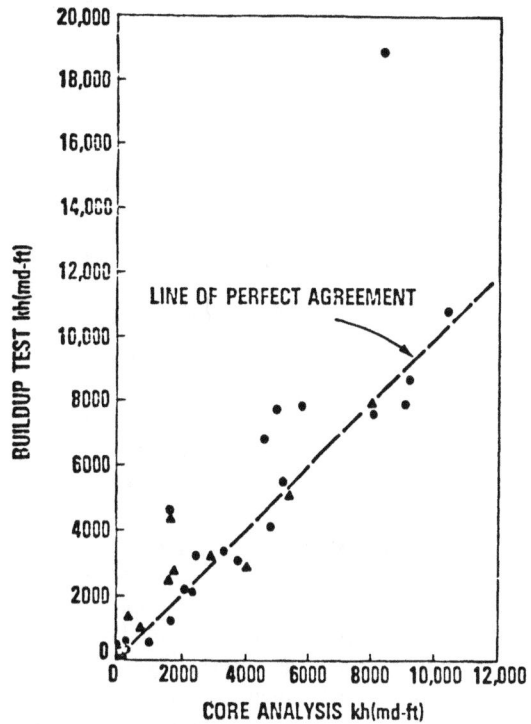

Fig. 7.127—Comparison of core analysis and buildup kh, Jay-LEC fields.[27]

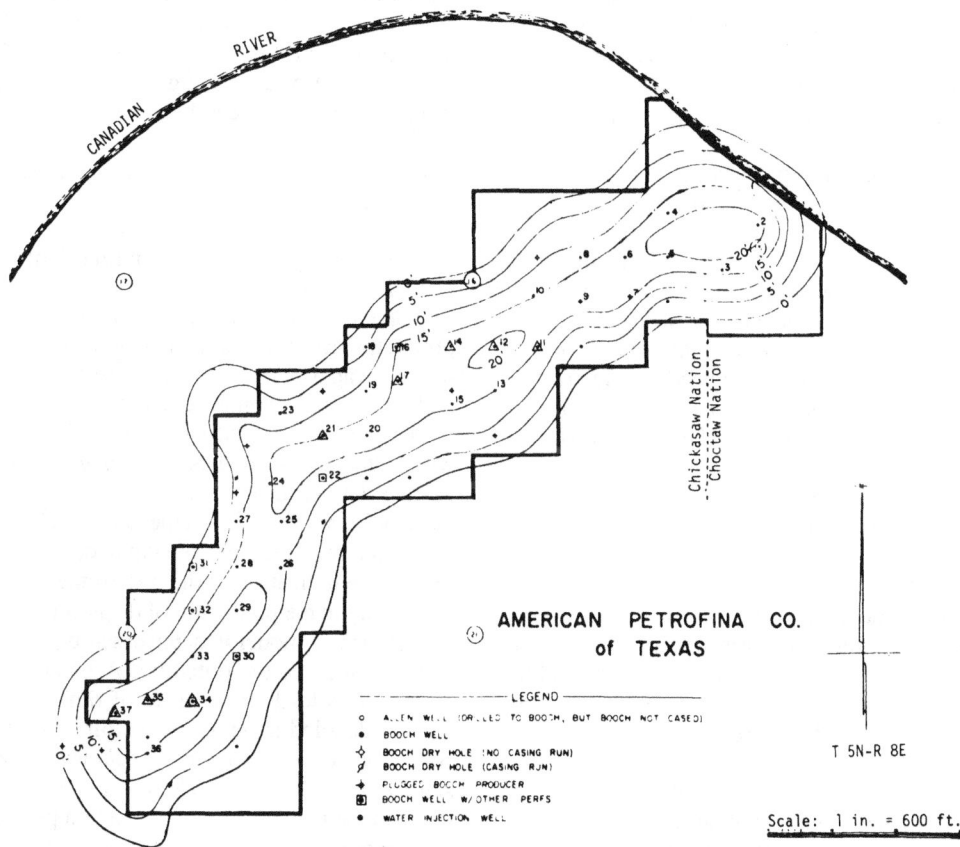

Fig. 7.128—Isopachous map, Booch sand reservoir.[96]

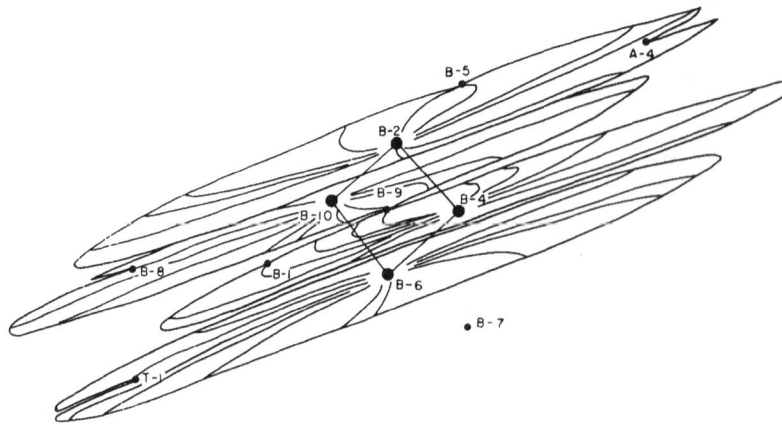

Fig. 7.129—Calculated position of flood front as of March 1957.[97]

bility is important. Vertical permeability may also be evaluated with the repeat formation tester.[100] Directional permeability may be identified also through analysis of a set of oriented cores. Normally, oriented cores are obtained to confirm directional permeability trends that have been discovered by other methods. Greenkorn *et al.*[101] show how oriented cores can be studied to deduce directional permeability trends.

Assignment of rock and fluid properties throughout a reservoir often has subjective elements that depend on the collective experience and judgment of the engineer/geologist team. This is an area where research and development are under way to develop better quantitative tools to assign properties spatially with limited data.

Integration Studies. Integration studies are the last step in developing a reservoir description. In this stage, data and results of rock, framework, and reservoir quality studies are integrated to obtain a three-dimensional description of the reservoir. This description is often the input to mathematical models simulating primary production. A match of pressure/production history during primary depletion serves as verification of the reservoir description or indicates areas where additional revision of the reservoir description is needed. Frequently, this is an iterative process in which geologic maps may be adjusted, reservoir data re-evaluated, or new data obtained. *Integration studies are aimed at evaluating the consistency of reservoir descriptions with known production performance.* An example showing the results of integration studies was presented in Chap. 6, where we illustrated how numerical simulators were used to evaluate waterflooding in the West Seminole field. Details of the simulation and geological considerations are found in Refs. 102 and 103.

Several activities must be completed to prepare data for mathematical simulation. Maps of reservoir PV, hydrocarbon PV, and permeability thickness usually are required in simulation studies. These maps can be constructed by cross contouring two parameters, such as porosity and thickness. Cross contours often are prepared by the overlaying of porosity and thickness isopachs on a light table and the construction of ϕh contours from values at the intersections. Transparencies may be used instead of a light table. Cross contouring also can be accomplished with computer mapping programs. Fig. 7.130 is a computer-generated ϕh map of the pilot area in the El Dorado field in Kansas corresponding to net thickness and porosity maps in Figs. 7.120 through 7.125. Other parameters can be crossplotted in the same manner. Fig. 7.131 is a permeability-index/thickness contour map from the same field. Fig. 7.132 shows the distribution of oil in the area.

The development of a reservoir description following the methodology presented in Table 7.19 requires extensive cooperation of both engineers and geologists. Fig. 7.133 summarizes the interplay between geologists and engineers in each study area as the reservoir description is developed.

Evolution of the Reservoir Description. A reservoir description is rarely completed at one time. The description evolves as more information becomes available at discrete times in the life of a reservoir. Fig. 7.134 illustrates the evolution of the zonation of the Swan Hills unit over a period of 19 years as engineers and geologists sought to reconcile differences between performance and predictions.[63]

There are several reasons why reservoir descriptions evolve over a period of time. Time and cost are two factors that enter into decision-making. Operators have different perceptions of the minimum information needed to develop a reservoir. When a reservoir is discovered, principal concerns are the size of the deposit, saturation, production rates, initial reservoir pressure, and presence of a gas cap and/or aquifer. These factors are needed to make an initial economic evaluation.

Design of waterfloods and other EOR processes is done with mathematical models of varying degrees of complexity and data requirements. There are tradeoffs between the sophistication of the mathematical model and the reservoir data required for the input data. Table 7.25 is a qualitative indication of the requirements of models used commonly in the prediction of displacement performance. The choice of model is determined by the data available and the value of the outcome. A mismatch of data and model often leads to a waste of time and money.

How detailed should a reservoir description be for a waterflood design? *History has shown that reservoirs are seldom over-described.* Therefore, a design team should develop the best reservoir description possible, considering the potential value of the waterflood, before a reservoir model is selected. The reservoir description then becomes a principal factor in selecting the model.

Fig. 7.130—Porosity/thickness map, El Dorado pilot area, KS.[90]

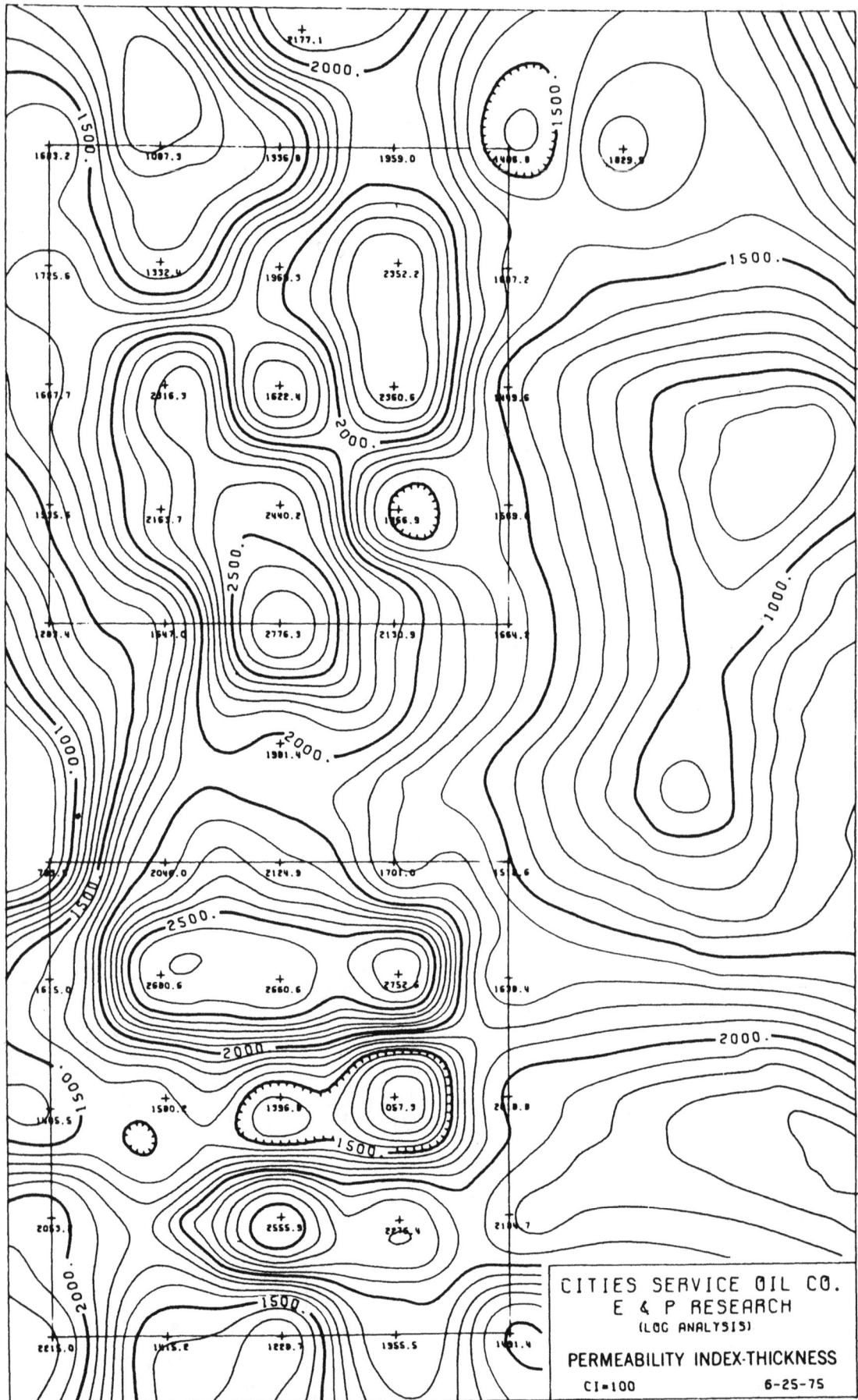

Fig. 7.131—Permeability index/thickness map, El Dorado pilot area, KS.[90]

Fig. 7.132—Distribution of oil in pilot area, El Dorado field, KS. [90]

TYPE OF GEOLOGIC ACTIVITY

ROCK STUDIES
- LITHOLOGY
- DEPOSITIONAL ORIGIN
- RESERVOIR ROCK TYPES

FRAMEWORK STUDIES
- STRUCTURE
- CONTINUITY
- GROSS THICKNESS TRENDS

RESERVOIR QUALITY STUDIES
- QUALITY PROFILES
- RESERVOIR ZONATION
- NET THICKNESS TRENDS

INTEGRATION STUDIES
- PORE VOLUME
- TRANSMISSIBILITY

EXAMPLES OF INTERPLAY OF EFFORT

CORE ANALYSIS

WELL TESTING

PRESSURE PRODUCTION HISTORY MATCH

Fig. 7.133—Interplay of geological and engineering activities in the development of a reservoir description. [21]

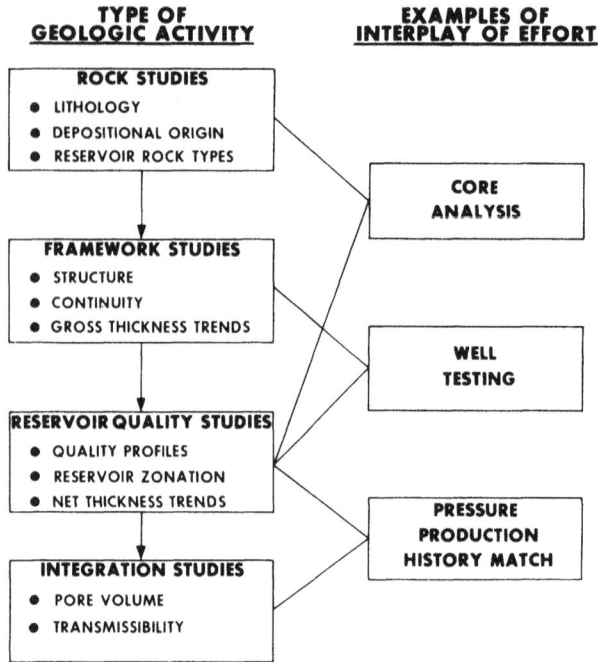

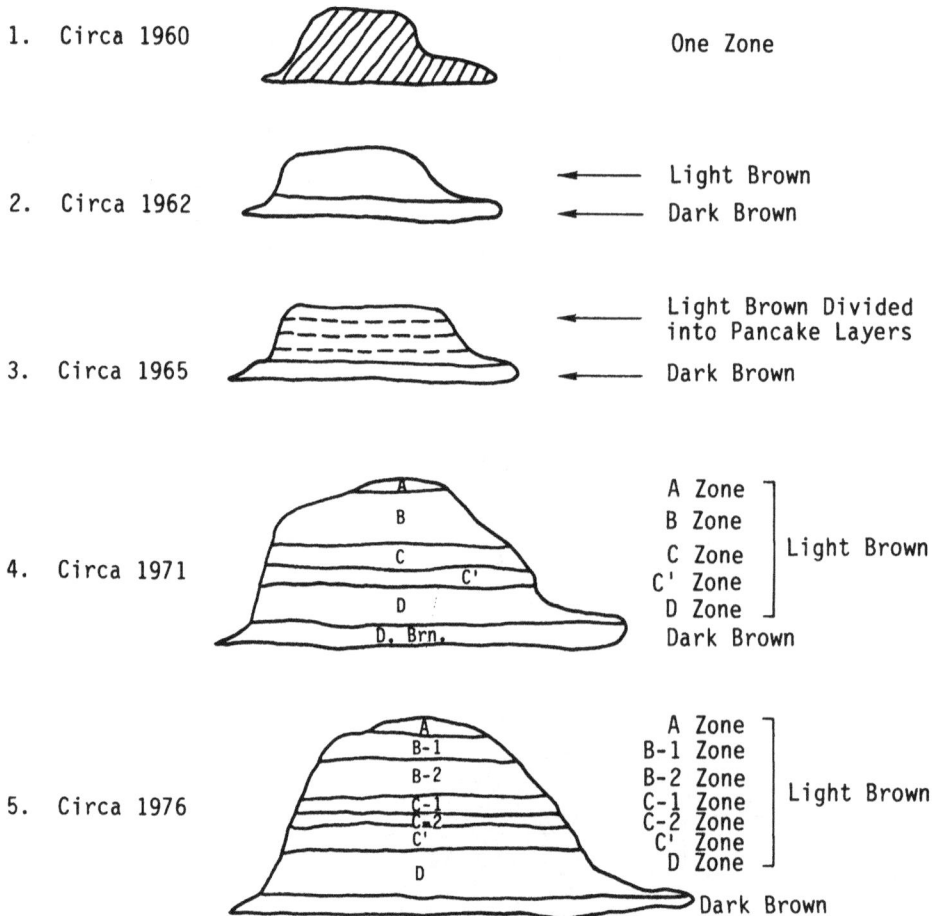

The examples presented in this chapter represent a small fraction of reservoir studies in the literature. Our emphasis was to introduce a methodology that can lead to a reservoir description consistent with known reservoir performance. Refs. 104 through 130 describe other reservoir studies.

TABLE 7.25—EVALUATION OF DATA REQUIREMENTS AND CAPABILITIES OF RESERVOIR SIMULATORS*

	Models and Constraints			
	Requirements			
Calculation Method	Data	Time	Cost	Proposed Value
Correlations	Minimal	Minimal	Minimal	Minimal
Stream tube	Reasonable	Short term	Low	Low-high
2D black oil	Major	Long term	High	High
3D black oil	Major	Long term	High	High
Compositional simulator	Massive	Long term	Very high	High

*Personal communication with H.M. Staggs, ARCO Oil and Gas Co., 1983.

1. Circa 1960   One Zone

2. Circa 1962   ← Light Brown
                          ← Dark Brown

3. Circa 1965   ← Light Brown Divided into Pancake Layers
                          ← Dark Brown

4. Circa 1971
    A
    B
    C
    C'
    D
    D. Brn.

    A Zone
    B Zone
    C Zone   Light Brown
    C' Zone
    D Zone
    Dark Brown

5. Circa 1976
    A
    B-1
    B-2
    C-1
    C-2
    C'
    D

    A Zone
    B-1 Zone
    B-2 Zone
    C-1 Zone   Light Brown
    C-2 Zone
    C' Zone
    D Zone
    Dark Brown

Fig. 7.134—Changes in engineering concepts of pay distribution, Swan Hills formation. [63]

References

1. Levorsen, A.I.: *Geology of Petroleum,* second edition, Wolf, Free-man and Co., San Francisco (1967).
2. Minor, H.E. and Hanna, M.A.: "East Texas Field, Rusk, Cherokee, Smith, Gregg, and Upshur Counties, Texas," *Stratigraphic Type Oil Fields,* AAPG, Tulsa, OK (1941) 625-26.
3. Rumble, R.C., Spain, H.H., and Stamm, H.E.: "A Reservoir Analyzer Study of the Woodbine Basin," *Trans.,* AIME (1951) **192,** 331-40.
4. Gruy, H.J.: "Thirty Years of Proration in the East Texas Field," *J. Pet. Tech.* (June 1962) 577-82.
5. Salathiel, R.A.: "Oil Recovery by Surface Film Drainage in Mixed-Wettability Rocks," *J. Pet. Tech.* (Oct. 1973) 1216-24; *Trans.,* AIME, **255.**
6. DesBrisay, C.L. and Daniel, E.L.: "Supplemental Recovery Development of The Intisar "A" and "D" Reef Fields, Libyan Arab Republic," *J. Pet. Tech.* (July 1972) 785-96.
7. Wayhan, D.A. and McCaleb, J.A.: "Elk Basin Madison Heterogeneity—Its Influence on Performance," *J. Pet. Tech.* (Feb. 1969) 153-59.
8. McCaleb, J.A.: "The Role of Geology in Reservoir Exploitation," notes prepared for AAPG Petroleum Reservoir Fundamentals School (1982).
9. Ghauri, W.K., Osborne, W.K., and Magnuson, W.L.: "Changing Concepts in Carbonate Waterflooding—West Texas Denver Unit Project—An Illustrative Example," *J. Pet. Tech.* (June 1974) 595-606.
10. Ghauri, W.K.: "Production Technology Experience in a Large Carbonate Waterflood, Denver Unit, Wasson San Andres Field," *J. Pet. Tech.* (Sept. 1980) 1493-1502.
11. Shirer, J.A., Ainsworth, W.J., and White, R.A.: "Selection of a Waterflood Pattern for the Jay-Little Escambia Creek Fields," paper SPE 4978 presented at the 1974 Annual Meeting, Houston, Oct. 6-9.
12. Shirer, J.A.: "Jay-Lec Waterflood Pattern Performs Successfully," paper SPE 5534 presented at the 1975 SPE Annual Technical Conference and Exhibition, Dallas, Sept. 28-Oct. 1.
13. Langston, E.P.: "Field Application of Pressure Buildup Tests, Jay-Little Escambia Creek Fields," paper SPE 6199 presented at the 1976 SPE Annual Technical Conference and Exhibition, New Orleans, Oct. 3-6.
14. Shirer, J.A., Langston, E.P., and Strong, R.B.: "Application of Field-Wide Conventional Coring in the Jay-Little Escambia Creek Unit," *J. Pet. Tech.* (Dec. 1978) 1774-80.
15. Langston, E.P., Shirer, J.A., and Nelson, D.E.: "Innovative Reservoir Management—Key to Highly Successful Jay/LEC Waterflood," *J. Pet. Tech.* (May 1981) 783-91.
16. Jardine, D. *et al.*: "Distribution and Continuity of Carbonate Reservoirs," *J. Pet. Tech.* (July 1977) 873-85.
17. Flewitt, W.E.: "Refined Reservoir Description Maximizes Petroleum Recovery," *Oil and Gas J.* (Jan. 5, 1976) 111-16.
18. Moore, R.C.: "Meaning of Facies," *Geol. Soc. Amer. Mem.* (1949) **39,** 1-34.
19. Diehl, A.L.: "The Role of Geology in Enhanced Oil Recovery," *J. Cdn. Pet. Tech.* (April-June 1977) 17-21.
20. Harris, D.G. and Hewitt, C.H.: "Synergism in Reservoir Management—The Geologic Perspective," *J. Pet. Tech.* (July 1977) 761-70.
21. Harris, D.G.: "The Role of Geology in Reservoir Simulation Studies," *J. Pet. Tech.* (May 1975) 625-32.
22. Ebanks, W.J. Jr.: "Geology in Enhanced Oil Recovery," *Proc.,* Second Tertiary Oil Recovery Conference, U. of Kansas, Lawrence (1977) 92-103.
23. LeBlanc, R.J. Sr.: "Distribution and Continuity of Sandstone Reservoirs—Part 1," *J. Pet. Tech.* (July 1977) 776-92.
24. LeBlanc, R.J. Sr.: "Distribution and Continuity of Sandstone Reservoirs—Part 2," *J. Pet. Tech.* (July 1977) 793-804.
25. Hewitt, C.H. and Morgan, J.T.: "The Fry In Situ Combustion Test—Reservoir Characteristics," *J. Pet. Tech.* (March 1965) 337-42; *Trans.,* AIME, **234.**
26. Sneider, R.M., Tinker, C.N., and Meckel, L.D.: "Deltaic Environment Reservoir Types and Their Characteristics," *J. Pet. Tech.* (Nov. 1978) 1538-46.
27. Sneider, R.M., Tinker, C.N., and Richardson, J.G.: "Reservoir Geology of Sandstones," SPE Short Course notes presented at the 1982 SPE Annual Technical Conference and Exhibition, New Orleans, Sept. 26-29.
28. Beard, D.C. and Weyl, P.K.: "Influence of Texture on Porosity and Permeability of Unconsolidated Sand," *Bull.,* AAPG (1973) **57,** 349-69.
29. Pettijohn, F.J., Potter, P.E., and Siever, R.: *Sand and Sandstone,* Springer-Verlag, New York City (1972).
30. Allen, J.R.L.: "A Review of the Origin and Characteristics of Recent Alluvial Sediments," *Sedimentology* (1965) **5,** 191.
31. Blatt, H., Middleton, G., and Murray, R.: *Origin of Sedimentary Rocks,* Prentice Hall Inc., Englewood Cliffs, NJ (1980).
32. Weller, J.M.: "The Geology of Edmonson County," *Ky. Geol. Survey* (1927) **24,** VI, 199-208.
33. Strachan, C.G.: "Pre-Pennsylvanian Channeling in Western Kentucky Units—Connection with Oil Accumulation," *Tulsa Geol. Soc. Digest* (1935) 35-40.
34. Charles, H.H.: "Bush City Oil Field, Anderson County, Kansas," *Stratigraphic Type Oil Fields,* AAPG, Tulsa, OK (1941) 43-56.
35. Ebanks, W.J. Jr. and Weber, J.F.: "Development of a Shallow Heavy Oil Deposit in Missouri," *Oil and Gas J.* (Sept. 27, 1982) 222-34.
36. Weber, K.J.: "Sedimentological Aspects of Oil Fields in the Niger Delta," *Geol. en Mijnbouw* (1971) **50,** 559-76.
37. Bernard, H.A. *et al.*: "Recent Sediments of Southeast Texas," U. of Texas Bureau of Economic Geology, Austin (July 1970).
38. Sneider, R.M. *et al.*: "Predicting Reservoir Rock Geometry and Continuity in Pennsylvanian Reservoirs, Elk City Field, Oklahoma," *J. Pet. Tech.* (July 1977) 851-66.
39. Davies, D.K. and Berg, R.R.: "Sedimentary Characteristics of Muddy Barrier-Bar Reservoir and Lagoonal Trap at Bell Creek Field," paper presented at the 20th Annual Conference of Montana Geologic Society, Eastern Montana Symposium (1969) 97-105.
40. Normark, S.R.: "Fan Valleys, Channels and Depositional Lobes on Modern Submarine Fans: Characteristics for Recognition of Sandy Turbidite Environments," *Bull.,* AAPG (1978) **62,** 912-31.
41. Mutti, E. and Ricci Lucchi, R.: "Le Torbiditi dell'Appennino Settentrionale: Introduzione alla Analisi di Facies," *Mem. Soc. Geol. Italia* (1972) **11,** 161-99; *Intl. Geol. Rev.* (1978) **20,** 125-66.
42. Walker, R.G.: "Deep-Water Sandstone Facies and Ancient Submarine Fans: Models for Exploration for Stratigraphic Traps," *Bull.,* AAPG (1978) **62,** 932-66.
43. Greensmith, J.T.: *Petrology of the Sedimentary Rocks,* sixth edition, George Allen and Unwin/Thomas Murby, Boston (1978).
44. Bouma, A.H.: *Sedimentology of Some Flysch Deposits,* Elsevier Scientific Pub. Co., Amsterdam (1962) 48-54.
45. Walker, R.G.: "Turbidite Sedimentary Structures and Their Relationship to Proximal and Distal Depositional Environments," *J. Sed. Petrology* (1967) **35,** 25-43.
46. Walton, E.K.: "The Sequence of Internal Structures in Turbidites," *Scottish J. Geol.* (1967) **3,** 306-17.
47. Middleton, G.V.: "Experiments on Density and Turbidity Currents, III," *Cdn. J. Earth Sci.* (1967) **4,** 475-505.
48. Klein, G. deV.: *Sandstone Depositional Models for Exploration for Fossil Fuels,* second edition, Burgess Publishing Co., Minneapolis (1980).
49. Hsu, K.J.: "Studies of Ventura Field, California, 1: Facies Geometry and Genesis of Lower Miocene Turbidites," *Bull.,* AAPG (1977) **61,** 137-68.
50. Wilson, J.L.: *Carbonate Facies in Geologic History,* Springer-Verlag, New York City (1975).
51. Dickey, P.A.: *Petroleum Development Geology,* second edition, PennWell Publishing Co., Tulsa (1981).
52. Wilson, J.L.: "Depositional Facies Across Carbonate Shelf Margins," *Trans.,* Gulf Coast Assoc. Geological Soc. (1970) **20,** 229-33.
53. Wilson, J.L.: "Characteristics of Carbonate Platform Margins," *Bull.,* Amer. Assoc. Pet. Geologists (1974) **58,** 810-24.
54. Folk, R.L.: "Practical Petrographic Classification of Limestones," *Bull.,* AAPG (1959) **43,** 1-38.
55. Folk, R.L.: "Special Subdivision of Limestone Types," *Classification of Carbonate Rocks,* W.E. Ham (ed.), AAPG, Mem. 1 (1962) 62-84.
56. Asquith, G.B.: *Subsurface Carbonate Depositional Models: A Concise Review,* PennWell Publishing Co., Tulsa (1979).
57. Dunham, R.J.: "Classification of Carbonate Rocks According to Depositional Texture," *Classification of Carbonate Rocks,* W.E. Ham (ed.), AAPG, Mem. 1 (1962) 108-21.

58. George, C.J. and Stiles, L.H.: "Improved Techniques for Evaluating Carbonate Waterfloods in West Texas," *J. Pet. Tech.* (Nov. 1978) 1547-54.

59. Silver, B.A. and Todd, R.G.: "Permian Cyclic Strata, Northern Midland and Delaware Basin, West Texas and Southeastern New Mexico," *Bull.,* Amer. Assoc. Pet. Geologists (1969) **53**, 2223-51.

60. Meissner, F.F.: "Cyclic Sedimentation in Middle Permian Strata of the Permian Basin, West Texas and New Mexico," *Cyclic Sedimentation in the Permian Basin,* second edition, West Texas Geol. Society, Midland (1972) 203-32.

61. Rieke, H.H. *et al.*: "Classification and Performance of Carbonate Reservoirs," *Oil and Gas Production From Carbonate Rocks,* Elsevier Scientific Pub. Co., New York City (1972) **6**.

62. Hemphill, C.R., Smith, R.I., and Szabo, F.: "Geology of Beaverhill Lake Reefs, Swan Hills Area, Alberta," *Geology of Giant Petroleum Fields,* M.T. Halbouty (ed.), AAPG, Mem. 14 (1970) 50-90.

63. Sears, J.R.: "Improved Reservoir Description: Swan Hills Unit, No. 1," *J. Cdn. Pet. Tech.* (Jan.-March 1978) 43-50.

64. Choquette, P.W. and Pray, L.C.: "Geological Nomenclature and Classification of Porosity in Sedimentary Carbonates," *Bull.,* AAPG (1970) **54**, 207-50.

65. Jodry, R.L.: "Pore Geometry of Carbonate Rocks (Basic Geologic Concepts)," *Oil and Gas Production from Carbonate Rocks,* G.V. Chilingar, R.W. Mannon, and H.H. Rieke III (eds.), Elsevier Scientific Pub. Co., New York City (1972) 35-82.

66. Chilingar, G.V., Bissell, H.J., and Wolf, K.H.: "Diagenesis of Carbonate Rocks," *Diagenesis in Sediments,* G. Larsen and G.V. Chilingar (eds.), Elsevier Scientific Pub. Co., Amsterdam (1967) 179-322.

67. Felsenthal, M. and Ferrell, H.H.: "Fluid Flow in Carbonate Reservoirs," *Oil and Gas Production from Carbonate Rocks,* G.V. Chilingar, R.W. Mannon, and H.H. Rieke III (eds.), Elsevier Scientific Pub. Co., New York City (1972) 83-137.

68. Ebanks, W.J.: "Geology in Enhanced Oil Recovery," *Reservoir Sedimentology and Synergy,* R.W. Tillman and K. Weber (eds.), Soc. Econ. Miner. and Paleon., Special Pub. No. 36, in press.

69. Langston, J.R. and Chin, G.E.: "Rainbow Member Facies and Related Reservoir Properties, Rainbow Lake, Alberta," *Bull.,* AAPG (1968) **152**, 1925-55.

70. Stadler, P.J.: "Influence of Crystallographic Habit and Aggregate Structure and Authigenic Clay Minerals on Sandstone Permeability," *Geol. en Mijnbouw* (1973) **52**, 217-20.

71. Ebanks, W.J. Jr., Euwer, R.M., and N.-Zeller, D.E.: "Mississippian Combination Trap, Bindley Field, Hodgeman County, Kansas," *Bull.,* AAPG (1977) **61**, 309-30.

72. *Well Logging and Interpretation Techniques,* Dresser, Atlas, Dresser Ind. Inc. (1982).

73. Elkins, L.F.: "Evaluation in Determination of Residual Oil Saturation," Interstate Oil Compact Commission, Oklahoma City (1978) 177-90.

74. Doveton, J.H.: *Prolog,* Kansas Geological Survey, Lawrence (1982).

75. Weber, K.J. *et al.*: "Simulation of Water Injection in a Barrier-Bar-Type, Oil-Rim Reservoir in Nigeria," *J. Pet. Tech.* (Nov. 1978) 1555-65.

76. Weber, K.J.: "Influence of Common Sedimentary Structures on Fluid Flow in Reservoir Models," *J. Pet. Tech.* (March 1982) 665-72.

77. Haldorsen, H.H. and Lake, L.W.: "A New Approach to Shale Management in Field-Scale Models," *Soc. Pet. Eng. J.* (Aug. 1984) 447-57.

78. Amyx, J.W., Bass, D.M. Jr., and Whiting, R.L.: *Petroleum Reservoir Engineering,* McGraw-Hill Book Co. Inc., New York City (1960) 142-74.

79. Tinker, G.E.: "Design and Operating Parameters That Affect Waterflood Performance in Michigan," *J. Pet. Tech.* (Oct. 1983) 1884-92.

80. Hearn, C.L. *et al.*: "Geological Factors Influencing Reservoir Performance of the Hartzog Draw Field, Wyoming," *J. Pet. Tech.* (Aug. 1984) 1335-45.

81. Richardson, J.G. *et al.*: "The Effect of Small, Discontinuous Shales on Oil Recovery," *J. Pet. Tech.* (Nov. 1978) 1531-37.

82. Zeito, G.A.: "Interbedding of Shale Breaks and Reservoir Heterogeneities," *J. Pet. Tech.* (Oct. 1965) 1223-28; *Trans.,* AIME, **234**.

83. Verrien, J.P., Courand, G., and Montadert, L.: "Applications of Production Geology Methods to Reservoir Characteristics Analysis From Outcrop Observations," *Proc.,* Seventh World Pet. Cong., Mexico (1967) 425-46.

84. Wadman, D.H., Lamprecht, D.E., and Mrosovsky, I.: "Joint Geological Engineering Analysis of the Sadlerochit Reservoir, Prudhoe Bay Field," *J. Pet. Tech.* (July 1979) 933-40.

85. Johnson, C.R., Greenkorn, R.A., and Woods, E.G.: "Pulse-Testing: A New Method for Describing Reservoir Flow Properties Between Wells," *J. Pet. Tech.* (Dec. 1966) 1599-1604; *Trans.,* AIME, **237**.

86. McKinley, R.M., Vela, S., and Carlton, L.A.: "A Field Application of Pulse-Testing for Detailed Reservoir Description," *J. Pet. Tech.* (March 1968) 313-21; *Trans.,* AIME, **243**.

87. Pierce, A.E.: "Case History: Waterflood Performance Predicted by Pulse Testing," *J. Pet. Tech.* (Aug. 1977) 914-18.

88. Earlougher, R.C. Jr.: *Advances in Well Test Analysis,* Monograph Series, SPE, Richardson, TX (1977) **5**.

89. Stiles, L.H.: "Optimizing Waterflood Recovery in a Mature Waterflood, The Fullerton Clearfork Unit," paper SPE 6198 presented at the 1976 SPE Annual Technical Conference and Exhibition, New Orleans, Oct. 3-6.

90. "El Dorado Micellar-Polymer Demonstration Project," first annual report, BERC/TPR-75/1, Natl. Technical Information Service (Oct. 1975).

91. Sewell, B.W.: "The Carter Pressure Core Barrel," *Drill. and Prod. Prac.,* API (1939) 69.

92. "Dowdco Sponge Coring," Dowdco, Midland, TX (1983).

93. Willmon, G.J.: "A Study of Displacement Efficiency in the Redwater Field," *J. Pet. Tech.* (April 1967) 449-56.

94. Testerman, J.D.: "A Statistical Reservoir-Zonation Technique," *J. Pet. Tech.* (Aug. 1962) 889-93; *Trans.,* AIME, **225**.

95. Hagen, K.: "Mapping of Surface Joints on Air Photos Can Help Understand Waterflood Performance at North Burbank Unit, Osage and Kay Counties, Oklahoma," MS thesis, U. of Tulsa, Tulsa, OK.

96. Lane, B.B.: "Determining a Proper Flow Pattern From a Three-Well Pilot in a Channel Sand," *J. Pet. Tech.* (Feb. 1971) 195-201.

97. Barfield, E.C., Jordan, J.K., and Moore, W.D.: "An Analysis of Large-Scale Flooding in the Fractured Spraberry Trend Area Reservoir," *J. Pet. Tech.* (April 1959) 15-19.

98. Hirasaki, G.J.: "Pulse Tests and Other Early Transient Pressure Analyses for In-Situ Estimation of Vertical Permeability," *Soc. Pet. Eng. J.* (Feb. 1974), 75-91; *Trans.,* AIME, **257**.

99. Falade, G.K. and Brigham, W.E.: "The Analysis of Single-Well Pulse Tests in a Finite-Acting Slab Reservoir," paper SPE 5055B presented at the 1974 SPE Annual Meeting, Houston, Oct. 6-9.

100. Bishlawi, M.: "Pressure Studies Optimize Montrose Field Production," *Oil and Gas J.* (Jan. 26, 1981) 177-87.

101. Greenkorn, R.A., Johnson, C.R., and Shallenberger, L.K.: "Directional Permeability of Heterogeneous Anisotropic Porous Media," *Soc. Pet. Eng. J.* (June 1964) 124-32; *Trans.,* AIME, **231**.

102. Barrett, D.D., Harpole, K.J., and Zaaza, M.W.: "Reservoir Data Pays Off: West Seminole San Andres Unit, Gaines County, Texas," paper SPE 6738 presented at the 1977 SPE Annual Technical Conference and Exhibition, Denver, Oct. 9-12.

103. Harpole, K.J.: "Improved Reservoir Characterization—A Key to Future Reservoir Management for the West Seminole San Andres Unit," *J. Pet. Tech.* (Nov. 1980) 2009-19.

104. Ader, J.C.: "Stratification Testing Results in Revised Concept of Reservoir Drive Mechanism, University Block 13, Ellenberger Field," *J. Pet. Tech.* (Aug. 1980) 1452-58.

105. Behm, E.J. Jr. and Ebanks, W.J. Jr.: "Comprehensive Geological and Reservoir Engineering Evaluation of the Lower San Andres Dolomite Reservoir, Mallet Lease, Slaughter Field, Hockley County, Texas," paper SPE 12015 presented at the 1983 SPE Annual Technical Conference and Exhibition, San Francisco, Oct. 5-8.

106. Bradford, R.N. and Hrkel, E.J.: "Evaluation of a Reservoir Simulation Study: West Civit Dykemann Sand Unit," *J. Pet. Tech.* (Dec. 1979) 1599-604.

107. Buchanan, R. and Hoogteyling, L.: "Auk Field Development: A Case History Illustrating the Need for a Flexible Plan," *J. Pet. Tech.* (Oct. 1979) 1305-12.

108. Burwell, R.B. and Hadlow, R.E.: "Reservoir Management of Blackjack Creek Field," paper SPE 6195 presented at the 1976 SPE Annual Technical Conference and Exhibition, New Orleans, Oct. 3-6.

109. Chakravorty, S.K., Brown, P.R., and Endsin, N.: "A Review of Waterflood Performance in Garrington Cardium A and B Pools, Unit No. 2," *J. Pet. Tech.* (June 1978) 869-76.

110. Chauvin, A.L. *et al.*: "Development Planning for the Statfjord Field Using Three Dimensional and Areal Simulation," paper SPE 8334 presented at the 1979 SPE Annual Technical Conference and Exhibition, Las Vegas, Sept. 23-26.

111. Cordiner, F.S. and Livingston, A.R.: "Tensleep Reservoir Study, Oregon Basin Field, Wyoming—Engineering Plans for Development and Operations, South Dome," *J. Pet. Tech.* (July 1977) 897-902.

112. Craig, F.F. Jr. *et al.*: "Optimized Recovery Through Continuing Interdisciplinary Cooperation," *J. Pet. Tech.* (July 1977) 755-60.

113. Driscoll, V.J.: "Recovery Optimization Through Infill Drilling—Concepts, Analysis, and Field Results," paper SPE 4977 presented at the 1974 SPE Annual Meeting, Houston, Oct. 6-9.

114. Emmett, W.R., Beaver, K.W., and McCaleb, J.A.: "Little Buffalo Basin Tensleep Heterogeneity—Its Influence on Drilling and Secondary Recovery," *J. Pet. Tech.* (Feb. 1971) 161-68.

115. Freedman, R. and Studlick, J.R.J.: "How a Texas Heavy Oil Deposit Was Evaluated," *Oil and Gas J.* (Nov. 30, 1981) 63-76.

116. Grantz, R.E.: "Waterflood History Match Study of the Torchlight Tensleep Micellar Pilot," *J. Pet. Tech.* (Sept. 1981) 1599—1605.

117. Guidroz, G.M.: "E.T. O'Daniel Project—A Successful Spraberry Flood," *J. Pet. Tech.* (Sept. 1967) 1137-40.

118. Harris, G.A.: "History of Development of the Tijerina-Canales-Blucher (Zone 21-B) Field High Pressure Extension Area," paper SPE 5538 presented at the 1975 Annual Technical Conference and Exhibition, Dallas, Sept. 28-Oct. 1.

119. Hewitt, C.H.: "How Geology Can Help Engineer Your Reservoirs," *Oil and Gas J.* (Nov. 14, 1966) 171-78.

120. Irwin, R.A., Tucker, C.W., and Schwartz, H.E. Jr.: "A Case History of the Postle Area—Computer Production Control and Reservoir Simulation," *J. Pet. Tech.* (July 1972) 775-84.

121. Kendall, G.H.: "Importance of Reservoir Description in Evaluating In-Situ Recovery Methods for Cold Lake Heavy Oil, Part 1—Reservoir Description," *J. Cdn. Pet. Tech.* (Jan.-March 1977) 41-47.

122. Knutson, C.F. *et al.*: "Characterization of the San Miguel Sandstone by a Coordinated Logging and Coring Program," *J. Pet. Tech.* (July 1961) 425-32.

123. Lang, R.V.: "Performance Simulation of the Snipe Lake Beaverhill Lake A Pool—A Geologically Complex Reservoir," *J. Cdn. Pet. Tech.* (April-June 1977) 68-90.

124. McLeod, J.G.F.: "Successful Injection Pattern Alternation, Pembina J Lease, Alberta," *J. Cdn. Pet. Tech.* (Jan.-March 1978) 51-55.

125. Morgan, J.T., Cordiner, F.S., and Livingston, A.R.: "Tensleep Reservoir Study, Oregon Basin Field, Wyoming—Reservoir Characteristics," *J. Pet. Tech.* (July 1977) 886-96.

126. Roy, M.G. *et al.*: "Waterflood Redevelopment Prior to Future Tertiary Attempts—A Case History," paper SPE 6460 presented at the 1977 SPE Oklahoma Regional Meeting, Oklahoma City, Feb. 20-22.

127. Thakur, G.G. *et al.*: "Engineering Studies of G-1, G-2 and G-3 Reservoirs, Meren Field, Nigeria," *J. Pet. Tech.* (April 1982) 721-32.

128. Thompson, J.K.: "Formation Analysis and Reservoir Performance Study—Tensleep Reservoir, Hamilton Dome Field," *J. Pet. Tech.* (Feb. 1981) 245-53.

129. Trantham, J.C., Threlkeld, C.B., and Patterson, H.L. Jr.: "Reservoir Description for a Surfactant/Polymer Pilot in a Fractured, Oil-Wet Reservoir—North Burbank Unit Tract 97," *J. Pet. Tech.* (Sept. 1980) 1647-56.

130. Webb, M.G.: "Monarch Sandstone: Reservoir Description in Support of a Steam Flood, Section 26C, Midway-Sunset Field, California," *J. Cdn. Pet. Tech.* (Oct.-Dec. 1978) 31-40.

SI Metric Conversion Factors

$$
\begin{array}{rl}
\text{acre} \times 4.046\ 873 & \text{E}+03 = \text{m}^2 \\
\text{bbl} \times 1.589\ 873 & \text{E}-01 = \text{m}^3 \\
\text{cu ft} \times 2.831\ 685 & \text{E}-02 = \text{m}^3 \\
\text{ft} \times 3.048^* & \text{E}-01 = \text{m} \\
\text{in.} \times 2.54^* & \text{E}+00 = \text{cm} \\
\text{mile} \times 1.609\ 344^* & \text{E}+00 = \text{km} \\
\text{psi} \times 6.894\ 757 & \text{E}+00 = \text{kPa}
\end{array}
$$

*Conversion factor is exact.

Appendix A

Computer Programs

Most of the solution techniques presented in Chaps. 3 through 6 are suited for programming on a digital computer. The required programming skills are well within those expected of engineering students. Development of programs for waterflood calculations serves two purposes. First, a thorough understanding of a method is required to write a satisfactory program. Second, it is possible to study waterflood performance over a wide range of parameters without the tedium of many hand calculations.

The author advocates the development of computer programs as part of a waterflood course. To do this within reasonable time constraints, providing selected "canned" programs is helpful. Five programs are included in this Appendix and are summarized as follows.

SWF. This is a function subprogram that computes the breakthrough saturation and the derivative of the fractional flow curve at breakthrough for both mobile and immobile initial water saturations. Generalized relative permeability functions presented in Example 3.5 are used.

PROP. This subroutine subprogram computes k_{ro}, k_{rw}, f_w, and f'_{Sw} from the generalized relative permeability functions in Example 3.5 for a specific value of S_w when μ_o and μ_w are given.

APVIS. This subroutine subprogram computes the average apparent viscosity for linear displacement following the procedures described in Sec. 3.6.

CGM. This subroutine subprogram computes $Q_i^*/Q_{i}^{*}{}_{bt}$ for the Craig *et al.* model.

SWFW. This function subprogram computes the breakthrough saturation and derivative of the fractional flow at breakthrough when k_{rw} and k_{ro} are expressed as continuous functions of water saturation. The Wegstein method is used to find S_{wf}.

The programs are written in FORTRAN 66 and have been used extensively on the Honeywell 6600. A brief description of each program is given in the following section.

Program Descriptions

SWF. SWF is a function subprogram that computes the breakthrough saturation and f_{Swf} for the generalized form of the relative permeability curves given by Eqs. 3.143 and 3.144:

$$k_{ro} = \alpha_1 (1 - S_{wD})^m$$

and

$$k_{rw} = \alpha_2 S_{wD},$$

where

$$S_{wD} = \frac{S_w - S_{iw}}{1 - S_{or} - S_{iw}}.$$

The program finds the value of S_{wf} that satisfies Eq. A-1.

$$\frac{f_{Swf} - f_{Swi}}{S_{wf} - S_{wi}} = \left(\frac{\partial f_w}{\partial S_w} \right)_{S_{wf}}, \quad \dots\dots\dots\dots\dots (A-1)$$

where S_{wi} is the initial water saturation and f_{Swi} is the fractional flow at the initial water saturation. By substituting Eq. 3.147 into Eq. A-1, Eq. A-2 is obtained.

$$\frac{f_{Swf} - f_{Swi}}{S_{wf} - S_{wi}}$$

$$= \frac{AB[n S_{wDf}^{n-1} (-S_{wDf})^m + m S_{wDf}^n (1 - S_{wDf})^{m-1}]}{[S_{wDf}^n - A(1 - S_{wDf})^m]^2},$$

$$\dots\dots\dots\dots\dots\dots\dots\dots\dots (A-2)$$

where

$$A = \frac{\alpha_1 \mu_w}{\alpha_2 \mu_o}$$

and

$$B = \frac{1}{1 - S_{or} - S_{iw}}.$$

Eq. 3.146 expresses f_w as a function of S_w so f_{Swf} is given by Eq. A-3.

$$f_{Swf} = \frac{S_{wDf}^n}{S_{wDf}^n + A(1 - S_{wDf})^m}. \quad \dots\dots\dots\dots (A-3)$$

Substitution of Eq. A-3 for f_{Swf} in Eq. A-2 yields a complex equation in which S_{wDf} is the only unknown. The value of S_{wDf} is found by Newton's method.[1]

SWF is accessed by the FORTRAN statement given below.

SWBT = SWF (VISO, VISW, SWC, SOR, FPSWF).

When this statement is executed, the parameters ALPHA1, ALPHA2, M, N, SWC, SWI, SOR, VISO, and VISW are read into the subprogram in an interactive mode. The arguments VISO, VISW, SWC, SOR, and FPSWF are returned to the calling program for use there. The parameters ALPHA1, ALPHA2, M, N, SOR1, and SWC1 are placed in a labeled common named PDATA. Values of these parameters are available for another program having the same labeled common. M and N are real variables.

Table A-1 is a sample of the interactive input and output following a call to SWF in a FORTRAN program. SWF checks the value of S_{wf} by computing f_{Swf} with Eqs. A-1 through A-3. SWF also computes the endpoint mobility ratio and the mobility ratio based on the average water saturation behind the flood front. A listing of SWF is listed in Table A-2.

PROP. PROP is a subroutine subprogram that computes k_{ro}, k_{rw}, f_w, and f'_{Sw} from Eqs. 3.143, 3.144, 3.146, and 3.147 when S_w and water and oil viscosities are given:

$$k_{ro} = \alpha_1 (1 - S_{wD})^m,$$

$$k_{rw} = \alpha_2 S_{wD}^n,$$

$$f_w = \frac{S_{wD}}{S_{wD}^n + A(1 - S_{wD})^m},$$

and

$$\frac{\partial f_w}{\partial S_w} = \frac{AB[nS_{wD}^{n-1}(1 - S_{wD})^m + mS_{wD}^n(1 - S_{wD})^{m-1}]}{[S_{wD}^n + A(1 - S_{wD})^m]^2},$$

where

$$A = \frac{\alpha_1 \mu_w}{\alpha_2 \mu_o},$$

$$B = \frac{1}{1 - S_{or} - S_{iw}}, \text{ and}$$

$$S_{wD} = \frac{S_w - S_{iw}}{1 - S_{or} - S_{iw}}.$$

PROP is accessed through the following FORTRAN call statement:

CALL PROP (VISO, VISW, SW1, KRO1, KRW1, FW1, FPSW1).

A call to this subroutine gives values for only the value of SW1 specified in the calling sequence. Thus, PROP must be called for every saturation where values of k_{ro}, k_{rw}, f_w and f'_{Sw} are desired and is used in conjunction with SWF. The values of ALPHA1, ALPHA2, M, N, SOR, and SWC are made available to PROP through the common labeled PDATA. Thus SWF must be executed before PROP is called. The variables KRO1 and KRW1 are real. A listing of PROP is given in Table A-3.

APVIS. APVIS is a subroutine subprogram that computes the average apparent viscosity for a linear system with the methods described in Sec. 3.6. The average apparent viscosity is defined by Eq. 3.130:

$$\overline{\lambda_2^{-1}} = \frac{\displaystyle\int_0^{f'_{Sw_2}} \lambda_r^{-1} df'_{Sw}}{f'_{Sw_2}},$$

where

$$\lambda_r^{-1} = \frac{1}{\dfrac{k_{ro}}{\mu_o} + \dfrac{k_{rw}}{\mu_w}}.$$

TABLE A-1—SAMPLE INTERACTIVE INPUT AND OUTPUT (INTERACTIVE INPUT VALUES IN ITALIC)

INITIAL WATER SATURATION
= 0.363

CONNATE (INTERSTITIAL) WATER SATURATION
= 0.363

RESIDUAL OIL SATURATION
= 0.205

KRO = ALPHA1*(1. – SWD)**M

SPECIFY COEFFICIENT (ALPHA1) OF OIL RELATIVE PERMEABILITY
= 1.0

SPECIFY EXPONENT (M) OF OIL RELATIVE PERM RELATIONSHIP
= 2.56

KRW = ALPHA2*(SWD)**N

SPECIFY COEFFICIENT (ALPHA2) OF WATER RELATIVE PERMEABILITY
= 0.78

SPECIFY EXPONENT (N) OF WATER RELATIVE PERM RELATIONSHIP
= 3.72

OIL VISCOSITY IN CP
= 2.0

WATER VISCOSITY IN CP
= 1.0

MOBILITY RATIO BASED ON ENDPOINTS OF RELATIVE PERMEABILITY CURVES = 0.15600000E 01

SDMAX = 0.99999999E 00   DELSD = 0.20408163E – 01

SWDF = 0.69645134E 00   FOR MOBILITY RATIO = 0.15600000E 01

FWSF = 0.89576355E 00

SWBT = 0.66386698E 00

FRACTIONAL FLOW IS 0. AT INITIAL WATER SATURATION = 0.36300000E 00

FPRIME FROM COMPUTED TANGENT
FPRIME = 0.29772744E 01

FPRIME FROM FW CURVE 0.29772746E 01

FPSWW2 = 0.29772744E 01

The integral in Eq. 3.130 is solved by the trapezoidal rule applied to values of f'_{Sw} from $f'_{Sw} = 0$ to f'_{Sw2}. The subroutine is called as follows:

CALL APVIS (VISW, VISO, KRO, KRW, FPSW, NI, AVIS).

Dummy variables KRO, KRW, FPSW, and AVIS are 1D arrays with dimension NI. KRO and KRW must be real.
It is assumed that the arrays are ordered such that

$$\left.\begin{array}{l} \text{KRO(1)} \\ \text{KRW(1)} \\ \text{FPSW(1)} \end{array}\right\} \text{ correspond to } S_{wf}$$

and

$$\left.\begin{array}{l} \text{KRO(NI)} \\ \text{KRW(NI)} \\ \text{FPSW(NI)} \end{array}\right\} \text{ correspond to } 1 - S_{or}.$$

The arrays KRO, KRW, and FPSW must be filled with the appropriate values when APVIS is called. When execution is completed, the array AVIS will contain the average apparent viscosities corresponding to the saturations

TABLE A-2—LISTING OF SWF

```
10C        PROGRAM TO COMPUTE FRACTIONAL FLOW INFORMATION FOR
20C        RELATIVE PERMEABILITY CURVES INCLUDING MOBILE CONNATE WATER
40C
50C        KRO = ALPHA1*(1.-SWD)**M
60C
70C        KRW = ALPHA2*SWD**N
80C
90C        GPW 11-2-80
100C
110C       TAKEN FROM GSW2SW2
120C
130C       REVISION GPW 10-14-81
140C       FUNCTION SWF(VISO,VISW,SWC,SOR,FPSWF)
150        COMMON/PDATA/ALPHA1,ALPHA2,M,N,SWC1,SOR1
160        DIMENSION SW(101),KRO(101),KRW(101),FW(101),FP(101)
170        DIMENSION SWAVG(10)
180        REAL KRO,KRW,KO,KW,MOCGM,MOBEND
190        Y(SW1) = (SW1-SWC)*B
200        KW(Y) = ALPHA2*Y**N
210        KO(Y) = ALPHA1*(1.-Y)**M
220        G2(SWD1) = A*(B1(SWD1) + B2(SWD1))/B3(SWD1)**2
230        B1(SWD1) = N*SWD1**(N-1)*(1.-SWD1)**M
240        B2(SWD1) = M*SWD1**N*(1.-SWD1)**(M-1)
250        B3(SWD1) = SWD1**N + A*(1.-SWD1)**M
260        G2P(SWD1) = -2.*G2(SWD1)*C1(SWD1)/B3(SWD1)
270&          + A*B1(SWD1)*(C2(SWD1) + C3(SWD1))/B3(SWD1)**2
280        C1(SWD1) = N*SWD1**(N-1) + A*M*(1.-SWD1)**(M-1)
290        C2(SWD1) = ((N-1)/SWD1 + M/(1.-SWD1))*B1(SWD1)
300        C3(SWD1) = (M/(1.-SWD1) + M*(M-1)*SWD1/(N*(1.-SWD1)**2))*B1(SWD1)
310        G3(SWD1) = G1(SWD1)-G2(SWD1)
320        G3P(SWD1) = G1P(SWD1)-G2P(SWD1)
330        G1(SWD1) = (FWF(SWD1)-FWF(SWDI))/(SWDI-SWDI)
340        G1P(SWD1) = (G2(SWD1)-G1(SWD1))/(SWD1-SWDI)
350        FWF(SWD1) = SWD1**N/(SWD1**N + A*(1.-SWD1)**M)
360        SWF1(SWD1) = SWD1/B + SWC
370C
380        PRINT,"INITIAL WATER SATURATION"
390        READ,SWI
400        PRINT,"CONNATE (INTERSTITIAL) WATER SATURATION"
410        READ,SWC
420C
430        PRINT,"RESIDUAL OIL SATURATION"
440        READ,SOR
450        B = 1./(1.-SOR-SWC)
460        SWDI = Y(SWI)
470C
480C
490C
500        PRINT,"KRO = ALPHA1*(1.-SWD)**M"
510C
520        PRINT,"SPECIFY COEFFICIENT (ALPHA1) OF OIL RELATIVE
           PERMEABILITY"
530        READ,ALPHA1
540C
550C
560C
570        PRINT,"SPECIFY EXPONENT (M) OF OIL RELATIVE PERM RELATIONSHIP"
590        READ,M
600C
610C
620
630        PRINT,"KRW = ALPHA2*(SWD)**N"
640C
650C
660
670
690        PRINT,"SPECIFY COEFFICIENT (ALPHA2) OF WATER RELATIVE
           PERMEABILITY"
700        READ,ALPHA2
710C
720C
730
740        PRINT,"SPECIFY EXPONENT (N) OF WATER RELATIVE PERM
           RELATIONSHIP"
```

TABLE A-2—LISTING OF SWF (continued)

```
760C
770         READ,N
780C
790
800
820         PRINT,"OIL VISCOSITY IN CP"
830         READ,VISO
840C
850         PRINT,"WATER VISCOSITY IN CP"
860         READ,VISW
870C
880         A = ALPHA1*VISW/(ALPHA2*VISO)
890         SWDF = SQRT(A/(1. + A))
900         MOBEND = 1./A
910         PRINT,"MOBILITY RATIO BASED ON ENDPOINTS OF RELATIVE
            PERMEABILITY CURVES = ",MOBEND
920C        COMPUTE BREAKTHROUGH SATURATION FOR GENERAL
930C        FORM OF RELATIVE PERMEABILITY CURVES BY NEWTONIAN ITERATION
940         X1 = SWDI
950         SDMAX = Y(1. - SOR)
960         DSD = (SDMAX-SWDI)/49
970C        PRINT,"SDMAX = ",SDMAX,"DELSD = ",DSD
980         SWX = X1 + DSD
990         G3X = G3(SWX)
1000        DO 10 I = 1,100
1010        SWX = SWX + DSD
1020        G3XP1 = G3(SWX)
1030        PRINT,SWX,G3X,G3XP1
1040        IF ((G3X.LT.0.).AND.G3XP1.GT.0.) GO TO 90
1050        G3X = G3XP1
1060        GO TO 10
1070    90  SWX = SWX-DSD
1080        DSD = DSD/2.
1090        IF (ABS(G3X).LE.1.E-07) GO TO 20
1100    10  CONTINUE
1110        PRINT,"NO CONVERGENCE IN 100 ITERATIONS"
1120        STOP
1130    20  SWDF = SWX
1140C       PRINT,"SWDF = ",SWDF," FOR MOBILITY RATIO = ",MOBEND
1150        FWSF = FWF(SWDF)
1160C
1170        PRINT,"FWSF = ",FWSF
1180        SWF = SWF1(SWDF)
1190        PRINT,"SWF = ",SWF
1200        FSWF = FWF(SWDF)
1210        PRINT,"FPRIME FROM FW CURVE"
1220        FPSWF = B*G2(SWDF)
1230        PRINT,FPSWF
1240        FPSWW2 = B*G1(SWDF)
1250        PRINT,"FPSWW2 = ",FPSWW2
1260C
1270        RETURN
1280        END
```

TABLE A-3—LISTING OF PROP

```
10C       SUBPROGRAM TO COMPUTE KRO,KRW,FW,FPSW
20C
30C       GPW10-14-81
40C
50        SUBROUTINE PROP(VISO,VISW,SW1,KRO1,KRW1,FW1,FPSW1)
60        REAL KRO1,KRW1,KO,KW,M,N
70        COMMON /PDATA/ALPHA1,ALPHA2,M,N,SWC1,SOR1
80        Y(SW1) = (SW1-SWC1)*B
90        KW(Y) = ALPHA2*Y**N
100       KO(Y) = ALPHA1*(1.-Y)**M
110       G2(SWD1) = A*(B1(SWD1) + B2(SWD1))/B3(SWD1)**2
120       B1(SWD1) = N*SWD1**(N*-1)(1.-SWD1)**M
130       B2(SWD1) = M*SWD1**N*(1.-SWD1)**(M-1)
140       B3(SWD1) = SWD1**N + A*(1.-SWD1)**M
145       FWF(SWD1) = SWD1**N/(SWD1**N + A*(1.-SWD1)**M)
150       B51./(1.-SOR1-SWC1)
160       A = ALPHA1*VISW/(ALPHA2*VISO)
170       SWD = Y(SW1)
175       IF (ABS(SWD-1.0).LE.9.E-07) GO TO 10
180       KRO1 = KO(SWD)
190       KRW1 = KW(SWD)
200       FW1 = FWF(SWD)
210       FPSW1 = B*G2(SWD)
220       RETURN
222   10  KR01 = 0.
223       KRW1 = KW(SWD)
224       FW1 = 1.0
225       FPSW1 = 0.
226       RETURN
230       END
```

$S_w = S_{wf}$ to $1-S_{or}$. APVIS does not compute values of the average apparent viscosity for $S_w < S_{wf}$. APVIS is called only once per problem.

A listing of APVIS is given in Table A-4.

CGM. CGM is a subroutine subprogram that computes the value of Q_i^*/Q_{ibt}^* from Eqs. 4.19 and 4.20:

$$\frac{Q_i^*}{Q_{ibt}^*} = 1 + a_1 e^{-a_1}[Ei(a_2) - Ei(a_1)],$$

where

$$a_1 = 3.65\, E_{Abt},$$

$$a_2 = a_1 + \ln(W_i/W_{ibt}), \text{ and}$$

$$W_{ibt} \le W_i \le W_{i100},$$

and

$$Ei(x) = 0.57721577 + \ln x + \sum_{n=1}^{\infty} x^n/(nn!).$$

TABLE A-4—LISTING OF APVIS

```
0010C     SUBROUTINE TO DETERMINE THE AVERAGE APPARENT VISCOSITY
0020C     FOR A LINEAR SYSTEM WHEN THE FRACTIONAL FLOW DATA ARE
          AVAILABLE
0030C
0040C     GPW 10-14-81
0050C
0060      SUBROUTINE APVIS(VISW,VISO,KRO,KRW,FPSW,N,AVIS)
0070      REAL KRO(N),KRW(N)
0080      DIMENSION FPSW(N),AVIS(N),VISAP(100),FPSW1(100),VISAP1(100)
0090      DO 10 I = 1,N
0100  10  VISAP(I) = 1./(KRO(I.)/VISO + KRW(I)/VISW)
0110C
0120C     INVERT ORDER OF FPSW ARRAY TO EASE INTEGRATION
0130C
0140      DO 15 J = 1,N
0150      VISAP1(J) = VISAP(N-J+1)
0160  15  FPSW1(J) = FPSW(N-J+1)
0170      SUM = 0.
0180      DO 20 I = 2,N
0190      SUM = SUM + (FPSW1(I) - FPSW1(I-1))*(VISAP1(I) + VISAP1(I-1))*0.5
0200      J = N-I+1
0210  20  AVIS(J) = SUM/FPSW1(I)
0220      AVIS(N) = VISW/KRW(N)
0230      RETURN
0240      END
```

TABLE A-5—LISTING OF CGM

```
1100        SUBROUTINE CGM
            (WI,WIBT,EABT,QIQIBT)
1110        A1 = EABT/0.274
1120        E1 = EI(A1)
1130        WIWIBT = WI/WIBT
1140        A2 = A1 + ALOG(WIWIBT)
1150        QIQIBT = 1. + A1*EXP(-A1)*(EI(A2) - E1)
1160C   10  PRINT:I,WIWIBT,QIQIBT
1170        RETURN
1180        END
1190C       PROGRAM TO COMPUTE EI(X) CHECKED
            OUT AGAINST TABLES
1200C       OF JAHNKE AND EDME FOR X = 1,2,,--
            -10 GPW 6-21-79
1210        FUNCTION EI(X)
1220        DIMENSION TERM (100)
1230        GAMMA = 0.57721557
1240        TERM(1) = X
1250        SUM = TERM(1)
1260        DO 10 I = 2,100
1270        TERM(I) = TERM(I-1)*X*(I-1)/I**2
1280        IF (TERM(I).LT.1.E-8) GO TO 20
1290    10  SUM = SUM + TERM(I)
1300    20  EI = GAMMA + ALOG(X) + SUM
1310        RETURN
1320        END
```

TABLE A-6—SAMPLE INPUT/OUTPUT—SWFW (INTERACTIVE INPUT VALUES IN ITALICS)

INITIAL WATER SATURATION
= *0.248*

CONNATE (INTERSTITIAL) WATER SATURATION
= *0.248*

RESIDUAL OIL SATURATION
= *0.233*

POROSITY OF RESERVOIR-FRACTION
= *0.281*

OIL VISCOSITY IN CP
= *5.2*

WATER VISCOSITY IN CP
= *1.0*

FWSWI = 0.   AT SWI = 0.24800000E 00

CONVERGENCE IN   4 ITERATIONS VALUE OF SWF = 0.65303014 E 00

SWF = 0.65303014E 00   FOR ENDPOINT MOBILITY RATIO = 0.67753928E 00

FSWF = 0.84426913E 00

FPRIME FROM COMPUTED TANGENT = 0.20844600E 01

FPRIME FROM FW CURVE = 0.20844407E 01

As noted in Chap. 4, Eq. 4.20 is an infinite series representation for $Ei(x)$ that is accurate to 10^{-8} (Chap. 4, Ref. 12.)

When CGM is called in a FORTRAN statement, the value of Q_i^*/Q_{ibt}^* is computed and returned as the variable QIQIBT.

CALL CGM (WI, WIBT, EABT, QIQIBT).

Values of WI, WIBT, and EABT must be supplied as arguments. A listing of CGM is given in Table A-5.

SWFW. SWFW is a function subprogram that calculates the breakthrough saturation when the relative permeability relationships as a function of S_w are represented by analytic expressions. The expressions must have continuous first derivatives.

The method used to find f'_{Swf} is applicable to such analytical expressions as those obtained by Honarpour et al. (Chap. 3, Ref. 34), which are summarized in Problem 3.22. The version of SWFW presented in this Appendix uses one set of correlations from Honarpour et al.

The breakthrough saturation is the value of S_{wf}, which satisfies the relationship in Eq. A-1.

$$\frac{f_{Swf} - f_{wi}}{S_{wf} - S_{wi}} = \left(\frac{\partial f_w}{\partial S_w}\right)_{S_{wf}},$$

because

$$f_w = \frac{1}{1 + \left(\dfrac{k_{ro}}{\mu_o}\right)\left(\dfrac{\mu_w}{k_{rw}}\right)} \quad \dots\dots\dots\dots\dots (A-4)$$

and

$$\frac{\partial f_w}{\partial S_w} = \frac{\mu_w f_w}{\mu_o k_{rw}} \left[\frac{\partial k_{ro}}{\partial S_w} + \frac{k_{ro}}{k_{rw}}\left(\frac{\partial k_{rw}}{\partial S_w}\right)\right]. \quad \dots\dots (A-5)$$

Inspection of Eq. A-5 shows that f_{Sw} varies only with S_w when μ_w and μ_o are constant. Thus, solution of Eq. A-1 for S_{wf} amounts to a root-finding process in which the value of S_{wf} is obtained by iterative solution. The Wegstein method[2] of finding roots is used in this program. In the Wegstein method, Eq. A-1 is written as

$$S_w = S_{wi} + \frac{f_w - f_{wi}}{\left(\dfrac{\partial f_w}{\partial S_w}\right)}. \quad \dots\dots\dots\dots\dots\dots\dots (A-6)$$

Eq. A-6 is the equation of a straight line that intersects the fractional flow curve at two points when a tangent is drawn from S_{wi}. The points are S_{wi} and S_{wf}. Because there are two roots to Eq. A-6, it is necessary to be close to the value of S_{wf} to converge on S_{wf}.

The program included in this Appendix was developed with the following expressions for relative permeability as a function of saturation developed by Honarpour et al.

$$k_{rw} = 0.035388 \frac{(S_w - S_{iw})}{1 - S_{iw} - S_{or}} - 0.010874$$

$$\times \left[\frac{S_w - S_{iw}}{1 - S_{or} - S_{iw}}\right]^{2.9} + 0.56556(S_w)^{3.6}(S_w - S_{iw})$$

$$\dots\dots\dots\dots\dots\dots\dots\dots\dots\dots (A-7)$$

TABLE A-7—LISTING OF SWFW

```
10C          PROGRAM TO COMPUTE FRACTIONAL FLOW INFORMATION FOR
20C          RELATIVE PERMEABILITY CURVES INCLUDING MOBILE CONNATE
             WATER
25           FUNCTION SWFW(VISO,VISW,SWI,SWC,SOR,FPSWF)
30C
40C          GENERAL ROOT-FINDING PROGRAM
50C          TO COMPUTE SWC WHEN
60C          KRW AND KRO ARE GIVEN
70C          BY ANALYTICAL EXPRESSIONS
80C          GPW
90C
100C
110          DIMENSION SWA(101),SW(101),KRO(101),KRW(101),FW(101),FP(101)
120          DIMENSION SWAVG(10)
130          REAL KRO,KRW,KO,KW,M,N,MOBEND
140          KW(SW1) = 0.035388*(B(SW1))/A-0.010874*B(SW1)**2.9/A**2.9 +
150&            0.56556*(SW1)**3.6*B(SW1)
160          DKRW(SW1) = 0.035388/A-0.031535*B(SW1)**1.9/A*2.9
170&            + 0.56556*((3.6)*B(SW1)*SW1**2.6 + SW1**3.6)
180          KO(SW1) = 0.76067*B1(SW1)**1.8*B2(SW1)**2 + 2.6318*POROS*
190&            (1 - SOR)*(1.-SW1-SOR)
200          DKRO(SW1) = -1.*(1.369206*B1(SW1)**0.8((1.-SWC)*(1.-SOR))*
210&            B2(SW1)**2 + 1.52134*(B1(SW1))**1.8*B2(SW1)/
220&            (1.-SOR-SWC) + 2.6318*POROS*(1-SOR)
230          B(SW1) = SW1-SWC
240          B1(SW1) = ((1.-SW1)/(1.-SWC - SOR)/(1.-SOR)
250          B2(SW1) = (1.-SW1-SOR)/(1.-SWC-SOR)
260          FWF(SW1) = 1./(1. + KO(SW1)*VISW/(KW(SW1)*VISO))
270          DFWDSW(SW1) = FWF(SW1)*FWF(SW1)*VISW/VISO/KW(SW1)*
280&            (KO(SW1)/KW(SW1)*DKRW(SW1)-DKRO(SW1))
290          G1(SW1) = SWI + (FWF(SW1)-FWSWI)/DFWDSW(SW1)
300C
310          PRINT,"INITIAL WATER SATURATION"
320          READ,SWI
330          FWSWI = 0.
340          PRINT,"CONNATE (INTERSTITIAL) WATER SATURATION"
350          READ,SWC
360C
370          PRINT,"RESIDUAL OIL SATURATION"
380          READ,SOR
390          PRINT,"POROSITY OF RESERVOIR—FRACTION"
400          READ,POROS
410          A = 1.-SWC-SOR
420          PRINT,"OIL VISCOSITY IN CP"
430          READ,VISO
440C
450          PRINT,"WATER VISCOSITY IN CP"
460          READ,VISW
470          MOBEND = KW(1.-SOR)*VISO/(KO(SWC)*VISW)
480C
490C         IF(SWI.GT.SWC) FWSWI = FWF(SWI)
500          PRINT,"FWSWI = ",FWSWI," AT SWI = ",SWI
510C           COMPUTE BREAKTHROUGH SATURATION BY WEGSTEIN METHOD
520C           THERE ARE TWO ROOTS TO THE EQUATION SW = G1(SW)
530C         SW = SWC AND SW = SWF
540C
550C         MUST FIND INTERVAL WHERE SW IS APPROXIMATELY = SWF BEFORE
             ROOT FINDING
560C         TECHNIQUE IS USED. OTHERWISE, ROOT AT SW = SWC MAY BE
             FOUND
```

TABLE A-7—LISTING OF SWFW (continued)

```
570C
580          DELS = (1–SOR–SWI)/50.
590          DO 100 I = 1,50
600          SW(I) = SWI + DELS*I
610    100   SWA(I) = G1(SW(I))
620          INDEX = 1
630          IF(SWA(1).LT.SW(1)) INDEX = 2
640          GO TO (200,400), INDEX
650C
660C             G1(SW) > SW NEAR SWC
670C
680    200   DO 250 I = 1,50
690          IF(SWA(I).LE.SW(I)) GO TO 500
700    250   CONTINUE
710          PRINT,"NO INTERSECTION FOUND"
720          RETURN
730C
740c             G1(SW) < SW NEAR SWC
750C
760    400   DO 350 I = 1,50
770          IF(SWA(I).GE.SW(I)) GO TO 500
780    350   CONTINUE
790          PRINT,"NO INTERSECTION FOUND"
800          STOP
810C
820C             SWF IS NOW BRACKETED BY SW(I) AND SW(I–1)
830C
840    500   SWMAX = SW(I)
850          SWMIN = SW(I–1)
860          SWO = SWMIN
870C         PRINT,SWMIN,SWMAX
880          NC = 0
890          DO 10 IT = 1,100
900C         PRINT,IT,SWO,DKRO(SWO),DKRW(SWO)
910      5   SWW = G1(SWO)
920C            PRINT,SWW,DFWDSW(SWO),FWF(SWO)
930      7   CALL WEG(SWO,SWW,1,NC)
940C         PRINT,NC,SWO,SWW
950          GO TO (20,10),NC
960     10   CONTINUE
970          PRINT,"NO CONVERGENCE IN 100 ITERATIONS"
980          STOP
990     20   PRINT,"CONVERGENCE IN" ,IT, "ITERATIONS","VALUE OF
             SWF = ",SWW
1000         FSWF = FWF(SWW)
1010         SWF = SWW
1020         PRINT,"SWF = ",SWF, "FOR ENDPOINT MOBILITY RATIO = ",MOBEND
1030C
1040         PRINT,"FSWF = ",FSWF
1050         FPRIME = (FSWF–FWSWI)/(SWF–SWI)
1060         PRINT,"FPRIME FROM COMPUTED TANGENT = ",FPRIME
1070         FPSWF = DFWDSW(SWF)
1080         PRINT,"FPRIME FROM FW CURVE = ",FPSWF
1090         RETURN
1100         END
```

and

$$k_{ro} = 0.76067 \left[\frac{\left(\dfrac{S_o}{1-S_{iw}} \right) - S_{or}}{1 - S_{or}} \right]^{1.8}$$

$$\left(\frac{S_o - S_{or}}{1 - S_{iw} - S_{or}} \right)^{2.0} + 2.6318\phi(1-S_{or})(S_o - S_{or}).$$

$$\dots \dots \dots \dots \dots \dots \dots \dots (A-8)$$

SWFW is accessed using the FORTRAN statement below:

SWBT = SWFW (VISO, VISW, SWI, SWC, SOR, FPSWF).

Values of μ_o, μ_w, ϕ, S_{wi}, S_{iw}, and S_{or} are read into the program interactively. VISO, VISW, SWI, SWC, SOR, and FPSWF are returned to the calling program as arguments.

Table A-6 shows the output from a sample run.

A listing of the program is given in Table A-7. Table A-8 is a listing of Subroutine WEGS from Ref. 3 that is used in SWFW to carry out the Wegstein method.

References

1. Carnahan, B., Luther, H.A., and Wilkes, J.O.: *Applied Numerical Methods*, John Wiley and Sons Inc., New York City (1969) 171.

TABLE A-8—LISTING OF SUBROUTINE WEGS

```
0010C
0020C        WEGSTEIN METHOD TO COMPUTE
             ROOT OF AN ALGEBRAIC FUNCTION
0030C        source,"MODELING AND SIMULATION
             IN CHEMICAL ENGINEERING"
0040C        ROGER G.E. FRANKS, WILEY (1972)
0050C
0060C        GPW 6/26/84
0070C
0080         SUBROUTINE WEG(X,Y,NR,NC)
0090         DIMENSION XA(100),YA(100)
100          IF (ABS((X-Y)/(X+Y)).LT..10^-6) GO TO 6
0110         IF (NC.LE.1) GO TO 5
0120         XT = (XA(NR)*Y-YA(NR)*X)/(XA(NR)-
             X+Y-YA(NR))
0130         XA(NR) = X
0140         YA(NR) = Y
0150         X = XT
0160         RETURN
0170    5    XA(NR) = X
0180         YA(NR) = Y
0190         X = Y
0200         NC = 2
0210         RETURN
0220    6    X = Y
0230         NC = 1
0240         RETURN
0250         END
```

2. Wegstein, J.H.: "Accelerating Convergence of Iterative Processes," *Commun. Assoc. Computing Machinery* (June 1958) **1**, No. 6, 9–13.
3. Franks, R.G.E.: *Modeling and Simulation in Chemical Engineering*, John Wiley and Sons Inc., New York City (1972) 26–27.

Appendix B

Streamtube Parameters for Pattern Floods, Higgins-Leighton Streamtube Model

The use of streamtube models to predict waterflood performance in five-spot pattern floods was introduced in Chap. 4 and illustrated in waterflood design in Chap. 6. Waterflood performance of direct line drive, staggered line drive, and seven-spot patterns may also be predicted with the Higgins-Leighton program (Chap. 4, Ref. 18) with minor adjustments. Appendix B contains streamtube parameters for these patterns. Shape factors were obtained from Ref. 1. Other information was obtained by personal communication.[*]

[*]Personal communication with A.J. Leighton (1984).

TABLE B-1—DIRECT LINE-DRIVE PATTERN ($d/a = 1$), FOUR SYMMETRIC ELEMENTS/INJECTION WELL[1]

Cell No.	Direct Line-Drive Channel Number				
	1	2	3	4	5
1	19.604	14.155	14.167	11.424	13.960
2	1.761	0.843	1.037	1.019	0.986
3	0.786	0.491	0.666	0.570	0.623
4	0.510	0.332	0.455	0.446	0.438
5	0.324	0.259	0.348	0.335	0.350
6	0.244	0.228	0.326	0.228	0.290
7	0.174	0.221	0.297	0.264	0.266
8	0.146	0.175	0.267	0.249	0.251
9	0.151	0.169	0.247	0.209	0.210
10	0.165	0.175	0.229	0.196	0.186
11	0.193	0.177	0.223	0.196	0.173
12	0.239	0.182	0.221	0.194	0.170
13	0.312	0.197	0.215	0.191	0.167
14	0.414	0.218	0.203	0.185	0.167
15	0.530	0.232	0.187	0.179	0.164
16	0.640	0.245	0.181	0.169	0.149
17	0.779	0.266	0.181	0.169	0.144
18	0.908	0.281	0.190	0.190	0.155
19	0.931	0.287	0.205	0.225	0.171
20	0.933	0.288	0.209	0.231	0.173
21	0.933	0.288	0.209	0.231	0.173
22	0.931	0.287	0.205	0.225	0.171
23	0.908	0.281	0.190	0.190	0.155
24	0.779	0.266	0.181	0.169	0.144
25	0.640	0.245	0.181	0.169	0.149
26	0.530	0.232	0.187	0.179	0.164
27	0.414	0.218	0.203	0.185	0.167
28	0.312	0.197	0.215	0.191	0.167
29	0.239	0.182	0.221	0.194	0.170
30	0.193	0.177	0.223	0.196	0.173
31	0.165	0.175	0.229	0.196	0.186
32	0.151	0.169	0.247	0.209	0.210
33	0.146	0.175	0.267	0.249	0.251
34	0.174	0.221	0.297	0.264	0.266
35	0.244	0.228	0.326	0.288	0.290
36	0.324	0.259	0.348	0.335	0.350
37	0.510	0.332	0.455	0.446	0.438
38	0.786	0.491	0.666	0.570	0.623
39	1.761	0.843	1.037	1.019	0.986
40	19.604	14.155	14.167	11.424	13.960
Volume, fraction	0.2275	0.2798	0.1854	0.1610	0.1464
Angles, degrees					
Cells 1 and 40	17.9	16.6	17.6	24.8	13.1
CON1*	4.508				
CON2*	0.3562				

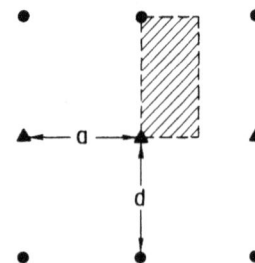

$$\frac{d}{r_w} = 688.86$$

*CON1 and CON2 are flow-rate and PV parameters for the Higgins-Leighton program (Chap. 4, Ref. 18).

TABLE B-2—STAGGERED LINE-DRIVE PATTERN, LENGTH/WIDTH = 3.0, FOUR SYMMETRIC ELEMENTS/INJECTION WELL[1]

Cell No.	Staggered Line-Drive Channel Number					
	1	2	3	4	5	6
1	15.650	15.395	14.648	15.311	15.008	16.884
2	1.430	1.283	1.333	1.281	1.261	1.444
3	1.004	0.779	0.852	0.588	0.735	0.603
4	0.827	0.629	0.720	0.575	0.509	0.372
5	0.724	0.535	0.583	0.433	0.425	0.241
6	0.648	0.465	0.570	0.369	0.399	0.180
7	0.614	0.430	0.521	0.349	0.338	0.126
8	0.589	0.423	0.498	0.326	0.357	0.112
9	0.564	0.414	0.465	0.313	0.362	0.109
10	0.538	0.406	0.450	0.311	0.363	0.113
11	0.521	0.404	0.448	0.309	0.364	0.140
12	0.516	0.402	0.439	0.313	0.372	0.156
13	0.513	0.402	0.405	0.327	0.379	0.170
14	0.512	0.401	0.403	0.342	0.383	0.183
15	0.520	0.398	0.401	0.355	0.384	0.215
16	0.499	0.398	0.392	0.358	0.385	0.252
17	0.479	0.397	0.380	0.360	0.385	0.296
18	0.465	0.396	0.375	0.360	0.386	0.336
19	0.455	0.394	0.374	0.365	0.387	0.376
20	0.428	0.392	0.373	0.371	0.388	0.405
21	0.405	0.388	0.371	0.373	0.392	0.428
22	0.376	0.387	0.365	0.374	0.394	0.455
23	0.336	0.386	0.360	0.375	0.396	0.465
24	0.296	0.385	0.360	0.380	0.397	0.479
25	0.252	0.385	0.358	0.392	0.398	0.499
26	0.215	0.384	0.355	0.401	0.398	0.511
27	0.183	0.383	0.342	0.403	0.401	0.512
28	0.170	0.379	0.327	0.405	0.402	0.513
29	0.156	0.372	0.313	0.439	0.402	0.516
30	0.140	0.364	0.309	0.448	0.404	0.521
31	0.113	0.363	0.311	0.450	0.406	0.538
32	0.109	0.362	0.313	0.465	0.414	0.564
33	0.112	0.357	0.326	0.498	0.423	0.589
34	0.126	0.338	0.349	0.521	0.430	0.614
35	0.180	0.399	0.369	0.570	0.465	0.648
36	0.241	0.425	0.433	0.583	0.535	0.724
37	0.372	0.509	0.575	0.720	0.629	0.827
38	0.603	0.735	0.588	0.852	0.779	1.004
39	1.144	1.261	1.281	1.333	1.283	1.430
40	16.884	15.008	15.311	14.648	15.395	15.650
Volume, fraction	0.1965	0.1565	0.1470	0.1470	0.1565	0.1965
Angles, degrees						
Cell 1	14.6	13.8	14.1	16.6	15.8	15.1
Cell 40	15.1	15.8	16.6	14.1	13.8	14.6
CON1*	4.508					
CON2*	0.217					

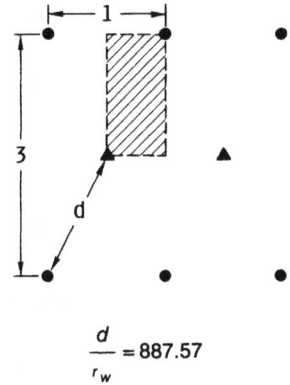

$$\frac{d}{r_w} = 887.57$$

*CON1 and CON2 are flow-rate and PV parameters for the Higgins-Leighton program (Chap. 4, Ref. 18).

TABLE B-3—SEVEN-SPOT PATTERN, 12 SYMMETRIC ELEMENTS/INJECTION WELL [1]

Cell No.	Seven-Spot Channel Number		
	1	2	3
1	12.952	7.885	7.321
2	1.549	3.085	3.224
3	0.602	0.590	0.846
4	0.387	0.391	0.349
5	0.277	0.335	0.328
6	0.224	0.233	0.246
7	0.187	0.218	0.220
8	0.157	0.204	0.199
9	0.124	0.185	0.172
10	0.100	0.166	0.169
11	0.085	0.148	0.152
12	0.082	0.146	0.141
13	0.080	0.139	0.139
14	0.069	0.129	0.129
15	0.056	0.134	0.132
16	0.051	0.135	0.136
17	0.050	0.137	0.138
18	0.058	0.138	0.139
19	0.074	0.140	0.140
20	0.113	0.142	0.142
21	0.116	0.148	0.143
22	0.123	0.150	0.146
23	0.128	0.153	0.155
24	0.132	0.172	0.167
25	0.144	0.188	0.180
26	0.159	0.193	0.185
27	0.172	0.200	0.195
28	0.206	0.222	0.213
29	0.236	0.239	0.234
30	0.256	0.257	0.252
31	0.300	0.299	0.281
32	0.323	0.317	0.312
33	0.393	0.349	0.351
34	0.427	0.403	0.407
35	0.546	0.486	0.491
36	0.639	0.598	0.608
37	0.847	0.815	0.742
38	1.234	1.092	1.022
39	2.070	3.793	4.330
40	26.358	21.040	19.708
Volume, fraction	0.4265	0.3001	0.2734
Angles, degrees			
Cell 1	17.6	21.1	21.3
Cell 40	9.4	10.2	10.4
CON1*	13.527		
CON2*	0.4626		

$$\frac{d}{r_w} = 1047.12$$

*CON1 and CON2 are flow-rate and PV parameters for the Higgins-Leighton program (Chap. 4, Ref. 18).

Reference

1. Higgins, R.V., Boley, D.W., and Leighton, A.J.: "Aids to Forecasting the Performance of Water Floods," *J. Pet. Tech.* (Sept. 1964) 1076–82; *Trans.*, AIME, **231**.

Appendix C

Commonly Used Equations in Alternative Units

Most of the equations developed in this text do not assume a specific set of units. Thus it is possible to perform calculations in field units or SI metric units by maintaining consistent units. There are always unit problems involving computation of flow rates because conventional units are mixed. Field units (barrels, darcies, centipoises, pounds per square inch, feet, days) were selected for the major flow equations and most examples in this text. Appendix C contains selected equations expressed in alternative units. For all equations except one, the alternative set of units is the SPE Metric Standard.[1]

$$N_{ca} = \frac{v\mu_w}{\sigma_{ow}}, \quad \dots\dots\dots\dots\dots(2.20)$$

which is multiplied by 3.528×10^{-6} when expressed in field units.

$$f_w = \frac{1}{1+\left(\dfrac{k_o}{k_w}\right)\left(\dfrac{\mu_w}{\mu_o}\right)}$$

$$+ \frac{\dfrac{k_o A}{\mu_o q_t}\left[\dfrac{\partial P_c}{\partial x}+(\rho_o-\rho_w)g\sin\alpha\right]}{1+\left(\dfrac{k_o}{k_w}\right)\left(\dfrac{\mu_w}{\mu_o}\right)}, \quad \dots\dots(3.73)$$

which becomes

$$f_w = \frac{1}{1+\left(\dfrac{k_o}{k_w}\right)\left(\dfrac{\mu_w}{\mu_o}\right)}$$

$$+ \frac{\dfrac{0.864\times10^{-4}k_o A}{\mu_o q_t}\left[\dfrac{\partial P_c}{\partial x}+0.001(\rho_o-\rho_w)g\sin\alpha\right]}{1+\left(\dfrac{k_o}{k_w}\right)\left(\dfrac{\mu_w}{\mu_o}\right)}$$

when expressed in SI metric values.

$$q_t = \frac{1.127 k_b A(p_i-p_p)}{\lambda^{-1}L}, \quad \dots\dots\dots\dots(3.126)$$

where 1.127 becomes 0.864×10^{-4} when expressed in SI metric values.

$$i = \frac{3.541 kh(\Delta p)}{\mu\left(\ln\dfrac{d}{r_w}-0.619\right)}, \quad \dots\dots\dots(4.32 \text{ and } 6.3)$$

where 3.541 becomes 2.714×10^{-3} when expressed in SI metric values.

$$G = \frac{(4.9\times10^{-4})kk_{rw}A(\rho_w-\rho_o)\sin\alpha}{\mu_w q_t}, \quad \dots\dots(5.72)$$

where 4.9×10^{-4} becomes 8.64×10^{-8} when expressed in SI metric values.

$$q_c = \frac{(4.9\times10^{-4})kk_{rw}A(\rho_w-\rho_o)\sin\alpha}{\mu_w(M-1)}, \quad \dots(5.75)$$

where 4.9×10^{-4} becomes 8.64×10^{-8} when expressed in SI metric values.

$$i = \frac{3.541 kh(\Delta p)}{\mu\left(\ln\dfrac{a}{r_w}+1.571\dfrac{d}{a}-1.838\right)}, \quad \dots\dots\dots(6.1)$$

where 3.541 becomes 2.714×10^{-4} when expressed in SI metric values.

$$i = \frac{3.541 kh(\Delta p)}{\mu\left(\ln\dfrac{a}{r_w}+1.571\dfrac{d}{a}-1.838\right)}, \quad \dots\dots(6.2)$$

where 3.541 becomes 2.714×10^{-4} when expressed in SI metric values.

$$i = \frac{4.72 kh(\Delta p)}{\mu\left(\ln\dfrac{d}{r_w}-0.569\right)}, \quad \dots\dots\dots\dots(6.4)$$

where 4.72 becomes 3.618×10^{-4} when expressed in SI metric values.

$$i = \frac{3.541 kh(\Delta p)_{i,c}}{\dfrac{1+R}{2+R}\left(\ln\dfrac{d}{r_w}-0.272\right)\mu}, \quad \dots\dots\dots(6.5)$$

where 3.541 becomes 2.714×10^{-4} when expressed in SI metric values.

$$i = \frac{7.082 kh(\Delta p)_{i,s}}{\left[\dfrac{3+R}{2+R}\left(\ln\dfrac{d}{r_w}-0.272\right)-\dfrac{0.693}{2+R}\right]\mu}, \quad \dots(6.6)$$

where 7.082 becomes 5.429×10^{-4} when expressed in SI metric values.

Reference

1. *The SI Metric System of Units and SPE Metric Standard*, SPE, Richardson, TX (1982).

Author Index

Subject Index

A

Absolute permeability, 24, 48, 66, 105, 150, 163, 176, 282
Acidization, 192
Acoustic log, 234, 273
Aeolian deposits, 243, 279
Afiesere field, 251
Air permeability, 272, 273
Air/water interface, 9
Alabama, 234, 235
Alaska, 18
Alberta, 19, 236, 263–267, 269, 287
Alcohols, low IFT's systems containing, 5
Allochemical rocks, 260
Alluvial channel, 281–283
Alluvial fans, 243, 279
Alluvial sand patterns, 243
American Petrofina Co. of Texas, 293
Amoco Production Co., 231
Amott wettability test, 19, 46
Anadarko basin, 281
APVIS computer subprogram, 100, 303, 304, 307
Aquifer, 2, 209, 225, 236, 240, 277, 294
Arbuckle reservoirs, 2
Archie's equation, 287
Areal displacement, 111–113, 129, 130
Areal sweep efficiency, 114, 115, 119, 133, 192, 194, 205, 210, 211, 218, 227
Argentina, 18
Authigenic clay cement, 272
Autochthonous reef rocks, 260
Average apparent mobilities, 123
Average apparent viscosity, 70–75, 83, 89, 100, 102–105, 203, 303, 307
Average water saturation (volumetric), 32, 64, 65, 68, 76, 80–82, 88, 96, 97, 115, 119, 122, 147, 149–152, 154, 156, 176, 202, 204, 210, 221, 303

B

Background:
  Development of waterflooding, 1, 2
  Introduction, 1
  Primary production, 2–4
  References, 4
Backreef, 257, 264, 272
Bandera sandstone, 35, 36
Barrier bar, 251, 254, 281–283
Barrier island sands, 251–253, 279
Base permeability, 21–23, 56, 66, 70, 75, 83, 96, 99, 105–107, 202, 203
Beaver Hill Lake platform, 263–265
Bell creek field, 253, 254
Berea sandstone, 11–16, 24, 27, 28, 35, 36, 41, 43, 44, 46–48, 57, 169
Bindley field, 273
Biohermal facies or reefs, 257, 260, 263–265, 268, 269
Biostromal facies or reefs, 257, 260, 264, 265, 268, 269
Biota, 258, 259
Black-oil reservoir simulator, 129, 298
Block diagrams, 244, 249
Boise sandstone, 47
Booch sand unit, 291, 293
Bottomwater drive, 99, 175, 225, 230, 239, 240
Bouma sequence, 255
Boundary effect absorbers, 58
Bradford field, 1, 2
Bradford sandstone, 8, 241

C

Braided streams, 243, 277, 279
Breakthrough (water), 37, 40, 59, 60, 63, 64, 66–74, 77–80, 84, 87–89, 91, 99, 100, 103, 107, 114–120, 122–125, 127, 133, 142, 143, 153, 156, 160, 161, 174, 183, 194, 195, 211, 229, 293, 303
Breakthrough saturation, 75, 76, 82, 89, 98, 100, 117, 123, 151, 153, 303, 306, 308, 309
Brine displacement by gas, 63
Bryans Hill sandstone, 36
Buckley-Leverett equation, 61, 129, 147
Buckley-Leverett model, 59, 60, 147
Buckley-Leverett solution, 63, 64, 93, 94, 107, 146, 155, 161, 167
Burbank sand, 9
Bush city pool, 244

Calcarenite, 261, 264, 269, 270
Calcite surface, 8, 9
Calculation of:
  Average apparent viscosities, 72–74
  Displacement performance, 32, 78, 79, 120, 125, 127, 143, 151, 153, 154, 166, 180
  q_t and time, 74
  Waterflood, 303
California, 4, 6, 19, 82, 186, 257
Canada, 18, 19, 236, 263–268, 287
Canadian Devonian reef, 263, 264
Capillary end effects, 23, 37, 57–59, 80, 89–92, 97, 158
Capillary equilibrium, 45, 46, 57, 58, 163–165, 184, 185, 288
Capillary forces, 35–38, 40–43, 90, 92–94, 111, 114, 129, 130, 139, 146, 148–154, 158, 160–167, 169, 174, 185, 214, 215
Capillary number, 42–44, 48, 49
Capillary number correlations, 42, 43, 48
Capillary pressure, 9, 10, 57–59, 62, 77, 97, 98, 148–150, 163, 184, 288
Capillary pressure curves, 14–16, 21, 43, 56–59, 91, 92, 148–150, 159, 163, 185, 284, 285, 288
Capillary pressure data, 13–16, 43, 45, 162, 163
Capillary pressure discontinuity, 57
Capillary-pressure distribution, 148–150, 184, 185
Capillary pressure drop, 39
Capillary pressure gradients, 59, 91–93, 144, 169
Capillary pressure/saturation relationship, 91, 145
Carbonate allochems, 261
Carbonate (shelf) bank, 260, 264, 268, 269
Carbonate platforms, 257–261, 263
Carbonate ramps, 257, 258, 260
Carbonate reefs, 263, 264
Carbonate reservoirs, 257–269
Carbonate rocks, classification by depositional texture, 261
Carbonate sediments, 260, 261, 268
Carson Creek field, 264
Cementation, 242, 249, 253, 268
CGM computer subprogram, 132, 303, 307, 308
Chain rule of differentiation, 158
Channel flow, 20, 21
Channeling, 162, 167

Cherokee shale, 244
Chesney-Hegberg pilot area, 288, 290
Circle flood, 1
CO_2 oil recovery process, 170
Colorado, 18
Compaction, 242, 249
Computer generated contour map of net pay, 286
Computer mapping programs, 294
Computer models for design of water-floods:
  Reservoir simulators, 214–217
  Streamtube models, 212–214
Computer program, 117, 118, 132, 171, 177, 212–214, 287, 303–311
Conductance, 119, 196
Condutance ratio, 119, 120, 194
Confidence level, 174, 175
Coning, 235
Connate water, 10, 33, 34, 68, 101, 155, 287, 288, 305, 309
Connate water saturation, 16, 25, 33, 59, 77, 85, 106, 134, 148, 179, 189,. 271, 287–289, 304, 305, 308, 309
Conservation of mass law, 53–55, 60
Constant-pressure-drop displacement, 82, 101, 102, 119, 120, 125, 128, 140, 177–181, 221
Constant-rate displacement, 82–85
Contact angle, 7–9, 17–20, 23, 25–28, 30, 39–41, 44, 86
Continental environment, 242–245
Continuity:
  Equation, 53–55, 105
  In mud-delta system, 252
  In sand-delta system, 251
  Of shale intercalations as function of depositional environment, 279
Contour maps, 271, 278, 286–288
Core permeability, 234, 275, 276
Core porosity, 234, 274, 275
Correlation cross section:
  Bradford sandstone, 241
  Elk City field, 281
  Judy Creek field, 236, 237
  Robinson sandstone, 241
Correlation of:
  Actual pay with apparent pay, 284
  Areal sweep efficiency at and after breakthrough, 114, 115
  Buildup results with core analyses, 291
  Capillary number with residual oil saturation, 42–44
  Capillary pressure, 97
  Conductance ratio for five-spot patterns—miscible displacement, 120, 194
  Connate water saturation with porosity, 287–289
  Core porosity with neutron porosity, 275
  Displacement performance for nine-spot flood, 210
  Displacement performance with permeability variation, 73
  Flood stabilization, 89, 90
  Grain size and porosity, 275
  Grain size with resistivity, 275
  High- and low-permeability zones, 234
  Net pay with actual pay, 285
  Oil saturation at water breakthrough, 36, 37
  Permeability data, empirical, 172, 173
  Permeability/saturation, 21
  Relative permeability data for porous rock, 24–31, 103, 107